www.wadsworth.com

wadsworth.com is the World Wide Web site for Wadsworth and is your direct source to dozens of online resources.

At *wadsworth.com* you can find out about supplements, demonstration software, and student resources. You can also send email to many of our authors and preview new publications and exciting new technologies.

wadsworth.com
Changing the way the world learns®

The Kaleidoscope of Gender

Prisms, Patterns, and Possibilities

JOAN Z. SPADE
State University of New York at Brockport

CATHERINE G. VALENTINE
Nazareth College

Australia • Canada • Mexico • Singapore • Spain
United Kingdom • United States

Sociology Editor: *Robert Jucha*
Editorial Assistant: *Melissa Walter*
Marketing Manager: *Matthew Wright*
Marketing Assistant: *Matthew Silverstein*
Advertising Project Manager: *Linda Yip*
Project Manager, Editorial Production: *Andy Marinkovich*
Print/Media Buyer: *Karen Hunt*
Permissions Editor: *Sarah Harkrader*
Production Service: *Buuji, Inc.*
Copy Editor: *Alan DeNiro, Buuji, Inc.*
Cover Designer: *Roger Knox*
Compositor: *Buuji, Inc.*
Printer: *Webcom*

Printed in Canada
2 3 4 5 6 7 07 06 05 04

Library of Congress Control Number:
2003100338

ISBN 0-534-57584-6

Wadsworth/Thomson Learning
10 Davis Drive
Belmont, CA 94002-3098
USA

Asia
Thomson Learning
5 Shenton Way #01-01
UIC Building
Singapore 068808

Australia/New Zealand
Thomson Learning
102 Dodds Street
Southbank, Victoria 3006
Australia

Canada
Nelson
1120 Birchmount Road
Toronto, Ontario M1K 5G4
Canada

Europe/Middle East/Africa
Thomson Learning
High Holborn House
50/51 Bedford Row
London WC1R 4LR
United Kingdom

Latin America
Thomson Learning
Seneca, 53
Colonia Polanco
11560 Mexico D.F.
Mexico

Spain/Portugal
Paraninfo
Calle/Magallanes, 25
28015 Madrid, Spain

To Kay's spouse, Paul J. Burgett,
for sharing love, knowledge, and adventure

and

To Joan's children, their spouses, and grandchildren,
for having the strength to seek out new ways of doing gender

Contents

Preface

The Kaleidoscope of Gender: Prisms, Patterns, and Possibilities provides an overview of the cutting edge literature and theoretical frameworks in the sociology of gender and related fields for understanding the social construction of gender. Although not ignoring classical contributions to gender research, this book focuses on where the field is moving and the changing paradigms and approaches to gender studies. *The Kaleidoscope of Gender* uses the metaphor of a kaleidoscope and three themes—prisms, patterns, and possibilities—to unify topic areas. It focuses on the prisms through which gender is shaped, the patterns which gender takes, and the possibilities for social change through a deeper understanding of ourselves and our relationships with others, both locally and globally.

The book begins by looking at gender and other social prisms that define gendered experiences across the spectrum of daily lives. We conceptualize prisms as social categories of difference and inequality that shape the way gender is defined and practiced, including culture, race/ethnicity, social class, sexuality, age, and ability/disability. Different as individuals' experiences might be, there are patterns to gendered experiences. The second section of the book examines these patterns across a multitude of areas of daily life. The last section of the book explores possibilities for change. Basic to the view of gender as a social construction is the potential for social change. Students will read that gender transformation has and can occur, and learn that it is possible to alter the genderscape. Because prisms, patterns, and possibilities themselves intersect, the framework for this book is fluid, interweaving topics and emphasizing the complexity and everchanging nature of gender.

We had multiple goals in mind as we developed this book:

1. Creating a book of readings that is accessible, timely, and stimulating in a text whose structure and content incorporate a fluid framework with gender presented as an emergent, evolving, complex pattern, not one fixed in traditional categories and topics;
2. Selecting articles that creatively and clearly explicate what gender is and is not, and what it means to say that gender is socially constructed by incorporating provocative illustrations and solid scientific evidence of the malleability of gender and the role of individuals, groups, and social institutions in the daily performance and transformation of gender practices and patterns;
3. Including readings that untangle and clarify the intricate ways in which gender is embedded in and defined by the prisms of culture, race/ethnicity, class, sexuality, age, ability/disability, and cultural patterns of identities, groups, and institutions;
4. Integrating articles with a cross-cultural focus to illustrate that gender is a continuum of categories, patterns, and expressions whose relevance is contextual and continuously shifting, and that gender inequality is not a universal, natural social pattern, but rather one of many systems of oppression, none of which can be generalized;
5. Assembling articles that offer students useful cognitive and emotional tools for making sense of the shifting and contradictory genderscape they inhabit, its personal relevance, its implications for relationships both locally and globally, and possibilities for change.

The reading selections include theoretical, classical, and review articles; however, the emphasis is on contemporary contributions to the field. We include classical and theoretical arguments throughout the book to provide a broader framework with which to understand gender. Chapter introductions and the introduction to the book contextualize the literature in gender, introduce the readings in the chapters and illustrate how they relate to analyses of gender, and develop the kaleidoscope metaphor as a tool for viewing gender. Introductions and questions for consideration precede each reading to help students to focus on and grasp the key points of the selections. Additionally, each section ends with questions for students to consider, the answers to which can be found using InfoTrac® College Edition or specified Web sites.

It is possible to use this book alone, as a supplement to a text, or in combination with other articles or monographs. It is designed for undergraduate audiences and has been tested successfully in four 300-level classes. The readings are appropriate for a variety of courses focusing on the study of gender such as sociology of gender, gender and social change, and women's studies. The book may be used in departments of sociology, anthropology, psychology, and women's studies.

We would like to thank those reviewers whose valuable suggestions and comments helped us to develop this book: Leslie Miller-Bernal, Wells College;

Janet C. Bogdan, LeMoyne College; Susan Chase, University of Tulsa; Catherine Connolly, University of Wyoming; Charlotte Dunham, Texas Tech University; Sarah Gatson, Texas A&M University; Kathryn Hausbeck, University of Nevada Las Vegas; Robert Heasley, Indiana University of Pennsylvania; Melissa Hippard, Colorado State University; Jocelyn Hollander, University of Oregon; Jennifer Lois, Western Washington University; Lauren J. Pivnick, Monroe Community College; Judith Warner, Texas A&M International; Matt Wray, University of Nevada Las Vegas; and Anna Zajicek.

Finally, we would like to thank the students in Kay's fall 2001 and 2002 Gender and Society classes and Joan's spring 2002 and 2003 Gender and Social Change classes for their insightful feedback on the material and for not complaining too much about the "draft" form of the book. Joan would like to thank Gloria Condoluci, sociology department secretary at SUNY Brockport, for her assistance with the project and her many colleagues who gave suggestions and comments along the way, including Ronnie Steinberg and Michael Ames. Kay would like to thank Kate Keck, work-study student in Women's Studies, and the Nazareth College support staff who provided important assistance throughout the project.

ABOUT THE AUTHORS

Joan Z. Spade is professor and chair of sociology at the State University of New York at Brockport. She received her Ph.D. from State University of New York at Buffalo and her B.A. from State Unviersity of New York at Geneseo. In addition to gender, Joan teaches courses on education, family, research methods, and statistics. She has published articles on gender, including work on rape culture in college fraternities and research on work and family, including orientations towards work. She has also coedited two books on education and has published articles on education, including research on tracking and a piece on gender and education. Joan is active in Sociologists for Women in Society, Eastern Sociological Society, and the American Sociological Association.

Catherine (Kay) G. Valentine is professor of sociology and director of women's studies at Nazareth College in Rochester, New York. She received her Ph.D. from Syracuse University and her B.A. from the State University of New York at Albany. Kay teaches a wide range of courses, such as sociology of gender, senior seminar in sociology, introduction to women's studies, sociology of the arts, and human sexuality. Her publications include articles on teaching sociology, women's bodies and emotions, gender and qualitative research, and the sociology of art museums. She is the founding director of women's studies at Nazareth College and long-time member of Sociologists for Women in Society. She has served as president of the New York State Sociological Association. Kay and her spouse, Paul J. Burgett, University of Rochester dean and vice president, are devotees of the arts and world travel.

Introduction

This book is an invitation to you, the reader, to enter the fascinating and challenging world of gender studies. Gender is briefly defined as the meanings, practices, and relations of femininity and masculinity that people create as they go about their daily lives in different social settings. Although we discuss gender throughout this book, it is a very complex term to understand. Chapter 1 provides a more detailed discussion of what gender is and how it is related to biological maleness and femaleness. In thinking about the complexity of the meaning of gender from a sociological viewpoint, we find the metaphor of a kaleidoscope useful.

THE KALEIDOSCOPE OF GENDER

A real kaleidoscope is a tube containing an arrangement of mirrors or prisms that produces different images and patterns. When you look through the eyepiece of a kaleidoscope, light is typically reflected from the mirrors or prisms through object cells containing glass pieces, seashells, and the like to create ever-changing patterns of design and color (Baker, 1999). In this book, we use the kaleidoscope metaphor to help us grasp the complex and ever-changing meaning and practice of gender as it interacts with other social prisms—such as race, age, sexuality, and social class—to create complex patterns of identities and relationships. Three themes emerge from the metaphor of the kaleidoscope: prisms, patterns, and possibilities.

Part One of the book focuses on prisms. A prism in a kaleidoscope is an arrangement of mirrors that refracts or disperses light into a spectrum of patterns (Baker, 1999). We use the term prism to refer to socially constructed categories of difference and inequality through which our lives are reflected or shaped into patterns of daily experiences. In addition to gender, we define socially constructed categories such as race, age, physical ability, social class, and sexual orientation as social prisms. Culture is also conceptualized as a prism in this book, as we examine how gender is shaped across cultures. The concept of social prisms helps us to understand that gender is not a universal or static entity, but rather it is continuously created within the social parameters of individual and group life. Looking at the interactions of the prism of gender with other social prisms helps us to see the bigger picture—gender practices and meanings are a montage of intertwined social divisions and connections that both pull us apart and bring us together.

Part Two of the book examines the patterns of gendered expressions and experiences created by the interaction of multiple prisms of difference and inequality. Patterns are regularized, prepackaged ways of thinking, feeling, and acting in society. Gendered patterns are present in almost all aspects of daily life. In the United States, examples of gendered patterns include the association of the color pink with girls and blue with boys, and the disproportionate numbers of female nurses and male engineers (see Table 7.1 in this book). However, these patterns of gender are experienced and expressed in different ways depending on the other social prisms that shape our identities and life chances. For example, as you take a closer look at who is an engineer and who is a nurse (as discussed in Chapter 7), you will note that it is predominately white men who are engineers and white women who are nurses. Consequently, the patterns of gender are a result of the complex interaction of multiple prisms.

Part Three of the book concerns possibilities for gender change. Just as the wonder of the kaleidoscope lies in the ever-evolving patterns it creates, gendered patterns are always in flux. Each life and the world we live in can be understood as a kaleidoscope of unfolding growth and continual change (Baker, 1999). This dynamic aspect of the kaleidoscope metaphor represents the opportunity we have, individually and collectively, to transform gendered patterns that are harmful to women and men. Although the theme of gender change is prominent throughout this book, it is addressed specifically in Chapter 10 and the Epilogue.

One caveat must be presented before we take you through the kaleidoscope of gender. A metaphor is a figure of speech in which a word ordinarily used to refer to one thing is applied to better understand another thing. A metaphor should not be taken literally. It does not directly represent reality. We use the metaphor of the kaleidoscope as an analytical tool to aid us in grasping the complexity, ambiguity, and fluidity of gender. However, unlike the prisms in a real kaleidoscope, the meaning and experience of social prisms (e.g., gender, race, social class, and culture) are socially constructed and change in response to the patterns in the larger society. Thus, although the prisms of a real kaleidoscope are static, the prisms of the gender kaleidoscope are shaped by the patterns of society.

As you step into the world of gender studies, you'll need to develop a capacity to see what is hidden by the cultural blinders that we all wear at least some of the time. This capacity to see into the complexities of human relationships and group life has been called sociological imagination or, to be hip, sociological radar. It is a capacity that is finely honed by practice and training both in and out of the classroom. A sociological perspective enables us to see through the cultural smokescreens that conceal the patterns, meanings, and dynamics of our relationships.

GENDER STEREOTYPES

What will the sociological perspective help you to understand about gender that you don't already know? It will, for example, help you to debunk "gender stereotypes," which are rigid, oversimplified, exaggerated beliefs about femininity and masculinity that misrepresent most women and men (Walters, 1999). Let's work through an illustration by analyzing one gender stereotype that many people in American society believe—women talk more than men (Wood, 1999; Anderson and Leaper, 1998).

Social scientific research is helpful in documenting whether women actually talk more than men, or whether this belief is just another "gender stereotype." Social scientists study the interactions of men and women in an array of settings and count how often men speak compared to women. They almost always find that, on average, men talk more in mixed-gender groups (Wood, 1999). Researchers also find that men interrupt more and tend to ignore topics brought up by women (Wood, 1999; Anderson and Leaper, 1998). In and of itself, these are important findings—the stereotype turns reality on its head.

So, why does the stereotype continue to exist? First, we might ask how it is that people believe something to be real, such as the stereotype that women talk more than men, when, in general, it isn't true? Part of the answer lies in the fact that culture, defined as the way of life of a group of people, shapes what we experience as reality (see Chapter 3 for a more detailed discussion). As Allan Johnson (1997) aptly puts it, "Living in a culture is somewhat like participating in the magician's magic because all the while we think we're paying attention to what's 'really' happening, alternative realities unfold without even occurring to us" (55). In other words, we don't usually reflect on our own culture; we are mystified by it without much awareness of its bewildering effect on us. The power of beliefs, including gender beliefs, is quite awesome. By providing stereotypes along which perceptions proceed, beliefs shape reality.

A second question we need to ask about gender stereotypes is: what is their purpose? For example, do they set men against women and contribute to the persistence of a system of inequality that disadvantages women? Certainly, the stereotype that many Americans hold of women as big talkers is not a positive one. The stereotype doesn't assume that women are assertive, articulate, or captivating speakers. Instead it tends to depict women's talk as trivial gossip or irritating nagging. In other words, the stereotype devalues women's talk while, at the same time, elevating men's talk as thoughtful and worthy of our attention. One of the consequences of this stereotype is that both men and women take men's talk more seriously (Wood, 1999). This pattern is reflected in the fact that the voice of authority in many areas of American culture, such as television and politics, is almost always a male voice. The message that is communicated is clear—women are less important than men. In other words, gender stereotypes help to legitimize role and power differences between men and women.

However, stereotypical images of men and women are not universal in their application because they are complicated by the kaleidoscopic nature of people's lives. Prisms or social categories such as race/ethnicity, social class, and age intersect with gender to produce stereotypes that differ in symbolic meaning and functioning. For example, the prisms of gender and race interact for African-American and Hispanic men, who are stereotyped as less intelligent but more sexual than white men, as you will read about in Mirande's (Chapter 2) and Ferguson's (Chapter 4) readings in this book. These variations in gender stereotypes act as controlling images that maintain complex systems of domination and subordination in which some individuals and groups are dehumanized and disadvantaged in relationship to others (see Collins and other readings in Chapter 2).

THE EVOLUTION OF THE CONCEPT OF GENDER

Just a few decades ago, social scientists reduced gender to a set of two "sex" roles—male/masculine and female/feminine—that they believed were tied to innate personality characteristics and to biological sex characteristics such as hormones and reproductive functions (Kimmel, 2000; Tavris, 1992). For example, women were thought to be naturally more nurturing because of their capacity to bear children, and men were seen as natural leaders because they were unencumbered by pregnancy and childcare. This definition of masculine and feminine was one-dimensional, relatively static, and ethnocentric, and it is not supported by biological, psychological, sociological or anthropological research.

The definition of gender expanded as social scientists conducted research that questioned the simplicity and accuracy of the "sex roles" perspective. First, social scientists debunked the notion that biological sex characteristics cause differences in men's and women's behaviors (Tavris, 1992). For example, testosterone does not cause aggression in men (see Sapolsky in Chapter 1), and the menstrual cycle does not cause women to be more "emotional" than men (Tavris, 1992).

Second, social scientific research demonstrated that men and women are far more physically, cognitively, and emotionally alike than different. Then and now, what we perceive as natural differences are actually rooted in the asymmetrical and unequal life experiences, resources, and power of women compared to men (Tavris, 1992). For example, it was, and for many still is, a commonly held belief that biological sex is related to physical ability and that women are athletically inferior to men. That belief has been challenged by the outcomes of a recent series of legal interventions that opened the doors to the world of competitive sport to girls and women. Once legislation such as Title IX was implemented, the expectation that women could not be athletes began to change as girls and young women received the same training and support for athletic pursuits as men. Not surprisingly, the gap in physical strength and skills between women and men decreased dramatically. Today, women athletes regularly break records and perform physical feats thought impossible for women just a few decades ago (see Berlage in Chapter 7).

Third, social scientists documented patterns of gender inequality within the economy, family, religion, and other social institutions that benefit men as a group. To illustrate, in the 1970s when researchers began studying gender inequality, they found that women made between 60 and 70 cents for every dollar men made. Things are not much better today. In 2000, the median salary for women was 72 percent of that of men (U.S. Census, 2001).

Fourth, social scientists documented the fact that gender is "humanly-made" (Schwalbe, 2001). The theory that argues that gender is a human invention is called social constructionism. Understanding gender as a social construction means seeing the social processes by which gender is defined into existence and maintained or changed by human actions and interactions. Social constructionism will be discussed in more detail later in this chapter and throughout the book.

One of the most important arguments in support of the idea that gender is socially constructed is derived from cross-cultural studies. The variations and fluidity in the definitions and expressions of gender across cultures illustrate that the American gender system is not universal. For example, some cultures have more than two genders (see Nanda in Chapter 1). Other cultures define men and women as similar, not different (see Helliwell in Chapter 3). Still others view gender as flowing and changing across the life span (Herdt, 1997).

Fifth, as social scientists examine gender variations through the prism of culture, their research has challenged the notion that masculinity and femininity are defined and experienced in the same way by all people, even within the subcultures of a particular society. For example, the meaning of femininity in orthodox, American religious subcultures is not the same as femininity outside those communities (Rose, 2001). The differences are expressed in a variety of ways, including the clothing of women. Typically, orthodox religious women adhere to modesty rules in dress, covering their heads, arms, and legs.

Elaborating on the idea of multiple masculinities and femininities, the Australian sociologist, Robert Connell, coined the terms "hegemonic masculinity" and "emphasized femininity" to understand the relations between and among masculinities and femininities in patriarchal societies; that is, societies that are dominated by men. According to Connell (1987), hegemonic masculinity is the idealized pattern of masculinity in patriarchal societies, while emphasized femininity is the vision of femininity that is held up as the model of womanhood in those societies. In Connell's definition, hegemonic masculinity is about domination. Key features of hegemonic masculinity include the subordination of women, the marginalization of gay men, and the celebration of toughness and competitiveness (Connell, 2000). Emphasized femininity, in contrast, is about women's subordination, with its key features being sociability, compliance with men's sexual and ego desires, and acceptance of marriage and childcare (Connell, 1987).

Interestingly, according to Connell, hegemonic masculinity and emphasized femininity are not necessarily the most common gender patterns. They are, however, the versions of manhood and womanhood against which other patterns of masculinity and femininity are measured and found wanting (Kimmel, 2000). For example, hegemonic masculinity produces marginalized masculinities, which are, according to Connell (2000), characteristic of exploited groups such as racial and ethnic minorities. These marginalized forms of masculinity may share features with hegemonic masculinity, such as "toughness," but are socially debased (see Mirande in Chapter 2 and Ferguson in Chapter 4).

In patriarchal societies, the culturally idealized form of femininity, emphasized femininity, is produced in relation to male dominance and violence. Emphasized femininity insists on compliance, nurturance, and empathy as ideals of womanhood to which all women should subscribe (Connell, 1987). Connell does not use the term "hegemonic" to refer to emphasized femininity because, he argues, emphasized femininity is always subordinated to masculinity. Other versions of femininity are constructed in the context of patriarchy and in relation to emphasized femininity. As examples, Connell (1987) points to the resistant and marginalized patterns of femininity in the experiences of lesbians, prostitutes, so-called spinsters, and women who do highly masculinized work such as manual labor. Connell's work is important in helping us to understand that multiple masculinities and femininities are associated with complex systems of domination and subordination across and within gender categories.

Sixth, a major factor associated with multiple masculinities and femininities is the relationship of gender to other social categories of difference and inequality. Allan Johnson (2001) points out that

> categories that define privilege exist all at once and in relation to one another. People never see me solely in terms of my race, for example, or my gender. Like everyone else's, my place in the social world is a package deal—white, male, heterosexual, middle-aged, married, . . . —and that's the way it is all the time. . . . It makes no sense to talk about the effect of being in one of these categories—say, white—without also looking at the others and how they're related to it (53).

Seeing gender through multiple social prisms is critical, but it is not a simple task, as you will discover in the readings in Chapters 2 and 3. We need be aware of how other social prisms alter life experiences and chances. For example, although an upper-class, African-American woman is privileged by her social class category, she will face obstacles related to her race and gender. Or, consider the situation of a middle-class white man who is gay; he might lose some of the privilege attached to his class and race because of his sexual orientation.

Think about what happens when we look at the difference in men's and women's incomes when we include the prisms of race and ethnicity. As compared to the 72 cent difference in men's and women's incomes mentioned earlier, white non-Hispanic women who graduated from college and worked full-time, year around in 2000 made 69.9 cents for every dollar their male counterparts made (using the median income for both; see U.S. Census, 2001). The situation for African-American women is slightly better at 84.3 cents for every dollar African-American men made; however, this is because African-American, college-educated men who work full-time, year around earn considerably less than their white counterparts (median income $45,068 compared to $55,896 for white men, whereas African-American, college-educated women earn $39,122 compared to their white counterparts who earn $38,017; see U.S. Census, 2001). The ratio of female-to-male earnings for Hispanics is also better than that for non-Hispanic whites, but Hispanics earn less than either whites or African-Americans ($42,828 for men and $32,097 for women; see U.S. Census, 2001). We will explore many other examples of patterns of gender inequities that are a consequence of the interaction of multiple social prisms in this book. We also include theoretical frameworks to help you understand why these inequalities occur and what we need to do to change them.

THEORIES OF GENDER

Numerous theories explain how and why gender differentiation and inequality exist in most societies. We cannot describe them all, but we will briefly mention some of the theoretical approaches you will encounter in this book. We describe how theories explain gendered behavior at different, interrelated levels of social life: individual, situational/interactional, and institutional/societal.

Theories help us better understand gender; however, we want to emphasize that theories are not reality. Theories are explanations that exist in a conceptual world and allow us to make predictions and generalizations about human behavior. As such, we encourage you to apply more than one theory to explain the same gender phenomenon. As with the definition of gender, theories explaining why gender differences and inequalities exist in many societies have evolved over time, and continue to evolve. Many of the readings in this book expand upon or criticize the theories that follow. A primary reason for criticism is that many of these theories ignore the interaction among gender, race, ethnicity, class and other social categories, a point we will return to later.

Individual At the individual level, social psychologists and sociologists study the social categories and stereotypes that individuals use to identify themselves and label others. Gendered expectations, including stereotypes, are incorporated into how we define ourselves, and shape how we act and react as well as how others perceive us. Considerable research has studied how boys and girls (later researchers also included adults) are socialized into gender identities and practices (see Chapter 4 and the selection on Halloween costumes by Nelson in Chapter 5). While ultimately gendered identities are the result of larger societal and cultural forces, they are deeply felt, and individuals act upon them. Gender ideals and images are translated into self-identities (Howard and Alamilla, 2001).

Using theories that focus on the individual, how might we explain that women are less likely to become physicians than men? We would do so by concentrating on how people define themselves and what they believe they are capable of accomplishing. For example, in the early 1970s, Joan's daughter was playing hospital with her young friends. One doll was "very sick," so Joan suggested that one of the children play at being a doctor (all three were pretending to be nurses). Her daughter's reaction was adamant: "Girls can't be doctors!" She was most likely responding to cultural messages from toys, books, and the mass media of that era that depicted doctors as men. These cultural images were more powerful in shaping her gender expectations than reality because, at that time, her daughter's pediatrician was a woman. Fortunately, our identities and expectations can change; socialization is not indoctrination. Although Joan's daughter is not a doctor, she is now practicing medicine as a physician assistant. Furthermore, in 1999, 23.4 percent of physicians were women, compared to 7.7 percent in 1970 (American Medical Association, 2001). Gender socialization does not produce static outcomes for individuals or societies.

Interactional Of course, socialization does not happen in a vacuum, but occurs as part of social interaction. We do not become gendered alone. Theories such as symbolic interactionism and social constructionism help us to understand how developing a self-identity is a social process that involves incorporating the views of others into our own perspectives. According to West and Zimmerman (1987), gender is "done" in interaction with others in specific situations. They argue that the very process of interaction involves the presentation of a gendered self, the responses of the persons we are interacting with, and our reactions to their anticipated or actual responses. As such, gender is an ongoing activity that is carried out in interaction with other people, and people vary their gender presentations as they move from situation to situation. For example, an aggressive macho attitude might work on the football field, but not generally in an interview for a job. Lots of makeup, a strapless dress, and stiletto heels are okay at a formal dance, but not on a physician in the operating room of a hospital. The point is that sociologically speaking, people make gender happen through what they do and don't do in relation to others, and what they think is appropriate for different situations (see Lucal in Chapter 1).

Institutional Some theorists use social constructionism to explain gender at the institutional level. Acker (1992) argues that institutions are gendered because "gender is present in the processes, practices, images and ideologies, and distributions of power in the various sectors of social life" (567). Institutions maintain gender inequality through processes that exclude women. There are many examples of these processes, including outright exclusion as in the Roman Catholic priesthood, subtle exclusion through linguistic conventions that privilege masculinity (e.g., chairman, mankind), and the interactional exclusion of women, such as men dominating decision making processes (West and Zimmerman, 1987).

Like Acker, Judith Lorber (1994, 2001) also combines a social constructionist perspective with the concept of social institutions. She argues that gender be viewed "as a society-wide institution that is built into all major organizations of society": family, schools, mass media, and so on (2001:180). By viewing gender as an institution, Lorber sees it as a basis for inequality in society because it is through gender that resources, power, and privilege are distributed. Gender, according to Lorber, is "a process of creating distinguishable social statuses for the assignment of rights and responsibilities" (1994:32). As such, the structures of American society are built upon and reinforce the idea that women and men are distinct and unequal human types (Lorber, 1994). Without the socially constructed idea of clear differences between men and women, the entire social system of inequality and power would, according to Lorber, be in jeopardy.

Macro-level Theories Other frameworks that have historically examined gender at the macro-societal levels are conflict and functionalist theories. These theories have gone through many transformations since first proposed around the turn of the twentieth century. Scholars at that time were trying to sort out massive changes in society resulting from the industrial and democratic revolutions. These scholars sought answers in the structure of society.

Functionalism Functionalism attempts to understand how all parts of a society (e.g., institutions such as family, education, economy, and the polity or state) fit together to form a smoothly running social system. According to this theoretical paradigm, parts of society tend to complement each other to create social stability (Durkheim, 1933). Translated into gender relationships, Talcott Parsons and Robert Bales (1955), writing after World War II, saw distinct and separate gender roles in the heterosexual nuclear family as a functional adaptation to a modern, complex society. Women were thought to be more "functional" if they were socialized and aspired to raise children. And men were thought to be more "functional" if they were socialized and aspired to support their children and wives. These theorists argued that such a system would be beneficial because all societies must "replace" members if they are to survive. According to Parsons and Bales (1955), it is not functional if both women and men perform the same tasks. If men and women performed similar tasks in society, both would have to be socialized to desire a broader spectrum of activities, thus creating a greater demand on the social system and much more confusion in interpersonal relationships within families and workplaces. Functionalists are less concerned about workplace and other structural inequities between men and women, because such inequities assure that women will be more likely to stay home and care for children, an important function in society.

Conflict Theories Karl Marx and later conflict theorists, however, did not see social systems as functional or benign. Instead, Marx and his colleague Friedrich Engels described industrial societies as systems of oppression in which one group, the dominant social class, uses its control of economic resources to oppress the working class. The economic resources of those in control are obtained through profits gained from exploiting the labor of subordinate groups. Marx and Engels predicted that the tension between the "haves" and the "have nots" would result in an underlying conflict between these two groups. Most early Marxist theories focused on social class oppression; however, Engels (1942, 1970) wrote an important essay on the oppression of women as the earliest example of oppression of one group by another.

Feminist Theories Feminist theorists expanded upon the ideas of Marx and Engels, turning attention to the causes of women's oppression. One group, socialist feminists, continued to emphasize the role of capitalism in interaction with a patriarchal family structure as the basis for the exploitation of women. These theorists argue that economic and power benefits accrue to men who dominate women in capitalist societies. Another group, radical feminists, argues that patriarchy—the domination of men over women—is the fundamental form of oppression of women. Both socialist and radical feminists call for far-reaching changes in all institutional arrangements and cultural forms, including the dismantling of systems of oppression such as sexism, racism, and classism; by replacing capitalism with socialism; developing more egalitarian family systems; and other structural changes (e.g., Bart & Moran, 1993; Daly, 1978; Dworkin, 1987; MacKinnon, 1989).

Not all feminist theorists call for deep, structural and cultural changes. Liberal feminists are inclined to work toward a more equitable form of democratic capitalism. They argue that policies such as Title IX and affirmative action laws opened up opportunities for women in education and increased the number of women professionals, such as physicians. These feminists strive to achieve gender equality by removing barriers to women's freedom of choice and equal participation in all realms of life, eradicating sexist stereotypes, and guaranteeing equal access and treatment for women in both public and private arenas (e.g., Reskin & Roos, 1990; Schwartz, 1994; Vannoy Hiller, & Philliber, 1989; Weitzman, 1985; Steinberg, 1982).

Postmodernism In contrast to the theories just discussed, postmodernism focuses on the way knowledge about gender is constructed, not on explaining gender relationships themselves. To postmodernists, knowledge is never absolute. It is always situated in a social reality that is specific to an historical time period. Postmodernism is based on the idea that it is impossible for anyone to see the world without presuppositions. From a postmodernist perspective, gender is socially constructed through discourses, which are the "series of stories" that we use to explain our world (Andersen, 2004). Postmodernists attempt to "deconstuct" the discourses or stories used to support a group's beliefs about gender (Andersen, 2004; Lorber 2001). For example, Jane Flax argues that in order to fully understand gender in Western cultures, we must deconstruct the meanings in Western religious, scientific, and other discourses relative to "biology/sex/gender/nature" (in Lorber, 2001:199). As you will come to understand from the readings in Chapters 1 and 3 (e.g., Nanda and Helliwell), the association between sex and gender in Western scientific (e.g., theories and texts) and nonscientific (e.g., film, newspapers, media) discourses is not shared in other cultural contexts. Thus, for postmodernists, gender is a product of the discourses within particular social contexts that define and explain gender.

Intersectional Theories A major shortcoming with many of the theoretical perspectives just described is their failure to recognize how gender interacts with other social categories or prisms of difference and inequality within societies, including race/ethnicity, social class, sexuality, age, and ability/disability (see Chapter 2). A growing number of social scientists are responding to the problem of incorporating multiple social categories in their research by developing a new form of analysis, called multiracial theory or intersectional analysis, which is referred to as prismatic analysis in this book. Chapter 2 explores these theories of how gender interacts with other prisms of difference and inequality to create complex patterns. Without an appreciation of the interactions of socially constructed categories of difference and inequality, or what we call prisms, we end up with not only an incomplete, but also an *inaccurate* explanation of gender.

As you read through the articles in this book, consider the basis for the authors' arguments. What observations, data, or works of other social science researchers do these authors use to support their claims? Use a critical eye to examine the evidence as you reconsider the assumptions about gender that guide your life.

THE KALEIDOSCOPE OF GENDER: PRISMS, PATTERNS, AND POSSIBILITIES

Before beginning the readings that take us through the kaleidoscope of gender, let us briefly review the three themes that shape the book's structure: prisms, patterns, and possibilities.

Part One: Prisms Chapter 1 explores the meanings of the pivotal prism, gender, and its relationship to biological sex. Chapter 2 presents an array of prisms or socially constructed categories that interact with gender in many human societies, including race/ethnicity, social class, sexuality, age, and ability/disability. Chapter 3 focuses on the prism of culture/nationality, which alters the meaning and practice of gender in surprising ways.

Part Two: Patterns The prisms of the kaleidoscope create an array of patterned expressions and experiences of femininity and masculinity. Part Two of this book examines some of these patterns. We look at how people learn, internalize, and "do" gender (Chapter 4); how gender is exploited by capitalism (Chapter 5); how gender acts upon bodies, sexualities, and emotions (Chapter 6); how gendered patterns are reproduced and modified in work and play (Chapter 7); how gender is created and transformed in our intimate relationships (Chapter 8); and how conformity to patterns of gender is enforced and maintained (Chapter 9).

Part Three: Possibilities In much the same way as the colors and patterns of kaleidoscopic images flow, gendered patterns and meanings are inherently changeable. Chapter 10 examines the shifting sands of the genderscape and reminds us of the many possibilities for change. Finally, in the Epilogue, we examine changes we have seen and encourage you to envision future changes.

We use the metaphor of the gender kaleidoscope to discover what is going on under the surface of a society whose way of life we don't often penetrate in a non-defensive, disciplined, and deep fashion. In doing so, we will expose a reality that is astonishing in its complexity, ambiguity, and fluidity. With the kaleidoscope, you never know what's coming next. Come along with us as we begin the adventure of looking through the kaleidoscope of gender.

REFERENCES

Acker, Joan. (1992). From sex roles to gendered institutions. *Contemporary Sociology* 21(5): 565–570.

American Medical Association. (2001). Personal call with Tom Pasko, co-editor, taken from Physican Characteristics and Distribution in the United States, 2001–2002 Edition.

Andersen, Margaret L. (2003). Thinking about women: Sociological perspectives on sex and gender (6th Ed.). Boston: Allyn & Bacon.

Anderson, Kristin J. and Campbell Leaper. (1998). Meta-analysis of gender effects on conversational interruption: Who, what, when, where, and how. *Sex Roles* 39(3–4): 225–52.

Baker, Cozy. (1999). *Kaleidoscopes: Wonders of wonder.* Lafayette, CA: C & T Publishing.

Bart, Pauline B. and Eileen Geil Moran, Eds. (1993). *Violence against women: The bloody footprints.* Newbury Park, CA: Sage.

Connell, R. W. (2000). *The men and the boys.* Berkeley: University of California Press.

———. (1987). *Gender and power: Society, the person and sexual politics.* Stanford, CA: University of Stanford Press.

Daly, Mary. (1978). *Gyn/ecology, the metaethics of radical feminism.* Boston: Beacon Press.

Durkheim, Emile. (1933). *The division of labor in society.* Glencoe, IL: Free Press.

Dworkin, Andrea. (1987). *Intercourse.* New York: Free Press.

Engels, Friederich. (1942, 1970). *Origin of the family, private property, and the state.* New York: International Publishers Company, Inc.

Herdt, Gilbert. (1997). *Same sex, different cultures.* Boulder, CO: Westview Press.

Hill Collins, Patricia. (1990). *Black feminist thought: Knowledge, consciousness, and the politics of empowerment.* New York: Routledge.

Howard, Judith A. and Ramina M. Alamilla. (2001). Gender and identity. In *Gender mosaics: Social perspectives,* Dana Vannoy (Ed.), pp. 54–64. Los Angeles: Roxbury Publishing Company.

Johnson, Allan. (2001). *Privilege, power, and difference.* Mountain View, CA: Mayfield Publishing.

———. (1997). *The gender knot.* Philadelphia: Temple University Press.

Kimmel, Michael S. (2000). *The gendered society.* New York: Oxford University Press.

Lorber, Judith. (2001). *Gender inequality: Feminist theories and politics.* Los Angeles, CA: Roxbury Publishing Company.

———. (1994). *Paradoxes of gender.* New Haven, CT: Yale University Press.

MacKinnon, Catherine A. (1989). *Toward a feminist theory of the state.* Cambridge, MA: Harvard University Press.

Parsons, Talcott and Robert F. Bales. (1955). *Family, socialization, and interaction process.* Glencoe, IL: Free Press.

Reskin, Barbara F. and Patricia A. Roos. (1990). *Job queues, gender queues: Explaining women's inroads into male occupations.* Philadelphia: Temple University Press.

Rose, Dawn Robinson. (2001). Gender and Judaism. In *Gender mosaics: Social perspectives,* Dana Vannoy (Ed.), pp. 415–424. Los Angeles: Roxbury Publishing.

Schwalbe, Michael. (2001). *The sociologically examined life: Pieces of the conversation.* 2nd ed. Mountain View, CA: Mayfield Publishing.

Schwartz, Pepper. (1994). *Love between equals: How peer marriage really works.* New York: Free Press.

Steinberg, Ronnie J. (1982). *Wages and hours: Labor and reform in twentieth-century America.* New Brunswick, NJ: Rutgers University Press.

Tavris, Carol. (1992). *The mismeasure of woman.* New York: Simon & Schuster.

U.S. Census. (2001). Annual Demographic Survey, March Supplement. A Joint Project Between the Bureau of Labor Statistics and the Bureau of the Census. http://www.bls.census.gov. Downloaded 6/6/02.

Vannoy Hiller, Dana and William W. Philliber. (1989). *Equal partners: Successful women in marriage.* Newbury Park, CA: Sage Publications.

Walters, Suzanna Danuta (1999). Sex, text, and context: (in)between feminism and cultural studies. In *Revisioning gender,* Myra Marx Ferree, Judith Lorber, Beth B. Hess (Eds.), pp. 193–257. Thousand Oaks, CA: Sage.

Weitzman, Lenore J. (1985). *The divorce revolution: The unexpected social and economic consequences for women and children in America.* New York: Free Press.

West, Candace and Don H. Zimmerman. (1987). Doing gender. *Gender & Society* 1(2): 125–151.

Wood, Julia T. (1999). *Gendered lives: Communication, gender, and culture,* 3rd ed. Belmont, CA: Wadsworth.

CHAPTER 1

The Prism of Gender

In the metaphorical kaleidoscope of our book, gender is the pivotal prism. It is central to the intricate patterning of social life, and encompasses power relations, the division of labor, symbolic forms, and emotional relations (Connell, 2000). The shape and texture of people's lives are affected in profound ways by the prism of gender as it operates in their social worlds. Indeed, our ways of thinking about and experiencing gender, and the related category of sex, originate in our society.

As we noted in the Introduction, gender is very complex. In part, the complexity of the prism of gender in North American culture derives from the fact that it is characterized by a marked contradiction between people's beliefs about gender and real behavior. Our real behavior is far more flexible, adaptable, and malleable than our beliefs would have it. To put it another way, contrary to the stereotypes of masculinity and femininity, there are no gender certainties or absolutes. Real people behave in both feminine and masculine ways as they respond to situational demands and contingencies (Glick and Fiske, 1999; Tavris, 1992).

Two questions are addressed in this chapter to help us think more clearly about the complexity of gender: (1) how does Western culture condition us to think about gender, especially in relation to sex? and (2) how does social scientific research challenge Western beliefs about gender and sex?

Western Beliefs About Gender and Sex Most people in Western cultures grow up learning that there are two and only two sexes, male and female, and two and only two genders, feminine and masculine (Bem, 1993). We are taught that a real woman is feminine, a real man is masculine, and that any deviation or variation is strange or unnatural. Most people also learn that femininity and masculinity flow from biological sex characteristics (e.g., hormones, secondary sex characteristics, external and internal genitalia). We are taught that testosterone, a beard, big muscles, and a penis make a man, while estrogen, breasts, hairless legs, and a vagina make a woman. Many of us never question what we have learned about sex and gender, so we go through life assuming that gender is a relatively simple matter: a person who wears lipstick, high heel shoes, and a skirt is a feminine female, while a person who plays rugby, belches in public, and walks with a swagger is a masculine male (Lorber, 1994).

The readings we've selected for this chapter reflect a growing body of social scientific research that challenges and alters the Western model of sex and gender. Overall, the readings are critical of the American tendency to explain virtually every human behavior in individual and biological terms. As Jodi O'Brien (1999) points out, Americans tend to assume that answers to the complex workings of social relationships can be found in the bodies and psyches of individuals, not in culture or in the interaction between bodies and the environment.

Americans habitually overemphasize biology and underestimate the power of social facts to explain sex and gender (O'Brien, 1999). For instance, Americans tend to equate aggression with biological maleness and vulnerability with femaleness; natural facility in physics with masculinity and natural facility in childcare with femininity; and lace and ribbons with girlness and rough and tumble play with boyness (Glick and Fiske, 1999). The notions of natural sex and gender difference, duality, and even opposition and inequality permeate our thinking, color our labeling of people and things in our environment, and affect our practical actions (Bem, 1993).

We refer to the American tendency to assume that biological sex differences cause gender differences as "the pink and blue syndrome." This syndrome is deeply lodged in our minds and feelings and reinforced through everyday talk, performance, and experience. It's everywhere. Any place, object, discourse, or practice can be gendered. Children's birthday cards come in pink and blue. Authors of popular books assert that men and women are from different planets. People love PMS and alpha male jokes. In "The Pink Dragon Is Female" (see Chapter 5), Adie Nelson's research reveals that even children's fantasy costumes tend to be gendered as masculine and feminine. The "pink and blue syndrome" is so embedded within our culture, and consequently within individual patterns of thinking and feeling, that most of us cannot remember when we learned gender stereotypes and expectations or came to think about sex and gender as natural, immutable, and fixed. It all seems so simple and natural. Or is it?

What is gender? What is sex? How are gender and sex related? Why do most people in our society believe in the "pink and blue syndrome"? Why do so many of us attribute one set of talents, temperaments, skills, and behaviors to women and another set to men? These are the kinds of questions social scientists in sociology, anthropology, psychology, and other disciplines have been asking and researching for almost fifty years. Thanks to decades of good work by an array of scientists, we now understand that gender and sex are not so simple. Social scientists have discovered that the gender landscape is complicated, shifting, and contradictory. Among the beliefs that have been called into question by research are:

the notion that there are two and only two sexes and, consequently, two and only two genders; the assumption that men and women are the same everywhere and all the time; and the belief that biological factors cause the "pink and blue syndrome."

Using Our Sociological Radar Before we look at how social scientists answer questions such as "What is gender?", let's do a little research of our own. Try the following: relax, turn on your sociological radar, and examine yourself and the people you know carefully. Do all the men you know fit the ideal of masculinity all the time, in all relationships, and in all situations? Do all the women in your life consistently behave in stereotypical feminine fashion? Do you always fit into one as opposed to the other culturally approved gender box? Or are most of the people you know capable of "doing" both masculinity and femininity, depending on the interactional context? Our guess is that none of the people we know are aggressive all the time, nurturing all the time, sweet and submissive all the time, or strong and silent all the time. Thankfully, we are complex and creative. We stretch and grow and develop as we meet the challenges, constraints and opportunities of different and new situations and life circumstances. Men can do mothering; women can "take care of business." Real people are not stereotypes.

Yet even in the face of real gender fluidity and complexity, the belief in gender dichotomy and opposition continues to dominate almost every aspect of the social worlds we inhabit. For example, recent research shows that even though men's and women's roles have changed and blended, the tendency of Americans to categorize and stereotype people based on the simple male/female dichotomy persists (Glick and Fiske, 1999). As Glick and Fiske (1999) put it, "we typically categorize people by sex effortlessly, even nonconsciously, with diverse and profound effects on social interactions" (368). To reiterate, many Americans perceive humankind as divided into mutually exclusive, nonoverlapping groups: males/masculine/men and females/feminine/women (Bem, 1993). The culturally created image of gender is nonkaleidoscopic: no spontaneity, no ambiguity, no complexity, no diversity, no surprises, no elasticity, and no unfolding growth.

Social Scientific Understandings of Sex and Gender Modern social science offers us a very different image of gender. It opens the door to the richness and diversity of human experience, and it resists the tendency to reduce human behavior to single factors. Research shows that the behavior of real women and men depends on time and place, and context and situation, not on fixed gender differences (Lorber, 1994; Tavris, 1992). For example, just a few decades ago in the United States, cheerleading was a men's sport because it was considered too rigorous for women (Dowling, 2000), women were thought to lack the cognitive and emotional "stuff" to pilot flights into space, and medicine and law were viewed as too intellectually demanding for women. As Carol Tavris (1992) says, research demonstrates that perceived gender differences turn out to be a matter of "now you see them, now you don't" (288).

If we expand our sociological examination of gender to include cultures outside North America, the real-life fluidity of gender comes fully alive. (See Chapter 3 for detailed discussion.) In some cultures (for example, the Aka hunter-gatherers) fathers as well as mothers suckle infants (Hewlett, 2001). In other cultures, such as the Agta Negritos hunter-gatherers, women as well as men are the hunters (Estioko-Griffin and Griffin, 2001). As Serena Nanda notes in her reading in this chapter, there was extraordinary gender diversity expressed in complex sex/gender systems in many precontact Native American societies.

Context, which includes everything in the environment of a person's life such as work, family, social class, race and more, is the real source of gender definitions and practices. Gender is flexible, and "in its elasticity it stretches and unfolds in manifold ways" so that depending upon its contexts, including the life progress of individuals, we see it and experience it differently" (Sorenson, 2000:203). Most of us "do" both masculinity and femininity, and what we do is situationally dependent and institutionally constrained.

Let's use sociological radar again and call upon the work of social scientists to help us think more precisely and "objectively" about what gender and sex are. It has become somewhat commonplace to distinguish between gender and sex in the following way: sex is a biological fact, meaning that it is noncultural, static, scientifically measurable and unproblematic, while gender is a cultural attribute, a means by which people are taught who they are, how to behave, and what their roles will be (Sorenson, 2000). *However,* this mode of distinguishing between sex and gender has come under criticism, largely because new studies have begun to reveal the cultural dimensions of sex itself. That is, the physical characteristics of sex cannot be separated from the cultural milieu in which they are labeled and given meaning. For example, Robert Sapolsky's reading debunks the widely held myth that testosterone causes males to be more aggressive and domineering than females. He ends his article by stating firmly that "our behavioral biology is usually meaningless outside the context of the social factors and environment in which it occurs." In other words, the relationship between biology and behavior is reciprocal, both inseparable and intertwined (Yoder, 2003).

Sex, as it turns out, is not a clear cut matter of DNA, chromosomes, external genitalia, and the like, factors which produce two and only two sexes—females and males. First, there is considerable biological variation. Sex is not fixed in two categories. There is overlap. For example, all humans have estrogen, prolactin, and testosterone, but in varying and changing levels (Abrams, 2002). Think about this. In our society, people tend to associate breasts and related phenomena, such as breast cancer and lactation, with women. However, men have breasts. Indeed, some men have bigger breasts than some women, some men lactate, and some men get breast cancer. Also, in our society, people associate facial hair with men. What's the real story? All women have facial hair and some have more of it than some men. Indeed, recent hormonal and genetic studies (e.g., Beale, 2001; Abrams, 2002) are revealing that, biologically, women and men are far more similar than distinct.

Second, geneticists have identified at least five chromosomal types (XX, XY, and three intersex types), or what some researchers have come to call five sexes (Fausto-Sterling, 2000). Sharon E. Preves's reading offers exciting insights into the meanings and consequences of contemporary and historical responses to individuals who are intersexed.

Third, biological sex characteristics can and do change in interaction with environmental factors. For example, the average age of onset of first menstruation has changed rather dramatically in countries such as the United States. In the early nineteenth century, first menstruation typically happened at age fifteen or sixteen. Today, the average age of menarche is twelve. Why? The answer lies in vast changes in the socioeconomic environment, for example better diets and a decline in infectious diseases, which boosted the general health of young women and, in turn, altered their body timetables (Brumberg, 1997).

Biology is complicated business, and that should come as no surprise. The more we learn about biology, the more elusive and complex sex becomes. What seemed so obvious—two, opposite sexes—turns out to be an oversimplification. Humans are not unambiguous, clearly demarcated, biologically distinct, nonoverlapping, invariant groups of males and females.

So what is gender? First, gender is not sex. Biological sex characteristics do not cause specific gender behaviors or activities. As discussed earlier, biological sex is virtually meaningless outside the social context in which it develops and is expressed (Yoder, 2003). Second, gender is not an essential identity. It "does not have a locus nor does it take a particular form" (Sorenson, 2000:202). In other words, individuals do not possess a clearly defined gender that is the same everywhere and all the time. At this point, you may be thinking, what in the world are the authors saying? We are saying that gender is a human invention, a means by which people are sorted (in our society, into two genders), a basic aspect of how our society organizes itself and allocates resources (e.g., certain tasks assigned to people called women and other tasks to those termed men), and a fundamental ingredient in how individuals understand themselves and others ("I feel feminine." "He's manly." "You're androgynous."). In her clever reading in this chapter, Kate Bornstein plays with models of gender—a pacifier, a circle, a square, a pretzel, a pyramid—to underscore the illusory, yet powerful, nature of our binary gender system.

One of the fascinating aspects of gender is the extent to which it is negotiable and dynamic. In effect, masculinity and femininity exist because people believe that women and men are distinct groups and, most importantly, because people "do gender," day in and day out. The reading by Betsy Lucal illustrates vividly how gender is a matter of attribution and enactment. Some social scientists call gender a performance, others term it a masquerade. The terms performance and masquerade emphasize that it is through the ways in which we present ourselves in our daily encounters with others that gender is created and recreated.

We even do gender by ourselves and sometimes quite self-consciously. Have you ever tried to make yourself look and act more masculine or feminine? What is involved in "putting on" femininity or masculinity? Consider transvestism and cross-dressing. "Cross-dressers know that successfully being a man or a woman simply means convincing others that you are what you appear to be. Just ask Ru Paul, who seems to float almost effortlessly between the two" (Kimmel, 2000:104). Most people are not entertainers like Ru Paul who move, almost effortlessly, between gender categories. Most people have deeply learned gender, and most people come to view the gender box they inhabit as natural or the way things should be. Moving too far outside the walls of the box is very scary and, of course, if one steps far enough outside the box, the price one may have to pay—ridicule, ostracism, violence—may simply be too high.

You may be wondering why we have not used the term "role," as in "gender role," to describe "doing gender." The problem with the concept of roles is that typical social roles, such as those of teacher, student, doctor, or nurse, involve situated positions and identities. However, gender, like race, is a status and identity that cuts across many situations and institutional arenas. In other words gender does not "appear and disappear from one situation to another" (Van Ausdale and Feagin, 2001:32). People are always doing gender. They rarely let their guard down. In part, this is a consequence of the pressures that other people exert on us to "do gender" no matter the situation in which we find ourselves. Even if an individual would like to "give up gender," others will define and interact with that individual in gendered terms. If you were a physician, you could "leave your professional role behind you" when you left the hospital or office and went shopping or vacationing. Gender is a different story. Could you leave gender at the office and go shopping or vacationing? What would that look like, and what would it take to make it happen?

So far, we have explored gender as a product of our interactions with others. It is something we do, not something we inherit. Gender is also built into the larger world we inhabit, including its institutions, images and symbols, organizations, and material objects. For example, jobs, wages, and hierarchies of dominance and subordination in workplaces are gendered. Even after decades of substantial increase in women's workforce participation, occupations continue to be allocated by gender (e.g., secretaries are overwhelmingly women; men dominate construction work) and a wage gap between men and women persists (Bose and Whaley, 2001; Steinberg, 2001; Introduction to the book; Introduction to Chapter 7). In addition, men are still more likely to be bosses and women to be bossed. The symbols and images with which we are surrounded and by which we communicate are another part of our society's gender story. Our language speaks of difference and opposition in phrases such as "the opposite sex" and the absence of any words, except awkward medical terms (e.g., transsexual) or epithets (e.g., pervert), to refer to sex and gender variants. The swirl of gendered images is almost overwhelming. Blatant gender stereotypes still dominate TV, film, magazines, and billboards (Lont, 2001). Gender is also articulated, reinforced, and transformed through material objects and locales (Sorenson, 2000). Shoes are gendered, body adornments are gendered, public restrooms are gendered, weapons are gendered, ships are gendered, wrapping paper is gendered, and deodorants are gendered. The list is endless. The point is that these locales and objects are transformed into a medium for gender to operate within (Sorenson, 2000). They make gender seem "real," and they give it material consequences (Sorenson, 2000:82).

In short, social scientific research underscores the complexity of the prism of gender and demonstrates how gender is constructed at multiple, interacting levels of society. Gender cannot be reduced to one level: individual, interactional, or institutional. We are literally and figuratively immersed in a gendered world, a world in which difference, opposition, and inequality are the culturally defined themes. And yet that world is kaleidoscopic in nature. The lesson of the kaleidoscope is that "nothing in life is immune to change" (Baker, 1999:29). Reality is in flux; you never know what's coming next. The metaphor of the kaleidoscope reminds us to keep seeking the shifting meanings as well as the recurring patterns of gender (Baker, 1999).

We live in an interesting time of kaleidoscopic change. Old patterns of gender difference and inequality keep reappearing, often in new guises, while new patterns of convergence, equality, and self-realization have emerged. Social science research is vital in helping us to stay focused on understanding the prism of gender as changeable and responding to its context, as a social dialogue about societal membership and conventions, and "as the outcome of how individuals are made to understand their differences and similarities" (Sorenson, 2000:203–4). With that focus in mind, we can more clearly and critically explore our gendered society.

REFERENCES

Abrams, Douglas Carlton. (2002). "Father nature: The making of a modern dad." *Psychology today.* March/April, pp. 38–42.

Baker, Cozy. (1999). *Kaleidoscopes: Wonders of wonder.* Lafayette, CA: C&T Publishing.

Beale, Bob. (2001). "The sexes: New insights into the X and Y chromosomes." *The scientist* 15(15):18. http://www.the-scientist.com/yr2001/jul/research1_010723.html. Retrieved 7/23/2001.

Bem, Sandra L. (1993). *The lenses of gender.* New Haven, CT: Yale University Press.

Bose, Christine E. and Rachel Bridges Whaley. (2001). "Sex segregation in the U.S. labor force." In *Gender mosaics;* D. Vannoy (Ed.), pp. 228–239. Los Angeles: Roxbury Publishing.

Brumberg, Joan Jactobs. (1997). *The body project.* New York: Vintage Books.

Connell, R. W. (2000). *The men and the boys.* Berkeley: University of California Press.

Dowling, Colette. (2000). *The frailty myth.* New York: Random House.

Epstein, Cynthia Fuchs. (1988). *Deceptive distinctions.* New Haven, CT: Yale University Press.

Estioko-Griffin, Agnes and P. Bion Griffin. (2001). "Woman the hunter: The Agta." In *Gender in cross-cultural perspective.* C. Brettell and C. Sargent (Eds.), pp. 238–239. (3rd ed.). Upper Saddle River, NJ: Prentice Hall.

Fausto-Sterling, Anne. (2000). *Sexing the body: Gender politics and the construction of sexuality.* New York: Basic Books.

Glick, Peter and Susan T. Fiske. (1999). "Gender, power dynamics, and social interaction." In *Revisioning gender.* M. Ferree, J. Lorber, and B. Hess (Eds.), pp. 365–398. Thousand Oaks, CA: Sage.

Hewlett, Barry S. (2001). "The cultural nexus of Aka father-infant bonding." In *Gender in cross-cultural perspective.* C. Brettell and C. Sargent (Eds.), pp. 45–46. Third Edition. Upper Saddle River, NJ: Prentice Hall.

Kimmel, Michael S. (2000). *The gendered society.* New York: Oxford University Press.

Lont, Cynthia M. (2001). "The influence of the media on gender images." In D. Vannoy (Ed.). *Gender mosaics.* Los Angeles, CA: Roxbury Publishing.

Lorber, Judith. (1994). *Paradoxes of gender.* New Haven, CT: Yale University Press.

O'Brien, Jodi. (1999). *Social prisms: Reflections on the everyday myths and paradoxes.* Thousand Oaks, CA: Pine Forge Press.

Sorenson, Marie Louise Stig. (2000). *Gender archaeology.* Cambridge, England: Polity Press.

Steinberg, Ronnie J. (2001). "How sex gets into your paycheck and how to get it out: The gender gap in pay and comparable worth." In D. Vannoy (Ed.), *Gender mosaics.* Los Angeles: Roxbury Publishing.

Tavris, Carol. (1992). *The mismeasure of woman.* New York: Simon & Schuster.

Van Ausdale, Debra and Joe R. Feagin. (2001). *The first r: How children learn race and racism.* Lanham, MD: Rowman & Littlefield Publishers.

Yoder, Janice D. (2003). *Women and gender: Transforming psychology.* Upper Saddle River, NJ: Prentice Hall.

1

Who's on Top?

BY KATE BORNSTEIN

Kate Bornstein, a transgender performance artist and writer, identifies several models for conceptualizing how the power of the American gender system constrains individual expression and to structure hierarchies of privilege and power. At the foundation of her analysis is the cultural concept of "real men" and "real women," which she argues is a measuring stick against which real people are evaluated and mostly come up short.

1. What are the key elements of Bornstein's "Gender/Identity/Power Pyramid"?
2. Where do you fit in the Gender/Identity/Power hierarchy and why?
3. Why does Bornstein argue that being at the bottom of the pyramid is a valuable experience?

SAFETY FIRST

If real men and real women are in fact social constructs, that means they're constructed of *something.* I've been looking more and more closely at gender, and I think I've got a better idea of its components. . . . But, this is tricky stuff, and we need to proceed safely.

One of the first tenets of safety is awareness. Ask anyone who works with hir hands: you need to know where you are, where your tools are, what they're capable of doing, how they're capable of hurting you if they're misused. So the first step in safety around gender play would be to look at what gender is, what it's made up of, what tools we use to perform man, woman, or whatever it is we're performing. If we *haven't* looked very closely at whatever comprises our gender, we may change something about ourselves that we truly value.

HOW DO WE LOOK AT GENDER?

Since gender itself can't be seen, we tend to rely on models and metaphors. There are quite a few models these days for gender, none of which I've found adequate to describe the deathgrip gender-as-system has on us both personally and culturally.

- There's the old *binary* model: these are two completely opposite creatures, and only two, who have nothing to do with each other. It doesn't work. That particular metaphor leaves me and a whole lot of people out of the picture. Maybe you, too?
- There's a *yin/yang* model that looks like this: black and white, being opposites, form a circle with each other, and each half contains a portion of the other. That might be a real good metaphor for principles like "active" and "passive," but we still don't

know what comprises "black" and what comprises "white" when it comes to *gender.*

- The idea of a *continuum* is currently coming into vogue among those who study gender, as well as with some transgender activists. This is a better metaphor, allowing as it does for a wide range of genders along a pole, with *man/male* on one end, and *woman/female* on the other. I don't like it for two reasons: the man/male part, and the woman/female part. Seriously, why hold those two as fixed points by which we define ourselves, when we can't for sure say what the two fixed points are made up of?
- Several Native-American nations have described gender as a *circle,* and anyone can be anywhere on the circle of gender expression. That's about the closest I can agree with. It does away with any idea of a binary, but I still don't find it satisfying, because again there is no clearly defined marker on that circle against which I can measure myself.

These days, people are coming up with new, truly creative systems to describe or delineate gender. One such schematic can be found in the book *The Apartheid of Sex,* by Martine Rothblatt. Ze's come up with a fascinating and seductive way to determine gender, using a metaphor of colors. Rothblatt isolates what ze refers to as three basic elements of sexual identity (hir words), and assigns each of them a basic color. Hir three basic elements are: ". . . activeness (or aggression), passiveness (or nurturing), and eroticism (or sex drive)." The idea is that as each of the basic elements shift in proportion to the others, then the resulting color combination will change, providing a unique representation of that gender. A truly innovative theory. My opinion is that while Rothblatt's color wheel may in fact be an excellent way to construct our genders in a world that accepts hir three criteria as essential to gender (and why not? they are very loving criteria), we need something that illustrates the destructive construct of gender within the dominant culture (which seems to have more criteria than Rothblatt's three areas) more clearly, if for no other reason than that we might begin to dismantle it. *Then* we can look at the possibility of mindfully constructing a very beautiful gender system, using Rothblatt's model.

The gender-as-color model, while possibly something to aim for in the future, is missing an intermediary model, as it does not reflect the current world that's driven by binary thinking. Rothblatt's gentle vision does not express that which we need to first overcome: a world driven by greed, acquisition, and the very human need to belong to some exclusive (and excluding) group. I wanted to come up with a visual representation of gender *the way it is* in the world today, something beyond man/woman in a world that says there's no such thing as "beyond."

GENDER IS A PACIFIER! NAH. CLOSE, BUT NO CIGAR

The bipolar gender system serves as a kind of safe harbor for most of us, and I'm definitely including myself in that, even though I don't personally identify as either a man or a woman, because I walk though this world *appearing* to be a woman for the most part. I *pass* as a woman. I can do that. And I do it because it allows me to rest for a moment. I use my passing times as moments when I don't have to fight the good fight against gender tyranny. It's a safe harbor from all the jeers and oppression that attend gender transgression. But I'm reminded of the text I read once on a rather smarmy Hallmark-type poster: "Ships are safe in harbor, but that's not what ships are built for."

All right, then . . . how can we look at gender?

GENDER IS A CIRCLE! NO, NO . . . IT'S A SQUARE! NO, NO . . .

I like pretzels.

THAT'S IT! GENDER IS A PRETZEL!

No, no . . . gender isn't a pretzel.

IT'S NOT?

Well, maybe gender is a pretzel, but that's not the model I want to use. I *really* like pretzels. I eat boxes of them, the real salty hard sourdough variety from Pennsylvania, if you ever wanna send me some. And I was eating pretzels one day, and I was reading the back of the box, and lo and behold, there was the US Food and Drug Administration's Basic Food Group Pyramid. It was a true *eureka* moment, let me tell you. . . .

The really good stuff [in the pyramid] is at the bottom: grains, complex (interesting word) carbohydrates, stuff we're supposed to eat a lot of, stuff that's good for us. Fruits and veggies come next on the pyramid; they're very important. So are dairy and meats, but less so. And at the very tippy-top, there are fats, oils, and sugars. Bad bad bad . . . a little goes a lonnnnnng way, and they're not all that good for you, right? Well, my *eureka* moment with that box of pretzels was simply that gender is like that pyramid.

Works like this. We're accustomed to defining gender by some sort of biological component, be that hormones or chromosomes or genitalia or reproductive ability. We've defined gender by biology. Okay, that's old. It's too simple for today's complex world. Some "forward thinkers" have said, ahhhhhh, but that's *sex,* and sex is biology, but gender is psychology. That sort of thinking still posits something (sex) that's biologically essential. Right idea, wrong solution. Let's get off that merry-go-round once and for all. . . .

Some theorists are now saying that sociological factors have a lot to do with gender, and I think this is closer to the mark. If gender is a social construct, which includes respectful nods to biology, physiology, and psychology, then let's develop a model of gender that *demonstrates* that system.

The hallmark of today's *two*-gender system is that the preferred gender, the privileged gender, the gender that goes home with all the cookies, is labeled male. In sociological terms, though, we need to be more specific. The easy thing would be to say "It's all men." But it's not, not if this pyramid image reproduces the way gender structures hierarchies in the world today.

Examining the food-group pyramid we've got this wide base at the bottom: grains and carbohydrates. We're supposed to eat a lot of these things, they form the basis of our healthy diet. They're good for us. Working up the food group pyramid, we require fewer and fewer of these foods in our diet, until we get to the very top: sugars and oils. Don't eat too much, they're not all that good for you. Okay, I'm going to do some metaphor stretching. What if the pyramid represented humanity as it's living in our Western or Western-influenced cultures today, and the height of the pyramid was a function of power? The higher up on this pyramid you are, the more power you have to do whatever it is you want to do, and the more access you have to things like wealth, care, protection from harm and wrongdoing. And what if the very topmost point on the pyramid represented some perfect identity, some perfect gender, that we've all been taught to be, be like, or be liked by. All of a sudden, this pyramid idea of gender makes sense, and we have the representation of a graduated gender system that reflects power and privilege as it exists in the world. If the breadth of the pyramid reflects quantity of people, the model also interestingly enough seems to reflect relative numbers of people *belonging* to the different levels of this graduated gender system. The higher up we go on the pyramid, the fewer people there are belonging to that preferred gender.

The higher up we go, those fewer people have more and more power. . . .

THE CASE FOR A PERFECT GENDER

This phenomenon of graduated perfection in gender is easy to spot; it gets back to the troublesome concept of "real men" and "real women." In terms of gender, there will be in any group of men some who are going to be more "real" as men than others. Similarly, in any group of women, some are going to be more "real" as women than others.

For example, when I was growing up, I was a boy. In any mixed crowd of girls and boys, I was one of the guys. But, and this is very important, when I was with a group of boys only, I wasn't a boy—I was a Jew. Because of our age, none of us were "men." We were all boys. Is "boy" another gender? I think so. And among those boys, was I less male simply because of my ethnicity, my religion? Yup, I was less male.

I would even go so far as to say Jewish men are a different gender than Christian men, and that's the way I see it, but *it's not a bad thing!* It's just a fact. It's how Jews are perceived within the larger culture, and so it has some cultural weight. No doubt, some people are going to think I'm saying that Jewish men are *lesser men* than Anglo-Americans. No, no, no. That's an old anti-Semitic argument that survives to this day, but it's not how I see it—and it's certainly not limited to Jewish men. Moving up or down on this pyramid representation of the gender system is a function of *power,* not a function of humanity. What I'm saying is that there's a *difference;* and that by some standard, not mine, Jews are *judged* by those differences to be *less,* as are Asians, African-Americans, Hispanics . . . the list goes on and on, and even includes *most* Anglos (when you get into areas of age, class, education, appearance, social polish, etc., etc., etc.).

I'm not saying that each of these categories necessarily considers *themselves* to be less. I'm taking the point of view from the top of the pyramid, right? From that point of view there's an attitude of perfection, and everyone else is less, or Other. Everyone else is less perfectly gendered. What I'm thinking is that different kinds of men might as well be tagged as different *genders,* different ways of expressing oneself within some sort of male range, none of which measures up to the cultural ideal: the perfect gender. Wanting to be *considered* a "real man" by impossible standards keeps most men in the position of *supporting* the impossible standards. . . .

Your Gender is HOW Long?

What makes me curious is why anyone *cares* what a real man is or isn't. Why is that so important? What is it about the classification "man," or the category "woman" for that matter, that makes us so enraged (and we *do* get enraged) when someone accuses us of not being a "real" one or the other? *Those* are some valuable questions to ask, and once they're finally raised, we can begin to topple the system that's been keeping us bound up in living most of our lives running around in some hamster wheel, failing to measure up.

The Top of the Heap

I wanted to nail down this perfect gender, so that I'd know who or what it is I have to watch out for. Contrary to the laws of physics, I started building my gender pyramid at the top. I looked for the folks who have most of the power and wealth, the folks who claim to be the ideal, the *very* few people who can actually hold themselves up as REAL MEN in the world. More than that, I wanted to see if there was anyone who himself was not conflicted by that designation. Some guy who's got all the confidence in the world, and the power to back it up. I put him at the top of the pyramid because there *are* so few of him, and frankly, they're like the oils and sugar: take them in moderation. They have no nutritional value, they're not good for the culture. And we'll get to see exactly what they look like in just a bit. But first, let's see how *you* measure up to the idea of some perfect gender. . . .

THE GENDER/IDENTITY/ POWER SYSTEM

We're starting to define "gender" as a hierarchical dynamic masquerading behind and playing itself out through each of only two socially privileged monogendered identities. The power of this kind of gender perfection would be in direct proportion to the power granted those who can stake legitimate claims to those identities. The power is derived from the very invisibility of the gender/identity hierarchy. This makes gender, identity, and power each functions of each other, inextricably woven into the web of our culture beneath an attractive tapestry called the bipolar gender system. . . .

So what does the dude at the top of this pyramid look like? Remember, the height of the pyramid measures the amount of power a person wields in the world, and the breadth of the pyramid measures the number of people who wield that much power. I'm thinking the guy at the top looks like this:

- white
- citizen of the USA
- Protestant-defined Christian
- middle-aged
- middle- to upper-class
- heterosexual
- monogamous, monofidelitous
- able-bodied
- tall, trim, and reasonably muscled
- attractive, according to cultural standards
- right-handed
- well-educated
- well-mannered
- professional or executive level
- politically conservative
- capitalist
- self-defining and self-measuring
- physically healthy, with access to health care
- in possession of all rights available under the law
- free and safe access to all private and public areas as allowed by the law
- property-owning
- binary-oriented
- logical (linear thinking)
- uses power over others
- possessing a well-formed, above-average-length penis, a pair of reasonably matched testicles, and at least an average sperm count
- parent of more of the same

The reason I chose these categories out of the many possible is that each of these factors contributes to the amount of power a person currently wields in the world. If the gender/identity/power pyramid is to work as a metaphor, than each of these factors must truly be a measurement of not only how much power a person wields, but also how much of a real man or a real woman that person might be; how close to being the perfect gender that person is.

Each of these components or qualifications can, of course, be further qualified. In any group of people who meet these criteria, there will always be someone who is taller, more educated, has more or "better" sons, and is healthier, etc., etc., etc. . . . so the pyramid reaches its point with *some* hypothetical person at the very top who's "better" (more culturally valuable, has more power) than everyone else. Conversely, as the qualities defining this perfect, unnamed gender identity drop away from an individual, that individual's gender identity shifts itself downward to the bottom of the pyramid where there are more and more people with less and less power.

It stuns me that most everything in the culture forwards this ideal gender identity and its exclusivity. Nearly everything in the culture pushes us to:

- *be* some perfect gender (impossible for most of us)

- be *like* that gender (possible for a very few people)
- or be *liked by* that gender (possible to many, but not all people)

What's more, the further removed we are from the qualities expressed by the top of the pyramid, the less and less our gender is perceived as *real.* For example, if our genitals are in any way anomalous to the prescribed genitals for our gender, that obviously makes us unreal men or women, right? Similarly, if we're in our late teens or early twenties, we're told we're not-quite-men and not-quite-women; we're told we'll grow into that. . . .

THE CASE FOR A PERFECT IDENTITY?

I'm toying with the idea of putting the gender/identity/power pyramid into three dimensions, calling it simply an identity/power pyramid. It's tempting to call the top of this pyramid, the very tip-top, the Perfect Identity. Looking down from the top, from the viewpoint of this perfectly identified individual, each side of the pyramid can be defined by some aspect of classification *by the standards of the top's own claimed perfection.* That is, how do the folks at the very top see the rest of us, and does the very top of the pyramid reflect a possible common source of oppression for many if not most oppressed groups?

For example, the two-dimensional side we've been looking at is gender. Another side *could* be race. Another side *could* be age, or class, or religious beliefs. There are so many ways to classify people, but the top of this pyramid just might remain the same: the Perfect Identity. At the top we'd have the Perfect Gender *and* the Perfect Race *and* the Perfect Class. So, the culturally agreed-upon standards of perfection just might all converge into one identity that's got the bulk of the power in the world, and *that* identity relies on its granted perfection from each of the classifications that support it.

The posited "perfect identity," this powerful oppressive force made up of the composite perfections of all systems of classifications, has a lot of names today. Feminists call it MAN. Jews have called it GENTILE. African-American activists call it WHITE. Bisexuals, lesbians, and gays call it STRAIGHT. Transgendered folks are beginning to call it GENDERED. In this binary-slanted world, we keep naming our oppressor (some person or group who has more power than us and is using that power to withhold access, resources, or wealth) in terms of some convenient opposite. On the other side of the fence, we have a tendency to call *our* gender or identity the "good" gender or the "good" identity. "Transgendered is better than traditionally gendered because blah blah blah."

We have to knock that off, all of that good-and-bad way of thinking. It's a tactic of the privileged to name others by using themselves as a yardstick. We need to realize that no single attribute gives a person enough power to oppress us. No *single* quality of identity resulting in a privileged status gives a person enough power to keep the rest of us in thrall.

Just something to think about, that's all. Try it out for yourself. See what it's like to devise other faces to this pyramid, call them race, age, class, whatever. . . see if it holds up for you. . . .

IT'S JUST A METAPHOR

The benefit of any agreed-upon metaphor is that it gives people a framework around which to develop theories and question concepts. A good metaphor for a good metaphor might be a coatrack on which we get to hang our favorite ideas. The *danger* of an agreed-upon metaphor, including this pyramid model of the gender/identity/power system, is that it might inhibit the development of newer, more inclusive metaphors.

So while I think there are a lot of ways you *could* break the remainder of the pyramid into little components (like the top half of the pyramid is all the people with penises, and the top part of *that* is all the people who are white *and* have penises), I don't think that's the point. The point is that as soon as we fall away to any degree from the established, privileged norms, we start to become a less-than-perfect gender; our anxiety to attain that perfection and its attendant power increases as we continue to find ourselves short in any of its defining categories. . . .

A group member who abandons the group-sanctioned identity for any reason can raise questions within the group about the *value* of their chosen identity; and that can result in the group labeling the playful or questioning one as outcast.

It's all a matter of degree.

A Simple Scale of Cultural Rejection

- If we *question* gender, we might be considered eccentric.
- If we play with our gender presentation, we might provoke hostility.
- If we cross the sacred line of the biological imperative and alter ::gasp:: our genitals, we are in effect banished from the culture, unless we hide very well indeed.

How does this scale relate to the broader concept of identity? Very nearly one to one.

- If we simply *question* a basic identity, we might be considered eccentric.
- If we play with our presentation of some basic identity, we might provoke hostility.
- If we cross the sacred line of the bottom-line qualifier for some basic identity, we are in effect banished from the culture, unless we hide very well indeed. . . .

It's Lonely at the Bottom (and it's still worth it!)

Outcast, Unclean, Outlaw . . . all terms for one identity not too many people want, especially if there's no nearby group of similar outcasts in which we might claim membership. For me, it boils down to loneliness, and that gets back to an even more basic drive: connection with others. Doing any kind of gender play can result in some degree of outcast status. It can result in a pretty horrible loneliness, and part of preparing for playing with or even simply questioning gender is to prepare oneself to deal with that loneliness.

No, I don't know how to prepare anyone to deal with loneliness. The best I can do right now is let you know it's going to happen. I think dealing with loneliness, coming to terms with it, and ultimately embracing alone-ness is a journey we all need to make on our own.

I'm learning the old difference between lonely and alone.

Lonely, for me, means I want to belong to some crowd, and I don't belong. *Alone,* on the other hand, is the same sense of apartness, but without the overwhelming desire to *belong.* . . .

Alone is the way I try to experience life. But there's always a tug: I've always wanted to be included. I've always been tempted to settle myself into one identity and say to the group, "Hey! Now I'm one of you, now can I spend time with you?" I like companionship. I like hanging out with folks. I just don't want to lock myself into an identity in order to do it, and what I've done is move on when I no longer have any room to shift and grow.

How about you? . . .

GENDER IS PURE AND SIMPLE WHEN *YOU'RE* PURE AND SIMPLE, BUT WHO EVER REALLY IS?

Playing with *any* identity can be scary, but our fear seems proportional to how valuable the identity is to our existence. Pure identities (or identities that pass as pure) are valuable things. They're valuable to those who have them, because there's a sense that *someplace* will always be *home,* a space with others who claim similar pure identities. And our pure identities are valuable to others. We become easier to deal with. Other people know who we are. So we begin to lean into an identity, we support our lives on some identity, and when we or someone else starts to mess with it, then all of a sudden we've got something, this identity, to lose, and we get very protective not only of our own identity, but of the purity of that identity as a membership requirement for others. This might be how identity politics does itself in. We need to get past this. . . .

So Okay, Would You Sell Your *Identity* for a Million Dollars?

What value do you place on the factors that make up your identity? Would you sell part of your identity for, say, a cool hundred thousand?

What's your identity worth to you?

I wonder about all our identities, all the ways we say to ourselves "This is me, not you." They're so valuable to us, these markers, that despite knowing their shortcomings, we cling to them like children to a raft in the middle of a stormy lake. We need some sort of security, don't we? And gender is one identity, by reason of the agreed-upon bipolar system, to which we find it extremely easy to cling. Well, cling we do. But is that necessarily a good thing?

In this next exercise, let's find out in more detail what you *like* about the various components of your gender identity. That way, going into some changes that might allow you to better express yourself, you'll be aware of what you might lose, what you don't *have* to lose, and what you may have to give up.

1. How does being a member of your race enhance or detract from what you consider to be your gender?
2. In terms of your gender, what do you enjoy about being a member of the ethnic group you are?
3. How does being as old or as young as you are affect how you enjoy or dislike your gender?
4. Are there any rights, privileges, or maybe any freedom you get from being a gendered member of the class you belong to?
5. Being the gender you are presenting, does the kind of work you do give you anything particular that enhances that gender presentation, either materially or socially?
6. Is there something that gives you pride in your gender that you get from being as educated (or not) as you are?
7. Do your religious, philosophical, or spiritual beliefs make it good to be the gender you are?
8. Are your genitals a source of pleasure for you just the way they are? How?
9. Is there something about your body type, shape, abilities, or disabilities that you really like?
10. Is there some sort of security or safety you get out of defining yourself in terms of the gender or genders you are romantically attracted to?
11. What benefits might you get by naming yourself something based on what you like to do sexually?
12. What does simply being the gender you were assigned at birth give you?
13. Is there something you enjoy about being defined by or defining yourself by something you used to be? (e.g., an ex-nun, a former woman, a widow)
14. How do your politics or your political beliefs make being the gender you are pleasing to you?
15. For someone of your gender, is how much economic power you do or don't wield particularly pleasing or disappointing to you?
16. What is it you might not want to give up that you get from defining yourself by your physical health, or lack thereof?
17. What do you (or have you) like(d) about being known as Other?
18. What sort of comfort or security do you get from defining yourself by your relationship to someone else (familial or otherwise)?

19. Is there anything about being defined by others as something you don't consider yourself to be, that you would miss if people no longer did that?
20. What perks do you get from being on the outside of some group or identity that won't have you as a member?
21. Is there something about your particular astrological sign or some other identity-system category you fit in that reinforces the way you see your gender?
22. What gendered benefits do you derive by reason of your membership in some club, group, party, or organization?
23. In addition to those listed above, what benefits or perks do you get from other identities or self-descriptives you use for yourself?
24. In addition to those listed above, what privileges or comforts do you enjoy by reason of any other identities or descriptives that others define you by?

IN TERMS OF GENDER, UPWARDLY MOBILE IS A CONTRADICTION

People continue daily to buck different parts of the gender/identity/power system, with varying degrees of success. There's a genuine need to claim more rights and access, a very genuine need to free ourselves from what we perceive is holding us back, limiting our self-expression, or locking us out of a circle of people who have more privileges at the expense of many others. Unfortunately, I think, this need is translated into a move *upward* on the gender/identity/power pyramid, as if being squeezed into some teeny tiny pure and simple and perfect identity with no hope of change would be a fun thing. Rather than move upward in hopes of achieving one's fair share of the power and access, why not simply dismantle the pyramid itself? . . .

2

Sexing the Intersexed

An Analysis of Sociocultural Responses to Intersexuality

BY SHARON E. PREVES

The author takes us through the historical and contemporary responses of Westerners to individuals who are born intersexed. Prior to modern stigmatization and medicalization of intersexuality, hermaphrodites were regarded as akin to a third sex. For example, in twelfth-century European theological and medical writings, sex was defined as continuous, not binary, while in seventeenth-century France, hermaphrodites were permitted to marry a person "opposite" of "their predominant sex." By the late eighteenth century, the medical establishment was gaining authority over definition and treatment of the body. Today, dominant medical models view intersexuals as deviant and requiring medical intervention to be turned into "real females or males."

1. What percentage of children are born with intersexed bodies? Generally speaking, does intersexuality pose a danger to the physiological and psychological well-being of individuals with intersexed bodies?
2. What is current medical practice with respect to intersexed newborns, and what is the rationale for this practice?
3. What is the intersex social movement, and why has it arisen?

I explore here the social construction of gender in North America through an analysis of contemporary and historical responses to infants who are born genitally ambiguous, or intersexed (hermaphroditic).[1] Bodies that are sexually ambiguous challenge prevailing binary understandings of sex and gender. Individuals who are intersexed have bodies that are quite literally queer or "culturally unintelligible" (Butler 1993, 2). That is, their bodies do not conform to an overarching and largely unexamined social expectation that all humans belong to one of two clearly delineated sex categories, female or male (Wilson 1998).

While being born with ambiguous sexual organs indeed problematizes binary understandings of sex and gender, the majority of intersexed children do not require medical intervention for their physiological health (Diamond and Sigmundson 1997b; Dreger 1998b; Kessler 1998).[2] Nevertheless, the majority of these infants are medically assigned a definitive sex, undergoing surgery and hormone treatments to "correct" their variation from the anatomies expected by the designations of female and male. The impetus to control intersexual "deviance" stems from cultural tendencies toward gender binarism, homophobia, and fear of difference (Butler 1990, 1993).

According to Erving Goffman (1963), the great rewards associated with "normalcy" will motivate individuals to attempt to pass as normal, whether they are or not. The case of intersexuality provides a poignant example of normative expectations remaining unfulfilled. When presented with someone or something that disrupts our basic understanding of the human body, such as intersexuality, we are presented with a case of what Goffman would label as *stigma*. Prevailing sex assignment theory suggests the psychological necessity of correcting intersexual "deviance" to preclude such stigma (Money 1968, 1991a, 1995; Money and Ehrhardt 1972).

Current medical interventions on intersexed bodies rest on the fundamental assumption that without the medical alteration of genitals to aid in unambiguous sex/gender assignment, intersexuals will live a life of alienation and despair. There is limited empirical evidence to demonstrate this and some recent research and activism that contradict this mode of thinking.[3] As is the case with most medical conditions, there are no mandatory or legislative standards of care for intersex conditions, only the work of individual scholars and researchers to guide physicians on medical protocol. . . .

A CHANGING FIELD: INTERSEX MEDICAL MANAGEMENT

Although we lack conclusive frequency data on the prevalence of intersex, current estimates based on a review of recent medical literature suggest that approximately 1 or 2 per 2,000 children are born with bodies considered appropriate for sex assignment surgery (Fausto-Sterling 2000) and that nearly 2 percent are born with chromosomes or other nongenital features that could be considered intersexed (Blackless et al. 2000).[4] Prior research in this area has been predominantly biomedical, with an emphasis on the etiology of intersexed conditions and the protocol for and success of medical intervention. In recent years the scope and focus of this field have been changing rapidly and significantly, especially since the emergence and activism of several intersex advocacy groups in the 1990s and the emergence of research results that contradict prior findings regarding the success of sex assignment.[5] Prior to this activism and research, the medical "sexing" of intersexuals remained largely unchallenged. . . .

"TRUE SEX" AND OTHER CURIOUS NOTIONS: TENETS OF MEDICAL SEX ASSIGNMENT

The primary concern of intersex medicalization is the swift assignment of a genitally appropriate gender for an infant.[6] Families typically remain marginal in the decision-making process regarding evaluation and treatment, while the medical team retains nearly exclusive control over the situation (Lee 1994; Kessler 1998). For example, a common method of handling an ambiguous birth involves telling the parent(s) at the time of delivery that the child's genitals are not yet "fully developed" and then whisking the infant away for myriad medical diagnostic procedures to ascertain the "most appropriate" gender assignment (Schwartz Cowan 1992; Kessler 1998). Most urban hospitals consult with members of their own gender ambiguity teams. Typically, these teams comprise urologists, pediatricians, pediatric endocrinologists, surgeons, and, at times, social workers or psychologists (Kessler 1998, 1990). Doctors often do not inform parents that the sex of their child is in question. Instead, they proceed to "uncover" an infant's "true sex," revealing it for all the world to behold. According to Ellen Hyun-Ju Lee, "Genital ambiguity is presented as 'hiding' and underlying sex, yet to be 'discovered' by the physicians" (1994, 17). Indeed, serious ethical implications are raised by physicians' failure to fully disclose the sexual ambiguity of the intersexed infant.[7]

Medical discussions about sex assignment revolve around infants' genitals, chromosomes, hormones, and gonads, with special emphasis given to the potential for appearance of genital, gender, and heterosexual normalcy in adulthood. The focus on genital appearance and size is so great, in fact, that normative ranges for infant clitoral and penile size have been codified. The range for medically acceptable clitoral size is between 0–0.9 centimeters, that is, three-eighths of an inch. Any phallus larger than 0.9 centimeters is considered too large and therefore unacceptable by Western clitoral standards. Thus, according to current medical standards, the overbearing clitoris must be "receded" or "trimmed back" despite potential loss of sexual function or other possible iatrogenic consequences. Conversely, to be considered a penis within this model, an organ must be at least 1 inch long, that is, at least 2.5 centimeters in length (Kessler 1997–98, 1998).[8]

The vast majority of intersex children are sexed as female. According to Patricia Donohoe, David Powell, and Mary Lee, "the decision to raise a child with male pseudohermaphroditism [XY sex chromosomes with ambiguous or female genitalia] as male or female is dictated entirely by the size of the phallus" (1991, 537). In other words, if an intersexed child has a Y chromosome, its "phalloclit" is examined for its social potential (read future size and sexual adequacy) as a penis.[9] Conversely, if a child has no Y chromosome, or a small or "inadequate" penis, the child is sexed as female. There is also great concern that a child with a male sex assignment is able to urinate from the tip of his penis while standing.[10] Thus, a child's genotypic makeup may be male, but unless the medical team deems the infant's phallus to be of adequate size, capable of "proper" urination while standing, and likely to pass as (hetero)sexually "normal," the child will likely be surgically and hormonally constructed as female (Donohoe, Powell, and Lee 1991; Kessler 1998; Slaughenhoupt and Van Savage 1999). According to a widely cited article from the *Johns Hopkins Magazine,* which includes interviews with leading intersex specialists Gary Berkovitz, John Gearheart, and Claude Migeon, "In truth, the choice of gender still often comes down to what the external genitals look like. Doctors who work with children with ambiguous genitalia sometimes put it this way, 'You can make a hole [vagina] but you can't build a pole [penis].' Surgeons can decrease the size of a phallus and create a vagina, but constructing a penis that will grow as the child grows is another matter" (Hendricks 1993, 15). . . .

The emphasis on infants' phallic size as a correlate of adult phallic capacity is questionable as the normative ranges of infant phallic and clitoral size have been shown to be unrelated to adult genital size (Lee 1994). In addition, initial studies on boys raised with micropenises show that if their small penises are left intact, these boys fare quite well socially and sexually—far better than if they had feminizing genitoplasty.[11]

THE SOCIAL CONSTRUCTION OF DISEASE AND CURE

In the 1970s, medical sociologists expanded their scope to include the study of natural human experiences labeled as "medically problematic." Irving Zola first labeled this phenomenon *medicalization* (Bell 1990). The term *medicalization* refers to viewing a natural phenomenon in a medical framework where the medical view is seen as the authoritative, if not hegemonic, view (Conrad and Schneider 1992). Once a phenomenon is seen through this medical lens, medical treatments may seem logical. An example of the medicalization of intersex was evident in a recent episode of the popular television hospital drama *Chicago Hope.* In this episode, following the birth of a healthy intersexed infant, the delivering physician immediately consulted a pediatric endocrinologist and urologist to ascertain the baby's sex. The medical team treated the infant's sexual ambiguity as a medical emergency and, in the end, opted for a female sex assignment and genitoplasty due to the small size of the infant's phallus. The necessity of surgical intervention was never discussed, it was simply presumed.

Medicalization both increases the range of social phenomena linked to medicine and those interpreted through the concepts of disease and treatment (Crawford 1980; Conrad and Schneider 1992). Zola (1990) notes the spread of medicalese into the language and thinking of the lay person. For example, when referring to newborns that are intersexed, clinicians typically describe an ambiguous phalloclit as an hypertrophied (enlarged) clitoris or a micropenis and use scientific instruments to measure and classify the ambiguous anatomy in question (Kessler 1997–98). This is especially notable when intersexuals talk about their own experiences using medical jargon, referring to their "conditions" or "endocrine disorders." The use of medical jargon reflects the widespread acceptance of a medical paradigm, which makes it difficult for lay persons to question medical opinion or authority.

The disparity in status between patient and doctor is furthered by the elevation of doctors to privileged "knowers" of medical treatments and cures, making patients more dependent upon the specialized knowledge of their doctors. In this way, the physician is seen as healer while the patient and family are relegated to a dependent position and in need of a professional cure (Fisher and Groce 1985). This shift in power of sexing from the family and intersexed individual her/himself to medical committees on gender ambiguity is evident in the history of hermaphroditic treatment.[12] In Alice Dreger's words, "By consulting with medical men, hermaphrodites supplied an acknowledgement of the medical men's authority, confirmation that the medical men were indeed the just and trustworthy arbiters of pathology and identity" (1995a, 57). In this way, as Goffman (1959, 1963, 1982) and others have demonstrated, the meaning of sexual ambiguity is created collectively by all the participants in social interaction and social discourse. It is certainly not just the "medical men" who desire the simplicity of gender binarism. Persons who are sexually ambiguous and their families (may) also desire some semblance of normalcy in terms of sociocultural expectations of sex and gender (Dreger 1995a). For example, prior to the medicalization of hermaphroditism in the West, intersexuals attempted to pass as "normal" through the use of observable signifiers of gender such as clothing and accessories, occupation, and sex/gender of partner. Most members of the population continue to use socially recognized gender signifiers as a means of identity expression in contemporary times. . . .

A HISTORICAL ANALYSIS OF INTERSEX MEDICALIZATION

. . . Cary Nederman and Jacqui True (1996) review twelfth-century theological and medical writings that characterize sex as continuous rather than binary. Unlike later, twentieth-century contentions that an intermediate intersexed state is not truly possible (Dreger 1995b; Nederman and True 1996), it appears that, during the twelfth century in Western Europe hermaphrodites were regarded as a discrete third sex.[13] This idea was based on Galen's second-century theory claiming that there were gendered differences in body temperature, especially in the reproductive and sexual organs ([170 C.E.] 1916).[14] According to this theory, sex distinctions were based on a continuum of heat, with males being internally hotter than females, thus creating the impetus for male external reproductive organs and female, "colder," internal organs.[15] In Thomas Laqueur's (1990) analysis, this differential temperature theory actually provided the basis for a one-sex conceptual model, with females being seen as the inverse of males. (That is, the vagina was viewed as an inverted penis, the uterus as a scrotum, the fallopian tubes as seminal vesicles, the ovaries as internal testicles, and so on.)

In the early seventeenth century, scientific thought about apparent females masculinizing at puberty was associated with Galen's temperature model. The esteemed seventeenth-century surgeon Ambroise Pare viewed an excessive amount of internal heat as the cause of this female pubescent masculinization. According to Pare (1634), this heat was typically brought on by the activities of children, such as jumping and playing roughly, which then led to the "pushing out" and transition of internal female organs into external male organs. Laqueur disagrees with both Pare and Galen. He claims that Pare's theory is based on "reading" male anatomy onto female bodies and that Galen viewed men not only as "hotter" than women but also as superior to them.

The widely read and highly acclaimed pseudo-Galenic *De Spermate* further defined sex differentiation in reproduction. According to *De Spermate,* both the male "seed" and the female uterus played active roles in determining the sex of the offspring. The uterus and testicles were seen as divided into hot (right) and cold (left) sections, with both having a mysterious indeterminate middle section. When male "seed" was planted in the warmer, right section of the uterus, the baby was a boy. A girl was produced by implantation in the colder, left side of the uterus. If the "seed" planted itself in the midsection, a hermaphroditic baby was produced (Cadden 1993; Nederman and True 1996). In this sense, with the uterus having a separate, "neutral" chamber for nurturing hermaphroditic fetuses, hermaphroditism was conceptualized as a natural, if not expected, state (Jones and Stallybrass 1991; Moscucci 1991; Nederman and True 1996). Further evidence of the definition of three distinct sexual categories comes from the late-twelfth-century Italian civil jurist Portius Azo, who wrote in his *Summa Institutionum* (1610): "There is another division between human beings, namely that some are male, others are female, others are hermaphrodites" (quoted in Nederman and True 1996, 512).[16] Although attitudes toward hermaphroditism varied widely in twelfth-century Europe, from viewing hermaphrodites as monsters to seeing hermaphroditism as natural, hermaphroditism was seen as a separate, third sex category (Nederman and True 1996).

Since this period precedes the separation of church and state, the writings of theologians had great influence. On the moral regulations applied to the hermaphrodites of the twelfth century, the Parisian, Peter the Chanter, wrote:

> The church allows a hermaphrodite—that is, someone with the organs of both sexes, capable of either active or passive functions—to use the organ by which (s)he is most aroused or the one to which (s)he is most susceptible. If (s)he is more active, (s)he may wed as a man, but if (s)he is more passive, (s)he may marry as a woman. If, however, (s)he should fail with one organ, the use of the other can never be permitted, but (s)he must be perpetually celibate to avoid any similarity to the role inversion of sodomy, which is detested by God. (Quoted in Nederman and True 1996, 510–11)[17]

This widely accepted tolerance of hermaphrodite choice in her/his sex/gender/sexuality stands in striking contrast to the rigidity of contemporary sex assignment.[18] It appears that although physiological ambiguity was allowed to persist in premedicalized times, the forms in which sexual identity could be displayed or expressed were culturally mandated.

In contemporary and historic times, legal concerns related to sex ambiguity are located within a system of gendered rights and obligations. In their normative breaching, "Hermaphrodites highlight the privilege differential between male and female precisely because they cannot participate neatly in it" (Epstein 1990, 124). Throughout history, family, church, and state have exercised control over overt expressions of gender such as choice of occupation, gender of marital/sexual partner, and type of clothing as a means to distinguish between women and men and to decrease the sex/gender/sexual ambiguity presented by hermaphrodites (Pagliassotti 1993). For example, in the mid-twelfth century, a person's ability to serve as a legal witness depended on her/his predominant gender. As women were excluded from providing court testimony, voting privileges, and property rights, hermaphrodites who presented as more female than male were also precluded from exercising the legal rights accorded to males at the time (Nederman and True 1996).

Early surgical attempts to solder sex, such as lowering abdominal testicles, appeared in the beginning of the nineteenth century (Pagliassotti 1993). A primary motive for the social insistence upon outward displays of gender clarity was fear of homosexuality, or hermaphroditism of the soul (Hekma 1994), a threat that was present in the sexually ambiguous (or, quite literally, *bisexual*) body of the hermaphrodite. By appearing, outwardly, to be of the "other" sex, it was feared hermaphrodites would tempt heterosexual partners into homosexual relations (Dreger 1995a).

The legal motivation for making precise sex distinctions was, and is, grounded in a morally based attempt to preserve the heterosexual institution of marriage, which is predicated on sex difference.[19] By the end of the eighteenth century, "The sex of husband and wife was beset by rules and recommendations. The marriage relation was the most intense focus of constraints; it was spoken of more than anything else" (Foucault 1978, 37).

Regardless of the time period, demarcations for lawful marital unions are precise even when legal definitions of sex are lacking. According to Roger Ormrod, "To constitute a valid marriage the parties must be of different sexes, for the simple reason that is what the word [marriage] means" (1972, 85).[20] Although overt legal discourse surrounding the validity of marital unions concentrates on sex (as in genitals, gonads, and chromosomes), the underlying motive for the insistence upon "opposite" sex wedlock appears to be social insurance against sodomy. This is evidenced by legal clauses relating to the traditional penile-vaginal heterosexual consummation of marriage vows. Ormrod (1972) and Donna Hawley (1977) note that an inability to consummate a marriage in this manner provides legal grounds for annulment.

Additional reasons for requiring legal registration and classification of sex at birth include the prevention of fraud; restriction for the carrying out of sex-specific rights, duties, and obligations; and the preservation of morality and family life (otherwise known as the prevention of homosexual relations) (Capron and D'Avino 1981). In Alexander Capron and Richard D'Avino's words, "To enforce the prohibition (and, incidentally, public sexual displays) between members of the same sex, society needs a legal means of classifying all individuals by sex" (1981, 220). In the same article, the authors reiterate the importance of maintaining a stable sex once it is proclaimed: "The state's insistence on a dual classification [of sex] is usually accompanied by a requirement of *permanency* in the designation" (Capron and D'Avino 1981, 221). In fact, transgressions of sex/gender stability have been met historically by punishments as severe as death.

In early seventeenth-century France, hermaphrodites were allowed to marry a person "opposite" of their predominant sex. Once a hermaphrodite made this type of visible gender choice, they were morally and legally expected to adhere to that decision—to uphold the ascribed social status in gender.[21] Echoing Peter the Chanter's statement above, if a hermaphrodite turned against their sex/gender decision in a sexual relationship, they were charged with sodomy, publicly whipped, hanged, and/or burned (Jones and Stallybrass 1991; Trumbach 1994; Nederman and True 1996). In contrast with twelfth-century allowance of social gender choice, in eighteenth- and nineteenth-century western Europe, female pseudohermaphrodites who gained access to male privilege through outwardly male displays such as marriage to women and performing "male" occupations were often charged with fraud or usurping male privilege and subjected to public punishment such as whipping, hanging, and burning (Epstein 1990; Jones and Stallybrass 1991; Trumbach 1994).

IDENTITY POLITICS AND THE INTERSEX SOCIAL MOVEMENT

In her 1993 article dealing with sexual categorization, Anne Fausto-Sterling stated that female and male sex categories exclude intersexuals. Using a taxonomic system established by British doctors George F. Blacker and T. W. P. Lawrence, she posed a five-sex system comprised of categories for female, female pseudohermaphrodite, true hermaphrodite, male pseudohermaphrodite, and male.[22] Kessler (1990, 1997–98, 1998) suggests that instead of seeing intersexed genitals as aberrant, we should respond to them as one variety of genital possibilities. Foucault went as far as to question whether we really needed to bother with establishing one's "true sex" (1980).

Recently, transgender activists have been quite visibly traversing, bridging, and blurring categories of sex, gender, and sexual attraction (Stone 1991; Bornstein 1994; Feinberg 1996). According to self-identified transactivist Leslie Feinberg, "Trans*gender* people traverse, bridge, or blur the boundary of the *gender expression* they were assigned at birth" (1996, x; emphases in original).[23] Persons who identify as transgendered (i.e., persons who move across or beyond the standard binary categories of gender) have garnered significant attention in the 1990s as they have mobilized and created a social movement of identity politics. In this work transgendered individuals are participating in "social movements which seek to alter the self conceptions and societal conceptions of their participants" (Anspach 1979, 765). Rather than attempting to adapt and conform to normative expectations of gender and sexual expression, transactivists seek to normalize transgenderism.

While traditional transsexual hormonal and surgical interventions conformed to gender binarism, the recent transgender movement challenged the necessity of matching one's genitals to one's gender, effecting a paradigm shift in the way transgender identities are conceptualized (Bockting 1995, 1997a, 1997b). Today, a number of individuals born male and female are living differently gendered lives without the aid of hormonal or surgical treatments and with less emphasis on passing as standard women and men.[24] Many transsexuals are choosing to live with a blended gender identity that blurs the boundaries of male and female. . . .

As I have noted, many individuals are now questioning the ethics and effectiveness of medical sex assignment procedures. From a biological perspective, Diamond and Sigmundson emphasize the importance of biological factors over postnatal socialization.[25] From a social constructionist perspective, Kessler (1998), Dreger (1998c), and Fausto-Sterling (2000) argue that intersex ought to be demedicalized as it is not in itself pathological. Rather, the pathology lies in the social system and its strict adherence to gender binarism. They contend that in a culture where childbirth is treated medically, intersex variation is seen within a medical framework—that is, as a disease requiring medical intervention. In addition, many individuals are questioning the ethics and effectiveness of current North American medical intervention on intersexed children,[26] and intersex individuals are openly expressing their dissatisfaction with this medical intervention.[27] Their dissatisfaction centers on the following issues: (1) certain medical interventions, from a strictly physical point of view, are not necessary for survival;[28] (2) as a result of medical intervention, both sexual and psychological satisfaction and functioning are often impaired; and (3) for intersex individuals the lack of open discussion of their intersex status results in feelings of shame and isolation. Evidence of their discontent is abundant in the writings and personal stories of intersexuals (Horowitz 1995). For example, in Morgan Holmes's words, "Not that I would necessarily have kept my phalloclit. . . . But I would have liked to have been able to choose for myself. I would have liked to have grown up in the body I was born with. . . . But physically, someone else made the decision of what and who I would always 'be' before I even knew who and what I 'was.' . . . [The doctors] used surgical force to make my appearance coincide with the medical and social standards of a 'normal' female body, thereby attempting to permanently jettison any trace of intersexuality" (1994, 29–30). . . .

Despite the medical aim to erase or hide sexual ambiguity, some intersexuals are finding each other through their own activism and are attempting to reclaim their difference as prideful. In fact, in the last twelve years intersexuals have been implementing their own networks of support and avenues for social change at a rapid pace. Here I provide a brief overview of a few of the significant developments on this front.[29]

The Turner's Syndrome Society, founded in Minneapolis in 1987, was the first-known support group for persons with atypical sex differentiation. This organization was founded by women who have Turner's syndrome and currently serves thousands of members nationally as a medical information clearinghouse.[30] A year later, in 1988, the mother of a girl with androgen insensitivity syndrome (AIS) founded the U.K.-based AIS Support Group.[31] At the time of this writing, the group has chapters in five countries and a widely distributed newsletter called *ALIAS.* In 1989 the mother of a child with Klinefelter's syndrome founded the U.S.-based K. S. & Associates.[32] This organization now serves over one thousand families.

In 1993 intersex activist and scholar Cheryl Chase founded the Intersex Society of North America (ISNA). Chase initially announced ISNA's existence by publishing a letter to the editor in response to Fausto-Sterling's article, "The Five Sexes" in *Sciences* (Chase 1993). In this letter, Chase listed a post office box for ISNA, which soon began filling with mail from other intersexuals around the world. The Intersex Society of North America published the first issue of its newsletter, *Hermaphrodites with Attitude,* in the winter of 1994, started bimonthly support groups in January 1995, and went on-line with an Internet Web site in January 1996 (Chase 1997, 1998b). In addition to the support it offers to intersexuals and their families, ISNA's mission is to destigmatize intersexuality and to legitimate genital variability (Kessler 1997–98). In a political move, the ISNA linked its objective of preventing genital surgery to the emotional and political movement against female genital mutilation. Since the October 1996 U.S. federal legislation banning female genital mutilation, ISNA has been lobbying against "intersex genital mutilation" (Chase 1997, 1998b; Harvey 1999). The society has approximately 1,500 members.

In 1995 the mother of an intersexed child formed the Ambiguous Genitalia Support Network (AGSN), an organization that fosters pen pal relationships among parents of intersexed children. In 1996, another mother founded the Hermaphrodite Education and Listening Post (HELP). This group offers peer support and medical information to its members. Both AGSN and HELP were started by mothers who were dissatisfied with intersex medical protocol. Both had children who at birth had male chromosomes and gonads and genitals deemed inadequate to be male. Each mother countered medical advice for feminizing surgery and is raising her young son as a boy. More recently, in 1998, an intersexed woman formed the U.S.- and Christian-based Intersex Support Group International. In support of its opposition to intersex medicalization, this group cites biblical passages referring to the sanctity of all God's creations.

Although the objectives of the above groups differ, all provide information and support to their members. The Intersex Society of North America has been most visible and vocal in its mission and significantly influenced scholars' suggestions for clinical change (Diamond and Sigmundson 1997a; Kipnis and Diamond 1998). The ISNA's model for treatment is as follows: (1) avoid surgery unless there is imminent medical risk to the infant; (2) provide in-depth counseling for the entire family of the infant; (3) insist that this counseling be given by qualified mental health professionals well versed in issues relevant to sex therapy and sexology; (4) provide age-appropriate information about medical care, diagnoses, and support patient autonomy, allowing intersexuals full knowledge of their situation and access to medical records; (5) provide peer support to intersexuals by connecting them to one of the many available support networks; and (6) provide access to surgical and/or hormonal medical technologies only after attaining the informed consent of the intersexual her/himself.[33]

Social Movements and Identity Formation

I would be remiss in not commenting on the particular sociocultural context within which this new intersex identity is emerging. First, the intersex social movement of the very late twentieth and early twenty-first centuries is largely the accomplishment of prior activists, and particularly gay, lesbian, bisexual, and transgender (GLBT) individuals. The great strides these activists have made in the last three decades have furthered GLBT human rights, identity politics movements, and the development of queer theory. The increasing visibility of GLBT persons has clearly provided an increased level of social tolerance for those with nonbinary genders and sexualities, ultimately paving the way for the intersex movement to move to center stage. In addition, both queer theory and GLBT activism parallel the politics of prior difference movements, such as the disability rights movement, in which paradigmatic understandings of disability shifted focus from moral, to medical, to minority concepts of difference (Mona 1998). The intersex movement is also quite reliant technologically upon electronic media and communications, including the Internet and electronic mail (Turner 1999). . . .

IMPLICATIONS FOR CLINICAL REFORM

Physicians are faced with a difficult dilemma: they are taught to improve the lives of their patients. Certainly surgery is warranted in those rare intersex cases where intervention is mandatory (as when elimination of feces or urine is not possible without such intervention). However, the vast majority of intersex conditions are simply not life threatening (Diamond and Sigmundson 1997a; Kessler 1998; Wilson and Reiner 1998). Given the injunction "do no harm" and the lack of data conclusively demonstrating the efficacy of more cosmetically based medical interventions, clinicians should opt for the least invasive treatment procedures and not conduct any irreversible surgical or hormonal intervention without the patient's direct consent.[34]

Perpetuating treatments that seem to be ineffective and even harmful raises serious ethical questions. Viktor Gecas notes that autonomous control over one's own existence is exceedingly important in the development of a positive self-concept: "It is the quality of the individual-environment interaction, primarily with regard to the opportunities it provides for engaging in efficacious action . . . that continues to be the major condition for self-efficacy throughout a person's life" (1989, 300). Why this reluctance to discontinue treatment when there is doubt as to its effectiveness? Resistance to clinical reform is evident in the continued use of ineffective treatments, especially use of various aversion or "reparative" therapies for GLBT individuals whose families hope for their "heterosexual conversions" or "shifts." According to Paulo Freire (1970), persons in positions of power, such as medical authorities, may disregard the misgivings and mobilization efforts of those they serve because supporting their dissent would only undermine their efforts to perpetuate and uphold the status quo. We can see this trend among intersex clinicians who regard intersex activists as "radical zealots" representing perhaps only 3 percent of the entire surgically altered intersex population (Yronwode 1999).

As Thomas Kuhn (1962) and Freire (1970) demonstrate, paradigmatic shifts are hard won, especially when attempted by disempowered or marginal populations. In order to affect change, open communication and collaboration must occur to ensure that those operating within the prevailing model have the capacity to be affected by the personal experiences and efforts of those seeking change. It is through such personal interaction, when both parties approach each other with compassion, humility, and hope, that change is possible. Both sides must be flexible enough to change their thinking and to compromise when necessary (Freire 1970; Bockting, Rosser, and Coleman 1999). As it is the patients who must live with the results of medical treatments, the doctors who serve them should take their experiences, perspectives, and desires into flail consideration.

As a result of the highly vocal criticisms of intersexuals, clinicians are beginning to reform their practices and to become far less eager advocates of surgical intervention on intersexed infants and children (Schober 1998b; Wilson and Reiner 1998; Nussbaum 1999).[35] These changes have been made by physicians who have had personal interaction with adult intersexuals who are critical of medicalization because they have suffered sexual dysfunction due to iatrogenic nerve damage and incorrect gender role assignment (Kipnis and Diamond 1998; Schober 1998a; Wilson and Reiner 1998).

CONCLUSIONS

Contemporary intersex activism emerged as a reaction to the broad societal marginalization experienced by intersexed individuals, with a specific critique directed at disempowering medical interventions. Before the existence of intersex support and advocacy organizations, intersexuals were dependent on their physicians for access to related resources. Even though they may be disappointed with the information they receive from their physicians, intersexed individuals turn to medical authorities for assistance with their medicalized condition rather than to typical avenues of social support (Berkun 1986). Cleo Berkun uses menopause as an example of this phenomenon: "Because of the perceptions of the private, medical, and moral nature of their [menopausal] thoughts, they did not turn to their usual sources of support and information but, rather, sought 'factual' information from physicians with which they were not satisfied" (1986, 381). Autonomy is so critical to one's self-concept that Edward Jones et al. claim that "the self-esteem of stigmatized individuals will increase to the extent that the individuals come to view themselves as other than helpless, dependent, and worthless" (1984, 136).

Several individuals and groups have already proposed new clinical guidelines to protect the rights and autonomy of intersexed individuals.[36] Through the development of alternative and compassionate understandings of intersex with less social focus on genitals or gender, the pejorative meanings associated with it may, perhaps, begin to be cast off. In order to transform social discourse in this way, far more consideration must be given to the impact of small-scale social movements and social support networks as vital agents of social change. . . .

NOTES

1. I view the terms *intersex* and *hermaphrodite* synonymously, despite their divergent historical context. When speaking in contemporary terms, I use the more contemporary term *intersex*. However, when using historical references, I use the more popular pre-twentieth-century term *hermaphrodite*. Interestingly, the term *intersex* emerged in the late nineteenth century and was used not only when referring to hermaphrodites but to homosexuals as well (Epstein 1990). For example, a woman who sexually desired other women was often labeled *intersexed*. Preceding the emergence of the term *intersex*, the label *hermaphrodite* was also used in relation to homosexuals in the seventeenth and eighteenth centuries (Pagliassotti 1993; Hekma 1994; Trumbach 1994).
2. While the majority of intersex conditions are found to be physiologically benign, some conditions do require surgical or hormonal intervention for reasons of physiological health. Most notably, this occurs in cases where elimination of urine and feces is rendered difficult due to physiological complications or, in rare cases of salt-wasting congenital adrenal hyperplasia, where hormone therapy is required to regulate the endocrine system (Diamond and Sigmundson 1994b; Kessler 1998; Wilson and Reiner 1998).
3. Several sources reveal this recent challenge. See, e.g., Kessler 1998; Dreger 1999; and Preves 2000.
4. Additional estimates note the frequency of intersex as comprising approximately 1–4 percent of all births (Edgerton 1964; Fiedler 1978; Money 1989). These estimates vary widely, depending on one's definition of intersex (Dreger 1998c; Kessler 1998). For example, some low estimates reflect acceptance of the traditional definition of *true hermaphroditism*, which accounts only for the rare occurrence of mixed gonadal tissue (i.e., the presence of ovarian and testicular tissue in the same body). Other researchers include children born with *pseudo hermaphroditism*, which typically presents in a child with internal gonads that are consistent with the karyotype (typically XX or XY) and external genitals that are incongruent with internal gonads and chromosomes. Finally, other researchers may also include chromosomal variations such as those found in Turner's syndrome (45, XO) and

Klinefelter's syndrome (e.g., 47, XXY). Blackless et al. 2000 suggest that the total frequency of nongenital intersex (intersex chromosomes or nongenital body parts) is much higher than 1 in 2000 and that working with a more inclusive definition of intersex would yield frequency estimates closer to 1 or 2 per 100.

5. See, e.g., Van Seters and Slob 1988; Reilly and Woodhouse 1989; Schober 1998a; Bin Abbas et al. 1999.
6. See Voorhess 1982; Donohoe, Powell, and Lee 1991; Horowitz and Glassberg 1992; Meyers-Seifer and Charest 1992.
7. See Kemp et al. 1996; Natarajan 1996; Kessler 1998; Wilson and Reiner 1998.
8. It is interesting to note that newborn penile size charts were first published in the 1960s. In striking contrast, newborn clitoral size charts were not published until the late 1980s (Lee 1994).
9. I use Holmes's (1994) term *phalloclit* to convey the ambiguity and bipotential of the genital tubercle to become clitoris or penis.
10. See Money 1985; Donohoe et al. 1991; Lee 1994; Kessler 1998.
11. Van Seters and Slob 1988; Reilly and Woodhouse 1989; Schober 1998b; Bin Abbas et al. 1999.
12. Foucault 1978, 1980; Epstein 1990; Pagliassotti 1993; Dreger 1995b.
13. Note that in the late nineteenth century, medical understanding regarded *hermaphrodite* as meaning male and female, whereas early use of the word *intersex* denoted in between male and female (Dreger 1998c).
14. Galenists and Aristotelians disagreed on the point of women's contribution to the formation of the embryo. According to Aristotle, females did not contribute "seed" or "sperm" to their offspring; Galen thought females actively provided "seed" to their progeny (Cadden 1993).
15. Aristotle valued "male" heat over "female" cold and viewed females' lack of heat as a sign of inferiority and even deformity (Cadden 1993).
16. Although Azo wrote this text in the late 1100s, it was not published until early in the seventeenth century.
17. The translation of Peter the Chanter (Peter Cantor) used here by Nederman and True (1996) is Boswell's (1980). Boswell's translation is from Peter Cantor's original Vulgate, with some Latin and Greek notation.
18. For evidence of historic tolerance of hermaphrodites, see Pare (1634) 1968; Epstein 1990; Jones and Stallybrass 1991; Pagliassotti 1993; Nederman and True 1996.
19. Ormrod 1972; Hawley 1977; Capron and D'Avino 1981; Greenberg 1999.
20. This discussion is, of course, relevant to current legislative discourse regarding the legalization of homosexual marriage and the Defense of Marriage Act (Greenberg 1999). The recent Vermont Supreme Court ruling requiring the state of Vermont to provide the benefits of marriage to same-sex couples is a good example.
21. Gender is made ever more important because it is performed or realized through primary social institutions including economic, religious, and familial spheres.
22. Dreger's historical research shows that British medical doctors George F. Blacker and T. W. P. Lawrence originated this five-sex system in 1896 (see Dreger 1998a).
23. To my knowledge, *transactivist* is Feinberg's term (1996). S/he uses it to refer to transgender activists.
24. Bolin 1994; Rothblatt 1995; Bockting 1997b; Devor 1997.
25. Diamond 1996b, 1997, 1999; and Diamond and Sigmundson 1997a.
26. Chase and Coventry 1997–98; Dreger 1998a, 1998b, 1998c; Kipnis and Diamond 1998; Preves in press.
27. Coventry 1998; Holmes 1998; Kaldera 1998; Chase 1999.
28. Some rare cases of intersexuality do require medical intervention. This is especially true of salt-wasting forms of congenital adrenal hyperplasia (CAH) or when infants have difficulty voiding urine or feces (Diamond and Sigmundson 1997b; Kessler 1998; Wilson and Reiner 1998).
29. I am indebted for much of this history to Chase 1998b.
30. Turner's syndrome is the most common form of "female" chromosomal variation where the typical karyotype is 45, XO, meaning that one sex chromosome is missing. A fetus with a 45,YO karyotype is not sustainable and will die (Grumbach and Conte 1998). Individuals with Turner's syndrome typically develop unambiguous female genitalia yet have underdeveloped breasts, uteruses, and vaginas. Some have testicular tissue and primitive gonadal "streak" tissue. They commonly do not develop secondary sex characteristics, are very short (dwarfism), have a webbed neck, a "shield chest," short fingers and toes, renal disorders, heart and circulatory problems, and are prone to deafness and mental retardation (Money 1968; Money and Ehrhardt 1972; Glanze et al. 1996; Grumbach and Conte 1998).
31. In androgen insensitivity syndrome (AIS), individuals have a male-typical 46, XY karyotype but lack a key androgen (male hormone) receptor that incapacitates the ability, fetally onward, to respond to the androgens produced in normal amounts. This results in a feminization of the external genitalia

and, typically, abdominal testicles. Some individuals with AIS are completely insensitive to androgen, and some only partially. Most individuals with AIS are sexed as female unless virilization is only slightly affected by insensitivity to androgen, in which case male sex assignment would prevail (Quigley et al. 1995; Diamond and Sigmundson 1997a). At puberty, individuals with AIS respond to the normal levels of estrogen produced by their bodies and develop breasts. Individuals with AIS typically develop very little, if any, body hair, and are tall and lean. (It is indeed paradoxical that appearance of AIS "boys" is consistent with contemporary Western ideals of female beauty.) Individuals with AIS are sterile, but many do adopt children (Josso 1981; Glanze et al. 1986; Kupfer et al. 1992; Grumbach and Conte 1998).

32. Klinefelter's syndrome is a type of chromosomal variation in which a "male" child has a karyotype with more than one X chromosome, such as a 47, XXY (or 48, XXYY, 48, XXXY, or 49, XXXXY). Genital ambiguity is not present, but testes may be small and firm, and breast development (gynecomastia) is common. Secondary sex characteristic development is limited, and these men are almost always sterile (Money 1968; Money and Ehrhardt 1972; Glanze, Anderson, and Anderson 1996; Grumbach and Conte 1998).
33. Chase 1998a, 1998c, 1999; see the ISNA Web site at http://www.isna.org.
34. By "patient," I mean only the person on whom a procedure will be performed, not parent(s) or guardian(s).
35. This shift was notable in pediatric urologist Yuri Reinberg's (1999) grand rounds presentation at the University of Minnesota Medical School. During his presentation and the discussion following it, several noted proponents of intersex surgery spoke of their newfound reluctance to perform genital operations on infants and children due to adult intersexuals' critiques of sexual dysfunction and inappropriate sex assignment.
36. Diamond and Sigmundson 1997b; Chase 1998c; Kipnis and Diamond 1998; Wilson and Reiner 1998.

REFERENCES

ALIAS. The biannual newsletter of the Androgen Insensitivity Syndrome Support Group. Available on-line at <http://www.medhelp.org/www/ais>.

Anspach, Renee R. 1979. "From Stigma to Identity Politics: Political Activism among the Physically Disabled and Former Mental Patients." *Social Science and Medicine* 13A:765–73.

Azo, Portius. 1610. *Summa Institutionem*. Venice.

Bell, Susan. 1990. "Sociological Perspectives on the Medicalization of Menopause." *Annals of the New York Academy of Sciences* 592:173–78.

Bem, Sandra Lipsitz. 1995. "Dismantling Gender Polarization and Compulsory Heterosexuality: Should We Turn the Volume Down or Up?" *Journal of Sex Research* 32(4):329–34.

Berkun, Cleo S. 1986. "On Behalf of Women over 40: Understanding the Importance of the Menopause." *Social Work* 31(5):378–84.

Bin-Abbas, Bassam S., Felix A. Conte, Melvin M. Grumbach, and Selna L. Kaplan. 1999. "Congenital Hypogonadotropic Hypogonadism and Micropenis: Effect of Testosterone Treatment on Adult Penile Size—Why Sex Reversal Is Not Indicated." *Journal of Pediatrics* 134(5):579–83.

Blackless, Melanie, Anthony Charuvastra, Amanda Derryck, Anne Fausto-Sterling, Karl Lauzanne, and Ellen Lee. 2000. "How Sexually Dimorphic Are We?" *American Journal of Human Biology* 12(2):151–66.

Bockting, Walter O. 1995. "Transgender Coming Out: Gender Revolution?" Paper presented at the annual meeting of the Society for the Scientific Study of Sexuality, San Francisco, November 9–12.

———. 1997a. "The Assessment and Treatment of Gender Dysphoria." *Directions in Clinical and Counseling Psychology* 7(11):1–23.

———. 1997b. "Transgender Coming Out: Implications for the Clinical Management of Gender Dysphoria." In *Gender Blending*, ed. Bonnie Bullough, Vern Bullough, and James Elias, 48–52. Amherst, Mass.: Prometheus.

Bockting, Walter O., Simon Rosser, and Eli Coleman. 1999. "Transgender HIV Prevention: Community Involvement and Empowerment." *International Journal of Transgenderism* 3(1–2). Available on-line at <http://www.symposion.com/ijt/hiv_risk/bockting.htm>.

Bolin, Anne. 1994. "Transcending and Transgendering: Male-to-Female Transsexuals, Dichotomy and Diversity." In *Third Sex, Third Gender: Beyond Sexual Dimorphism in Culture and History*, ed. Gilbert Herdt, 447–86. New York: Zone.

Bornstein, Kate. 1994. *Gender Outlaw: On Men, Women, and the Rest of Us*. New York: Routledge.

Boswell, John. 1980. *Christianity, Social Tolerance, and Homosexuality: Gay People in Western Europe from the Beginning of the Christian Era to the Fourteenth Century*. Chicago: University of Chicago Press.

Butler, Judith. 1990. *Gender Trouble: Feminism and the Subversion of Identity*. New York: Routledge.

———. 1993. *Bodies That Matter: On the Discursive Limits of "Sex."* New York: Routledge.

Cadden, Joan. 1993. *Meanings of Sex Differences in the Middle Ages: Medicine, Science, and Culture.* New York: Cambridge University Press.

Capron, Alexander Morgan, and Richard D'Avino. 1981. "Legal Implications of Intersexuality." In *Pediatric and Adolescent Endocrinology: Special Issue, the Intersex Child* 8:218–27.

Chanter, Peter the. *De Vitio Sodomitico (On Sodomy).* Translated in Boswell 1980.

Chase, Cheryl. 1993. "Letters from Readers." *Sciences,* July–August, 3.

———. 1997. "Making Media: An Intersex Perspective." *Images* (Fall): 22–25.

———. 1998a. "Affronting Reason." In *Looking Queer: Body Image and Gay Identity in Lesbian, Bisexual, Gay, and Transgender Communities,* ed. Dawn Atkins, 205–19. New York: Harrington Park.

———. 1998b. "Hermaphrodites with Attitude: Mapping the Emergence of Intersex Political Activism." *GLQ: A Journal of Lesbian and Gay Studies* 4(2):189–211.

———. 1998c. "Surgical Progress Is Not the Answer to Intersexuality." *Journal of Clinical Ethics* 9(4):385–92.

———. 1999. "Rethinking Treatment for Ambiguous Genitalia." *Pediatric Nursing* 25(4):451–55.

Chase, Cheryl, and Martha Coventry, eds. 1997–98. "Intersex Awakening," special issue of *Chrysalis: The Journal of Transgressive Gender Identities* 2(5):1–56.

Chicago Hope. 1996. "Episode on Intersex." Columbia Broadcasting System, April 4.

Conrad, Peter, and Joseph W. Schneider. 1992. *Deviance and Medicalization: From Badness to Sickness.* 2d ed. Philadelphia: Temple University Press.

Coventry, Martha. 1997–98. "Finding the Words." In Chase and Coventry 1997–98, 27–29.

———. 1998. "The Tyranny of the Esthetic: Surgery's Most Intimate Violation." *On the Issues: The Progressive Woman's Quarterly* 7(3):16–20, 60–61.

Cowley, Geoffrey. 1997. "Gender Limbo." *Newsweek,* May 19, 64–67.

Crawford, Robert. 1980. "Healthism and the Medicalization of Everyday Life." *International Journal of Health Services* 10(3):365–87.

Devor, Holly. 1997. *FTM: Female-to-Male Transsexuals in Society.* Bloomington: Indiana University Press.

Diamond, Milton. 1982. "Sexual Identity: Monozygotic Twins Reared in Discordant Sex Roles and a BBC Follow-Up." *Archives of Sexual Behavior* 11(2): 181–85.

———. 1996. "Prenatal Disposition and the Clinical Management of Some Pediatric Conditions." *Journal of Sex and Marital Therapy* 22(3):139–47.

———. 1997. "Sexual Identity and Sexual Orientation in Children with Traumatized or Ambiguous Genitalia." *Journal of Sex Research* 34(2):199–211.

Diamond, Milton, and Keith Sigmundson. 1997a. "Management of Intersexuality: Guidelines for Dealing with Persons with Ambiguous Genitalia." *Archives of Pediatric Adolescent Medicine* 151 (October): 1046–50.

———. 1997b. "Sex Reassignment at Birth: Long-term Review and Clinical Implications." *Archives of Pediatric and Adolescent Medicine* 150 (March): 298–304.

Donohoe, Patricia K., David M. Powell, and Mary M. Lee. 1991. "Clinical Management of Intersex Abnormalities." *Current Problems in Surgery* 28(8):513–79.

Dreger, Alice Domurat. 1995a. "Doubtful Sex: Cases and Concepts of Hermaphroditism in France and Britain, 1868–1915." Ph.D. dissertation, Indiana University.

———. 1995b. "Doubtful Sex: The Fate of the Hermaphrodite in Victorian Medicine." *Victorian Studies* (Spring): 335–70.

———. 1998a. "A History of Intersexuality: From the Age of Gonads to the Age of Consent." *Journal of Clinical Ethics* 9(4):345–55.

———. 1998b. "'Ambiguous Sex'—or Ambivalent Medicine? Ethical Issues in the Treatment of Intersexuality." *Hastings Center Report* 28(3):24–36.

———. 1998c. *Hermaphrodites and the Medical Invention of Sex.* Cambridge, Mass.: Harvard University Press.

———, ed. 1999. *Intersex in the Age of Ethics.* Hagerstown, Md.: University Publishing Group.

Edgerton, Robert. 1964. "Pokot Intersexuality: An East African Example of the Resolution of Sexual Incongruity." *American Anthropologist* 66(6):1288–99.

Epstein, Julia. 1990. "Either/or—Neither/Both: Sexual Ambiguity and the Ideology of Gender." *Genders* 7 (Spring): 99–142.

Fausto-Sterling, Anne. 1985. *Myths of Gender.* New York: Basic.

———. 1993. "The Five Sexes: Why Male and Female Are Not Enough." *Sciences* 33(2):20–25.

———. 1996. "How to Build a Man." In *Science and Homosexualities,* ed. Vernon A. Rosario, 219–25. New York: Routledge.

———. 2000. *Sexing the Body: Gender Politics and the Construction of Sexuality.* New York: Basic.

Feinberg, Leslie. 1996. *Transgender Warriors: Making History from Joan of Arc to Ru-Paul.* Boston: Beacon.

Fiedler, Leslie. 1978. *Freaks: Myths and Images of the Secret Self.* New York: Anchor.

Fisher, Sue, and Stephen B. Groce. 1985. "Doctor-Patient Negotiation of Cultural Assumptions." *Sociology of Health & Illness* 7(3):342–74.

Flaubert, Gustave. (1887) 1965. *Madame Bovary.* Trans. Paul de Man. New York: Norton.

Foucault, Michel. 1970. *The Order of Things: An Archaeology of the Human Sciences.* New York: Vintage.

Freire, Paulo. 1970. *Pedagogy of the Oppressed.* New York: Continuum.

Galen. (170 C.E.) 1916. *On the Natural Faculties.* Ed. and trans. Arthur John Brock. Loeb Classical Library. Cambridge, Mass.: Harvard University Press; London: William Heinmann.

Gecas, Viktor. 1989. "The Social Psychology of Self-Efficacy." *Annual Review of Sociology* 15: 291–316.

———. 1991. "The Self-Concept as a Basis for a Theory of Motivation." In *The Self-Society Dynamic,* ed. Judith A. Howard and Peter L. Callero, 171–85. Cambridge: Cambridge University Press.

Glanze, Walter D., Kenneth N. Anderson, and Lois E. Anderson. 1996. *The Signet Mosby Medical Encyclopedia.* Rev. ed. New York: Signet.

Goffman, Erving. 1959. *The Presentation of Self in Everyday Life.* New York: Anchor.

———. 1963. *Stigma: Notes on the Management of Spoiled Identity.* Englewood Cliffs, N.J.: Prentice-Hall.

———. 1982. *Interaction Ritual.* New York: Pantheon.

Greenberg, Julie A. 1999. "Defining Male and Female: Intersexuality and the Collision between Law and Biology." *Arizona Law Review* 41(2):265–328.

Harvey, Kay. 1999. "A Mother's Dilemma." *St. Paul Pioneer Press,* March 4, 1F.

Hawley, Donna Lee. 1977. "The Legal Problems of Sex Determination." *Alberta Law Review* 15:122–41.

Hekma, Gert. 1994. "A Female Soul in a Male Body: Sexual Inversion as Gender Inversion in Nineteenth-Century Sexology." In *Third Sex, Third Gender: Beyond Sexual Dimorphism in Culture and History,* ed. Gilbert Herdt, 213–39. New York: Zone.

Hendricks, Melissa. 1993. "Is It a Boy or a Girl?" *Johns Hopkins Magazine,* November, 10–16.

Hermaphrodites with Attitude. The quarterly publication of the Intersex Society of North America. Available on-line at <http://www.isna.org>.

Holmes, Morgan M. 1994. "Medical Politics and Cultural Imperatives: Intersexual Identities beyond Pathology and Erasure." Master's thesis, York University.

———. 1995. "Queer Cut Bodies: Intersexuality & Homophobia in Medical Practice." Paper presented at the fifth annual National Lesbian, Gay, and Bisexual Graduate Student Queer Frontiers Conference, University of Southern California, Los Angeles.

———. 1998. "In(to) Visibility: Intersexuality in the Field of Queer." In *Looking Queer: Body Image and Gay Identity in Lesbian, Bisexual, Gay, and Transgender Communities,* ed. Dawn Atkins, 221–26. New York: Harrington Park.

Horowitz, Mark, and Kenneth I. Glassberg. 1992. "Ambiguous Genitalia: Diagnosis, Evaluation, and Treatment." *Urologic Radiology* 14(4):306–18.

Horowitz, Sarah. 1995. "The Middle Sex." *San Francisco Weekly,* February 1, 11–13.

Jacobs, Sue-Ellen, Wesley Thomas, and Sabine Lang, eds. 1997. Introduction to their *Two Spirit People: Native American Gender Identity, Sexuality, and Spirituality.* Chicago: University of Illinois Press.

Jones, Ann Rosalind, and Peter Stallybrass. 1991. "Fetishizing Gender: Constructing the Hermaphrodite in Renaissance Europe." In *Body Guards: The Cultural Politics of Gender Ambiguity,* ed. Julia Epstein and Kristina Straub, 80–111. New York: Routledge.

Jones, Edward E., Amerigo Farina, Albert H. Hastorf, Hazel Markus, Dale T. Miller, and Robert A. Scott. 1984. *Social Stigma: The Psychology of Marked Relationships.* New York: W. H. Freeman.

Josso, Nathalie. 1981. "Physiology of Sex Differentiation: A Guide to the Understanding and Management of the Intersex Child." In "The Intersex Child," special issue of *Pediatric and Adolescent Endocrinology,* 8:1–13.

Kaldera, Raven. 1998. "Agdistis' Children: Living Bi-Gendered in a Single-Gendered World." In *Looking Queer: Body Image and Gay Identity in Lesbian, Bisexual, Gay, and Transgender Communities,* ed. Dawn Atkins, 227–32. New York: Harrington Park.

Kemp, B. Diane, Sherri A. Groveman, Anonymous, H. Deni Tako, Karl M. Irwin, Anita Natarajan, and Patrick Sullivan. 1996. "Sex, Lies and Androgen Insensitivity Syndrome." *Canadian Medical Association Journal* 154(12):1829–33.

Kessler, Suzanne J. 1990. "The Medical Construction of Gender: Case Management of Intersexed Infants." *Signs* 16(1):3–26.

———. 1997–98. "Meanings of Genital Variability." In Chase and Coventry 1997–98, 33–38.

———. 1998. *Lessons from the Intersexed.* New Brunswick, N.J.: Rutgers University Press.

Kessler, Suzanne J., and Wendy McKenna. 1978. *Gender: An Ethnomethodological Approach.* Chicago: University of Chicago Press.

Kuhn, Thomas S. 1962. *The Structure of Scientific Revolutions.* Chicago: University of Chicago Press.

Lee, Ellen Hyun-Ju. 1994. "Producing Sex: An Interdisciplinary Perspective of Sex Assignment Decisions for Intersexuals." Senior thesis, Brown University.

Martin, Emily. 1987. *The Woman in the Body: A Cultural Analysis of Reproduction.* Boston: Beacon.

Meyers-Seifer, Cynthia H., and Nancy J. Charest. 1992. "Diagnosis and Management of Patients with Ambiguous Genitalia." *Seminars in Perinatology* 16(5): 332–39.

Mona, Linda. 1998. "Cognitive Adaption Styles and Sexual Self-Esteem as Predictors of Sexual and Psychological Adjustment Following Spinal Cord Injury." Paper presented at the annual meeting of the Society for the Scientific Study of Sexuality, Los Angeles, November 11–15.

Money, John. 1975. "Ablatio Penis: Normal Male Infant Sex-Reassignment as a Girl." *Archives of Sexual Behavior* 4(1):65–71.

———. 1989. *The Geraldo Rivera Show.* "Hermaphrodites: The Sexually Unfinished." National Broadcasting Company, July 27.

———. 1991b. "Serendipities on the Sexological Pathway to Research in Gender Identity and Sex Research." *Journal of Psychology and Human Sexuality* 4(1): 101–13.

———. 1995. *Gendermaps: Social Constructionism, Feminism, and Sexosophical History.* New York: Continuum.

Moscucci, Ornella. 1991. "Hermaphroditism and Sex Difference: The Construction of Gender in Victorian England." In *Science and Sensibility: Gender and Scientific Enquiry, 1780–1945,* ed. Marina Benjamin, 174–99. Williston, Fla.: Basil Blackwell.

Natarajan, Anita. 1996. "Medical Ethics and Truth Telling in the Case of Androgen Insensitivity." *Canadian Medical Association Journal* 154(4):568–70.

Nederman, Cary J., and Jacqui True. 1996. "The Third Sex: The Idea of the Hermaphrodite in Twelfth-Century Europe." *Journal of the History of Sexuality* 6(4):497–517.

Nussbaum, Emily. 1999. "The Sex That Dare Not Speak Its Name." *Lingua Franca: The Review of Academic Life* 9(4):42–51.

Open Secret: The First Question. 1980. Produced by P. Williams and M. Smith. BBC television production.

Ormrod, Roger. 1972. "The Medico-Legal Aspects of Sex Determination." *Medico-Legal Journal* 40 (3):78–88.

Pagliassotti, Druann. 1993. "On the Discursive Construction of Sex and Gender." *Communication Research* 20(3):472–93.

Pare, Ambroise. (1634) 1968. *The Collected Works of Ambroise Pare.* Trans. Thomas Johnson. New York: Milford House.

Preves, Sharon E. 1998. "For the Sake of the Children: Destigmatizing Intersexuality." *Journal of Clinical Ethics: Special Issue on Intersex* 9(4):411–20.

———. 1999. "Sexing the Intersexed: Lived Experiences in Socio-Cultural Context." Ph.D. dissertation, University of Minnesota.

———. In press. *Sexing the Intersexed: Stories of Medicalization and the Toll of Gender Conformity.* New Brunswick, N.J.: Rutgers University Press.

Quigley, Charmaine A., A. De Bellis, K. B. Merschke, M. K. El-Awady, E. M. Wilson, and Frank S. French. 1995. "Androgen Receptor Defects: Historical, Clinical and Molecular Perspectives." *Endocrine Review* 16(3):271–321.

Reilly, Justine M., and C. R. J. Woodhouse. 1989. "Small Penis and the Male Sexual Role." *Journal of Urology* 142 (August): 569–72.

Reinberg, Yuri. 1999. "Evaluation and Treatment of Children Born with Ambiguous Genitalia: Current Controversies in the Field of Sexual Reassignment." Grand Rounds Presentation, University of Minnesota Medical School, Minneapolis, January 27.

Rosario, Vernon A. 1997. Review of *Changing Sex: Transsexualism, Technology, and the Idea of Gender,* by Bernice Hausman. Configurations 5:243–46.

Rothblatt, Martine. 1995. *The Apartheid of Sex: A Manifesto on the Freedom of Gender.* New York: Crown.

Schober, Justine Murat. 1998a. "Feminizing Genitoplasty for Intersex." In *Pediatric Surgery and Urology: Long-Term Outcomes,* ed. M. D. Stringer, K. T. Oldham, P. D. E. Mouriquand, and E. R. Howard, 549–58. Philadelphia: W. B. Saunders.

———. 1998b. "A Surgeon's Response to the Intersex Controversy." *Journal of Clinical Ethics* 9(4):393–97.

Slaughenhoupt, Bruce L., and John G. Van Savage. 1999. "The Child with Ambiguous Genitalia." *Infectious Urology* 12(4):113–18.

Stone, Sandy. 1991. "The *Empire* Strikes Back." In *Body Guards: The Cultural Politics of Gender Ambiguity,* ed. Julia Epstein and Kristina Straub, 280–304. New York: Routledge.

Triea, Kiira. 1997. "As a Former Intersexed Patient." Letter in *Time,* April 14, 19.

Trumbach, Randolph. 1994. "London's Sapphists: From Three Sexes to Four Genders in the Making of Modern Culture." In *Third Sex, Third Gender: Beyond Sexual Dimorphism in Culture and History,* ed. Gilbert Herdt, 111–36. New York: Zone.

Turner, Stephanie S. 1999. "Intersex Identities: Locating New Intersections of Sex and Gender." *Gender and Society* 13(4):457–79.

Van Seters, A. P., and A. K. Slob. 1988. "Mutually Gratifying Heterosexual Relationship with Micropenis of Husband." *Journal of Sex and Marital Therapy* 14(2):98–107.

Voorhess, Martin L. 1982. "Normal and Abnormal Sexual Development." In *Core Textbook of Pediatrics,* ed. Robert Kaye, Frank A. Oski, and Lewis A. Barness, 214–37. Philadelphia: J. B. Lippincott.

Walcutt, Heidi. 1997–98. "Time for a Change." In Chase and Coventry 1997–98, 25–26.

Wilson, Bruce E., and William G. Reiner. 1998. "Management of Intersex: A Shifting Paradigm." *Journal of Clinical Ethics* 9(4):360–69.

Wilson, Elizabeth A. 1998. *Neural Geographies: Feminism and the Microstructure of Cognition.* New York: Routledge.

Young, Diony. 1982. *Changing Childbirth: Family Birth in the Hospital.* Rochester, N.Y.: Childbirth Graphics.

Yronwode, Althaea. 1999. "Intersex Individuals Dispute Wisdom of Surgery on Infants." *Synapse* 43(22):3–5.

Zola, Irving K. 1990. "Medicine as an Institution of Social Control." In *The Sociology of Health and Illness: Critical Perspectives,* ed. Peter Conrad and Rochelle Kern, 398–408. New York: St. Martin's.

3

The Trouble with Testosterone

BY ROBERT SAPOLSKY

Robert Sapolsky is professor of biology and neurology at Stanford University and a research associate with the Institute of Primate Research, National Museum of Kenya. He is the author of The Trouble with Testosterone, Why Zebras Don't Get Ulcers *and, most recently,* A Primate's Memoir. *Sapolsky has lived as a member of a baboon troop in Kenya, conducting cutting-edge research on these beautiful and complex primates. In this article, he uses his keen wit and scientific understanding to debunk the widely held myth that testosterone causes aggression in males.*

1. Why do many Americans want to believe that biological factors, such as hormones, are the basis of gender differences and inequalities?
2. Sapolsky says that hormones have a "permissive effect." What does "permissive effect" mean in terms of the relationship between testosterone and aggression?
3. How does research on testosterone, male monkeys, and spotted hyenas help one to grasp the role of social factors and environment in behavioral biology?

Face it, we all do it. We all believe in certain stereotypes about certain minorities. The stereotypes are typically pejorative and usually false. But every now and then, they are true. I write apologetically as a member of a minority about which the stereotypes are indeed true. I am male. We males account for less than 50 percent of the population, yet we generate an incredibly disproportionate percentage of the violence. Whether it is something as primal as having an ax fight in an Amazonian clearing or as detached as using computer-guided aircraft to strafe a village, something as condemned as assaulting a cripple or as glorified as killing someone wearing the wrong uniform, if it is violent, males excel at it.

Why should that be? We all think we know the answer. A dozen millennia ago or so, an adventurous soul managed to lop off a surly bull's testicles and thus invented behavioral endocrinology. It is unclear from the historical records whether this individual received either a grant or tenure as a result of this experiment, but it certainly generated an influential finding—something or other comes out of the testes that helps to make males such aggressive pains in the ass.

That something or other is testosterone.[1] The hormone binds to specialized receptors in muscles and causes those cells to enlarge. It binds to similar receptors in laryngeal cells and gives rise to operatic basses. It causes other secondary sexual characteristics, makes for relatively unhealthy blood vessels, alters biochemical events in the liver too dizzying to even contemplate, has a profound impact, no doubt, on the workings of cells in big toes. And it seeps into the brain, where it binds to those same "androgen" receptors and influences behavior in a way highly relevant to understanding aggression.

What evidence links testosterone with aggression? Some pretty obvious stuff. Males tend to have higher testosterone levels in their circulation than do females (one wild exception to that will be discussed later) and to be more aggressive. Times of life when males are swimming in testosterone (for example, after reaching puberty) correspond to when aggression peaks. Among numerous species, testes are mothballed most of the year, kicking into action and pouring out testosterone only during a very circumscribed mating season—precisely the time when male–male aggression soars.

Impressive, but these are only correlative data, testosterone repeatedly being on the scene with no alibi when some aggression has occurred. The proof comes with the knife, the performance of what is euphemistically known as a "subtraction" experiment. Remove the source of testosterone in species after species and levels of aggression typically plummet. Reinstate normal testosterone levels afterward with injections of synthetic testosterone, and aggression returns.

To an endocrinologist, the subtraction and replacement paradigm represents pretty damning proof: this hormone is involved. "Normal testosterone levels appear to be a prerequisite for normative levels of aggressive behavior" is the sort of catchy, hummable phrase that the textbooks would use. That probably explains why you shouldn't mess with a bull moose during rutting season. But that's not why a lot of people want to understand this sliver of science. Does the action of this hormone tell us anything about *individual* differences in levels of aggression, anything about why some males, some human males, are exceptionally violent? Among an array of males—human or otherwise—are the highest testosterone levels found in the most aggressive individuals?

Generate some extreme differences and that is precisely what you see. Castrate some of the well-paid study subjects, inject others with enough testosterone to quadruple the normal human levels, and the high-testosterone males are overwhelmingly likely to be the more aggressive ones. However, that doesn't tell us much about the real world. Now do something more subtle by studying the normative variability in testosterone—in other words, don't manipulate anything, just see what everyone's natural levels are like—and high levels of testosterone and high levels of aggression still tend to go together. This would seem to seal the case—interindividual differences in levels of aggression among normal individuals are probably driven by differences in levels of testosterone. But this turns out to be wrong.

Okay, suppose you note a correlation between levels of aggression and levels of testosterone among these normal males. This could be because *(a)* testosterone elevates aggression; *(b)* aggression elevates testosterone secretion; *(c)* neither causes the other. There's a huge bias to assume option *a,* while *b* is the answer. Study after study has shown that when you examine testosterone levels when males are first placed together in the social group, testosterone levels predict nothing about who is going to be aggressive. The subsequent behavioral differences drive the hormonal changes, rather than the other way around.

Because of a strong bias among certain scientists, it has taken forever to convince them of this point. Behavioral endocrinologists study what behavior and hormones have to do with each other. How do you study behavior? You get yourself a notebook and a stopwatch and a pair of binoculars. How do you measure the hormones? You need a gazillion-dollar machine, you muck around with radiation and chemicals, wear a lab coat, maybe even goggles—the whole nine yards. Which toys would you rather get for Christmas? Which facet of science are you going to believe in more? Because the endocrine aspects of the business are more high-tech, more reductive, there is the bias to think that it is somehow more scientific, more powerful. This is a classic case of what is often called physics envy, the disease among scientists where the behavioral biologists fear their discipline lacks the rigor of physiology, the physiologists wish for the techniques of the biochemists, the biochemists covet the clarity of the answers revealed by the molecular biologists, all the way down until you get to the physicists, who confer only with God.[2] Hormones seem to many to be more real, more substantive, than the ephemera of behavior, so when a correlation occurs, it must be because hormones regulate behavior, not the other way around.

As I said, it takes a lot of work to cure people of that physics envy, and to see that interindividual differences in testosterone levels don't predict subsequent differences in aggressive behavior among individuals. Similarly, fluctuations in testosterone levels within one individual over time do not predict subsequent changes in the levels of aggression in that one individual—get a hiccup in testosterone secretion one afternoon and that's not when the guy goes postal.

Look at our confusing state: normal levels of testosterone are a prerequisite for normal levels of aggression, yet changing the amount of testosterone in someone's bloodstream within the normal range doesn't alter his subsequent levels of aggressive behavior. This is where, like clockwork, the students suddenly start coming to office hours in a panic, asking whether they missed something in their lecture notes.

Yes, it's going to be on the final, and it's one of the more subtle points in endocrinology—what is referred to as a hormone having a "permissive effect." Remove someone's testes and, as noted, the frequency of aggressive behavior is likely to plummet. Reinstate precastration levels of testosterone by injecting that hormone, and precastration levels of aggression typically return. Fair enough. Now this time, castrate an individual and restore testosterone levels to only 20 percent of normal and . . . amazingly, normal precastration levels of aggression come back. Castrate and now generate twice the testosterone levels from before castration—and the same level of aggressive behavior returns. You need some testosterone around for normal aggressive behavior—zero levels after castration, and down it usually goes; quadruple it (the sort of range generated in weight lifters abusing anabolic steroids), and aggression typically increases. But anywhere from roughly 20 percent of normal to twice normal and it's all the same; the brain can't distinguish among this wide range of basically normal values.

We seem to have figured out a couple of things by now. First, knowing the differences in the levels of testosterone in the circulation of a bunch of males will not help you much in figuring out who is going to be aggressive. Second, the subtraction and reinstatement data seem to indicate that, nevertheless, in a broad sort of way, testosterone causes aggressive behavior. But that turns out not to be true either, and the implications of this are lost on most people the first thirty times you tell them about it. Which is why you'd better tell them about it thirty-one times, because it is the most important point of this piece.

Round up some male monkeys. Put them in a group together, and give them plenty of time to sort out where they stand with each other—affiliative friendships, grudges and dislikes. Give them enough time to form a dominance hierarchy, a linear ranking system of numbers 1 through 5. This is the hierarchical sort of system where number 3, for example, can pass his day throwing around his weight with numbers 4 and 5, ripping off their monkey chow, forcing them to relinquish the best spots to sit in, but, at the same time, remembering to deal with numbers 1 and 2 with shit-eating obsequiousness.

Hierarchy in place, it's time to do your experiment. Take that third-ranking monkey and give him some testosterone. None of this within-the-normal-range stuff. Inject a ton of it into him, way higher than what you normally see in a rhesus monkey; give him enough testosterone to grow antlers and a beard on every neuron in his brain. And, no surprise, when you then check the behavioral data, it turns out that he will probably be participating in more aggressive interactions than before.

So even though small fluctuations in the levels of the hormone don't seem to matter much, testosterone still causes aggression. But that would be wrong. Check out number 3 more closely. Is he now raining aggressive terror on any and all in the group, frothing in an androgenic glaze of indiscriminate violence? Not at all. He's still judiciously kowtowing to numbers 1 and 2, but has simply become a total bastard to numbers 4 and 5. This is critical: testosterone isn't *causing* aggression, it's *exaggerating* the aggression that's already there.

Another example just to show we're serious. There's a part of your brain that probably has lots to do with aggression, a region called the amygdala.[3] Sitting right near it is the Grand Central Station of emotion-related activity in your brain, the hypothalamus. The amygdala communicates with the hypothalamus by way of a cable of neuronal connections called the stria terminalis. No more jargon, I promise. The amygdala has its influence on aggression via that pathway, with bursts of electrical excitation called action potentials that ripple down the stria terminals, putting the hypothalamus in a pissy mood.

Once again, do your hormonal intervention; flood the area with testosterone. You can do that by injecting the hormone into the bloodstream, where it eventually makes its way to this part of the brain. Or you can be elegant and surgically microinject the stuff directly into this brain region. Six of one, half a dozen of the other. The key thing is what doesn't happen next. Does testosterone now cause there to be action potentials surging down the stria terminalis? Does it turn on that pathway? Not at all. If and only if the amygdala is sending an aggression-provoking volley of action potentials down the stria terminalis, testosterone increases the rate of such action potentials by shortening the resting time between them. It's not turning on the pathway, it's increasing the volume of signaling if it is already turned on. It's not causing aggression, it's exaggerating the preexisting pattern of it, exaggerating the response to environmental triggers of aggression.

This transcends issues of testosterone and aggression. In every generation, it is the duty of behavioral biologists to try to teach this critical point, one that seems a maddening cliché once you get it. You take that hoary old dichotomy between nature and nurture, between biological influences and environmental influences, between intrinsic factors and extrinsic ones, and, the vast majority of the time, regardless of which behavior you are thinking about and what underlying biology you are studying, the dichotomy is a sham. No biology. No environment. Just the interaction between the two.

Do you want to know how important environment and experience are in understanding testosterone and aggression? Look back at how the effects of castration were discussed earlier. There were statements like "Remove the source of testosterone in species after species and levels of aggression typically plummet." Not "Remove the source . . . and aggression always goes to zero." On the average it declines, but rarely to zero, and not at all in some individuals. And the more social experience an individual had being aggressive prior to castration, the more likely that behavior persists sans *cojones.* Social conditioning can more than make up for the hormone.

Another example, one from one of the stranger corners of the animal kingdom: If you want your assumptions about the nature of boy beasts and girl beasts challenged, check out the spotted hyena. These animals are fast becoming the darlings of endocrinologists, sociobiologists, gynecologists, and tabloid writers. Why? Because they have a wild sex-reversal system—females are more muscular and more aggressive than males and are socially dominant over them, rare traits in the mammalian world. And get this: females secrete more of certain testosterone-related hormones than the males do, producing the muscles, the aggression (and, as a reason for much of the gawking interest in these animals, wildly masculinized private parts that make it supremely difficult to tell the sex of a hyena). So this appears to be a strong vote for the causative powers of high androgen levels in aggression and social dominance. But that's not the whole answer. High up in the hills above the University of California at Berkeley is the world's largest colony of spotted hyenas, massive bone-crunching beasts who fight with each other for the chance to have their ears scratched by Laurence Frank, the zoologist who brought them over as infants from Kenya. Various scientists are studying their sex-reversal system. The female hyenas are bigger and more muscular than the males and have the same weirdo genitals and elevated androgen levels that their female cousins do back in the savannah. Everything is in place except . . . the social system is completely different from that in the wild. Despite being stoked on androgens, there is a very significant delay in the time it takes for the females to begin socially dominating the males—they're growing up without the established social system to learn from.

When people first grasp the extent to which biology has something to do with behavior, even subtle, complex, human behavior, there is often an initial evangelical enthusiasm of the convert, a massive placing of faith in the biological components of the story. And this enthusiasm is typically of a fairly reductive type—because of physics envy, because reductionism is so impressive, because it would be so nice if there were a single gene or hormone or neurotransmitter or part of the brain that was *it,* the cause, the explanation of everything. And the trouble with testosterone is that people tend to think this way in an arena that really matters.

This is no mere academic concern. We are a fine species with some potential. Yet we are racked by sickening amounts of violence. Unless we are hermits, we feel the threat of it, often as a daily shadow. And regardless of where we hide, should our leaders push the button, we will all be lost in a final global violence. But as we try to understand and wrestle with this feature of our sociality, it is critical to remember the limits of the biology. Testosterone is never going to tell us much about the suburban teenager who, in his after-school chess club, has developed a particularly aggressive style with his bishops. And it certainly isn't going to tell us much about the teenager in some inner-city hellhole who has taken to mugging people. "Testosterone equals aggression" is inadequate for those who would offer a simple solution to the violent male—just decrease levels of those pesky steroids. And "testosterone equals aggression" is certainly inadequate for those who would offer a simple excuse: Boys will be boys and certain things in nature are inevitable. Violence is more complex than a single hormone. This is endocrinology for the bleeding heart liberal—our behavioral biology is usually meaningless outside the context of the social factors and environment in which it occurs.

NOTES

1. Testosterone is one of a family of related hormones, collectively known as "androgens" or "anabolic steroids." They all are secreted from the testes or are the result of a modification of testosterone, they all have a similar chemical structure, and they all do roughly similar things. Nonetheless, androgen mavens spend entire careers studying the important differences in the actions of different androgens. I am going to throw that subtlety to the wind and, for the sake of simplification that will horrify many, will refer throughout to all of these related hormones as "testosterone."
2. An example of physics envy in action. Recently, a zoologist friend had obtained blood samples from the carnivores that he studies and wanted some hormones in the sample assays in my lab. Although inexperienced with the technique, he offered to help in any way possible. I felt hesitant asking him to do anything tedious but, so long as he had offered, tentatively said, "Well, if you don't mind some unspeakable drudgery, you could number about a thousand assay vials." And this scientist, whose superb work has graced the most prestigious science journals in the world, cheerfully answered, "That's okay, how often do I get to do *real* science, working with test tubes?"
3. And no one has shown that differences in the size or shape of the amygdala, or differences in the numbers of neurons in it, can begin to predict differences in normal levels of aggression. Same punch line as with testosterone.

FURTHER READING

For a good general review of the subject, see E. Monaghan and S. Glickman, "Hormones and Aggressive Behavior," in J. Becker, M. Breedlove, and D. Crews, eds., *Behavioral Endocrinology* (Cambridge, Mass.: MIT Press, 1992), 261.

This also has an overview of the hyena social system, as Glickman heads the study of the Berkeley hyenas. For technical papers on the acquisition of the female dominance in hyenas, see S. Jenks, M. Weldele, L. Frank, and S. Glickman, "Acquisition of Matrilineal Rank in Captive Spotted Hyenas: Emergence of a Natural Social System in, Peer-Reared Animals and Their Offspring," *Animal Behavior* 50 (1995): 893; and L. Frank, S. Glickman, and C. Zabel, "Ontogeny of Female Dominance in the Spotted Hyaena: Perspectives from Nature and Captivity," in P. Jewell and G. Maloiy, eds., "The Biology of Large African Mammals in Their Environment," *Symposium of the Zoological Society of London* 61 (1989): 127.

I have emphasized that while testosterone levels in the normal range do not have much to do with aggression, a massive elevation of exposure, as would be seen in anabolic steroid abusers, does usually increase aggression. For a recent study in which even elevating into that range (approximately five times normal level) still had no effect on mood or behavior, see S. Bhasin, T Storer, N. Berman, and colleagues, "The Effects of Supraphysiologic Doses of Testosterone on Muscle Size and Strength in Normal Men," *New England Journal of Medicine* 335 (1996): 1.

The study showing that raising testosterone levels in the middle-ranking monkey exaggerates preexisting patterns of aggression can be found in A. Dixson and J. Herbert, "Testosterone, Aggressive Behavior and Dominance Rank in Captive Adult Male Talapoin Monkeys *(Miopithecus talapoin)," Phsiology and Behavior* 18 (1977): 539. For the demonstration that testosterone shortens the resting period between action potentials in neurons, see K. Kendrick and R. Drewert, "Testosterone Reduces Refractory Period of Stria Terminalis Neurons in the Rat Brain," *Science* 204 (1979): 877.

4

What It Means to Be Gendered Me

BY BETSY LUCAL

Sociologist Betsy Lucal describes the rigidity of the American binary gender system and the consequences for people who do not fit by analyzing the challenges she faces in the course of her daily experience of negotiating the boundaries of our gendered society. Since her physical appearance does not clearly define her as a woman, she must navigate a world in which some people interact with her as though she is a man. Through analysis of her own story, Lucal demonstrates how gender is something we do, rather than something we are.

1. Why does Lucal argue that we cannot escape "doing gender"?
2. How does Lucal negotiate "not fitting" into the American two-and-only-two gender structure?
3. Have you ever experienced a mismatch between your gender-identity and the gender that others perceive you to be? If so, how did you feel and respond?

I understood the concept of "doing gender" (West and Zimmerman 1987) long before I became a sociologist. I have been living with the consequences of inappropriate "gender display" (Goffman 1976; West and Zimmerman 1987) for as long as I can remember.

My daily experiences are a testament to the rigidity of gender in our society, to the real implications of "two and only two" when it comes to sex and gender categories (Garfinkel 1967; Kessler and McKenna 1978). Each day, I experience the consequences that our gender system has for my identity and interactions. I am a woman who has been called "Sir" so many times that I no longer even hesitate to assume that it is being directed at me. I am a woman whose use of public rest rooms regularly causes reactions ranging from confused stares to confrontations over what a man is doing in the women's room. I regularly enact a variety of practices either to minimize the need for others to know my gender or to deal with their misattributions.

I am the embodiment of Lorber's (1994) ostensibly paradoxical assertion that the "gender bending" I engage in actually might serve to preserve and perpetuate gender categories. As a feminist who sees gender rebellion as a significant part of her contribution to the dismantling of sexism, I find this possibility disheartening.

In this article, I, examine how my experiences both support and contradict Lorber's (1994) argument using my own experiences to illustrate and reflect on the social construction of gender. My analysis offers a discussion of the consequences of gender for people who do not follow the rules as well as an examination of the possible implications of the existence of people like me for the gender system itself. Ultimately, I show how life on the boundaries of gender affects me and how my life, and the lives of others who make similar decisions about their participation in the gender system, has the potential to subvert gender.

Because this article analyzes my experiences as a woman who often is mistaken for a man, my focus is on the social construction of gender for women. My assumption is that, given the gendered nature of the gendering process itself, men's experiences of this phenomenon might well be different from women's.

THE SOCIAL CONSTRUCTION OF GENDER

It is now widely accepted that gender is a social construction, that sex and gender are distinct, and that gender is something all of us "do." This conceptualization of gender can be traced to Garfinkel's (1967) ethnomethodological study of "Agnes."[1] In this analysis, Garfinkel examined the issues facing a male who wished to pass as, and eventually become, a woman. Unlike individuals who perform gender in culturally expected ways, Agnes could not take her gender for granted and always was in danger of failing to pass as a woman (Zimmerman 1992).

This approach was extended by Kessler and McKenna (1978) and codified in the classic "Doing Gender" by West and Zimmerman (1987). The social constructionist approach has been developed most notably by Lorber (1994, 1996). Similar theoretical strains have developed outside of sociology, such as work by Butler (1990) and Weston (1996). . . .

Given our cultural rules for identifying gender (i.e., that there are only two and that masculinity is assumed in the absence of evidence to the contrary), a person who does not do gender appropriately is placed not into a third category but rather into the one with which her or his gender display seems most closely to fit; that is, if a man appears to be a woman, then he will be categorized as "woman," not as something else. Even if a person does not want to do gender or would like to do a gender other than the two recognized by our society, other people will, in effect, do gender for that person by placing her or him in one and only one of the two available categories. We cannot escape doing gender or, more specifically, doing one of two genders. (There are exceptions in limited contexts such as people doing "drag" [Butler 1990; Lorber 1994].)

People who follow the norms of gender can take their genders for granted. Kessler and McKenna asserted, "Few people besides transsexuals think of their gender as anything other than 'naturally' obvious"; they believe that the risks of not being taken for the gender intended "are minimal for nontranssexuals" (1978, 126). However, such an assertion overlooks the experiences of people such as those women Devor (1989) calls "gender blenders" and those people Lorber (1994) refers to as "gender benders." As West and Zimmerman (1987) pointed out, we all are held accountable for, and might be called on to account for, our genders.

People who, for whatever reasons, do not adhere to the rules, risk gender misattribution and any interactional consequences that might result from this misidentification. What are the consequences of misattribution for social interaction? When must misattribution be minimized? What will one do to minimize such mistakes? In this article, I explore these and related questions using my biography.

For me, the social processes and structures of gender mean that, in the context of our culture, my appearance will be read as masculine. Given the common conflation of sex and gender, I will be assumed to be a male. Because of the two-and-only-two genders rule, I will be classified, perhaps more often than not, as a man—not as an atypical woman, not as a genderless person. I must be one gender or the other; I cannot be neither, nor can I be both. This norm has a variety of mundane and serious consequences for my everyday existence. Like Myhre (1995), I have found that the choice not to participate in femininity is not one made frivolously.

My experiences as a woman who does not do femininity illustrate a paradox of our two-and-only-two gender system. Lorber argued that "bending gender rules and passing between genders does not erode but rather preserves gender boundaries" (1994, 21). Although people who engage in these behaviors and appearances do "demonstrate the social constructedness of sex, sexuality, and gender" (Lorber 1994, 96), they do not actually disrupt gender. Devor made a similar point: "When gender blending females refused to mark themselves by publicly displaying sufficient femininity to be recognized as women, they were in no way challenging patriarchal gender assumptions" (1989, 142). As the following discussion shows, I have found that my own experiences both support and challenge this argument. Before detailing these experiences, I explain my use of my self as data.

MY SELF AS DATA

This analysis is based on my experiences as a person whose appearance and gender/sex are not, in the eyes of many people, congruent. How did my experiences become my data? I began my research "unwittingly" (Krieger 1991). This article is a product of "opportunistic research" in that I am using my "unique biography, life experiences, and/or situational familiarity to understand and explain social life" (Riemer 1988, 121; see also Riemer 1977). It is an analysis of "unplanned personal experience" that is, experiences that were not part of a research project but instead are part of my daily encounters (Reinharz 1992).

This work also is, at least to some extent, an example of Richardson's (1994) notion of writing as a method of inquiry. As a sociologist who specializes in gender, the more I learned, the more I realized that my life could serve as a case study. As I examined my experiences, I found out things—about my experiences and about theory—that I did not know when I started (Richardson 1994).

It also is useful, I think, to consider my analysis an application of Mills's (1959) "sociological imagination." Mills (1959) and Berger (1963) wrote about the importance of seeing the general in the particular. This means that general social patterns can be discerned in the behaviors of particular individuals. In this article, I am examining portions of my biography, situated in U.S. society during the 1990s, to understand the "personal troubles" my gender produces in the context of a two-and-only-two gender system. I am not attempting to generalize my experiences; rather, I am trying to use them to examine and reflect on the processes and structure of gender in our society.

Because my analysis is based on my memories and perceptions of events, it is limited by my ability to recall events and by my interpretation of those events. However, I am not claiming that my experiences provide the truth about gender and how it works. I am claiming that the biography of a person who lives on the margins of our gender system can provide theoretical insights into the processes and social structure of gender. Therefore, after describing my experiences, I examine how they illustrate and extend, as well as contradict, other work on the social construction of gender.

GENDERED ME

Each day, I negotiate the boundaries of gender. Each day, I face the possibility that someone will attribute the "wrong" gender to me based on my physical appearance.

I am six feet tall and large-boned. I have had short hair for most of my life. For the past several years, I have worn a crew cut or flat top. I do not shave or otherwise remove hair from my body (e.g., no eyebrow plucking). I do not wear dresses, skirts, high heels, or makeup. My only jewelry is a class ring, a "men's" watch (my wrists are too large for a "women's" watch), two small earrings (gold hoops, both in my left ear), and (occasionally) a necklace. I wear jeans or shorts, T-shirts, sweaters, polo/golf shirts, button-down collar shirts, and tennis shoes or boots. The jeans are "women's" (I do have hips) but do not look particularly "feminine." The rest of the outer garments are from men's departments. I prefer baggy clothes, so the fact that I have "womanly" breasts often is not obvious (I do not wear a bra). Sometimes, I wear a baseball cap or some other type of hat. I also am white and relatively young (30 years old).[2]

My gender display—what others interpret as my presented identity—regularly leads to the misattribution of my gender. An incongruity exists between my gender self-identity and the gender that others perceive. In my encounters with people I do not know, I sometimes conclude, based on our interactions, that they think I am a man. This does not mean that other people do not think I am a man, just that I have no way of knowing what they think without interacting with them.

Living with It

I have no illusions or delusions about my appearance. I know that my appearance is likely to be read as "masculine" (and male) and that how I see myself is socially irrelevant. Given our two-and-only-two gender structure, I must live with the consequences of my appearance. These consequences fall into two categories: issues of identity and issues of interaction.

My most common experience is being called "Sir" or being referred to by some other masculine linguistic marker (e.g., "he," "man"). This has happened for years, for as long as I can remember, when having encounters with people I do not know.[3] Once, in fact, the same worker at a fast-food restaurant called me "Ma'am" when she took my order and "Sir" when she gave it to me.

Using my credit cards sometimes is a challenge. Some clerks subtly indicate their disbelief, looking from the card to me and back at the card and checking my signature carefully. Others challenge my use of the card, asking whose it is or demanding identification. One cashier asked to see my driver's license and then asked me whether I was the son of the cardholder. Another clerk told me that my signature on the receipt "had better match" the one on the card. Presumably, this was her way of letting me know that she was not convinced it was my credit card.

My identity as a woman also is called into question when I try to use women-only spaces. Encounters in public rest rooms are an adventure. I have been told countless times that "This is the ladies' room." Other women say nothing to me, but their stares and conversations with others let me know what they think. I will hear them say, for example, "There was a man in there. " I also get stares when I enter a locker room. However, it seems that women are less concerned about my presence, there, perhaps because, given that it is a space for changing clothes, showering, and so forth, they will be able to make sure that I am really a woman. Dressing rooms in department stores also are problematic spaces. I remember shopping with my sister once and being offered a chair outside the room when I began to accompany her into the dressing room.

Women who believe that I am a man do not want me in women-only spaces. For example, one woman would not enter the rest room until I came out, and others have told me that I am in the wrong place. They also might not want to encounter me while they are alone. For example, seeing me walking at night when they are alone might be scary.[4]

I, on the other hand, am not afraid to walk alone, day or night. I do not worry that I will be subjected to the public harassment that many women endure (Gardner 1995). I am not a clear target for a potential rapist. I rely on the fact that a potential attacker would not want to attack a big man by mistake. This is not to say that men never are attacked, just that they are not viewed, and often do not view themselves, as being vulnerable to attack.

Being perceived as a man has made me privy to male-male interactional styles of which most women are not aware. I found out, quite by accident, that many men greet, or acknowledge, people (mostly other men) who make eye contact with them with a single nod. For example, I found that when I walked. down the halls of my brother's all-male dormitory making eye contact, men nodded their greetings at me. Oddly enough, these same men did not greet my brother,

I had to tell him about making eye contact and nodding as a greeting ritual. Apparently, in this case I was doing masculinity better than he was!

I also believe that I am treated differently, for example, in auto parts stores (staffed almost exclusively by men in most cases) because of the assumption that I am a man. Workers there assume that I know what I need and that my questions are legitimate requests for information. I suspect that I am treated more fairly than a feminine-appearing woman would be. I have not been able to test this proposition. However, Devor's participants did report "being treated more respectfully" (1989, 132) in such situations.

There is, however, a negative side to being assumed to be a man by other men. Once, a friend and I were driving in her car when a man failed to stop at an intersection and nearly crashed into us. As we drove away, I mouthed "stop sign" to him. When we both stopped our cars at the next intersection, he got out of his car and came up to the passenger side of the car, where I was sitting. He yelled obscenities at us and pounded and spit on the car window. Luckily, the windows were closed. I do not think he would have done that if he thought I was a woman. This was the first time I realized that one of the implications of being seen as a man was that I might be called on to defend myself from physical aggression from other men who felt challenged by me. This was a sobering and somewhat frightening thought.

Recently, I was verbally accosted by an older man who did not like where I had parked my car. As I walked down the street to work, he shouted that I should park at the university rather than on a side street nearby. I responded that it was a public street and that I could park there if I chose. He continued to yell, but the only thing I caught was the last part of what he said: "Your tires are going to get cut!" Based on my appearance that day—I was dressed casually and carrying a backpack, and I had my hat on backward—I believe he thought that I was a young male student rather than a female professor. I do not think he would have yelled at a person he thought to be a woman—and perhaps especially not a woman professor.

Given the presumption of heterosexuality that is part of our system of gender, my interactions with women who assume that I am a man also can be viewed from that perspective. For example, once my brother and I were shopping when we were "hit on" by two young women. The encounter ended before I realized what had happened. It was only when we walked away that I told him that I was pretty certain that they had thought both of us were men. A more common experience is realizing that when I am seen in public with one of my women friends, we are likely to be read as a heterosexual dyad. It is likely that if I were to walk through a shopping mall holding hands with a woman, no one would look twice, not because of their open-mindedness toward lesbian couples but rather because of their assumption that I was the male half of a straight couple. Recently, when walking through a mall with a friend and her infant, my observations of others' responses to us led me to believe that many of them assumed that we were a family on an outing, that is, that I was her partner and the father of the child.

Dealing with It

Although I now accept that being mistaken for a man will be a part of my life so long as I choose not to participate in femininity, there have been times when I consciously have tried to appear more feminine. I did this for a while when I was an undergraduate and again recently when I was on the academic job market. The first time, I let my hair grow nearly down to my shoulders and had it permed. I also grew long fingernails and wore nail polish. Much to my chagrin, even then one of my professors, who did not know my name, insistently referred to me in his kinship examples as "the son." Perhaps my first act on the way to my current stance was to point out to this man, politely and after class, that I was a woman.

More recently, I again let my hair grow out for several months, although I did not alter other aspects of my appearance. Once my hair was about two and a half inches long (from its original quarter inch), I realized, based on my encounters with strangers, that I had more or less passed back into the category of "woman." Then, when I returned to wearing a flat top, people again responded to me as if I were a man.

Because of my appearance, much of my negotiation of interactions with strangers involves attempts to anticipate their reactions to me. I need to assess whether they will be likely to assume that I am a man and whether that actually matters in the context of our encounters. Many times, my gender really is irrelevant, and it is just annoying to be misidentified. Other times, particularly when my appearance is coupled with something that identifies me by name (e.g., a check or credit card) without a photo, I might need to do something to ensure that my identity is not questioned. As a result of my experiences, I have developed some techniques to deal with gender misattribution.

In general, in unfamiliar public places, I avoid using the rest room because I know that it is a place where there is a high likelihood of misattribution and where misattribution is socially important. If I must use a public rest room, I try to make myself look as nonthreatening as possible. I do not wear a hat, and I try to rearrange my clothing to make my breasts more obvious. Here, I am trying to use my secondary sex characteristics to make my gender more obvious rather than the usual use of gender to make sex obvious. While in the rest room, I never make eye contact, and I get in and out as quickly as possible. Going in with a woman friend also is helpful; her presence legitimizes my own. People are less likely to think I am entering a space where I do not belong when I am with someone who looks like she does belong.[5]

To those women who verbally challenge my presence in the rest room, I reply, "I know," usually in an annoyed tone. When they stare or talk about me to the women they are with, I simply get out as quickly as possible. In general, I do not wait for someone I am with because there is too much chance of an unpleasant encounter.

I stopped trying on clothes before purchasing them a few years ago because my presence in the changing areas was met with stares and whispers. Exceptions are stores where the dressing rooms are completely private, where there are individual stalls rather than a room with stalls separated by curtains, or where business is slow and no one else is trying on clothes. If I am trying on a garment clearly intended for a woman, then I usually can do so without hassle. I guess the attendants assume that I must be a woman if I have, for example, a women's bathing suit in my hand. But usually, I think it is easier for me to try the clothes on at home and return them, if necessary, rather than risk creating a scene. Similarly, when I am with another woman who is trying on clothes, I just wait outside.

My strategy with credit cards and checks is to anticipate wariness on a clerk's part. When I sense that there is some doubt or when they challenge me, I say, "It's my card." I generally respond courteously to requests for photo ID, realizing that these might be routine checks because of concerns about increasingly widespread fraud. But for the clerk who asked for ID and still did not think it was my card, I had a stronger reaction. When she said that she was sorry for embarrassing me, I told her that I was not embarrassed but that she should be. I also am particularly careful to make sure that my signature is consistent with the back of the card. Faced with such situations, I feel somewhat nervous about signing my name—which, of course, makes me worry that my signature will look different from how it should.

Another strategy I have been experimenting with is wearing nail polish in the dark bright colors currently fashionable. I try to do this when I travel by plane. Given more stringent travel regulations, one always must present a photo ID. But my experiences have shown that my driver's license is not necessarily convincing. Nail polish might be. I also flash my polished nails when I enter airport rest rooms, hoping that they will provide a clue that I am indeed in the right place.

There are other cases in which the issues are less those of identity than of all the norms of interaction that, in our society, are gendered. My most common response to misattribution actually is to appear to ignore it, that is, to go on with the interaction as if nothing out of the ordinary has happened. Unless I feel that there is a good reason to establish my correct gender, I assume the identity others impose on me for the sake of smooth interaction. For example, if someone is selling me a movie ticket, then there is no reason to make sure that the person has accurately discerned my gender. Similarly, if it is clear that the person using "Sir" is talking to me, then I simply respond as appropriate. I accept the designation because it is irrelevant to the situation. It takes enough effort to be alert for misattributions and to decide which of them matter; responding to each one would take more energy than it is worth.

Sometimes, if our interaction involves conversation, my first verbal response is enough to let the other person know that I am actually a woman and not a man. My voice apparently is "feminine" enough to shift people's attributions to the other category. I know when this has happened by the apologies that usually accompany the mistake. I usually respond to the apologies by saying something like "No problem" and/or "It happens all the time." Sometimes, a misattributor will offer an account for the mistake, for example, saying that it was my hair or that they were not being very observant.

These experiences with gender and misattribution provide some theoretical insights into contemporary Western understandings. of gender and into the social structure of gender in contemporary society. Although there are a number of ways in which my experiences confirm the work of others, there also are some ways in which my experiences suggest other interpretations and conclusions.

WHAT DOES IT MEAN?

Gender is pervasive in our society. I cannot choose not to participate in it. Even if I try not to do gender, other people will do it for me. That is, given our two-and-only-two rule, they must attribute one of two genders to me. Still, although I cannot choose not to participate in gender, I can choose not to participate in femininity (as I have), at least with respect to physical appearance.

That is where the problems begin. Without the decorations of femininity, I do not look like a woman. That is, I do not look like what many people's commonsense understanding of gender tells them a woman looks like. How I see myself, even how I might wish others would see me, is socially irrelevant. It is the gender that I *appear* to be (my "perceived gender") that is most relevant to my social identity and interactions with others. The major consequence of this fact is that I must be continually aware of which gender I "give off" as well as which gender I "give" (Goffman 1959).

Because my gender self-identity is "not displayed obviously, immediately, and consistently" (Devor 1989, 58), I am somewhat of a failure in social terms with respect to gender. Causing people to be uncertain or wrong about one's gender is a violation of taken-for-granted rules that leads to embarrassment and discomfort; it means that something has gone wrong with the interaction (Garfinkel 1967; Kessler and McKenna 1978). This means that my nonresponse to misattribution is the more socially appropriate response; I am allowing others to maintain face (Goffman 1959, 1967). By not calling attention to their mistakes, I uphold their images of themselves as competent social actors. I also maintain my own image as competent by letting them assume that I am the gender I appear to them to be.

But I still have discreditable status; I carry a stigma (Goffman 1963). Because I have failed to participate appropriately in the creation of meaning with respect to gender (Devor 1989), I can be called on to account for my appearance. If discredited, I show myself to be an incompetent social actor. I am the one not following the rules, and I will pay the price for not providing people with the appropriate cues for placing me in the gender category to which I really belong.

I do think that it is, in many cases, safer to be read as a man than as some sort of deviant woman. "Man" is an acceptable category; it fits properly into people's gender worldview. Passing as a man often is the path of least resistance" (Devor 1989; Johnson 1997). For example, in situations where gender does not matter, letting people take me as a man is easier than correcting them.

Conversely, as Butler noted, "We regularly punish those who fail to do their gender right" (1990, 140). Feinberg maintained, "Masculine girls and women face terrible condemnation and brutality—including sexual violence—for crossing the boundary of what is 'acceptable' female expression" (1996, 114). People are more likely to harass me when they perceive me to be a woman who looks like a man. For example, when a group of teenagers realized that I was not a man because one of their mothers identified me correctly, they began to make derogatory comments when I passed them. One asked, for example, "Does she have a penis?"

Because of the assumption that a "masculine" woman is a lesbian, there is the risk of homophobic reactions (Gardner 1995; Lucal 1997). Perhaps surprisingly, I find that I am much more likely to be taken for a man than for a lesbian, at least based on my interactions with people and their reactions to me. This might be because people are less likely to reveal that they have taken me for a lesbian because it is less relevant to an encounter or because they believe this would be unacceptable. But I think it is more likely a product of the strength of our two-and-only-two system. I give enough masculine cues that I am seen not as a deviant woman but rather as a man, at least in most cases. The problem seems not to be that people are uncertain about my gender, which might lead them to conclude that I was a lesbian once they realized I was a woman. Rather, I seem to fit easily into a gender category—just not the one with which I identify.

In fact, because men represent the dominant gender in our society, being mistaken for a man can protect me from other types of gendered harassment. Because men can move around in public spaces safely (at least relative to women), a "masculine" woman also can enjoy this freedom (Devor 1989).

On the other hand, my use of particular spaces—those designated as for women only—may be challenged. Feinberg provided an intriguing analysis of the public rest room experience. She characterized women's reactions to a masculine person in a public rest room as "an example of genderphobia" (1996, 117), viewing such women as policing gender boundaries rather than believing that there really is a man in the women's rest room. She argued that women who truly believed that there was a man in their midst would react differently. Although this is an interesting perspective on her experiences, my experiences do not lead to the same conclusion.[6] Enough people have said to me that "This is the ladies' room" or have said to their companions that "There was a man in there" that I take their reactions at face value.

Still, if the two-and-only-two gender system is to be maintained, participants must be involved in policing the categories and their attendant identities and spaces. Even if policing boundaries is not explicitly intended, boundary maintenance is the effect of such responses to people's gender displays.

Boundaries and margins are an important component of both my experiences of gender and our theoretical understanding of gendering processes. I am, in effect both woman and not-woman. As a woman who often is a social man but who also is a woman living in a patriarchal society, I am in a unique position to see and act. I sometimes receive privileges usually limited to men, and I sometimes am oppressed by my status as a deviant woman. I am, in a sense, an outsider-within (Collins 1991). Positioned on the boundaries of gender categories, I have developed a consciousness that I hope will prove transformative (Anzaldua 1987).

In fact, one of the reasons why I decided to continue my nonparticipation in femininity was that my sociological training suggested that this could be one of my contributions to the eventual dismantling of patriarchal gender constructs. It would be my way of making the personal political. I accepted being taken for a man as the price I would pay to help subvert patriarchy. I believed that all of the inconveniences I was enduring meant that I actually was doing something to bring down the gender structures that entangled all of us.

Then, I read Lorber's (1994) *Paradoxes of Gender* and found out, much to my dismay, that I might not actually be challenging gender after all. Because of the way in which doing gender works in our two-and-only-two system, gender displays are simply read as evidence of one of the two categories. Therefore, gender bending, blending, and passing between the categories do not question the categories themselves. If one's social gender and personal (true) gender do not correspond, then this is irrelevant unless someone notices the lack of congruence.

This reality brings me to a paradox of my experiences. First, not only do others assume that I am one gender or the other, but I also insist that I *really am* a member of one of the two gender categories. That is, I am female; I self-identify as a woman. I do not claim to be some other gender or to have no gender at all. I simply place myself in the wrong category according to stereotypes and cultural standards; the gender I present, or that some people perceive me to be presenting, is inconsistent with the gender with which I identify myself as well as with the gender I could be "proven" to be. Socially, I display the wrong gender; personally, I identify as the proper gender.

Second, although I ultimately would like to see the destruction of our current gender structure, I am not to the point of personally abandoning gender. Right now, I do not want people to see me as genderless as much as I want them to see me as a woman. That is, I would like to expand the category of "woman" to include people like me. I, too, am deeply embedded in our gender system, even though I do not play by many of its rules. For me, as for most people in our society, gender is a substantial part of my personal identity (Howard and Hollander 1997). Socially, the problem is that I do not present a gender display that is consistently read as feminine. In fact, I consciously do not participate in the trappings of femininity. However, I do identify myself as a woman, not as a man or as someone outside of the two-and-only-two categories.

Yet, I do believe, as Lorber (1994) does, that the purpose of gender, as it currently is constructed, is to oppress women. Lorber analyzed gender as a "process of creating distinguishable social statuses for the assignment of rights and responsibilities" that ends up putting women in a devalued and oppressed position (1994, 32). As Martin put it, "Bodies that clearly delineate gender status facilitate the maintenance of the gender hierarchy" (1998, 495).

For society, gender means difference (Lorber 1994). The erosion of the boundaries would problematize that structure. Therefore, for gender to operate as it currently does, the category "woman" be expanded to include people like me. The maintenance of the gender structure is dependent on the creation of a few categories that are mutually exclusive, the members of which are as different as possible (Lorber 1994). It is the clarity of the boundaries between the categories that allows gender to be used to assign rights and responsibilities as well as resources and rewards.

It is that part of gender—what it is used for—that is most problematic. Indeed, is it not *patriarchal*—or, even more specifically, *hetero-patriarchal*—constructions of gender that are actually the problem? It is not the differences between men and women, or the categories themselves, so much as the meanings ascribed to the categories and, even more important, the hierarchical nature of gender under patriarchy that is the problem (Johnson 1997). Therefore, I am rebelling not against my femaleness or even my womanhood; instead, I am protesting contemporary constructions of femininity and, at least indirectly, masculinity under patriarchy. We do not, in fact, know what gender would look like if it were not constructed around heterosexuality in the context of patriarchy.

Although it is possible that the end of patriarchy would mean the end of gender, it is at least conceivable that something like what we now call gender could exist in a postpatriarchal future. The two-and-only-two categorization might well disappear, there being no hierarchy for it to justify. But I do not think that we should make the assumption that gender and patriarchy are synonymous.

Theoretically, this analysis points to some similarities and differences between the work of Lorber (1994) and the works of Butler (1990), Goffman (1976, 1977), and West and Zimmerman (1987). Lorber (1994) conceptualized gender as social structure, whereas the others focused more on the interactive and processual nature of gender. Butler (1990) and Goffman (1976, 1977) view gender as a performance, and West and Zimmerman (1987) examined it as something all of us do. One result of this difference in approach is that in Lorber's (1994) work, gender comes across as something that we are caught in—something that, despite any attempts to the contrary, we cannot break out of. This conclusion is particularly apparent in Lorber's argument that gender rebellion, in the context of our two-and-only-two system, ends up supporting what it purports to subvert. Yet, my own experiences suggest an alternative possibility that is more in line with the view of gender offered by West and Zimmerman (1987): If gender is a product of interaction, and if it is produced in a particular context, then it can be changed if we change our performances. However, the effects of a performance linger, and gender ends up being institutionalized. It is institutionalized, in our society, in a way that perpetuates inequality, as Lorber's (1994) work shows. So, it seems that a combination of these two approaches is needed.

In fact, Lorber's (1994) work seems to suggest that effective gender rebellion requires a more blatant approach—bearded men in dresses, perhaps, or more active, responses to misattribution. For example, if I corrected every person who called me "Sir," and if I insisted on my right to be addressed appropriately and granted access to women-only spaces, then perhaps I could start to break down gender norms. If I asserted my right to use public facilities without being harassed, and if I challenged each person who gave me "the look," then perhaps I would be contributing to the demise of gender as we know it. It seems that the key would be to provide visible evidence of the nonmutual exclusivity of the categories. Would *this* break down the patriarchal components of gender? Perhaps it would, but it also would be exhausting.

Perhaps there is another possibility. In a recent book, *The Gender Knot,* Johnson (1997) argued that when it comes to gender and patriarchy, most of us follow the paths of least resistance; we "go along to get along," allowing our actions to be shaped by the gender system. Collectively, our actions help patriarchy maintain and perpetuate a system of oppression and privilege. Thus, by withdrawing our support from this system by choosing paths of greater resistance, we can start to chip away at it. Many people participate in gender because they cannot imagine any alternatives. In my classroom, and in my interactions and encounters with strangers, my presence can make it difficult for people not to see that there *are* other paths. In other words, following from West and Zimmerman (1987), I can subvert gender by doing it differently.

For example, I think it is true that my existence does not have an effect on strangers who assume that I am a man and never learn otherwise. For them, I do uphold the two-and-only-two system. But there are other cases in which my existence can have an effect. For example, when people initially take me for a man but then find out that I actually am a woman, at least for that moment, the naturalness of gender may be called into question. In these cases, my presence can provoke a "category crisis" (Garber 1992, 16) because it challenges the sex/gender binary system.

The subversive potential of my gender might be strongest in my classrooms. When I teach about the sociology of gender, my students can see me as the embodiment of the social construction of gender. Not all of my students have transformative experiences as a result of taking a course with me; there is the chance that some of them see me as a "freak" or as an exception. Still, after listening to stories about my experiences with gender and reading literature on the subject, many students begin to see how and why gender is a social product. I can disentangle sex, gender, and sexuality in the contemporary United States for them. Students can begin to see the connection between biographical experiences and the structure of society. As one of my students noted, I clearly live the material I am teaching. If that helps me to get my point across, then perhaps I am subverting the binary gender system after all. Although my gendered presence and my way of doing gender might make others—and sometimes even me—uncomfortable, no one ever said that dismantling patriarchy was going to be easy.

NOTES

1. Ethnomethodology has been described as "the study of commonsense practical reasoning" (Collins 1988, 274). It examines how people make sense of their everyday experiences. Ethnomethodology is particularly useful in studying gender because it helps to uncover the assumptions on which our understandings of sex and gender are based.
2. I obviously have left much out by not examining my gendered experiences in the context of race, age, class, sexuality, region, and so forth. Such a project clearly is more complex. As Weston pointed out gender presentations are complicated by other statuses of their presenters: "What it takes to kick a person over into another gendered category can differ with race, class, religion, and time" (1996, 168). Furthermore, I am well aware that my whiteness allows me to assume that my experiences are simply a product of gender (see, e.g., hooks 1981; Lucal 1996; Spelman 1988; West and Fenstermaker 1995). For now, suffice it to say that it is my privileged position on some of these axes and my more disadvantaged position on others that combine to delineate my overall experience.
3. In fact, such experiences are not always limited to encounters with strangers. My grandmother, who does not see me often, twice has mistaken me for either my brother-in-law or some unknown man.
4. My experiences in rest rooms and other public spaces might be very different if I were, say, African American rather than white. Given the stereotypes of African American men, I think that white women would react very differently to encountering me (see, e.g., Staples [1986] 1993).
5. I also have noticed that there are certain types of rest rooms in which I will not be verbally challenged; the higher the social status of the place, the less likely I will be harassed. For example, when I go to the theater, I might get stared at, but my presence never has been challenged.
6. An anonymous reviewer offered one possible explanation for this. Women see women's rest rooms as their space; they feel safe, and even empowered, there. Instead of fearing men in such space, they might instead pose a threat to any man who might intrude. Their invulnerability in this situation is, of course, not physically based but rather socially constructed. I thank the reviewer for this suggestion.

REFERENCES

Anzaldua, G. 1987. *Borderlands/La Frontera.* San Francisco: Aunt Lute Books.

Berger, P. 1963. *Invitation to sociology.* New York: Anchor.

Bordo, S. 1993. *Unbearable weight.* Berkeley: University of California Press.

Butler, J. 1990. *Gender trouble.* New York: Routledge.

Collins, P. H. 1991. *Black feminist thought.* New York: Routledge.

Collins, R. 1988. *Theoretical sociology.* San Diego: Harcourt Brace Jovanovich.

Devor, H. 1989. *Gender blending: Confronting the limits of duality.* Bloomington: Indiana University Press.

Feinberg, L. 1996. *Transgender warriors.* Boston: Beacon.

Garber, M. 1992. *Vested interests: Cross-dressing and cultural anxiety.* New York: HarperPerennial.

Gardner, C. B. 1995. *Passing by: Gender and public harassment.* Berkeley: University of California.

Garfinkel, H. 1967. *Studies in ethnomethodology.* Englewood Cliffs, NJ: Prentice Hall.

Goffman, E. 1959. *The presentation of self in everyday life.* Garden City, NY. Doubleday.

———. 1963. *Stigma.* Englewood Cliffs, NJ: Prentice Hall.

———. 1967. *Interaction ritual.* New York: Anchor/Doubleday.

———. 1976. Gender display. *Studies in the Anthropology of Visual Communication* 3:69–77.

———.1977. The arrangement between the sexes. *Theory and Society* 4:301–31.

hooks, b. 1981. *Ain't I a woman: Black women and feminism.* Boston: South End Press.

Howard, J. A., and J. Hollander. 1997. *Gendered situations, gendered selves.* Thousand Oaks, CA: Sage.

Kessler, S. J., and W. McKenna. 1978. *Gender: An ethnomethodological approach.* New York: John Wiley.
Krieger, S. 1991. *Social science and the self.* New Brunswick, NJ: Rutgers University Press.
Johnson, A. G. 1997. *The gender knot: Unraveling our patriarchal legacy.* Philadelphia: Temple University Press.
Lorber, J. 1994. *Paradoxes of gender.* New Haven, CT. Yale University Press.
———. 1996. Beyond the binaries: Depolarizing the categories of sex, sexuality, and gender. *Sociological Inquiry* 66:143–59.
Lucal, B. 1996. Oppression and privilege: Toward a relational conceptualization of race. *Teaching Sociology* 24:245–55.
———. 1997. "Hey, this is the ladies' room!": Gender misattribution and public harassment. *Perspectives on Social Problems* 9:43–57.
Martin, K. A. 1998. Becoming a gendered body: Practices of preschools. *American Sociological Review* 63:494–511.
Mills, C. W. 1959. *The sociological imagination.* London: Oxford University Press.
Myhre, J. R. M. 1995. One bad hair day too many, or the hairstory of an androgynous young feminist. In *Listen up: Voices from the next feminist generation,* edited by B. Findlen. Seattle, WA: Seal Press.
Reinharz, S. 1992. *Feminist methods in social research.* New York: Oxford University Press.
Richardson, L. 1994. Writing: A method of inquiry. In *Handbook of Qualitative Research,* edited by N. K. Denzin and Y. S. Lincoln. Thousand Oaks, CA: Sage.
Riemer, J. W. 1977. Varieties of opportunistic research. *Urban Life* 5:467–77.
———. 1988. Work and self. In *Personal sociology,* edited by P. C. Higgins and J. M. Johnson. New York: Praeger.
Spelman, E. V. 1988. *Inessential woman: Problems of exclusion in feminist thought.* Boston: Beacon.
Staples, B. 1993. Just walk on by. In *Experiencing race, class, and gender in the United States,* edited by V. Cyrus. Mountain View, CA: Mayfield. (Originally published 1986)
West, C., and S. Fenstermaker. 1995. Doing difference. *Gender & Society* 9:8–37.
West, C., and D. H. Zimmerman. 1987. Doing gender. *Gender & Society* 1:125–51.
Zimmerman, D. H. 1992. They were all doing gender, but they weren't all passing: Comment on Rogers. *Gender & Society* 6:192–98.

5

Multiple Genders among North American Indians

BY SERENA NANDA

The anthropologist, Serena Nanda, is widely known for her ethnography of India's Hijaras, entitled Neither Man nor Woman. *This article is from her new book on multiple sex/gender systems around the world. Nanda's analysis of multiple genders among Native North Americans is rich and detailed. As you read this piece, consider the long-term consequences of the failure of European colonists and early anthropologists to get beyond their ethnocentric assumptions so that they could understand and respect the gender diversity of North American Indian cultures.*

1. Why does Serena Nanda use the term "gender variants" instead of "two-spirit" and "berdache"?
2. What was the relationship between sexual orientation and gender status among American Indians whose cultures included more than two sex/gender categories? How about hermaphroditism and gender status?
3. Why was there often an association between spiritual power and gender variance in Native American cultures?

The early encounters between Europeans and Indian societies in the New World, in the fifteenth through the seventeenth centuries, brought together cultures with very different sex/gender systems. The Spanish explorers, coming from a society where sodomy was a heinous crime, were filled with contempt and outrage when they recorded the presence of men in American Indian societies who performed the work of women, dressed like women, and had sexual relations with men (Lang 1996; Roscoe in 1995).

Europeans labelled these men "berdache," a term originally derived from an Arabic word meaning male prostitute. As such, this term is inappropriate and insulting, and I use it here only to indicate the history of European (mis)understanding of American Indian sex/gender diversity. The term berdache focused attention on the sexuality associated with mixed gender roles, which the Europeans identified, incorrectly, with the "unnatural" and sinful practice of sodomy in their own societies. In their ethnocentrism, the early European explorers and colonists were unable to see beyond their own sex/gender systems and thus did not understand the multiple sex/gender systems they encountered in the Americas. They also largely overlooked the specialized and spiritual functions of many of these alternative sex/gender roles and the positive value attached to them in many American Indian societies.

By the late-nineteenth and early-twentieth centuries, some anthropologists included accounts of North American Indian sex/gender diversity in their ethnographies. They attempted to explain the berdache from various functional perspectives, that is, in terms of the contributions these sex/gender roles made to social structure or culture. These accounts, though less contemptuous than earlier ones, nevertheless largely retained the emphasis on berdache sexuality. The berdache was defined as a form of "institutionalized homosexuality," which served as a social niche for individuals whose personality and sexual orientation did not match the definition of masculinity in their societies, or as a "way out" of the masculine or warrior role for "cowardly" or "failed" men (see Callender and Kochems 1983).

Anthropological accounts increasingly paid more attention to the association of the berdache with shamanism and spiritual powers and also noted that mixed gender roles were often central and highly valued in American Indian cultures, rather than marginal and deviant. These accounts were, nevertheless, also ethnocentric in misidentifying indigenous gender diversity with European concepts of homosexuality, transvestism, or hermaphroditism, which continued to distort their indigenous meanings.

In American Indian societies, the European homosexual/heterosexual dichotomy was not culturally relevant and the European labeling of the berdache as homosexuals resulted from their own cultural emphasis on sexuality as a central, even defining, aspect of gender and on sodomy as an abnormal practice and/or a sin. While berdache in many American Indian societies did engage in sexual relations and even married persons of the same sex, this was not central to their alternative gender role. Another overemphasis resulting from European ethnocentrism was the identification of berdache as *transvestites.* Although berdache often cross-dressed, transvestism was not consistent within or across societies. European descriptions of berdache as *hermaphrodites* were also inaccurate. Considering the variation in alternative sex/gender roles in native North America, a working definition may be useful: the berdache in the anthropological literature refers to people who partly or completely take on aspects of the culturally defined role of the other sex and who are classified neither as women nor men, but as genders of their own (see Callender and Kochems 1983:443). It is important to note here that berdache thus refers to variant gender roles, rather than a complete crossing over to an opposite gender role.

In the past twenty-five years there have been important shifts in perspectives on sex/gender diversity among American Indians and anthropologists, both Indian and non-Indian (Jacobs, Thomas, and Lang 1997:Introduction). Most current research rejects institutionalized homosexuality as an adequate explanation of American Indian gender diversity, emphasizing the importance of occupation rather than sexuality as its central feature. Contemporary ethnography views multiple sex/gender roles as a normative part of American Indian sex/gender systems, rather than as a marginal or deviant part (Albers 1989:134; Jacobs et al. 1997; Lang 1998). A new emphasis on the variety of alternative sex/gender roles in North America undercuts the earlier treatment of the berdache as a unitary phenomenon across North (and South) America (Callender and Kochems 1983; Jacobs et al. 1997; Lang 1998; Roscoe 1998). Current research also emphasizes the integrated and often highly valued position of gender variant persons and the association of sex/gender diversity with spiritual power (Roscoe 1996; Williams 1992).

A change in terminology has also taken place. Berdache generally has been rejected, but there is no unanimous agreement on what should replace it. One widely accepted suggestion is the term *two-spirit* (Jacobs et al. 1997; Lang 1998), a term coined in 1990 by urban American Indian gays and lesbians. Two-spirit has the advantage of conveying the spiritual nature of gender variance as viewed by gay, lesbian, and transgendered American Indians and also the spirituality associated with traditional American Indian gender variance, but the cultural continuity suggested by two-spirit is in fact a subject of debate. Another problem is that two-spirit emphasizes the Euro-American gender construction of only two genders. Thus, I use the more culturally neutral term, variant genders (or gender variants) and specific indigenous terms wherever possible.

DISTRIBUTION AND CHARACTERISTICS OF VARIANT SEX/GENDER ROLES

Multiple sex/gender systems were found in many, though not all, American Indian societies. Male gender variant roles (variant gender roles assumed by biological males) are documented for 110 to 150 societies. These roles occurred most frequently in the region extending from California to the Mississippi Valley and upper-Great Lakes, the Plains and the Prairies, the Southwest, and to a lesser extent along the Northwest Coast tribes. With few exceptions, gender variance is not historically documented for eastern North America, though it may have existed prior to European invasion and disappeared before it could be recorded historically (Callender and Kochems 1983; Fulton and Anderson 1992).

There were many variations in North American Indian gender diversity. American Indian cultures included three or four genders: men, women, male variants, and female variants (biological females who by engaging in male activities were reclassified as to gender). Gender variant roles differed in the criteria by which they were defined; the degree of their integration into the society; the norms governing their behavior; the way the role was acknowledged publicly or sanctioned; how others were expected to behave toward gender variant persons; the degree to which a gender changer was expected to adopt the role of the opposite sex or was limited in doing so; the power, sacred or secular, that was attributed to them; and the path to recruitment.

In spite of this variety, however, there were also some common or widespread features: transvestism, cross-gender occupation, same sex (but different gender) sexuality, some culturally normative and acknowledged process for recruitment to the role, special language and ritual roles, and associations with spiritual power.

TRANSVESTISM

The degree to which male and female gender variants were permitted to wear the clothing of the other sex varied. Transvestism was often associated with gender variance but was not equally important in all societies. Male gender variants frequently adopted women's dress and hairstyles partially or completely, and female gender variants partially adopted the clothing of men; sometimes, however, transvestism was prohibited. The choice of clothing was sometimes an individual matter and gender variants might mix their clothing and their accoutrements. For example, a female gender variant might wear a woman's dress but carry (male) weapons. Dress was also sometimes situationally determined: a male gender variant would have to wear men's clothing while engaging in warfare but might wear women's clothes at other times. Similarly, female gender variants might wear women's clothing when gathering (women's work), but male clothing when hunting (men's work) (Callender and Kochems 1983:447). Among the Navajo, a male gender variant, *nádleeh,* would adopt almost all aspects of a woman's dress, work, language and behavior; the Mohave male gender variant, called *alyha,* was at the extreme end of the cross-gender continuum in imitating female physiology as well as transvestism. . . . Repression of visible forms of gender diversity, and ultimately the almost total decline of transvestism, were a direct result of American prohibitions against it.

OCCUPATION

Contemporary analysis emphasizes occupational aspects of American Indian gender variance as a central feature. Most frequently a boy's interest in the implements and activities of women and a girl's interest in the tools of male occupations signaled an individual's wish to undertake a gender variant role (Callender and Kochems 1983:447; Whitehead 1981). In hunting societies, for example, female gender variance was signaled by a girl rejecting the domestic activities associated with women and participating in playing and hunting with boys. In the arctic and subarctic, particularly, this was sometimes encouraged by a girl's parents if there were not enough boys to provide the family with food (Lang 1998). Male gender variants were frequently considered especially skilled and industrious in women's crafts and domestic work (though not in agriculture, where this was a man's task) (Roscoe 1991; 1996). Female gender crossers sometimes won reputations as superior hunters and warriors.

Male gender variants' households were often more prosperous than others, sometimes because they were hired by whites. In their own societies the excellence of male gender variants' craftwork was sometimes ascribed to a supernatural sanction for their gender transformation (Callender and Kochems 1983:448). Female gender variants opted out of motherhood, so were not encumbered by caring for children, which may explain their success as hunters or warriors. In some societies, gender variants could engage in both men's and women's work, and this, too, accounted for their increased wealth. Another source of income was payment for the special social activities of gender variants due to their intermediate gender status, such as acting as go-betweens in marriage. Through their diverse occupations, then, gender variants were often central rather than marginal in their societies.

Early anthropological explanations of male gender variant roles as a niche for a "failed" or cowardly man who wished to avoid warfare or other aspects of the masculine role are no longer widely accepted. To begin with, masculinity was not associated with warrior status in all American Indian cultures. In some societies, male gender variants were warriors and in many others, men who rejected the warrior role did not become gender variants. Sometimes male gender variants did not go to war because of cultural prohibitions against their using symbols of maleness, for example, the prohibition against their using the bow among the Illinois. Where male gender variants did not fight, they sometimes had other important roles in warfare, like treating the wounded, carrying supplies for the war party, or directing postbattle ceremonials (Callender and Kochems 1983:449). In a few societies male gender variants become outstanding warriors, such as Finds Them and Kills Them, a Crow Indian who performed daring feats of bravery while fighting with the United States Army against the Crow's traditional enemies, the Lakota Sioux (Roscoe 1998:23).

GENDER VARIANCE AND SEXUALITY

Generally, sexuality was not central in defining gender status among American Indians. But in any case, the assumption by European observers that gender variants were homosexuals meant they did not take much trouble to investigate or record information on this topic. In some American Indian societies same-sex sexual desire/practice did figure significantly in the definition of gender variant roles; in others it did not (Callender and Kochems 1983:449). Some early reports noted specifically that male gender variants lived with and/or had sexual relations with women as well as men; in other societies they were reported as having sexual relations only with men, and in still other societies, of having no sexual relationships at all (Lang 1998:189–95).

The bisexual orientation of some gender variant persons may have been a culturally accepted expression of their gender variance. It may have resulted from an individual's life experiences, such as the age at which he or she entered the gender variant role, and/or it may have been one aspect of the general freedom of sexual expression in many American Indian societies. While male and female gender variants most frequently had sexual relations with, or married, persons of the same biological sex as themselves, these relationships were not considered homosexual in the contemporary Western understanding of that term. In a multiple gender system the partners would be of the same sex but different genders, and homogender, rather than homosexual, practices bore the brunt of negative cultural sanctions. The sexual partners of gender variants were never considered gender variants themselves.

The Navajo are a good example (Thomas 1997). The Navajo have four genders; in addition to man and woman there are two gender variants: masculine female-bodied nádleeh and feminine male-bodied nádleeh. A sexual relationship between a female nádleeh and a woman or a sexual relationship between a male-bodied nádleeh and a man were not stigmatized because these persons were of different genders, although of the same biological sex. However, a sexual relationship between two women, two men, two female-bodied nádleeh or two male-bodied nádleeh, was considered homosexual, and even incestual, and was strongly disapproved of.

The relation of sexuality to valiant sex/gender roles across North America suggests that sexual relations between gender variants and persons of the same biological sex were a result rather than a cause of gender variance. Sexual relationships between a man and a male gender variant were accepted in most American Indian societies, though not in all, and appear to have been negatively sanctioned only when it interfered with child-producing heterosexual marriages. Gender variants' sexual relationships varied from casual and wide-ranging (Europeans used the term promiscuous), to stable, and sometimes even involved life-long marriages. In some societies, however, male gender variants were not permitted to engage in long-term relationships with men, either in or out of wedlock. In many cases, gender variants were reported as living alone.

There are some practical reasons why a man might desire sexual relations with a (male) gender variant: in some societies taboos on sexual relations with menstruating or pregnant women restricted opportunities for sexual intercourse; in other societies, sexual relations with a gender variant person were exempt from punishment for extramarital affairs; in still other societies, for example, among the Navajo, some gender variants were considered especially lucky and a man might hope to vicariously partake of this quality by having sexual relations with them (Lang 1998:349).

BIOLOGICAL SEX AND GENDER TRANSFORMATIONS

European observers often confused gender variants with hermaphrodites. Some American Indian societies explicitly distinguished hermaphrodites from gender variants and treated them differently; others assigned gender variant persons and hermaphrodites to the same alternative gender status. With the exception of the Navajo, in most American Indian societies biological sex (or the intersexedness of the hermaphrodite) was not the criterion for a gender variant role, nor were the individuals who occupied gender variant roles anatomically abnormal. The Navajo distinguished between the intersexed and the alternatively gendered, but treated them similarly, though not exactly the same (Thomas 1997; Hill 1935).

And even as the traditional Navajo sex/gender system had biological sex as its starting point, it was only a starting point, and Navajo nádleeh were distinguished by sex-linked behaviors, such as body language, clothing, ceremonial roles, speech style, and work. Feminine, male bodied nádleeh might engage in women's activities such as cooking, weaving, household tasks, and making pottery. Masculine, female-bodied nádleeh, unlike other female-bodied persons, avoided childbirth; today they are associated with male occupational roles such as construction or firefighting (although ordinary women also sometimes engage in these occupations). Traditionally, female-bodied nádleeh had specific roles in Navajo ceremonials.

Thus, even where hermaphrodites occupied a special gender variant role, American Indian gender variance was defined more by cultural than biological criteria. In one recorded case of an interview with and physical examination of a gender variant male, the previously mentioned Finds Them and Kills Them, his genitals were found to be completely normal (Roscoe 1998).

If American Indian gender variants were not generally hermaphrodites, or conceptualized as such, neither were they conceptualized as transsexuals. Gender transformations among gender variants were recognized as only a partial transformation, and the gender variant was not thought of as having become a person of the opposite sex/gender. Rather, gender variant roles were autonomous gender roles that combined the characteristics of men and women and had some unique features of their own. This was sometimes symbolically recognized: among the Zuni a male gender variant was buried in women's dress but men's trousers on the men's side of the graveyard (Parsons quoted in Callender and Kochems 1983:454; Roscoe 1991:124, 145). Male gender variants were neither men—by virtue of their chosen occupations, dress, demeanor, and possibly sexuality—nor women, because of their anatomy and their inability to bear children. Only among the Mohave do we find the extreme imitation of women's physiological processes related to reproduction and the claims to have female sexual organs—both of which were ridiculed within Mohave society. But even here, where informants reported that female gender variants did not menstruate, this did not make them culturally men. Rather it was the mixed quality of gender variant status that was culturally elaborated in native North America, and this was the source of supernatural powers sometimes attributed to them.

SACRED POWER

The association between the spiritual power and gender variance occurred in most, if not all, Native American societies. Even where, as previously noted, recruitment to the role was occasioned by a child's interest in occupational activities of the opposite sex, supernatural sanction, frequently appearing in visions or dreams, was also involved. Where this occurred, as it did mainly in the Prairie and Plains societies, the visions involved female supernatural figures, often the moon. Among the Omaha, for example, the moon appeared in a dream holding a burden strap—a symbol of female work—in one hand, and a bow—a symbol of male work—in the other. When the male dreamer reached for the bow, the moon forced him to take the burden strap (Whitehead 1981). Among the Mohave, a child's choice of male or female implements heralding gender variant status was sometimes prefigured by a dream that was. believed to come to an embryo in the womb (Devereux 1937).

Sometimes, by virtue of the power associated with their gender ambiguity, gender variants were ritual adepts and curers, or had special ritual functions (Callender and Kochems 1983:453, Lang 1998). Gender variants did not always have important sacred roles in native North America, however. Where feminine qualities were associated with these roles, male gender variants might become spiritual leaders or healers, but where these roles were associated with male qualities they were not entered into by male gender variants. Among the Plains Indians, with their emphasis on the vision as a source of supernatural power, male gender variants were regarded as holy persons, but in California Indian societies, this was not the case and in some American Indian societies gender variants were specifically excluded from religious roles (Lang 1998:167). Sometimes it was the individual personality of the gender variant rather than his/her gender variance itself, that resulted in occupying sacred roles (see Commentary following Callender and Kochems 1983). Nevertheless, the importance of sacred power was so widely associated with sex/gender diversity in native North America that it is generally agreed to be an important explanation of the frequency of gender diversity in this region of the world.

In spite of cultural differences, some significant similarities among American Indian societies are particularly consistent with multigender systems and the positive value placed on sex/gender diversity (Lang 1996). One of these similarities is a cosmology (system of religious beliefs and way of seeing the world) in which transformation and ambiguity are recurring themes. Thus a person who contains both masculine and feminine qualities or one who is transformed from the sex/gender assigned at birth into a different gender in later life manifests some of the many kinds of transformations and ambiguities that are possible, not only for humans, but for animals and objects in the natural environment. Indeed, in many American Indian cultures, sex/gender ambiguity, lack of sexual differentiation, and sex/gender transformations play an important part in the story of creation. American Indian cosmology may not be "the cause" of sex/gender diversity but it certainly (as in India) provides a hospitable context for it (Lang 1996:187). . . .

As a result of Euro-American repression and the growing assimilation of Euro-American sex/gender ideologies, both female and male gender variant roles among American Indians largely disappeared by the 1930s, as the reservation system was well under way. And yet, its echoes may remain. The current academic interest in American Indian multigender roles, and particularly the testimony of contemporary two-spirits, remind us that alternatives are possible and that understanding American Indian sex/gender diversity in the past and present makes a significant contribution to understandings of sex/gender diversity in the larger society.

REFERENCES

Albers, Patricia C. 1989. "From Illusion to Illumination: Anthropological Studies of American Indian Women." In *Gender and Anthropology: Critical Reviews for Research and Teaching,* edited by Sandra Morgen. Washington, DC: American Anthropological Association.

Callender, Charles, and Lee M. Kochems. 1983. "The North American Berdache." *Current Anthropology* 24 (4): 443–56 (Commentary, pp. 456–70).

Fulton, Robert, and Steven W. Anderson. 1992. "The Amerindian 'Man-Woman': Gender, Liminality, and Cultural Continuity." *Current Anthropology* 33 (5): 603–10.

Hill, Willard W. 1935. "The Status of the Hermaphrodite and Transvestite in Navaho Culture." *American Anthropologist* 37:273–79.

Jacobs, Sue-Ellen, Wesley Thomas, and Sabine Lang, eds. 1997. *Two-Spirit People: Native American Gender Identity, Sexuality, and Spirituality.* Urbana and Chicago: University of Illinois.

Lang, Sabine. 1996. "There Is More than Just Men and Women: Gender Variance in North America." In *Gender Reversals and Gender Culture,* edited by Sabrina Petra Ramet, pp. 183–96. London and New York: Routledge

———. 1998. *Men as Women, Women as Men: Changing Gender in Native American Cultures.* Trans. from the German by John L. Vantine. Austin: University of Texas Press.

Roscoe, Will, and Stephen O. Murray, eds. 1998. *Boy-Wives and Female-Husbands: Studies in African Homosexualities.* New York: St. Martins.

Thomas, Wesley. 1997. "Navajo Cultural Constructions of Gender and Sexuality." In *Two-Spirit People: Native American Gender Identity, Sexuality, and Spirituality,* edited by Sue-Ellen Jacobs, Wesley Thomas, and Sabine Lang, pp. 156–73. Urbana and Chicago: University of Illinois.

Williams, Walter. 1992. *The Spirit and the Flesh: Sexual Diversity in American Indian Culture.* Boston: Beacon.

CHAPTER 2

The Interaction of Gender with Other Socially Constructed Prisms

After considering what gender is and isn't, we are going to complicate things a bit by looking at how a variety of other socially constructed categories of difference and inequality such as race, ethnicity, social class, religion, age, sexual orientation and identification, and ability/disability shape gender. As is the case with prisms in a kaleidoscope, the interaction of gender with other social prisms creates complex patterns of identity and relationships for people across groups and situations. Because there are so many different social prisms that interact with gender in daily life, we can only discuss a few in this chapter; however, other social categories are explored throughout this book. The articles we have selected for this chapter illustrate two key arguments. First, gender is a complex and multifaceted array of experiences and meanings that cannot be understood without considering the social context within which they are situated. Second, variations in the meaning and display of gender are related to different levels of prestige, privilege, and power associated with membership in other socially constructed categories of difference and inequality.

PRIVILEGE

In our daily lives, there usually isn't enough time or opportunity to consider how the interaction of multiple social categories to which we belong affects beliefs, behaviors, and life chances. In particular, we are discouraged from critically examining our culture, as will be discussed in Chapter 3. People who occupy positions of privilege often do not notice how their privileged social positions influence them. In the United States, privilege is associated with white skin color, masculinity, wealth, heterosexuality, youth, able-bodiedness, and so on. It seems normal to those of us who occupy privileged positions and those we interact with, that our positions of privilege be deferred to, allowing us to move more freely in society. Peggy McIntosh is a pioneer in examining the hidden and unearned benefits of privilege. Her article in this chapter compares white privilege with male privilege and helps us to develop a better understanding of what it means to be privileged. McIntosh illustrates that persons who have "unearned advantages" often do not understand how their privilege is a function of the disempowerment of others. For example, male privilege seems "normal" and white privilege seems "natural."

The struggle for women's rights also has seen the effects of privilege, with white women dominating this struggle historically. The privilege of race and social class created a view of woman as a universal category. Many women of color stood up for women's rights; however, they did so in response to a universal definition of womanhood derived from privilege (hooks, 1981). In 1867, former slave Sojourner Truth, in response to a white man who felt women were more delicate than men, told of the exertion and toil of her work as a slave and asked "ain't I a woman" (Guy-Scheftall, 1995). In the late 1970s, women of color including Audre Lorde (1982), bell hooks (1981), Angela Davis (1981), Gloria Anzaldua (1987), and others spoke out against white privilege within the women's movement. These women, recognizing that the issues facing women of color were *not* always the same as those of white women, carried on a battle to make African-American women visible in the second wave of feminism. For example, while white feminists were fighting for the right to abortion, African-American women were fighting other laws and sterilization practices that denied them the right to control their own fertility. Women of color, including Patricia Hill Collins (this chapter) and others, some of which are listed in this introduction, continue to challenge the white-dominated definitions of gender and fight to include in the analysis of gender an understanding of the experience of domination and privilege of all women.

UNDERSTANDING THE INTERACTION OF GENDER WITH OTHER CATEGORIES OF DIFFERENCE AND INEQUALITY

Throughout this chapter introduction, you will read about social scientists and social activists who attempt to understand the interactions between "interlocking oppressions." Social scientists develop theories, with their primary focus on explanation, whereas social activists explore the topic of interlocking oppressions from the perspective of initiating social change (see the reading by Barbara Ryan in this chapter for a discussion of the latter). Although the goals of explanation and change are rarely separated in feminist research, they reflect different emphases (Hill Collins, 1990). As such, these two different agendas shape attempts to understand the interaction of gender with multiple social prisms of difference and inequality.

The effort to include the perspectives and experiences of *all* women in understanding gender is complicated. Previous theories had to be expanded to include the interaction of gender with other social categories of difference and privilege. We group these efforts into three different approaches. The earliest approach was to treat each social category of difference and inequality as if it was separate and nonoverlapping. A second, more recent approach is to add up the different social categories an individual belongs to and summarize the effects of the social categories of privilege and power. The third and newest approach attempts to understand the simultaneous interaction of gender with all other categories of difference and inequality. These three approaches are described in more detail in the following paragraphs.

Separate and Different Approach Deborah King (1988) describes the earliest approach as the "race-sex analogy." She characterizes this approach as one in which oppressions related to race are compared to those related to gender, but each is seen as a separate influence. King quotes Elizabeth Cady Stanton, who in 1860 stated that "Prejudice against color, of which we hear so much, is no stronger than that against sex" (43). This approach, the "race-sex analogy," continued well into the late twentieth century and can get in the way of a deeper understanding of the complexity of gender. For example, in the race-sex analogy, gender is assumed to have the same effect for African-American and white women, while race is defined as the same experience for African-American men and women. In fact, the reading by Alfredo Mirandé in this chapter, which explores Mexican and Latino men's definitions of "macho," illustrates how we cannot assume that gender will mean the same thing to different men from the same society. The reality is that we cannot draw a line down our bodies separating gender from other social categories to which we belong. For example, African-American females cannot always be certain that the discrimination they face is due to race or gender or both. As individuals, we are complex combinations of multiple social identities. Separating the effects of multiple social prisms does not always make sense.

King (1988) argues that attempting to determine which "ism" is most oppressive and most important to overcome (e.g., sexism, racism, or classism) does not address the actual situations of individuals. This approach pits the interests of each subordinated group against the others and asks individuals to choose one group identity over others. For example, must poor, African-American women decide which group will best address their situations in society: groups fighting racial inequality, gender inequality, or class inequality? The situations of poor, African-American women are more complex than a single-issue approach can address.

The race-sex analogy of treating one "ism" at a time has been criticized because, although some needs and experiences of oppressed people are included, others are ignored. For instance, Patricia Hill Collins (1990) describes the position of African-American women as that of "outsiders within" the feminist movement and in relations with women in general. She and others have criticized the women's movement for leaving out of its agenda and awareness the experiences and needs of African-American women (e.g., King, 1988). This focus on one "ism" or another—the formation of social action groups around one category of difference and inequality—is called identity politics. In her reading in this chapter, Barbara Ryan describes why identity politics did not and cannot meet the needs of people whose identities encompass multiple social categories of difference and inequality.

Additive Approach The second approach used by theorists to understand how multiple social prisms interact at the level of individual life examines the effects of multiple social categories in an additive model. In this approach, the effects of race, ethnicity, class, and other social prisms are added together as static, equal parts of a whole (King, 1988). Returning to the earlier example of poor, African-American women, the strategy is one of adding up the effects of racism, sexism, and classism to equal what is termed "triple jeopardy." If that same woman was also a lesbian, according to the additive model her situation would be that of "quadruple jeopardy." Although this approach takes into account multiple social identities in understanding oppression, King and others, including Yen Espiritu whose work appears in this chapter, reject it as too simplistic. We cannot simply add up the complex inequalities across social categories of difference and inequality because the weight of each social category varies based upon individual situations. McIntosh in her reading in this chapter argues that privileges associated with membership in particular social categories interact to create "interlocking oppressions" whose implications and meanings shift across time and situations. For example, for African-American women in some situations, their gender will be more salient, while in other situations their race will be more salient.

Interaction Approach The third approach to understanding the social and personal consequences of membership in multiple socially constructed categories is called multiracial theory. Various terms are used by multiracial theorists to describe "interlocking oppressions," including: intersectional analysis (Baca Zinn and Thornton Dill, 1996), interrelated (Weber, 2001), simultaneous (Weber, 2001; Hill Collins, 1990), multiplicative or multiple jeopardy (King, 1988), matrix of domination (Hill Collins, 1990), and relational (Baca Zinn and Thornton Dill, 1996; Baca Zinn, Hondagneu-Sotelo, and Messner, 2001). For the purposes of this book, we describe this approach as consisting of "prismatic" interactions, which occur when socially constructed categories of difference and inequality interact with other categories in the patterns of individuals' lives. A brief discussion of some of these different models for explaining "interlocking oppressions" is useful in deepening our understanding of the complex interactions of membership in multiple social prisms.

King (1988) discusses the concept of multiple jeopardy, in which she refers not only "to several, simultaneous oppressions, but to the multiplicative relationships among them as well" (47). As a result, different socially constructed categories of difference and inequality fold into individual identities; not in an additive way, but in a way in which the total construction of an individual's identity incorporates the relationship of the identities to each other. King's model includes both multiple social identities and situational factors to understand individual differences. For example, being a submissive woman might matter more in certain religious groups where women have more restricted roles, while race or class may be less salient in that situation because the latter social prisms are likely to be similar across the religious group.

Patricia Hill Collins (1990), on the other hand, conceptualizes oppressions as existing in a "matrix of domination" in which individuals not only experience but also resist multiple inequalities. Collins argues that domination and resistance can be found at three levels: personal, cultural, and institutional. Individuals with the most privilege and power—white, upper-class men—control dominant definitions of gender in this model. Hill Collins discusses how "white skin privilege" has limited white feminists' understandings of gender oppression to their own experiences and created considerable tension in the women's movement. Tensions occur on all three levels—personal, within and between groups, and at the level of institutions such as the women's movement itself—all of which maintain power differentials. In her article in this chapter, Collins seeks a deeper understanding of gender oppression by considering how oppressions related to sexism, racism, and classism operate simultaneously.

Baca Zinn, Hondagneu-Sotelo, and Messner (2001) emphasize that gender is relational. Focusing on gender as a process (Connell, 1987), they argue that "the meaning of *woman* is defined by the existence of women of different races and classes" (Baca Zinn, Hondagneu-Sotelo, and Messner 2001:174). Yen Espiritu's reading in this chapter looks at the interaction of gender with race and class for Asian-American men. In this article, social class and gender are defined as instrumental in determining the form of gender relationships in the home. Espiritu describes how many Asian-American men seek but do not achieve hegemonic masculinity. In his reading in this chapter, Alfredo Mirandé describes the responses of Hispanic men, a subordinated group, to the dominant group's social construction of "macho." Readings throughout this book illustrate how forms of masculinity and femininity are defined in relationship to dominant forms of womanhood and manhood, and vice versa.

As you can see, this third approach does not treat interlocking oppressions as strictly additive. For example, in Espiritu's research we find that power, particularly for those men with the least power, affects their interpersonal power in their relations with other men and women. Men whose power is minimal outside of the home relative to other men can seek power in their relationships by disempowering their wives and partners. The articles in this chapter illustrate how multiple socially constructed prisms interact to shape both the identities and opportunities of individuals. They also show how interpersonal relationships are intricately tied to the larger structures of society and, as described in the Mirandé reading, how gender is maintained across groups in society.

These efforts to understand gender through the lens of multiple social prisms of difference and inequality can be problematic. One concern is that gender could be reduced to what has been described as a continually changing quilt of life experiences (Connell, 1992; Baca Zinn, Hondagneu-Sotelo, and Messner, 2001). That is, if the third approach is taken to the extreme, gender is seen as a series of individual experiences and the approach could no longer be used as a tool for explaining patterns across groups, which would be meaningless in generating social action. Thus, the current challenge for researchers and theorists is to forge an explanation of the interaction of gender with other socially constructed prisms that both recognizes and reflects the experiences of individuals, while at the same time highlighting the patterns that occur across groups of individuals.

Since we live in a world that includes many socially constructed categories of difference and inequality, understanding the ways social prisms come together is critical for understanding gender. How we visualize the effects of multiple social prisms depends upon whether we seek social justice, theoretical understanding, or both. If multiple social categories are linked, then what are the mechanisms by which difference is created, supported, and changed? Are multiple identities interlocked as McIntosh describes them? Are they multiplied, as King suggests; added, as others suggest; or combined into a matrix, as Hill Collins suggests?

We raise these questions not to confuse you, but rather to challenge you to try to understand and explain gender beyond the "boxes" which Bornstein described in the first reading in this book. Bornstein used a pyramid to understand gender inequality, with the most powerful individuals forming a small group at the top. Bornstein, however, didn't differentiate between men and women, only by how much power each person had on this pyramid. Her solution to the fact that individuals hold membership in multiple socially constructed categories of difference and inequality was to add other sides to this pyramid of power to represent socially constructed categories of race and class.

Ryan takes a slightly different look at this question in her reading describing the struggle within the women's movement, as women and men formed coalitions to address the oppression of women, while also dealing with the multiple social identities that we all share. She discusses how identity politics exist in a world made up of multiple identities. Ryan's reading adds another wrinkle to the discussion, by asking whether attention to difference achieved by highlighting the multitude of identities of women within the women's movement might damage the effectiveness of what would otherwise be considered a single identity group—women. She notes how the women's movement has experienced divisiveness as individuals seek to associate with those groups that best reflect their identities (e.g., sexual orientation, ability, age, etc.). Although it may seem that divisiveness is bad for social organization, the alternative of not acting to demand change in oppressive practices can be even worse (Lorde, 1984).

PRISMATIC INTERACTIONS

We return to the metaphor of the kaleidoscope to help us sort out this question of how to deal with multiple social identities in explaining gender. Understanding the interaction of several socially constructed identities can be compared to the ray of light passing through the prisms of a kaleidoscope. Socially constructed categories serve as prisms that create life experiences. Although the kaleidoscope produces a flowing and constantly changing array of patterns, so too do we find individual life experiences that are unique and flowing. However, similar colors and patterns often reoccur in slightly different forms. Sometimes, when we look through a real kaleidoscope, we find a beautiful image in which blue is dominant. Although we may not be able to replicate the specific image, it would not be unusual for us to see another blue-dominant pattern. Gender differences emerge in a similar form; not as a single, fixed pattern, but as a dominant, broad pattern that encompasses many unique but similar patterns.

The prism metaphor offers an avenue for systematically envisioning what McIntosh calls "interlocking oppressions." We would argue that to fully understand this interaction of social influences, one must focus on power. The distribution of privilege and oppression is a function of power relations (Baca Zinn, Hondagneu-Sotelo, and Messner, 2001). All of the articles in this chapter focus on differences defined by power relations. It is difficult to explain the combined effects of multiple social prisms without focusing on power. We argue that the power one accrues from a combination of socially constructed categories explains the patterns created by these categories. However, one cannot add up the effects from each category one belongs to, as in the additive approach described earlier. Instead, one must understand that, like the prisms in the kaleidoscope, the power of any single socially constructed identity is related to all other categories. The final patterns that appear are based upon the combinations of power that shape the patterns. Individuals' life experiences take unique forms, as race, class, ethnicity, religion, age, ability/disability, body type, and other individual socially constructed characteristics are combined to create patterns that emerge across individual experiences. However, we cannot make statements about groups of individuals if we do not look across individual experiences to describe the patterns of social issues.

Consider your own social identities and the social categories to which you belong. How do they mold you at this time and how did they, or might they, mold your experience of gender at other times and under different circumstances? Consider other social prisms that Ryan discusses in her article, such as age, ability/disability, religion, and national identity. If you were to build your own kaleidoscope of gender, what prisms would you include? What prisms interact with gender to shape your life? Do these prisms create privilege or disadvantage for you? Think about how these socially constructed categories combine to create your life experiences and how they are supported by the social structure in which you live. Keep your answers in mind as you read these articles to gain a better understanding of the role prisms play in shaping gender and affecting your life.

REFERENCES

Anzaldúa, Gloria. (1987). *Borderlands lafrontera: The new mestiza*. San Francisco: Aunt Lute Book Company.

Baca Zinn, Maxinne and Bonnie Thornton Dill. (1996). Theorizing difference from multiracial feminism. *Feminist Studies* 22(2): 321–327.

Baca Zinn, Maxinne, Pierrette Hondagneu-Sotelo, and Michael A. Messner. (2001). Gender through the prism of difference. In *Race, class, and gender: An anthology*, 4th ed, Margaret L. Andersen and Patricia Hill Collins (Eds.), pp. 168–176. Belmont, CA: Wadsworth.

Connell, R. W. (1992). A very straight gay: Masculinity, homosexual experience, and the dynamics of gender. *American Sociological Review* 57: 735–751.

———. (1987). *Gender and power: Society, the person, and sexual politics.* Stanford, CA: Stanford University Press.

Davis, Angela Y. (1981). *Women, race, and class.* New York: Random House.

Guy-Scheftall, Beverly, Ed. (1995). *Words of fire: An anthology of African-American feminist thought.* New York: The New Press.

Hill Collins, Patricia. (1990). *Black feminist thought: Knowledge, consciousness, and the politics of empowerment.* New York: Routledge.

hooks, bell. (1981). *Ain't I a woman: Black women and feminism.* Boston: South End Press.

King, Deborah. (1988). Multiple jeopardy: The context of a Black feminist ideology. *Signs* 14(1): 42–72.

Lorde, Audre. (1984). *Sister outsider.* Trumansberg, NY: The Crossing Press.

———. (1982). *Zami, A new spelling of my name.* Trumansberg, NY: The Crossing Press.

Weber, Lynn. (2001). *Understanding race, class, gender, and sexuality: A conceptual framework.* Boston: McGraw-Hill.

6

White Privilege and Male Privilege

Unpacking the Invisible Knapsack

BY PEGGY MCINTOSH

Peggy McIntosh has been unpacking the knapsack of privilege for some time. In this early article she takes a hard look at what it means to the person who occupies a position of privilege. By using her own whiteness, she helps us to understand how we don't often see our own positions of privilege. She uses examples from her own life to illustrate how everything we do in our everyday lives is made easier when we occupy positions of privilege. This reading helps us to understand how the concept of privilege is related to gender and other systems of difference in our society.

1. What advantages do you get from occupying positions of privilege?
2. What does McIntosh mean by "privilege systems?"
3. How do "privilege systems" relate to the three levels of analysis discussed in the Introduction: individual, interactional/situational, and institutional/societal?

Through work to bring materials from Women's Studies into the rest of the curriculum, I have often noticed men's unwillingness to grant that they are over-privileged, even though they may grant that women are disadvantaged. They may say they will work to improve women's status, in the society, the university, or the curriculum, but they can't or won't support the idea of lessening men's. Denials which amount to taboos surround the subject of advantages which men gain from women's disadvantages. These denials protect male privilege from being fully acknowledged, lessened or ended.

Thinking through unacknowledged male privilege as a phenomenon, I realized that since hierarchies in our society are interlocking, there was most likely a phenomenon of white privilege which was similarly denied and protected. As a white person, I realized I had been taught about racism as something which puts others at a disadvantage, but had been taught not to see one of its corollary aspects, white privilege, which puts me at an advantage.

I think whites are carefully taught not to recognize white privilege, as males are taught not to recognize male privilege. So I have begun in an untutored way to ask what it is like to have white privilege. I have come to see white privilege as an invisible package of unearned assets which I can count on cashing in each day, but about which I was "meant" to remain oblivious. White privilege is like an invisible weightless knapsack of special provisions, maps, passports, codebooks, visas, clothes, tools and blank checks.

Describing white privilege makes one newly accountable. As we in Women's Studies work to reveal male privilege and ask men to give up some of their power, so one who writes about having white privilege must ask, "Having described it, what will I do to lessen or end it?"

After I realized the extent to which men work from a base of unacknowledged privilege, I understood that much of their oppressiveness was unconscious. Then I remembered the frequent charges from women of color that white women whom they encounter are oppressive. I began to understand why we are justly seen as oppressive, even when we don't see ourselves that way. I began to count the ways in which I enjoy unearned skin privilege and have been conditioned into oblivion about its existence.

My schooling gave me no training in seeing myself as an oppressor, as an unfairly advantaged person, or as a participant in a damaged culture. I was taught to see myself as an individual whose moral state depended on her individual moral will. My schooling followed the pattern my colleague Elizabeth Minnich has pointed out: whites are taught to think of their lives as morally neutral, normative, and average, and also ideal, so that when we work to benefit others, this is seen as work which will allow "them" to be more like "us."

I decided to try to work on myself at least by identifying some of the daily effects of white privilege in my life. I have chosen those conditions which I think in my case *attach somewhat more to skin-color privilege* than to class, religion, ethnic status, or geographical location, though of course all these other factors are intricately intertwined. As far as I can see, my African American co-workers, friends and acquaintances with whom I come into daily or frequent contact in this particular time, place, and line of work cannot count on most of these conditions.

1. I can if I wish arrange to be in the company of people of my race most of the time.
2. If I should need to move, I can be pretty sure of renting or purchasing housing in an area which I can afford and in which I would want to live.
3. I can be pretty sure that my neighbors in such a location will be neutral or pleasant to me.
4. I can go shopping alone most of the time, pretty well assured that I will not be followed or harassed.
5. I can turn on the television or open to the front page of the paper and see people of my race widely represented.
6. When I am told about our national heritage or about "civilization," I am shown that people of my color made it what it is.
7. I can be sure that my children will be given curricular materials that testify to the existence of their race.
8. If I want to, I can be pretty sure of finding a publisher for this piece on white privilege.
9. I can go into a music shop and count on finding the music of my race represented, into a supermarket and find the staple foods which fit with my cultural traditions, into a hairdresser's shop and find someone who can cut my hair.
10. Whether I use checks, credit cards, or cash, I can count on my skin color not to work against the appearance of financial reliability.
11. I can arrange to protect my children most of the time from people who might not like them.
12. I can swear, or dress in second hand clothes, or not answer letters, without having people attribute these choices to the bad morals, the poverty, or the illiteracy of my race.
13. I can speak in public to a powerful male group without putting my race on trial.
14. I can do well in a challenging situation without being called a credit to my race.
15. I am never asked to speak for all the people of my racial group.
16. I can remain oblivious of the language and customs of persons of color who constitute the world's majority without feeling in my culture any penalty for such oblivion.
17. I can criticize our government and talk about how much I fear its policies and

behavior without being seen as a cultural outsider.

18. I can be pretty sure that if I ask to talk to "the person in charge," I will be facing a person of my race.
19. If a traffic cop pulls me over or if the IRS audits my tax return, I can be sure I haven't been singled out because of my race.
20. I can easily buy posters, postcards, picture books, greeting cards, dolls, toys, and children's magazines featuring people of my race.
21. I can go home from most meetings of organizations I belong to feeling somewhat tied in, rather than isolated, out-of-place, outnumbered, unheard, held at a distance, or feared.
22. I can take a job with an affirmative action employer without having coworkers on the job suspect that I got it because of race.
23. I can choose public accommodation without fearing that people of my race cannot get in or will be mistreated in the places I have chosen.
24. I can be sure that if I need legal or medical help, my race will not work against me.
25. If my day, week, or year is going badly, I need not ask of each negative episode or situation whether it has racial overtones.
26. I can choose blemish cover or bandages in "flesh" color and have them more or less match my skin.

I repeatedly forgot each of the realizations on this list until I wrote it down. For me white privilege has turned out to be an elusive and fugitive subject. The pressure to avoid it is great, for in facing it I must give up the myth of meritocracy. If these things are true, this is not such a free country; one's life is not what one makes it; many doors open for certain people through no virtues of their own.

In unpacking this invisible knapsack of white privilege, I have listed conditions of daily experience which I once took for granted. Nor did I think of any of these perquisites as bad for the holder. I now think that we need a more finely differentiated taxonomy of privilege, for some of these varieties are only what one would want for everyone in a just society, and others give licence to be ignorant, oblivious, arrogant and destructive.

I see a pattern running through the matrix of white privilege, a pattern of assumptions which were passed on to me as a white person. There was one main piece of cultural turf; it was my own turf, and I was among those who could control the turf. *My skin color was an asset for any move I was educated to want to make.* I could think of myself as belonging in major ways, and of making social systems work for me. I could freely disparage, fear, neglect, or be oblivious to anything outside of the dominant cultural forms. Being of the main culture, I could also criticize it fairly freely.

In proportion as my racial group was being made confident, comfortable, and oblivious, other groups were likely being made inconfident, uncomfortable, and alienated. Whiteness protected me from many kinds of hostility, distress, and violence, which I was being subtly trained to visit in turn upon people of color.

For this reason, the word "privilege" now seems to me misleading. We usually think of privilege as being a favored state, whether earned or conferred by birth or luck. Yet some of the conditions I have described here work to systematically overempower certain groups. Such privilege simply *confers dominance* because of one's race or sex.

I want, then, to distinguish between earned strength and unearned power conferred systemically. Power from unearned privilege can look like strength when it is in fact permission to escape or to dominate. But not all of the privileges on my list are inevitably damaging. Some, like the expectation that neighbors will be decent to you, or that your race will not count against you in court, should be the norm in a just society. Others, like the privilege to ignore less powerful people, distort the humanity of the holders as well as the ignored groups.

We might at least start by distinguishing between positive advantages which we can work to spread, and negative types of advantages which unless rejected will always reinforce our present hierarchies. For example, the feeling that one belongs within the human circle, as Native Americans say, should not be seen as privilege for a few. Ideally it is an *unearned entitlement*. At present, since only a few have it, it is an *unearned advantage* for them. This paper results from a process of coming to see that some of the power which I originally saw as attendant on being a human being in the U.S. consisted in *unearned advantage* and *conferred dominance.*

I have met very few men who are truly distressed about systemic, unearned male advantage and conferred dominance. And so one question for me and others like me is whether we will be like them, or whether we will get truly distressed, even outraged, about unearned race advantage and conferred dominance and if so, what we will do to lessen them. In any case, we need to do more work in identifying how they actually affect our daily lives. Many, perhaps most, of our white students in the U.S. think that racism doesn't affect them because they are not people of color; they do not see "whiteness" as a racial identity. In addition, since race and sex are not the only advantaging systems at work, we need similarly to examine the daily experience of having age advantage, or ethnic advantage, or physical ability, or advantage related to nationality, religion, or sexual orientation.

Difficulties and dangers surrounding the task of finding parallels are many. Since racism, sexism, and heterosexism are not the same, the advantaging associated with them should not be seen as the same. In addition, it is hard to disentangle aspects of unearned advantage which rest more on social class, economic class, race, religion, sex and ethnic identity than on other factors. Still, all of the oppressions are interlocking, as the Combahee River Collective Statement of 1977 continues to remind us eloquently.

One factor seems clear about all of the interlocking oppressions. They take both active forms which we can see and embedded forms which as a member of the dominant group one is taught not to see. In my class and place, I did not see myself as a racist because I was taught to recognize racism only in individual acts of meanness by members of my group, never in invisible systems conferring unsought racial dominance on my group from birth.

Disapproving of the systems won't be enough to change them. I was taught to think that racism could end if white individuals changed their attitudes. [But] a "white" skin in the United States opens many doors for whites whether or not we approve of the way dominance has been conferred on us. Individual acts can palliate, but cannot end, these problems.

To redesign social systems we need first to acknowledge their colossal unseen dimensions. The silences and denials surrounding privilege are the key political tool here. They keep the thinking about equality or equity incomplete, protecting unearned advantage and conferred dominance by making these taboo subjects. Most talk by whites about equal opportunity seems to me now to be about equal opportunity to try to get into a position of dominance while denying that *systems* of dominance exist.

It seems to me that obliviousness about white advantage, like obliviousness about male advantage, is kept strongly inculturated in the United States so as to maintain the myth of meritocracy, the myth that democratic choice is equally available to all. Keeping most people unaware that freedom of confident action is there for just a small number of people props up those in power, and serves to keep power in the hands of the same groups that have most of it already.

Though systemic change takes many decades, there are pressing questions for me and I imagine for some others like me if we raise our daily consciousness on the perquisites of being light-skinned. What will we do with such knowledge? As we know from watching men, it is an open question whether we will choose to use unearned advantage to weaken hidden systems of advantage, and whether we will use any of our arbitrarily-awarded power to try to reconstruct power systems on a broader base.

7

Toward a New Vision

Race, Class, and Gender as Categories of Analysis and Connection

BY PATRICIA HILL COLLINS

Patricia Hill Collins is a sociologist and a leader in the efforts to explain and understand the interaction of gender with race and social class. She has worked toward an understanding of gender that moves beyond theories that narrowly apply to white, middle and upper-class women. This paper was an address to faculty attending a workshop on integrating race and social class into courses on gender. Although it is not intended for a student audience, it provides many ways that we can eliminate racism and sexism in our daily lives. She also challenges us to think more critically about the meaning of oppression and how we might apply it to social categories of difference and inequality.

1. What is wrong with an additive analysis of oppression?
2. What does she mean by institutional oppression?
3. How is Collins' discussion of privilege different from or similar to that of McIntosh in the previous article?

The true focus of revolutionary change is never merely the oppressive situations which we seek to escape, but that piece of the oppressor which is planted deep within each of us.

—AUDRE LORDE, *SISTER OUTSIDER,* 123

Audre Lorde's statement raises a troublesome issue for scholars and activists working for social change. While many of us have little difficulty assessing our own victimization within some major system of oppression, whether it be by race, social class, religion, sexual orientation, ethnicity, age or gender, we typically fail to see how our thoughts and actions uphold someone else's subordination. Thus, white feminists routinely point with confidence to their oppression as women but resist seeing how much their white skin privileges them. African-Americans who possess eloquent analyses of racism often persist in viewing poor White women as symbols of white power. The radical left fares little better. "If only people of color and women could see their true class interests," they argue, "class solidarity would eliminate racism and sexism." In essence, each group identifies the type of oppression with which it feels most comfortable as being fundamental and classifies all other types as being of lesser importance.

Oppression is full of such contradictions. Errors in political judgment that we make concerning how we teach our courses, what we tell our children, and which organizations are worthy of our time, talents and financial support flow smoothly from errors in theoretical analysis about the nature of oppression and activism. Once we realize that there are few pure victims or oppressors, and that each one of us derives varying amounts of penalty and privilege from the multiple systems of oppression that frame our lives, then we will be in a position to see the need for new ways of thought and action.

To get at that "piece of the oppressor which is planted deep within each of us," we need at least two things. First, we need new visions of what oppression is, new categories of analysis that are inclusive of race, class, and gender as distinctive yet interlocking structures of oppression. Adhering to a stance of comparing and ranking oppressions—the proverbial, "I'm more oppressed than you"—locks us all into a dangerous dance of competing for attention, resources, and theoretical supremacy. Instead, I suggest that we examine our different experiences within the more fundamental relationship of damnation and subordination. To focus on the particular arrangements that race or class or gender take in our time and place without seeing these structures as sometimes parallel and sometimes interlocking dimensions of the more fundamental relationship of domination and subordination may temporarily ease our consciences. But while such thinking may lead to short term social reforms, it is simply inadequate for the task of bringing about long term social transformation.

While race, class and gender as categories of analysis are essential in helping us understand the structural bases of domination and subordination, new ways of thinking that are not accompanied by new ways of acting offer incomplete prospects for change. To get at that "piece of the oppressor which is planted deep within each of us," we also need to change our daily behavior. Currently, we are all enmeshed in a complex web of problematic relationships that grant our mirror images full human subjectivity while stereotyping and objectifying those most different than ourselves. We often assume that the people we work with, teach, send our children to school with, and sit next to, will act and feel in prescribed ways because they belong to given race, social class or gender categories. These judgments by category must be replaced with fully human relationships that transcend the legitimate differences created by race, class and gender as categories of analysis. We require new categories of connection, new visions of what our relationships with one another can be.

Our task is immense. We must first recognize race, class and gender as interlocking categories of analysis that together cultivate profound differences in our personal biographies. But then we must transcend those very differences by reconceptualizing race, class and gender in order to create new categories of connection.

[This paper] addresses this need for new patterns of thought and action. I focus on two basic questions. First, how can we reconceptualize race, class and gender as categories of analysis? Second, how can we transcend the barriers created by our experiences with race, class and gender oppression in order to build the types of coalitions essential for social exchange? To address these questions I contend that we must acquire both new theories of how race, class and gender have shaped the experiences not just of women of color, but of all groups. Moreover, we must see the connections between these categories of analysis and the personal issues in our everyday lives. . . . As Audre Lorde points out, change starts with self, and relationships that we have with those around us must always be the primary site for social change.

HOW CAN WE RECONCEPTUALIZE RACE, CLASS AND GENDER AS CATEGORIES OF *ANALYSIS*?

To me, we must shift our discourse away from additive analyses of oppression (Spelman 1982; Collins 1989). Such approaches are typically based on two key premises. First, they depend on either/or, dichotomous thinking. Persons, things and ideas are conceptualized in terms of their opposites. For example, Black/White, man/woman, thought/feeling, and fact/opinion are defined in oppositional terms. Thought and feeling are not seen as two different and interconnected ways of approaching truth that can coexist in scholarship and teaching. Instead, feeling is defined as antithetical to reason, as its opposite. In spite of the fact that we all have "both/and" identities (I am both a college professor and a mother—I don't stop being a mother when I drop my child off at school, or forget everything I learned while scrubbing the toilet), we persist in trying to classify each other in either/or categories. I live each day as an African-American woman—a race/gender specific experience. And I am not alone. Everyone has a race/gender/class specific identity. Either/or, dichotomous thinking is especially troublesome when applied to theories of oppression because every individual must be classified as being either oppressed or not oppressed. The both/and position of simultaneously being oppressed and oppressor becomes conceptually impossible.

A second premise of additive analyses of oppression is that these dichotomous differences must be ranked. One side of the dichotomy is typically labeled dominant and the other subordinate. Thus, Whites rule Blacks, men are deemed superior to women, and reason is seen as being preferable to emotion. Applying this premise to discussions of oppression leads to the assumption that oppression can be quantified, and that some groups are oppressed more than others. I am frequently asked, "Which has been most oppressive to you, your status as a Black person or your status as a woman?" What I am really being asked to do is divide myself into little boxes and rank my various statuses. If I experience oppression as a both/and phenomenon, why should I analyze it any differently?

Additive analyses of oppression rest squarely on the twin pillars of either/or thinking and the necessity to quantify and rank all relationships in order to know where one stands. Such approaches typically see African-American women as being more oppressed than everyone else because the majority of Black women experience the negative effects of race, class and gender oppression simultaneously. In essence, if you add together separate oppressions, you are left with a grand oppression greater than the sum of its parts.

I am not denying that specific groups experience oppression more harshly than others—lynching is certainly objectively worse than being held up as a sex object. But we must be careful not to confuse this issue of the saliency of one type of oppression in people's lives with a theoretical stance positing the interlocking nature of oppression. Race, class and gender may all structure a situation but may not be equally visible and/or important in people's self-definitions. . . . This recognition that one category may have salience over another for a given time and place does not minimize the theoretical importance of assuming that race, class and gender as categories of analysis structure all relationships.

In order to move toward new visions of what oppression is, I think that we need to ask new questions. How are relationships of domination and subordination structured and maintained in the American political economy? How do race, class and gender function as parallel and interlocking systems that shape this basic relationship of domination and subordination? Questions such as these promise to move us away from futile theoretical struggles concerned with ranking oppressions and towards analyses that assume race, class and gender are all present in any given setting, even if one appears more visible and salient than the others. Our task becomes redefined as one of reconceptualizing oppression by uncovering the connections among race, class and gender as categories of analysis.

1. INSTITUTIONAL DIMENSION OF OPPRESSION

Sandra Harding's contention that gender oppression is structured along three main dimensions—the institutional, the symbolic, and the individual—offers a useful model for a more comprehensive analysis encompassing race, class and gender oppression (Harding 1989). Systemic relationships of domination and subordination structured through social institutions such as schools, businesses, hospitals, the work place, and government agencies represent the institutional dimension of oppression. Racism, sexism and elitism all have concrete institutional locations. Even though the workings of the institutional dimension of oppression are often obscured with ideologies claiming equality of opportunity, in actuality, race, class and gender place Asian-American women, Native American men, White men, African-American women, and other groups in distinct institutional niches with varying degrees of penalty and privilege.

Even though I realize that many . . . would not share this assumption, let us assume that the institutions of American society discriminate, whether by design or by accident. While many of us are familiar with how race, gender and class operate separately to structure inequality, I want to focus on how these three systems interlock in structuring the institutional dimension of oppression. To get at the interlocking nature of race, class and gender, I want you to think about the antebellum plantation as a guiding metaphor for a variety of American social institutions. Even though slavery is typically analyzed as a racist institution, and occasionally as a class institution, I suggest that slavery was a race, class, gender specific institution. Removing any one piece from our analysis diminishes our understanding of the true nature of relations of domination and subordination under slavery.

Slavery was a profoundly patriarchal institution. It rested on the dual tenets of White male authority and White male property, a joining of the political and the economic within the institution of the family. Heterosexism was assumed and all Whites were expected to marry. Control over affluent White women's sexuality remained key to slavery's survival because property was to be passed on to the legitimate heirs of the slave owner. Ensuring affluent White women's virginity and chastity was deeply intertwined with maintenance of property relations.

Under slavery, we see varying levels of institutional protection given to affluent White women, working class and poor White women, and enslaved African women. Poor White women enjoyed few of the protections held out to their upper class sisters. Moreover, the devalued status of Black women was key in keeping all White women in their assigned places. Controlling Black women's fertility was also key to the continuation of slavery, for children born to slave mothers themselves were slaves.

African-American women shared the devalued status of chattel with their husbands, fathers and sons. Racism stripped Blacks as a group of legal rights, education, and control over their own persons. African-Americans could be whipped, branded, sold, or killed, not because they were poor, or because they were women, but because they were Black. Racism ensured that Blacks would continue to serve Whites and suffer economic exploitation at the hands of all Whites.

So we have a very interesting chain of command on the plantation—the affluent White master as the reigning patriarch, his White wife helpmate to serve him, help him manage his property and bring up his heirs, his faithful servants whose production and reproduction were tied to the requirements of the capitalist political economy, and largely propertyless, working class White men and women watching from afar. In essence, the foundations for the contemporary roles of elite White women, poor Black women, working class White men, and a series of other groups can be seen in stark relief in this fundamental American social institution. While Blacks experienced the most harsh treatment under slavery, and thus made slavery clearly visible as a racist institution, race, class and gender interlocked in structuring slavery's systemic organization of domination and subordination.

Even today, the plantation remains a compelling metaphor for institutional oppression. Certainly the actual conditions of oppression are not as severe now as they were then. To argue, as some do, that things have not changed all that much denigrates the achievements of those who struggled for social change before us. But the basic relationships among Black men, Black women, elite White women, elite White men, working class White men and working class White women as groups remain essentially intact.

A brief analysis of key American social institutions most controlled by elite White men should convince us of the interlocking nature of race, class and gender in structuring the institutional dimension of oppression. For example, if you are from an American college or university, is your campus a modern plantation? Who controls your university's political economy? Are elite White men over represented among the upper administrators and trustees controlling your university's finances and policies? Are elite White men being joined by growing numbers of elite White women helpmates? What kinds of people are in your classrooms grooming the next generation who will occupy these and other decision-making positions? Who are the support staff that produce the mass mailings, order the supplies, fix the leaky pipes? Do African-Americans, Hispanics or other people of color form the majority of the invisible workers who feed you, wash your dishes, and clean up your offices and libraries after everyone else has gone home? . . .

2. THE SYMBOLIC DIMENSION OF OPPRESSION

Widespread, societally-sanctioned ideologies used to justify relations of domination and subordination comprise the symbolic dimension of oppression. Central to this process is the use of stereotypical or controlling images of diverse race, class and gender groups. In order to assess the power of this dimension of oppression, I want you to make a list, either on paper or in your head, of "masculine" and "feminine" characteristics. If your list is anything like that compiled by most people, it reflects sonic variation of the following:

Masculine	*Feminine*
aggressive	passive
leader	follower
rational	emotional
strong	weak
intellectual	physical

Not only does this list reflect either/or dichotomous thinking and the need to rank both sides of the dichotomy, but ask yourself exactly which men and women you had in mind when compiling these characteristics. This list applies almost exclusively to middle class White men and women. The allegedly "masculine" qualities that you probably listed are only acceptable when exhibited by elite White men, or when used by Black and Hispanic men against each other or against women of color. Aggressive Black and Hispanic men are seen as dangerous, not powerful, and are often penalized when they exhibit any of the allegedly "masculine" characteristics. Working class and poor White men fare slightly better and are also denied the allegedly "masculine" symbols of leadership, intellectual, competence, and human rationality. Women of color and working class and poor White women are also not represented on this list, for they have never had the luxury of being "ladies." What appear to be universal categories representing all men and women instead are unmasked as being applicable to only a small group.

It is important to see how the symbolic images applied to different race, class and gender groups interact in maintaining systems of domination and subordination. If I were to ask you to repeat the same assignment, only this time, by making separate lists for Black men, Black women, Hispanic women and Hispanic men, I suspect that your gender symbolism would be quite different. In comparing all of the lists, you might begin to see the interdependence of symbols applied to all groups. For example, the elevated images ofWhite womanhood need devalued images of Black womanhood in order to maintain credibility.

While the above exercise reveals the interlocking nature of race, class and gender in structuring the symbolic dimension of oppression, part of its importance lies in demonstrating how race, class and gender pervade a wide range of what appears to be universal language. Attending to diversity . . . in our daily lives provides a new angle of vision on interpretations of reality thought to be natural, normal and "true." Moreover, viewing images of masculinity and femininity as universal gender symbolism, rather than as symbolic images that are race, class and gender specific, renders the experiences of people of color and of non-privileged White women and men invisible. One way to dehumanize an individual or a group is to deny the reality of their experiences. So when we refuse to deal with race or class because they do not appear to be directly relevant to gender, we are actually becoming part of some one else's problem.

Assuming that everyone is affected differently by the same interlocking set of symbolic images allows us to move forward toward new analyses. Women of color and White women have different relationships to White male authority and this difference explains the distinct gender symbolism applied to both groups. Black women encounter controlling images such as the mammy, the matriarch, the mule and the whore, that encourage others to reject us as fully human people. Ironically, the negative nature of these, images simultaneously encourages us to reject them. In contrast, White women are offered seductive images, those that promise to reward them for supporting the status quo. And yet seductive images can be equally controlling. Consider, for example, the views of Nancy White, a 73-year old Black woman, concerning images of rejection and seduction:

> My mother used to say that the black woman is the white man's mule and the white woman is his dog. Now, she said that to say this: we do the heavy work and get beat whether we do it well or not. But the white woman is closer to the master and he pats them on the head and lets them sleep in the house, but he ain't gon' treat neither one like he was dealing with a person. (Gwalatney, 148)

Both sets of images stimulate particular political stances. By broadening the analysis beyond the confines of race, we can see the varying levels of rejection and seduction available to each of us due to our race, class and gender identity. Each of us lives with an allotted portion of institutional privilege and penalty, and with varying levels of rejection and seduction inherent in the symbolic images applied to us. This is the context in which we make our choices. Taken together, the institutional and symbolic dimensions of oppression create a structural backdrop against which all of us live our lives.

3. THE INDIVIDUAL DIMENSION OF OPPRESSION

Whether we benefit or not, we all live within institutions that reproduce race, class and gender oppression. Even if we never have any contact with members of other race, class and gender groups, we all encounter images of these groups and are exposed to the symbolic meanings attached to those images. On this dimension of oppression, our individual biographies vary tremendously. As a result of our institutional and symbolic statuses, all of our choices become political acts.

Each of us must come to terms with the multiple ways in which race, class and gender as categories of analysis frame our individual biographies. I have lived my entire life as an African-American woman from a working class family and this basic fact has had a profound impact on my personal biography. Imagine how different your life might be if you had been born Black, or White, or poor, or of a different race/class/gender group than the one with which you are most familiar. The institutional treatment you would have received and the symbolic meanings attached to your very existence might differ dramatically from what you now consider to be natural, normal and part of everyday life. You might be the same, but your personal biography might have been quite different.

I believe that each of us carries around the cumulative effect of our lives within multiple structures of oppression. If you want to see how much you have been affected by this whole thing, I ask you one simple question—who are your close friends? Who are the people with whom you can share your hopes, dreams, vulnerabilities, fears and victories? Do they look like you? If they are all the same, circumstance may be the cause. For the first seven years of my life I saw only low income Black people. My friends from those years reflected the composition of my community. But now that I am an adult, can the defense of circumstance explain the patterns of people that I trust as my friends and colleagues? When given other alternatives, if my friends and colleagues reflect the homogeneity of one race, class and gender group, then these categories of analysis have indeed become barriers to connection.

I am not suggesting that people are doomed to follow the paths laid out for them by race, class and gender as categories of analysis. While these three structures certainly frame my opportunity structure, I as an individual always have the choice of accepting things as they are, or trying to change them. As Nikki Giovanni points out, "we've got to live in the real world. If we don't like the world we're living in, change it. And if we can't change it, we change ourselves. We can do something" (Tate 1983, 68). While a piece of the oppressor may be planted deep within each of us, we each have the choice of accepting that piece or challenging it as part of the "true focus of revolutionary change." . . .

Since I opened with the words of Audre Lorde, it seems appropriate to close with another of her ideas . . . :

> Each of us is called upon to take a stand. So in these days ahead, as we examine ourselves and each other, our works, our fears, our differences, our sisterhood and survivals, I urge you to tackle what is most difficult for us all, self-scrutiny of our complacencies, the idea that since each of us believes she is on the side of right, she need not examine her position. (1985)

I urge you to examine your position.

REFERENCES

Collins, Patricia Hill. 1989. "The Social Construction of Black Feminist Thought." *Signs*. Summer 1989.

Gwalatney, John Langston. 1980. *Drylongso: A Self-Portrait of Black America*. New York: Vintage.

Harding, Sandra. 1986. *The Science Question in Feminism*. Ithaca, New York: Cornell University Press.

Lorde, Audre. 1984. *Sister Outsider*. Trumansberg, New York: The Crossing Press.

———. 1985. "Sisterhood and Survival." Keynote address, Conference on the Black Woman Writer and the Diaspora, Michigan State University.

Spelman, Elizabeth. 1982. "Theories of Race and Gender: The Erasure of Black Women." *Quest* 5: 36–32.

Tate, Claudia, ed. 1983. *Black Women Writers at Work*. New York: Continuum.

8

Race, Gender, Class in the Lives of Asian Americans

BY YEN L. ESPIRITU

Espiritu's description of Asian Americans as being multiply disadvantaged helps us to understand the difficulty of viewing gender through multiple prisms. In trying to understand the situation of disadvantage for Asian-American men and women, we can see that looking at gender alone confounds rather than clarifies the reality of their daily lives. Asian men, who may be privileged as men, may not be privileged as Asians in their work places. Espiritu suggests that these men may exert their maleness at home, using their power as men against their wives.

1. How does the use of mutually exclusive binaries (as Yen Espiritu calls them, or dichotomous categories as referred to in other readings) and dualism work against a deeper understanding of the interactions of gender with other social categories?
2. Why does Espiritu argue against an additive model when trying to understand the intersection of gender with other social categories of privilege and power?
3. How is Espiritu's argument different from and similar to that of Hill Collins in the previous reading?

Societies tend to organize themselves around sets of mutually exclusive binaries: white or black, man or woman, professional or laborer, citizen or alien. In the United States, this binary construction of difference—of privileging and empowering the first term and reducing and disempowering the second—structures and maintains race, gender, and class privilege and power (Lowe, 1991; Grozs, 1994). Thus, white/male/professional/citizen constitutes the norm against which black/female/laborer/alien is defined (Okihiro, 1995). Normed on this white, male, bourgeois hierarchy, working class immigrant women of color are subordinated and suppressed (Mohanty, 1991). There is also another kind of dualism, one that treats race, gender, and class as mutually exclusive categories. White feminist scholars engage in this either/or dichotomous thinking when they assert that gender oppression transcends divisions among women created by race, social class and other forms of difference. Similarly, men of color rely on dualism when they insist that the system of racial oppression takes precedence over that of gender oppression.

By privileging either race or gender or class instead of recognizing their interconnections, this dichotomous stance marginalizes the experiences of those who are multiply disadvantaged (Crenshaw, 1990). As a multiply disadvantaged people, Asians in the Untied States complicate either/or definitions and categories and carve out for themselves a "third space" as "neither/nor" and as "both/and" (Kim, 1993). Because of their racial ambiguity, Asian Americans have been constructed historically to be both "like black" and "like white," as well as *neither* black nor white.

Similarly, Asian women have been both hyperfemininized and masculinized, and Asian men have been both hypermasculinized and feminized. And in social class and cultural terms, Asian Americans have been cast both as the "inassimilable alien" and the "model minority" (Okihiro, 1995). Their ambiguous, middling positions maintain systems of privilege and power but also threaten and destabilize these constructs of hierarchies. This essay discusses how Asian Americans, as racialized "others" who occupy a "third" position, both disrupt and conform to the hegemonic dualism of race, gender, and class.

The problems of race, gender, and class are closely intertwined in the lives of Asian American men and women. It is racial and class oppression against "yellows" that restricts their material lives, (re)defines their gender roles, and provides material for degrading and exaggerated sexual representations of Asian men and women in U.S. popular culture. Asian Americans have always, but particularly since the 1960s, resisted race, class, and gender exploitation through political, economic, and cultural activism. As a result, the objectification of Asian Americans as exotic aliens who are different from, and inferior to, white Americans has never been absolute.

On the other hand, in demanding legitimacy, some Asian Americans have adopted the either/or dichotomies of the dominant patriarchal structure, "unwittingly upholding the criteria of those whom they assail" (Cheung, 1990). Lisa Lowe (1991:31) argued that "in accepting the binary terms ('white' and 'nonwhite,' or 'majority' and 'minority') . . . , we forget that these binary schemes are not neutral descriptions." For example, men who have been historically devalued are likely to take their rage and frustration out on those closest to them (Lipsitz, 1988:204–205; Crenshaw, 1990:185–189). Having been forced into "feminine" subject positions, some Asian American men seek to reassert their masculinity by physically and emotionally abusing those who are even more powerless: the women and children in their families. In particular, men's inability to earn a family wage and subsequent reliance on their wives' income undermines severely their sense of well-being. Though it is useful to view male tyranny within the context of racial inequality and class exploitation, it is equally important to note that this aggression is informed by Eurocentric gender ideology, particularly its emphasis on oppositional dichotomous sex roles. Because these Asian American men can see only race oppression, and no gender domination, they are unable, or unwilling, to view themselves as both oppressed and oppressor. This dichotomous stance has led to the marginalization of Asian American women and their needs. Concerned with recuperating their identities as men and as Americans, some Asian American political and cultural workers have subordinated feminism to nationalist concerns. From this limited standpoint, Asian American feminists who expose Asian American sexism are cast as "anti-ethnic," criticized for undermining group solidarity, and charged with exaggerating the community's patriarchal structure to please the larger society. In an analysis of the display of machismo among Mexican immigrant men, Pierrette Hondagneu-Sotelo (1994:193–94) characterized these men's behaviors as "personally and collectively constructed performances of masculine gender display . . . [which] should be distinguished from structurally constituted positions of power." In other words, these displays of male prowess are indicators of "marginalized subordinated masculinities."

The racist debasement of Asian men makes it difficult for Asian American women to balance the need to expose the problems of male privilege with the desire to unite with men to contest the overarching racial ideology that confines them both. As Asian American women negotiate this difficult feat, they, like men, tend to subscribe to either/or dichotomous thinking. They do so when they adopt the fixed masculinist Asian American identity, even when it marginalizes their positions, or when they privilege women's concerns over men's or over concerns about other forms of inequality. Both of these positions advance the dichotomous stance of man and woman, gender or race or class, without recognizing the "complex *relationality* that shapes our social and political lives" (Mohanty, 1991:13). Finally, Asian American women enforce Eurocentric gender ideology when they accept the objectification and feminization of Asian men and the parallel construction of white men as the most desirable sexual and marital partners.

Traditional white feminists likewise succumb to binary definitions and categories when they insist on the primacy of gender, thereby dismissing racism and other structures of oppression. The feminist mandate for gender solidarity accounts only for hierarchies between men and women and ignores power differentials among women, among men, and between white women and men of color. This exclusive focus on gender makes it difficult for white women to see the web of multiple oppressions that constrain the lives of most women of color, thus limiting the potential bonding among all women. Furthermore, it bars them from recognizing the oppression of men of color: the fact that there are men, and not only women, who have been "feminized" and the fact that white, middle class women hold cultural power and class power over certain groups of men (Cheung, 1990:245–46; Wiegman, 1991:311).

In sum, Asian American men, Asian American women, and white women unwittingly comply with the ideologies of racialized patriarchy. Asian American men fulfill traditional definitions of manhood when they conflate might and masculinity and sweep aside the needs and well-being of Asian American women. Asian American women accept these racialized gender ideologies when they submit to white and Asian men or when they subordinate racial, class, or men's concerns to feminism. And white women advance a hierarchical agenda when they fail to see that the experiences of white women, women of color, and men of color are connected in systematic ways.

BEYOND DUALISM: CONSTRUCTING AN "IMAGINED COMMUNITY"

As a multiply marginalized group, Asian Americans pose a fundamental problem to the binary oppositions that structure and maintain privilege and power in the United States (Okihiro, 1995). The conditions of their lives challenge the naturalism of this dualism and reveal how multiple structures of difference and disempowerment reinforce one another. In other words, they show how race, gender, and class, as categories of difference, do not parallel but instead intersect and confirm each other (Wiegman, 1991:311). The task for feminist theory, then, is to develop paradigms that articulate the complicity among these categories of oppression, that strengthen the alliance between gender and ethnic studies, and that reach out not only to women, but perhaps also to men, of color (Cheung, 1990:245).

A central task in feminist scholarship is to expose and dismantle the stereotypes that traditionally have provided ideological justifications for women's subordination. However, ideologies of manhood and womanhood have as much to do with class and race as they have to do with sex. Class and gender intersect when the culture of patriarchy, which assigns men to the public sphere and women to the private sphere, makes it possible for capitalists to exploit and profit from the labor of both men and women. Because patriarchy mandates that men be the breadwinners, it pressures them to work in the capitalist wage market, even in jobs that are low paying, physically punishing, and without opportunities for upward mobility. In this sense, the sexual division of labor within the family produces a steady supply of male labor for the benefits of capital. The culture of patriarchy is also responsible for the capitalist exploitation of women. The assumption that women are not the main income earners in their families, and therefore can afford to work for less, provides ideological justification for employers to hire women at lower wages and in poorer working conditions than exist for men (Hossfeld, 1994:74). On the other hand, in however limited a way, wage employment does allow women to challenge the confines and dictates of traditional patriarchal social relations. It affords women some opportunities to leave the confines of the home, delay marriage and childbearing, develop new social networks, and exercise more personal independence (Lim, 1083:83). As such, wage labor both oppresses and liberates women, exploiting them as workers but also strengthening their claims against patriarchal authority (Okihiro, 1994:91). But this potential liberation is limited. As Linda Y.C. Lim (1983:88) pointed out, because capitalist employment and exploitation of female labor is based on patriarchal exploitation, "The elimination of these conditions may well bring about an elimination of the jobs themselves."

U.S. capital also profits from racism. In the pre–World War II era, white men were considered "free labor" and could have a variety of jobs in the industrialized economic sector, whereas Asian Men were racialized as "coolie labor" and confined to nonunionized, degrading low paying jobs in the agricultural and service sectors. Asian immigrants faced a special disability: They could not become citizens and thus were a completely disfranchised group. As noncitizens, Asian immigrants were subjected to especially onerous working conditions compared to other workers, including longer hours, lower wages, more physically demanding labor, and more dangerous tasks. The alien, and thus rights-deprived, status of Asian immigrants increased the ability of capital to control them; it also allowed employers to use the cheapness of Asian labor to undermine and discipline the white small producers and white workers (Bonacich, 1984:165–66).

The post-1965 Asian immigrant group, though much more differentiated along social class lines, is still racialized and exploited. In all occupational sectors, Asian American men and women fare worse than their white counterparts. Unskilled and semiskilled Asian immigrant labor is relegated to the lower-paying job brackets of racially segregated industries. Due to their gender, race, and noncitizen status, Asian immigrant women fare the worst because they are seen as being the most desperate for work at any wage (Hossfeld, 1994:75). The highly educated, on the other hand, encounter institutionalized economic and cultural racism that restricts their economic mobility. In sum, capitalist exploitation of Asians has been possible mainly because Asian labor had already been categorized by a racist society as being worth less than white worker's labor. This racial hierarchy then confirms the "manhood" of white men while rendering Asian men impotent.

Racist economic exploitation of Asian Americans has had gender implications. Due to the men's inability to earn a family wage, Asian American women have had to engage in paid labor to make up the income discrepancies. In other words, the racialized exploitation of Asian American men has historically been the context for the entry of Asian American women into the labor force. Access to wage work and relative economic independence, in turn, has given women solid ground for questioning their subordination. But progress has been slow and uneven. In some instances, more egalitarian divisions of labor and control of domestic resources have emerged. In others, men's loss of status in the public and domestic spheres has placed severe pressures on the traditional family, leading at times to resentment, verbal or physical abuse, and divorce.

Moreover, Asian women's ability to transform traditional patriarchy is often constrained by their social-structural location in the dominant society. The articulation between the processes of gender discrimination, racial discrimination of (presumed or actual) immigrant workers, and capitalist exploitation makes their position particularly vulnerable. Constrained by these overlapping categories of oppression, Asian American women may accept certain components of the traditional patriarchal system to have a strong and intact family—an important source of support to sustain them in the work world (Glenn, 1986; Kibria, 1990). Indeed, in this hostile environment, the act of maintaining families is itself a form of resistance. Finally, women's economic resources have remained too meager for them to maintain their economic independence from men. Therefore, some Asian American women may choose to preserve the traditional family system, though in a tempered form, because they value the promise of male economic protection. As Evelyn Nakano Glenn (1986:218) pointed out, for Asian Americans, the family has been "simultaneously a unity, bound by interdependence in the fight for survival, and a segmented institution in which men and women struggled over power, resources, and labor."

To recognize the interconnections of race, gender, and class is also to recognize that the conditions of our lives are connected to and shaped by the conditions of others' lives. Thus men are privileged precisely because women are not, and whites are advantaged precisely because people of color are disadvantaged. In other words, both people of color and white people live racially structured lives, both women's and men's lives are shaped by their gender, and all of our lives are influenced by the dictates of the patriarchal economy of U.S. society (Wiegman, 1991:311; Frankenberg, 1993:1). But the intersections among these categories of oppression mean that there are also hierarchies among women and men and that some women hold cultural and economic power over certain groups of men. On the other hand, the "intersecting, contradictory, and cross-category functioning of U.S. culture" (Wiegman, 1991:331) also presents opportunities for transforming the existing hierarchical structure. If Asian men have been "feminized" in the United States, then they can best attest to and fight against patriarchal oppression that has long denied all women male privilege. If white women recognize that ideologies of womanhood have as much to do with race and class as they have to do with sex, then they can better work with, and not for, women (and men) of color. And if men and women of all social classes understand how capitalism distorts and diminishes all peoples' lives, then they will be more apt to struggle together for a more equitable economic system. Thus, to name the categories of oppression and to identify their interconnections is also to explore, forge, and fortify cross-gender, cross-racial, and cross-class alliances. It is to construct what Chandra Mohanty (1991:4) called an "imagined community": a community that is bounded not only by color, race, or class but crucially by a shared struggle against all pervasive and systemic forms of domination.

BIBLIOGRAPHY

Bonacich, E. 1984. Asian Labor in the Development of Hawaii and California. Pp. 130–185 in *Asian Workers in the United States before World War II,* edited by L. Cheng & E. Bonacich. Berkeley: University of California Press.

Cheung, K.K. 1990. "The Woman Warrior Versus the Chinaman Pacific: Must a Chinese American Critic Choose Between Feminism and Heroism? Pp. 234–251 in *Conflicts in Feminism,* edited by M. Hirsch and E.P. Keller. New York: Routledge.

Crenshaw, L. 1989. "Demarginalizing the Intersection of Race and Sex: A Black Feminist Critique of Antidiscrimination Doctrine, Feminist Theory and Antiracist Politics." Pp. 139–167 in University of Chicago Legal Forum: Feminism in the Law: Theory, Practice and Criticism. Chicago: University Press.

Frankenberg, R. 1993. *White Women, Race, Matters: The Social Construction of Whiteness.* Minneapolis, University of Minnesota Press.

Glenn, E.N. 1986. *Issei, Nisei, War Bride: Three Generations of Japanese American Women at Domestic Work.* Philadelphia: Temple University Press.

Grosz, E. 1994. *Volatile Bodies: Toward a Corporeal Feminism.* Bloomington: Indiana University Press.

Hondagneu-Sotelo, P. 1994. *Gendered Transition: Mexican Experiences in Immigration.* Berkeley, University of California Press.

Hossfeld, K.J. 1994. "Hiring Immigrant Women: Silicon Valley's Simple Formula." Pp. 65–93 in *Women of Color in US Society,* edited by M. Baca Zinn & B. T. Dill. Philadelphia: Temple University Press.

Kim, E. 1993. Preface. Pp. vii–xiv in *Charlie Chan Is Dead: An Anthology of Contemporary Asian American Fiction,* edited by J. Hagedoen. New York: Penguin.

Lim, L.Y.C. 1983. Capitalism, Imperialism, and Patriarchy: The Dilemma of Third World Women Workers in Multinational Factories. Pp. 70–91 in *Women, Men and the International Division of Labor,* edited by J. Nash and M. P. Fernandez-Kelly. Albany: State University of New York Press.

Lowe, L. 1991. Heterogeneity, Hybridity, Multiplicity: Marking Asian American Difference. *Diaspora* 1: 24–44.

Mohanty, C.T. 1991. "Cartographies of Struggle: Third World Women and the Politics of Feminism." Pp. 1–47 in *Third World Women and the Politics of Feminism,* edited by C.T. Mohanty, A. Russo, & L. Torres, Bloomington: University of Indiana Press.

Okihiro, G.Y. 1995 (November). *Reading Asian Bodies, Reading Anxieties.* Paper presented at the University of California, San Diego Ethnic Studies Colloquium, La Jolla.

9

"Macho"

Contemporary Conceptions

BY ALFREDO MIRANDÉ

Alfredo Mirandé examines how Anglo definitions of "macho" compare to its meaning in Mexican or Latino culture. Not only does he show how definitions of masculinity vary across cultures, but he helps us to understand the meaning of "macho" within Latino cultures. Using data from interviews with Latino men, he explores the negative, positive, and neutral meanings that are attributed to the word "macho." This reading provides a deeper understanding of definitions of masculinity, and variations of those definitions within the same culture, as well as how definitions of gender are embedded in cultures.

1. How do the definitions of "macho" differ between those who view it negatively and those who view it positively?
2. What factors contribute to the different definitions of "macho"?
3. How is the definition of masculinity in Mexican and Latino cultures linked to gendered roles in family, relationships, and other areas of life?

"MACHO": AN OVERVIEW

Mexican folklorist Vicente T. Mendoza suggested that the word "macho" was not widely used in Mexican songs, *corridos* (folk ballads), or popular culture until the 1940s (Mendoza 1962, 75–86). Use of the word was said to have gained in popularity after Avila Camacho became president. The word lent itself to use in *corridos* because "macho" rhymed with "Camacho."

While "macho" has traditionally been associated with Mexican or Latino culture, the word has recently been incorporated into American popular culture, so much so that it is now widely used to describe everything from rock stars and male sex symbols in television and film to burritos. When applied to entertainers, athletes, or other "superstars," the implied meaning is clearly a positive one that connotes strength, virility, masculinity, and sex appeal. But when applied to Mexicans or Latinos, "macho" remains imbued with such negative attributes as male dominance, patriarchy, authoritarianism, and spousal abuse. Although both meanings connote strength and power, the Anglo macho is clearly a much more positive and appealing symbol of manhood and masculinity. In short, under current usage the Mexican macho oppresses and coerces women, whereas his Anglo counterpart appears to attract and seduce them.

This chapter focuses on variations in perceptions and conceptions of the word "macho" held by Mexican and Latino men. Despite all that has been written and said about the cult of masculinity and the fact that male dominance has been assumed to be a key feature of Mexican and Latino culture, very little research exists to support this assumption. Until recently such generalizations were based on stereotypes, impressionistic evidence, or the observations of ethnographers such as Oscar Lewis (1960, 1961), Arthur Rubel (1966), and William Madsen (1973). These Anglo ethnographers were criticized by noted Chicano folklorist Américo Paredes (1977) for the persistent ignorance and insensitivity to Chicano language and culture that is reflected in their work. Paredes contended, for example, that although most anthropologists present themselves as politically liberal and fluent in Spanish, many are only minimally fluent and fail to grasp the nuance and complexity of Chicano language. There is, it seems, good reason to be leery of their findings and generalizations regarding not only gender roles but also all aspects of the Mexican/Latino experience.

Utilizing data obtained through qualitative open-ended questions, I look in this chapter at how Latino men themselves perceive the word "macho" and how they describe men who are considered *"muy machos."* Although all of the respondents were living in the United States at the time of the interviews, many were foreign-born and retained close ties with Mexican/Latino culture. Since they had been subjected to both Latino and American influences, I wondered whether they would continue to adhere to traditional Mexican definitions of "macho" or whether they had been influenced by contemporary American conceptions of the word.

Specifically, an attempt was made in the interviews to examine two polar views. The prevailing view in the social science literature of the Mexican macho is a negative one. This view holds that the origins of excessive masculine displays and the cult of masculinity in México and other Latino countries can be traced to the Spanish Conquest, as the powerless colonized man attempted to compensate for deep-seated feelings of inadequacy and inferiority by assuming a hypermasculine, aggressive, and domineering stance. There is a second and lesser-known view that is found in Mexican popular culture, particularly in film and music, one that reflects a more positive, perhaps idyllic, conception of Mexican culture and national character. Rather than focusing on violence and male dominance, this second view associates macho qualities with the evolution of a distinct code of ethics.

Un hombre que es macho is not hypermasculine or aggressive, and he does not disrespect or denigrate women. Machos, according to the positive view, adhere to a code of ethics that stresses humility, honor, respect of oneself and others, and courage. What may be most significant in this second view is that being "macho" is not manifested by such outward qualities as physical strength and virility but by such inner qualities as personal integrity, commitment, loyalty, and, most importantly, strength of character. . . . It is not clear how this code of ethics developed, but it may be linked to nationalist sentiments and Mexican resistance to colonization and foreign invasion. Historical figures such as Cuauhtémoc, *El Pipíla, Los Niños Héroes,* Villa, and Zapata would be macho according to this view. In music and film positive macho figures such as Pedro Infante, Jorge Negrete, and even Cantinflas are patriots, but mostly they are *muy hombres,* men who stand up against class and racial oppression and the exploitation of the poor by the rich.

Despite the apparent differences between the two views, both see the macho cult as integral to Mexican and Latino cultures. Although I did not formulate explicit hypotheses, I entered the field expecting that respondents would generally identify with the word "macho" and define it as a positive trait or quality in themselves and other persons. An additional informal hypothesis proposed was that men who had greater ties to Latino culture and the Spanish language would be more likely to identify and to have positive associations with the word. I expected, in other words, that respondents would be more likely to adhere to the positive view of macho.

FINDINGS: CONCEPTIONS OF MACHO

Respondents were first asked the following question: "What does the word 'macho' mean to you?" The interviewers were instructed to ask this and all other questions in a neutral tone, as we wanted the respondents to feel that we really were interested in what they thought. We stressed in the interviews that there were no "right" or "wrong" answers to any of the questions. This first question was then followed by a series of follow-up questions that included: "Can you give me an example (or examples) of someone you think is really macho?"; "What kinds of things do people who are really macho do?"; and "Can a woman be macha?"

One of the most striking findings is the extent to which the respondents were polarized in their views of macho. Most had very strong feelings; very few were neutral or indifferent toward the word. In fact, only 11 percent of the 105 respondents were classified as neutral by our judges. No less surprising is the fact that, contrary to my expectations, very few respondents viewed the word in a positive light. Only 31 percent of the men were positive in their views of macho, compared to 57 percent who were classified as negative. This means, in effect, that more than two-thirds of the respondents believed that the word "macho" had either negative or neutral connotations.

My expectation that those individuals with greater ties to Latino culture would be more likely to identify and to have positive associations with "macho" was also not supported by the data. Of the thirty-nine respondents who opted to be interviewed in Spanish, only 15 percent were seen as having a positive association with macho, whereas 74 percent were negative and 10 percent were neutral toward the term. In contrast, of the sixty-six interviewed in English, 41 percent were classified as positive, 47 percent as negative, and 12 percent as neutral toward the term.

Although negative views of the word "macho" were more prevalent than I had expected, the responses closely parallel the polar views of the word "macho" discussed earlier. Responses classified as "negative" by our judges are consistent with the "compensatory" or "deficit" model, which sees the emphasis on excessive masculinity among Mexicans and Latinos as an attempt to conceal pervasive feelings of inferiority among native men that resulted from the Conquest and the ensuing cultural, moral, and spiritual rape of the indigenous population. Those classified as "positive," similarly, are roughly consistent with an "ethical" model, which sees macho behavior as a positive, nationalist response to colonization, foreign intervention, and class exploitation.

Negative Conceptions of "Macho"

A number of consistent themes are found among the men who were classified as viewing the word "macho" in a negative light. Though I divide them into separate themes to facilitate the presentation of the findings, there is obviously considerable overlap between them.

Negative Theme 1: Synthetic/Exaggerated Masculinity A theme that was very prevalent in the responses is that machos are men who are insecure in themselves and need to prove their manhood. It was termed a "synthetic self-image," "exaggerated masculinity," "one who acts tough and is insecure in himself," and an "exaggerated form of manliness or super manliness." One respondent described a macho as

> one who acts "bad." One who acts tough and who is insecure of himself. I would say batos [dudes] who come out of the *pinta* [prison] seem to have a tendency to be insecure with themselves, and tend to put up a front. [They] talk loud, intimidate others, and disrespect the meaning of a man.

Another person described it as

> being a synthetic self-image that's devoid of content. . . . It's a sort of facade that people use to hide the lack of strong, positive personality traits. To me, it often implies a negative set of behaviors. . . . I have a number of cousins who fit that. I have an uncle who fits it. He refuses to have himself fixed even though he was constantly producing children out of wedlock.

Negative Theme 2: Male Dominance/Authoritarianism A second, related theme is that of male dominance, chauvinism, and the double standard for men and women. Within the family, the macho figure is viewed as authoritarian, especially relative to the wife. According to one respondent, "They insist on being the dominant one in the household. What they say is the rule. They treat women as inferior. They have a dual set of rules for women and men." Another respondent added:

> It's someone that completely dominates. There are no two ways about it; it's either his way or no way. My dad used to be a macho. He used to come into the house drunk, getting my mother out of bed, making her make food, making her cry.

A Spanish-speaker characterized the macho as follows:

> *Una persona negativa completamente. Es una persona que es irresponsable en una palabra. Que anda en las cantinas. Ese no es hombre. Si, conozco muchos de mi tierra; una docena. Toman, pelean. Llegan a la casa gritando y golpeando a la señora, gritando, cantando. Eso lo vi yo cuando era chavalillo y se me grabó. Yo nunca vi a mi papá que golpeara a mi mamá* (A completely negative person. In a word, it's a person who is irresponsible. Who is out in the taverns. That's not a man. Yes, I know many from my homeland; a dozen. They drink, fight. They come home yelling and hitting the wife, yelling, singing. I saw this as a child and it made a lasting impression on me. I never saw my father hit my mother).

Negative Theme 3: Violence/Aggressiveness A third, related theme is macho behavior manifested in expressions of violence, aggressiveness, and irresponsibility, both inside and outside the family. It is "someone that does not back down, especially if they fear they would lose face over the most trivial matters." Another person saw macho as the exaggeration of perceived masculine traits and gave the example of a fictional figure like Rambo and a real figure like former president Ronald Reagan. This person added that it was "anyone who has ever been in a war," and "it's usually associated with dogmatism, with violence, with not showing feelings." A Spanish-speaking man summarized it succinctly as *"el hombre que sale de su trabajo los viernes, va a la cantina, gasta el cheque, y llega a su casa gritando, pegándole a su esposa diciendo que él es el macho"* (the man who gets out of work on Friday, goes to a bar, spends his check, and comes home yelling and hitting his wife and telling her that he is the macho [i.e., man]). Still another felt that men who were macho did such things as "drinking to excess," and that associated with the word "macho" was "the notion of physical prowess or intimidation of others. A willingness to put themselves and others at risk, particularly physically. For those that are married, the notion of having women on the side."

One of our Spanish-speaking respondents mentioned an acquaintance who lost his family because he would not stop drinking. *"Él decía, 'La mujer se hizo para andar en la casa y yo pa' andar en las cantinas'"* (He used to say, "Woman was made to stay at home and I was made to stay in taverns"). This respondent also noted that men who are real machos tend not to support their families or tend to beat them, to get "dandied up," and to go out drinking. Another said that they "drink tequila" and "have women on their side kissing them."

Negative Theme 4: Self-Centeredness/ *Egoísmo* Closely related is the final theme, which views someone who is macho as being self-centered, selfish, and stubborn, a theme that is especially prevalent among respondents with close ties to México. Several men saw machismo as *un tipo de egoísmo* (a type of selfishness) and felt that it referred to a person who always wanted things done his way—*a la mía.* It is someone who wants to impose his will on others or wants to be right, whether he is right or not. It is viewed, for example, as

> *un tipo de egoísmo que nomás "lo mío" es bueno y nomás mis ideas son buenas. Como se dice, "Nomás mis chicharrones truenan." . . . Se apegan a lo que ellos creen. Todo lo que ellos dicen está correcto. Tratan que toda la gente entre a su manera de pensar y actuar, incluyendo hijos y familia* (a type of selfishness where only "mine" is good and only my ideas are worthwhile. As the saying goes, "Whatever I say goes." . . . They cling to their own beliefs. Everything they say is right. They try to get everyone, including children and family, to think and act the way they do).

Some respondents who elaborated on the "self-centeredness" or *egoísta* theme noted that some men will hit their wives "just to prove that they are machos," while others try to show that they "wear the pants" by not letting their wives go out. One person noted that some men believe that wives and daughters should not be permitted to cut their hair because long hair is considered "a sign of femininity," and another made reference to a young man who actually cut off a finger in order to prove his love to his sweetheart.

Because the word "macho" literally means a "he-mule" or a "he-goat," respondents often likened macho men to a dumb animal such as a mule, goat, or bull: "Somebody who's like a bull, or bullish"; "The man who is strong as though he were an animal"; "It's an ignorant person, like an animal, a donkey or mule"; and "It's a word that is outside of that which is human." One person described a macho as

> the husband of the mule that pulls the plow. A macho is a person who is dumb and uneducated. *Hay tienes a* [There you have] Macho Camacho [the boxer]. He's a wealthy man, but that doesn't make a smart man. I think he's dumb! . . . They're aggressive, and they're harmful, and insensitive.

Another respondent said, "Ignorant, is what it means to me, a fool. They're fools, man. They act bully type." Another similarly linked it to being "ignorant, dumb, stupid," noting that they "try to take advantage of their physical superiority over women and try to use that as a way of showing that they are right."

Given that these respondents viewed "macho" in a negative light, it is not surprising to find that most did not consider themselves macho. Only eight of the sixty men in this category reluctantly acknowledged that they were "somewhat" macho. One said, "Yes, sometimes when I drink, I get loud and stupid," and another, "Yes, to an extent because I have to be head-strong and bullish as a teacher."

Positive Conceptions of Macho: Courage, Honor, and Integrity

* * *

As was true of men who were classified as negative toward the word "macho," several themes were discernible among those classified as positive. And as with the negative themes, they are separate but overlapping. A few respondents indicated that it meant "masculine" or "manly" (*varonil*), a type of masculinity (*una forma de masculinidad*), or male. The overriding theme, however, linked machismo to internal qualities like courage, valor, honor, sincerity, respect, pride, humility, and responsibility. Some went so far as to identify a distinct code of ethics or a set of principles that they saw as being characteristic of machismo.

Positive Theme 1: Assertiveness/Standing Up for Rights A more specific subtheme is the association of machismo with being assertive, courageous, standing up for one's rights, or going "against the grain" relative to other persons. The following response is representative of this view:

> To me it means someone that's assertive, someone who stands up for his or her rights when challenged. . . . Ted Kennedy because of all the hell he's had to go through. I think I like [Senator] Feinstein. She takes the issues by the horns. . . . They paved their own destiny. They protect themselves and those that are close to them and attempt to control their environment versus the contrast.

It is interesting to note that this view of being macho can be androgynous. Several respondents mentioned women who exemplified "macho qualities" or indicated that these qualities may be found among either gender. . . .

Spanish-speaking respondent added:

> *En respecto a nuestra cultura es un hombre que defiende sus valores, en total lo fisico, lo emocional, lo psicológico. En cada mexicano hay cierto punto de macho. No es arrogante, no es egoísta excepto cuando tiene que defender sus valores. No es presumido* (Relative to our culture, it's a man that stands up for what he believes, physically, emotionally, and psychologically. Within every Mexican there is a certain sense of being macho. He is not arrogant, not egoistic, except when he has to defend his values. He is not conceited).

Positive Theme 2: Responsibility/Selflessness A second positive macho theme is responsibility, selflessness, and meeting obligations. In direct opposition to the negative macho who is irresponsible and selfish, the positive macho is seen as having a strong sense of responsibility and as being very concerned with the welfare and well-being of other persons. This second positive macho theme was described in a number of ways: "to meet your obligations"; "someone who shoulders responsibility"; "being responsible for your family"; "a person who fulfills the responsibility of his role . . . irrespective of the consequences"; "they make firm decisions . . . that take into consideration the well-being of others." According to one respondent,

> A macho personality for me would be a person that is understanding, that is caring, that is trustworthy. He is all of those things and practices them as well as teaches them, not only with family but overall. It encompasses his whole life.
>
> It would be a leader with compassion. The image we have of Pancho Villa. For the Americans it would be someone like Kennedy, as a strong person, but not because he was a womanizer.

Positive Theme 3: General Code of Ethics The third theme we identified embodies many of the same traits mentioned in the first and second themes, but it differs in that respondents appear to link machismo not just to such individual qualities as selflessness but to a general code of ethics or a set of principles. One respondent who was married to an Israeli woman offered a former defense minister of Israel as exemplifying macho qualities. He noted that

> It's a man responsible for actions, a man of his word. . . . I think a macho does not have to be a statesman, just a man that's known to stand by his friends and follow through. A man of action relative to goals that benefit others, not himself.

Another said that it means living up to one's principles to the point of almost being willing to die for them. . . .

Positive Theme 4: Sincerity/Respect The final positive theme overlaps somewhat with the others and is often subsumed under the code of ethics or principles. A number of respondents associated the word "macho" with such qualities as respect for oneself and others, acting with sincerity and respect, and being a man of your word. One of our interviewees said,

> *Macho significa una persona que cumple con su palabra y que es un hombre total. . . . Actúan con sinceridad y con respeto* (Macho means a person who backs up what he says and who is a complete man. . . . They act with sincerity and respect).

Another mentioned self-control and having a sense of oneself and the situation.

> Usually they are reserved. They have kind of an inner confidence, kind of like you know you're the fastest gun in town so you don't have to prove yourself. There's nothing to prove. A sense of self.

Still another emphasized that physical prowess by itself would not be sufficient to identify one as macho. Instead, "It would be activities that meet the challenge, require honor, and meet obligations." Finally, a respondent observed:

> Macho to me means that you understand your place in the world. That's not to say that you are the "he-man" as the popular conception says. It means you have respect for yourself, that you respect others.

Not surprisingly, all of the respondents who viewed machismo in a positive light either already considered themselves to have macho qualities or saw it as an ideal they hoped to attain.

Neutral Conceptions of Macho

Twelve respondents could not be clearly classified as positive or negative in their views of "macho." This so-called neutral category is somewhat of a residual one, however, because it includes not only men who were, in fact, neutral but also those who gave mixed signals and about whom the judges could not agree. One said that "macho" was just a word that didn't mean anything; another said that it applied to someone strong like a boxer or a wrestler, but he did not know anyone who was macho, and it was not clear whether he considered it to be a positive or negative trait. Others were either ambivalent or pointed to both positive and negative components of being macho. A street-wise young man in his mid-twenties, for example, indicated that

> The word macho to me means someone who won't take nothing from no one. Respects others, and expects a lot of respect from others. The person is willing to take any risks. . . . They always think they can do anything and everything. They don't take no shit from no one. They have a one-track mind. Never want to accept the fact that women can perform as well as men.

* * *

Another person observed that there were at least two meanings of the word—one, a brave person who is willing to defend his ideals and himself, and the other, a man who exaggerates his masculinity—but noted that "macho" was not a term that he used. Another respondent provided a complex answer that distinguished the denotative (i.e., macho) and connotative (i.e., machismo) meanings of the term. He used the word in both ways, differentiating between being macho or male, which is denotative, and machismo, which connotes male chauvinism. He considered himself to be macho but certainly not *machista.*

> *Ser macho es ser valiente o no tener miedo. La connotación que tiene mal sentido es poner los intereses del hombre adelante de los de la mujer o*

> *del resto de la familia. Representa egoísmo. . . . Macho significa varón, hombre, pero el machismo es una manera de pensar, y es negativo* (To be macho is to be brave or to not be afraid. The connotation that is negative is to put the interests of the man ahead of those of the woman or the rest of the family. It represents selfishness. . . . Macho means male, man, but machismo is a way of thinking, and it is negative).

Another person similarly distinguished between being macho and being *machista.*

> *Pues, en el sentido personal, significa el sexo masculino y lo difiere del sexo femenino. La palabra machismo existe solamente de bajo nivel cultural y significa un hombre valiant, borracho y pendenciero* (Well, in a personal sense, it means the masculine gender and it distinguishes it from the feminine. The word machismo exists only at a low cultural level and it means a brave man, a drunkard, and a hell-raiser).

Six of the twelve respondents who were classified as neutral considered themselves to be at least somewhat macho.

CONCLUSION

These data provide empirical support for two very different and conflicting models of masculinity. The compensatory model sees the cult of virility and the Mexican male's obsession with power and domination as futile attempts to mask feelings of inferiority, powerlessness, and failure, whereas the second perspective associates being macho with a code of ethics that organizes and gives meaning to behavior. The first model stresses external attributes such as strength, sexual prowess, and power; the second stresses internal qualities like honor, responsibility, respect, and courage.

Although the findings are not conclusive, they have important implications. First, and most importantly, the so-called Mexican/Latino masculine cult appears to be a more complex and diverse phenomenon than is commonly assumed. But the assumption that being macho is an important Mexican cultural value is seriously called into question by the findings. Most respondents did not define macho as a positive cultural or personal trait or see themselves as being macho. Only about one-third of the men in the sample viewed the word "macho" positively. If there is a cultural value placed on being macho, one would expect that those respondents with closer ties to Latino culture and the Spanish language would be more apt to identify and to have positive associations with macho, but the opposite tendency was found to be true. Respondents who preferred to be interviewed in English were much more likely to see macho positively and to identify with it, whereas the vast majority of those who elected to be interviewed in Spanish viewed it negatively.

A major flaw of previous conceptualizations has been their tendency to treat machismo as a unitary phenomenon. The findings presented here suggest that although Latino men tend to hold polar conceptions of macho, these conceptions may not be unrelated. In describing the term, one respondent observed that there was almost a continuum between a person who is responsible and one who is chauvinistic. If one looks more closely at the two models, moreover, it is clear that virtually every trait associated with a negative macho trait has its counterpart in a positive one. . . .

From this perspective much of what social scientists have termed "macho" behavior is not macho at all, but its antithesis. Rather than attempting to isolate a modal Mexican personality type or determining whether macho is a positive or a negative cultural trait, social scientists would be well served to see Mexican and Latino culture as revolving around certain focal concerns or key issues such as honor, pride, dignity, courage, responsibility, integrity, and strength of character. Individuals, in turn, are evaluated positively or negatively according to how well they are perceived to respond to these focal concerns.

But because being macho is ultimately an internal quality, those who seek to demonstrate outwardly that they are macho are caught in a double bind. A person who goes around holding his genitals, boasting about his manliness, or trying to prove how macho he is would not be considered macho by this definition. In the final analysis it is up to others to determine the extent to which a person lives up to these expectations and ideals.

It is also important to note that to a great extent, the positive internal qualities associated with the positive macho are not the exclusive domain of men but extend to either gender. One can use the same criteria in evaluating the behavior of women and employ parallel terminology such as *la hembra* (the female) and *hembrismo* (femaleness). *Una mujer que es una hembra* (a woman who is a real "female") is neither passive and submissive nor physically strong and assertive, for these are external qualities. Rather, *una hembra* is a person of strong character who has principles and is willing to defend them in the face of adversity. Thus, whereas the popular conception of the word "macho" refers to external male characteristics such as exaggerated masculinity or the cult of virility, the positive conception isolated here sees being macho as an internal, androgynous quality.

REFERENCES

Lewis, Oscar. 1960. *Tepoztlan*. New York: Holt, Rinehart & Winston.

———. 1961. *The Children of Sanchez*. New York: Random House.

Madsen, William. 1973. *The Mexican-Americans of South Texas*. 2d ed. New York: Holt, Rinehart & Winston.

Mendoza, Vicente T. 1962. *"El machismo en Mexico a traves de las canciones, corridos, y cantares."* In *Cuadernos del instituto nacional de antropologia III*, 75–86. Buenos Aires: Minsterio de Educación y Justicia.

Paredes, Américo. 1977. "On Ethnographic Work Among Minority Groups: A Folklorist's Perspective." *New Scholar* 6 (fall and spring): 1–33.

Rubel, Arthur J. 1966. *Across the Tracks: Mexican Americans in a Texas City*. Austin: University of Texas Press.

10

Identity Politics in the Women's Movement

BY BARBARA RYAN

Barbara Ryan, a sociologist, has studied social movements and feminism for over twenty years and has authored three books on the topic. This reading is the introduction to her latest book of the same title. She describes different perspectives, comparing postmodernists' emphasis on uniqueness to that of identity politics, which focuses on similarities across individuals' experiences. Activism is a central part of feminist thought; however, as we noted in the introduction to this chapter, activism for women's rights and equality is not a simple or straightforward issue. Ryan's discussion of the women's movement reflects the argument that gender is a complex concept representing multiple, socially constructed identities. It is these multiple social identities that have shaped and divided the women's movement.

1. What does Ryan mean by "identity politics" and how does this differ from a postmodernist perspective?
2. Are identity politics an impediment or facilitator of efforts to gain rights for women?
3. How are multiple social identities incorporated into the women's movement?

What is identity? Is identity recognition of a shared characteristic that enables a solidarity with members of a group? And does it conversely entail distance from those who lack this common feature? Or, is identity a social construction that ebbs and flows, is always in process, multilayered, and fragmented? What is gender? As a defining identity, does it fit the former or the latter conceptualization?

These questions are at the heart of understanding how identity politics affects social movements and the women's movement. The women's movement is a gender-focused movement. Yet, within the category of women there are other identities that work to keep women from recognizing gender commonalties. For instance, living in a largely segregated society, women of color feel a bonding with men of color that they do not usually feel with white women (Bell-Scott 1994; Collins 1990; Dill 1983; Fleming 1993; McKay 1993). Likewise, studies of lesbian feminist communities reveal the positive aspects of joining together to find acceptance and emotional support where it is lacking in the straight world (Franzen 1993; Kreiger 1982, 1983; Taylor and Whittier 1992).

Proponents of identity politics believe it important to affiliate with those who confront similar experiences based on social group characteristics. Members of an oppressed group may organize to change their situation, as well as their feelings of self-worth and place in the social structure. Hence, social characteristics that have been used to exclude certain groups have led to social movements organized by those groups to change their condition.

Critics of identity politics assert that it leads to further marginalization and that it prevents uniting with those who are working on similar issues but who differ in physical/social features (Gitlin 1993). Class, too, is left out of this analysis, as are differences within groups, which may have everyday practical consequences (Allison 1993). Indeed, Hall (1996:4–5) argues that unities based on essentialist identities are constructed within the dynamics of power and exclusion, "and thus are more the product of the making of difference and exclusion, than they are the sign of a naturally constituted unity." And problematically, as Grossberg (1996:88) points out, groups organized around their own model of repression often lack the capability of creating alliances with others.

CRITIQUES

Postmodern thought sees identity as a process rather than a fact or deterministic force. Yet, recognizable identity traits continue to draw people together and to provide them with support for attempting social change. This means identity and identity politics are serious contenders in the political process and social movement arenas.

Eric Hobsbawm (1996) points out that in the late 1960s the *International Encyclopedia of the Social Sciences* had no entry under identity. Thus, Hobsbawm sees identity politics as a recent phenomenon. He also sees it as a problematic category. First, he argues that a collective identity is defined against others and is based not on what their members have in common but, rather, on differences between them when, in fact, "we" may have little in common except not being the "others." Second, no one has only a single identity; yet, identity politics leads one to disclaim other identities. And finally, identities are not fixed—they depend on contexts, which can change. . . .

Todd Gitlin (1995), like Hobsbawm, calls for a Left politics based on class position. However, his analysis is somewhat different, fashioned on a contemporary and North American model. According to Gitlin, the most serious and negative identity politics is white men who fear identity gains will come at their expense. But he is also opposed to identity politics in general because he feels it cultivates unity only within special groups, and there is an obsession with difference leading to the "borders identity politics draws" (severson and stanhope 1998).

Rather than organizing to reduce inequalities between rich and poor, Gitlin argues that identity politics struggles to change the color of inequality. And, what we need to be doing, instead, is to tend mutualities. Identity politics, according to Gitlin, (1995:236) has failed to tend and, even worse, has left the centers of power uncontested.

Gitlin calls for a Left politics that includes everyone, a common—a cause of all. Although his argument is compelling in many ways, he barely mentions gender. Socialist feminists in the 1960s and 1970s explicitly pointed out how the inclusion of "all" in the Left of their day did not include them. When leftist writers in the late 1990s have little to say specifically to gender, are we to believe them? Does commonality leave women's "difference" out? Is leftist universalism like postmodern and deconstructionist analysis, wiping away all difference, even denying there is difference because there is no reality? Does identity politics as we know it in the women's movement leave the centers of power uncontested? These questions raised by leftist scholars present serious critiques of identity politics.

Feminist analysis also contains critiques of identity politics. Daphne Patai (1992), for instance, discusses the zealousness of feminist adherents to control thought and appearance, what she calls "ideological policing." Patai objects to the assumption that one's racial/ethnic identity is the same as one's views. Even more, she believes there has been a reversal of privilege, now residing with women of color, in which no white person can challenge their version of reality. These inclinations have led to concern for the "dogmatic turn" identity politics has presented to women's studies. In her experience, *Eurocentric* became a slur and teaching courses on other racial/ethnic groups was not accepted of a North American white.

Further, Patai (1992:B3) questions the ways identity politics gets used in a scarce job market, calling it "the fraud that accompanies familiar old ambitions dressed up in appropriate ideology." Patai feels distress that these tendencies have arisen and, even more, that they are not discussed. Instead, identity politics has led to silencing. She considers her writing on these issues to be a defense of feminism.

Others also point to problems within feminism, particularly the focus on personal experience, which may have isolated the women's movement from more general social change struggles. Often, rather than oppression's being fought in the wider society, struggles are being fought on local levels (Adams 1989). L. A. Kauffman (1990) takes a more nuanced look at identity politics, dividing it into political and nonpolitical frameworks. Kauffman dates the beginning of identity politics not with black women's challenge to sisterhood but to the civil rights movement of Martin Luther King and the Black Power movement, where activists called for a new collective identity to offset white imperialism. In turning to the women's movement, Kauffman credits Kate Millett's *Sexual Politics* (1970) with defining gendered power as politics—structured relationships whereby one group controls another. In the 1980s and 1990s, though, she fears identity politics has evolved into fragments where "the notion of solidarity, so central to any progressive politics" is lost (Kauffman 1990:76).

Kauffman (1990:78) makes the point that the increasing movement of self-transformation (as political change) leads to thinking that problems are attitudes rather than power differentials and vested interests. Like Patai, she see this leading to an emphasis on lifestyle (who one reads, what one eats or wears) rather than on the actions one takes. Still, Kauffman calls for using identity as entry to challenging institutions of power, and as politics intent upon both social and individual transformation.

A central issue of importance is whether difference has displaced inequality as a central concern of social movements. As Anne Phillips (1997) points out, an *injustice* perspective seeks to eliminate differentiation used against powerless groups, and *difference* perspectives are intent on highlighting these identities. She cites the dilemma between strategies that are meant to diminish the significance of gender and strategies that focus on the intrinsic worth of one's sex.

Hazel Cathy (1990) adds another perspective when she questions whether the emphasis on diversity in feminist thought and practices is a way to avoid the politics of race, even as it appears that race is being confronted. Similarly, the disjunction between inclusive feminism and the reality of the organizations that make up the women's movement raises the crucial point "on whose behalf" inclusive ideologies are meant (Leldner 2001).

IDENTITY CLAIMS

In spite of critical questions associated with identity politics, there are important rationales for the development of a politics of identity, beginning with *The Second Sex,* Simone de Beauvoir's (1953) classic work. Her introduction sets the tone, unveiling a gendered identity politics by calling women "the other." By this, she means that women have failed to identify themselves as a group because they are considered a part of man (the subject) and, thus, are not segregated into their own group, as are some racial and ethnic groups. They have no history or religion that is particularly their own. She calls for women to see themselves as a group in order to change their situation.

De Beauvoir's writings inspired a collective conscience of women, which laid the foundation for the reemergence in the 1960s of women's activism in their own behalf. Yet, by the early

1980s, writings by women of color spoke to the need to claim an identity of their own. They formulated a base for organizing around that identity, even if it separated women from one another. The Combahee River Collective, a group of black feminists and forerunner to this claim, began meeting in 1974. They issued the first statement on black feminism, twelve years after the contemporary women's movement emerged and many more years after the U.S. publication of *The Second Sex*. The statement combined gender and race identity. Black women proclaimed the task of combating simultaneous oppressions as theirs because other movements failed to acknowledge their specific oppression. They named what they were doing "identity politics" based on their conclusion that "the only people who care enough about us to work consistently for our liberation are us" (Combahee River Collective 1978:275).

How did white feminists react to this challenge? Some were angry or dismissive. Some had already reached this awareness. Others welcomed it. Still others struggled with their past and worked to become multicultural in their feminist thought and actions, even as this became a painful process of stripping away their own identity, deciding what to keep, what to eliminate, what to change (Pratt 1984). This difficult process involved the acknowledgment of another's existence while not denying one's own. For instance, Minnie Bruce Pratt (1984:73) describes her fears as she tries to understand herself in "relation to folks different from me, when there are discussions, conflicts about anti-Semitism and racism among women, criticisms, criticisms of me; when, for instance in a group discussion about race and class, I say I feel we have talked too much about race, not enough about class, and a woman of color asks me in anger and pain if I don't think her skin has something to do with class."

Part of the problem in understanding "other worlds" is that women grow up learning different gender roles. For example, in many Native American groups women are strong and valued (Allen 1995), whereas other women have had to work at developing that consciousness. And having acquired an ideology of strength and independence, what happens if you become disabled or when you grow old (Klein 1992)?

MULTIPLE IDENTITIES AND CHANGING IDENTITIES

An obvious complexity within the field of identity politics is the reality of multiple identities, including those we are in the process of becoming or losing. Audre Lorde, (1984a:41) who called upon women to speak—"your silence will not protect you"—used her life as an example of how we can rid ourselves of others' distortions by reclaiming all our identities so we can define them for ourselves.

As a forty-nine-year-old black lesbian socialist feminist, who was also a mother of two and part of an interracial couple, Lorde discussed her many group identities, including acquired identities that did not fit into acceptable society. This makes life difficult, and yet, she notes it is oppressed people who are expected to bridge the gap between their differences with more privileged groups. Lorde asserted that it is not the differences among us that separate us; it is the refusal to recognize the differences. An example she cites is the idea of "sisterhood." In a famous quote, she tells us: "Some problems we share as women, some we do not. You fear your children will grow up to join the patriarchy and testify against you, we fear our children will be dragged from a car and shot down in the street, and you will turn your backs upon the reasons they are dying" (Lorde 1984b:119).

In applying a wide lens, Lorde also talked about differences within black communities. Where racism is a living reality, differences within groups seem dangerous and suspect. The need for unity is often misnamed as a need for homogeneity, and a black feminist vision mistaken for betrayal. There is a refusal of some black women to recognize and protest against their oppression as women within the black community and of heterosexual women against lesbians, particularly among black women.

She urged women to identify with one another and develop new ways of being in this world and new ways for this world to be. For, as she tells us, "the master's tools will never dismantle the master's house" (Lorde 1984b:123). In line with Lorde's analysis of divisions within groups, Marilyn Frye (1992) makes clear that even in what appears to be a cohesive commonality (in this case a lesbian community) there are substantial differences that must be acknowledged and worked through.

The necessity for claiming more than one identity is also true for Chicanas (Nieto 1997), Asians (Shah 1994), and women in developing countries. Like African American women, members of these racial/cultural groups often find U.S. feminism's focus on male/female relations alienating because they, too, are working against multiple oppressions of gender, class, race, and nationalism. But they add a difference to the experiences of African American women, where much of identity politics has been focused, and that is invisibility, which, for them, is another form of oppression (Friedman 1995). Moreover, not all Third World women are women of color and not all women of color are Third World or poor. Hispanic women have reported experiencing racism through the rejection of black sisters because of being light skinned (Quintanales 1983), and class is a dividing agent among women of all nationalities and races.

Regardless of the identity issues that may divide them, feminists are concerned that in the rush to acknowledge and celebrate difference, the relations of power that create that difference are often ignored. Women of color, in particular, find the current popularity of diversity rhetoric all too often offers a decontextualized politics of difference, which turns out to be another way of preserving stratified social arrangements (Aguilar 1995).

In another vein, Carol Queen (1997) voices an unwelcome (at least for some) claim to feminist identity. Queen finds that sex radicals (regardless of race, class, ethnicity, or sexual orientation) have been silenced in the women's movement. She questions what it is that separates women who are opposed to sex work from those who do it, and why feminism does not take a more thoughtful look at this divide. She asks feminists to confront their "whorephobia" and agree to a dialogue, for she believes women have much to learn from sex workers.

The complexity of multiple identities is poignantly voiced by June Jordan (1985). From a vacation experience in the Bahamas she finds that, compared to the Bahamian people, she is a rich American woman. She is dismayed to find herself, as well as other black Americans (and whites), arguing prices on handmade items. Jordan uses this story to raise awareness of the complex interplay of race, class, and gender identity. She notes that she and the women workers are engaged in interactions that preclude seeing themselves as a united group of women. Jordan wonders how women are to connect with such different life circumstances, particularly when many women do not feel poor women's issues of poverty and crime are theirs. She asks, "Why aren't they everyone's?"

Jordan's story shows that race, class, and gender are not automatic paths of connection; there are differences within identities that have been imposed.

VOICES OF AFRICAN AMERICAN WOMEN

It was African American women in the early 1980s, more than any other group, who confronted the women's movement on identity politics issues. There was a desire for a more pluralistic approach to "sisterhood" that recognized similarities and differences among women (Dill 1983). Bernice McNair Barnett (1995:207) makes the interesting and telling remark that the barring of black women from the League of Women Voters in Montgomery, Alabama, showed that "it was white women, rather than black women, who placed their primary emphasis on race over gender." She also points out that the 1940s and 1950s were not a period of "doldrums" for women activists, as has been claimed for the women's movement (see Rupp and Taylor 1987). These were years of activism for black women in the civil rights movement, a movement dedicated to issues of freedom and equality.

Taking a different approach, Barbara Smith (1983) succinctly discusses the reasons feminism frightens black and Third World men and why they resist it. In her introduction to *Home Girls,* an early contribution to writings by black women, Smith shows why black women need a movement of their own. Revealing another perspective some ten years later, Ann duCille (1994) wonders if the effort to promote black women's lives has not gone too far. As a black woman who has long studied black women, she acknowledges having mixed feelings about this rise of "the occult of black womanhood." For instance, she questions the career-enhancing path women academics, white and black, have gained by claiming a "new" specialness for women of color and those who focus on them.

New questions are raised, such as looking at white middle-class women who are "housewives" to their husbands and the black working-class domestics they employ (Kaplan 1995). Both groups of women are in roles of serving others, but the white women exist with race and class privilege by means of their domestics. One conclusion, which can be drawn from this relationship, is that white women collude with the patriarchal/capitalist system that oppresses women. Another conclusion is that domestic workers enable white middle-class woman to avoid confronting their spouses about sharing household duties (Kaplan 1995:81). Moreover, household help releases the middle-class woman to become a woman of leisure or to have a career. The ways that women treat other women (using domestics as an example) may help explain why many black women stay away from the (white) feminist movement.

SEXUALITY AND SEXUALITIES

A second area of identity contestation centers on questions of sexual orientation and preference. One of the onerous aspects of heterosexual society is the normative expectation of appropriate sexual behavior that excludes homosexuality, bisexuality, sadomasochism, or transsexuality. Dichotomous thinking, rather than a continuum model of sexual identity, had long been critiqued within the feminist movement, yet it arose in the 1980s among lesbian feminists. These divisions revealed that there are exclusions and antagonisms among gays that differ from the full acceptance of sexual expression found in queer theory or, in the past, in the lesbian concept of the "woman-identified woman" (Radicalesbians 1970).

One challenge to agreed-upon thought was the deconstruction of commonly held views of sex workers, that is, to see them as workers deserving of workers' rights. Women in unions, armed with feminist ideologies, concretely address many of the issues of the women's movement—sexual harassment, maternity benefits, parental leave, and comparable worth (Chernow and Moir 1995). Yet, in the debates over prostitution and pornography, feminists who have argued for other women workers have not taken up these workers' cause (Alexander 1997).

There are differences within lesbianism, within feminism, and even within radical feminism. Eileen Bresnahan (2001) humorously relates an incident where the "original" radical feminists (with roots in the Left) collided with newer radical feminists, who were called cultural feminists. She laments the shift away from political process to lifestyle affirmation that she saw occurring in the mid-1970s. For her, this shift left an ambiguous meaning of radical feminism and was also a departure from the past, when being a radical feminist meant that one accepted definite agreed-upon principles of radical politics. Bresnahan explains her distrust of cultural feminism as the end product of her seriousness about feminist identity. She states that because "I'm a working-class woman who grew up in the 1950s and 1960s, the women's movement was the first time I took myself seriously and the first time I was taken seriously by others whom I could also respect." Thus, the challenge to agreed-upon thought was unwelcome in her mind and in her radical feminist circle.

What is a lesbian—who counts—is a continuing theme of sexuality inquiry. Divisions are found among lesbian feminists based on bisexuality, dress, associations, s/m practices, gender roles, and transgendered people. The 1970s woman-identified women represented a sisterhood against the patriarchy; 1990s lesbians aligned with gay men. Young lesbians have focused more on sex than political theory and often call themselves queer or "bad girls" rather than lesbians. These generational differences have led to clashes between lesbian feminism and queer ideologies. Problematically, these clashes have also raised charges of who is a real or fake lesbian, for example, what if you have a heterosexual past? The essentialism (true lesbian) of the 1970s is now confronted with more than one model for lesbian behavior. And even though this may seem confusing, as Vera Whisman (1993:58) says, "[T]he truth is, most of us sometimes feel incredibly queer, at other times indelibly female."

One highly contested issue that his created division among lesbian feminists is sadomasochism. Shane Phelan considers it a mistake for activists to get involved in arguments of this kind, an issue that is rooted in the identity politics of what feminism is. By this, she does not mean that identity politics should be abandoned rather, she asks that we be more careful in distinguishing "the sorts of identity issues that are vital to our growth and freedom from those that are not" (Phelan 1989:133).

Other, more "acceptable" divisions among lesbians have been identified as class age, and ideology. Trisha Franzen discovered that lesbian-feminist university students considered the butch/fem roles played by many working-class-bar lesbians to be tainted with heterosexuality. Thus, she argues that "sexuality is a problematic basis for political solidarity among women" (Franzen 1993:903). A similar dispute occurred at the Michigan Womyn's Music Festival over the admittance of transsexuals. In researching this issue, Joshua Garrison (1997:183) argues that identity requires difference and that building collective identities requires not simply pointing out commonalties but also marking off "who we are not." He finds these acts to be the boundary patrol of identity politics.

From these examples we can see that sexual identities and political affiliations often shift and are always contingent (Whisman 1993:58). Hence, we can no longer assume what the foundation of identical politics presumes; that is, the idea that identity groups, in this case lesbians, share an identity and therefore a politics. Even more pointed at the turn of the twenty-first century is to recognize that sexual identity, indeed all identities, are more provisional than most people realize. As Arlene Stein (1997) discussed in her research on ex-lesbians, there was a restructuring of the identity process based on situational factors that some feminists went through as they moved into lesbianism in the 1970s and out of it by the 1980s.

MORE DIVERSITIES— MORE IDENTITIES

While major divisions have arisen over issues of race and sexual diversity, there are other gender-plus identities that confront the women's movement. There is the issue of Jewish feminists and their place in the movement (Beck 1988), of age from the older woman's perspective (Macdonald 1995), as well as of the younger feminist viewpoint (Dietzel 1999; Heywood and Drake 1997; Looser and Kaplan 1997; Walker 1992). Inclusion itself has been questioned. For instance, Rosa María Pegueros (2001) reports on her own experience as a Latina activist in the National Organization for Woman (NOW). Achieving a high, visible position, she questions what that means for her. Is she a token, a traitor to her group, or an accepted member of a feminist elite (and does she want this)? What she has to say raises questions about the sincerity of inclusion some groups are promoting.

And, what of men? How does being a male feminist affect one's identity? There is the possibility that male feminism may be seen as a traitorous identity, indeed, traitorous perhaps to both men and women (Bettie 2001). Although it is self-evident that not all men are powerful, there is a danger in pointing out how men, too, are oppressed. This can seen as a denial of the history, and meaning of gender power relations. De-essentializing identity categories may be a necessary corrective to the conventional application of identity politics, but we must be careful not to become, then, an identity skeptic, refusing to recognize gender, race, and other identities (Bettie 2001).

Cutting across identity concerns is always the issue of class and class transformation. Moving from the working class to the middle class does not mean one has left all vestiges of one's background behind. Class has not been adequately explored in identity politics, perhaps because of the simultaneous desire to both reject and to retain this cultural identity. The challenge is to maintain a vigilant awareness of the inherent power these relations present while guarding against incorrect parallelisms, which can erase the political histories of difference (Bettie 2001).

The history of division within women's studies is legendary. Indeed, one could say that women's studies is itself identity politics (Perry 1995). Yet, in spite of the contentious debates over the category of women in the academy, the term *woman* has not been so starkly problematized in the larger society. Women of all races, ethnicity, sexual orientation (and preferences) are disadvantaged in a society that does not value women.

There are other divisions among women . . . Women in revolutions, prisons, and armed services, and those living in rural areas . . . What is their relationship to other women and to feminism? What about differences between single and married women? Or those with or without children?

Divisions are often magnified when we begin to talk of global feminism and organizing transnationally to unite women from around the world. In Yemen in 2000, the Women's Studies Program at San'a University was closed down and the director fled the country because of the use of the word *gender* (Abu-nasr 2000). In Kuwait, women continue to be told they are not to be allowed the vote. How can women join together in India, where women are divided by caste? In what ways can Muslim women organize when ideas of a constructed gender identity are considered a Western concept? There is fear that a transnational unity of women might raise a counterargument, and perhaps repression, from conservative and religious forces that have used biological determinism in order to maintain gender segregation.

At the same time, within countries, it must be recognized that the ideological and political realities women face limits the kinds of issues that can be raised. . . .

REFERENCES

Abu-nasr, Donna. 2000. "Gender Controversy Shuts Down Women's Studies Program in Yemen." *Philadelphia Inquirer,* May 21, p. A14.

Adams, Mary Louise. 1989. "There's No Place Like Home: On the Place of Identity in Feminist Politics." *Feminist Review* 31 (Spring):22–33.

Aguilar, Delia D. 1995. "What's Wrong with the 'F' Word?" In *Frontline Feminism,* edited by Karen Kahn. San Francisco: Aunt Lute Books.

Alexander, Priscilla. 1997. "Feminism, Sex Workers, and Human Rights." In *Whores and Other Feminists,* edited by Jill Nagle. New York: Routledge.

Allen, Paula Gunn. 1995. "Where I Come From God Is a Grandmother." In *Frontline Feminism,* edited by Karen Kahn. San Francisco: Aunt Lute Books.

Allison, Dorothy. 1993. "A Question of Class." Pp. 46–60 in *Sisters, Sexperts, Queers: Beyond the Lesbian Nation,* edited by Arlene Stein. New York: Plume Books.

Barnett, Bernice McNair. 1995. "Black Women's Collectivist Movement Organizations: Their Struggles during the 'Doldrums.'" In *Feminist Organizations: Harvest of the New Women's Movement,* edited by Myra Marx Ferree and Patricia Yancy Martin. Philadelphia: Temple University Press.

Beck, Evelyn Torton. 1988. "The Politics of Jewish Invisibility." *NWSA Journal* 1:93–102.

Bell-Scott, Patricia, ed. 1994. *Life Notes: Personal Writings by Contemporary Black Women.* New York: Norton.

Bettie, Julie. 2001. "Changing the Subject: Male Feminism, Class Identity, and the Politics of Location." Chap. 15 in Identity Politics in the Women's Movement by Barbara Ryan. New York: New York University Press.

Bresnahan, Eileen. 2001. "The Strange Case of Jackie East: When Identities Collide." Chap. 21 in Identity Politics in the Women's Movement by Barbara Ryan. New York: New York University Press.

Carby, Hazel. 1990. "The Politics of Difference." *Ms.* 1 (September): 84–85.

Chernow, Harneen, and Susan Moir. 1995. "Feminism and Labor: Building Alliances." In *Frontline Feminism,* edited by Karen Kahn. San Francisco: Aunt Lute Books.

Collins, Patricia Hill. 1990. *Black Feminist Thought: Knowledge, Consciousness, and the Politics of Empowerment.* Boston: Unwin Hyman.

Combahee River Collective. 1978. "A Black Feminist Statement." In *Capitalist Patriarchy and the Case for Socialist Feminist,* edited by Zillah R. Eisenstein. New York: Monthly Review Press. Also, chap. 7 in this volume.

de Beauvoir, Simone. 1953. *The Second Sex.* New York: Alfred A. Knopf.

Dietzel, Susanne. 1999. "Talking about My Generation." Unpublished paper.

Dill, Bonnie Thornton. 1983. "Race, Class and Gender: Prospects for an All-Inclusive Sisterhood." *Feminist Studies* 9:131–150.

duCille, Ann. 1994. "The Occult of True Black Womanhood: Critical Demeanor and Black Feminist Studies." *Signs: Journal of Women in Culture and Society* 19 (3):591–621.

Fleming, Cynthia Griggs. 1993. "Black Women Activists and the Student Nonviolent Coordinating Committee: The Case of Ruby Doris Smith Robinson." *Journal of Women's History* 4:64–82.

Franzen, Trisha. 1993. "Differences and Identities: Feminism and the Albuquerque Lesbian Community." *Signs: Journal of Women in Culture and Society* 18 (4):891–906.

Friedman, Susan Stanford. 1995. "Beyond White and Other: Relationality and Narratives of Race in Feminist Discourse." *Signs: Journal of Women in Culture and Society* 21 (1):112–149.

Frye, Marilyn. 1992. "Lesbian Community: Heterodox Congregation." In *Willful Virgin: Essays in Feminism, 1976–1992,* by Marilyn Frye. Freedom, CA: Crossing Press.

Gamson, Joshua. 1997. "Messages of Exclusion: Gender, Movements, and Symbolic Boundaries." *Gender & Society* 11 (April):178–199.

Gitlin, Todd. 1993. The Rise of 'Identity Politics': An Examination and a Critique." *Dissent* (Spring):172–177.

Gitlin, Todd. 1995. *The Twilight of Common Dreams: Why America Is Wracked by Culture Wars.* New York: Henry Holt & Co.

Grossberg, Lawrence. 1996. "Identity and Cultural Studies: Is That All There Is?" Pp. 87–107 in *Questions of Cultural Identity,* edited by Stuart Hall and Paul du Gay. Thousand Oaks, CA: Sage.

Hall, Stuart. 1996. "Introduction: Who Needs 'Identity'?" Pp. 1–17 in *Questions of Cultural Identity,* edited by Stuart Hall and Paul du Gay. Thousand Oaks, CA: Sage.

Heywood, Leslie, and Jennifer Drake, eds. 1997. *Third Wave Agenda: Being Feminist, Doing Feminism.* Minneapolis: University of Minnesota Press.

Hobsbawm, Eric. 1996. "Identity Politics and the Left." *New Left Review* 217:38–47.

Jordan, June. 1985. "Report from the Bahamas." In *On Call: Political Essays,* by June Jordan. Boston: South End Press.

Kaplan, Elaine Bell. 1995. "I Don't Do No Windows." In *Frontline Feminism,* edited by Karen Kahn. San Francisco: Aunt Lute Books.

Kauffman, L. A. 1990. "The Anti-Politics of Identity." *Socialist Review* 90:67–80.

Klein, Bonnie Sherr. 1992. "We Are Who You Are: Feminism and Disability." *Ms.* 3 (Nov/Dec):70–74.

Kreiger, Susan. 1982. "Lesbian Identity and Community: Recent Social Science Literature." *Signs: Journal of Women in Culture and Society* 8:91–108.

Kreiger, Susan. 1983. *The Mirror Dance: Identity in a Women's Community.* Philadelphia: Temple University Press.

Leidner, Robin. 2001. "On Whose Behalf? Feminist Ideology and Dilemmas of Constituency." Chap. 6 in Identity Politics in the Women's Movement by Barbara Ryan. New York: New York University Press.

Looser, Devoney, and E. Ann Kaplan, eds. 1997. *Generations: Academic Feminists in Dialogue.* Minneapolis: University of Minnesota Press.

Lorde, Audre. 1984a. "The Transformation of Silence into Language and Action." In *Sister Outsider: Essays and Speeches,* by Audre Lorde. Trumansburg, NY: Crossing Press.

Lorde, Audre. 1984b. "Age, Race, Class, and Sex: Women Redefining Difference." In *Sister Outsider: Essays and Speeches,* by Audre Lorde. Trumansburg, NY: Crossing Press.

Macdonald, Barbara. 1995. "An Open Letter to the Women's Movement." In *Frontline Feminism,* edited by Karen Kahn. San Francisco: Aunt Lute Books.

McKay, Nellie Y. 1993. "Acknowledging Differences: Can Women Find Unity Through Diversity?" Pp. 267–282 in *Theorizing Black Feminisms: The Visionary Pragmatism of Black Women,* edited by S. James and A. Busia. New York: Routledge.

Millett, Kate. 1970. *Sexual Politics.* New York: Doubleday.

Nieto, Consuelo. 1997. "The Chicana and the Woman's Rights Movement." In *Chicana Feminist Thought: The Basic Historical Writings,* edited by Alma M. Garcia. New York: Routledge.

Patai, Daphne. 1992. "The Struggle for Feminist Purity Threatens the Goals of Feminism." *Chronicle of Higher Education* (February 5):B1.

Pegueros, Rosa Maria. 2001. "Sharing Power: A Latina in NOW." Chap 28 in Identity Politics in the Women's Movement by Barbara Ryan. New York: New York University Press.

Perry, Ruth. 1995. "I Brake for Feminists: Debates and Divisions within Women's Studies." *Transformations* 17 (Spring):1–13.

Phelan, Shane. 1989. *Identity Politics: Lesbian Feminism and the Limits of Community.* Philadelphia: Temple University Press.

Phillips, Anne. 1997. "From Inequality to Difference: A Severe Case of Displacement." *New Left Review* 224 (July/Aug):143–153.

Pratt, Minnie Bruce. 1984. "Who Am I If I'm Not My Father's Daughter?" Earlier version of "Identity: Skin Blood Heart," in *Rebellion Essays, 1980–1991,* by Bruce Pratt. Ithaca: Firebrand Books.

Queen, Carol. 1997. "Sex Radical Politics, Sex-Positive feminist Thought, and Whore Stigma." In *Whores and Other Feminists,* edited by Jill Nagle. New York: Routledge.

Quintanales, Mirtha. 1983. "I Paid Very Hard for My Immigrant Ignorance." In *This Bridge Called My Back: Writings by Radical Women of Color,* edited by Cherríe Moraga and Gloria Anzaldúa. New York: Kitchen Table: Women of Color Press.

Rupp, Leila J., and Verta Taylor. 1987. *Survival in the Doldrums: The American Women's Rights Movement, 1945 to the 1960s.* New York: Oxford University Press.

severson, kristin, and victoria stanhope. 1998. "Identity Politics and Progress: Don't Fence Me In (or Out)." *off our backs* 18 (4):12–13.

Shah, Sonia. 1994. "Presenting the Blue Goddess: Toward a Bicultural Asian-American Feminist Agenda." In *The State of Asian America: Activism and Resistance in the 1990s,* edited by Karin Aguilar-San Juan. Boston: South End Press.

Smith, Barbara. 1983. "Introduction to *Home Girls.*" In *Home Girls,* edited by Barbara Smith. New York: Kitchen Table: Women of Color Press.

Stein, Arlene. 1997. *Sex and Sensibility: Stories of a Lesbian Generation.* Berkeley: University of California Press.

Taylor, Verta, and Nancy Whittier. 1992. "Collective Identity in Social Movement Communities: Lesbian Feminist Mobilization." Pp. 104–129 in *Frontiers in Social Movement Theory,* edited by Aldon Morris and Carol McClurg Mueller. New Haven: Yale University Press.

Walker, Rebecca. 1992. "Becoming the Third Wave." *Ms.* (Jan/Feb):39–41

Whisman, Vera. 1993. "Identity Crisis: Who Is a Lesbian, Anyway?" In *Sisters, Sexperts, Queers: Beyond the Lesbian Nation,* edited by Arlene Stein. New York: Plume Books.

CHAPTER 3

Gender and the Prism of Culture

Now that we've introduced you to the ways in which the American gender system interacts with, and is modified by, a complex set of categories of difference and inequality, we turn to an exploration of the ways in which the prism of culture shapes gender definitions and arrangements. Generations of researchers in the social sciences, especially anthropology and sociology, have opened our eyes to the array of "genderscapes" around the globe. When we look through our kaleidoscope at the interaction between the prisms of gender and culture, we see different patterns that blur, blend, and are cast into a variety of culturally gendered configurations (Baker, 1999).

What Is Culture? Before we examine the intersection of gender and culture, it is important to define culture. At a very general level, the term "culture" is regarded by anthropologists as "the notion of human consciousness" and its products, such as systems of thought and technologies (Ortner, 1996:26). More specifically, culture consists of the beliefs, practices, and material objects that are created and shared within a group of people, thus constituting their way of life (Stone and McKee, 1999). The anthropological view of culture makes it clear that, without culture, human experience would have little shape or meaning. A group's culture provides members with the assumptions and expectations on which their social interaction is built and in which their identities are forged. It also makes groups distinct from each other.

Indeed different cultural groups create very different realities. For example in many cultures, such as the Sambia of New Guinea, people do not recognize categories or concepts that are equivalent to the meanings of the modern Western notion of homosexuality, and yet some members of those cultures regularly engage in same-gender sex (Herdt, 1997). Not only do different groups of people produce different cultures, the cultures they produce are dynamic. That is, people continually generate and alter culture both as individuals and as members of particular groups; as a result, all cultures undergo change as people evaluate, resist, and challenge beliefs and practices (Stone and McKee, 1999). To illustrate, in the United States today, many cultural categories are controversial and disputed, including race, sexual orientation, and gender. Consequently, those categories are undergoing major alterations in meaning and practice.

The prisms of gender and culture are inextricably intertwined. That is, people construct specific gender beliefs and practices in relation to particular cultural traditions and societal conditions. Cultures are gendered in distinctive ways and gender systems, in turn, shape both material and symbolic cultural products (see, for example, Chapter 5). As you will discover, the crosscultural analyses of gender presented in this chapter provide critical support for the social constructionist argument that gender is a situated, negotiated, contested, and changing set of practices and understandings.

Let's begin our exploration of the intersections of gender and culture with a set of observations about gender in different cultures. Do you know that there are cultures in which more than two genders operate or cultures in which individuals can move from one gender category to another without being stigmatized? If you traveled from country to country around the world, you would find cultures in which men are gentle, soft-spoken, and modest; and cultures in which women are viewed as strong and take on roles that are labeled masculine in the United States. Although we hear news about extreme forms of oppression of women in some places in the world (e.g., bride-burning in parts of India), there are other places where women and men live in relative harmony and equality. Also, there are cultural groups in which the social prisms of difference and inequality that operate in the United States (e.g., social class, race, sexual orientation) are minimal, inconsequential, or nonexistent (see readings by Christine Helliwell and Maria Lepowsky).

The Problem of Bias in Crosscultural Research If you find any of these observations unsettling or even shocking, then you have probably tapped into the problem of bias in crosscultural studies. One of the great challenges of crosscultural research is learning to transcend one's own cultural biases to be able to value and understand another culture in and of itself. It takes practice, conscious commitment, and self-awareness to get outside of one's own cultural box. After all, seeing what our culture wants us to see is precisely what socialization is about. Not only do cultural blinders make it difficult for us to see what gender is and how it is configured and reconfigured by various social prisms within our own culture, they make it even more challenging to grasp the profoundly different ways in which people in many other cultures think about and organize gender relations.

We tend to "see what we believe," which means that we are likely to deny gender patterns that vary from our own cultural experience and/or to misinterpret patterns that are different from our own. For example, the Europeans who first explored and colonized Africa were horrified by the ways in which African forms of gender and sexuality diverged from their own. They had no framework in which to understand warrior women, such as Nzinga of the Ndongo kingdom of the Mbundu (Murray and Roscoe, 1998). Nzinga was king of her people, dressed as man, and was surrounded by a harem of young men who dressed as women and were her "wives" (Murray and Roscoe, 1998). However, her behavior was a product of her culture, one in which people perceived gender as situational and symbolic, thus allowing for alternative genders (Murray and Roscoe, 1998).

It is a challenge for Westerners to control their cultural biases when confronted by cultures in which "races," "social classes," "sexual orientations," and other American modes of categorizing humans do not exist. For example, it's easy for people, including researchers, who are steeped in Western cultural traditions, to engage in ethnocentrism (i.e., the belief that the ideas and practices of one's own group are the standard and that divergent cultures are substandard or inferior) by imposing Western views of homosexuality on non–Western, same-gender sexual relations. The ethnocentrism of Westerners leads them to assume that sexual relations take the same form and have the same motivations and meanings everywhere in the world (Herdt, 1997). In reality, the spectrum of sexual cultures is immense.

The wide-ranging research of sociologists and anthropologists shows us the peculiarities of our gender system and opens us up to worlds in which gender is defined, organized, and experienced in ways that are very different from the American system. We can transcend ethnocentrism by developing our sociological radar to penetrate the myths and facades of our own culture and expose its underlying tensions, contradictions, and ambiguities. One of the enduring values of challenging one's own perspective on the world is that one can see anew one's own way of life, and understand more deeply lifeways, including genderscapes, in other places in the world.

The readings in this chapter will introduce you to some of the variety in gender beliefs and practices across cultures and illustrate four of the most important findings of crosscultural research on gender: (1) there is no universal definition or experience of gender, no core masculinity or femininity; (2) the American binary gender system is not universal; (3) gender inequality, specifically the dominance of men over women, is not the rule everywhere in the world; and (4) gender arrangements, whatever they may be, are socially constructed and, thus, ever-evolving.

There Is No Universal Definition or Experience of Gender Although people in virtually every contemporary culture perceive at least some differences between women and men, and assign different tasks and responsibilities to people based on gender categories, these differences vary both from culture to culture and within cultures. There is no unified ideal or definition of masculinity or femininity across cultures. In some cultures, such as the Ju/'hoansi of Namibia and Botswana, women and men alike can become powerful and respected healers, while in others, such as the United States today, powerful healing roles are dominated by men (Bonvillain, 2001). Among the seminomadic, pastoral Tuareg of the Sahara and the Sahel, women have considerable economic independence as livestock owners, herders, gardeners, and leathersmiths, while in other cultural groups, such as in formerly Taliban-controlled Afghanistan, women were severely restricted to household labor and economic dependence on men (Rasmussen, 2001).

The readings in this chapter highlight some of the extraordinary crosscultural differences in beliefs about men and women and in the tasks and rights assigned to them. They offer insights into how gender is shaped by a number of factors including ideology, participation in economic production, and control over sexuality and reproduction across cultures. For example, in the article with the titillating title, "It's Only A Penis," Christine Helliwell provides an account of the Gerai of Borneo, a cultural group in which rape does not occur. Helliwell argues that the Gerai belief in the biological sameness of women and men is a key to understanding their rape-free society. Her research offers an important account of how assumptions about human biology, in this case femaleness and maleness, are culturally shaped.

The American Binary Gender System Is Not Universal Simply said, the two-sex (male or female), two-gender (feminine or masculine) and two sexual orientation (homosexual or heterosexual) system of Western culture is not a universal mode of categorization. As you know from reading Serena Nanda's article on gender variants in Native North America (Chapter 1), the two-spirit role was widespread and accepted in many American Indian tribes. Gilbert Herdt (1997), an expert on the anthropology of sexual orientation and gender, points out that the two-spirit role reached a high point in its cultural elaboration among the Mojave Indians who "sanctioned both male (alyha) and female (hwame) two-spirit roles, each of which had its own distinctive social positions and worldviews" (92).

Two readings in this chapter address cultural traditions in which gender, sex, and sexuality are not framed in binary terms. The first is a poem by a Vietnamese woman that speaks to a tradition of acceptance of intersex people. The second is an article by anthropologists, Wairimu Ngaruiya Njambi and William E. O'Brien, on the meanings of woman-woman marriage among the Gikuyu of Kenya. The authors tell the stories of Gikuyu women in contemporary Kenya who participate in a tradition of woman-woman marriage that coexists with woman-man marriage. This reading gives us insight into a culture in which the American prism of binary gender and sexual orientation is virtually meaningless.

Gender Inequality Is Not the Rule Everywhere Gender and power go together but not in only one way. The relationship of power to gender in human groups varies from extreme male dominance to relative equality between women and men. Most societies are organized so that men, in general, have greater access to and control over valued resources such as wealth, authority, and prestige. At the extreme are intensely patriarchal societies, such as traditional China and India, in which women were dominated by men in multiple contexts and relationships. In traditional China, for example, sons were preferred, female infanticide was common, divorce could only be initiated by husbands, restrictions on girls and women were embodied in the mutilating practice of foot binding, and the suicide rate among young wives—who typically endured extreme isolation and hardship—was higher than any other age and gender category (Bonvillain, 2001).

The United States also has a history of gender relations in which white men as a group have had power over women as a group (and over men of color). For many decades, men's power was overt and legal. For example, in the nineteenth century, husbands were legally empowered to beat wives, women did not have voting rights, and women were legally excluded from many occupations (Stone and McKee, 1999). Today gender inequality takes more covert and subtle forms. For example, women earn less, on average, than do men of equal educational and occupational level; women are far more likely to be sexually objectified; and women are more likely to shoulder the burden of a double day of work inside and outside home (Coltrane and Adams, 2001; Chapters 7 and 8).

Understanding the relationship between power and gender requires us to use our sharpest sociological radar. To start, it is important to understand that power does not reside in individuals per se. For example, neither presidents nor bosses have power in a vacuum. They always need the support of personnel and special resources such as media, weapons, and money. Power is a group phenomenon, and it exists only so long as a powerful group, its ruling principles, and its control over resources are sustained (Kimmel, 2000).

In addition, not all members of an empowered group have the same amount of power. In the United States and similar societies, male power benefits some men more than others. In fact, many individual men do not occupy formal positions of power, and many do not feel powerful in their everyday lives (Johnson, 2001; Kimmel, 2000; see the Espiritu reading in Chapter 2 and the reading by Pyke in Chapter 8). Yet, major institutions and organizations (e.g., government, big business, the military, the mass media) in the United States are gendered masculine, with controlling positions in those arenas dominated by men, but not just any men (Johnson, 2001). Controlling positions are overwhelmingly held by white men who typically come from privileged backgrounds and whose lives appear to conform to a straight and narrow path (Johnson, 2001; Kimmel, 2000). As we learned in Chapter 2, the relationship of gender to power in a nation such as the United States is complicated by interactions among structures of domination and subordination such as race, social class, and sexual orientation.

Not all societies are as highly and intricately stratified by gender, race, social class, and other social categories of privilege and power as is the United States. Many cultural groups organize relationships in ways that give most or all adults access to similar rights, prestige, decision-making authority, and autonomy in their households and communities. Traditional Agta, Ju/'hoansi, and Iroquois societies are good examples. In other cultural groups, such as the precontact Tlingit and Haida of the Canadian Pacific coastal region, relations among people were based on their position in economic and status hierarchies, yet egalitarian valuation of women and men prevailed (Bonvillain, 2001).

The point is that humans do not inevitably create inequalities out of perceived differences. Thus, even though there is at least some degree of division of labor by gender in every cultural group today, differences in men's and women's work do not inexorably lead to patriarchal relations in which men monopolize high-status positions in important institutions and women are relegated to a restricted world of low-status activities and tasks. Several readings in this chapter examine cultural groups in which egalitarianism is the way of life. Maria Lepowsky's ethnography of Vanatinai social relations provides us with a model of a society in which the principles of personal autonomy and freedom of choice prevail. The gender ideology of the Vanatinai is egalitarian, and their belief in equality manifests itself in daily life. For example, women as well as men own and inherit land and other valuables. Women choose their own marriage partners and lovers, and they divorce at will. Any individual on Vanatinai may try to become a leader by demonstrating superior knowledge and skill.

Gender inequality is not the rule everywhere. Male dominance, patriarchy, gender inequality—whatever term one uses—is not the inevitable state of human relations. Additionally, patriarchy itself is not unitary. Patriarchy does not assume a particular shape, and it does not mean that women have no control or influence in their communities. Even in the midst of patriarchy, women and men may create identities and relationships that allow for autonomy and independence. Sally Cole's chapter portrait of a Portuguese fisherwoman named Maria illustrates this point. Portugal has a history of male dominance, but male dominance is not complete. In Maria's community, many women move with ease between women's work and men's work. Maria herself is an economically independent fisherwoman who owns her own boat, a proud mother, and a household manager. In her community, she is regarded as both woman and man.

Gender Arrangements Are Ever-Evolving The crosscultural story of gender takes us back to the metaphor of the kaleidoscope. Life is an ongoing, neverending process of change from one pattern to another. We can never go back to "the way things were" at some earlier moment in time, nor can we predict exactly how the future will unfold. This is, of course, the story of gender around the world. One of the major sources of change in gender meanings and practices across cultures has been culture contact (Sorenson, 2000). Among the most well-documented accounts of such change have been those that demonstrate how Western gender biases were imposed on people whose gender beliefs and arrangements varied from Western assumptions and practices. For example, Native American multiple gender systems were actively, and sometimes violently, discouraged by European colonists (Herdt, 1997; see reading by Cantu in Chapter 10). Today expressions of globalization—and the increasing economic, cultural, political, and environmental interconnectedness and interdependence of societies worldwide—raise the problem of the development of a world order, including a gender order, dominated by Western, especially American, cultural values and patterns (Held, McGrew, Goldbatt, and Perraton, 1999; see Chapter 10 for further discussion). How do women and men in cultures around the world experience globalized culture contact? How will Americans respond to increasing knowledge of cultures very different from their own?

Culture contact is by no means the only source of changing gender arrangements (see Chapter 10 for detailed discussion of gender change). The forces of change are many and complex, and they have resulted in both tendencies toward rigid, hierarchical gender relations and toward gender flexibility and equality, depending upon the specific cultural context and forces of change experienced by particular groups of people. In all of this, there is one fact: people are not bound by any set of gender beliefs and practices. Culture change is inevitable and so is change in the genderscape.

REFERENCES

Baker, Cozy. (1999). *Kaleidoscopes: Wonders of wonder.* Lafayette, CA: C&T Publishing.

Bonvillain, Nancy. (2001). *Women and men: Cultural constructs of gender.* Third Edition. Upper Saddle River, NJ: Prentice Hall.

Coltrane, Scott and Michele Adams. (2001). "Men, women, and housework." In *Gender mosaics,* D. Vannoy (Ed.), pp. 145–154. Los Angeles, CA: Roxbury Publishing.

Held, David, Anthony McGrew, David Goldblatt, and Jonathan Perraton. (1999) *Global transformations.* Stanford, CA: Stanford University Press.

Herdt, Gilbert. (1997). *Same sex, different cultures.* Boulder, CO: Westview Press.

Johnson, Allan G. (2001). *Privilege, power, and difference.* Mountain View, CA: Mayfield Publishing.

Kimmel, Michael. (2000). *The gendered society.* New York: Oxford University Press.

Murray, Stephen O. and Will Roscoe. (1998). *Boy-wives and female husbands: Studies of African homosexualities.* New York: St. Martin's Press.

O'Brien, Jodi. (1999). *Social prisms.* Thousand Oaks, CA: Pine Forge Press.

Ortner, Sherry B. (1996). *Making gender: The politics and erotics of culture.* Boston, MA: Beacon Press.

Rasmussen, Susan. (2001). "Pastoral nomadism and gender." In *Gender in cross-cultural perspective,* Caroline B. Brettell and Carolyn F. Sargent (Eds.), pp. 280–293. Upper Saddle River, NJ: Prentice Hall.

Sorenson, Marie Louise Stig. (2000). *Gender archeology.* Cambridge, England: Polity Press.

Stone, Linda and Nancy P. McKee. (1999). *Gender and culture in America.* Upper Saddle River, NJ: Prentice Hall.

11

Girl without a Sex

BY HO XUAN HUONG

The following is a poem by a Vietnamese woman. You may need to look up words such as "maidenhead," "ovule," and "anther" in order to fully interpret it. Note carefully the cultural point of view about intersex expressed in the final stanza.

1. What does the "it" refer to in the third stanza?
2. Why does the poet view the "girl's" intersex status as "fine" and "okay"?
3. How might participants in the intersex movement described in Preves's article in Chapter 1 interpret this poem?

Did the fairy midwives have a falling out
and somehow misplace her maidenhead?

The little father mouse squeaking about, doesn't care,
nor the mother honeybee buzzing along, fat with pollen.

Can anyone tell whether it's ovule or anther?
Can anyone say if it's stem or bud?

Well, fine. It's really okay. Since her whole life
she'll never have to hear "daughter-in-law!"

12

"It's Only a Penis"

Rape, Feminism, and Difference

BY CHRISTINE HELLIWELL

Anthropologist Christine Helliwell provides a challenging account of a cultural group, the Gerai of Indonesia, in which rape does not occur. She links the freedom from rape among the Gerai people to peculiar aspects of the relatively egalitarian nature of their gender relations. Helliwell's research questions many gender beliefs held by members of Western cultures today.

1. How are men's and women's sexual organs conceptualized among the Gerai, and what are the consequences for Gerai understandings of sexual intercourse?
2. Genitalia do not determine identity in Gerai. What does?
3. What does Helliwell mean when she states that "rape imposes difference as much as it is produced by difference" (812)?

In 1985 and 1986 I carried out anthropological fieldwork in the Dayak community of Gerai in Indonesian Borneo. One night in September 1985, a man of the village climbed through a window into the freestanding house where a widow lived with her elderly mother, younger (unmarried) sister, and young children. The widow awoke, in darkness, to feel the man inside her mosquito net, gripping her shoulder while he climbed under the blanket that covered her and her youngest child as they slept (her older children slept on mattresses nearby). He was whispering, "be quiet, be quiet!" She responded by sitting up in bed and pushing him violently, so that he stumbled backward, became entangled with her mosquito net, and then, finally free, moved across the floor toward the window. In the meantime, the woman climbed from her bed and pursued him, shouting his name several times as she did so. His hurried exit through the window, with his clothes now in considerable disarray, was accompanied by a stream of abuse from the woman and by excited interrogations from wakened neighbors in adjoining houses.

I awoke the following morning to raucous laughter on the longhouse verandah outside my apartment where a group of elderly women gathered regularly to thresh, winnow, and pound rice. They were recounting this tale loudly, and with enormous enjoyment, to all in the immediate vicinity. As I came out of my door, one was engaged in mimicking the man climbing out the window, sarong falling down, genitals askew. Those others working or lounging near her on the verandah—both men and women—shrieked with laughter.

When told the story, I was shocked and appalled. An unknown man had tried to climb into the bed of a woman in the dead, dark of night? I knew what this was called: attempted rape. The woman had seen the man and recognized him (so had others in the village, wakened by her shouting). I knew what he deserved: the full weight of the law. My own fears about being a single woman alone in a strange place, sleeping in a dwelling that could not be secured at night, bubbled to the surface. My feminist sentiments poured out. "How can you laugh?" I asked my women friends; "this is a very bad thing that he has tried to do." But my outage simply served to fuel the hilarity. "No, not bad," said one of the old women (a particular friend of mine), "simply stupid."

I felt vindicated in my response when, two hours later, the woman herself came onto the verandah to share betel nut and tobacco and to broadcast the story. Her anger was palpable, and she shouted for all to hear her determination to exact a compensation payment from the man. Thinking to obtain information about local women's responses to rape, I began to question her. Had she been frightened? I asked. Of course she had—Wouldn't I feel frightened if I awoke in the dark to find an unknown person inside my mosquito net? Wouldn't I be angry? Why then, I asked, hadn't she taken the opportunity, while he was entangled in her mosquito net, to kick him hard or to hit him with one of the many wooden implements near at hand? She looked shocked. Why would she do that? she asked—after all, he hadn't hurt her. No, but he had wanted to, I replied. She looked at me with puzzlement. Not able to find a local word for *rape* in my vocabulary, I scrabbled to explain myself: "He was trying to have sex with you." I said, "although you didn't want to. He was trying to hurt you." She looked at me, more with pity than with puzzlement now, although both were mixed in her expression. "Tin [Christine], it's only a penis" she said. "How can a penis hurt anyone?"

RAPE, FEMINISM, AND DIFFERENCE

A central feature of many feminist writings about rape in the past twenty years is their concern to eschew the view of rape as a natural function of male biology and to stress instead its bases in society and culture. It is curious, then, that so much of this work talks of rape in terms that suggest—either implicitly or explicitly—that it is a universal practice. To take only several examples: Pauline Bart and Patricia O'Brien tell us that "every female from nine months to ninety years is at risk" (1985, 1); Anna Clark argues that "all women know the paralyzing fear of walking down a dark street at night. . . . It seems to be a fact of life that the fear of rape imposes a curfew on our movements" (1987, 1); Catharine MacKinnon claims that "sexuality is central to women's definition and forced sex is central to sexuality," so "rape is indigenous, not exceptional, to women's social condition" (1989b, 172) and "all women live all the time under the shadow of the threat of sexual abuse" (1989a, 340); Lee Madigan and Nancy Gamble write of "the global terrorism of rape" (1991, 21–2); and Susan Brison asserts that "the fact that all women's lives are restricted by sexual violence is indisputable" (1993, 17). . . . This is particularly puzzling given that Peggy Reeves Sanday, for one, long ago demonstrated that while rape occurs widely throughout the world, it is by no means a human universal: some societies can indeed be classified as rape free (1981).

There are two general reasons for this universalization of rape among Western feminists. The first of these has to do with the understanding of the practice as horrific by most women in Western societies. In these settings, rape is seen as "a fate worse than, or tantamount to, death" (S. Marcus 1992, 387): a shattering of identity that, for instance, left one North American survivor feeling "not quite sure whether I had died and the world went on without me, or whether I was alive in a totally alien world" (Brison 1993, 10). . . .

A second, equally deep-seated reason for the feminist tendency to universalize rape stems from Western feminism's emphasis on difference between men and women and from its consequent linking of rape and difference. Two types of difference are involved here. The first of these is difference in social status and power; thus rape is linked quite explicitly, in contemporary feminist accounts, to patriarchal social forms. Indeed, this focus on rape as stemming from difference in social position is what distinguishes feminist from other kinds of accounts of rape (see Ellis 1989, 10). In this view, inequality between men and women is linked to men's desire to possess, subjugate, and control women, with rape constituting a central means by which the freedom of women is limited and their continued submission to men ensured. Since many feminists continue to believe that patriarchy is universal—or, at the very least, to feel deeply ambivalent on this point—there is a tendency among us to believe that rape, too, is universal.[1]

However, the view of women as everywhere oppressed by men has been extensively critiqued within the anthropological literature. A number of anthropologists have argued that in some societies, while men and women may perform different roles and occupy different spaces, they are nevertheless equal in value, status, and power.[2] . . .

But there is a second type of difference between men and women that also, albeit largely implicitly, underlies the assumption that rape is universal, and it is the linkage between this type of difference and the treatment of rape in feminist accounts with which I am largely concerned in this article. I refer to the assumption by most Western feminists writing on rape that men and women have different bodies and, more specifically, different genitalia: that they are, in other words differently sexed. Furthermore, it is taken for granted in most feminist accounts that these differences render the former biologically, or "naturally," capable of penetrating and therefore brutalizing the latter and render the latter "naturally" able to be brutalized. . . . Rape of women by men is thus assumed to be universal because the same "biological" bodily differences between men and women are believed to exist everywhere.

Unfortunately, the assumption that preexisting bodily difference between men and women underlies rape has blinded feminists writing on the subject to the ways the practice of rape itself creates and inscribes such difference. This seems particularly true in contemporary Western societies where the relationship between rape and bodily/genital dimorphism appears to be an extremely intimate one. Judith Butler (1990, 1993) has argued (following Foucault 1978) that the Western emphasis on sexual difference is a product of the heterosexualization of desire within Western societies over the past few centuries, which "requires and institutes the production of discrete and asymmetrical oppositions between 'feminine' and 'masculine' where these are understood as expressive attributes of 'male' and 'female'" (1990, 17).[3] The practice of rape in Western contexts can only properly be understood with reference to this heterosexual matrix, to the division of humankind into two distinct—and in many respects opposed—types of body (and hence types of person).[4] While it is certainly the case that rape is linked in contemporary Western societies to disparities of power and status between men and women, it is the particular discursive form that those disparities take—their elaboration in terms of the discourse of sex—that gives rape its particular meaning and power in these contexts.

Sharon Marcus has already argued convincingly that the act of rape "feminizes" women in Western settings, so that "the entire female body comes to be symbolized by the vagina, itself conceived of as a delicate, perhaps inevitably damaged and pained inner space" (1992, 398). I would argue further that the *practice* of rape in these settings—both its possibility and its actualization—not only feminizes women but masculinizes men as well.[5] This masculinizing character of rape is very clear in, for instance, Sanday's ethnography of fraternity gang rape in North American universities (1990b) and, in particular, in material on rape among male prison inmates. In the eyes of these rapists the act of rape marks them as "real men" and marks their victims as not men, that is, as feminine.[6] In this iconography, the "masculine" body (along with the "masculine" psyche), is viewed as hard, penetrative, and aggressive, in contrast to the soft, vulnerable, and violable "feminine" sexuality and psyche. Rape both reproduces and marks the pronounced sexual polarity found in these societies.

Western understandings of gender difference have almost invariably started from the presumption of a presocial bodily difference between men and women ("male" and "female") that is then somehow acted on by society to produce gender. In particular, the possession of either male genitals or female genitals is understood by most Westerners to be not only the primary marker of gender identity but, indeed, the underlying cause of that identity. . . .

I seek to do two things in this article. First, in providing an account of a community in which rape does not occur, I aim to give the lie to the widespread assumption that rape is universal and thus to invite Western feminists to interrogate the basis of our own tendency to take its universality for granted.[7] The fundamental question is this: Why does a woman of Gerai see a penis as lacking the power to harm her, while I, a white Australian/New Zealand woman, am so ready to see it as having the capacity to defile, to humiliate, to subjugate and, ultimately, to destroy me?

Second, by exploring understandings of sex and gender in a community that stresses identity, rather than difference, between men and women (including men's and women's bodies), I aim to demonstrate that Western beliefs in the "sexed" character of bodies are not "natural" in basis but, rather, are a component of specifically Western gendering and sexual regimes. And since the practice of rape in Western societies is profoundly linked to these beliefs, I will suggest that it is an inseparable part of such regimes. This is not to say that the practice of rape is always linked to the kind of heterosexual regime found in the West; even the most cursory glance at any list of societies in which the practice occurs indicates that this is not so.[8] But it is to point out that we will be able to understand rape only ever in a purely localized sense, in the context of the local discourses and practices that are both constitutive of and constituted by it. In drawing out the implications of the Gerai stress on identity between men and women for Gerai gender and sexual relations, I hope to point out some of the possible implications of the Western emphasis on gender difference for Western gender and sexual relations—including the practice of rape.

GENDER, SEX, AND PROCREATION IN GERAI

Gerai is a Dayak community of some seven hundred people in the Indonesian province of Kalimantan Barat (West Borneo).[9] In the twenty months I spent in the community, I heard of no cases of either sexual assault or attempted sexual assault (and since this is a community in which privacy as we understand it in the West is almost nonexistent—in which surveillance by neighbors is at a very high level [see Helliwell 1996]—I would certainly have heard of any such cases had they occurred). In addition, when I questioned men and women about sexual assault, responses ranged from puzzlement to outright incredulity to horror.

While relations between men and women in Gerai can be classified as relatively egalitarian in many respects, both men and women nevertheless say that men are "higher" than women (Helliwell 1995, 364). This is especially the case in the context of formal community-wide functions such as village meetings and moots to settle legal disputes. While women are not required to remain silent on such occasions, their voices carry less authority than those of men, and, indeed, legal experts in the community (all men) told me that a woman's evidence in a moot is worth seven-tenths of a man's (see also Tsing 1990). In addition, a husband is granted a degree of formal authority over his wife that she does not have over him; thus a wife's disobedience of her husband is theoretically a punishable offense under *adat,* or local law. I have noted elsewhere that Gerai people stress the ideal of *diri,* literally meaning "standing" or "to stand," according to which each rice group should take primary responsibility for itself in all spheres of life and make its own decisions on matters concerning its members (Helliwell 1995). It is on the basis of their capacity to stand that rice groups within the community are ranked against one another. The capacity to stand is predicated primarily on the ability to produce rice surpluses: yet, significantly, although men and women work equally at rice-field work, it is only men who occasionally are individually described as standing. As in some other societies in the same region (Ilongot, Wana), Gerai people link men's higher status to their greater bravery.[10] This greater bravery is demonstrated, they say, by the fact that it is men who *mampat* (cut down the large trees to make a rice field), who burn off the rice field to prepare for planting, and who enter deep primary jungle in search of game and jungle products such as aloe wood—all notoriously dangerous forms of work.

This greater status and authority does not, however, find expression in the practice of rape, as many feminist writings on the subject seem to suggest that it should. This is because the Gerai view of men as "higher" than women, although equated with certain kinds of increased potency vis-à-vis the world at large, does not translate into a conception of that potency as attached to and manifest through the penis—of men's genitals as able to brutalize women's genitals.

Shelly Errington has pointed our that a feature of many of the societies of insular Southeast Asia is a stress on sameness, even identity, between men and women (1990, 35, 39), in contrast to the Western stress on difference between the passive "feminine" object and the active, aggressive "masculine" subject.[11] Gerai understandings of gender fit Errington's model very well. In Gerai, men and women are not understood as fundamentally different types of persons: there is no sense of a dichotomized masculinity and femininity. Rather, men and women are seen to have the same kinds of capacities and proclivities, but with respect to some, men are seen as "more so" and with respect to others, women are seen as "more so." Men are said to be braver and more knowledgeable about local law (*adat*), while women are said to be more persistent and more enduring. All of these qualities are valued. Crucially, in terms of the central quality of nurturance (perhaps the most valued quality in Gerai), which is very strongly marked as feminine among Westerners, Gerai people see no difference between men and women. As one (female) member of the community put it to me: "We all must nurture because we all need."[12] The capacity both to nurture and to need, particularly as expressed through the cultivation of rice as a member of a rice group, is central to Gerai conceptions of personhood: rice is the source of life, and its (shared) production humanizes and socializes individuals (Helliwell, forthcoming). Women and men have identical claims to personhood based on their equal contributions to rice production (there is no notion that women are somehow diminished as persons even though they may be seen as less "high"). As in Strathern's account of Hagen (1988), the perceived mutuality of rice-field work in Gerai renders inoperable any notion of either men or women as autonomous individual subjects.

It is also important to note that while men's bravery is linked to a notion of their greater physical strength, it is not equated with aggression—aggression is not valued in most Gerai contexts.[13] As a Gerai man put it to me, the wise man is the one "who fights when he has to, and runs away when he can"; such avoidance of violence does not mark a man as lacking in bravery. . . . While it is recognized that a man will sometimes need to fight—and skill and courage in fighting are valued—aggression and hotheadedness are ridiculed as the hallmarks of a lazy and incompetent man. In fact, physical violence between adults is uncommon in Gerai, and all of the cases that I did witness or hear about were extremely mild.[14] Doubtless the absence of rape in the community is linked to this devaluing of aggression in general. However, unlike a range of other forms of violence (slapping, beating with a fist, beating with an implement, knifing, premeditated killing, etc.), rape is not named as an offense and accorded a set punishment under traditional Gerai law. In addition, unlike these other forms of violence, rape is something that people in the community find almost impossible to comprehend ("How would he be able to do such a thing?" one woman asked when I struggled to explain the concept of a man attempting to put his penis into her against her will). Clearly, then, more is involved in the absence of rape in Gerai than a simple absence of violence in general.

Central to all of the narratives that Gerai people tell about themselves and their community is the notion of a "comfortable life": the achievement of this kind of life marks the person and the household as being of value and constitutes the norm to which all Gerai people aspire. Significantly, the content of such a life is seen as identical for both men and women: it is marked by the production of bountiful rice harvests each year and the successful raising of a number of healthy children to maturity. The core values and aspirations of men and women are thus identical; of the many life histories that I collected while in the community—all of which are organized around this central image—it is virtually impossible to tell those of men from those of women. Two points are significant in this respect. First, a "comfortable life" is predicated on the notion of a partnership between a man and a woman (a conjugal pair). This is because while men and women are seen to have the same basic skills and capacities, men are seen to be "better" at certain kinds of work and women to be 'better" at other kinds. Second, and closely related to this, the Gerai notion of men's and women's work does not constitute a rigid division of labor: both men and women say that theoretically women can perform all of the work routinely carried out by men, and men can perform all of the work routinely carried out by women. However, men are much better at men's work, and women are much better at women's work. Again, what we have here is a stress on *identity* between men and women at the expense of radical difference.

This stress on identity extends into Gerai bodily and sexual discourses. A number of people (both men and women) assured me that men sometimes menstruate; in addition, menstrual blood is not understood to be polluting, in contrast to how it is seen in many societies that stress more strongly the difference between men and women. While pregnancy and childbirth are spoken of as "women's work," many Gerai people claim that under certain circumstances men are also able to carry out this work—but, they say, women are "better" at it and so normally undertake it. In line with this claim, I collected a Gerai myth concerning a lazy woman who was reluctant to take on the work of pregnancy and childbirth. Her husband instead made for himself a lidded container out of bark, wood, and rattan ("like a betel nut container"), which he attached around his waist beneath his loincloth and in which he carried the growing fetus until it was ready to be born. On one occasion when I was watching a group of Gerai men cut up a boar, one, remembering an earlier conversation about the capacity of men to give birth, pointed to a growth in the boar's body cavity and said with much disapproving shaking of the head: "Look at this. He wants to carry his child. He's stupid." In addition, several times I saw fathers push their nipples into the mouths of young children to quieten them; while none of these fathers claimed to be able to produce milk, people nevertheless claimed that some men in the community were able to lactate, a phenomenon also attested to in myth. Men and women are thought to produce the same genital fluid, and this is linked in complex ways to the capacity of both to menstruate. All of these examples demonstrate the community's stress on bodily identity between men and women.

Furthermore, in Gerai, men's and women's sexual organs are explicitly conceptualized as the same. This sexual identity became particularly clear when I asked several people who had been to school (and hence were used to putting pencil to paper) to draw men's and women's respective organs for me: in all cases, the basic structure and form of each were the same. One informant, endeavoring to convince me of this sameness, likened both to wooden and bark containers for holding valuables (these vary in size but have the same basic conical shape, narrower at the base and wider at the top). In all of these discussions, it was reiterated that the major difference between men's and women's organs is their location: inside the body (women) and outside the body (men).[15]

In fact, when I pressed people on this point, they invariably explained that it makes no sense to distinguish between men's and women's genitalia themselves; rather, it is location that distinguishes between penis and vulva."[16]

Heterosexuality constitutes the normative sexual activity in the community and, indeed, I was unable to obtain any information about homosexual practices during my time there. In line with the stress on sameness, sexual intercourse between a man and a woman in Gerai is understood as an equal coming together of fluids, pleasures, and life forces. The same stress also underlies beliefs about conception. Gerai people believe that repeated acts of intercourse between the same two people are necessary for conception, since this "prepares" the womb for pregnancy. The fetus is deemed to be created through the mingling of equal quantities of fluids and forces from both partners. Again, what is seen as important here is not the fusion of two different types of bodies (male and female) as in Western understandings; rather, Gerai people say, it is the similarity of the two bodies that allows procreation to occur. As someone put it to me bluntly: "If they were not the same, how could the fluids blend? It's like coconut oil and water: they can't mix!"

What needs to be stressed here is that both sexual intercourse and conception are viewed as involving a mingling of similar bodily fluids, forces, and so on, rather than as the penetration of one body by another with a parallel propulsion of substances from one (male) body only into the other, very different (female) one. Nor is there anything in Gerai understandings that equates with the Western notion of conception as involving an aggressive active male cell (the sperm) seeking out and penetrating a passive, immobile female cell (the egg) (Martin 1991). What Gerai accounts of both sexual intercourse and conception stress are tropes of identity, mingling, balance, and reciprocity. In this context it is worth noting that many Gerai people were puzzled by the idea of gender-specific "medicine to prevent contraception—such as the injectable or oral contraceptives promoted by state-run health clinics in the area. Many believed that, because both partners play the same role in conception, it should not matter whether husband or wife received such medicine (and indeed, I knew of cases where husbands had taken oral contraceptives meant for their wives). This suggests that such contraceptive regimes also serve (like the practice of rape) to reinscribe sex difference between men and women (see also Tsing 1993, 104–20). . . .

While Gerai people stress sameness over difference between men and women, they do, nevertheless, see them as being different in one important respect: their life forces are, they say, oriented differently ("they face different ways," it was explained to me). This different orientation means that women are "better" at certain kinds of work and men are "better" at other kinds of work—particularly with respect to rice-field work. Gerai people conceive of the work of clearing large trees for a new rice field as the definitive man's work and regard the work of selecting and storing the rice seed for the following year's planting—which is correlated in fundamental ways with the process of giving birth—as the definitive woman's work. Because women are perceived to lack appropriate skills with respect to the first, and men are perceived to lack appropriate skills with respect to the second, Gerai people say that to be viable a household must contain both adult males and adult females. And since a "comfortable life" is marked by success in production not only of rice but also of children, the truly viable household must contain at least one conjugal pair. The work of both husband and wife is seen as necessary for the adequate nurturance of the child and successful rearing to adulthood (both of which depend on the successful cultivation of rice). Two women or two men would not be able to produce adequately for a child since they would not be able to produce consistently successful rice harvests; while such a household might be able to select seed, clear a rice field, and so grow rice in some rudimentary fashion, its lack of expertise at one of these tasks would render it perennially poor and its children perennially unhealthy, Gerai people say. . . .

Gender difference in Gerai, then, is not predicated on the character of one's body, and especially of one's genitalia as in many Western contexts. Rather, it is understood as constituted in the differential capacity to perform certain kinds of work, a capacity assigned long before one's bodily being takes shape.[17] In this respect it is important to note that Gerai ontology rests on a belief in predestination, in things being as they should (see Helliwell 1995). In this understanding, any individual's *semongan'* is linked in multifarious and unknowable ways to the cosmic order, to the "life" of the universe as a whole. Thus the new fetus is predestined to become someone "fitted" to carry out either men's work or women's work as part of the maintenance of a universal balance. Bodies with the appropriate characteristics—internal or external genitalia, presence or absence of breasts, and so on—then develop in line with this prior destiny. At first sight this may not seem enormously different from Western conceptions of gender, but the difference is in fact profound. While, for Westerners, genitalia, as significant of one's role in the procreative process, are absolutely fundamental in determining ones identity, in Gerai the work that one performs is seen as fundamental, and genitalia along with other bodily characteristics, are relegated to a kind of secondary, derivative function.

Gerai understandings of gender were made quite clear through circumstances surrounding my own gender classification while in the community. Gerai people remained very uncertain about my gender for some time after I arrived in the community because (as they later told me) "I did not . . . walk like a woman, with arms held out from the body and hips slightly swaying; I was "brave" trekking from village to village through the jungle on my own; I had bony kneecaps; I did not know how to tie a sarong in the appropriate way for women; I could not distinguish different varieties of rice from one another; I did not wear earrings; I had short hair; I was tall" (Helliwell 1993, 260). This was despite the fact that people in the community knew from my first few days with them both that I had breasts (this was obvious when the sarong that I wore clung to my body while I bathed in the river) and that I had a vulva rather than a penis and testicles (this was obvious from my trips to defecate or urinate in the small stream used for that purpose, when literally dozens of people would line the banks to observe whether I performed these functions differently from them). As someone said to me at a later point, "Yes, I saw that you had a vulva, but I thought that Western men might be different."

My eventual, more definitive classification as a woman occurred largely fortuitously. My initial research proposal focused on the creation of subjectivity and sociality through work and accordingly, as soon as I arrived in the community, I began accompanying people to work in the rice fields. Once I had negotiated a longhouse apartment of my own in which to live (several weeks after arrival), I also found myself, in concert with all other households in the community, preparing and cooking rice at least twice daily. These activities rapidly led to a quest for information concerning rice itself, particularly concerning the different strains, how they are cultivated, and what they are used for. As I learned to distinguish types of rice and their uses, I became more and more of a woman (as I realized later), since this knowledge—including the magic that goes with it—is understood by Gerai people as foundational to femininity. However, while people eventually took to referring to me as a woman, for many in the community my gender identity remained deeply ambiguous, partly because so many of my characteristics and behaviors were more like those of a man than a woman, but also, and more importantly, because I never achieved anything approaching the level of knowledge concerning rice-seed selection held by even a girl child in Gerai.

In fact, Gerai people talk of two kinds of work as defining a woman: the selection and storage of rice seed and the bearing of children.[18] But the first of these is viewed as prior, logically as well as chronologically. People are quite clear that in the womb either "someone who can cut down the large trees for a ricefield is made, or someone who can select and store rice." When I asked if it was not more important whether or not someone could bear a child, it was pointed out to me that many women do not bear children (there is a high rate of infertility in the community), but all women have the knowledge to select and store rice seed. In fact, at the level of the rice group the two activities of "growing" rice and "growing" children are inseparable: a rice group produces rice in order to raise healthy children, and it produces children so that they can in turn produce the rice that will sustain the group once their parents are old and frail (Helliwell, forthcoming). For this reason, any Gerai couple unable to give birth to a child of their own will adopt one, usually from a group related by kinship. The two activities of growing rice and growing children are constantly talked about together, and the same imagery is used to describe the development of a woman's pregnancy and the development of rice grains on the plant. . . .

Gerai, then, lacks the stress on bodily—and especially genital—dimorphism that most feminist accounts of rape assume. Indeed, the reproductive organs themselves are not seen as "sexed." In a sense it is problematic even to use the English categories *woman* and *man* when writing of this community, since these terms are saturated with assumptions concerning the priority of biological (read, bodily) difference. In the Gerai context, it would be more accurate to deal with the categories of, on the one hand, "those responsible for rice selection and storage" and, on the other, "those responsible for cutting down the large trees to make a ricefield." There is no discursive space in Gerai for the distinction between an active, aggressive, penetrating male sexual organ (and sexuality) and a passive, vulnerable, female one. Indeed, sexual intercourse in Gerai is understood by both men and women to stem from mutual "need" on the part of the two partners; without such need, people say, sexual intercourse cannot occur, because the requisite balance is lacking. Since, as I have describe at length elsewhere (Helliwell, forthcoming), a relationship of "needing" is always reciprocal (it is almost inconceivable, in Gerai terms, to need someone who does not need you in return, and the consequences of unreciprocated needing are dire for both individual and rice group), the sexual act is understood as preeminently mutual in its character, including in its initiation. The idea of having sex with someone who does not need you to have sex with them—and so the idea of coercing someone into sex—is thus almost unthinkable to Gerai people. In addition, informants asserted that any such action would destroy the individual's spiritual balance and that of his or her rice group and bring calamity to the group as a whole.[19]

In this context, a Gerai man's astonished and horrified question "How can a penis be taken into a vagina if a woman doesn't want it?" has a meaning very different from that of the same statement uttered by a man in the West. In the West, notions of radical difference between men and women—incorporating representations of normative male sexuality as active and aggressive, normative female sexuality as passive and vulnerable, and human relationships (including acts of sexual intercourse) as occurring between independent, potentially hostile, agents—would render such a statement at best naive, at worst misogynist. In Gerai, however, the stress on identity between men and women and on the sexual act as predicated on mutuality validates such a statement as one of straightforward incomprehension (and it should be noted that I heard similar statements from women). In the Gerai context, the penis, or male genitalia in general, is not admired, feared, or envied. . . . In fact, Gerai people see men's sexual organs as more vulnerable than women's for the simple reason that they are outside the body, while women's are inside. This reflects Gerai understandings of "inside" as representing safety and belonging, while "outside" is a place of strangers and danger, and it is linked to the notion of men as braver than women.[20] In addition, Gerai people say, because the penis is "taken into" another body, it is theoretically at greater risk during the sexual act than the vagina. This contrasts, again, quite markedly with Western understandings, where women's sexual organs are constantly depicted as more vulnerable during the sexual act—as liable to be hurt, despoiled, and so on (some men's anxieties about *vagina dentata* not withstanding). In Gerai a penis is "only a penis": neither a marker of dimorphism between men and women in general nor, in its essence, any different from a vagina.

CONCLUSIONS

The Gerai case suggests that, in some contexts at least, the practice of rape is linked to sexual dimorphism and, indeed, that in these contexts discourses of rape (including the act of rape itself) reinscribe such dimorphism. While the normative sexual practice in Gerai is heterosexual (between men and women), it is not accompanied by a heterosexual regulatory regime in the sense meant by Foucault (1978) in his discussion of the creation of sex as part of the heterosexualization of desire in the West, nor is it part of what Butler terms "the heterosexual matrix" (Butler 1990, 1993). The notion of "heterosexualization" as used by these thinkers refers to far more than the simple establishment of sexual relations between men and women as the normative ideal; it denotes the entire governmental regime that accompanies this normative ideal in Western contexts. Gerai stresses sameness between men and women more than difference, and such difference as occurs is based on the kinds of work people perform. Although this process certainly naturalizes a division between certain kinds of tasks—and the capacity to perform those tasks effectively—clearly, it does not involve sex or sexed bodies in the way Westerners normally understand those terms—as a naturalized difference between bodies (located primarily in the genitals) that translates into two profoundly different types of person. In this context, sexual assault by a man on a woman is almost unthinkable (both by women and by men).

With this background, I return now to the case with which I began this article—and, particularly, to the great differences between my response to this case and that of the Gerai woman concerned. On the basis of my own cultural assumptions concerning the differences—and particularly the different sexual characters—of men and women, I am inclined (as this case showed me) to read any attempt by a man to climb into a woman's bed in the night without her explicit consent as necessarily carrying the threat of sexual coercion and brutalization. This constant threat has been inscribed onto my body as part of the Western cultural process whereby I was "girled" (to use Butler's felicitous term [1993, 7]), or created as a gendered being in a context where male and female sexualities are perceived as penetrative and aggressive and as vulnerable and self-protective, respectively. The Gerai woman, in contrast, has no fear of coerced sexual intercourse when awakened in the dark by a man. She has no such fear because in the Gerai context "girling" involves the inscription of sexual sameness, of a belief that women's sexuality and bodies are no less aggressive and no more vulnerable than men's.

In fact, in the case in question, the intruding man did expect to have intercourse with the woman.[21] He claimed that the woman had already agreed to this through her acceptance of his initiatory gifts of soap.[22] The woman, however, while privately agreeing that she had accepted such gifts, claimed that no formal agreement had yet been reached. Her anger, then, did not stem from any belief that the man had attempted to sexually coerce her ("How would he be able to do such a thing?"). Because the term "to be quiet" is often used as a euphemism for sexual intercourse in Gerai, she saw the man's exhortation that she "be quiet" as simply an invitation to engage in sex with him, rather than the implicit threat that I read it to be.[23] Instead, her anger stemmed from her conviction that the correct protocols had not been followed, that the man ought to have spoken with her rather than taking her acceptance of the soap as an unequivocal expression of assent. She was, as she put it, letting him know that "you have sexual relations together when you talk together. Sexual relations cannot be quiet."[24]

Yet, this should not be taken to mean that the practice of rape is simply a product of discourse: that brutality toward women is restricted to societies containing particular, dimorphic representations of male and female sexuality and that we simply need to change the discourse in order to eradicate such practices.[25] Nor is it to suggest that a society in which rape is unthinkable is for that reason to be preferred to Western societies. To adopt such a position would be still to view the entire world through a sexualized Western lens. There are, in fact, horrific things that may be done to women in places such as Gerai—things that are no less appalling in their implications for the fact that they do not involve the sexualized brutality of rape. In Gerai, for instance, while a woman does not fear rape, she does fear an enemy's bewitchment of her rice seed (the core of her gendered identity in this context) and the subsequent failure of the seed to sprout, resulting in hunger and illness for herself and her rice group. . . . Gerai women live constantly with the fear of this bewitchment (much as Western women live with the fear of rape), and even talking of it (always in whispers) reduces them to a state of terror.[26] The fact that this kind of attack can be carried out on a woman by either a woman or a man, and that it strikes not at her alone but at her rice group as a whole, marks it as belonging to a very different gendering regime from that which operates in the West. But it is no less horrific in its implications for that.

In order to understand the practice of rape in countries like Australia and the United States, then—and so to work effectively for its eradication there—feminists in these countries must begin to relinquish some of our most ingrained presumptions concerning difference between men and women and, particularly, concerning men's genitalia and sexuality as inherently brutalizing and penetrative and women's genitalia and sexuality as inherently vulnerable and subject to brutalization. Instead, we must begin to explore the ways rape itself *produces* such experiences of masculinity and femininity and so inscribes sexual difference onto our bodies. In a recent article, Moira Gatens asks of other feminists, "Why concede to the penis the power to push us around, destroy our integrity, 'scribble on us,' invade our borders and boundaries, and . . . occupy us in our (always already) conquered 'privacy'?" (1996, 43). This article echoes her lament. The tendency among many Western feminists writing on rape to accept as a seeming fact of nature the normative Western iconography of sexual difference leads them to reproduce (albeit unwittingly) the very discursive framework of Western rapists themselves, with their talk of "tools" and "holes," the very discursive framework in which rape is possible and which it reinscribes. For rape imposes difference as much as it is produced by difference. In fact, the highly racialized character of rape in many Western contexts suggests that the practice serves to police not simply sexual boundaries but racial ones as well. This is hardly surprising, given the history of the present "heterosexual matrix" in the West: as Stoler (1989, 1995) has demonstrated, the process of heterosexualization went hand-in-hand with that of colonialism. As a result, in contemporary Western settings sexual othering is inextricably entangled with racial othering. Unfortunately, in universalizing rape, many Western feminists risk naturalizing these othering processes and so contributing to a perpetuation of the very practices they seek to eradicate.

NOTES

1. Among "radical" feminists such as Andrea Dworkin and Catharine MacKinnon this belief reaches its most extreme version, in which all sexual intercourse between a man and a woman is viewed as akin to rape (Dworkin 1987; MacKinnon 1989a, 1989b).
2. Leacock 1978 and Bell 1983 are well-known examples. Sanday 1990a and Marcus 1992 are more recent examples, on Minangkabau and Turkish society, respectively.
3. See Laqueur 1990 for a historical account of this process.

4. On the equation of body and person within Western (especially feminist) thought, see Moore 1994.
5. See Plaza 1980: "[Rape] is very sexual in the sense that [it] is frequently a sexual activity, but especially in the sense that it opposes men and women: it is *social sexing* which is latent in rape. . . . Rape is sexual essentially because it rests on the very social difference between the sexes" (31).
6. The material on male prison inmates is particularly revealing in this respect. As an article by Stephen Donaldson, a former prisoner and the president of the U.S. advocacy group Stop Prisoner Rape, makes clear, "hooking up" with another prisoner is the best way for a prisoner to avoid sexual assaults, particularly gang rapes. Hooking up involves entering a sexual liaison with a senior partner ("jocker," "man," "pitcher," "daddy") in exchange for protection. In this arrangement, the rules are clear: the junior partner gives up his autonomy and comes under the authority of the senior partner; he is often expected by the senior partner to be as feminine in appearance and behaviour as possible," including shaving his legs, growing long hair, using a feminine nickname, and performing work perceived as feminine (laundry, cell cleaning, giving backrubs. etc.) (Donaldson 1996, 17, 20). See also the extract from Jack Abbott's prison letters in Halperin 1993 (424–25).
7. While I am primarily concerned here with the feminist literature (believing that it contains by far the most useful and insightful work on rape), it needs to be noted that many other (nonfeminist) writers also believe rape to be universal. See, e.g., Ellis 1989; Palmer 1989.
8. For listings of "rape-prone" societies, see Minturn, Grosse, and Haider 1969; Sanday 1981.
9. I carried out anthropological fieldwork in Gerai from March 1985 to February 1986 and from June 1986 to January 1987. The fieldwork was funded by an Australian National University Ph.D. scholarship and carried out under the sponsorship of Lembaga Ilmu Pengetahuan Indonesia. At the time that I was conducting my research a number of phenomena were beginning to have an impact on the community—these had the potential to effect massive changes in the areas of life discussed in this article. These phenomena included the arrival of a Malaysian timber company in the Gerai region and the increasing frequency of visits by Malay, Bugis, Chinese, and Batak timber workers to the community; the arrival of two American fundamentalist Protestant missionary families to live and proselytize in the community; and the establishment of a Catholic primary school in Gerai, resulting in a growing tendency among parents to send their children (both male and female) to attend Catholic secondary school in a large coastal town several days' journey away.
10. On the Ilongot, see Rosaldo 1980a, on the Wana, see Atkinson 1990.
11. The Wana, as described by Jane Atkinson (1990), provide an excellent example of a society that emphasizes sameness. Emily Martin points out that the explicit Western opposition between the "natures" of men and women is assumed to occur even at the level of the cell, with biologists commonly speaking of the egg as passive and immobile and the sperm as active and aggressive even though recent research indicates that these descriptions are erroneous and that they have led biologists to misunderstand the fertilization process (1991). See also Lloyd 1984 for an excellent account of how (often latent) conceptions of men and women as having opposed characteristics are entrenched in the history of Western philosophical thought.
12. The nurture-need dynamic (that I elsewhere refer to as the "need-share dynamic") is central to Gerai sociality. Need for others is expressed through nurturing them; such expression is the primary mark of a "good" as opposed to a "bad" person. See Helliwell (forthcoming) for a detailed discussion.
13. In this respect, Gerai is very different from, e.g., Australia or the United States, where, as Michelle Rosaldo has pointed out, aggression is linked to success, and women's constitution as lacking aggression is thus an important element of their subordination (1980b, 416; see also Myers 1988, 600).
14. See Helliwell 1996, 142–43, for an example of a "violent" altercation between husband and wife.
15. I have noted elsewhere that the inside-outside distinction is a central one within this culture (Helliwell 1996).
16. While the Gerai stress on the sameness of men's and women's sexual organs seems, on the face of it, to be very similar to the situation in Renaissance Europe as described by Laqueur 1990, it is profoundly different in at least one respect: in Gerai, women's organs are not seen as emasculated versions of men's—"female penises"—as they were in Renaissance Europe. This is clearly linked to the fact that, in Gerai, as we have already seen, *people* is not synonymous with *men,* and women are not relegated to positions of emasculation or abjection, as was the case in Renaissance Europe.
17. In this respect Gerai is similar to a number of other peoples in this region (e.g., Wana, Ilongot), for whom difference between men and women is also

seen as primarily a matter of the different kinds of work that each performs.

18. In Gerai, pregnancy and birth are seen not as semimystical "natural" processes, as they are for many Westerners, but simply as forms of work, linked very closely to the work of rice production.
19. Sanday 1986 makes a similar point about the absence of rape among the Minangkabau. See Helliwell (forthcoming) for a discussion of the different kinds of bad fate that can afflict a group through the actions of its individual members.
20. In Gerai, as in nearby Minangkabau (Sanday 1986), vulnerability is respected and valued rather than despised.
21. The man left the community on the night that this event occurred and went to stay for several months at a nearby timber camp. Community consensus—including the view of the woman concerned—was that he left because he was ashamed and distressed, not only as a result of having been sexually rejected by someone with whom he thought he had established a relationship but also because his adulterous behavior had become public, and he wished to avoid an airing of the details in a community moot. Consequently, I was unable to speak to him about the case. However, I did speak to several of his close male kin (including his married son), who put his point of view to me.
22. The woman in this particular case was considerably younger than the man (in fact, a member of the next generation). In such cases of considerable age disparity between sexual partners, the older partner (whether male or female) is expected to pay a fine in the form of small gifts to the younger partner, both to initiate the liaison and to enable its continuance. Such a fine rectifies any spiritual imbalance that may result from the age imbalance and hence makes it safe for the relationship to proceed. Contrary to standard Western assumptions, older women appear to pay such fines to younger men as often as older men pay them to younger women (although it was very difficult to obtain reliable data on this question, since most such liaisons are adulterous and therefore highly secretive). While not significant in terms of value (women usually receive such things as soap and shampoo, while men receive tobacco or cigarettes), these gifts are crucial in their role of "rebalancing" the relationship. It would be entirely erroneous to subsume this practice under the rubric of "prostitution."
23. Because Gerai adults usually sleep surrounded by their children, and with other adults less than a meter or two away (although the latter are usually inside different mosquito nets), sexual intercourse is almost always carried out very quietly.
24. In claiming that "sexual relations cannot be quiet," the woman was playing on the expression "be quiet" (meaning to have sexual intercourse) to make the point that while adulterous sex may need to be even "quieter" than legitimate sex, it should not be so "quiet" as to preclude dialogue between the two partners. Implicit here is the notion that in the absence of such dialogue, sex will lack the requisite mutuality.
25. Foucualt, e.g., once suggested (in a debate in French reprinted in *La Folie Encerclee* [see Plaza 1980]) that an effective way to deal with rape would be to decriminalize it in order to "desexualize" it. For feminist critiques of his suggestion, see Plaza 1980; de Lauretis 1987; Woodhull 1988.
26. Men fear a parallel form of bewitchment that causes death while engaged in the definitive "men's work" of cutting down large trees to make a rice field. Like women's death in childbirth, this is referred to as an "evil death" *(mati jat)* and is believed to involve the transformation of the man into an evil spirit.

REFERENCES

Atkinson, Jane Monnig. 1990. "How Gender Makes a Difference in Wana Society" In *Power and Difference: Gender in Island Southeast Asia,* ed. Jane Monnig Atkinson and Shelly Errington, 59–93. Stanford, Calif.: Stanford University Press.

Barry, Kathleen. 1995. *The Prostitution of Sexuality.* New York and London: New York University Press.

Bart, Pauline B., and Patricia H. O'Brien. 1985. *Stopping Rape: Successful Survival Strategies.* New York: Pergamon.

Bell, Diane. 1983. *Daughters of the Dreaming.* Melbourne: McPhee Gribble.

Bourque, Linda B. 1989. *Defining Rape.* Durham, N.C., and London: Duke University Press.

Brison, Susan J. 1993. "Surviving Sexual Violence: A Philosophical Perspective." *Journal of Social Philosophy* 24(1):5–22.

Butler, Judith. 1990. *Gender Trouble: Feminism and the Subversion of Identity.* New York and London: Routledge.

———. 1993. *Bodies That Matter: On the Discursive Limits of "Sex."* New York and London: Routledge.

Clark, Anna. 1987. *Women's Silence, Men's Violence: Sexual Assault in England, 1770–845.* London and New York: Pandora.

de Lauretis, Teresa. 1987. "The Violence of Rhetoric: Considerations on Representation and Gender." In her *Technologies of Gender: Essays on Theory, Film and Fiction,* 31–50. Bloomington and Indianapolis: Indiana University Press.

Dentan, Robert Knox. 1968. *The Semai: A Nonviolent People of Malaya.* New York: Holt, Rinehart & Winston.

———. 1978. "Notes on Childhood in a Nonviolent Context: The Semai Case (Malaysia)." In *Learning Non-Aggression: The Experience of Non-Literate Societies,* ed. Ashley Montagu, 94–143. New York: Oxford University Press.

Donaldson, Stephen. 1996. "The Deal behind Bars" *Harper's* (August): 17–20.

Dubinsky, Karen. 1993. *Improper Advances: Rape and Heterosexual Conflict in Ontario, 1880–1929.* Chicago and London: University of Chicago Press.

Dworkin, Andrea. 1987. *Intercourse.* London: Secker & Warburg.

Ellis, Lee. 1989. *Theories of Rape: Inquiries into the Causes of Sexual Aggression.* New York: Hemisphere.

Errington, Shelly. 1990. "Recasting Sex, Gender, and Power: A Theoretical and Regional Overview" In *Power and Difference: Gender in Island Southeast Asia,* ed. Jane Monnig Atkinson and Shelly Errington, 1–58. Stanford, Calif.: Stanford University Press.

Foucault, Michel. 1978. *The History of Sexuality.* Vol. 1, *An Introduction.* Harmondsworth: Penguin.

Gatens, Moira. 1996. "Sex, Contract, and Genealogy? *Journal of Political Philosophy* 4(1):29–44.

Geddes, W. R. 1957. *Nine Dayak Nights.* Melbourne and New York: Oxford University Press.

Gibson, Thomas. 1986. *Sacrifice and Sharing in the Philippine Highlands: Religion and Society among the Buid of Mindoro.* London and Dover: Athlone.

Gilman, Sander L. 1985. "Black Bodies, White Bodies: Toward an Iconography of Female Sexuality in Late Nineteenth-Century Art, Medicine, and Literature." In *"Race," Writing, and Difference,* ed. Henry Louis Gates, Jr., 223–40. Chicago and London: University of Chicago Press.

Gordon, Margaret T., and Stephanie Riger. 1989. *The Female Fear.* New York: Free Press.

Gregor, Thomas. 1990. "Male Dominance and Sexual Coercion." In *Cultural Psychology: Essays on Comparative Human Development,* ed. James W. Stigler, Richard A. Shweder, and Gilbert Herdt, 477–95. Cambridge: Cambridge University Press.

Griffin, Susan. 1986. *Rape: The Politics of Consciousness.* San Francisco: Harper & Row.

Halperin, David M. 1993. "Is There a History of Sexuality?" In *The Lesbian and Gay Studies Reader,* ed. Henry Abelove, Michele Barale, and David M. Halperin, 416–31. New York and London: Routledge.

Helliwell, Christine 1993. "Women in Asia: Anthropology and the Study of Women." In *Asia's Culture Mosaic,* ed. Grant Evans, 260–86. Singapore: Prentice Hall.

———. 1995. "Autonomy as Natural Equality: Inequality in 'Egalitarian' Societies." *Journal of the Royal Anthropological Institute* 1(2) :359–75.

———. 1996. "Space and Sociality in a Dayak Longhouse." In *Things as They Are: New Directions in Phenomenological Anthropology,* ed. Michael Jackson, 128–48. Bloomington and Indianapolis: Indiana University Press.

———. Forthcoming. *"Never Stand Alone": A Study of Borneo Sociality.* Williamsburg: Borneo Research Council.

Howell, Signe. 1989. *Society and Cosmos: Chewong of Peninsular Malaysia.* Chicago and London: University of Chicago Press.

Kilpatrick, Dean G., Benjamin E. Saunders, Lois J. Veronen, Connie L. Best, and Judith M. Von. 1987. "Criminal Victimization: Lifetime Prevalence, Reporting to Police, and Psychological Impact." *Crime and Delinquency* 33 (4) :479–89.

Koss, Mary P., and Mary R. Harvey. 1991. *The Rape Victim: Clinical and Community Interventions.* 2d ed. Newbury Park, Calif.: Sage.

Laqueur, Thomas. 1990. *Making Sex: Body and Gender from the Greeks to Freud.* Cambridge, Mass., and London: Harvard University Press.

Leacock, Eleanor. 1978. "Women's Status in Egalitarian Society: Implications for Social Evolution." *Current Anthropology* 19(2):247–75.

Lloyd, Genevieve. 1984. *The Man of Reason: "Male" and "Female" in Western Philosophy.* London: Methuen.

MacKinnon, Catharine A. 1989a. "Sexuality, Pornography, and Method: 'Pleasure under Patriarchy.'" *Ethics* 99: 314–46.

———. 1989b. *Toward a Feminist Theory of the State.* Cambridge, Mass., and London: Harvard University Press.

Madigan, Lee, and Nancy C. Gamble. 1991. *The Second Rape: Society's Continued Betrayal of the Victim.* New York: Lexington.

Marcus, Julie. 1992. *A World of Difference: Islam and Gender Hierarchy in Turkey.* Sydney: Allen & Unwin.

Marcus, Sharon. 1992. "Fighting Bodies, Fighting Words: A Theory and Politics of Rape Prevention." In *Feminists Theorize the Political,* ed. Judith Butler and Joan W. Scott, 385–403. New York and London: Routledge.

Martin, Emily 1991. "The Egg and the Sperm: How Science Has Constructed a Romance Based on Stereotypical Male-Female Roles." *Signs: Journal of Women in Culture and Society* 16(3):485–501.

McColgan, Aileen. 1996. *The Case for Taking the Date Out of Rape.* London: Pandora.

Minturn, Leigh, Martin Grosse, and Santoah Haider. 1969. "Cultural Patterning of Sexual Beliefs and Behaviour." *Ethnology* 8(3):301–18.

Mohanty, Chandra Talpade. 1991. "Under Western Eyes: Feminist Scholarship and Colonial Discourses." In *Third World Women and the Politics of Feminism,* ed. Chandra Talpade Mohanty, Ann Russo, and Lourdes Torres, 51–80. Bloomington and Indianapolis: Indiana University Press.

Moore, Henrietta L. 1994. *A Passion for Difference: Essays in Anthropology and Gender.* Cambridge and Oxford: Polity.

Myers, Fred R. 1988. "The Logic and Meaning of Anger among Pintupi Aborigines." *Man* 23 (4):589–610.

Palmer, Craig. 1989. "Is Rape a Cultural Universal? A Re-Examination of the Ethnographic Data." *Ethnology* 28(1):1–16.

Plaza, Monique. 1980. "Our Costs and Their Benefits." *m/f* 4:28–39.

Rosaldo, Michelle Z. 1980a. *Knowledge and Passion: Ilongot Notions of Self and Social Life.* Cambridge: Cambridge University Press.

———. 1980b. "The Use and Abuse of Anthropology: Reflections on Feminism and Cross-cultural Understanding." *Signs* 5(3):389–417.

Russell, Diana E. H. 1984. *Sexual Exploitation: Rape, Child Abuse, and Workplace Harassment.* Beverly Hills, Calif.: Sage.

Sanday, Peggy Reeves. 1981. "The Socio-Cultural Context of Rape: A Cross-Cultural Study." *Journal of Social Issues* 37(4):5–27.

———. 1986. "Rape and the Silencing of the Feminine." In *Rape,* ed. Sylvana Tomaselli and Roy Porter, 84–101. Oxford: Blackwell.

———. 1990a. "Androcentric and Matrifocal Gender Representations in Minangkabau Ideology." In *Beyond the Second Sex: New Directions in the Anthropology of Gender,* ed. Peggy Reeves Sanday and Ruth Gallagher Goodenough, 141–68. Philadelphia: University of Pennsylvania Press.

———. 1990b. *Fraternity Gang Rape: Sex, Brotherhood, and Privilege on Campus.* New York and London: New York University Press.

Stoler, Ann Laura. 1989. "Carnal Knowledge and Imperial Power: Gender, Race, and Morality in Colonial Asia." In *Gender at the Crossroads of Knowledge: Feminist Anthropology in the Postmodern Era,* ed. Micaela di Leonardo, 51–101. Berkeley and Los Angeles: University of California Press.

———. 1995. *Race and the Education of Desire: Foucault's* History of Sexuality *and the Colonial Order of Things.* Durham, N.C., and London: Duke University Press.

Strathern, Marilyn. 1987. "Conclusion." In *Dealing with Inequality: Analysing Gender Relations in Melanesia and Beyond,* ed. Marilyn Strathern, 278–302. Cambridge: Cambridge University Press.

———. 1988. *The Gender of the Gift: Problems with Women and Problems with Society in Melanesia.* Berkeley and Los Angeles: University of California Press.

Trumbach, Randolph. 1989. "Gender and the Homosexual Role in Modern Western Culture: The Eighteenth and Nineteenth Centuries Compared." In *Homosexuality, Which Homosexuality?* ed. Dennis Altman, 149–69. Amsterdam: An Dekker/Schorer; London: GMP.

———. 1993. "London's Sapphists: From Three Sexes to Four Genders in the Making of Modern Culture." In *Third Sex, Third Gender: Beyond Sexual Dimorphism in Culture and History,* ed. Gilbert Herdt, 111–36. New York: Zone.

Tsing, Anna Lowenhaupt. 1990. "Gender and Performance in Meratus Dispute Settlement." In *Power and Difference: Gender in Island Southeast Asia,* ed. Jane Monnig Atkinson and Shelly Errington, 95–125. Stanford, Calif.: Stanford University Press.

———. 1993. *In the Realm of the Diamond Queen: Marginality in an Out-of-the-Way Place.* Princeton, N.J.: Princeton University Press.

van der Meer, Theo. 1993. "Sodomy and the Pursuit of a Third Sex in the Early Modern Period." In *Third Sex, Third Gender: Beyond Sexual Dimorphism in Culture and History,* ed. Gilbert Herdt, 137–212. New York: Zone.

Woodhull, Winifred. 1988. "Sexuality, Power, and the Question of Rape." In *Feminism and Foucault: Reflections on Resistance,* ed. Irene Diamond and Lee Quinby, 167–76. Boston: Northeastern University Press.

Young, Iris Marion. 1990. "Throwing like a Girl: A Phenomenology of Feminine Body Comportment, Motility, and Spatiality." In her *Throwing like a Girl and Other Essays in Feminist Philosophy and Social Theory,* 141–59. Bloomington and Indianapolis: Indiana University Press.

13

Revisiting "Woman-Woman Marriage"

BY WAIRIMŨ NGARŨIYA NJAMBI AND WILLIAM E. O'BRIEN

Anthropologists Wairimũ Ngarũiya Njambi and William E. O'Brien offer an eye-opening account of the institution of woman-woman marriage as practiced in Kenya today. Based on in-depth interviews with Gikuyu women married to other women, the authors argue for an extensive revision of scholarship on woman-woman marriage that reflects the situated, complex, and empowering nature of this tradition.

1. What are the major shortcomings of past studies of woman-woman marriage?
2. How are woman-woman marriages regarded in the Gikuyu community today?
3 Why do the authors view woman-woman marriage as a mode of empowerment?

I ask myself, "What is it that women who are married to men have that I don't have? Is it land? I have land. Is it children? I have children. I don't have a man, but I have a woman who cares for me. I belong to her and she belongs to me. And I tell you, I don't have to worry about a man telling me what to do."

—CIRU, MARRIED TO NDUTA

INTRODUCTION

The practice of women marrying women is somewhat common in certain societies in West Africa, Southern Africa, East Africa, and the Sudan (O'Brien 1977). Yet, besides a total lack of discussion in the popular media, what is typically called woman-woman marriage is the subject of a very small body of academic literature.[1] Early scholarship is limited to the margins of several colonial-era ethnographies such as those of Evans-Pritchard, Herskovits, and Leakey. . . . More recent work remains equally marginal. Precious few writings address woman-woman marriage practices exclusively (e.g., Amadiume 1987; Burton 1979; Krige 1974; Oboler 1980); within others the subject remains little more than a footnote (e.g., Davis and Whitten 1987; Mackenzie 1990; Okonjo 1992). Since O'Brien's (1977) call for field research into woman-woman marriages more than two decades ago, there has been no study of Gĩkũyũ woman-woman marriages, and few studies anywhere else. Our study attempts to revive this dormant discourse in relation to the Gĩkũyũ.

Based on interviews with members of households containing woman-woman marriages, we attempt to provide images of this institution as practiced in central Kenya. Relying upon these women's voices, we present these Gĩkũyũ woman-woman marriages in relation to major themes in the literature.[2] . . . Our attention is on the ambiguities and flexibility inherent in women's decision to marry women. In addition, we point to the strong emotional bonds to one another expressed by these women, shedding critical light on the omissions of purely functionalist perceptions of woman-woman marriage relationships. We also challenge the generalized conceptualizations of women who initiate such marriages as "female husbands." That term, used by Leakey and virtually all other authors on the topic, regardless of cultural context, imposes a "male" characterization upon a situation where none necessarily exists. Emphasizing a term such as "female husband" prompts sex-role presumptions that do not fit these Gĩkũyũ women, who bristle at the implied male-identification regarding their roles.

This study is based on interviews with women in eight households in a small village in Murang´a District in central Kenya. This case study approach does not attempt to portray a generalized picture of woman-woman marriages, but relies upon the women's situated words to explain why they have married women, allowing them to present their own illuminating perspectives (see Smith 1987). . . .

The Gĩkũyũ are the largest ethnic group in Kenya, generally occupying the administrative unit of Central Province. "Kikuyuland," as it is commonly called, is bounded by Nairobi to the south and Mt. Kirinyaga (Mt. Kenya) to the north, the Rift Valley and Nyandarua Range (Aberdares) to the west, and the Mbeere Plain to the east. . . .

Most of the woman-woman marriage households in the study engaged in peasant farming for a living, dividing their agricultural production between cash crops and subsistence crops, a pattern typical of this rural setting. However, some of the women were engaged in other occupations including shop ownership, market trading of small commodities, and, in one case, *matatu* (mini bus) driving. The initiators of these relationships, who are called *ahikania,* were all landowners, and the households all had modest living standards similar to most others in the locality. Though the interviews took place in a rural setting, two of the subjects were residents of Nairobi, while another lived and worked in a nearby small urban center.

The majority of the *ahikania* were middle-aged at the time of marriage, and two were in their early 30s. All of the *ahiki,* the women who accepted the marriage offer, were between the ages of 20 and 30 when they were married. Education patterns of the subjects show that most of the initiators of the marriages were educated through the traditional Gĩkũyũ educational system of *githomo gia ugikuyu:* one had a high school education, one primary school. Almost all of the women who accepted the marriage offer had at least a primary school education. The wide range of age and education suggests to us that woman-woman marriage continues to be a relevant potential life-option for Gĩkũyũ women.

Kuhikania, the process of getting married, and *uhiki,* the marriage ceremony, takes place in the same manner for woman-woman marriages as with woman-man marriages. In fact, there is no separate term to differentiate a woman-woman marriage from a woman-man marriage. Even the term which describes the marriage initiator, *muhikania,* is used to describe a woman or a man.[3] As woman-woman marriages are not sanctioned by the various Christian churches in the region, *kuhikania* and *uhiki* continue to be performed through customary guidelines. The woman seeking a marriage partner, the *muhikania,* announces, either through a *kiama* (a customary civic organization) or through her own effort, her desire to find a marriage partner, or *muhiki.* Once the word is out, interested women go to visit, and once a suitable partner is found the *muhikania's* friends and family bring *ruracio* (gifts associated with *uhiki*) to those of the future wife and vice-versa. *Uhiki* takes place after this gift exchange and is performed with ceremonial blessings, termed *irathimo,* by elders of both families as the new wife moves into the *muhikania's* house.

WOMAN-WOMAN MARRIAGES AND FAMILY DEFINITIONS

While woman-woman marriage may be familiar to most anthropologists, at least in passing, the topic remains relatively obscure to most people outside Africa. In family studies discourse, the topic is pushed to the extreme margins by an historical fixation on western nuclear families as a universal ideal. This normative presumption of nuclearity makes it very difficult for particular non-western family forms, such as the woman-woman marriages in this study, to be evaluated as anything but bizarre novelties. As Skolnick and Skolnick argue:

> The assumption of universality has usually defined what is normal and natural both for research and therapy and has subtly influenced our thinking to regard deviations from the nuclear family as sick or perverse or immoral. (1989, 7)

. . . The Gĩkũyũ woman-woman marriages we studied challenge this thinking on all counts. Not only are the adults involved in these marriages of the same sex, but also there may be more than two, and the form of the family is not necessarily permanent (as an ideal) once a union is made, but may change periodically. Furthermore, men are often absent from such relationships, though they may be involved in married relationships as spouses of women who initiate woman-woman marriages.

One example of such a relationship in our study is Kũhĩ's household. In this complex case, Kũhĩ (a woman) and Huta (a man) were originally married to each other. Later, they decided together that Huta would marry a second woman, Kara, creating a polygynous marriage.[4] Later still, Kũhĩ entered into a woman-woman marriage with a woman named Wamba. Wamba came to that family as Kũhĩ's marriage partner, and to assist in raising the children of that household. In this particular case, Wamba could have a sexual relationship with Huta (whom she also informally regarded as a husband), and was not restricted from having sexual relationships with other men outside their household. Later in her life, while still married to Kũhĩ, Wamba married a woman named Wambũi. The result is that this single household contains four marriages: two woman-man marriages and two woman-woman marriages. Such complex relationships do not break any "rules," expectations, or ideals of woman-woman marriages, but are an accepted aspect of such relationships in Gĩkũyũ contexts. . . .

The idea of same sex relationships has spurred discussion of the sexuality of women in such marriages. A few texts imply that there may be sexual involvement in these marriages. Herskovits, for example, suggested that Dahomey woman-woman marriages sometimes involved sexual relations between the women (1937). Davis and Whitten go so far as to state that the main issue in explaining these relationships generally is over whether reasons for such partnerships are in fact "homoerotic" or strictly socio-economic (1987, 87). While sexuality was not directly discussed in our interviews, we can glean from the experience that this dichotomy makes little sense.[5]

In our Gĩkũyũ locale, women in these relationships did not talk about sexual involvement with one another, although some did indicate sharing the same bed at night. . . . Given the ambiguity in this Gĩkũyũ context, one might borrow Obbo's assertion regarding the Kamba of Kenya that while there may be no clear indication of sexual relations among women in these marriages, we simply cannot dismiss the possibility (1976). We agree with Carrier that this possibility has been too quickly dismissed by some authors, and suggest that the subject deserves more careful investigation (1980). At the same time, we question the assumption that sexual contact is the only factor that determines whether one should be considered as "homosexual" (see Martin 1992).

On the other side of the dichotomy, to suggest that such relationships are based solely on socio-economic factors like access to land and other resources or lineage ignores the close emotional ties experienced by these women. Such functionalist views have strongly influenced historical, and still-held stereotypes of African marriages generally. African family relations, compared to the privileged, western nuclear family form, are often portrayed as relatively primitive since they are presumed to be based on practical considerations alone, such as access to resources, as opposed to having a significant emotional aspect (e.g., Albert 1971; Ainsworth 1967; Beeson 1990; Kilbride and Kilbride 1990; Le Vine 1970). The women interviewed help undermine such rigid notions, demonstrating clear emotional commitment to the women they marry. For example, one participant, Nduta, proclaims her feelings for her *muka wakwa,* or co-wife, Cirũ:[6]

> No one dare to disturb my co-wife in any way, and especially knowing what I would do to them. No one dares point a finger at her. I tell her to proudly proclaim her belongingness to me, and I to her. . . . What I hate most is when people come to gossip to me about my co-wife's whereabouts or whom they have seen her with. I don't care as long as she is here for me now and even after I am gone. . . . Regardless of what she does, she is here because of me. Then why should I tell her what to do and what not to do. She is a free woman. And that is what I want her to be. So, when they come here to gossip, I tell them to leave her alone. She is mine and she is here on my property, not yours. . . . She who sincerely loved me and I loved back, let her stay mine. It is she who shall enshrine and take over this household when my time comes. (in interview)

In addition to expressing love (*wendo*) for Cirũ, Nduta also alludes to the fact that Cirũ is not restricted from having sexual encounters with men outside the woman-woman marriage relationship. Such liaisons, however, in no way undermine Cirũ's reciprocated love and appreciation for Nduta. In a separate interview, Cirũ, who has been married to Nduta for over 25 years, presents her deep feelings for her marriage partner:

> I know that some people do talk negatively about our marriage. Although honestly I have never caught anybody personally. But I ask myself, "What is it that women who are married to men have that I don't have? Is it land?" I have land. "Is it children?" I have children. I don't have a man, but I have a woman who cares for me. I belong to her and she belongs to me. And I tell you, I don't have to worry about a man telling me what to do. Here, I make all the decisions for myself. Nduta likes women who are able to stand on their own, like herself. I do what I want and the same goes for Nduta. Now I'm so used to being independent, and I like that a lot. I married Nduta because I knew we could live together well. She is a very wonderful woman with a kind heart. (in interview)[7]

While functionalist interpretations perceive African family relationships in terms of the purposes they serve in the functioning of a society, our interviewees highlight the complex and intertwined aspects of relationships that one would expect to find in a discussion of any committed, caring marriage partnership, undermining prevailing notions of the non-emotional African "Other."

One other point in the ideology of the nuclear family that remains strong, even among scholars, but is challenged by the woman-woman marriage data, is the alleged need for a father figure to maintain "functionality" (Cheal 1991). . . . The presence of a father is apparently not so important in many woman-woman marriages. During interviews, some women downplayed the importance of men in their households. Of the eight households in our study, six did not include permanent relationships with male partners. Among these six households, it seemed clear based on our interviews that male involvement with children, beyond procreation, was restricted, even identities of designated male genitors could not be revealed. Cirũ's comments support the view that males are viewed principally as friends and/or sex partners with no claim on children or property. What does she desire from men? Not much, apparently, except perhaps sex, and she can get that when she wants on her own terms:

> I have freedom to have sex with any man that I desire, for pleasure and for conceiving babies. And none of these men can ever settle here at our home or claim the children. They can't. They are not supposed to, and they know that very well. They come and go. (in interview)

Nduta's comments present the same lack of interest in having a man around as the ideal situation, expressing the independence provided by keeping men out of the household:

> We have no interest with a man who wants to stay in our home. We only want the *arume a mahutini* [men met in "the bush," a term for "male genitors"]—meaning those who are met only for temporary needs. *The meaning for this is for a woman to be independent enough so that she can make her own homestead shrine.* Cirũ sees also that I myself do not keep a man here. What for? To make me miserable? If I kept a man here who will then start asking me for money to buy alcohol, where would I find such money? No, I won't agree to live like that. It is better for one to look after oneself. It is better for one to look after oneself. (in interview, emphasis added)

Another case that downplays the importance of a male presence is that of Mbura, who had been married to a man, though he had died over 40 years ago. She was more recently married to a woman, Nimũ, who subsequently left after a couple of years. Mbura was later married to a woman named Kabura on the last day of this fieldwork. Mbura responds as well to the question of the place of men in the woman-woman marriage household, adding that, to her, men are not trustworthy, though she still appreciates their temporary presence:

> Men, even the good friends, know that they are not welcome here. They are here just for a visit and to leave. Whatever they come here to do, they must leave. They cannot be trusted. That is not good. One is given respect and that's all. (in interview)

Despite the fact that the other two households in the study *did* have men present as partners of one of the *ahikania,* or marriage initiators, the need for a "father figure," an ideal of most heterosexual nuclear families is clearly not a universal reality for all family situations.

BEYOND COMMON EXPLANATIONS

An overview of the literature on woman-woman marriages in African societies might tempt a reader to make three intertwined cross-cultural generalizations. The first generalization regards access to children. Sudarkasa suggests that the basis for woman-woman marriage, as with African marriages generally, is the desire "to acquire rights over a woman's childbearing capacity" ([1986] 1989, 155). That is, the woman who initiates a marriage seeks access to children that she herself does not have. Rights over childbearing capacity are often linked to a second general theme: that children are desired by such women as a means of transferring property through inheritance. . . . Connected to both general circumstances is the third common assertion that women's "barrenness" is a fundamental factor prompting woman-woman marriages. In fact, one of the most widely held general assumptions, as Burton points out, is that woman-woman marriages must involve women who cannot themselves have children (1979). . . .

Gĩkũyũ women in our relatively small study sample, living within a very proscribed spatial setting, expressed multiple and heterogeneous reasons for marrying women. . . . The women initiating these marriages pursued various objectives: companionship to appease loneliness, to be remembered after death, to have children to increase the vibrancy of the household, to fulfill social obligations in accordance with indigenous spiritual beliefs, and not least to avoid direct domination by male partners in a strongly patriarchal society, including men's control of both the women's behavior and household finances.

Our study does not deny the inability to bear children, inheritance, or lineage as partial explanations for some, or even many, Gĩkũyũ woman-woman marriages. Expressed reasons for marrying women in our study *did* often include the desire for the *muhikania* to have a child to inherit property and/or to perpetuate her family lineage. However, such explanations are never offered as the exclusive reasons, nor are they offered by all women. Such women appear to have much greater latitude in choosing how and why they participate in woman-woman marriages. For example, situations that defy Leakey's account include those in which women who are already married to men (who are still alive) and have their own children then initiate *uhiki,* or marriage, with a woman, as in the above described case of Kũhĩ (married to Wamba).

Mbura's explanation for *kuhikia,* or marrying a woman superficially resembles Leakey's account, since she expresses a desire for children that she herself cannot bear, as indicated in the following statement:

> I married Nimũ because I could never have children myself. I did not even give birth to children who later died, nor did I experience any miscarriage. I remained the way I came out of my mother's womb. And now I'm getting old and there is no way I can sit, think and decide to have a baby because my time is over, unless *Ngai's* [God's] miracle happens to me [she laughs].[8] I think a lot about how my husband left me and how I can't have a baby. That is why a cry of a baby makes me happy and sad at the same time. One has to realize how special a child is. . . . So, when I think about all these things: how I can't have a child, how my husband died and left me nothing, and how I have this illness, I ask *Ngai wenda mdathima na mutumia ungi* [God, please bless me with another woman]. . . . "Won't you please send that woman here to my home." Who knows, that woman might . . . give me a child. . . . Don't you see when I die I will be satisfied that I have left somebody in that home, who shall continue and revive that home? (in interview)

While she seems to portray a conventional account—marrying a woman to have a child to continue a lineage—Mbura's explanation is more complicated, indicating a desire for children beyond their role as inheritors of land and name. This is not to suggest, however, that lineage is not important in Mbura's decision to marry a woman. But the lineage she seeks to perpetuate is not necessarily her husband's, as Leakey and others would argue. Rather, Mbura is most interested in being remembered herself, as she indicates in the following statement:

> If I were to die even as we speak, that would be the end of it. I would be completely forgotten. No one would ever mention my name. That is simply because there would be no one to carry on my name. Since my husband died he is still remembered by many. But the key reason why he is still remembered is because of me. Someone may pass through here and demand to know "Whose home is that?" Then turn around and ask, "What about the next one?" One would reply, "Did you know so and so? This is his wife's home." Now do you see that the reason he is being remembered is because of me? Because I can be seen. But if I were to die, who will make me be remembered? . . . That is why the idea of marrying another woman came to me. Even now as we speak, if *Ngai* would bless me with another woman I would appreciate her.

Mbura continues, suggesting that companionship to appease loneliness is another strong motivation for marrying a woman:

> Let me tell you, I'm not the only one or the first one to marry a woman. And certainly, there are many others out there like me. I'm all alone just like that. No husband, no child. Just poor me. No one is here to keep me company or even to ask me "Did you sleep well?," except for occasional visits by some people like those you met here the other day. (in interview)

While Leakey's explanation may partly account for Mbura's case, Nduta's case clearly has emerged under a set of circumstances not fully considered by Leakey. First of all, Nduta's decision is the result of women's collaboration, namely between Nduta and her mother-in-law. Nduta married a man named Ndũngũ with whom she had three sons and a daughter. However, early in their marriage, her husband and their three sons were poisoned to death by some people in her husband's clan who wanted their land. After their deaths, Nduta's mother-in-law advised her to marry a woman as a way of protecting their family and land from male relatives who were trying to take her land, a sign of the tenuous hold that women have over land in Gĩkũyũ society (Mackenzie 1990). Rather than being victimized by men within their family, Nduta's case shows how women collaborate to look out for one another to protect women's interests:

> When a woman is left alone, she should not be frightened, but must be brave. You must make yourself a queen, otherwise, be a coward and everything you stand for will be taken away from you by those who are hungry for what you have. . . . If you were a woman, and you had properties, you will be the first one to be stolen from by the men who thought they were more important than women. So, she must act. . . . I had a lot of properties and if it were not for *karamu* [the "pen"] that cheated me out of many of them, I would still have a lot.[9] I lost many of them because I was a woman and I had no sons. So, my mother-in-law advised me to marry my own woman because all my people had been finished [i.e., killed] except for my daughter. And that is the piece of advice that I myself chose to follow. So I married her. When I married her [Cirũ], she said "It is better to live with a woman. I'm tired of men." I responded, "Is that so?! I love that." We became good friends and partners and thereafter I gave *ruracio* to her family. (in interview) . . .

Nduta's case is similar to Mackenzie's and Leakey's images of woman-woman marriage presented by those authors in that she had been married to a man who died and she had no sons (they died as well). However, upon marrying a woman after her husband's death, she asserts that she could have passed her land to her daughter, Ceke. Indeed, Ceke was given half of Nduta's land. While Nduta explained that she could have left all of her land to Ceke, she decided against doing so because she did not want to constrain her daughter with the social expectations that "staying at home" entails:

> . . . I didn't want my daughter, Ceke, to stay here. I gave her freedom to fly and land wherever she wanted. That is the same freedom that brought me here. So why would I want to hold her here? Women like to go far. They don't like to be held down at their birth home. (in interview)

While the issue of inheritance is important in Nduta's case, related to her difficult struggle as a woman to maintain control over land resources, Nduta adds an important dimension drawn from Gĩkũyũ mythology. This reason becomes clear when we hear Nduta, who is about 90 years old, speak of her dead sons who, she says, visit her in her sleep to thank her for marrying a woman:

> *Roho wa anake akwa makwrire* [the spirits of my dead sons] come to visit me to show

> appreciation for what I have done for them. One time they came and told me, "Thank you, mother for marrying Cirũ for us. We are very grateful for bringing us dead people back home again. We are grateful indeed. For that we will always be watching over you. Nothing will ever harm you. We will take care of you." And then I would say, "If I didn't marry Cirũ for them, who else would I have married her for?" Then the other day they came to tell me that I have got only five years to live; that I'm going to die soon [she laughs hard]. I said, "Is that so? Thanks a lot and may *Ngai* be praised!" That is fine for me. I need rest. (in interview)

Nduta's sons died long ago, very young, and had not been able to accomplish much in their lives. Some Gĩkũyũ still believe that if someone dies suddenly, his or her life activities can be carried out as if they are still alive so that their opportunities would not be denied. Thus, when their mother married Cirũ, she married her in the same way her sons would have married had they lived. In this sense, even though these sons were already dead, they feel quite at home because of Cirũ's presence.

While Nduta's and Mbura's cases push the limits of Leakey's narrow inheritance-focused account of woman-woman marriage, the case of Nduta's daughter, Ceke, falls largely outside the scope of his scenario. Ceke's decision to marry a woman appears to be heavily influenced by the example set by her mother, who acted as a role model. However, unlike her mother she was at the same time still married and living with her husband, Ngigĩ, together with her daughter, Wahu, along with Wahu's six children. Having grown attached to Wahu's children, Ceke was insecure about whether Wahu would move away with them, leaving Ceke in a household without children. Ceke's marriage to a woman (Ngware) was thus viewed as a way Ceke could have more children. Ceke's intention was that her wife, Ngware, would have children with her husband, Ngigĩ. After having a child, however, Ngware left the household. Ceke and Wahu (her daughter) then reached an agreement that the children would be welcome to remain with Ceke even if Wahu decides to leave:

> Although my daughter was living with me at the time, and had all these children that you see here, I did not know what to expect from her. I did not know whether one day I will wake up and find her gone with all her children that I personally have raised and who actually call me *maitu* [mother], or whether she had already made up her mind that she will never leave. I made that move of wanting to find out when my wife [Ngware] left us. After that, my husband and I made an agreement with Wahu that she will live with us permanently and that if she will ever feel like leaving, her children that we have raised as our own will be welcome to remain with us where they are already guaranteed good care as well as land settlement when they grow up. In any case, this is her land too, you know. Since we have got no other children, everything we have belongs to her and her children and to my other son borne by my wife before she left. (in interview)

While this example supports the general claim that women marry women to acquire rights over childbearing capacity (Sudarkasa [1986] 1989), Ceke's decision is not linked to property inheritance, "barrenness," or widowhood: the three essential criteria for a Gĩkũyũ woman-woman marriage, according to Leakey. Like Mbura, Ceke's strong desire for children was an important factor in her decision. The option of woman-woman marriage as a means to fulfill this desire was immediately apparent, given the influence and example of her mother, Nduta.

Finally, we have already alluded to the more overtly political motivations for marrying women expressed by some of our interviewees. The relative freedom from male control, which appears to be built into Gĩkũyũ woman-woman marriages, is expressed most forcefully by Cirũ and Nduta in previous quotations. Recall, for example, Nduta's conversation with her then wife-to-be, Cirũ, who commented, "I'm tired of men," to which Nduta responded, "I love that." And Nduta's comment about why she doesn't live with a man, stating "What for? To make me miserable?" Recall also these women's comments regarding the sexual freedom they find in these relationships. And finally, recall the opening quote in which Cirũ states that her woman-woman marriage allows her to avoid having "a man telling me what to do."

These examples demonstrate that flexibility, heterogeneity, and ambiguity appear as guiding principles in explaining such marriages, rather than being governed by somewhat rigid social rules, as the literature so often implies. However, contributions to the woman-woman marriage literature have continually, since the early-twentieth century, presented these relationships in functionalist terms. Cheal suggests that functionalist explanations continue to be perceived as having a "subterranean" influence on the study of families, describing such relationships in terms of "the ways in which they meet society's needs for the continuous replacement of its members" (1991, 4). Our alternative has been to present the institution of woman-woman marriage, at least in the Gĩkũyũ context, as a flexible option available to women within which they may pursue any number of interests: political, social, economic, and personal.

WHAT'S IN A NAME? RETHINKING THE "FEMALE HUSBAND"

Another area of concern for us in the literature is the unquestioned use of the term "female husband," the general term used to describe women who initiate woman-woman marriages. . . .

Not surprisingly, the major debate regarding the term "female husband" is over the male social traits often attributed to such women. Some have criticized the emphasis placed on gendered assumptions regarding sex-roles. For example, Krige suggests that one cannot assume that female husbands generally are taking on male roles (1974). Rather, one must carefully study sex roles in particular societies. For the Lovedu, Krige points out that numerous roles involve both males and females. Oyewumi, writing about the Yoruba, argues that local terms for both "husband" and "wife" are not gender-specific since both males and females can be husbands or wives (1994). As a result, as Burton (1979, 69) contends, the assumption that "husband" and "male" are automatically connected "confounds roles with people" since "husband" is a role that can be carried out by women as well as by men. Amadiume (1987), Burton (1979), Krige (1974), Oyewumi (1994), and Sudarkasa (1986) all suggest that in many societies, "masculinity" and "femininity" are not as clearly defined categories as they are in the West; presuming that "husband" automatically connotes "male" and that "wife" connotes "female" imposes western sex-role presumptions on other societies, ignoring local ambiguity regarding these roles (Sudarkasa [1986] 1989).

While our study supports views that women initiating marriages are not characterized as "male," we question the continued use of the term "female husband" to describe such women. Burton (1979), Krige (1974), and Sudarkasa (1986), while criticizing those who confuse social roles with genders, implicitly suggest that the term "female husband" is adequate and that the only task is to transform its connotative meaning.

We argue that the term "female husband" should be reconsidered on the grounds that the male connotation of "husband" cannot be so easily disposed of; just as the term "wife" conjures an association with "female," so does "husband" with "male." Especially in contexts where gender roles are ambiguous, this implicit association will easily mislead readers to impose western presumptions upon woman-woman marriages. Thus, in our view, efforts to theoretically disassociate gender from such role-centered terms—like "husband" and "wife" in this instance—imposed originally by western researchers in colonial contexts, will in a practical way continue to impose a male/female dichotomy. . . .

We acknowledge that there is nothing essential about the term "husband" that necessitates domination and control. But we also acknowledge, as does Oyewumi, that historically the term "husband" in most western contexts is normally associated with the role of "breadwinner," "decision-maker," and "head of household" (1994). We feel that the use of the term "female husband" serves to mask the relatively egalitarian woman-woman marriage relationships we encountered. . . .

The relative absence of domination, for example, is evident in the terms the women used to describe one another. The women interviewed never used the Gĩkũyũ term for "husband" (*muthuri*) to describe their partners. Instead, they consistently referred to each other using the terms *mutumia wakwa* and *muka wakwa,* which when used by these women translates as "co-wife," or *muiru wakwa,* which translates as "partner in marriage," indicating the mutual respect and relative equality between them. While most women in our study who initiate the marriages tended to be women with social influence and/or relatively greater material wealth, within the marriages both women interpreted their relationship as semiotically and materially equal.

Furthermore, women in our study rejected any male-association with their position of initiator of the marriage. None of the women interviewed indicated that they aspired to be like "males." What follows is Nduta's perception of herself in relation to the seeming "maleness" of her marital position:

> I stayed at Nairobi for three weeks at my daughter's house, and when I came back they were joyfully shouting "She is back!" And because I brought them bread just like other men who work in the city do around here, the children started shouting, "Here comes our *baba* [father]! Our *baba* has arrived! Our *baba* has arrived!" (she laughs). I called them *ndungana ici.*[10] "Who told you that I am your *baba?*" (she laughs again). So I asked them, "Is that what you see me as? I'm not your *baba.* But thank you for appreciating that I can also bring bread home." Therefore, even when you see me quarrel with them sometimes, I don't store those quarrels in my heart. I brought this family together not to destroy it but to care for it. (in interview)

Nduta is being teased by her children, who called her *baba,* because of the bread she brought from the city, just as men with urban jobs do when they visit their rural homes, not because her position as initiator of the marriage automatically connotes male characteristics.

As a tentative alternative to "female husband" we have been using the phrase "marriage initiator" to describe women in that position. However, we acknowledge that such description can be problematic, especially if it is used to focus more attention on the "initiator" at the expense of the agency of the one "initiated" into the marriage. We also acknowledge that descriptions of such concepts will differ from one culture to another.

AVENUES FOR FUTURE RESEARCH AND CONCLUSIONS

This article addresses the neglected topic of woman-woman marriages in Africa, relying upon Gĩkũyũ women to speak about issues that have lain dormant for a number of years. Our effort has been to challenge researchers on the topic to rethink the ways in which such relationships have been represented up to this point.

Future research must rely more heavily on the voices of Gĩkũyũ women to investigate this subject, not necessarily as a "better" and "authentic" way to tell these women's stories, but also as a constant reminder that these women have typically not had opportunities to speak and tell their own stories. Research must become sensitized to the idea that local voices relate the complexities, ambiguities, and heterogeneity involved in practices of woman-woman marriage, and in the analysis of how these practices take place and how the women involved perceive them. We do not suggest that because these Gĩkũyũ examples suggest flexibility, ambiguity, and heterogeneity that all African woman-woman marriages are the same. Rather, we raise the possibility that earlier explanations import assumptions that obscure different interpretations.

A number of issues regarding Gĩkũyũ woman-woman marriage remain. Our study did not investigate, for example, the prevalence of such marriages among the Gĩkũyũ. . . .

Another issue that needs further exploration is the emergence and transformation of Gĩkũyũ woman-woman marriages during the 500 year history of the Gĩkũyũ. Muriuki offers evidence that the matriarchal origins of Gĩkũyũ society had been superseded by patrilineal and patrilocal social and political organization by the mid-seventeenth century (1974). It is certainly not clear, but perhaps Gĩkũyũ woman-woman marriage is a remnant of a matriarchal past. While nothing has been written of origins, more recent twentieth-century social transformations have without doubt profoundly impacted practices of woman-woman marriage. As with other indigenous practices, Christian churches have severely and unfairly questioned the morality of woman-woman marriages, and have, in turn, shaped public opinion. For example, recent baptism guidelines from the Catholic Church in Kenya include their policy on, in their words, "women who 'marry' other women:"

> In regard to this traditional practice, the first step is to insist that this arrangement be given up completely and that meantime [sic] all those involved, plus any other persons directly responsible for the arrangement, be denied the sacraments. After the women have separated completely, each one will be helped separately and any infants will be baptised. (Kenya Catholic Bishops 1991, 21)

Such official condemnation impacts public perception by suggesting that such marriages represent an affront to Christian values.[11] . . .

Perhaps twentieth-century social changes, which have seriously dislocated, though not completely eliminated many indigenous institutions, fosters an ambivalence toward woman-woman marriage as a practice that is simultaneously acceptable, yet also incurs hostility. The acceptability of woman-woman marriages is evident in the fact that despite some hostility, these Gĩkũyũ woman-woman marriages are in no way secretive or hidden. All of the women in the study underwent a marriage ceremony to affirm their relationships, a ceremony no different than that for an opposite sex, indigenous (i.e., non-Christian) Gĩkũyũ marriage. Like other marriages, woman-woman marriages are facilitated by clan elders from both women's families (rather than priests or ministers), and involve an exchange of gifts between both families as well as dances and food. Such marriages are clearly not "underground" in any way.

Silence among feminists regarding the issue of woman-woman marriages is another issue. By now, it has been well documented that well-meaning feminists from western contexts have often represented "Third World women" in problematic ways. A common view is that of a linear women's emancipation, suggesting that societies have moved through evolutionary stages from women's oppression toward liberation, with western feminists having made the greatest progress and "Third World women" still mired in more overt forms of oppression. As a result, Third World women, a problematic category in itself, are often described by feminists "in terms of the underdevelopment, oppressive traditions, high illiteracy, rural and urban poverty, religious fanaticism, and "overpopulation" that appear to rule their lives in relation to those in the relatively liberated West (Mohanty 1991, 5). Such a linear view ignores what in many cases are long histories of women's empowerment and resistance, demonstrated here by woman-woman marriages.

By marrying women, these Gĩkũyũ women are clearly radically disrupting the male domination that operates in their everyday lives. Their stories may begin with land and struggles over material resources, but they are also stories of love, commitment, children, sexual freedom, vulnerability, and empowerment. The "implosion" of all these things makes these women's stories unique and all the more compelling to feminists who are constantly searching for unique practices of feminism that resemble, but are not engineered by, western feminism (Haraway 1997).

NOTES

1. Other terms include "woman-marriage" and "woman-to-woman marriage."
2. The names of interviewees have been changed to protect their identities.
3. Note that multiple Gĩkũyũ terms seem to describe the same concept. Choice of term depends upon the context in which the concept is employed. For example, while "marriage initiator" in one context is expressed as *muhikania,* the plural form of the concept is *ahikania.*
4. It is important to acknowledge that in most cases polygynous marriages among the Gĩkũyũ come as a result of negotiation between the first wives and husbands.
5. While the sexuality of the women involved in woman-woman marriages is clearly one of the most interesting unresolved issues on the topic, the Human Subjects Review Board reviewing the research proposal decided that the topic was too sensitive, and therefore declared such questions off limits.
6. Interviews were conducted in the Gĩkũyũ language and were translated by the primary author.
7. *Miario miuru,* or "negative talk," that is mentioned by Cirũ in this quotation points to fundamental changes that have occurred in Gĩkũyũ society over the course of the twentieth century with colonialist religious and educational training. These changes are reflected in complex local attitudes toward indigenous practices and are discussed briefly in the last section of this paper.
8. *Ngai* commonly translates as "God," although the Gĩkũyũ term carries no gendered connotation.
9. *Karamu,* or "the pen," refers to the use of title deeds (by those who could read and write—mainly men) that conferred private ownership of property since the 1960s. This private ownership was started under colonial rule and undermined (though it did not eliminate completely) more customary land tenure rules (Mackenzie 1990).
10. *Ndungana ici* is a derogatory term that translates most benignly as "You Stink!" However, the term also has sexual connotations, and is only used by elders to criticize misbehavior of younger people.
11. Related to condemnations of woman-woman marriage practices are official condemnations of homosexuality as expressed in recent homophobic statements by Presidents Moi of Kenya, Museveni of Uganda, and Mugabe of Zimbabwe, who referred to homosexuals and homosexual practices in terms such as "scourge," "abominable acts," and "lower than pigs and dogs" respectively. Some in Africa argue that the lack of local African terms for "homosexual" is evidence that homosexuality is foreign to Africa. However, in the Gĩkũyũ language there is no term for "heterosexual" either. Should this be taken as evidence that sexual relationships between women and men do not exist? Among the Gĩkũyũ, male-to-male sexual contact is traditionally prohibited; but prohibition suggests to us that such practices are already in place.

REFERENCES

Ainsworth, Mary D. Salter. 1967. *Infancy in Uganda: Infant Care and the Growth of Love.* Baltimore, MD: Johns Hopkins University.

Albert, Ethel M. 1971. "Women of Burundi: A Study of Social Values." In *Women of Tropical Africa,* ed. D. Paulme. Los Angeles: University of California Press.

Amadiume, Ifi. 1987. *Male Daughters and Female Husbands: Gender and Sex in an African Society.* London: Zed Press.

Beeson, R. W. 1990. "The Clinical Distribution of Family Systems." *International Journal of Contemporary Sociology* 27:89–127.

Burton, Clare. 1979. "Woman-Marriage in Africa: A Critical Study for Sex-Role Theory?" *Australian and New Zealand Journal of Sociology* 15(2):65–71.

Cheal, David. 1991. *Family and the State of Theory.* Toronto, Canada: University of Toronto Press.

Davis, D. L., and R. G. Whitten. 1987. "The Cross-Cultural Study of Human Sexuality." *Annual Review of Anthropology* 16:69–98.

Haraway, Donna. 1997. *Modest Witness@Second Millennium. FemaleMan Meets OncoMouse: Feminism and Technoscience.* New York: Routledge.

Kenya Catholic Bishops. 1991. *Guidelines for the Celebration of the Sacrament of Baptism for Infants and Special Cases.* Nairobi, Kenya: St. Paul Publications.

Kilbride, Philip Levey, and Janet Capriotti Kilbride. 1990. *Changing Family Life in East Africa: Women and Children at Risk.* University Park: The Pennsylvania State University Press.

Krige, Eileen Jensen. 1974. "Woman-Marriage, With Special Reference to the Lovedu—Its Significance for the Definition of Marriage." *Africa* 44:11–37.

Le Vine, R. 1970. "Personality and Change." In *The African Experience, Vol 1,* eds. J. N. Paden and E. W. Soja. Evanston, IL: Northwestern University Press.

Mackenzie, Fiona. 1990. "Gender and Land Rights in Murang'a District, Kenya." *The Journal of Peasant Studies* 17:609–43.

Mohanty, Chandra T. 1991. "Under Western Eyes: Feminist Scholarship and Colonial Discourses." In *Third World Women and the Politics of Feminism,* eds. C. T. Mohanty, A. Russo, and L. Torres, 51–80. Bloomington: Indiana University Press.

Muriuki, Godfrey. 1974. *A History of the Kikuyu 1500–1900.* New York: Oxford University Press.

Obbo, Christine. 1976. "Dominant Male Ideology and Female Options: Three East African Case Studies." *Africa* 46(4):371–89.

Oboler, Regina Smith. 1980. "Is the Female Husband a Man? Woman/Woman Marriage Among the Nandi of Kenya." *Ethnology* 19:69–88.

O'Brien, Denise. 1977. "Female Husbands in Southern Bantu Societies." In *Sexual Stratification: A Cross-Cultural View,* ed. A. Schlegel, 109–26. New York: Columbia University Press.

Okonjo, Kamene. 1992. "Aspects of Continuity and Change in Mate Selection Among the Igbo West of the Niger River." *Journal of Comparative Family Studies* 23:339–60.

Oyewumi, Oyeronke. 1994. "Inventing Gender: Questioning Gender in Precolonial Yorubaland." In *Problems in African History: The Precolonial Centuries,* eds. R. Collins et al., 244–50. New York: Marcus Wiener Publishing, Inc.

Skolnick, Arlene S., and Jerome H. Skolnick. 1989. "Introduction: Family in Transition." In *Families in Transition,* 6th ed., eds. A. S. Skolnick and J. H. Skolnick, 1–18. Boston: Scott, Foresman and Company.

Smith, Dorothy E. 1987. *The Everyday World As Problematic: A Feminist Sociology.* Boston: Northeastern University Press.

Sudarkasa, Niara. (1986) 1989. "'The Status of Women' in Indigenous African Societies." In *Feminist Frontiers II: Rethinking Sex, Gender, and Society,* eds. L. Richardson and V. Taylor, 152–58. New York: McGraw-Hill, Inc.

Talbot, Percy A. (1926) 1969. *The Peoples of Southern Nigeria.* London: Cass.

14

Gender and Power

BY MARIA ALEXANDRA LEPOWSKY

Maria Lepowsky is an anthropologist who lived among the Melanesian people of Vanatinai, a small, remote island near New Guinea, from 1977–1979, for two months in 1981, and again for three months in 1987. She chose Vanatinai, which literally means "motherland," because she wanted to do research in a place where "the status of women" is high. The egalitarianism of the Vanatinai challenges the Western belief in the universality of male dominance and female subordination.

1. What is the foundation of women's high status and gender equality among the people of Vanatinai?
2. What does gender equality mean on Vanatinai? Does it mean that women and men split everything 50–50? Are men and women interchangeable?
3. What are the similarities and differences between the egalitarianism of the Gerai people (depicted in Helliwell's article in this chapter) and that of the people of Vanatinai?

Vanatinai customs are generally egalitarian in both philosophy and practice. Women and men have equivalent rights to and control of the means of production, the products of their own labor, and the products of others. Both sexes have access to the symbolic capital of prestige, most visibly through participation in ceremonial exchange and mortuary ritual. Ideologies of male superiority or right of authority over women are notably absent, and ideologies of gender equivalence are clearly articulated. Multiple levels of gender ideologies are largely, but not entirely, congruent. Ideologies in turn are largely congruent with practice and individual actions in expressing gender equivalence, complementarity, and overlap.

There are nevertheless significant differences in social influence and prestige among persons. These are mutable, and they fluctuate over the lifetime of the individual. But Vanatinai social relations are egalitarian overall, and sexually egalitarian in particular, in that at each stage in the life cycle all persons, female and male, have equivalent autonomy and control over their own actions, opportunity to achieve both publicly and privately acknowledged influence and power over the actions of others, and access to valued goods, wealth, and prestige. The quality of generosity, highly valued in both sexes, is explicitly modeled after parental nurture. Women are not viewed as polluting or dangerous to themselves or others in their persons, bodily fluids, or sexuality.

Vanatinai sociality is organized around the principle of personal autonomy. There are no chiefs, and nobody has the right to tell another adult what to do. This philosophy also results in some extremely permissive childrearing and a strong degree of tolerance for the idiosyncrasies of other people's behavior. While working together, sharing, and generosity are admirable, they are strictly voluntary. The selfish and antisocial person might be ostracized, and others will not give to him or her. If kinfolk, in-laws, or neighbors disagree, even with a powerful and influential big man or big woman, they have the option, frequently taken, of moving to another hamlet where they have ties and can expect access to land for gardening and foraging. Land is communally held by matrilineages, but each person has multiple rights to request and be given space to make a garden on land held by others, such as the mother's father's matrilineage. Respect and tolerance for the will and idiosyncrasies of individuals is reinforced by fear of their potential knowledge of witchcraft or sorcery.

Anthropological discussions of women, men, and society over the last one hundred years have been framed largely in terms of "the status of women," presumably unvarying and shared by all women in all social situations. Male dominance and female subordination have thus until recently been perceived as easily identified and often as human universals. If women are indeed universally subordinate, this implies a universal primary cause: hence the search for a single underlying reason for male dominance and female subordination, either material or ideological.

More recent writings in feminist anthropology have stressed multiple and contested gender statuses and ideologies and the impacts of historical forces, variable and changing social contexts, and conflicting gender ideologies. Ambiguity and contradiction, both within and between levels of ideology and social practice, give both women and men room to assert their value and exercise power. Unlike in many cultures where men stress women's innate inferiority, gender relations on Vanatinai are not contested, or antagonistic: there are no male versus female ideologies which vary markedly or directly contradict each other. Vanatinai mythological motifs, beliefs about supernatural power, cultural ideals of the sexual division of labor and of the qualities inherent to men and women, and the customary freedoms and restrictions upon each sex at different points in the life course all provide ideological underpinnings of sexual equality.

Since the 1970s writings on the anthropology of women, in evaluating degrees of female power and influence, have frequently focused on the disparity between the "ideal" sex role pattern of a culture, often based on an ideology of male dominance, publicly proclaimed or enacted by men, and often by women as well, and the "real" one, manifested by the actual behavior of individuals. This approach seeks to uncover female social participation, overt or covert, official or unofficial, in key events and decisions and to learn how women negotiate their social positions. The focus on social and individual "action" or "practice" is prominent more generally in cultural anthropological theory of recent years. Feminist analyses of contradictions between gender ideologies of female inferiority and the realities of women's and men's daily lives—the actual balance of power in household and community—have helped to make this focus on the actual behavior of individuals a wider theoretical concern.[1]

In the Vanatinai case gender ideologies in their multiple levels and contexts emphasize the value of women and provide a mythological charter for the degree of personal autonomy and freedom of choice manifested in real women's lives. Gender ideologies are remarkably similar (though not completely, as I discuss below) as they are manifested situationally, in philosophical statements by women and men, in the ideal pattern of the sexual division of labor, in taboos and proscriptions, myth, cosmology, magic, ritual, the supernatural balance of power, and in the codifications of custom. Women are not characterized as weak or inferior. Women and men are valorized for the same qualities of strength, wisdom, and generosity. If possessed of these qualities an individual woman or man will act in ways which bring prestige not only to the actor but to the kin and residence groups to which she or he belongs.

Nevertheless, there is no single relationship between the sexes on Vanatinai. Power relations and relative influence vary with the individuals, sets of roles, situations, and historical moments involved. Gender ideologies embodied in myths, beliefs, prescriptions for role-appropriate behavior, and personal statements sometimes contradict each other or are contradicted by the behavior of individuals.

* * *

THE SEXUAL DIVISION OF LABOR

Vanatinai custom is characterized by a marked degree of overlap in the sexual division of labor between what men normally do and what women do. This kind of overlap has been suggested as a primary material basis of gender equality, with the mingling of the sexes in the tasks of daily life working against the rise of male dominance.

Still, sorcerers are almost all male. Witches have less social power on Vanatinai and are blamed for only a small fraction of deaths and misfortunes. Only men build houses or canoes or chop down large trees for construction or clearing garden lands. Women are forbidden by custom to hunt, fish, or make war with spears, although they may hunt for possum and monitor lizard by climbing trees or setting traps and catching them and use a variety of other fishing methods. Despite the suppression of warfare men retain greater control of the powers that come with violence or the coercive threat of violent death.

Some Vanatinai women perceive an inequity in the performance of domestic chores. Almost all adult women are "working wives," who come home tired in the evening, often carrying both a young child in their arms and a heavy basket of yams or other produce on their heads for distances of up to three miles. They sometimes complain to their husbands or to each other that, "We come home after working in the garden all day, and we still have to fetch water, look for firewood, do the cooking and cleaning up and look after the children while all men do is sit on the verandah and chew betel nut!" The men usually retort that these are the work of women. Here is an example of contested gender roles.[2]

Men are tender and loving to their children and often carry them around or take them along on their activities, but they do this only when they feel like it, and childcare is the primary responsibility of a mother, who must delegate it to an older sibling or a kinswoman if she cannot take care of the child herself. Women are also supposed to sweep the house and the hamlet ground every morning and to pick up pig excrement with a sago-bark "shovel" and a coconut-rib broom.

When speaking of their many responsibilities, women say, "Vanatinai women have to be very strong." Rosaldo (1974) sees participation of men in domestic activities such as childrearing and cleaning as a sign of high female status, as does Bacdayan (1977), an argument reminiscent of Western feminists' analyses of the politics of housework. Vanatinai men are less involved in these activities than are the Ilongot and Western Bontoc men of the Philippines whom Rosaldo and Bacdayan describe. While the roles of Vanatinai men and women overlap, they do not overlap completely.

Vanatinai is not a perfectly egalitarian society, either in terms of a lack of difference in the status and power of individuals or in the relations between men and women. Women in young and middle adulthood are likely to spend more time on childcare and supervision of gardens and less on building reputations as prominent transactors of ceremonial valuables. The average woman spends more of her time sweeping up the pig excrement that dots the hamlet from the unfenced domestic pigs wandering through it. The average man spends more time hunting wild boar in the rain forest with his spear (although some men do not like to hunt). His hunting is more highly valued and accorded more prestige by both sexes than her daily maintenance of hamlet cleanliness and household order. The sexual division of labor on Vanatinai is slightly asymmetrical, despite the tremendous overlap in the roles of men and women and the freedom that an individual of either sex has to spend more time on particular activities—gardening, foraging, fishing, caring for children, traveling in quest of ceremonial valuables—and to minimize others.

Yet the average Vanatinai woman owns many of the pigs she cleans up after, and she presents them publicly during mortuary rituals and exchanges them with other men and women for shell-disc necklaces, long axe blades of polished greenstone, and other valuables. She then gains status, prestige, and influence over the affairs of others, just as men do and as any adult does who chooses to make the effort to raise pigs, grow large yam gardens, and acquire and distribute ceremonial valuables. Women who achieve prominence and distribute wealth, and thus gain an enhanced ability to mobilize the labor of others, are highly respected by both sexes. An overview of the life course and the sexual division of labor on Vanatinai reveals a striking lack of cultural restrictions upon the autonomy of women as well as men and the openness of island society to a wide variety of lifestyles.

GENDER AND HISTORY

Since the arrival of the ancestors the permeable boundaries of the motherland's beaches have been swept by regional and global flows of persons, objects, and ideas over time and through space. These are the conditions in which Vanatinai gender relations as we see them today have been formed and transformed. We cannot consider them as part of a discretely bounded and wholly coherent social universe. They are localized and temporally specific artifacts that are components of ethnoscapes, ever changing landscapes of persons and beliefs, charged with the changing valences of complicated power relations (cf. Appadurai 1990, 1991). Vanatinai people explicitly recognize a basic unity of custom and language among the fruit of the motherland. At the same time they point to variations of belief and speech from one district on the island to another, explain that exchange partners from other islands regularly participate in the rituals that are the core of the way of the ancestors, and analyze the changes that have taken place, both before and after the coming of the Europeans, in their most crucial rituals.

Island customs are notably resilient and resistant to externally imposed changes. At the same time they are innovative and adaptive, working out new solutions to the problems posed by new forms of power and desires for new forms of wealth. Key changes in the Vanatinai region include increases and decreases in the frequency of warfare and raiding, the build-up of population (closely related to warfare patterns in the precolonial period), resulting scarcity of resources on smaller islands, new technologies and forms of wealth, the introduction of new forms of religious ideology with the advent of missionization, and the imposition of colonial and then national political authority upon previously autonomous island peoples.

Vanatinai has lost one source of gender asymmetry, the male opportunity in earlier generations to obtain power and influence through a reputation as a champion fighter, or asiara. In spite of major involvement by women in warfare and diplomacy, only men—a few younger men—could gain respect, fame, and power by becoming known as champion. . . .

Vanatinai warfare, as remembered today and as recorded in the 1880s by European visitors, was characterized by the male defense of people and property from raiders, not heroic attacks on others. In Vanatinai historical memory men killed other men to protect their families, just as they kill animals to nurture women and children and thus, paradoxically, to give life (cf. Brightman 1993b).

Both women and men are brought up to have assertive personalities. But physical violence against women—and men—is abhorred and occurs only rarely today. I have never heard of a case of rape. One of the last battles on the island took place as retaliation for a man's attack on his wife. Descendants mention the justifications of compensation and revenge but deplore the uncontrolled violence of earlier times. It is of course highly likely that male physical aggression was more highly valued in precolonial times, when warriors with spears defended land and people against attack and led raids against enemy districts. The admiration still evoked today by the word *asiara* is probably a muted echo of what it was.

The former male monopoly on warfare on Vanatinai probably did lead to certain kinds of influence within kin or hamlet groups being reserved for certain men, the asiara, in precolonial times. Some of the influence formerly wielded by the asiara, the feared warrior whose success and skill was based on his supernatural power, is now held by the (male) sorcerer.

The pacification imposed by British and Australian colonial authorities between 1888 and 1942 thus may have led to greater equality of opportunity for women to achieve renown and influence through taking prominent roles in ceremonial exchanges and mortuary ritual. There was no competing role of asiara for which they were not eligible. Women may also have benefited even more than men from the increased mobility for everyone made possible by pacification, a mobility that resulted in an expansion of interisland travel and exchange.

The Vanatinai case corroborates the importance of documenting local circumstances in evaluating the impact of historical changes upon gender ideologies and the material position of women. Contrary to Leacock (1978) and other writers, there is no automatic lowering of female autonomy in all aspects of the life of a small-scale society with the intrusion of colonialism or a centralized state. In some ways, in some places, women may gain advantages relative to men. At the same time everyone in the society may suffer the threat of violence, be coerced by outsiders, and lose autonomy and freedom of movement or behavior.

The impact on women's authority and influence of the absorption of Vanatinai into colonial and national polities has probably been negative overall. The new political and religious systems emphasize hierarchical authority controlled by distant male outsiders, with a few local adult males chosen to exercise control over the rest of the population as local government councillors, policemen and pastors. Colonial and national men coming to the island have expected to talk, in English, and delegate authority to other men, not bare-breasted women wearing coconut-leaf skirts. The new systems have their own ideologies of gender that more highly valorize men and that expect men and women to occupy spheres of activity that are more separate than is customary on Vanatinai. And the new systems directly oppose the Vanatinai ethic of egalitarian relations among autonomous individuals and exclude women from the new positions of authority.

Vanatinai people say that the prominent position of women in daily and ceremonial life is taubwaragha, the way of the ancestors. . . . Adhering to the way of the ancestors has for several generations often been a conscious act of resistance in the face of pressure from government, mission, and commercial interests. This resistance includes defending the power and influence of women: to own land and distribute goods, to acquire customary and magical knowledge, to travel, exchange, and host feasts, and to speak out forcefully in public about community concerns.

Why have Vanatinai people been successful to a large degree in their resistance and their cultural conservatism? Their physical circumstances have helped. They live on a remote island surrounded by treacherous reefs, a mountain guarded by large tracts of swampland. The island, though rich in resources, is hard for outsiders to exploit. Nobody could ever find the motherlode of the gold, transport and shipping of tropical commodities are difficult, and there have been few white residents in this century. Beginning a century ago, Vanatinai men and women could pan for gold if they wanted to trade for steel tools and other new needs and wants rather than migrate to sell their labor. Colonial suppression of warfare and the resulting efflorescence of interisland ceremonial exchange and of mortuary ritual provided enhanced opportunities for men and women such as the people of Vanatinai to find satisfaction in the way of the ancestors and the kinds of social relations it embodies. They were affected by colonial policies but far away from the controlling presence of colonial officers and missionaries.

MATERIAL AND IDEOLOGICAL BASES OF EQUALITY

Does equality or inequality, including between men and women, result from material or ideological causes? We cannot say whether an idea preceded or followed specific economic and social circumstances. Does the idea give rise to the act, or does the act generate an ideology that justifies it or mystifies it?

If they are congruent ideology and practice reinforce one another. And if multiple levels of ideology are in accord social forms are more likely to remain unchallenged and fundamentally unchanged. Where levels of ideology, or ideology and practice, are at odds, the circumstances of social life are more likely to be challenged by those who seek a reordering of social privileges justified according to an alternative interpretation of ideology. When social life embodies these kinds of contradictions, the categories of people in power—aristocrats, the rich, men—spend a great deal of energy maintaining their power. They protect their material resources, subdue the disenfranchised with public or private violence, coercion, and repression, and try to control public and private expressions of ideologies of political and religious power.

On Vanatinai, where there is no ideology of male dominance, the material conditions for gender equality are present. Women—and their brothers—control the means of production. Women own land, and they inherit land, pigs, and valuables from their mothers, their mothers' brothers, and sometimes from their fathers equally with men. They have the ultimate decision-making power over the distribution of staple foods that belong jointly to their kinsmen and that their kinsmen or husbands have helped labor to grow. They are integrated into the prestige economy, the ritualized exchanges of ceremonial valuables. Ideological expressions, such as the common saying that the woman is the owner of the garden, or the well-known myth of the first exchange between two female beings, validate material conditions.

I do not believe it would be possible to have a gender egalitarian society, where prevailing expressions of gender ideology were egalitarian or valorized both sexes to the same degree, without material control by women of land, means of subsistence, or wealth equivalent to that of men. This control would encompass anything from foraging rights, skills, tools, and practical and sacred knowledge to access to high-paying, prestigious jobs and the knowledge and connections it takes to get them. Equal control of the means of production, then, is one necessary precondition of gender equality. Vanatinai women's major disadvantage is their lack of access to a key tool instrumental in gaining power and prestige, the spear. Control of the means of production is potentially greater in a matrilineal society.

* * *

GENDER IDEOLOGIES AND PRACTICE IN DAILY LIFE

In Melanesian societies the power of knowing is privately owned and transmitted, often through ties of kinship, to heirs or younger supporters. It comes not simply from acquiring skills or the experience and the wisdom of mature years but is fundamentally a spiritual power that derives from ancestors and other spirit forces.

In gender-segregated societies, such as those that characterize most of Melanesia, this spiritual knowledge power is segregated as well into a male domain through male initiations or the institutions of men's houses or male religious cults. Most esoteric knowledge—and the power over others that derives from it—is available to Vanatinai women if they can find a kinsperson or someone else willing to teach it to them. There are neither exclusively male nor female collectivities on Vanatinai nor characteristically male versus female domains or patterns of sociality (cf. Strathern 1988:76).

Decisions taken collectively by Vanatinai women and men within one household, hamlet, or lineage are political ones that reverberate well beyond the local group, sometimes literally hundreds of miles beyond. A hundred years ago they included decisions of war and peace. Today they include the ritualized work of kinship, more particularly of the matrilineage, in mortuary ritual. Mortuary feasts, and the interisland and interhamlet exchanges of ceremonial valuables that support them, memorialize the marriages that tied three matrilineages together, that of the deceased, the deceased's father, and the widowed spouse. Honoring these ties of alliance, contracted by individuals but supported by their kin, and threatened by the dissolution of death, is the major work of island politics. . . .

The small scale, fluidity (cf. Collier and Rosaldo 1981), and mobility of social life on Vanatinai, especially in combination with matriliny, are conducive of egalitarian social relations between men and women and old and young. They promote an ethic of respect for the individual, which must be integrated with the ethic of cooperation essential for survival in a subsistence economy. People must work out conflict through face to face negotiation, or existing social ties will be broken by migration, divorce, or death through sorcery or witchcraft.

Women on Vanatinai are physically mobile, traveling with their families to live with their own kin and then the kin of their spouse, making journeys in quest of valuables, and attending mortuary feasts. They are said to have traveled for these reasons even in precolonial times when the threat of attack was a constant danger. The generally greater physical mobility of men in human societies is a significant factor in sexual asymmetries of power, as it is men who generally negotiate and regulate relationships with outside groups (cf. Ardener 1975:6).

Vanatinai women's mobility is not restricted by ideology or by taboo, and women build their own far-ranging personal networks of social relationships. Links in these networks may be activated as needed by the woman to the benefit of her kin or hamlet group. Women are confined little by taboos or community pressures. They travel, choose their own marriage partners or lovers, divorce at will, or develop reputations as wealthy and generous individuals active in exchange.

BIG MEN, BIG WOMEN, AND CHIEFS

Vanatinai giagia, male and female, match Sahlins's (1963) classic description of the Melanesian big man, except that the role of gia is gender-blind. There has been renewed interest among anthropologists in recent years in the big man form of political authority.[3] The Vanatinai case of the female and male giagia offers an intriguing perspective. . . .

Any individual on Vanatinai, male or female, may try to become known as a gia by choosing to exert the extra effort to go beyond the minimum contributions to the mortuary feasts expected of every adult. He or she accumulates ceremonial valuables and other goods both in order to give them away in acts of public generosity and to honor obligations to exchange partners from the local area as well as distant islands. There may be more than one gia in a particular hamlet, or even household, or there may be none. A woman may have considerably more prestige and influence than her husband because of her reputation for acquiring and redistributing valuables. While there are more men than women who are extremely active in exchange, there are some women who are far more active than the majority of men.

Giagia of either sex are only leaders in temporary circumstances and if others wish to follow, as when they host a feast, lead an exchange expedition, or organize the planting of a communal yam garden. Decisions are made by consensus, and the giagia of both sexes influence others through their powers of persuasion, their reputations for ability, and their knowledge, both of beneficial magic and ritual and of sorcery or witchcraft. . . .

On Vanatinai power and influence over the actions of others are gained by achievement and demonstrated superior knowledge and skill, whether in the realm of gardening, exchange, healing, or sorcery. Those who accumulate a surplus of resources are expected to be generous and share with their neighbors or face the threat of the sorcery or witchcraft of the envious. Both women and men are free to build their careers through exchange. On the other hand both women and men are free *not* to strive toward renown as giagia but to work for their own families or simply to mind their own business. They can also achieve the respect of their peers, if they seek it at all, as loving parents, responsible and hard-working lineage mates and affines, good gardeners, hunters, or fishers, or skilled healers, carvers, or weavers.

Mead (1935) observes that societies vary in the degree to which "temperament types" or "approved social personalities" considered suitable for each sex or a particular age category differ from each other. On Vanatinai there is wide variation in temperament and behavior among islanders of the same sex and age. The large amount of overlap between the roles of men and women on Vanatinai leads to a great deal of role flexibility, allowing both individual men and women the freedom to specialize in the activities they personally enjoy, value, are good at performing, or feel like doing at a particular time. There is considerable freedom of choice in shaping individual lifestyles.

An ethic of personal autonomy, one not restricted to the powerful, is a key precondition of social equality. Every individual on Vanatinai from the smallest child to an aged man or woman possesses a large degree of autonomy. Idiosyncrasies of personality and character are generally tolerated and respected. When you ask why someone does or does not do something, your friends will say, emphatically and expressively, "We [inclusive we: you and I both] don't know," "It is something of theirs" [their way], or, "She doesn't want to." Islanders say that it is not possible to know why a person behaves a certain way or what thoughts generate an action. Persisting in a demand to "know" publicly the thoughts of others is dangerous, threatening, and invasive. Vanatinai people share, in part, the perspectives identified with postmodern discussions of the limits of ethnographic representation: it is impossible to know another person's thoughts or feelings. If you try they are likely to deceive you to protect their own privacy or their own interests. Your knowing is unique to you. It is your private property that you transmit only at your own volition, as when you teach magical spells to a daughter or sister's son.[4]

The prevailing social sanction is also individualistic: the threat of somebody else's sorcery or witchcraft if you do not do what they want or if you arouse envy or jealousy. But Vanatinai cultural ideologies stress the strength of individual will in the face of the coercive pressures of custom, threat of sorcery, and demands to share. This leads to a Melanesian paradox: the ethic of personal autonomy is in direct conflict to the ethic of giving and sharing so highly valued on Vanatinai, as in most Melanesian cultures. Nobody can make you share, short of stealing from you or killing you if you refuse them. You have to want to give: your nurture, your labor, your valuables, and your person. This is where persuasion comes in. It comes from the pressure of other people, the force of shame, and magical seduction made potent by supernatural agency. Vanatinai custom supplies a final, persuasive argument to resolve this paradox: by giving, you not only strengthen your lineage and build its good name, you make yourself richer and more powerful by placing others in your debt.

What can people in other parts of the world learn from the principles of sexual equality in Vanatinai custom and philosophy? Small scale facilitates Vanatinai people's emphasis on face-to-face negotiations of interpersonal conflicts without the delegation of political authority to a small group of middle-aged male elites. It also leaves room for an ethic of respect for the will of the individual regardless of age or sex. A culture that is egalitarian and nonhierarchical overall is more likely to have egalitarian relations between men and women.

Males and females on Vanatinai have equivalent autonomy at each life cycle stage. As adults they have similar opportunities to influence the actions of others. There is a large amount of overlap between the roles and activities of women and men, with women occupying public, prestige-generating roles. Women share control of the production and the distribution of valued goods, and they inherit property. Women as well as men participate in the exchange of valuables, they organize feasts, they officiate at important rituals such as those for yam planting or healing, they counsel their kinfolk, they speak out and are listened to in public meetings, they possess valuable magical knowledge, and they work side by side in most subsistence activities. Women's role as nurturing parent is highly valued and is the dominant metaphor for the generous men and women who gain renown and influence over others by accumulating and then giving away valuable goods.

But these same characteristics of respect for individual autonomy, role overlap, and public participation of women in key subsistence and prestige domains of social life are also possible in large-scale industrial and agricultural societies. The Vanatinai example suggests that sexual equality is facilitated by an overall ethic of respect for and equal treatment of all categories of individuals, the decentralization of political power, and inclusion of all categories of persons (for example, women and ethnic minorities) in public positions of authority and influence. It requires greater role overlap through increased integration of the workforce, increased control by women and minorities of valued goods—property, income, and educational credentials—and increased recognition of the social value of parental care. The example of Vanatinai shows that the subjugation of women by men is not a human universal, and it is not inevitable. Sex role patterns and gender ideologies are closely related to overall social systems of power and prestige. Where these systems stress personal autonomy and egalitarian social relations among all adults, minimizing the formal authority of one person over another, gender equality is possible.

NOTES

1. See, for example, Rogers (1975) and Collier and Rosaldo (1981) on ideal versus real gender relations. Ortner (1984) summarizes approaches to practice; cf. Bourdieu (1977).
2. Another example is a young woman who complained at feast when the men did their ritual bathing upstream from the women. Women and men generally bathe in the same pools, either in family or same-sex groups.
3. The appropriateness of using the big man institution to define Melanesia versus a Polynesia characterized by chiefdoms, the relationship of big men to social equality, rank, and stratification, and the interactions of this form of leadership with colonialism and modernization are central issues in recent anthropological writings on big men (e.g., Brown 1987, Godelier 1986, Sahlins 1989, A. Strathern 1987, Thomas 1989, Lederman 1991). I discuss the implications of the Vanatinai case of the giagia at greater length in Lepowsky (1990b).
4. See, for example, Clifford (1983), Clifford and Marcus (1986), and Marcus and Fischer (1986) on representations. In this book I have followed my own cultural premises and not those of Vanatinai by publicly attributing thoughts, motives, and feelings to others and by trying to find the shapes in a mass of chaotic and sometimes contradictory statements and actions. But my Vanatinai friends say, characteristically, that my writing is "something of mine"—my business.

REFERENCES

Bourdieu, Pierre. 1977. *Outline of a Theory of Practice.* T. R. Nice. Cambridge: Cambridge University Press.

Brown, Paula. 1987. "New Men and Big Men: Emerging Social Stratification in the Third World, A Case Study from the New Guinea Highlands." *Ethnology* 26:87–106.

Clifford, James. 1983. "On Ethnographic Authority." *Representations* 1:118–146.

Clifford, James, and George Marcus, eds. 1986. *Writing Culture: The Poetics and Politics of Ethnography.* Berkeley: University of California Press.

Collier, Jane, and Michelle Rosaldo. 1981. "Politics and Gender in Simple Societies." In Sherry Ortner and Harriet Whitehead, eds., *Sexual Meanings: The Cultural Construction of Gender and Sexuality.* Cambridge: Cambridge University Press.

Gailey, Christine. 1980. "Putting Down Sisters and Wives: Tongan Women and Colonization." In Mona Etienne and Eleanor Leacock, eds., *Women and Colonization.* New York: Bergin/Praeger.

Godelier, Maurice. 1986. *The Making of Great Men: Male Domination and Power Among the New Guinea Baruya.* Cambridge: Cambridge University Press.

Kan, Sergei. 1989. *Symbolic Immortality: The Tlingit Potlatch of the Nineteenth Century.* Washington, D.C.: Smithsonian Institution Press.

Lederman, Rena. 1991. "'Interests' in Exchange: Increment, Equivalence, and the Limits of Bigmanship." In Maurice Godelier and Marilyn Strathern, eds., *Big Men and Great Men: Personifications of Power in Melanesia.* Cambridge: Cambridge University Press.

Lepowsky, Maria. 1990. "Big Men, Big Women, and Cultural Autonomy." *Ethnology* 29(10):35–50.

Linnekin, Jocelyn. 1990. *Sacred Queens and Women of Consequence: Rank, Gender, and Colonialism in the Hawaiian Islands.* Ann Arbor: University of Michigan Press.

Marcus, George, and Michael Fischer, eds. 1986. *Anthropology as Cultural Critique: An Experimental Moment in the Human Sciences.* Chicago: University of Chicago Press.

Ortner, Sherry. 1984. "Theory in Anthropology Since the Sixties." *Comparative Studies in Society and History* 26(1):126–166.

Rogers, Susan Carol. 1975. "Female Forms of Power and the Myth of Male Dominance: A Model of Female/Male Interaction in Peasant Society." *American Ethnologist* 2:727–756.

Sahlins, Marshall. 1989. "Comment: The Force of Ethnology: Origins and Significance of the Melanesia/Polynesia Division." *Current Anthropology* 30:36–37.

Silverblatt, Irene. 1987. *Moon, Sun, and Witches: Gender Ideologies and Class in Inca and Colonial Peru.* Princeton: Princeton University Press.

Strathern, Marilyn. 1987. "Introduction." In Marilyn Strathern, ed., *Dealing with Inequality: Analysing Gender Relations in Melanesia and Beyond.* Cambridge: Cambridge University Press.

Thomas, Nicholas. 1989. "The Force of Ethnology: Origins and Significance of the Melanesia/Polynesia Division." *Current Anthropology* 30:27–34.

15

Maria, a Portuguese Fisherwoman

BY SALLY COLE

The author's brief account of the life of Maria, a Portuguese fisherwoman, challenges the assumption that women in patriarchal cultures are uniformly oppressed. Maria's life story also questions the widely held beliefs that women are not fit to do men's work and that women cannot combine mothering and work outside the home without men's assistance.

1. What do the fisherwomen like Maria mean when they say they fish "like men?"
2. How does Maria constitute herself as an autonomous person?
3. What is the relationship between women and property in Maria's community?

"I was at the same time dona de casa (housewife) and pescador (fisherman)."

"In my childhood girls used to collect seaweed both from the beach with a hand net (ganha-pão) and from boats, using a type of rake *(gan-chola)*. It was also common for them to accompany relatives fishing. When only ten years old I began to accompany neighbors when they went fishing. When I was fourteen I took out my license and I continued to fish on boats owned by neighbors. These men are all dead now, but it was they who taught me this work.

"I married when I was only twenty years old and I think this was too young. My husband was a pescador from a neighboring parish. He came to live with me and my mother and grandmother and took up fishing in Vila Chã. I continued fishing whenever I could, and after my daughters were born I left them in the care of my mother in order to go out on the sea. I also worked at the seaweed harvest, often going out alone in the boat to collect seaweed.

"From the beginning my husband was selfish. He never helped me with my work but would instead go off to attend to his own affairs *(a vida dele)*. I married too young. We had two daughters and, when I was pregnant with the third, my husband emigrated to Brazil. He was gone for almost four years, during which time I heard nothing from him and he sent no money. I decided to go to Brazil to find him. In 1955 I went by ship with my sister-in-law who was going to join her husband, my brother, in Brazil. I found my husband involved in a life of women and drink *(amigas e bebidas)*, and after a few months I returned to Vila Chã alone. I wanted to make my life in Vila Chã and I missed my daughters. I took up full-time fishing and harvested seaweed when I wasn't fishing, and in this way I supported my mother and my children. In 1961, I bought a boat of my own and took out my skipper's license.

"I like my profession but I fished because I was forced to. My marriage had become difficult. My husband went away to Brazil, leaving me in the street *(deixando-me na rua)* with three children, and I had to face life on my own. Fishing was not as productive then as it is now *(a pesca não dava o que dá agora),* and the life of a pescadeira was difficult. But I had to turn to what I knew. First I fished in a boat belonging to another pescador, and then for eighteen years I owned and fished in my own boat *Três Marias.* About fourteen years ago I managed to buy a small house, which little by little I have fixed up, and that's where I live now.

"Although in recent years I have been the only woman skipper *(mestre),* there have been no difficulties for me at all, because I know my profession very well—as well as any of my comrades *(camaradas).* Men used to like to fish with me because they knew I was strong. C., one of my partners, used to say that I was stronger than he. Fishing holds no secrets for me and, besides, I think that women have the right to face life beside men *(acho que as mulheres têm o direito de enfrentar a vida ao lado dos homens).* What suits men, suits women *(O que serve para os homens serve para as mulheres).* I am respected by everyone, men and women. I have many friends, and when the weather prevents fishing we all stay here on the beach working on the nets and conversing. I have always enjoyed my work on the sea. I was never one who liked to stay at home *(Nunca gostei de estar em casa).*

"When my daughters were small I used to be at sea day and night—whenever there were fish. They stayed at home with my mother. Later, when they were older and I was fishing, my daughters assisted my mother harvesting seaweed, and in this way they contributed to the maintenance *(o sustento)* of the household. As soon as I returned from fishing I would start the housework. You see, I was at the same time housewife and fisherman *(Olhe, eu era ao mesmo tempo dona de casa e pescador).*

"I retired in 1979. I sold my boat and I gave my fishing gear to my son-in-law. I sold my boat to a fisherman in Matosinhos because I could not bear to see it anymore here on the beach. In 1982 I bought a piece of land in [Lugar do] Facho, and two of my daughters are building a duplex there now. My third daughter and her husband and three children live with me in my house. I have helped all of my daughters establish their households. I have been very good to them. Now that I am old and my heart is not good, they are looking after me. When I returned from Brazil leaving my husband there, I could have found another man to live with. I couldn't marry but I could have lived with another man. But I never wanted to do that because, if things didn't work out with us, I worried that he would take it out on my daughters because they were nothing to him. I preferred to have my daughters *(Antes quis as minhas filhas).*

"Recently, my husband has begun writing to me from Brazil. He wants to return to Portugal and he wants me to take him back. He needs someone to care for him now in his old age. But I won't take him back. It's not right at all *(Não tem jeito nenhum).* I liked him once but that's all over now. The best part of the life of a couple is passed. I'm not interested in his returning. I'm not an object to be put away and then picked up, dusted off, and used again *(Não sou um objecto p'ra deixar e depois retornar e limpar e usar mais uma vez).* I am not an object. I am a human being. I have the right to be a human being, don't you think? I managed to make a life for myself here, but he has arranged nothing for himself there, nothing *(não arranjou nada lá, nadinha).* He arranged nothing here, but he also arranged nothing there. He has never done anything for me or for my daughters and now he wants to come back. Who does he think he is? I am not crazy *(Não sou tola).* He has no right whatsoever *(Ele não tem direito nenhum).*"

Maria's narrative illustrates women's economic autonomy and women's strategies in the maritime households. Maria could support herself, her mother, and her three children without the help of her husband or of any other man because fishing was locally perceived to be an occupation suitable for women as well as for men. Having been left by her husband "in the street" with nothing, Maria recognized the importance of owning property. Maria bought her own boat and gear and later a house and a piece of land. She solidified her economic security by investing in property and in her relationship with her daughters. When she made the decision to raise her daughters on her own, Maria consciously invested in her relationship with her daughters to provide for herself in her old age. And, having supported herself and her children, she is clear about her rights as a person and as a woman. Maria has worked side by side with men all her life; she has earned the respect of the pescadores of Vila Chã, and she will not now subordinate herself to a man. This is why, now that her estranged husband is trying to claim her services to care for him in his old age, she is refusing. Her reasons are found in the strong self-image she developed through a life of hard work and independence. As she sees it: "I am not an object to be put away and then picked up, dusted off, and used again. I am a human being."

Maria's story further confirms the existence of a local understanding that such vocations as those of fisherman or skipper were masculine roles but could be filled by women as well as men. Maria's case illustrates how gender, rather than being static or predetermined, is on the contrary negotiated through social roles. Maria has drawn on existing gender ideas in local society and manipulated them. She adopted a masculine self-image: she is the only woman of her generation who wears trousers. Maria wears the typical dress of a maritime woman—a skirt, apron, slippers, wool socks, a head scarf, and a wool shawl—but she also wears trousers under her skirt. She is large and muscular and walks with a masculine stride. Maria speaks with a man's self-assurance and lack of restraint (although it must be said that women of maritime households generally are outspoken and assertive). Maria's presentation of self and her assumption of masculine rights (exemplified in her refusal to take back her spouse) suggest that she has negotiated a masculine gender identity or perhaps a third-gender identity in the local context. Certainly, her self-perception and behavior do not resemble the stereotype of women in the code of honor and shame.

Thus, in Vila Chã, there existed cultural definitions of men's work and women's work: men fished and women worked onshore. The social reality of the division of labor, however, was that both men and women fished. Women who fished described themselves as "like men," but, as Lucília explained, "there wasn't anything special" about women fishing. Fishing was socially constructed as a masculine role, but either sex could, and did, do the work. Women like Maria . . . manipulated social constructions of gender in order to maximize their economic autonomy, that is, their decision-making authority, access to resources, and control of their own labor. Fishing "like men" was one strategy that some women employed.

Topics for Further Examination

CHAPTER 1

- Look up the intersex movement, including the most recent research on intersex. Also, check out some of the following Web sites: http://www.tgforum.com/ and http://www.isn.org/.
- Locate research on gender bending in the arts (e.g., performance art, literature, music videos).
- Find research articles that demonstrate how gender is socially constructed in everyday life.

CHAPTER 2

- Using Infotrac, check out the most recent work of Patricia Hill Collins, Yen L. Espiritu, and Alfredo Mirandé. (Use parentheses around their names and ask for referred journals only.)
- Do a Web search using "feminist theory" and another category of difference and inequality, i.e., "feminist theory" and "race."
- Using the Web, locate information on those cited in the introduction to Chapter 2: Audre Lorde, bell hooks, Angela Davis, Gloria Anzaldua, Sojourner Truth. When doing so, try to find the names of others who challenged the whiteness of the women's movement.

CHAPTER 3

- Locate research on the Hijras of India and the Fa'afafines of Samoa.
- Find research articles on genital cutting and its meanings and consequences in different societies today. Check out the following Web site: http://www.fgmnetwork.org/.
- Look up studies of the egalitarian relationship structure of pre-contact Iroquoian societies of Northeastern North America.

CHAPTER 4

Learning and Doing Gender

We began this book by discussing the shaping of gender in Western and non–Western cultures. Part Two expands upon the idea of prisms by examining the patterns of gendered experiences that emerge from the interaction of gender with other socially constructed prisms. As multiple patterns are created by the refraction of light as it travels through a kaleidoscope containing prisms, so too are the patterns of individuals' life experiences influenced by gender and other social prisms discussed in Part One.

GENDERED PATTERNS

Social patterns are the center of social scientists' work. Schwalbe (1998), a sociologist, defines social patterns as "a regularity in the way the world works" (101). For example, driving down the "right" side of the street is a regularity people appreciate. You will read about different gendered patterns in Part Two, many of which are regularities you will find problematic because they deny the individuality of women and men. Clearly, there are exceptions to social patterns; however, these exceptions are in the details, not in the regularity of social behavior itself (Schwalbe, 1998).

A deeper understanding of how and why particular social patterns exist helps us to interpret our own behavior and the world around us. Gender, as we discussed in Part One, is not a singular pattern of masculinity or femininity that carries from one situation to another. Instead it is complex, multifaceted, and everchanging depending upon the social context, whom we are with, and where we are. Our behavior in almost all situations is framed within our knowledge of ideal gender—

hegemonic masculinity and emphasized femininity—as discussed in the Introduction.

Keep the concepts of hegemonic masculinity and emphasized femininity in mind as we examine social patterns of gender. To illustrate this, let's return to the stereotype discussed in the Introduction—that women talk more than men. We know from research that the real social pattern in mixed-gender groups is that men talk more, interrupt more, and change the topic more often than women (Wood, 1999; Anderson and Leaper, 1998). The stereotype, while trivializing women's talk and ignoring the dominance of men in mixed-gender groups, maintains the patterns of dominance and subordination associated with hegemonic masculinity and emphasized femininity, influencing women's as well as men's behaviors. In Chapter 5, Jean Kilbourne describes how girls often receive encouragement to be silent, use a nice voice, and not talk too much. Later, as they grow older and join mixed-gender groups at work or play, women's voices are often ignored and women are subordinated as they monitor what they say and how often they talk, and check to make sure they are not dominating the conversation. By examining how these idealized versions of masculinity and femininity pattern daily practices, we can understand better the patterns and meanings of our behavior and the behaviors of others.

Gendered patterns of belief and behavior influence us throughout our lives, in almost every activity in which we engage. Readings in Part Two describe gendered patterns in work and play (Chapter 7) and in daily intimate relationships with family and friends (Chapter 8). We also explore how gendered patterns affect our bodies, sexualities, and emotions (Chapter 6), and how patterns of dominance, control, and violence enforce gender patterns (Chapter 9). The patterns that emerge from the gender kaleidoscope are not unique experiences in individual lives; they occur in many people's lives. Institutions and groups enforce certain types of gendered relationships in the home, workplace, and daily life as described in the readings throughout Part Two. These patterns overlap and reinforce gender differences and inequalities. For example, gender discrimination in wages affects families' decisions about parenting roles and relationships. Since most men still earn more than most women, the choices of families who wish to break away from traditional gender patterns are limited (see Chapters 7 and 8).

LEARNING GENDER

We begin this part of the book by examining the processes by which we acquire self-perceptions and behaviors that fit our culture's patterns of masculinities and femininities (Chapters 4 and 5). The readings in this chapter emphasize that, regardless of our inability or unwillingness to attain idealized femininity and masculinity, almost everyone in a culture learns what idealized gender is and organizes their lives around those expectations. The term sociologists use to describe how we learn gender is "gender socialization," and sociologists approach it from a variety of different perspectives (Coltrane, 1998).

Socialization is the process of teaching new members of a society the ways of the larger culture. Socialization takes place in all interactions and situations, with families and schools typically having formal responsibility for socializing new members in Western societies. Early attempts to explain gender socialization gave little attention to the response of individuals to agents of socialization such as parents, peers, and teachers. There was an underlying assumption in this early perspective that individuals were blank tablets (tabulae rasae) upon which the cultural definitions of gender and other appropriate behaviors were written. This perspective assumed that, as individuals developed, they took on a gender identity appropriate to their biological sex category (Howard and Alamilla, 2001).

Social scientists now realize that individuals are not blank tablets; gender socialization isn't just something that is "done" to us. Theorists now describe socialization into gender as a series of complex and dynamic processes. Individuals create, as well as respond to, social stimuli (Howard and Alamilla, 2001; Carlton-Ford and Houston, 2001). For example, the Urla and Swedlund reading in Chapter 5 discusses children's reactions to and interpretations of gender-specific toys such as Barbie. Moreover, socialization doesn't simply end after childhood. Socialization is a process that lasts throughout one's life, from birth to death (Bush and Simmons, 1981). Throughout our lives, we assess cues around us and behave as situations dictate; gender is a key factor in determining what is appropriate.

SOCIALIZING CHILDREN

There are many explanations for why children gravitate toward gender-appropriate behavior. It is not just family members who teach children to behave as "good boys" or "good girls." Almost every person a child comes into contact with, and virtually all aspects of a child's material world (e.g., toys, clothing), reinforce gender. It is not long before most children come to understand that they are "boys" or "girls" and segregate themselves accordingly. Adults also help segregate children by gender, such as teachers who separate children into gender-segregated spaces in lunch lines or playground areas (Sadker and Sadker, 1994; Thorne, 1993).

Most children quickly understand the gender-appropriate message directed toward them and behave accordingly. Although not all boys are dominant and not all girls are subordinate, studies in a variety of areas find that most white boys tend toward active and aggressive behaviors, while most white girls tend to be quieter and focus on relationships. These patterns have been documented in schools and in play (e.g., Sadker and Sadker, 1994; Thorne, 1993; Ferguson, in this chapter). The consequences for gender-appropriate behavior are considerable. Gender-appropriate behavior is related to lower self-confidence and self-esteem for girls (e.g., Spade, 2001; Eder, 1995; Orenstein, 1994) whereas boys are taught to "mask" their feelings and compete with everyone for control, thus isolating themselves and ignoring their own feelings (e.g., Connell, 2000; Pollack, 2000; Messner, 1992).

The dominant pattern of gender expectations, the "pink and blue syndrome" described in Chapter 1, begins at birth. Once external genital identification takes place, immediate expectations for masculine and feminine behavior follow. Exclamations of "he's going to be a great baseball (or football or soccer) player" and "she's so cute" are accompanied by gifts of little sleepers in pink or blue with gender-appropriate decorations. Try as we might, it is very difficult to find gender-neutral clothing for children (see Nelson in Chapter 5). These expectations, and the way we treat young children, reinforce idealized gender constructions of dominance and subordination.

SCHOOLS AND SOCIALIZATION

Schools reinforce separate spheres for boys and girls (Orenstein, 1994; Sadker and Sadker, 1994; Thorne, 1993). Considerable research by the American Association of University Women (1992, 1998, 1999) documents how schools "short-change" girls. Schools are social institutions that maintain patterns of power and dominance. In her reading, Dorothy Smith argues that schools are primary vehicles to organize and perpetuate inequality by gender as well as race and class. Smith's emphasis is on the structure of schools and how it serves to limit young women's agency. Bernice Sandler, in another reading in this chapter, lists many patterns in schools and organizations that put and keep girls and women in their "place," thus maintaining hegemonic masculinity and emphasized femininity. She gives examples of patterns of interaction in which supervisors or teachers and peers silence women and girls in classrooms or at work. In essence these adults tell them that what they say is not important. By allowing boys and men to interrupt and cut off women and girls, Sandler argues that we teach them dominance, a pattern that extends far beyond the immediate situation.

In effect, the structure of society pronounces boys and girls as different and teaches them how to behave accordingly. For example, Ellen Goodman, in this chapter, provides a commentary on some of the "cues" recently directed toward college women. In Chapter 5 you will read more about how capitalist societies reinforce and maintain gender for children and adults. Television, music, books, clothing, and toys differentiate and prescribe appropriate behavior for girls and boys. For example, studies of children's books find some distinctive patterns that reinforce idealized forms of gender. One study of children's readers from the early 1970s found that while Jane looked on, Dick did exciting and interesting things. Stories about boys outnumbered those about girls by 5 to 2 in that study, with boys engaged in adventures. Any adventures or discoveries on the part of girls were attributed to an accident or luck (Women on Words and Images, 1974). This research was the beginning of careful analyses of how the books American children read socialize children into gender patterns. Also, a recent study of award-winning books from 1995 to 1999 found that although boys and girls were equally represented as main characters, portrayals of male characters were likely to be dominant, while female characters were likely to be subordinate. Although other evidence also supported the depiction of traditional gender patterns, on the plus side the researchers found that girls and women are more likely to be portrayed in gender-atypical roles in many recent children's books (Gooden and

Gooden, 2001). Unfortunately, boys and men continue to be depicted in gender-typical roles in these recent, award-winning children's books (Gooden and Gooden, 2001).

However, not all boys and men are allowed to be dominant across settings (Eder, 1995). Ann Arnett Ferguson, in this chapter, describes how schools discourage African-American boys from claiming their blackness and masculinity. Although white boys may be allowed to be "rambunctious" and disrespectful, African-American boys are punished more severely than white peers when they "act out." Girls also exist within a hierarchy of relationships (Eder, 1995). Girls from racial, ethnic, economically disadvantaged, or other subordinated groups must fight even harder to succeed under multiple systems of domination and inequality in schools (Chapter 2). Bettie (2002) compared the paths to success for upwardly mobile white and Mexican high school girls and found some similarities in gender experiences at home and school that facilitated mobility, such as participation in sports. There were also differences in their experiences because race was always salient for the Mexican-American girls. However, Bettie (2002) believes that achieving upward mobility may be easier for these Mexican girls than their brothers because it is easier for them to transgress gender boundaries. Their brothers, on the other hand, feel pressure to "engage in the rituals of proving masculinity" (Bettie, 2002:419; Mirande, Chapter 2), which often lead to trouble similar to that described by Ferguson in this chapter.

Bettie's (2002) study emphasizes the fact that multiple social prisms of difference and inequality create an array of patterns, which would not be possible if gender socialization were universal. Individuals' lives are constructed around many factors, including gender. Barajas and Pierce (2001) described how successful Latina college students had to find paths by which they could succeed in college while at the same time maintaining a positive sense of their own ethnic identity. These patterns of interaction across social prisms emerge around the same goal—the maintenance of hegemonic masculinity and emphasized femininity. The process of gender socialization is rooted in the principle that people are not equal and that the socially constructed categories of difference and inequality (gender, race, ethnicity, class, religion, age, culture, etc.) are legitimate.

SPORTS AND SOCIALIZATION

Sports, particularly organized sports, provide other examples of how institutionalized activities reinforce the gender identities that children learn. Boys learn the meaning of competition and success, including the idea that winning *is* everything (e.g., Messner, 1992). Girls, on the other hand, often are found on the edge of the playing field, or on the sides of the playgrounds, watching the boys (Thorne, 1993). Yet not all children play in the same ways. Goodwin (1990) finds that children from urban, lower-class, high-density neighborhoods, where households are closer together, are more likely to play in mixed-gender and mixed-age groups. In suburban middle-class households, which are farther apart than urban households, parents are more likely to drive their children to sporting activities or houses to play with same-gender, same-age peers. The consequences of social class and place of residence are that lower-class children are more comfortable with their sexuality as they enter preadolescence and are less likely to gender segregate in schools (Goodwin, 1990).

GENDER TRANSGRESSIONS

Children learn to display gender-appropriate behavior; however, there are times when children step out of gender-appropriate zones as described by C. Shawn McGuffey and B. Lindsey Rich in this chapter. The patterns they found are similar to the trend found in children's books of the late 1990s, that girls and women are more likely to transgress and do masculine things than boys and men are to participate in feminine activities. McGuffey and Rich's reading helps us to understand that hegemonic masculinity is complex and constantly negotiated, even among children at play. They find that girls who transgress into the "boys' zone" may eventually be respected by their male playmates if they are good at conventionally male activities such as playing baseball. Boys, however, are harassed and teased when they try to participate in any activity associated with girls. By denying boys access to girls' activity, the dominance of hegemonic masculinity is maintained, even when boys are ridiculed because they "throw like a girl."

As you can see, learning gender is complicated. Clearly, gender is something that we "do" as well as learn, and in doing gender we are responding to structured expectations from institutions in society. Every time we enter a new social situation, we look around for cues and guides to determine how to behave in a gender-appropriate manner. In some situations, we might interpret gender cues as calling for a high degree of gender conformity, while in other situations, the clues allow us to be more flexible. We create gender as well as respond to expectations for it.

DOING MASCULINITY AND FEMININITY THROUGHOUT OUR LIVES

Most men have learned to "do" the behaviors that maintain hegemonic masculinity, while at the same time suppress feelings and behaviors that might make them seem feminine (Connell, 1987). Elizabeth Gilbert, a white woman journalist, helps us to understand how adult Americans define and enact masculinity and femininity. In her article in this chapter, she tells how she learned to see the world through the eyes of men by becoming a "drag king." She became a man in appearance, physical posturing, and cognitive and emotional approaches to others. In doing so, she discovered how hegemonic masculinity limits one's identity, including how she felt about herself and interacted with others. The ease with which she was able to "become a man" points to the relational nature of gender. That is, by learning how to be a woman she also learned how the "opposite" gender, men, must act.

As you can see, hegemonic masculinity is maintained in a hierarchy that is realized by only a few men, with everyone else subordinated to them—women, poor white men, men of color, gay men, and men from devalued ethnic and religious groups. Furthermore, this domination is not always one-on-one, but can be institutionalized in the structure of the situation, as Ferguson illustrates for African-American boys and as Smith and Sandler argue is the case for girls in schools. As you read these and other articles in this chapter, you will see that gender is not something that we learn once in one setting, such as an inoculation or shot for rabies. Instead, we learn to do gender over time in virtually everything we undertake.

Moreover, learning to do gender is complicated by the other prisms that interact in our lives. Recall the lessons from Chapter 2 and remember that gender does not stand alone, but rather is reflected in other social identities. The last reading in this chapter, by Leora Tanenbaum, illustrates the intersection of gender with sexuality. Just as race was used to reinforce status hierarchies in the reading by Ferguson on African-American boys, sexuality is also used to reinforce and control individuals' behaviors. By determining what is "right" and what is not, sexuality keeps most individuals neatly subordinated within the framework of hegemonic masculinity and emphasized femininity.

It is not easy to separate the learning and doing of gender from other patterns. As you read selections in other chapters in this part of the book, you will be able to see the influence of social processes and institutions on how we learn and do gender across all aspects of our lives. Before you start to read, ask yourself how you learned gender and how well you do it. Not succeeding at doing gender is normal. That is, if we all felt comfortable with ourselves, no one would be striving for idealized forms of gender—hegemonic masculinity or emphasized femininity. Imagine a world in which we all felt comfortable with who we are! As you read through the rest of this book, ask yourself why that world doesn't exist.

REFERENCES

American Association of University Women. (1999). *Voices of a generation: Teenage girls on sex, school, and self.* Washington, DC: American Association of University Women Educational Foundation.

———. (1998). *Gender gaps: Where schools still fail our children.* Washington, DC: American Association of University Women Educational Foundation.

———. (1992). *How schools shortchange girls.* Washington, DC: American Association of University Women Educational Foundation.

Anderson, Kristin J. and Campbell Leaper. (1998). Meta-analysis of gender effects on conversational interruption: Who, what, when, where, and how. *Sex Roles* 39(3–4): 225–52.

Barajas, Heidi Lasley and Jennifer L. Pierce. (2001). The significance of race and gender in school success among Latinas and Latinos in college. *Gender & Society* 15(5): 859–878.

Bettie, Julie. (2002). Exceptions to the rule: Upwardly mobile white and Mexican American high school girls. *Gender & Society* 16(3): 403–422.

Bush, Diane Mitsch and Roberta G. Simmons. (1981). Socialization processes over the life course. In *Social psychology: Sociological perspectives,* Morris Rosenberg and Ralph H. Turner (eds.), pp. 133–164. New York: Basic Books.

Carlton-Ford, Steve and Paula V. Houston. (2001). Children's experience of gender: Habitus and field. In *Gender mosaics: Societal perspectives,* Dana Vannoy (ed.), pp. 65–74. Los Angeles: Roxbury Publishing Company.

Coltrane, Scott. (1998). *Gender and families.* Thousand Oaks, CA: Pine Forge Press.

Connell, R. W. (2000). *The men and the boys.* Berkeley: University of California Press.

———. (1987). *Gender and power: Society, the person, and sexual politics.* Stanford, CA: Stanford University Press.

Eder, Donna. (1995). *School talk: Gender and adolescent culture.* New Brunswick, NJ: Rutgers University Press.

Gooden, Angela M. and Mark A. Gooden. (2001). Gender representation in notable children's picture books: 1995–1999. *Sex Roles* 45(1/2): 89–101.

Goodwin, M. H. (1990). *He-said-she-said: Talk as social organization among black children.* Bloomington: Indiana University Press.

Howard, Judith A. and Ramira M. Alamilla. (2001). Gender and identity. In *Gender mosaics: Societal perspectives,* Dana Vannoy (ed.), pp. 54–64. Los Angeles: Roxbury Publishing Company.

Messner, Michael A. (1992). *Power at play: Sports and masculinity.* Boston: Beacon Press.

Orenstein, Peggy. (1994). *School girls: Young women, self-esteem, and the confidence gap.* New York: Anchor Books.

Pollack, William S. (2000). *Real boys' voices.* New York: Penguin Putnam.

Sadker, David and Myra Sadker. (1994). *Failing at fairness: How our schools cheat girls.* New York: Simon & Schuster.

Schalbe, Michael. (1998). *The sociologically examined life: Pieces of the conversation.* Mountain View, CA: Mayfield Publishing Company.

Spade, Joan Z. (2001). Gender and education in the United States. In *Gender mosaics: Societal perspectives,* Dana Vannoy (ed.), pp. 85–93. Los Angeles: Roxbury Publishing Company.

Thorne, Barrie. (1993). *Gender play: Girls and boys in school.* New Brunswick, NJ: Rutgers University Press.

Women on Words and Images. (1974). Look Jane look. See sex stereotypes. In *And Jill came tumbling after: Sexism in American education,* Judity Stacey, Susan Bereaud, and Joan Daniels (eds.), pp. 159–177. New York: Dell Publishing Co., Inc.

Wood, Julia T. (1999). *Gendered lives: Communication, gender, and culture* (3rd ed.). Belmont, CA: Wadsworth.

16

Playing in the Gender Transgression Zone

Race, Class, and Hegemonic Masculinity in Middle Childhood

BY C. SHAWN MCGUFFEY AND B. LINDSAY RICH

C. Shawn McGuffey and B. Lindsay Rich use their sociological radar to examine the ways 5- to 12-year-olds from various racial and ethnic backgrounds create and maintain gender patterns in a summer day camp. The first author was a counselor at this camp and spent over nine weeks during the summer of 1996 observing children's play. He kept daily logs and met weekly with the second author to discuss his observations. He also interviewed 22 children from the camp and six parents to get a deeper understanding of the ways children constructed gendered meanings and activities. Their conceptualization of the gender transgression zone helps us to understand how gendered boundaries are maintained and violated by young children.

1. What do the authors mean by the "organization of homosocial status systems"?
2. How easy is it for boys and girls across racial and class backgrounds to enter the gender transgression zone?
3. How does hegemonic masculinity combine with boys' racial and class backgrounds to affect their ability to transgress gender?

By now, R. W. Connell's concept of "hegemonic masculinity" has wide currency among students of gender.[1] The concept implies that there is a predominant way of doing gender relations (typically by men and boys, but not necessarily limited to men and boys) that enforces the gender order status quo: It elevates the general social status of masculine over feminine qualities and privileges some masculine qualities over others. The notion that "masculinities" and "femininities" exist and can be interrogated as negotiated realities allows us to further our understanding about the larger gender order in which they are embedded.

We want to caution, however, against the temptation to overgeneralize the concept of hegemonic masculinity. To do so runs the risk of glossing the modalities, both historical and social-spatial (in terms of class, ethnoracial, sexual, and age variations), in which hegemonic masculinity emerges. We believe that hegemonic masculinity, while having general qualities as a form of social power, may take on many valences and nuances, depending on the social setting and the social actors involved. Connell (1987, 1995, 36–37) is himself careful to make the sorts of qualifications we make here while similarly claiming the general analytic utility of the concept. We agree with advocates of the concept that it indeed gives us great theoretical leverage and explanatory power toward clarifying and refining how and why men's dominance works at higher levels of social organization, perhaps even at the global level (Connell 1995; Hawkesworth 1997). In this article, we provide evidence of how hegemonic masculinity is manifest in middle childhood play and used to re-create a gender order among children wherein the larger social relations of men's dominance are learned, employed, reinforced, and potentially changed. Specifically, we present and discuss the results of a preliminary participant observation study of microlevel processes of gender boundary negotiation in middle childhood (ages 5–12).

Providing empirical evidence about the ways in which boys and girls negotiate gender relations within specific social contexts can further understanding about why gender relations take the forms they do in childhood. Using the concept of hegemonic masculinity as a heuristic tool, we decided to focus on how gender relations—specifically, the enactment of masculine hegemony within these relations—were "done" (West and Zimmerman 1987). We eschew the notion that men and women (or boys and girls) merely enact "sex roles" as handed-down scripts. Rather, while acknowledging structural gender socialization implied by the concept of role, we focus on the ways in which the relations between girls and boys are negotiated (Connell 1987, 1995; Messner 1998).

* * *

[W]hen we refer to specific children in this study, we designate ethnoracial, gender, and age distinctions as follows: W = White, B = Black, A = Asian, B = Boy, G = Girl, and a number representing the age of the child. For example, an African American girl who is seven years old will be represented as (BG7). If a child is of another racial category other than white, Black, or Asian, his or her specific classification will be marked accordingly.

ORGANIZATION OF HOMOSOCIAL STATUS SYSTEMS

When examined as two separate social groups, boys and girls organize themselves differently based on distinct systems of valuing. We must reiterate that masculinity and femininity are not bipolar or opposites but are rather "separate and relatively independent dimensions" (Absi-Semaan, Crombie, and Freeman 1993, 188). Gender is a social construction that is constantly being modified as individuals mature. What may be gender appropriate at one stage in life may be gender inappropriate at a later stage. Boys, for example, are free to touch each other affectionately in early middle childhood, but this is subsequently stigmatized, with a few exceptions (such as victory celebrations). As "independent dimensions," one can develop a clearer view of masculinity and femininity by studying how they differ in context to intragender (homosocial) relations and then how they interact in intergender (heterosocial) relations. Homosocial relationships—nonsexual attractions held by members of the same sex—define how heterosocial relationships are maintained. Thus, it is essential to understand how boys and girls organize themselves within each homosocial group to understand how they negotiate boundaries between the two (Bird 1996).[2]

Structural Formation of Boys in Middle Childhood

Boys in middle childhood organize themselves in a definite hierarchical structure in which the high-status boys decide what is acceptable and valued—that which is hegemonically masculine—and what is not. A boy's rank in the hierarchy is chiefly determined by his athletic ability. Researchers have identified sports as a central focus in boys' development (Fine 1992; Messner 1992, 1994). Boys in this context were observed using words such as *captain, leader,* and various other ranking references, even when they were not playing sports. Messner (1994, 209) explains the attraction of sports in hegemonic masculinity as a result of young males finding the "rulebound structure of games and sports to be a psychologically 'safe' place in which [they] can get (nonintimate) connection with others within a context that maintains clear boundaries, distance, and separation from others." Sharon Bird (1996) identifies three characteristics in maintaining hegemonic masculinity: emotional detachment, competitiveness, and the sexual objectification of women, in which masculinity is thought of as different from and better than femininity. As another essential feature of hegemonic masculinity, we want to add to these characteristics the ability to draw attention to one's self. Because hegemony is sustained publicly, being able to attract positive attention to one's self is vital. The recognition a boy receives from his public performance of masculinity allows him to maintain his high status and/or increase his rank in the hierarchy.

Conflicts and disagreements in the boys' hierarchy are resolved by name-calling and teasing, physical aggression, and exclusion from the group. These forms of aggression structure and maintain the hierarchy by subordinating alternate propositions and identities that threaten hegemonic masculinity. Although direct and physical aggression are the most physically damaging, the fear of being exiled from the group is the most devastating since the hierarchy confirms masculinity and self-worth for many young boys. According to Kaufman (1995, 16), the basis for a hegemonic masculinity is "unconsciously rooted before the age of six" and "is reinforced as the child develops." Lower-status boys adhere to the hegemonic rules as established by the top boys even if they do not receive any direct benefits from the hierarchy within the homosocial context. The overwhelming majority of boys support hegemonic masculinity in relation to subordinated masculinities and femininities because it not only gives boys power over an entire sex (i.e., girls), but it also gives them the opportunity to acquire power over members of their own sex. This helps maintain the hierarchical frame by always giving boys—even low-status boys—status and power over others. Connell (1987, 183) states that hegemonic masculinity "is always constructed in relation to various subordinated masculinities as well as in relation to women." Connell (1995, 79) describes this panmasculine privilege over girls and women as the "patriarchal dividend." Hegemonic masculinity is publicly used to sustain the power of high-status boys over subordinate boys and boys over girls.

Emotional detachment, competitiveness, and attention arousal could be witnessed in any game of basketball. High-status boys in our study generally performed the best and always distinguished themselves after scoring points. Three high-status boys demonstrate this particularly well. After scoring, Adam[3] (WB11) usually jumped in the air, fist in hand, and shouted either, "In your face!" or "You can't handle this!" Brian's (BB11) style consisted of a little dance followed by, "It's all good and it's all me!" Darrel (BB11) also had a shuffle he performed and ended his routine with, "You can't handle my flow!" or "Pay attention and take notes on how a real 'G' [man] does it." These three are also the most aggressive, often times running over their own teammates. By constantly displaying their athletic superiority, these high-status boys are validating their position and maintaining separation from lower-status boys. Most boys usually did some "attention getting" as well when they scored.

The sexual objectification of women can easily be seen in boys' homosocial interactions. Sexually degrading remarks by boys about women and girls at the pool were common; harassment by young boys occasionally occurred. In one instance at a nearby swimming pool, an adolescent girl, approximately 16 years old, was on her stomach sunbathing with the top portion of her bikini unfastened. Adam (WB11)—the highest-ranked boy—walked over to the young lady and asked if he could put some tanning lotion on her back. After she refused his offer, Adam—with a group of boys urging him on—poured cold water on her back, causing her to instinctually raise up and reveal her breasts. While he was being disciplined, the other boys cheered him on, and Adam smiled with pride. In "The Dirty Play of Little Boys," Gary Fine (1992, 137) argues that "given the reality that many talkers have not reached puberty, we can assume that their sexual interests are more social than physiological. Boys wish to convince their peers that they are sexually mature, active, and knowledgeable" and, we might add, definitely heterosexual.

Despite the fact that there were definite racial and class differences in the boys' hierarchy, these factors had surprisingly little consequence for rankings in the power structure. Black and/or economically disadvantaged boys were just as likely to hold high positions of authority as their white and/or middle-class counterparts. Though a white middle-class boy (Adam) was the highest-ranked youth in the boys' hierarchy, two poor Black youths (Brian and Darrel) held the second and third positions in the hierarchy. Furthermore, when Adam went on a two-week vacation with his family, Darrel surpassed Brian and assumed the alpha position in the boys' social order. Nonetheless, upon his return, Adam reasserted his dominance in the group.

Structural Formation of Girls in Middle Childhood

Girls' homosocial organizational forms are distinct from boys'. The tendency toward a single hierarchy, for example, is quite rare. Social aggression (e.g., isolating a member of the group) is used to mark boundaries of femininity. These boundaries do not seem to involve a singular notion of hegemonic femininity with which to subordinate other forms or to heighten public notice of a higher-status femininity. Girls' boundaries are less defined than boys'. The girls in our study generally organized themselves in small groups ranging from two to four individuals. These groups, nonetheless, usually had one girl who was of higher status than the other girls in the clique. The highest-status girl was generally the one considered the most sociable and the most admired by others in the immediate clique as well as others in the camp. Much as Luria and Thorne (1994, 52) observed, the girls were connected by shifting alliances. Girls deal with personal conflicts by way of exclusion from the group and social manipulation. Social manipulation includes gossiping, friendship bartering, and indirectly turning the group against an individual. Contrary to the findings of many sociologists and anthropologists who only characterize aggression in physical aspects, we—like Kaj Bjorkqvist (1994) in "Sex Differences in Physical, Verbal, and Indirect Aggression"—found that girls display just as much aggression as boys but in different ways. When Elaine (WG8), for example, would not share her candy with her best friends Brandi (WG8) and Darlene (WG7), Brandi and Darlene proclaimed that Elaine could no longer be their friend. Elaine then joined another group of girls. This is quite representative of what happens when one girl is excluded from a clique. To get back at Brandi and Darlene, Elaine told her new "best friends" that Brandi liked Kevin (biracial B11) and that Darlene urinated on herself earlier that day. This soon spread throughout the camp, and Darlene and Brandi were teased for the rest of the day, causing them to cry. As Bjorkqvist (1994, 180) suggests, there is no reason to believe that girls are any less aggressive than boys. In fact, social manipulation may be more damaging than physical aggression because though physical wounds heal, gossip and group exclusion can persist eternally (or at least until the end of summer).

Unlike the boys who perform or comply with a predominant form of masculinity, no such form of hegemonic femininity was observed. Connell (1987) explains the lack of a hegemonic form of femininity as the result of the collective subordination of women to the men's homosocial hierarchy. According to Connell, since power rests in the men's (boys') sphere, there is no reason to form power relations over other women (girls). Hence, "no pressure is set up to negate or subordinate other forms of femininity in the way hegemonic masculinity must negate other masculinities" (Connell 1987, 187).

Girls were inclined to gather in different groups, or cliques, reflective of various ways to define femininity; they gathered with those who defined their girlhood on the same terms. Just as Ann Beutel and Margaret Marini (1995, 436) discovered in their work, "Gender and Values," girls in our study also formed girl cliques "characterized by greater emotional intimacy, self-disclosure, and supportiveness." Intimacy helps ensure faithfulness to the group. All the girl groups—regardless of racial makeup, socioeconomic background, or age differences—had an idea of being "nice," which enhanced clique solidarity. Various girls were asked to give a definition of what it meant to be nice: "Nice just means, you know, helping each other out" (BG12); "Nice just means doing the right thing" (BG7); "Nice means . . . getting along" (WG11). Despite this notion of being nice, however, being nice in one group may be seen as being mean in another. In some groups, for example, it was considered nice for one girl to ask another if the former could have some of the other's chips at lunch. In others, though, this was considered rude; the nice, or proper, conduct was to wait until one was offered some chips. Nice was relative to the particular group. Being nice among the girls observed in this study generally entailed sharing, the aversion of physical and direct aggression, and the avoidance of selfish acts.

The organization of African American girls was somewhat unique. As mentioned, campers were divided into four age groups. In each age group, there were no more than four or five Black girls. Within these age groups, African American girls had the same structural patterns as Caucasians—small cliques of two to four in which a person may drift from clique to clique at a given time. There were no problems with the Black girls mixing with the white girls in age groups or organized activities.

During times when the campers were not restricted to specific groups (e.g., snack time, most field trips, at the pool, group games, and free time), however, the preponderance of African American girls gravitated to each other, despite age differences. This differs from previous research that notes that children in middle childhood associate with near-aged members (Absi-Semaan, Crombie, and Freeman 1993; Andersen 1993; Beutel and Marini 1995; Block 1984; Curran and Renzetti 1992; Luria and Thorne 1994). The first author also visited another camp with similar demographics and observed a similar lack of age segregation among African American girls. African American girls formed larger groups and occupied more space than white girls.

In general, Black girls were more assertive and therefore less likely to be bothered by boys. A loose hierarchy formed in which the older girls made most of the decisions for the younger ones in the group. This hierarchy was by no means hegemonic as in the boys' hierarchy. Rather, this hierarchy used a communal approach to decision making, with the older girls working to facilitate activities for the group. This process was illustrated every day as this group of girls decided which activity they would participate in at free time. The oldest girls—Brittany (BG11), Alexia (BG12), and Melanie (BG11)—would give options such as arts and crafts, checkers, basketball, and jump roping for the group to choose from. After considering all the options—taking into account what they had played the day before, the time left to participate in the activity, and the consensus of the group—the older girls indirectly shifted the focus to a particular activity that seldom received objection from the younger girls in the clique. . . .

Moreover, whereas the boys displayed little class segregation, the girls were clearly marked by class affiliations. Girls usually formed groups with other girls from their neighborhood. Most of the girls in a clique knew each other as neighbors or schoolmates. Even when girls switched groups or bartered for friendship, they often did so along class lines. This was especially evident in unstructured activities in which children could freely choose to associate with whomever they wanted (e.g., snack time and free time). The data here suggest that class and racial distinctions are more salient for girls than boys in middle childhood.

THE GENDER TRANSGRESSION ZONE

How do boys and girls negotiate boundaries in the GTZ? This area of activity—where boys and girls conduct heterosocial relations in hopes of either expanding or maintaining current gender boundaries in child culture—is where gender transgression takes place. A boy playing hand-clapping games (e.g., patty cake) or a girl completing an obstacle course that is designed to determine one's "manliness" are instances of transgression that occur in this zone. . . .

HEGEMONIC MASCULINITY IN THE GENDER TRANSGRESSION ZONE

Boys spend the majority of their time trying to maintain current gender boundaries. It is through the enforcement of gender boundaries that boys construct their social status. High-status boys are especially concerned with gender maintenance because they have the most to lose. By maintaining gender boundaries, top boys secure resources for themselves—such as playing area, social prestige/status, and power. The social prestige procured by high-status boys causes lower-status boys and girls to grant deference to high-ranked boys. If a high-ranked boy insults a lower-status boy or interrupts girls' activities, he is much less likely to be socially sanctioned by boys or girls. The position of lower-status boys in the hierarchy prevents them from challenging the higher-ranked boy's authority, while the collective subordination of girls to boys inhibits much dissension from girls. Connell (1987, 187) would likely suggest that girls' deference to high-status boys is an adaptive strategy to "the global dominance of heterosexual men."

To young children, "masculinity is power" (Kaufman 1995, 16). As a social construction, then, masculinity is maintained through a hegemonic process that excludes femininity and alternate masculinities. Hence, in the GTZ, boys seldom accept deviant boys or girls. Just as boys actively participate in the maintenance of the hegemonic hierarchy by using name-calling, physical aggression, and exclusion to handle personal conflicts, these same tactics are used to handle gender transgressors. The GTZ, then, is where hegemonic masculinity flexes its social muscle.

Boys Patrolling Boys in the GTZ

High-status boys maximize the influence of hegemonic masculinity and minimize gender transgressors by identifying social deviants and labeling them as outcasts. A continuous process occurs of homosocial patrolling and stigmatizing anomalies. Boys who deviate are routinely chastised for their aberrant behavior. Two examples of this process are particularly obvious. Joseph (WB7) is a seven-year-old who is recognized as a "cry baby." He is not very coordinated and gets along better with girls than boys. Because Joseph is so young, he is not directly affected by the full scrutiny of the solidified form of hegemonic masculinity. His age still allows him the luxury of displaying certain behaviors (e.g., crying) that are discredited in subsequent stages of middle childhood. Although the first author did not observe any kids in Joseph's own age group calling Joseph names, many older boys figure that he will "probably be gay when he grows up," as stated by Daniel (WB10). Fewer of the older boys associate with Joseph during free time, and he is not allowed around the older boys as are some of the other more "hegemonically correct" younger boys.

Phillip (WB10) was rejected by all the boys, which, in turn, aided in the maintenance of hegemonic masculinity. Phillip acted rather feminine and looked feminine as well. He lacked coordination, was small in stature, and had shoulder-length hair. Phillip often played with girls and preferred stereotypically feminine activities (e.g., jump rope). It was not uncommon to hear him being referred to as a faggot, fag, or gay. He was the ultimate pariah in the boys' sphere. He was constantly rejected from all circles of boys but got along quite fine with girls. His untouchable status was exemplified clearly in two instances. First, during a game of trains and tunnels—which requires partners linking arms—all players voluntarily paired up with same-sex companions except Phillip. As parents came to pick up their children, however, cross-gendered pairs began to form. This caused little disruption. However, there came a point when a hegemonically masculine boy, Sean (WB9), should have paired up with Phillip. Upon finding out who his new partner would be, Sean violently rejected Phillip. Sean was told that if he did not accept Phillip as his partner, he would have to sit out the rest of the game. Sean screamed, "I don't care if I have to sit out the whole summer 'cause I'm not going to let that faggot touch me!" In another situation during an arts and crafts activity, Phillip finished early. When kids finished early, the staff usually asked them to help an individual who was having problems. Usually everyone accepted help. However, when Phillip attempted to assist Markus (WB9), Markus rejected him harshly. Nonetheless, Markus did accept help from Karen (WG10). Phillip threatened a boy's masculinity because Phillip had been labeled homosexual; receiving help from a girl in this particular area is nonthreatening. If Joseph's behavior continues, we expect that he will experience the same harsh rejections that Phillip received. By stigmatizing Joseph and rejecting Phillip, homophobia emerges as a cautionary tale in the GTZ that deters other boys from deviating from the norm out of fear of rejection.

The boys in our case study used Joseph and Phillip to represent what would happen to other boys who transgressed the bounds of hegemonic masculinity. If a boy started slipping from gender-appropriate activities, then other boys would simply associate him with one of the two pariahs, Joseph or Phillip, or call him a fag to get him back in the hegemonic group. The boys devalue homosexuality; the threat of being labeled gay is used as a control mechanism to keep boys conforming to the norms of hegemonic masculinity. Gregory Lehne (1992, 389) says that the fear of being labeled gay "is a threat used by societies and individuals to enforce social conformity in the male role, and maintain social control . . . used in many ways to encourage certain types of male behavior and to define the limits of 'acceptable' masculinity." Talk of faggots and gays is also used to help define a boy's own masculinity.[4] By negatively talking about gays and excluding members who are presumed homosexual, individual boys are defining their own heterosexuality, while collectively they are endorsing hegemonic masculinity. Because most of these boys are not sexually mature or knowledgeable, many do not have an accurate conception of homosexuality (or, for that matter, heterosexuality) at this age. Gay bashing is another way boys can separate themselves from gender-deviant behavior.

Boys Patrolling Girls in the GTZ

Just as it is important for boys to patrol their own sex, it is equally important for boys to monitor the activities of girls and to keep them out of the boys' domain. If girls entered the boys' sphere in substantial numbers, the hegemonic hierarchy would be jeopardized. Girls who enter the boys' realm, therefore, are made to feel inadequate by the boys. The few girls who do succeed in the boys' sphere, nevertheless, are either marginalized or adopted into boys' middle-childhood culture (masculinized). Marginalization or masculinization depends on the girl's overall athletic prowess and emotional detachment while in the boys' sphere. This is illustrated by the following incident.

During one of the camp field trips, campers went to a university athletic training center. During the tennis rotation, Adrianne (WG10) was put with three boys. Adrianne, who took tennis lessons, was ignored by the boys. While the boys were arguing over the proper way to hit a backhand, Adrianne sat quietly on the sideline. When one boy finally asked her if she knew how to hit a backhand, she shook her head no. The first author knew this was incorrect because Adrianne had explained to him the proper way to hit a backhand earlier that day. Therefore, the first author asked Adrianne why she responded no. She replied, "When you're with boys, sometimes it's better to pretend like you don't know stuff because they're going to ignore you or tell you you're wrong."

Marginalization also occurs when girls meet some, but not all, of the requirements of hegemonic masculinity. The group of African American girls, for example, was marginalized. Many were just as assertive, and two were more athletic than some boys of high status. Yet, these girls remained marginal, retaining too many feminine characteristics, such as expressive acts of emotion when comforting teammates when they performed poorly in an activity. When a group of boys was asked why they did not associate with these girls who were more athletic than many of the boys in the hierarchy, Adam replied, "They're just different. I don't know about them. That whole group of them are just different. They're all weird."

Girl masculinization occurs when boys dissociate a girl from her feminine gender. The best example of a girl being adopted into a hegemonic masculine identity is Patricia (WG11). She is very athletic and can outplay many boys in basketball, the game that seemed to most signify one's masculinity at this site.[5] She also remained emotionally detached while interacting with boys. One time at the playground, Adam (WB11) created an obstacle course that he contended proved whether or not one was a "man." Some of the "manhood" tests were very dangerous—such as balancing on the rails of a high overhang—and had to be stopped. Each boy who completed a task successfully received applause and high fives. Those who did not complete successfully were laughed at because, according to the other boys, they were not "men." There was one catch to this test of masculinity—Patricia. She completed the numerous tasks faster and better than many of the boys. She did not get the screams of jubilation and high fives as did the other boys at first. As she proved her "manhood," however, she began to be accepted by the boys. By the end of the tests, Patricia was proclaimed a "man." About eight weeks later, when the first author asked a group of boys why Patricia was accepted as a member of their group, Adam, the apparent spokesman for the hierarchy, said, "Well, Patricia is not really a girl. Technically she is, but not really. I mean, come on, she acts like a boy most of the time. She even passed the 'manhood' test, remember?" Though this reveals Patricia's acceptance into the boys' hierarchy, she had to forfeit her feminine gender. As Thorne (1994) recognizes, girls who successfully transgress into the boys' activities under boys' terms do not challenge stereotypical gender norms. Hence, Patricia's participation in boys' activities "does little to challenge existing arrangements" (Thorne 1994, 133).

Hegemonic masculinity in middle childhood maintains itself in regards to girls in the GTZ. Girls are not welcome into the boys' sphere, which occupies more space. If girls partially meet standards, they are marginalized and thought of as "weird." In a way, they are almost degendered. Girls who fit all hegemonic requirements (tomboys) are conceptualized as masculine, or a boy/man. This reasoning is especially disturbing because masculinity is not only maintaining and defining itself, but it is also defining femininity.

FEMININITY IN THE GENDER TRANSGRESSION ZONE

As previously stated, girls find various forms of femininity acceptable, despite how different the form may be from their own. With this in mind, one can understand that while some girls do not challenge gender boundaries, those who do are not stigmatized by other girls. To test this observation, a group of girls (who were stereotypically "gender appropriate") were asked during lunch one day their views about the behavior of various girls who transgressed into the boys' sphere. Speaking of Patricia (WG11)—the girl who was proclaimed a "man"—Melissa (WG11) said, "She's pretty nice," and Lucia (WG9) added,

"Yeah, she's pretty cool. . . . She just likes to do different stuff. There's nothing wrong with that." They were then asked about the various members of the Black girl clique, and Melissa responded, "They're nice to [us]." When the girls were specifically asked if there was anything wrong with the way these gender transgressors behaved, Melissa and Lucia simply said no. Even Robin (WG10), who was not completely comfortable with the actions of these transgressors, replied, "I guess not. They just have their own way of acting. I just don't think it's very lady-like acting." As one can see, gender transgression is virtually accepted among even gender-traditional girls. Nevertheless, girls deal with clique deviants—those who are not "nice" relative to the clique's definition—just as they handle personal conflicts: exclusion from a particular group and social manipulation.

Girls Patrolling Girls in the GTZ

As girls get older, they recognize the higher value that society puts on masculine traits as well as the resources accumulated in the boys' sphere. Girls also see masculinity as power (Connell 1987). With increasing encouragement from the larger society (parents, teachers, and other pro-feminist role models), many girls attempt to access these resources as they mature. It should be noted, however, that high-status girls in these small groups also have social power in their cliques. The highest-ranked girl largely dictated who was gossiped about and who would be banished from the group. Yet, girls' resources were limited in comparison to their masculine-gendered playmates' because their resources did not extend much further than their small clique. Interestingly, girls who dare to participate in the boys' realm not only avoid stigmatization from most girls but are often praised by other girls if they succeed in the boys' sphere. For the most part, girls only receive restrictions from the prime agents of hegemonic masculinity at play—boys. Though girls' relations generally consist of small, intimate groupings when dealing with each other, large group affiliation and support seem to be the gender strategy when girls transgress onto traditional boys' turf. This was observed frequently throughout the summer.

Whenever a girl beat a boy in an athletic event, girls, as a collectivity, cheered them on despite age differences. During a Connect Four contest, Travis (WB9), the champion, was bragging about winning—especially when he beat girls. He would say, "It only takes me two minutes to beat girls," and "Girls aren't a challenge." This changed, however, when Corisa (BG6) started to play. Corisa beat Travis four times in a row. Girls of all ages rallied behind Corisa. For the duration of the day, girls praised Corisa, and some even introduced their parents to Corisa in admiration. One introduction went as follows: "Mommy, this is Corisa. She beats boys in Connect Four."

The best example of group solidarity in resistance to boys' dominance was provided one day when leaving the swimming pool. Molly (WG9)—whose eyes were irritated by chlorine and was basically walking to the locker room with her eyes closed—accidentally entered the boys' locker room while the campers were changing. Many of the boys laughed at her and ridiculed Molly for her mistake. Brian (BB11) said, "She just wanted to look at our private stuff," and Thomas (WB12) called her a "slut." Molly started to cry. Girls, however, came to Molly's rescue. While Molly's immediate clique comforted her, the other girls scared off boys who attempted to harass Molly for the rest of the day. Crysta (BG9) and Brittany (BG11) were the most effective protectors. This was surprising because even though Molly and Crysta were in the same age group, they did not get along, and Brittany—who is in the oldest group—to the best of our knowledge, had never even talked to Molly. As a gender strategy, girls—regardless of age, class, or racial differences—united together to combat the dominance of boys.

Girls Patrolling Boys in the GTZ

Without a uniform or constant form of femininity, girls were more lenient to both girls and boys when either ventured into the GTZ. Girls accepted Joseph and Phillip, both gender-deviant boys, into all their activities without a problem.

These boys, nevertheless, had to adhere to the same principles of "niceness" as did the girls. If the boys did not, they were punished in the same manner as girls—exclusion and social manipulation. When Joseph (WB7) did not share his "Now and Later" candy during lunch one day with the group of girls he was eating with, he soon found himself eating alone and the subject of much gossip in the girls' sphere.

"ALPHAS RULE! OTHERS DROOL!" OR HOW HIGH-STATUS BOYS DIRECT CHANGE IN THE GTZ

The top-ranked boys in the hierarchy direct the actions of all the boys who aspire to hegemonic masculinity (or are, at least, complicit with it). High-status boys are primarily concerned with maintaining gender boundaries to retain status and all the luxuries that are a result of being hegemonically masculine. Dominant boys make decisions for the group and can manipulate the other boys to sustain high status and its privileges. Examples of status privileges include being picked first for teams, getting first dibs on other people's lunches, being allowed to cut in line, and being freed by other males during prison ball—a game similar to dodge ball—with no reciprocal obligation to free low- or middle-status boys.

High-status boys have the unique power of negotiating gender boundaries by accepting, denying, or altering gender codes. The power of high-status boys to alter gender boundaries was strikingly borne out by a series of events that, for weeks, redefined a feminine gender-stereotyped activity, hand-clapping games, into a hegemonically masculine one. . . .

Here is how the defeminization occurred. One day, right before the closing of the camp, Adam—the highest-ranked boy—was the only boy left waiting for his mother to pick him up. Four girls remained as well and were performing the *Rockin' Robin* hand-clapping routine. When one of the girls left, one of the three remaining girls asked Adam if he would like to learn the routine. He angrily replied, "No, that's girly stuff." Having been a camp counselor for three years, the first author knows every clapping routine from *Bo Bo See Aut In Totin* to *Miss Susie's Steamboat*. He, therefore, volunteered. The girls were amazed that he knew so many of what they referred to as "their" games. After a while, only two girls remained and *Rockin' Robin* requires four participants. Surprisingly, Adam asked to learn. Before he left, Adam had learned the sequence and was having a good time.

We believe that Adam transgressed for three reasons. One, all the other boys were gone, so there were no relevant or important (to him) witnesses to his transgression. Thorne (1994, 54) repeatedly states that witnesses hinder gender deviance: "Teasing makes cross-gender interaction risky, increases social distance between girls and boys, and has the effect of making and policing gender boundaries." Second, Adam saw the first author participate freely in an activity that was previously reserved for girls. Third, as the highest-ranked boy, Adam has a certain degree of freedom that allows him to transgress with little stigmatization. Thorne asserts that the highest-status boy in a hierarchy has "extensive social leeway" (p. 123) since his masculinity is rarely questioned. The next day, Adam was seen perfecting the routine he learned the day before. Many of the boys looked curiously and questioned why Adam was partaking in such an activity. Soon after, other boys started playing, and boys and girls were interacting heterosocially in what was formerly defined as a "girls-only" activity. Cross-gendered hand-clapping games continued for the rest of the summer and remained an area in which both girls and boys could come together. Defeminization occurred because Adam—the highest-status boy—set the standard and affirmed this type of entertainment as acceptable for boys. This incident supports our view that high-status boys control gender negotiations by showing that gender boundaries can be modified if someone of high status changes the standard of hegemonic masculinity.

To make hand-clapping more masculine, nonetheless, the first author documented boys changing the verses of the most popular hand-clapping game, *Rockin' Robin,* to further defeminize the activity. One of the original verses is "All the little birdies on J-Bird Street like to hear the robin go tweet, tweet, tweet." The boys changed this to "All the little birdies on J-Bird Street like to hear robin say eat my meat!" About a month later, the first author discovered another altered verse from the boys. They changed "Brother's in jail waiting for bail" to "Brother's in jail raising hell!" Since these verses were not condoned at the camp—though we are sure the children used them out of the hearing distance of counselors—girls cleverly modified one of the profane verses by singing, "Brother's in jail raising H-E- double hockey sticks!" This, too, the boys picked up and started applying as their own. Hand-clapping games moved from the girls' sphere to the GTZ. Defeminization of hand clapping exposes the constant fluctuation and restructuring of gender norms in childhood play.

Boys in middle childhood organize themselves in a definite hierarchy that is run by high-status boys in accordance with the hegemonic form of masculinity that they embody and police in the GTZ. Boys are not accepting of deviant boys or girls. Gender deviants are handled by teasing and name-calling, marginalization and exclusion from the group, and physical aggression. High-status boys, though, have the unique power to negotiate gender boundaries by either accepting, denying, or altering gender codes. Girls who enter the boys' realm are made to feel inadequate by the boys. Those girls who do succeed in the boys' sphere, nevertheless, are either marginalized from or masculinized into boys' middle-childhood culture, They are forced to leave their femininity behind if they want to cross the border fully. Therefore, no feminization of hegemonic masculinity is allowed. As can be seen in the hand-clapping phenomenon, the redefinition entails defeminization.

★ ★ ★

NOTES

1. To the best of our knowledge, this concept was first set out in his book *Gender and Power* (Connell 1987, 183–88).
2. Bird (1996) explains how homosocial interactions maintain gender boundaries among adult men. Beutel and Marini (1995) discuss the contrasting value systems of males and females.
3. The names of children in this study are pseudonyms.
4. This may be part and parcel of what McCreary (1994) refers to as the universal avoidance of femininity: Homophobia may be a rejection of the "abnormality" of being attracted to boys (i.e., being "girlish").
5. At the other site that the first author visited, football was the most masculinizing athletic activity.

REFERENCES

Absi-Semaan, N., G. Crombie, and C. Freeman. 1993. Masculinity and femininity in middle childhood: Development and factor analyses. *Sex Roles* 28 (3/4): 187–206.

Andersen, Margaret L. 1993. *Thinking about women.* New York: Macmillan.

Beutel, Ann M., and Margaret M. Marini. 1995. Gender and values. *American Sociological Review* 60 (3): 436–38.

Bird, Sharon R. 1996. Welcome to the men's club: Homosociality and the maintenance of hegemonic masculinity. *Gender & Society* 10 (2): 120–32.

Bjorkqvist, Kaj. 1994. Sex differences in physical, verbal, and indirect aggression: A review of recent research. *Sex Roles* 30 (3/4): 177–88.

———. 1995. *Masculinities* Berkeley: University of California Press.

Connell, R. W. 1987. *Gender & power.* Stanford, CA: Stanford University Press.

———. 1995. *Masculinities.* Berkeley: University of California Press.

Curran, Daniel J., and Claire M. Renzetti. 1992. *Women, men, and society,* 2d ed. Needham Heights, MA: Allyn & Bacon.

Fine, Gary Alan. 1992. The dirty play of little boys. In *Men's lives,* edited by Michael S. Kimmel and Michael A. Messner. New York: Macmillan.

Hawkesworth, Mary. 1997. Confounding gender. *Signs: Journal of Women in Culture and Society* 22 (3): 649–86.

Kaufman, Michael. 1995. The construction of masculinity and the triad of men's violence. In *Men's lives,*

edited by Michael Kimmel and Michael Messner. New York: Macmillan.
Lehne, Gregory K. 1992. Homophobia among men: Supporting and defining the male role. In *Men's lives,* edited by Michael S. Kimmel and Michael Messner. New York: Macmillan.
Luria, Zella, and Barrie Thorne. 1994. Sexuality and gender in children's daily worlds. In *Sociology: Windows on society,* edited by John W. Heeren and Marylee Mason. Los Angeles: Roxbury.
McCreary, Donald R. 1994. The male role and avoiding femininity, *Sex Roles* 31 (9): 517–32.
Messner, Michael A. 1992. *Power at play: Sports and the problem of masculinity.* Boston: Beacon.
———. 1994. The meaning of success: The athletic experience and the development of male identity. In *Sociology: Windows on society,* edited by John W. Heeren and Marylee Mason. Los Angeles: Roxbury.
———. 1998. The limits of "the male sex role": An analysis of the men's liberation and men's rights movements discourse. *Gender & Society* 12 (3): 255–76.
Thorne, Barrie. 1994. *Gender play: Girls and boys in school.* New Brunswick, NJ: Rutgers University Press.
West, Candace, and Don H. Zimmerman. 1987. Doing gender. *Gender & Society* 1 (2): 125–51.

17

Schooling for Inequality

BY DOROTHY E. SMITH

Dorothy Smith is a Canadian sociologist whose appointment within a college of education makes her views of gender inequality in schools particularly valuable. Smith describes gender inequality in classrooms, examining how classroom discourse and practices silence girls. In addition, she provides a framework for achieving gender equity in our schools.

1. How does the discourse in classrooms disadvantage girls?
2. What does Smith mean by, "agency as constituted socially"?
3. What are one or two changes you would make in schools and colleges to stop replicating assymetrical gender relations?

The topic of schooling as an institution productive of inequities—of gender, as well as race and class—has never been, as I believe it should be, a major issue for feminism.

For more than twenty years now I've worked in a sociology department in an institute for studies in education. In the early years of my work here, I was active with women teachers in their associations, helping to build women's organization. Yet, talking with activists in other areas, I found a profound disinterest in, a turning away from, issues concerning girls and women in the school system. Although feminism has grown and developed among educators, the inequalities produced by the school system have never become a central topic for feminist thought and debate.[1]

Unlike at the university level—where members of a feminist intelligentsia are intimately involved, where much work has been done on the situation of women in academic life, where feminist pedagogy is debated, and where we have access to the rich resources of women's studies—the school system is strikingly well insulated from initiatives originating in the public discourse of the intelligentsia and strikingly effective at preventing localized grassroots initiatives from generalizing throughout. I'd like to see that change.

For me, starting from women's standpoint means that inquiry must begin in the everyday/everynight actualities of people's experience; it means problematizing the objectified institutional order of large-scale corporations, of schooling and health care, of the professions, and of the academic, cultural, and scientific discourses, including the mass media. The institutional order puts people to work in particular local settings, coordinating their work translocally, largely through the medium of texts (print or electronic). The texts integral to the social organization of the institutional order are complemented by technologies or disciplined practices that produce standardized local states of affairs or events corresponding to the standardized texts. The institutional order has in a sense extracted organization from the direct connectedness of people's everyday/everynight activities and built specialized and differentiated relations that connect the multiple settings of people's work. It has become independent of particular individuals; individuals participate in it through the forms of agency and subjectivity that it establishes.

Postmodern theory has contributed the notion that a subject, or subject position, is constituted in discourse rather than being a property of persons. Similarly, I think of agency as constituted socially; being at work in the institutional order doesn't automatically accord agency (Smith 1998). Some years back, Marilee Reimer (1988) described the way the executive work of senior secretaries in a government office in Ontario was recognized only as *delegated* from their boss. They acted but were not agents. A worker on an assembly line in an automotive plant is governed by the operation of the line; her or his movements may be strictly prescribed. She or he has no agency in that corporation. A Hispanic man on his way to pick up his father after work is stopped and hassled by police, who find nothing incriminating but search and impound his car anyway. He is advised by a friend who is a cop to pay the fine and let it go, even though it means that he will have a record. He has no sense of himself/is not recognized as an agent within the judicial process. When I give a lecture in a large university on feminist issues, although the majority of the audience are women, my interlocutors are almost exclusively male. Patricia Hill Collins (1998, 3–4) tells of teaching a second-grade class of African-American children whose experiences were silenced by the standard curriculum. Lew Dunn, a grassroots environmental activist, describes his own lack of agency: "I was intimidated by government officials. I wouldn't challenge them. I was fearful of them, to be honest with you. I feared making a mistake, saying the wrong thing. I didn't think I knew enough to challenge them" (quoted in Szasz 1994, 95). Not everyone can take for granted the capacity of agent within the institutional order.

Teaching courses in gender equity in the classroom, I have come to think that schools are an integral part of the institutional processes for the differential allocation of agency. At the outset of this phase of the women's movement, a major emphasis was on voice, silencing, exclusion. It is still a major issue among nonwhite feminists. Way back, Mary Ellman (1968) described a distinction, which she saw as both obvious and unnoticed, between women and men in intellectual matters. A man's body (today, we would say, "a white man's body") gives credibility to his utterance, whereas a woman's takes it away. If we are puzzled by the persistence of gender and race inequality in the higher reaches of corporations, the state, and intellectual activity, we should look, I think, toward relationships and groupings formed in the school system.

Schools reproduce the social organization of inequality at multiple levels. At the level of the school system as a whole, differences of class as income level appear in the segregated private schools and colleges and in the public school system by the economic and racial character of school districts. Within the school, class, race, and gender emerge as dynamic and exclusionary groupings formative in students' identities and associations (Olsen 1997). To state it very simply, some students learn that their own voices have authority, that they count and should be heard; others learn their lack. Some learn that they belong to groups that have agency in society and that they can count on being recognized as such. This forming of groups is more than the "socialization" of individuals; these are ways of relating that are projected and perpetuated beyond school.

Research on gender and schooling shows a persistent replication of gender relations that develop over time as exclusive gender groupings marked by the privileging of male voices and male activity in the classroom, playground, sportsfield, and hallway.

One of the earliest observers of this dynamic, if not *the* earliest, is Raphaela Best, who described the formation of exclusive groups among boys who defined their masculinity and its attendant privileges as antithetical to what was attributed to girls (1983). Barrie Thorne's (1994) observations in elementary schools show similar patterns on playgrounds and in classrooms, partially overlaid by, yet powerfully present in, the order of the classroom. Myra Sadker and David Sadker (1994) explore how teachers' interactions with students in the classroom contribute actively to male dominance of classroom activity (this isn't news for feminist educators who've been talking and writing about this phenomenon for twenty years or so). Alison Lee writes of an Australian high school geography class: "The most lasting impression I have of this classroom is of boys' voices . . . of male voices physically swamping girls' . . . [the boys'] voices were often loud, the physiological difference combining with the classroom spatial arrangements and their apparent sense of freedom to produce their voices in ways which asserted their presence fairly effectively. . . . There was a marked absence of girls' voices, despite their physical presence in the room" (1996, 72–73). Psychological terminology gives us the concept of low self-esteem, said to be endemic among high school girls. Peggy Orenstein records the explanation of a girl whose story, voted the best in class, featured a boy as the central character: "It was an adventure; it wouldn't be right if you used a girl" (1994, 15). In a study that asked girls and young women to evaluate their schooling from a feminist standpoint, one young woman said, "I would have trouble asking a male classmate for . . . help [with an assignment] because I would feel, even if he wasn't judging, I would feel he was judging. I would feel judged" (the same informant described a math class in which boys held up for ridicule the test results of girls who had not done well) (Smith, McCoy, and Bourne 1995, 17).

I collect such examples as indicators of social processes that reproduce circles of exclusion from agency within the institutional order. What takes shape in school is a child's membership in a collectivity that projects her relationships with others into her adult future. Paul Willis's study (1977) of young men "learning to labour" in an English secondary school describes a dynamic interplay within the school that impels the "lads" into a future of unskilled labor. We are only just beginning to explore dynamics built into the school system that organize such exclusions. The insulation of the school system that I identified at the outset of this essay allows the institutional order to deny agency to people who do not share the interests and experiences it embeds. These school dynamics are not, of course, part of the curriculum or intended in the professional training of teachers (and, in any case, individual teachers are not as powerful within the school system as students, parents, and the media imagine). Nonetheless, such dynamics are a profound impairment of the democratic process in our societies and merit feminist attention and debate as we enter this new millennium.

NOTE

1. However, useful work has been done by organizations such as the American Association of University Women (1997).

REFERENCES

American Association of University Women. 1997. *How Schools Shortchange Girls: The AAUW Report, a Study of Major Findings on Girls and Education*. New York: Marlowe.

Best, Raphaela. 1983. *We've All Got Scars: What Boys and Girls Learn in Elementary School*. Bloomington: Indiana University Press.

Collins, Patricia Hill. 1998. *Fighting Words: Black Women and the Search for Justice*. Minneapolis: University of Minnesota Press.

Ellman, Mary. 1968. *Thinking about Women*. New York: Harcourt Brace Jovanovich.

Lee, Alison. 1996. *Gender, Literacy, Curriculum: Re-writing School Geography*. London: Taylor & Francis.

Olsen, Laurie. 1997. *Made in America: Immigrant Children in Our Public Schools*. New York: New Press.

Orenstein, Peggy. 1994. *School Girls: Young Women, Self-Esteem, and the Confidence Gap*. New York: Doubleday, with the American Association of University Women.

Reimer, Marilee. 1988. "The Social Organization of the Labour Process: A Case Study of the Documentary Management of Clerical Labour in the Public Service" Ph.D. dissertation, University of Toronto.

Sadker, Myra, and David Sadker. 1994. *Failing at Fairness: How America's Schools Cheat Girls*. New York: Macmillan.

Smith, Dorothy E. 1998. *Writing the Social: Critique, Theory and Investigations*. Toronto: University of Toronto Press.

Smith, Dorothy E., Liza McCoy, and Paula Bourne. 1995. "Girls and Schooling: Their Own Critique." Gender and Schooling Papers, no. 2. Toronto: Centre for Women's Studies in Education, University of Toronto.

Szasz, Andrew. 1994. *Ecopopulism: Toxic Waste and the Movement for Environmental Justice*. Minneapolis: University of Minnesota Press.

Thorne, Barry. 1994. *Gender Play: Girls and Boys in School*. New Brunswick, N.J.: Rutgers University Press.

Willis, Patti E. 1977. *Learning to Labour: How Working Class Kids Get Working Class Jobs*. London: Saxon House.

18

The Chilly Climate

Subtle Ways in Which Women Are Often Treated Differently at Work and in Classrooms

BY BERNICE R. SANDLER

Bernice Sandler is a long-time public advocate for women and girls in education. Her work on the "chilly climate" for girls in classrooms is widely known and respected. This piece from her Web site provides detailed descriptions of the behaviors that Smith, in the previous article, would argue deny girls and women agency in our schools. Unlike other readings in this book, this is a listing of behaviors rather than an article. This list describes behaviors that create a hostile environment for women and girls in school and at work.

1. What gender behaviors have you observed in classrooms or at work sites?
2. Is ignoring women and girls less or more harmful than hostile behavior toward them?
3. What recommendations would you suggest to make schools and work more equitable spaces?

The word "women" as used here includes all women. However, for women of color, disabled women, lesbians and older women these behaviors may be exacerbated and these women may experience other forms of differential behavior as well. Additionally, other "outsiders" such as men of color, persons for whom English is a second language, and those from working class backgrounds often experience many of the same behaviors described here.

Most of the behaviors are what has been described as "microinequities," a term coined by Mary Rowe of Massachusetts Institute of Technology. They describe the small everyday inequities through which individuals are often treated differently because of their gender, race, age, or other "outsider" status. Taken by itself, a microinequity may have a minuscule effect, if it has any at all, and is typically not noticed by the person it happens to or by the person who asserts it. Yet when these behaviors occur again and again, and especially if they are not noticed or understood, they often have a damaging cumulative effect, creating an environment that is indeed chilly—an environment that dampens women's self-esteem, confidence, aspirations and their participation.

Because overt behaviors are more easily recognized, they have generally been omitted from this article. Those that are included here are the types of behaviors that are typically minimized by the person engaging in the behavior. Some of the behaviors below may fit in more than one category.

Behaviors that Communicate Lower Expectations for Women

Asking women easier, more factual questions, men the harder, open-ended ones that require critical thinking.

Grouping women in ways which indicate they have less status or are less capable.

Doubting women's work and accomplishments: "Did you really do that without any help from someone else?"

Expecting less of women in the future.

Calling males "men" and women "girls" or "gals" which implies that women are not as serious or as capable as men.

Yielding to the Influence of Internalized Stereotypes

Using examples that reflect stereotypes.

Addressing women in ways that reinforce stereotypes and social roles rather than intellectual ones, for instance, calling women "honey."

Focusing on a woman's appearance, personal qualities and relationships rather than on her accomplishments: "I'd like you to meet our new charming colleague" rather than "I'd like you to meet the new hot-shot we just hired."

Judging women by their physical appearance and downgrading those who are not "attractive."

Describing women by their physical characteristics, such as a "blonde."

Using a different vocabulary to describe similar behavior or accomplishments, such as "angry man" but " bitchy woman."

Expressing stereotypes that discourage women from pursuing professional careers, such as "Women are naturally more caring and men are naturally more aggressive."

Assigning classroom tasks according to stereotyped roles. Women are assigned to be the note-takers.

Falling back on disparaging stereotyped words when angry or annoyed with females: "Look here, sweetie," and "Don't talk back to me, little girl."

Excluding Women from Participation in Meetings and Conversations

Ignoring women while recognizing men, even when women clearly volunteer to participate by raising their hands.

Addressing a group as if there were no women present: "When you were a boy . . ."

Interrupting women more than men or allowing their peers to interrupt them. Women may be more vulnerable when interrupted—they may not participate again for the rest of a meeting.

Treating Men and Women Differently When Their Behavior or Achievements Are the Same

Treating women who ask extensive questions as trouble-makers and men as interested and bright.

Believing that women who ask for information don't know the materials, but that men who ask are smart, inquisitive and involved.

Viewing marriage and parental status differently for men and women—as disadvantages for women and advantages for men.

Attributing women's achievements to something other than their abilities, such as good luck, affirmative action, beauty, or having "slept their way to the top."

Frowning when women speak (male and female students may also do this). Men and women alike may be less reinforcing when women speak.

Judging women who speak tentatively as being less competent or knowledgeable.

Giving Women Less Attention and Intellectual Encouragement

Making less eye contact with women.

Nodding and gesturing more and paying more attention in general to men than to women when they speak.

Responding more to men's comments by making additional comments, coaching, and asking questions, and responding more often to women with "uh-huh."

Calling on males more frequently in meetings and in conversations.

Calling males by name more frequently.

Coaching men but not women: "Tell me more about that."

Waiting longer for a man to respond to a question than a woman, before going on to another person.

Crediting men's comments to their owner or "author" ("As Bill said . . .") but not giving authorship or ownership to women. Sometimes a comment made by a woman is later credited to a male.

Giving men more detailed instructions for a task.

Giving women less feedback—less criticism, less help and less praise. (This is one of the critical ways in which women and men are treated differently.)

Being more concerned about men's behavior than that of women's, such as worrying about a male who doesn't participate but not being concerned about women who do not.

Giving women less encouragement to take on harder tasks.

Engaging in more informal conversation with men than with women.

Discouraging Women through Politeness

Using some forms of politeness that shift the focus from intellectual activities to social behavior: "I like to see the girls' smiling faces."

Males may perform hands-on tasks for women (as when helping them with a computer task) under the guise of being helpful, thereby depriving women of the experience and communicating lower expectations for them.

Faculty members may be excessively kind and paternalistic or maternalistic in trying to be helpful and hold women to a lower standard.

Men may tell a group that they are refraining from telling certain jokes or using certain words because there are "ladies" present.

(True courtesy and respect does not patronize, trivialize or depersonalize another person's abilities and talents, nor do they disappear when a woman acts in a way that deviates from gender stereotypes.)

Singling Out Women

Singling out women and other groups such as people of color: "What do you women think about this?"

Males are more likely to touch women than other men. If touch is being used to reassure or indicate friendliness, males are being excluded. Touch is often associated with power; frequently the message transmitted by a touch conveys a "power play."

Defining Women by Their Sexuality

Relating to women in a sexual manner—sexual comments about or toward specific women or women in general, such as discussing appearance or physical attributes or using sexual humor.

Valuing and praising women for their physical appearance, not for their intellectual ability.

Devaluing or ignoring comments made by women perceived as "unfeminine" or believed to be lesbian or bisexual.

Using the words "lesbian" and "bisexual" as pejorative terms, especially when women raise women's issues.

Engaging in sexually harassing behaviors or allowing others to do so.

Overt Hostile Behavior toward Women

Ridiculing or making denigrating remarks about women's issues, or making light of issues such as sexual harassment and sexual assault.

Discouraging women from conducting research on women's issues.

Calling women names if they are interested in women's issues or protest sexism.

Making sexist remarks about women in general or about specific women.

Using Humor in a Hostile Manner

Engaging in negative body language or behavior (for example, men rolling their eyeballs) when women speak.

Hissing or ridiculing women who raise women's issues.

Denigrating or ridiculing women or engaging in other rude behaviors that express hostility to women.

Telling sexist or sexual jokes which denigrate women.

Not taking women's comments or their work seriously.

Devaluation

Devaluation is often used as a partial explanation or rationale for differential treatment.

Gender affects our view of someone's competence. What is viewed as male is usually seen as more important than that associated with women.

Perceptual bias is not uncommon. For instance, a woman's success, such as getting into a prestigious program, is said to result from "luck" or "affirmative action" while a man's similar success will be attributed to talent.

Women's issues may be devalued, as well as women's ways of speaking.

Devaluation and Power

It is the power difference between men and women that gives value to or devalues whatever differences exist.

Stereotypes which reinforce differences are maintained precisely because they reinforce power and privilege. Behaviors which are valued such as competitive, status-seeking behavior, are behaviors that reinforce privilege. Males may assert power and expect to be treated more favorably than females.

19

College Gender Gap Stirs Old Bias

BY ELLEN GOODMAN

Ellen Goodman is a national columnist writing for the Boston Globe. *In this column, she describes the concern expressed in 2002 that college-educated women would no longer be able to find a mate. She compares it to an earlier discussion of the effects of college on women in 1873.*

1. What is the situation for women in higher education today and what was it in 1873?
2. What was the problem in 1873 and how is it similar to the debate today?
3. Is the problem today real or imagined? What are the consequences for college women today?

I know that educated women have always made some people nervous.

In 1873, when less than 15 percent of the college students were female, Harvard's Edward Clarke explained scientifically how expanding a woman's brain would make her uterus shrink.

Now it's 2002.

The fall semester opens with reports that 57 percent of all bachelor's degrees will go to women. And—surprise—it's being heralded as a national crisis.

It seems that everyone in higher education is majoring in the same subject: the gender gap. The pop quiz has three questions: Are women winning? Are men losing? Is good news for the goose bad news for the gander?

Before you answer, a few facts. There are now about 8.3 million women and 6.4 million men in college. The gap is the latest phase in a long gradual trend that began after the Vietnam War, when men had a very strong incentive to stay in school—college or combat. In the last couple of decades, the number of men with bachelor's degrees has risen, but the number of women has increased much further and faster.

Jacqueline King at the American Council on Education calls this "more of a success story for women than failure for men." Of course, "success" also has its limits. The gender gap reverses by the time you get to doctorates and goes into full-tilt retreat down the tenure track.

But even when we are counting bachelor's degrees, gender is just the sexy part of the statistics. The real educational gap has a lot more to do with race and class.

For openers, there is no gap between traditional-age white male and female students. The real differences appear among minorities. In fact, most of the overall gap is due to a huge increase in the number of minority women. Among African Americans, two women now get degrees for every man.

As for income, we know that low-income students of every ethnic group are less likely to go to college. But in turns out that more low-income women than men still find their way onto campus.

This has a lot to do with the job market, says King. An undecided low-income male may look around and see a job that looks decent by an 18-year-old's standard, she says. But a woman "can't get a job in the traditional female worlds of health care or office work without some post secondary school." Indeed, the good/bad news is that a woman still needs a college degree to match the income of a man with a high school diploma.

I don't want to minimize a real problem. We *should* worry about young people left behind in an economy that is increasingly dependent on education. And it's worth asking why low-income and minority men are less successful academically than other populations.

But if there's a crisis, it's not a man versus woman thing. As Jacqueline Woods, head of the American Association of University Women, puts it, "They're saying that women are replacing men and isn't this alarming. Those people are playing a zero-sum game and I refuse to play."

You don't need a bachelor's degree to pick up a subtext in the crisis reporting: Somehow or other, women are upsetting the natural order of things. Even, or especially, the marital order. As Thomas Mortenson at the Pell Institute told one reporter, "There's 170,000 more bachelor's degrees awarded to women than men. That's 170,000 women that will not be able to find a college-educated man to marry."

Hmmm. Was anyone alarmed when there weren't enough educated women for the bachelors with bachelor's degrees? College women hear this: "You are on the brink of destruction . . . Beware!!" Get the degree and you face a shrinking population of marriageable men.

Let's see now, is that better or worse than a shrinking uterus?

20

Bad Boys

Public Schools in the Making of Black Masculinity

BY ANN ARNETT FERGUSON

Ann Arnett Ferguson, a sociologist, used participant observation methods to gather data on twenty boys in an urban intermediate school (grades 4 through 6) located in a medium size city on the West Coast. Over half the students at this school were African American, with one-third white and the remaining classified as Asian American, Hispanic or "other." The staff and teachers in this school, however, were predominately white and female. She studied intensely 20 African-American boys, ten of whom she called "schoolboys"—as they had been identified by the school personnel as "doing well." The other ten she called "troublemakers" because they were identified as children who'd get into trouble. The Punishing Room or jailhouse is predominant in this reading; it is a place where the school sent disobedient children. She provides for us a close look at the dynamics of race, gender, and class in the classroom, focusing on young African-American males.

1. What are the dynamics of the relationships between the troublemakers and their white teachers?
2. What mechanisms do the troublemakers use to establish and maintain their masculinity and black identity?
3. Does the application of rules and punishments to African-American boys, as compared to white boys, change the way gender is viewed for these children? That is, do teachers use stereotypes to characterize the misbehavior of African-American boys that are different from those applied to white boys?

★ ★ ★

[F]or African American children the conditions of schooling are not simply tedious; they are also replete with symbolical forms of violence. Troublemakers are conscious of the fact that school adults have labeled them as problems, social and educational misfits; that what they bring from home and neighborhood—family structure and history, forms of verbal and nonverbal expression, neighborhood lore and experiences—has little or even deficit value. The convergence of the routine with the harsh, exclusionary ambience of school calls forth a more intensive mode of identity work. My concern now is to hone in on this work through an examination of the relationship between trouble and masculinity; of the specific circular relationship between risky, rule-breaking behavior, getting in trouble, and the experience of being and becoming male. Making a name for yourself through identity work and self-performance, even if the consequence is punishment, becomes a highly charged necessity given the conditions of school for the Troublemakers.

MAKING A NAME FOR YOURSELF: TRANSGRESSIVE ACTS AND GENDER PERFORMANCE

Though girls as well as boys infringe the rules, the overwhelming majority of violations in every single category, from misbehavior to obscenity, are by males. In a disturbing tautology, transgressive behavior is that which constitutes masculinity. Consequently, African American males in the very act of identification, of signifying masculinity, are likely to be breaking rules.

* * *

African American boys at Rosa Parks School use three key constitutive strategies of masculinity in the embrace of the masculine "we" as a mode of self-expression. These strategies speak to and about power. The first is that of heterosexual power, always marked as male. . . . The second involves classroom performances that engage and disrupt the normal direction of the flow of power. The third strategy involves practices of "fighting." All three invoke a "process of iterability, a regularized and constrained repetition of norms," in doing gender; constitute masculinity as a natural, essential corporeal style; and involve imaginary, fantasmatic identifications.[1]

These three strategies often lead to trouble, but by engaging them a boy can also make a name for himself as a real boy, the Good Bad Boy of a national fantasy. All three illustrate and underline the way that normative male practices take on a different, more sinister inflection when carried out by African American boys. Race makes a significant difference both in the form of the performance as well as its meaning for the audience of adult authority figures and children for whom it is played.

Heterosexual Power

One group of transgressions specifically involves behavior that expresses sexual curiosity and attraction. These offenses are designated as "personal violations" and given more serious punishment. Inscribed in these interactions are social meanings about relations of power between the sexes as well as assumptions about male and female differences at the level of the physical and biological as well as the representational. It is assumed that females are sexually passive, unlikely to be initiators of sexual passes, while males are naturally active sexual actors with strong sexual drives. Another assumption is that the feminine is a contaminated, stigmatizing category in the sex/gender hierarchy.

Typically, personnel violations involved physical touching of a heterosexual nature where males were the "perpetrators" and females the "victims." A few examples from the school files remind us of some of the "normal" displays of sexual interest at this age.

- Boy was cited with "chasing a girl down the hall" [punishment: two days in the Jailhouse].
- Boy pulled a female classmate's pants down during recess [punishment: one and a half days in the Jailhouse].
- Boy got in trouble for, "touching girl on private parts. She did not like" [punishment: a day in the Jailhouse].
- Boy was cited for "forcing girl's hand between his legs" [punishment: two and a half days in the Jailhouse].

In one highly revealing case, a male was cast as the "victim" when he was verbally assaulted by another boy who called him a girl. The teacher described the "insult" and her response to it on the referral form in these words:

> During the lesson, Jonas called Ahmed a girl and said he wasn't staying after school for detention because "S" [another boy] had done the same thing. Since that didn't make it ok for anyone to speak this way I am requesting an hour of detention for Jonas. I have no knowledge of "S" saying so in my presence.

This form of insult is not unusual. When boys want to show supreme contempt for another boy they call him a girl or liken his behavior to female behavior. What is more troubling is that adults capitulate in this stigmatization. The female teacher takes for granted that a comment in which a boy is called a girl is a symbolic attack, sufficiently derogatory to merit punishment. All the participants in the classroom exchange witness the uncritical acknowledgment of adult authority to a gender order of female debasement.

Of course, this is not news to them. Boys and girls understand the meaning of being male and being female in the field of power; the binary opposition of male/female is always one that expresses a norm, maleness, and its constitutive outside, femaleness. In a conversation with a group of boys, one of them asserted and then was supported by others that "a boy can be a girl, but a girl can never be a boy." Boys can be teased, controlled, punished by being accused of being "a girl." A boy faces the degradation of "being sissified," being unmanned, transferred to the degraded category of female. Girls can be teased about being a tomboy. But this is not the same. To take on qualities of being male is the access to and performance of power. So females must now fashion themselves in terms of male qualities to partake of that power. Enactments of masculinity signal value, superiority, power.

★ ★ ★

African American males have historically been constructed as hypersexualized within the national imagination. Compounding this is the process of the adultification of their behavior. Intimations of sexuality on their part, especially when directed toward girls who are bused in—white girls from middle-class families—are dealt with as grave transgressions with serious consequences.

Power Reversals: Class Acts

Performance is a routine part of classroom work. Students are called upon to perform in classes by teachers to show off their prowess or demonstrate their ineptitude or lack of preparation. They are required to read passages aloud, for example, before a highly critical audience of their peers. This display is teacher initiated and reflects the official curricula; they are command performances with well-scripted roles, predictable in the outcome of who has and gets respect, who is in control, who succeeds, who fails.

Another kind of performance is the spontaneous outbreaks initiated by the pupils generally defined under the category of "disruption" by the school. These encompass a variety of actions that punctuate and disrupt the order of the day. During the school year about two-thirds of these violations were initiated by boys and a third by girls. Here are some examples from the discipline files of girls being "disruptive":

- Disruptive in class—laughing, provoking others to join her. Purposely writing wrong answers, being very sassy, demanding everyone's attention.
- Constantly talking; interrupting; crumpling paper after paper; loud.

Some examples of boys' disruption:

- Constant noise, indian whoops, face hiccups, rapping.
- Chanting during quiet time—didn't clean up during art [punishment: detention].
- Joking, shouting out, uncooperative, disruptive during lesson.

From the perspective of kids, what the school characterizes as "disruption" on the referral slips is often a form of performance of the self: comedy, drama, melodrama become moments for self-expression and display. Disruption adds some lively spice to the school day; it injects laughter, drama, excitement, a delicious unpredictability to the classroom routine through spontaneous, improvisational outbursts that add flavor to the bland events.

In spite of its improvisational appearance, most performance is highly ritualized with its own script, timing, and roles. Teachers as well as students engage in the ritual and play their parts. Some kids are regular star performers. Other kids are audience. However, when a substitute is in charge of the class and the risk of being marked as a troublemaker is mimimal, even the most timid kids "act up." These rituals circulate important extracurricular knowledge about relations of power.

These dramatic moments are sites for the presentation of a potent masculine presence in the classroom. The Good Bad Boy of our expectations engages power, takes risks, makes the class laugh, and the teacher smile. Performances mark boundaries of "essential difference"—risk taking, brinkmanship. The open and public defiance of the teacher in order to get a laugh, make things happen, take center stage, be admired, is a resource for doing masculinity.

These acts are especially meaningful for those children who have already been marginalized as outside of the community of "good," hardworking students. For the boys already labeled as troublemakers, taking control of the spotlight and turning it on oneself so that one can shine, highlights, for a change, one's strengths and talents. Already caught in the limelight, these kids put on a stirring performance.

Reggie, one of the Troublemakers, prides himself on being witty and sharp, a talented performer. He aspires to two careers: one is becoming a Supreme Court justice, the other an actor. He had recently played the role of Caliban in the school production of *The Tempest* . . .

Here is one official school activity where Reggie gets to show off something that he is "good at." He is also proud to point out that this is not just a role in any play, but one in a play by Shakespeare. Here his own reward, which is not just doing something that he is good at, but doing it publicly so that he can receive the attention and respect of adults and peers, coincides with the school's educational agenda of creating an interest in Shakespeare among children.

Reggie also plays for an audience in the classroom, where he gets in trouble for disruption. He describes one of the moments for me embellished with a comic imitation of the teacher's female voice and his own swaggering demeanor as he tells the story . . .

This performance, like others I witnessed, is a strategy for positioning oneself in the center of the room in a face-off with the teacher, the most powerful person up to that moment. Fundamental to the performance is engagement with power; authority is teased, challenged, even occasionally toppled from its secure heights for brief moments. Children-generated theatrics allow the teasing challenge of adult power that can expose its chinks and weaknesses. The staged moments heighten tension, test limits, vent emotions, perform acts of courage. . . .

* * *

So why are the black kids "more open" in their confrontations with power? Why not be really "smart" and adopt a style of masculinity that allows them to engage in these rituals that spice the school day and help pass time, but carry less risk of trouble because it is within certain mutually understood limits?

These rituals are not merely a way to pass time, but are also a site for constituting a gendered racial subjectivity. For African American boys, the performance of masculinity invokes cultural conventions of speech performance that draws on a black repertoire. Verbal performance is an important medium for black males to establish a reputation, make a name for yourself, and achieve status.[2] . . .

Oral performance has a special significance in black culture for the expression of masculinity. Harper points out that verbal performance functions as an identifying marker for masculinity only when it is delivered in the vernacular and that "a too-evident facility in white idiom can quickly identify one as a white-identified uncle Tom who must also be therefore weak, effeminate, and probably a fag."[3] Though the speech performances that I witnessed were not always delivered in the strict vernacular, the nonverbal, bodily component accompanying it was always delivered in a manner that was the flashy, boldly flamboyant popular style essential to a good performance. The body language and spoken idiom openly engage power in a provocative competitive way. To be indirect, "sly," would not be performing masculinity.

This nonstandard mode of self-representation epitomizes the very form the school seeks to exclude and eradicate. It is a masculine enactment of defiance played in a black key that is bound for punishment. Moreover, the process of adultification translates the encounter from a simple verbal clash with an impertinent child into one interpreted as an intimidating threat.

Though few white girls in the school were referred to the office for disruptive behavior, a significant number of African American girls staged performances, talked back to teachers, challenged authority, and were punished. But there was a difference with the cultural framing of their enactments and those of the boys. . . . Boys expect to get attention. Girls vie for attention too, but it is perceived as illegitimate behavior. As the teacher described it in the referral form, the girl is "demanding attention." The prevailing cultural framework denies her the rights for dramatic public display.

Male and female classroom performance is different in another respect. Girls are not rewarded with the same kind of applause or recognition by peers or by teachers. Their performance is sidelined; it is not given center stage. Teachers are more likely to "turn a blind eye" to such a display rather than call attention to it, for girls are seen as individuals who operate in cliques at most and are unlikely to foment insurrection in the room. Neither the moral nor the pragmatic principle prods teachers to take action. The behavior is not taken seriously; it is rated as "sassy" rather than symptomatic of a more dangerous disorder. In some classrooms, in fact, risk taking and "feistiness" on the part of girls is subtly encouraged given the prevailing belief that what they need is to become more visible, more assertive in the classroom. The notion is that signs of self-assertion on their part should be encouraged rather than squelched.

Disruptive acts have a complex, multifaceted set of meanings for the male Troublemakers themselves. Performance as an expression of black masculinity is a production of a powerful subjectivity to be reckoned with, to be applauded; respect and ovation are in a context where none is forthcoming. The boys' anger and frustration as well as fear motivate the challenge to authority. Troublemakers act and speak out as stigmatized outsiders.

Ritual Performances of Masculinity: Fighting

Each year a substantial number of kids at Rosa Parks get into trouble for fighting. It is the most frequent offense for which they are referred to the Punishing Room. Significantly, the vast majority of the offenders are African American males.[4]

The school has an official position on fighting: it is the wrong way to handle any situation, at any time, no matter what. Schools have good reasons for banning fights: kids can get hurt and when fights happen they sully the atmosphere of order, making the school seem like a place of danger, of violence.

The prescribed routine for schoolchildren to handle situations that might turn into a fight is to tell an adult who is then supposed to take care of the problem. This routine ignores the unofficial masculine code that if someone hits you, you should solve the problem yourself rather than showing weakness and calling an adult to intervene. However, it is expected that girls with a problem will seek out an adult for assistance. Girls are assumed to be physically weaker, less aggressive, more vulnerable, more needy of self-protection; they must attach themselves to adult (or male) power to survive. . . .

Referrals of males to the Punishing Room, therefore, are cases where the unofficial masculine code for problem resolution has prevailed. Telling an adult is anathema to these youth. According to their own codes, the act of "telling" is dangerous for a number of reasons. The most practical of these sees it as a statement to the "whole world" that you are unable to deal with a situation on your own—to take care of yourself—an admission that can have disastrous ramifications when adult authority is absent. This is evident from the stance of a Troublemaker who questions the practical application of the official code by invoking knowledge of the proper male response when one is "attacked" that is shared with the male student specialist charged with enforcing the regulation: "I said, 'Mr. B., if somebody came up and hit you, what would you do?' 'Well,' he says, 'We're not talking about me right now, see.' That's the kind of attitude they have. It's all like on you."

Another reason mentioned by boys for not relying on a teacher to take care of a fight situation is that adults are not seen as having any real power to effectively change the relations among kids:

> If someone keep messing with you, like if someone just keep on and you tell them to leave you alone, then you tell the teacher. The teacher can't do anything about it because, see, she can't hit you or nothing. Only thing she can do is tell them to stop. But then he keep on doing it. You have no choice but to hit 'em. You already told him once to stop.

This belief extends to a distrust of authority figures by these young offenders. The assumption that all the children see authority figures as teachers, police, psychologists as acting on their behalf and trust they will act fairly may be true of middle- and upper-class children brought up to expect protection from authority figures in society. This is not the case with many of the children at the school. Their mistrust of authority is rooted in the historical and locally grounded knowledge of power relations that come from living in a largely black and impoverished neighborhood.

Fighting becomes, therefore, a powerful spectacle through which to explore trouble as a site for the construction of manhood. The practice takes place along a continuum that ranges from play—spontaneous outbreaks of pummeling and wrestling in fun, ritualistic play that shows off "cool" moves seen on video games, on TV, or in movies—to serious, angry socking, punching, fistfighting. A description of some of these activities and an analysis of what they mean provides the opportunity for us to delve under the surface of the ritualized, discrete acts that make up a socially recognizable fight event into the psychic, emotional, sensuous aspects of gender performativity. The circular, interactive flow between fantasmatic images, internal psychological processes, and physical acts suggest the dynamics of attachment of masculine identification.

Fighting is one of the social practices that add tension, drama, and spice to the routine of the school day. Pushing, grabbing, shoving, kicking, karate chopping, wrestling, fistfighting engage the body and the mind. Fighting is about play and games, about anger, and pain, about hurt feelings, about "messing around." To the spectator, a fight can look like serious combat, yet when the combatants are separated by an adult, they claim, "We were only playing." In fact, a single fight event can move along the continuum from play to serious blows in a matter of seconds. As one of the boys explained, "You get hurt and you lose your temper."

Fighting is typically treated as synonymous with "aggression" or "violence," terms that already encode the moral, definitional frame that obscures the contradictory ways that the practice, in all its manifestations, is used in our society. We, as good citizens, can distance ourselves from aggressive and violent behavior. "Violence" as discourse constructs "fighting" as pathological, symptomatic of asocial, dangerous tendencies, even though the practice of "fighting" and the discourses that constitute this practice as "normal" are in fact taken for granted as ritualized resources for "doing" masculinity in the contemporary United States.

The word *fighting* encompasses the "normal" as well as the pathological. It allows the range of meanings that the children, specifically the boys whom I interviewed and observed, as well as some of the girls, bring to the practice. One experience that it is open to is the sensuous, highly charged embodied experience before, during, and after fighting; the elating experience of "losing oneself" that I heard described in fight stories.

War Stories

I began thinking about fights soon after I started interviews with the Troublemakers and heard "fight stories." Unlike the impoverished and reluctantly told accounts of the school day, these stories were vivid, elaborate descriptions of bodies, mental states, and turbulent emotional feelings. They were stirring, memorable moments in the tedious school routine.

Horace described a fight with an older boy who had kept picking on him. He told me about the incident as he was explaining how he had broken a finger one day when we were trading "broken bones" stories.

> When I broke this finger right here it really hurted. I hit somebody in the face. It was Charles. I hit him in the face. You know the cafeteria and how you walk down to go to the cafeteria. Right there. That's where it happened. Charles picked me up and put me on the wall, slapped me on the wall, and dropped me. It hurt. It hurt bad. I got mad because he used to be messing with me for a long time so I just swung as hard as I could, closed my eyes, and just *pow,* hit him in the face. But I did like a roundhouse swing instead of doing it straight and it got the index finger of my right hand. So it was right there, started right here, and all around this part [he is showing me the back of his hand] it hurt. It was swollen. Oooh! It was like this! But Charles, he got hurt too. The next day I came to school I had a cast on my finger and he had a bandage on his ear. It was kinda funny, we just looked at each other and smiled.

The thing that most surprised and intrigued me about Horace's story was that he specifically recalled seeing Charles the next day and that they had looked at each other and smiled. Was this a glance of recognition, of humor, of recollection of something pleasing, of all those things? The memory of the exchanged smile derailed my initial assumption that fighting was purely instrumental. This original formulation said that boys fight because they have to fight in order to protect themselves from getting beaten up on the playground. Fighting from this instrumental perspective is a purely survival practice. Boys do fight to stave off the need to fight in the future; to stop the harassment from other boys on the playground and in the streets. However, this explains only a small group of boys who live in certain environments; it relegates fighting to the realm of the poor, the deviant, the delinquent; the pathological. This position fails to address these physical clashes as the central normative practice in the preparation of bodies, of mental stances, of self-reference for manhood and as the most effective form of conflict resolution in the realm of popular culture and international relations.

I listened closely to the stories to try to make sense of behavior that was so outside of my own experience, yet so familiar a part of the landscape of physical fear and vulnerability that I as a female walked around with every day. I asked school adults about their own memories of school and fighting. I was not surprised to find that few women seemed to recall physical fights at school, though they had many stories of boys who teased them or girlfriends whom they were always "fighting" with. This resonated with my own experience. I was struck, however, by the fact that all of the men whom I talked to had had to position themselves in some way with regard to fighting. I was also struck that several of these men framed the memory of fighting in their past as a significant learning experience.

* * *

As I explored the meaning of fighting I began to wonder how I, as female, had come to be shaped so fighting was not a part of my own corporeal or mental repertoire. A conversation with my brother reminded me of a long forgotten self that could fight, physically, ruthlessly, inflict hurt, cause tears. "We were always fighting," he recalled. "You used to beat me up." Memories of these encounters came back. I am standing with a tuft of my brother's hair in my hand, furious tears in my eyes. Full of hate for him. Kicking, scratching, socking, feeling no pain. Where had this physical power gone? I became "ladylike" repressing my anger, limiting my physical contact to shows of affection, fearful. I wondered about the meaning of being female in a society in which to be female is to be always conscious of men's physical power and to consciously chart one's everyday routines to avoid becoming a victim of this power, but to never learn the bodily and mental pleasure of fighting back.

Bodily Preparations: Pain and Pleasure

Fighting is first and foremost a bodily practice. I think about fighting and physical closeness as I stand observing the playground at recess noticing a group of three boys, bodies entangled, arms and legs flailing. In another area, two boys are standing locked closely in a wrestling embrace. Children seem to gravitate toward physical contact with each other. For boys, a close, enraptured body contact is only legitimate when they are positioned as in a fight. It is shocking that this bodily closeness between boys would be frowned on, discouraged if it were read as affection. Even boys who never get in trouble for "fighting" can be seen engaging each other through the posturing and miming, the grappling of playfight encounters.

This play can lead to "real" fights. The thin line between play and anger is crossed as bodies become vulnerable, hurt, and tempers are lost. One of the white boys in the school who was in trouble for fighting describes the progression this way:

> Well we were messing with each other and when it went too far, he started hitting me and then I hit him back and then it just got into a fight. It was sorta like a game between me, him and Thomas. How I would get on Thomas's back an—he's a big guy—and Stephen would try to hit me and I would wanta hit him back. So when Thomas left it sorta continued and I forgot which one of us wanted to stop—but one of us wanted to stop and the other one wouldn't.

Fighting is about testing and proving your bodily power over another person, both to yourself and to others through the ability to "hurt" someone as well as to experience "hurt."

* * *

Pain is an integral part of fighting. Sometimes it is the reason for lashing out in anger. . . .

Fighting is a mechanism for preparing masculinized bodies through the playful exercise of bodily moves and postures and the routinized rehearsal of sequences and chains of stances of readiness, attack, and defense. Here it is crucial to emphasize that while many boys in the school never ever engage in an actual physical fight with another boy or girl during school hours, the majority engage in some form of body enactments of fantasized "fight" scenarios. They have observed boys and men on TV, in the movies, in video games, on the street, in the playground adopting these stances.

These drills simultaneously prepare and cultivate the mental states in which corporeal styles are grounded. So for instance, boys are initiated into the protocol of enduring physical pain and mental anguish—"like a man"—through early and small infusions of the toxic substance itself in play fights. The practice of fighting is the site for a hot-wiring together of physical pain and pleasure, as components of masculinity as play and bodily hurt inevitably coincide.

Consequently, it also engages powerful emotions. . . .

One of the white boys in the school who had gotten in trouble for fighting described his thoughts and feelings preceding a fight and the moment of "just going black" in a loss of self. . . .

Fighting is a practice, like sports, that is so symbolically "masculine" that expressions of emotion or behavior that might call one's manhood into question are allowed without danger of jeopardizing one's manliness. Even crying is a permissible expression of "masculinity" under these circumstances. . . .

Fighting in school is a space in which boys can feel free to do emotional work.[5] In a social practice that is so incontrovertibly coded as masculine, behaviors marked as feminine, such as crying, can be called upon as powerful wellsprings for action.

One of the questions that I asked all the boys about fighting came out of my own ignorance. My query was posed in terms of identity work around the winning and losing of fights. Did you ever win a fight? Did you ever lose a fight? How did you feel when you lost? How did you feel when you won? I found the answers slippery, unexpected, contradictory. I had anticipated that winning would be described in proud and boastful ways, as success stories. But there seemed to be a surprising reluctance to embellish victory. I learned that I was missing the point by posing the question the way I had in terms of winning and losing. Trey enlightened me when he explained that what was at stake was not winning or losing per se but in learning about the self:

> I won a lot of fights. You know you won when they start crying and stuff or when they stop and leave. I lost fights. Then you feel a little okay. At least you lost. I mean like you ain't goin' win every fight. At least you fought back instead of just standing there and letting them hit you. . . .

Proving yourself to others is like a game, a kind of competition:

> Me and Leslie used to fight because we used to be the biggest boys, but now we don't care anymore. We used to get friends and try and fight each other. I fought him at Baldwin school all the time. We stopped about the fifth grade [the previous year]. Just got tired, I guess.

Standing and proving yourself today can be insurance against future harassment in the yard as you make a name for yourself through readiness to fight: "Like if somebody put their hands on you, then you have to, you have to hit them back. Because otherwise you going be beat up on for the rest of your life." . . .

In constructing the self through fight stories, it is not admirable to represent oneself as the aggressor or initiator in a fight. All the boys whom I talked to about fighting presented themselves as responding to a physical attack that had to be answered in a decisive way. No one presented himself as a "bully," though I knew that Horace had that reputation. Yet he told me that "only fights I been in is if they hit me first."

There are, however, times when it is legitimate to be the initiator. When verbal provocation is sufficient. This is when "family" has been insulted. Talking about "your momma" is tantamount to throwing down the gauntlet . . .

The boys talked about how they learned to fight. How one learns to fight and what one learns about the meaning of fighting—why fight, to fight or not to fight—involved both racial identity and class positioning. Ricky and Duane, two of the Schoolboys, have been enrolled by their parents in martial arts classes. Fighting remains a necessary accoutrement of masculinity that is "schooled," not a "natural" acquisition of doing. As such, it becomes a marker of higher class position. Fighting takes place in an institutionalized arena rather than spontaneously in just any setting. The mind seems to control the body here, rather than vice versa.

Horace, on the other hand, like the majority of boys with whom I talked, explained that he had learned to fight through observation and practice:

> I watched people. Like when I was younger, like I used to look up to people. I still do. I look up to people and they knew how to fight so I just watched them. I just like saw people fight on TV, you know. Boxing and stuff.

Another boy told me that he thought kids learned to fight "probably from theirselves. Like their mom probably say, if somebody hit you, hit them back." This advice about proper behavior is grounded in the socialization practices that are brought into school as ways of responding to confrontations.

Gender Practice and Identification

Fighting acts reproduce notions of essentially different gendered natures and the forms in which this "difference" is grounded. Though class makes some difference in when, how, and under what conditions it takes place, fighting is the hegemonic representation of masculinity. Inscribed in the male body—whether individual males fight or not, abjure fighting or not—is the potential for this unleashing of physical power. By the same token, fighting for girls is considered an aberration, something to be explained.

Girls do get in fights at school. Boys asserted that girls can fight, even that "sometimes they get in fights easier. Because they got more attitude." Indeed, girls do make a name for themselves this way. One of the girls at Rosa Parks was in trouble several times during the school year for fighting. Most of her scraps were with the boys who liked to tease her because she was very tall for her age. This, however, was not assumed to be reflective of her "femaleness" but of her individuality. Mr. Sobers, for example, when I asked him about her, made a point of this singularity rather than explaining her in terms of race, class, or gender: "Oh, Stephanie is just Stephanie."

Fighting is not a means of "doing gender" for girls. They do not use physical clashes as a way to relate to each other in play. Girls did not practice "cool" moves or engage in play fights with each other. They used other strategies for making the day go by, such as the chain of stories about other children, the "he said, she said," which can build up to a more physical confrontation.[6] More often it leads to injured feelings, the isolation and ostracism of individuals, and the regrouping of friendships. On the playground at Rosa Parks, girls were more likely to interact physically with boys than with other girls. They often initiated encounters with boys to play chase games by pushing, prodding, hitting, or bumping into them.

Through male fighting we can see how gender difference is grounded in a compulsory and violently enforced heterosexuality. The interaction involves the convergence of the desire for physical and emotional closeness with another, the anxiety over presenting a convincing performance of a declarative act of identification, and the risk of ostracism or punishment. Boys from an early age learn that affectionate public physical contact such as an embrace with those who are seen as most like oneself, other males, is taboo. For them, a physical embrace, the close intertwining of bodies is culturally permissible only in the act of the rituals of the fight.[7] Thus the fulfillment of desire for physical intimacy, for body contact, can most safely be accomplished publicly through the apparent or actual infliction and experience of bodily pain. A desire for closeness, for identification with a reflection of oneself, can be achieved through an act that beckons and embraces using apparently threatening and hostile gestures. In a revealing story of the constraining boundaries of male self-expression, Mac an Ghaill recounts how the public exchange of flowers between two males in a high school was regarded by personnel as more unnatural, reprehensible, and threatening than the physical violence of the fight that the gesture provoked.[8]

Most men don't have to actually fight; they can participate in the inscription of power on male bodies through watching. Fighting acts are a major form of entertainment in our society. From popular cultural figures on television, screen, in video games, boxing and wrestling matches, the use of fists and agile feet deeply encode the hegemonic representation of masculinity. Even as the pantheon of cultural superheroes real and fantasy, such as Mike Tyson, Dennis Rodman, Schwarzenegger's "Terminator" and Stallone's "Rambo," are supplemented by cultural representations of the "New Man" who is a more fitting partner for the stronger images of the liberated woman—Kevin Costner's Robin Hood or Keaton's Batman, for instance—these more "sensitive" heroes are also still skilled and courageous physical fighters. They become "real men" because they can, when inevitably called upon to do so, physically vanquish the villain and save the female "victim." The fight scene/shootout between hero and villain continues to be the most enduring convention of climax and resolution in film and television. Violence remains the most predictable way of males resolving conflict and problems in the popular culture as well as in world events.

The presence of spectators is a key element. The performance of fighting in settings such as the playground, the boxing ring, the movie theater, the sports arena is not only rousing entertainment for an audience, but a reinscription of an abstract masculine power. This performance is affirmed by ardent spectators, mostly men but some women, who consume the ritualistic enactment of raw, body power. Video games are an excellent example of how even males who avoid physical aggressive behavior in their own personal life symbolically perform a violent masculinity in order to play the game at all.[9]

Fighting is the emblematic ritual performance of male power. Participation in this ritual for boys and for men is not an expression of deviant, antisocial behavior but is profoundly normative, a thoroughly social performance. Though it is officially frowned on as a means of resolving personal problems, it is in fact culturally applauded as a way of settling differences among men. As boys mock fight through imitating martial arts movements with or without an audience, playfight with peers, play video games, they bodily and psychically inhabit male power through fantasy and imagination.

For Troublemakers, who are already sidelined as academic failures, one route to making a name for yourself, for expressing "normalcy," competency, and humanity, is through this identification with physical power. Once again, a sense of anger and frustration born of marginalization in school intensifies the nature of these performances.

Race makes a difference in how physical power is constituted and perceived. African American boys draw on a specific repertoire of racial images as well as the lived experience and popular knowledge from the world outside the school. Most of the black boys live in an environment where being mentally and physically prepared to stand up for oneself through words and deeds is crucial. However, there is another reason specifically grounded in the history and evolution of race relations in the United States. Up to the 1960s, physical violence wielded by whites in the form of individuals, mobs, or the state was the instrument used to police the racial order; demonstrations of male privilege or assertions of rights on the part of black men was the cause for brutal retaliation. The prevailing wisdom in black communities was that in order to survive males had to be carefully taught to mask any show of power in confrontations with whites. With the emergence of Black Power as an ideology and a practice, the right of black men to stand up for their manhood and their racial pride through physical force was asserted. This was the right to have the physical privileges of white men. This "right" is inculcated into young black males in family and community, many of whom are taught, "Don't let anyone take advantage of you. If someone hits you, you hit them back." First blows are not always physical, but sometimes symbolic; racial epithets are violent attacks. A physical response is especially likely on the part of the Troublemakers, who have a heightened racial consciousness.

Simultaneously, this manifestation of physicality is the very material presence that the school seeks to exclude: black males are already seen as embodying the violence and aggression that will drive away "desirable" families and their children. Fighting on the part of black boys is more visible as a problem, so it is viewed with extreme concern and responded to more swiftly and harshly. Once again, the process of adultification of black male behavior frames the act as symptomatic of dangerous tendencies. The Troublemakers, who have already been labeled as bound for jail, have little to lose and everything to gain in using this form of rule breaking as a way of making a name for themselves, gaining recognition through performances of masculinity.

NOTES

1. Judith Butler, *Bodies That Matter: On the Discursive Limits of "Sex"*(New York: Routledge, 1993), 95.
2. Geneva Smitherman, *Talkin and Testifyin: Language of Black America* (Detroit: Wayne State University Press, 1977); Lawrence Levine, *Black Culture and Black Consciousness: Afro-American Folk Thought from Slavery to Freedom* (New York: Oxford University Press, 1977); Philip Brian Harper, *Are We Not Men? Masculine Anxiety and the Problem of African-American Identity* (New York: Oxford University Press, 1996); Keith Gilyard, *Voices of the Self: A Study of Language Competence* (Detroit: Wayne State University Press, 1991).
3. Harper, *Are We Not Men?* 11.
4. One-quarter of the 1,252 referrals to the Punishing Room were for fighting; four-fifths of the incidents involved boys, nine out of ten of whom were African Americans. All except three of the girls who were in fights were black.
5. Arlie Russell Hochschild, *The Managed Heart: Commercialization of Human Feeling* (Berkeley and Los Angeles: University of California Press, 1983). Hochschild explores the feeling rules that guide and govern our own emotional displays as well as how we interpret the emotional expression of others.
6. For a full discussion of this strategy see Marjorie Harness Goodwin, *He-Said–She Said: Talk as a Social Organization among Black Children* (Bloomington: Indiana University Press, 1990).
7. In the U.S. context, we see passionate public embraces between males in certain high-contact team sports such as football, basketball, and soccer in moments of great emotion. It is less likely to be witnessed in sports such as tennis or golf where team camaraderie cannot develop or where the masculinity of the participants is not so indubitably demonstrated.
8. Mairtin Mac an Ghaill, *The Making of Men: Masculinities, Sexualities, and Schooling* (Buckingham, England: Open University Press, 1994), 1.
9. R. W. Connell, "Teaching the Boys: New Research on Masculinity and Gender Strategies for Schools," *Teachers College Record* 98, no. 2 (1996).

21

My Life as a Man

BY ELIZABETH GILBERT

Journalist Elizabeth Gilbert wrote this article for GQ *in 2001. She became a "man" with the help of Diane Torr, a performance artist who runs workshops designed to help women experience masculinity by becoming men. Gilbert learned that it takes more to become a man than simply bandaging her breasts and wearing a birdseed penis. This article helps us to understand the complex emotional and interactional basis upon which we all "do gender," as well as the close relationship between doing femininity and doing masculinity.*

1. How does Gilbert become a "man"?
2. What effect does this transformation have on the way she interacts with and looks at women?
3. What effect does this transformation have on her long-term relationship with her husband and others she interacts with?

* * *

The first time I was ever mistaken for a boy, I was 6 years old. I was at the county fair with my beautiful older sister, who had the long blond tresses one typically associates with storybook princesses. I had short messy hair, and I had scabs all over my body from falling out of trees. My beautiful sister ordered a snow cone. The lady at the booth asked, "Doesn't your little brother want one, too?"

I was mortified. I cried all day.

The last time I was mistaken for a boy was only a few weeks ago. I was eating in a Denny's with my husband, and the waitress said, "You fellas want some more coffee?"

This time I didn't cry. It didn't even bother me, because I've grown accustomed to people making the mistake. Frankly, I can understand why they do. I'm afraid I'm not the most feminine creature on the planet. I don't exactly wish to hint that Janet Reno and I were separated at birth, but I do wear my hair short, I am tall, I have broad shoulders and a strong jaw, and I have never really understood the principles of cosmetics. In many cultures, this would make me a man already. In some very primitive cultures, this would actually make me a king.

But sometime after the Denny's incident, I decided, *Ah, to hell with it. If you can't beat 'em, join 'em.* What would it take, I began to wonder, for me to actually transform into a man? To live that way for an entire week? To try to fool everyone? . . .

Fortunately, I have plenty of male friends who rally to my assistance, all eager to see me become the best man I can possibly be. And they all have wise counsel to offer about exactly How to Be a Guy:

"Interrupt people with impunity from now on," says Reggie. "Curse recklessly. And never apologize."

"Never talk about your feelings," says Scott. "Only talk about your accomplishments."

"The minute the conversation turns from something that directly involves you," says Bill, "let your mind wander and start looking around the room to see if there's anything nearby you can have sex with."

"If you need to win an argument," says David, "just repeat the last thing the guy you're fighting with said to you, but say it much louder."

So I'm thinking about all this, and I'm realizing that I already do all this stuff. I always win arguments, I'm shamefully slow to apologize, I can't imagine how I could possibly curse any more than I already goddamn do, I've spent the better part of my life looking around to see what's available to have sex with, I can't shut up about my accomplishments, and I'm probably interrupting you right this moment.

Another one of my friends warns, "You do this story, people are gonna talk. People might think you're gay." Aside from honestly not caring what people think, I'm not worried about this possibility at all. I'm worried about something else entirely: that this transformation thing might be *too* easy for me to pull off.

What I'm afraid I'll learn is that I'm *already* a man.

My real coach in this endeavor, though, is a woman. Her name is Diane Torr. Diane is a performance artist who has made her life's work the exploration of gender transformation. As a famous drag king, she has been turning herself into a man for twenty years. She is also known for running workshops wherein groups of women gather and become men for a day.

I call Diane and explain my goal, which is not merely to dress up in some silly costume but to genuinely pass as male and to stay in character for a week.

"That's a tough goal," Diane says, sounding dubious. "It's one thing to play with gender for the afternoon, but really putting yourself out there in the world as a man takes a lot of balls, so to speak. . . ."

Diane agrees to give me a private workshop on Monday. She tells me to spend the weekend preparing for my male life and buying new clothes. Before hanging up, I ask Diane a question I never thought I would ever have to ask anybody:

"What should I bring in terms of genitalia?"

This is when she informs me of the ingredients for my penis.

"Of course," I say calmly.

I write *birdseed* on my hand, underline it twice and make a mental note to stay away from the aviary next week.

I SPEND THE WEEKEND INVENTING MY CHARACTER.

One thing is immediately clear: I will have to be younger. I'm 31 years old, and I look it, but with my smooth skin, I will look boyish as a man. So I decide I will be 21 years old for the first time in a decade.

As for my character, I decide to keep it simple and become Luke Gilbert—a midwestern kid new to the city, whose entire background is cribbed from my husband, whose life I know as well as my own.

Luke is bright but a slacker. He really doesn't give a damn about his clothes, for instance. Believe me, I know—I'm the one who shopped for Luke all weekend. By Sunday night, Luke owns several pairs of boring Dockers in various shades of khaki, which he wears baggy. He has Adidas sneakers. He has some boxy short-sleeve buttondown shirts in brown plaids. He has a corduroy jacket, a bike messenger's bag, a few baseball caps and clean underwear. He also has, I'm sorry to report, a really skinny neck.

I haven't even met Luke yet, but I'm beginning to get the feeling he's a real friggin' geek.

THE TRANSFORMATION BEGINS PAINLESSLY ENOUGH.

It starts with my hair. Rayya, my regular hairdresser, spends the morning undoing all her work of the past months—darkening out my brightest blond highlights, making me drab, brownish, inconsequential; chopping off my sassy Dixie Chick pixie locks and leaving me with a blunt cut.

"Don't wash it all week," Rayya advises. "Get good and greasy; you'll look more like a guy."

Once the hair is done, Diane Torr gets to work on me. She moves like a pro, quick and competent. Together we stuff my condom ("This is the arts-and-crafts portion of the workshop!"), and Diane helps me insert it into my Calvins. She asks if I want my penis to favor the left or right side. Being a traditionalist, I select the right. Diane adjusts me and backs away; I look down and there it is—my semierect penis, bulging slightly against my briefs. I cannot stop staring at it and don't mind saying that it freaks me out to no end. Then she tries to hide my breasts. To be perfectly honest, my breasts are embarrassingly easy to make disappear. Diane expertly binds them down with wide Ace bandages. Breathing isn't easy, but my chest looks pretty flat now—in fact, with a men's undershirt on, I almost look as if I have well-developed pectoral muscles.

But my ass? Ah, here we encounter a more troublesome situation. I don't want to boast, but I have a big, fat, round ass. You could lop off huge chunks of my ass, make a nice osso buco out of it, serve it up to a family of four and still eat the left-overs for a week. This is a woman's ass, unmistakably. But once I'm fully in costume, I turn around before the mirror and see that I'm going to be OK. The baggy, low-slung pants are good ass camouflage, and the boxy plaid shirt completely eliminates any sign of my waist, so I don't have that girlie hourglass thing happening. I'm a little pear-shaped, perhaps, but let us not kid ourselves, people. There are pear-shaped men out there, walking among us every day.

Then Diane starts on my facial transformation. She has brought crepe hair—thin ropes of artificial hair in various colors, which she trims down to a pile of golden brown stubble. I elect, in homage to Tom Waits, to go with just a small soul patch, a minigoatee, right under my bottom lip. Diane dabs my face with spirit gum—a kind of skin-friendly rubber cement—and presses the hair onto me. It makes for a shockingly good effect. I suggest sideburns, too, and we apply these, making me look like every 21-year-old male art student I've ever seen. Then we muss up and darken my eyebrows. A light shadow of brown under my nose gives me a hint of a mustache. When I look in the mirror, I can't stop laughing. *I am a goddamn man, man!*

Well, more or less.

Diane looks me over critically. "Your jaw is good. Your height is good. But you should stop laughing. It makes you look too friendly, too accessible, too feminine." I stop laughing. She stares at me. "Let's see your walk."

I head across the floor, hands in my pockets.

"Not bad," Diane says, impressed.

Well, I've been practicing. I'm borrowing my walk from Tim Goodwin, a guy I went to high school with. Tim was short and slight but an amazing basketball player (we all called him "Tim *God*win"), and he had an athletic, kneeknocking strut that was very cool. There's also a slouch involved in this walk. But it's—and this is hard to explain—a *stiff* slouch. Years of yoga have made me really limber, but as Luke, I need to drop that ease of motion with my body, because men are not nearly as physically free as women. Watch the way a man turns his head: His whole upper torso turns with it. Unless he's a dancer or a baseball pitcher, he's probably operating his entire body on a ram-rod, unyielding axis. On the other hand, watch the way a woman drinks from a bottle. She'll probably tilt her whole head back to accommodate the object, whereas a man would probably hold his neck stiff, tilting the bottle at a sharp angle, making the bottle accommodate *him*. Being a man, it seems, is sometimes just about not budging.

Diane goes on to coach my voice, telling me to lower the timbre and narrow the range. She warns me against making statements that come out as questions, which women do constantly (such as when you ask a woman where she grew up and she replies, "Just outside Cleveland?"). But I don't do that begging-for-approval voice anyway, so this is no problem. As I'd suspected, in fact, all this turning-male stuff is coming too easily to me.

But then Diane says, "Your eyes are going to be the real problem. They're too animated, too bright. When you look at people, you're still too engaged and interested. You need to lose that sparkle, because it's giving you away."

The rest of the afternoon, she's on me about my eyes. She says I'm too flirtatious with my eyes, too encouraging, too appreciative, too attentive, too *available.* I need to intercept all those behaviors, Diane says, and erase them. Because all that stuff is "shorthand for girl." Girls typically flirt and engage and appreciate and attend; men typically don't. It's too generous for men to give themselves away in such a manner. Too dangerous, even. Granted, there are men in this world who are engaging, attentive and sparkly eyed, but Luke Gilbert cannot be one of them. Luke Gilbert's looks are so on the border of being feminine already that I can't afford to express any behavior that is "shorthand for girl," or my cover is blown. I can only emit the most stereotypical masculine code, not wanting to offer people even the faintest hint that I'm anything but a man.

Which means that gradually throughout Monday afternoon, I find myself shutting down my entire personality, one degree at a time. It's very similar to the way I had to shut down my range of physical expression, pulling in my gestures and stiffening up my body. Similarly, I must not budge emotionally. I feel as if I'm closing down a factory, silencing all the humming machines of my character, pulling shut the gates, sending home the workers. All my most animated and familiar facial expressions have to go, and with them go all my most animated and familiar emotions. Ultimately, I am left with only two options for expression—boredom and aggression. Only with boredom and aggression do I truly feel male. It's not a feeling I like at all, by the way. In fact, I am amazed by how much I don't like it. We've been laughing and joking and relating all morning, but slowly now, as I turn into Luke, I feel the whole room chill.

Toward the end of the afternoon, Diane gives me her best and most disturbing piece of advice.

"Don't look at the world from the surface of your eyeballs," she says. "All your feminine availability emanates from there. Set your gaze back in your head. Try to get the feeling that your gaze originates from two inches behind the surface of your eyeballs, from where your optic nerves begin in your brain. Keep it right there."

Immediately, I get what she's saying. I pull my gaze back. I don't know how I appear from the outside, but the internal effect is appalling. I feel—for the first time in my life—a dense barrier rise before my vision, keeping me at a palpable distance from the world, roping me off from the people in the room. I feel dead eyed. I feel like a reptile. I feel my whole face change, settling into a hard mask.

Everyone in the room steps back. Rayya, my hairdresser, whistles under her breath and says, "Whoa . . . you got the guy vibe happenin' now, Luke."

Slouching and bored, I mutter a stony thanks.

Diane finally takes me outside, and we stroll down the street together. She has dressed in drag, too. She's now Danny King—a pompous little man who works in a Pittsburgh department store. She seems perfectly at ease on the street, but I feel cagey and nervous out here in the broad daylight, certain that everyone in the world can see that my face is covered with fake hair and rubber cement and discomfort. The only thing that helps me feel even remotely relaxed is the basketball I'm loosely carrying under my arm—a prop so familiar to me in real life that it helps put me at ease in disguise. We head to a nearby basketball court. We have a small crowd following us—my hairdresser, the makeup artist, a photographer. Diane and I pose for photos under the hoop. I set my basketball down, and almost immediately, a young and muscular black guy comes over and scoops it off the pavement.

"Hey," he says to the crowd. "Whose basketball is this?"

Now, if you want to learn how to define your personal space as a man, you could do worse than take lessons from this guy. His every motion is offense and aggression. He leads with his chest and chin, and he's got a hard and cold set of eyes.

"I said, whose basketball is this?" he repeats, warning with his tone that he doesn't want to have to ask again.

"It's hers," says my hairdresser, pointing at me.

"Hers?" The young man looks at me and snorts in disgust. "What are you talkin' about, *hers*? That ain't no *her*. That's a *guy*."

My first gender victory!

But there's no time to celebrate this moment, because this aggressive and intimidating person needs to be dealt with. Now, here's the thing. Everyone on the court is intimidated by this guy, but I am not. In this tense moment, mind you, I have stopped thinking like Luke Gilbert; I'm back to thinking like Liz Gilbert. And Liz Gilbert always thinks she can manage men. I don't know if it's from years of tending bar, or if it's from living in lunatic-filled New York City, or if it's just a ridiculous (and dangerously naive) sense of personal safety, but I have always believed in my heart that I can disarm any man's aggression. I do it by paying close attention to the aggressive man's face and finding the right blend of flirtation, friendliness and confidence to put on my face to set him at ease, to remind him: *You don't wanna hurt me, you wanna like me.* I've done this a million times before. Which is why I'm looking at this scary guy and I'm thinking, *Give me thirty seconds with him and he'll be on my side.*

I step forward. I open up my whole face in a big smile and say teasingly, "Yeah, that's my basketball, man. Why, you wanna play? You think you can take me?"

"You don't know nothin' about this game," he says.

In my flirtiest possible voice, I say, "Oh, I know a *little* somethin' about this game. . . ."

The guy takes a menacing step forward, narrows his eyes and growls, "You don't know *shit* about this game."

This is when I snap to attention. This is when I realize I'm on the verge of getting my face punched. What the hell am I doing? This guy honestly thinks I'm a man! Therefore, my whole cute, tomboyish, I'm-just-one-of-the-guys act is not working. One-of-the-guys doesn't work when you actually *are* one of the guys. I have forgotten that I am Luke Gilbert—a little white loser on a basketball court who has just challenged and pissed off and *flirted* with an already volatile large black man. I have made a very bad choice here. I've only been on the job as a male for a few minutes, but it appears as though I'm about to earn myself a good old-fashioned New York City ass-kicking.

He takes another step forward and repeats, "You don't know shit about nothin'."

"You're right, man," I say. I drop my eyes from his. I lower my voice, collapse my posture, show my submission. I am a stray dog, backing away from a fight, head down, tail tucked. "Sorry, man. I was just kidding. I don't know anything about basketball."

"Yeah, that's right," says the guy, satisfied now that he has dominated me. "You don't know shit."

He drops the ball and walks away. My heart is slamming. I'm angry at my own carelessness and frightened by my newfound helplessness. Luke didn't know how to handle that guy on the court, and Luke almost got thrown a beating as a result (and would have deserved it, too—the moron). Realizing this makes me feel suddenly vulnerable, suddenly aware of how small I've become.

My hands, for instance, which have always seemed big and capable to me, suddenly appear rather dainty when I think of them as a man's hands. My arms, so sturdy only hours before, are now the thin arms of a weenie-boy. I've lost this comfortable feeling I've always carried through the world of being strong and brave. A five-foot-nine-inch, 140-pound woman can be a pretty tough character, after all. But a five-foot-nine-inch, 140-pound man? Kinda small, kinda wussy. . . .

* * *

My world-famously tolerant husband seems to have no trouble with my transformation at first. He unwinds my breast bandages every night before bed and listens with patience to my complaints about my itching beard. In the mornings before work, he binds up my breasts again and lends me his spice-scented deodorant so I can smell more masculine. We vie for mirror space in the bathroom as he shaves off his daily stubble and I apply mine. We eat our cereal together, I take my birth control pills, I pack my penis back into my slacks. . . .

It's all very domestic.

Still, by Wednesday morning, my husband confesses that he doesn't want to hang around with me in public anymore. Not as long as I'm Luke. It's not that he's grossed out by my physical transformation, or threatened by the sexual politics at play, or embarrassed by the possibility of exposure. It's simply this: He is deeply, emotionally unsettled by my new personality.

"I miss you," he says. "It's seriously depressing for me to be around you this way."

What's upsetting to Michael is that as a man, I can't give him what he has become accustomed to getting from me as a woman. And I'm not talking about sex. Sex can always be arranged, even this week. (Although I do make a point now of falling asleep immediately after it's over, just to stay in character.) What Michael hates is that I don't engage him anymore. As Luke, I don't laugh at my husband's jokes or ask him about his day. Hell, as Luke, I don't even have a husband—just another drinking buddy whose jokes and workday concerns I don't really care about. Michael, still seeing his wife under her goatee, keeps thinking I'm mad at him, or—worse—bored by him. But I can't attend to him on this, can't reassure him, or I risk coming across like a girl.

The thing is, I don't like Luke's personality any more than Michael does. As Luke, I feel completely and totally bound—and not just because of the tight bandage wrapped around my chest. I keep thinking back to my drag-king workshop, when Diane Torr talked about "intercepting learned feminine habits." She spoke of those learned feminine habits in slightly disparaging terms. Women, she said, are too attentive, too concerned about the feelings of others, too *available*. This idea of women as lost in empathy is certainly a standard tenet of feminism (Oprah calls it the Disease to Please), and, yes, there are many women who drown in their own overavailability. But I've never personally felt that attentiveness and engagement are liabilities. As a writer—indeed, as a *human being*—I think the most exciting way you can interact with this fantastic and capricious world is by being completely available to it. Peel me wide open; availability is my power. I would so much rather be vulnerable and experience existence than be strong and defend myself from it. And if that makes me a girlie-girl, then so be it—I'll be a goddamn girlie-girl.

Only, this week I'm not a girl at all. I'm Luke Gilbert. And poor Luke, I must say, is completely cut off from the human experience. The guy is looking at the world from a place two inches behind his eyeballs. No wonder my husband hates being around him. I'm not crazy about him myself.

* * *

[Wednesday], I'm walking home alone. Just ahead of me, a blond woman steps out of a bar, alone. She's screamingly sexy. She's got all the props—the long hair, the tiny skirt, the skimpy top, the wobbly stiletto heels, the eternal legs. I walk right behind this woman for several blocks and observe the tsunami she causes on 23rd Street in every man she passes—everyone has to react to her somehow. What amazes me, though, is how many of the men end up interacting with *me* after passing *her*. What happens is this: She saunters by, the guy stares at her in astonishment and then makes a comment about her to me because I'm the next man on the scene. So we have a little moment together, the guy and me, in which we share an experience. We get to bond. It's an ice-breaker for us.

The best is the older construction worker who checks out the babe, then raises his eyebrows at me and declares: "Fandango!"

"You said it!" I say, but when I walk on by, he seems a little disappointed that I haven't stuck around to talk more about it with him.

This kind of interaction happens more than a dozen times within three blocks. Until I start wondering whether this is actually the game.

Until I start suspecting that these guys maybe don't want to talk to the girl at all, that maybe they just desperately want to talk to *one another.*

Suddenly, I see this sexy woman in front of me as being just like sports; she's an excuse for men to try to talk to one another. She's like the Knicks, only prettier—a connection for people who otherwise cannot connect at all. It's a very big job, but I don't know if she even realizes she's doing it.

⋆ ⋆ ⋆

[Friday] night, taking a friend's advice, I go out drinking in the East Village, where seven out of ten young men look just like Luke Gilbert. I end up at a bar that is crawling with really cute pierced-nosed girls. I'm wondering whom I should try to pick up when an opportunity falls into my lap. A pretty red-haired girl in a black camisole walks into the bar alone. She has cool tattoos all over her arms. The bouncer says to her, "Hey, Darcy, where's your crowd tonight?"

"Everyone copped out," Darcy says. "I'm flying solo."

"So lemme buy you a drink," I call over from the bar.

"Rum and Coke," she says, and comes over to sit next to me.

Fandango!

We get to talking. Darcy's funny, friendly, from Tennessee. She tells me all about her roommate problems. She asks me about myself, but I don't share—Luke Gilbert is not available for sharing. Instead, I compliment Darcy on her pretty starfish necklace, which Darcy tells me was a gift from a childhood neighbor who was like a grandmother to her. I ask Darcy about her job, and she tells me she works for a publishing house that prints obscure journals with titles like *Catfish Enthusiast Monthly.*

"Damn, and here I just let my subscription to *Catfish Enthusiast Monthly* run out," I say, and she laughs. Darcy actually does that flirty thing girls do sometimes where they laugh and touch your arm and move closer toward you all at the same time. I know this move. I've been doing this move my whole life. And it is with this move and this touch and this laugh that I lose my desire to play this game anymore, because Darcy, I can tell, actually likes Luke Gilbert. Which is incredible, considering that Luke is a sullen, detached, stiff guy who can't make eye contact with the world. But she still likes him. This should feel like a victory, but all I feel like is a complete shitheel. Darcy is nice. And here I'm lying to her already.

Now I really *am* a guy.

"You know what, Darcy?" I say. "I have to go. I'm supposed to hook up with some friends for dinner."

She looks a little hurt. But not as hurt as she would look if, say, we dated for a month and then she found out the truth about me.

I give her a little kiss good-bye on the cheek.

"You're great," I tell her.

And then I'm done.

UNDOING IT ALL TAKES A FEW DAYS.

Rubbing alcohol gets the last of the spirit gum and fake hair off my face. I pluck my eyebrows and put on my softest bra (my skin has become chafed from days of binding and taping). I scatter my penis across the sidewalk for the pigeons. I make an appointment to get my hair lightened again. I go to yoga class and reawaken the idea of movement in my body. I cannot wait to get rid of this gender, which I have not enjoyed. But it's a tricky process, because I'm still walking like Luke, still standing like Luke, still thinking like Luke.

In fact, I don't really get my inner Liz back until the next weekend. It's not until the next Saturday night, when I am sitting at a bar on my own big fat ass, wearing my own girlie jeans, talking to an off-duty New York City fireman, that I really come back into myself. The fireman and I are both out with big groups, but somehow we peel off into our own private conversation. Which quickly gets serious. I ask him to tell me about the crucifix around his neck, and he says he's been leaning on God pretty hard this year. I want to know why. The fireman starts telling me about how his beloved father died this winter, and then his fiancée left him, and now the pressures of his work are starting to kill him, and there are times when he just wishes he could cry but he doesn't want people to see him like that. My guy friends are all playing darts in the corner, but I'm the one sitting here listening to this fireman tell me about how he never cries because his dad was such a hard-ass Irish cop, don'tcha know, because he was raised to hang so tough.

I'm looking right into this guy. I'm not touching him at all, but I'm giving him my entire self. He needs me right now, to tell all this to. He can have me. I've got my eyes locked on him, and I can feel how bad he wants to cry, and with my entire face I am telling this man: *Tell me everything.*

He says, "Maybe I was hard on her, maybe that's why she left me, but I was so worried about my father. . . ."

The fireman digs at his eye with a fist. I hand him a bar napkin. He blows his nose. He keeps talking. I keep listening. He can talk to me all night because I am unbound and I am wide-open. I'm open around the clock, open twenty-four hours a day; I never close. I'm really concerned for this guy, but I'm smiling while he spills his story because it feels so good to catch it. It feels so good to be myself again, to be open for business again—open once more for the rewarding and honest human business of complete *availability.*

22

Slut!

Growing Up Female with a Bad Reputation

BY LEORA TANENBAUM

Leora Tanenbaum uses her own life experiences to help us understand the complex and powerful link between gender and sexuality. In a world permeated by sexual images (as will be described in Chapter 6) and increased sexual activity, she argues that being sexually active is still stigmatized for girls and women. In this selection from her book of the same title, she also addresses responses to her book that she received from various sources. Her responses to these comments help us to understand what the situation is for young girls who are thought to be sexually active and how we might change it.

1. What does she mean by "slut-bashing"? Have you seen instances of it?
2. What does power have to do with the social construction of femininity and sexuality?
3. How does "slut-bashing" maintain traditional gender patterns and pit women against women?

Women living in the United States are fortunate indeed. Unlike women living in Muslim countries, who are beaten and murdered for the appearance of sexual impropriety, we enjoy enormous sexual freedom.[1] Yet even we are routinely evaluated and punished for our sexuality. In 1991, Karen Carter, a twenty-eight-year-old single mother, lost custody of her two-year-old daughter in a chain of events that began when she called a social service hot line to ask if it's normal to feel sexual arousal while breast feeding. Carter was charged with sexual abuse in the first degree, even though her daughter showed no signs of abuse; when she revealed in court that she had had a lifetime total of eight (adult male) lovers, her own lawyer referred to her "sexual promiscuity."[2] In 1993, when New Mexico reporter Tamar Stieber filed a sex discrimination lawsuit against the newspaper where she worked because she was earning substantially less than men in similar positions, defense attorneys deposed her former lover to ask him how often they'd had sex.[3] In the 1997 sexual-harassment lawsuits against Mitsubishi Motor Manufacturing, a company lawyer asked for the gynecological records of twenty-nine women employees charging harassment, and wanted the right to distribute them to company executives.[4] And in 1997 a North Carolina woman sued her husband's secretary for breaking up their nineteen-year-marriage and was awarded $1 million in damages by a jury. During the seven-day trial the secretary was described as a "matronly" woman who deliberately began wearing heavy makeup and short skirts in order to entice the husband into an affair.[5] . . .

In the realm of sexual choices we are light-years beyond the 1950s. Today a teenage girl can explore her sexuality without getting married, and most do. By age eighteen over half of all girls and nearly three quarters of all boys have had intercourse at least once.[6] Yet at the same time, a fifties-era attitude lingers: Teens today are fairly conservative about sex. A 1998 *New York Times*/CBS News poll of a thousand teens found that 53 percent of girls believe that sex before marriage is "always wrong," while 41 percent of boys agree.[7] Teens may be having sex, but they also look down on others, especially girls, who are sexually active. Despite the sexual revolution, despite three decades of feminism, despite the Pill, and despite legalized abortion, teenage girls today continue to be defined by their sexuality. The sexual double standard—and the division between "good" girls and "bad" or "slutty" ones—is alive and well. Some of the rules have changed, but the playing field is startlingly similar to that of the 1950s. . . .

In 1988, educators Janie Victoria Ward and Jill McLean Taylor surveyed Massachusetts teenagers across six different ethnic groups—black, white, Hispanic, Haitian, Vietnamese, and Portuguese—and found that the different groups upheld different sexual values. But one thing was universal: The sexual double standard. Regardless of race or ethnicity, "boys were generally allowed more freedom and were assumed to be more sexually active than girls." Ward and Taylor found that "sexual activity for adolescent males usually met cultural expectations and was generally accepted by adults and peers as part of normal male adolescence. . . . In general, women are often seen in terms of their sexual reputation rather than in terms of their personal characteristics."[8]

The double standard, we know, does not vaporize after high school. Sociologist Lillian Rubin surveyed six hundred students in eight colleges around the country in the late 1980s and found that 40 percent of the sexually active women said that they routinely understate their sexual experience because "my boyfriend wouldn't like it if he knew," "people wouldn't understand," and "I don't want him to think I'm a slut." Indeed, these women had reason to be concerned. When Rubin queried the men about what they expected of the women they might marry, over half said that they would not want to marry a woman who had been "around the block too many times," that they were looking for someone who didn't "sleep around," and that a woman who did was a "slut."[9]

Similarly when sex researcher Shere Hite surveyed over 2,500 college men and women, 92 percent of the men claimed that the double standard was unfair. Yet overwhelmingly they themselves upheld it. When asked, "If you met a woman you liked and wanted to date, but then found out she had had sex with ten to twenty men during the preceding year, would you still like her and take her seriously?," 65 percent of the men admitted that they would not take her seriously. At the same time only 5 percent said they would lose respect if a male friend had had sex with ten to twenty women in one year.[10]

Teenage girls who are called sluts today experience slut-bashing at its worst. Caught between the conflicting pressures to have sex and maintain a "good" reputation, they are damned when they do and damned when they don't. Boys and girls both are encouraged to have sex in the teen years—by their friends, magazines, and rock and rap lyrics—yet boys alone can get away with it. "There's no way that anyone who talks to girls thinks that there's a new sexual revolution out there for teenagers," sums up Deborah Tolman, a developmental psychologist at the Wellesley College Center for Research on Women. "It's the old system very much in place." It *is* the old system, but with a twist: Today's teenage girls have grown up after the feminist movement of the late 1960s and 1970s. They have been told their whole lives that they can, and should, do anything that boys do. But soon enough they discover that sexual equality has not arrived. Certain things continue to be the privilege of boys alone.

With this power imbalance, it's no wonder high school girls report feeling less comfortable with their sexual experiences than their male counterparts do. While 81 percent of adolescent boys say that "sex is a pleasurable experience," only 59 percent of girls feel the same way.[11] The statistical difference speaks volumes. Boys and girls both succumb to early sex due to peer and media pressures, but boys still get away with it while girls don't.

* * *

Most people who meet me for the first time are surprised by two things: that I am the type of person who would ever write a book with a title like *Slut!,* and that I was once known as a "slut." I have been described in print as "demure" and as "a petite brunette with wire-rimmed glasses"—code words for nice, shy, bookish. On the Oprah Winfrey show, I was presented as a nice, middle-class woman married to a nice, middle-class man. Over and over I am told, "But you're so clean-cut—you don't seem like a slut at all."

My point exactly: Any girl or woman can be labeled a "slut." Looks and attitude often have nothing to do with it.

Yet the word continues to evoke for most people an image of someone trampy and pathetic—the kind of girl or woman who wears short, tight, cleavage-enhancing clothes, always makes a beeline for the guy who enters the room, and can't string two sentences together without making a non sequitur. In short, she deserves to be called a "slut." . . .

"Slut" is, of course, a disturbing insult. But it is part of the vocabulary of adolescents—and adults—and a key word in the vocabulary of the sexual double standard. The severity of the word might offend some people, such as a racial or ethnic insult would, but refraining from using it in serious discussion serves only to reinforce its power. After all, "nigger" is a profoundly disturbing word, but can we have an honest conversation about racism without using it? I don't think so. Likewise, we must use the word "slut" and openly discuss its ramifications in order to eliminate the sexual double standard.

Below are some of the comments I've received from men and women in bookstores and radio call-in shows, and from television and radio show hosts. It's clear that most people are far more concerned with the sexuality of girls than with that of boys. My responses point out that females as well as males should be entitled to express their sexual desires. Hardly a radical concept, but it can stir up a lot of hostility.

Slut-bashing is a terrible thing, but let's face it: it affects only a small number of girls. Why write a whole book about it?

A reputation acquired in adolescence can damage a young woman's self-perception for years. She may become a target for other forms of harassment and even rape, since her peers see her as "easy" and therefore not entitled to say "no." She may become sexually active with a large number of partners (even if she had not been sexually active before her reputation). Or she may shut down her sexual side completely, wearing baggy clothes and being unable to allow a boyfriend to even kiss her.

It's true that most girls escape adolescence unscathed by slut-bashing. Nevertheless, just about every girl is affected by it. Every girl internalizes the message that sex is bad—because it can earn you a reputation. The result is that even years later, when she is safely out of adolescence, a woman may suffer from a serious hangup about sex and intimacy—even if she was not herself called a "slut." Second, the fear of being called a "slut" makes many girls unlikely to carry or use contraceptives, leading of course to the risk of pregnancy or disease.

Slut-bashing also affects boys. It fosters a culture of sexual entitlement that says that "easy" girls are expendable while only "good" girls deserve to be treated well. And that means that only some girls are treated with the respect that they all deserve.

You make it seem as if we're living in the 1950s. But this is the twenty-first century. Lots of girls are having sex; most of them are not called "sluts."

Of course, a girl today has many freedoms that her 1950s counterpart did not possess, including the license to sexually experiment before marriage. But even today, the prevailing attitude is that there is something wrong with the girl who behaves just as a boy does. Compare, for example, the fate of two recent movies about teenagers and the pursuit of sex, one involving girls, the other involving boys. *Coming Soon,* a witty comedy set in the world of wealthy teen Manhattanites and boasting a star-filled cast (Mia Farrow, Spalding Gray, Ryan O'Neal, Gaby Hoffman), follows a female high school senior who has never had an orgasm and wonders why her boyfriend leaves her sexually unfulfilled. She worries that there is something wrong with her; after all, her girlfriends report that *they* feel completely fulfilled. (It turns out they're lying.) The movie is actually far from raunchy. There is no nudity, and the raciest scene involves only the protagonist and a Jacuzzi. It garnered positive reviews from *Variety, The Hollywood Reporter,* and many prestigious film festivals.

Yet the Motion Picture Association of America gave Coming Soon the dreaded NC-17 rating, effectively barring it from theaters—until director Colette Burson agreed to cut several scenes that the MPAA deemed "lurid." Now it has been granted a "respectable" R rating (and has been released by a small distributor in a few theaters), but at the expense of a serious exploration of female sexuality. Burson explains that the MPAA "really didn't like the idea of girls and orgasm."

While Burson was busy tranquilizing *Coming Soon,* kids lined up at theaters across the country to see the vulgar antics of four teenaged boys desperate to lose their virginity in *American Pie.* In a weak nod to the notion that one's partner should enjoy sex too, one of the buddy-boy characters works hard to give his girlfriend an orgasm—not because he cares about her satisfaction, but because that will induce her to go "all the way" with him. The movie is far more sexually explicit than *Coming Soon* ever was and utterly contemptuous of girls (the tag line is "There's something about your first piece"), yet it merited an R rating. *American Pie* has grossed over a hundred million dollars in United States box-office receipts alone.

Put side by side, these two movies demonstrate that the idea of females exploring sex is taboo, while the idea of males exploring sex is an opportunity for slapstick and knowing guffaws. With this double standard in place, it's no wonder that any girl who asserts her sexual desire (or is presumed to) is treated like a freak. Her behavior is considered so deviant that it can't even be represented in the same theaters that screen bloody, ultra-violent films like *Reservoir Dogs* or incest-themed films like *Spanking the Monkey.* (For information about *Coming Soon,* go to *www.comingsoonmovie.com.*)

The sexual double standard was also in, full force in the summer of 1999 when a dozen Virginia junior high school girls were discovered to have engaged in oral sex throughout the school year during parties and at local parks. *The Washington Post* broke the story, which, was subsequently picked up by the Associated Press and reprinted in newspapers across the country. Parents, health educators, and guidance counselors weighed in with a loud chorus of condemnation. Certainly it was disturbing that kids so young were engaging in meaningless sexual encounters. But much more disturbing was that, first, the girls had been nothing more than sexual servicers to the boys; and second, that all of the censure was directed to the girls. It turns out that the school principal had called the parents of the girls to a special meeting to discuss the matter—but none of the boys or their parents was approached. Boys will be boys—but girls will be "sluts."

Girls today dress so provocatively, even to school, in skimpy outfits that expose a lot of flesh. They practically *invite* people to call them "sluts" and other names.

I have to admit that I am often appalled by some of the outfits I see young girls wearing these days: It's one thing for an adult woman to showcase her sexual appeal and a different thing entirely for an eighth grader to do likewise. But I don't blame the girls. On the contrary, I am sympathetic to them. These girls believe that if they attract a boyfriend and fall in love, their lives would be better and they would be happier. Sadly, many of these girls believe that their sexuality is the only power or appeal they have, and so they play it up to the hilt. They also feel competitive with other girls in a battle for the most desirable guys, so they feel the need to out-dress their peers. Dressing in sexy outfits, then, is both a strategy to obtain romance and a competition with other girls. But just because a girl dresses in a sexually provocative way doesn't mean that she is sexually promiscuous. In reality, she may not be any more sexually active than the prissy girl in tailored pants, loafers, and sweater set.

Some of your interviewees were called "sluts" even though they weren't sexually active at all. They were innocent victims. But the girls who were sexually promiscuous are a different story: they deserved what they got.

I don't believe that there should be a distinction between those who deserve a bad reputation and those who don't. Because frankly, I don't think that any girl deserves to be called a "slut." After all, boys who are sexually active are congratulated as studs.

Dividing "sluts" into the innocent and the guilty merely reinforces the sexual double standard. This is why, when I have been asked about my own sexual history—believe it or not, radio show hosts, aping Howard Stern— have felt perfectly comfortable quizzing me about the details of my sex life—I have refused to respond. Besides the fact that the answers are no one's business, they would serve only to buttonhole me as either "innocent" or "guilty," and I reject both categories.

* * *

If females practiced an ethic of sexual modesty, males would be more likely to treat them with respect.

Ideally, I think sex should be harnessed within a romantic relationship, but that ideal isn't possible or desirable for everybody. There are young women who perhaps would like to wait and initiate their first sexual encounter in a loving relationship, but for whatever reason, they have desire and want to act on it before they've met the "right" person—or they may never meet the "right" person. I worry that these young women are going to feel guilty and ashamed of their own sexual desire. I'm also concerned that they are going to make bad choices about who their mate is going to be, perhaps marrying too young. A sex drive is a natural appetite for males and females. If you say that females are innately modest, then you're also saying that a girl or woman who isn't modest is doing something unfemale and wrong.

In her book *A Return to Modesty,* Wendy Shalit argues that girls have to be "good" in order for boys to behave properly—to stop sexually harassing them. But I worry about the implications of being a "good" girl. Once you start characterizing some females as "good," you inevitably label others as "bad." And once you start thinking of some girls as "bad," in essence you are saying that those girls don't deserve to be treated with respect. The irony is that so many girls who are regarded as slutty aren't even particularly sexually active, and they are rarely more sexually active than their peers are. So the whole good girl/bad girl thing is a sham. Its purpose is to elevate some girls and to degrade others, and in the long run it hurts everyone. Boys will treat girls with respect, and loveless, casual sexual encounters will decrease, when we have one standard for both sexes—that is, when we have sexual equality. . . .

School sexual harassment lawsuits are getting out of hand. How can a school be monetarily responsible for sexual harassment? These lawsuits hurt everyone, since they take money away from education.

The Supreme Court ruled in May 1999 (*Davis v. Monroe County Board of Education*) that school districts receiving federal money can be liable for monetary damages if they fail to prevent severe, persistent sexual harassment among students. The ruling was in response to a case brought by the mother of a fifth-grade girl in rural Georgia. The girl, LaShonda Davis, was harassed by a male classmate who made repeated unwanted sexual advances over the course of five months. At least two teachers, as well as the principal, were aware of the incidents, but no disciplinary action was taken against the boy. Meanwhile, Davis's grades dropped and her father discovered that she had written a suicide note.

The ruling is important because it sends the message that schools must be vigilant in halting sexual harassment—which includes slut-bashing, a verbal form of sexual harassment. I agree that a school should be liable if the sexual harassment is severe and persistent and if the school is aware of the behavior but does not take steps to halt it. If, on the other hand, the school makes a good-faith effort to stop the behavior, then I don't believe it should be liable.

It's unfortunate that a ruling against a school results in a monetary loss, but it's also unfortunate that the threat of monetary payment is the most effective wake-up call to school administrators. As for the argument that these payments take money away from education, sexual harassment also impedes the ability of teachers to effectively educate and the ability of students to effectively learn.

* * *

Why are girls often worse than boys when it comes to slut-bashing?

All of us yearn for one arena in which we can wield power. For girls, this desire is often thwarted. After all, girls may get better grades—but boys, especially athletes, by and large receive more attention and congratulatory pats on the back from school administrators and teachers. Boys call out more in class and get away with it. They rule the playground. Many feel a sense of entitlement to grope girls' bodies. With these depressing realities, it's no wonder that many girls develop a sense of self-hatred. Sensing that femininity is devalued, they may feel, at some level, uncomfortable with being a girl, and therefore are reluctant to bond with other girls. Instead, they latch on to one small sphere of power they can call their own: the power to make or break reputations. Slut-bashing is a cheap and easy way to feel powerful. If you feel insecure or ashamed about your own sexual desires, all you have to do is call a girl a "slut" and suddenly you're the one who is "good" and on top of the social pecking order.

What can we do to stop slut-bashing?

Teachers must recognize that slut-bashing is a serious problem. Too often, they dismiss it as part of the normal fabric of adolescent life. But slut-bashing is a form of sexual harassment, and it is illegal under Title IX, which entitles students to a harassment-free education. If a teacher witnesses slut-bashing, she must make sure that it stops. She must confront the ringleader and other name-callers. Of course, teachers and school administrators shouldn't wait for slut-bashing to occur. They must create and publicize awareness through sexual harassment policies for their schools.

Parents should be open about sexuality with their kids—and that means being open about female sexuality as well as male sexuality. They should teach their daughters and sons that girls as well as boys have sexual feelings, and that sexual feelings are entirely normal. That way they won't have to pin their sexual anxieties on a scapegoat and then distance themselves from her.

But the most important thing that all of us need to work on is this: to stop calling or thinking of women as "sluts." Face it: At one time or another, many of us have called a woman a "slut." We see a woman who's getting away with something we wish we could get away with. What do we call her? A "slut."

We see a woman who dresses provocatively, and maybe we wish we had the guts to dress that way ourselves. What do we call her? A "slut."

If we think of a woman or girl as a "slut," it's like she's not one of us. She's one of *them*. She is other. "Slut," like any other derogatory label, is a shorthand for one who is different, strange—and not worth knowing or caring about. Unlike other insults, however, it carries a unique sting: the stigma of the out-of-control, trampy female. Most of us recognize that this stigma is unjust and unwarranted. Yet we have used the "slut" insult anyway: Our social conditioning runs too deep. We must will ourselves to be aware of the sexual double standard and of how we lapse into slut-bashing on an everyday level. If we become aware of our behavior, then we have the power to stop.

And never again be slut-bashers or self-bashers.

NOTES

1. In Jordan in 1993 a sixteen-year-old girl who had been raped by her older brother was killed by her family because, it was said, she had seduced him into sleeping with her. Kristen Golden, "Rana Husseini: A Voice for Justice," *Ms.*, July/August 1998, p. 36; Tali Edut, "Global Woman: Rana Husseini," *HUES*, Summer 1998, p. 41. In Afghanistan, where women must remain covered from head to toe in shrouds called *burqas*, the General Department for the Preservation of Virtue and Prevention of Vice beats women for wearing white socks or plastic sandals with no socks, attire that is said to provoke "impure thoughts" in men. John F. Burns, "Sex and the Afghan Woman: Islam's Straitjacket," *The New York Times*, August 29, 1997, p. A4. And in Turkey in 1998 five girls attempted suicide by eating rat poison and jumping into a water tank to avoid a forced virginity examination. An unmarried woman discovered not to be a virgin risks being beaten or killed. The virginity tests were carried out as the girls recovered in their hospital beds; when one girl did succeed in killing herself, her father had the exam performed on her corpse. Kelly Couturier, "Suicide Attempts Fuel Virginity Test Debate," *The Washington Post*, January 27, 1998, p. A18.
2. Lauri Umansky, "Breastfeeding in the 1990s: The Karen Carter Case and the Politics of Maternal Sexuality" in Molly Ladd-Taylor and Lauri Umansky, eds., *"Bad" Mothers: The Politics of Blame in Twentieth-Century America* (New York: New York University Press, 1998), pp. 299–309. Karen Carter is a pseudonym.
3. Tamar Stieber, "Viewpoint," *Glamour* August 1996, p. 138.
4. Stieber, p. 138.
5. Jon Jeter, "Woman Who Sued Ex-Husband's Mistress Is Awarded $1 Million," *The Washington Post*, August 7, 1997, p. A3.
6. *Sex and America's Teenagers* (New York and Washington: The Alan Guttmacher Institute, p. 20.
7. Laurie Goodstein with Marjorie Connelly, "Teen-Age Poll Finds a Turn to the Traditional," *The New York Times*, April 30, 1998, p. A20. The poll, of 1,048 teenagers ages thirteen to seventeen, was conducted by telephone in April 1998. The poll also found that only 18 percent of thirteen- to fifteen-year-olds said they had ever had sex, as against 38 percent of sixteen- and seventeen-year-olds.
8. Janie Victoria Ward and Jill McLean Taylor, "Sexuality Education for Immigrant and Minority Students: Developing a Culturally Appropriate Curriculum," in Janice M. Irvine, *Sexual Cultures and the Construction of Adolescent Identities* (Philadelphia: Temple University Press, 1994), p. 63.
9. Lillian B. Rubin, *Erotic Wars: What Happened to the Sexual Revolution?* (New York: HarperPerennial, 1991), p. 119.
10. Shere Hite, *Women and Love* (New York: St. Martin's Press, 1987), p. 205.
11. Tamar Lewin, "Boys Are More Comfortable With Sex Than Girls Are, Survey Finds," *The New York Times*, May 18, 1994.

CHAPTER 5

Buying and Selling Gender

In the video called *Adventures in the Gender Trade,* Kate Bornstein (see Chapter 1) looks into the camera and says, "Once you buy gender, you'll buy anything to keep it." Her observation goes to the heart of deep connections between economic processes and institutionalized patterns of gender difference, opposition, and inequality in contemporary society. Readings in this chapter examine the ways in which modern marketplace forces such as commercialization, commodification, and consumerism exploit and construct gender. However, before we explore the buying and selling of gender, we want to review briefly the major elements of contemporary American economic life—corporate capitalism—which form the framework for the packaging and delivery of gender to consumers.

Defining Corporate Capitalism Corporate capitalism is an economic system in which large, national and transnational corporations are the dominant forces. The basic goal of corporate capitalism is the same as it was when social scientists such as Karl Marx studied early capitalist economies: converting money into more money (Johnson, 2001). Corporate capitalists invest money in the production of all sorts of goods and services for the purpose of selling at a profit. Capitalism, as Gitlin (2001) observes, requires a consumerist way of life.

In today's society, corporate capitalism affects virtually every aspect of life—most Americans work for a corporate employer, whether a fast food chain or a bank, and virtually everyone buys the products and services of capitalist production (Johnson, 2001; Ritzer, 1999). Those goods and services include things we must have in order to live (e.g., food and shelter) and, most important for contemporary capitalism's survival and growth, things we have learned to want or desire (e. g., microwave ovens, televisions, cruises, fitness fashions, cosmetic surgery), even though we do not need them in order to live (Ritzer, 1999).

From an economic viewpoint, we are a nation of consumers, people who buy and use a dizzying array of objects and services conceived, designed, and sold to us by corporations. George Ritzer (1999), a leading analyst of American consumerism, observes that consumption plays such as big role in the lives of contemporary Americans that it has, in many respects, come to define our society. In fact, as Ritzer notes, Americans spend most of their available resources on consumer goods and services. Corporate, consumer capitalism depends on luring people into what he calls the "cathedrals of consumption," such as book superstores, shopping malls, theme parks, fast food restaurants, and casinos, where we will spend money to buy an array of goods and services.

Our consumption-driven economy counts on customers whose spending habits are relatively unrestrained and who view shopping as pleasurable. Indeed, Americans spend much more today than they did just forty years ago (Ritzer, 1999). Most of our available resources go to purchasing and consuming "stuff." Americans consume more of everything and more varieties of things than people in other nations. We are also more likely to go into debt than Americans of earlier generations and people in other nations today. Some social scientists (e.g, Schor, 1998) use the term "hyperconsumption" to describe what seems to be a growing American passion for and obsession with consumption.

Marketing Gender Gender is a fundamental element of the modern machinery of marketing. It is an obvious resource from which the creators and distributors of goods and services can draw ideas, images, and messages. The imagery of consumer culture thrives on gender difference and asymmetry. For example, consumer emblems of hyperfemininity and hypermasculinity, such as Barbie and GI Joe, stand in stark physical contrast to each other (Schiebinger, 2000). This is not happenstance. Barbie and GI Joe intentionally reinforce beliefs in essential differences between women and men. The exaggerated, gendered appearances of Barbie and GI Joe can also belong to adult consumers who have the resources to purchase new cosmetic surgeries, such as breast and calf implants, that literally inscribe beliefs about physical differences between women and men into their flesh (Sullivan, 2001). As Walters observes (2001 p. 289), turning difference into "an object of barter is perhaps the quintessentially American experience." Indeed, virtually every product and service, including the most functional, can be designed and consumed as masculine or feminine (e.g., deodorants, bicycles, greeting cards, wallpaper, cars, and hair styles).

Gender-coding of products and services is a common strategy employed by capitalist organizations to sell their wares. It is also integral to the processes by which gender is constructed, because it frames and structures gender practices. Let's look at the gender-coding of clothing to illustrate how consumer culture participates in the construction of gender through ordinary material forms. As the gender archeologist Sorenson (2000) observes, clothing is an ideal medium for the expression of a culture's gender beliefs because it is an extension of the body and an important element in identity and communication. No wonder corporate capitalists have cashed in on the business of fabricating gender through dress (Sorenson, 2000). Sorenson (2000) notes that simple observation of the clothing habits of people reveals a powerful pattern of "dressing gender" (124). Throughout life, she argues, the gender-coding of colors, patterns, decorations, fabrics, fastenings, trimmings, and other aspects of dress create and maintain differences between boys and girls and men and women.

Even when clothing designers and manufacturers create what appear to be "unisex" fashions (e.g., tuxedos for women), they incorporate just enough gendered elements (e.g., lacy trim or a revealing neckline) to insure that the culturally created gender categories—feminine and masculine—are not completely erased. Consider the lengths to which the fashion industry has gone to create dress that conveys a "serious yet feminine" business appearance for the increasing number of women in management and executive levels of the corporate world (Kimle and Damhorst, 1997). Contemplate the ferocity of the taboo against boys and men wearing skirts and dresses. Breaking the taboo (except on a few occasions such as Halloween) typically results in negative sanctions. The reading in this chapter by Adie Nelson examines the extent to which even fantasy dress for children ends up conforming to gender stereotypes.

Gender-coded clothing is one example of corporate exploitation of gender to sell all kinds of goods and services, including gender itself. Have we arrived at a moment in history when identities, including gender identity, are largely shaped within the dynamics of consumerism? Will we, as Bornstein observes, buy anything to keep up gender appearances? The readings in this chapter help us to answer these questions. They illuminate some of the key ways in which capitalist, consumer culture makes use of cultural definitions and stereotypes of gender to produce and sell goods and services. First, the reading by Jean Kilbourne examines the exploitation of gender by the key facilitators of consumption—advertising and the mass media (Featherstone, 1991; Kilbourne, 1999; Ritzer, 1999). The mass media (e.g., television and magazines) deliver potential consumers to advertisers whose job it is to persuade us to buy particular products and services (Kilbourne, 1999; Ritzer, 1999).

The advertising industry devotes itself to creating and keeping consumers in the marketplace, and it is very good at what it does. Today's advertisers use sophisticated strategies for hooking consumers. The strategies work because they link our deepest emotions and most beloved ideals to products and services and persuade us that identity and self-worth can be fashioned out of the things we buy (Featherstone, 1991). Advertisers transform gender into a commodity, and convince consumers that we can transform ourselves into more masculine men and more feminine women by buying particular products and services. Men are lured into buying cars that will make them feel like hypermasculine machines, and women are sold a wondrous array of cosmetic products and procedures that are supposed to turn them into drop-dead beauties.

Jacqueline Urla and Alan Swedlund's article expands upon Kilbourne's analysis of gender advertising and its effects on real people by exploring the story that Barbie doll, a well advertised and wildly popular toy that has become an icon, tells about femininity in consumer culture. They note that although Barbie's long, thin body and big breasts are remarkably unnatural, she stands as an ideal that has played itself out in the real body trends of *Playboy* magazine centerfolds and Miss

America contestants. The authors provide evidence that between 1959 and 1978, the average weight and hip size for women centerfolds and beauty contestants decreased steadily. A follow-up study for 1979–88 found the acceleration of this trend with "approximately 69 percent of Playboy centerfolds and 60 percent of Miss America contestants weighing in at 15 percent or more below their expected age and height category" (298). One lesson we might glean from this story is that a toy (Barbie) and real women (centerfolds and beauty contestants) are converging in a culture in which the bonds of beauty norms are narrowing and tightening their grip on both products and persons (Sullivan, 2001).

Any analysis of the marketing of femininity and masculinity has to take into account the ways in which the gendering of products and services is tightly linked to prisms of difference and inequality such as sexuality, race, age, and ability/disability. Consumer culture thrives, for example, on heterosexuality, whiteness, and youthfulness. Automobile advertisers market cars made for heterosexual romance and marriage. Liquor ads feature men and women in love (Kilbourne, 1999). Recent research on race and gender imagery in the most popular advertising medium, television, confirms the continuing dominance of images of white, affluent, young adults. "Virtually all forms of television marketing perpetuate images of White hegemonic masculinity and White feminine romantic fulfillment" (Coltrane and Messineo, 2000:386). In spite of what is called niche marketing or marketing to special audiences such as Latinos, gay men, and older Americans, commercial television imagery continues to rely on stereotypes of race, gender, age and the like (Coltrane and Messineo, 2000). Stereotypes sell.

Can You Buy In Without Selling Out? Second, readings in this chapter on popular culture images and messages explore the tension between creativity, resistance, and rebellion, on the one hand; and the lure and power of commercialization on the other. For example, the article by Rana Emerson raises a version of an old question: can you buy in, without selling out? Emerson documents the contradictory and stereotypical themes in the music videos of African-American women artists who must negotiate pressures to conform to stereotypes as they seek to create ways to express independence, self-reliance, and autonomy. This article raises several questions. Can we produce and consume the gendered products and services of corporate capitalism without wanting and trying to be just like Barbie or Madonna, the Marlboro Man or Eminem? Does corporate, commercial culture consume everything and everyone in its path, including the creators of countercultural forms?

The latter question is important. Consider the fact that "grunge," which began as antiestablishment fashion, became a national trend when companies such as Diesel and Urban Outfitters coopted and commercialized it (O'Brien, 1999). Then contemplate how commercial culture has cleverly exploited the women's movement by associating serious social issues and problems with trivial or dangerous products. "New Freedom" is a maxipad. "ERA" is a laundry detergent. Cigarette ads often portray smoking as a symbol of women's liberation (Kilbourne, 1999). Commercial culture is quite successful in enticing artists of all sorts to "sell out." For example, Madonna began her career as a rebel who dared to display a rounded belly. But, over time, she has been "normalized," as reflected in the transformation of her body to better fit celebrity appearance norms (Bordo, 1997).

The culture of the commodity is also successful in mainstreaming the unconventional by turning nonconformity into obedience to Madison Avenue (Harris, 2000). Analysts of the commodification of gayness have been especially sensitive to the potential problems posed by advertising's recent creation of a largely fictional identity of gay as "wealthy white man" with a lifestyle defined by hip fashion (Walters, 2001). What will happen if lesbian and gay male styles are increasingly drawn into mass-mediated, consumer culture? Will those modes of rebellion against the dominance of heterosexism lose their political clout? Will they become mere "symbolic forms of resistance, ineffectual strategies of rebellion" (Harris, 2000 p. xxiii)?

The Global Reach of American Gender Images and Ideals Third, several readings in this chapter, especially the piece by James Farrer, address the global reach of American culture. Transnational corporations are selling American popular culture and consumerism in countries around the world (Kilbourne, 1999; Ritzer, 1999). People across the globe are now regularly exposed to American images, icons, and ideals. For example, *Baywatch,* with its array of perfect (albeit cosmetically enhanced) male and female bodies, has been seen by more people in the world than any other television show (Kilbourne, 1999). American popular music and film celebrities dominate the world scene. Everyone knows Marilyn Monroe and James Dean, Tom Cruise and Julia Roberts.

You might ask, and quite legitimately, so what? The answer to that question is not a simple one, in part because cultural import-export relations are intricate. As Gitlin (2001) observes, "the cultural gates . . . swing both ways. American rhythm and blues influenced Jamaican ska, which evolved into raggae, which in turn was imported to the United States via Britain" (188). However, researchers have been able to document some troubling consequences of the global advantage of American commercial, consumer culture for the lifeways of people outside the United States. For example, social scientists (e.g., Connell, 1999; Herdt, 1997) are tracing how American definitions of sexual orientation are altering the modes of organization and perception of same-gender relations in some non-Western societies that have traditionally been more fluid and tolerant of sexual diversity than the United States (also Cantu, Chapter 10 in this book).

Scientists are also documenting the impact of American mass media images of femininity and masculinity on consumers in far corners of the world. The island country of Fiji is one such place. Researchers have discovered that as the young women of Fiji consume American television on a regular basis, eating disorders such as anorexia nervosa are being recorded for the first time. The ultra-thin images of girls and women that populate U.S. television ads and TV shows have become the measuring stick of femininity in a culture in which, previously, an ample, full body was the norm for women and men (Goode, 1999). The troubling consequences of the globalization of American consumer culture do not end with these examples. Consider the potential negative impact of idealized images of whiteness in a world in which most people are brown. Or how about the impact of America's negative images of older women and men on the people of cultures in which the elderly are revered?

Although corporate, capitalist economies provide many people with all the creature comforts they need and more, as well as making consumption entertaining and more accessible, there is a price to pay (Ritzer, 1999). This chapter explores one troubling aspect of corporate, consumer culture—the commodification and commercialization of gender.

A few final questions emerge from our analysis of patterns of gender in relationship to consumer capitalism. How can the individual develop an identity and self-worth that are not contingent upon and defined by a whirlwind of products and services? How do we avoid devolving into caricatures of stereotyped images of femininity and masculinity, whose needs and desires can only be met by gendered commodities? Is Kate Bornstein correct when she states that "Once you buy gender, you'll buy anything to keep it"? Or can we create and preserve alternative ways of life, even ways of life that undermine the oppression of dominant images and representations?

REFERENCES

Bordo, Susan. (1997). "Material girl: The effacements of postmodern culture." In *The gender/sexuality reader,* R. Lancaster and M. di Leonardo (Eds.), pp. 335-358. New York: Routledge.

Coltrane, Scott and Melianda Messineo. (2000). "The perpetuation of subtle prejudice: Race and gender imagery in 1990s television advertising." *Sex roles* 42:363-389.

Connell, R. W. (1999). "Making gendered people: Bodies, identities, sexualities." In *Revisioning gender,* M. Ferree, J. Lorber, and B. Hess (Eds.), pp. 449-471. Thousand Oaks, CA: Sage.

Featherstone, Mike. (1991). "The body in consumer culture." In *The body: Social process and cultural theory,* Featherstone, Hepworth, and Turner (Eds.), pp. 170-196. London: Sage.

Gitlin, Todd. (2001). *Media unlimited: How the torrent of images and sounds overwhelms our lives.* New York: Henry Holt and Company.

Goode, Erica. (1999). "Study finds TV alters Fiji girls' view of body." *New York Times,* May 20, p. A17.

Harris, Daniel. (2000). *Cute, quaint, hungry and romantic: The aesthetics of consumerism.* Cambridge, MA: Da Capo Press.

Herdt, Gilbert. (1997). *Same sex, different cultures.* Boulder, CO: Westview.

Johnson, Alan. (2001). *Privilege, power, and difference.* Mountain View, CA: Mayfield.

Kilbourne, Jean. (1999). *Can't buy my love.* New York: Simon & Schuster.

Kimle, Patricia A. and Mary Lynn Damhorst. (1997). "A grounded theory model of the ideal business image for women." *Symbolic Interaction* 20(1): 45-68.

Marenco, Susan, with Kate Bornstein. (1993). *Adventures in the gender trade: A case for diversity.* Filmakers Library.

O'Brien, Jodi. (1999). *Social prisms.* Thousand Oaks, CA: Pine Forge.

Ritzer, George. (1999). *Enchanting a disenchanted world.* Thousand Oaks, CA: Pine Forge.

Schiebinger, Londa. (2000). "Introduction." In *Feminism and the body,* L. Schiebinger (Ed.), pp. 1-21.

Schor, Juliet. (1998). *The overspent American.* New York: Basic Books.

Sorenson, Marie Louise Stig. (2000). *Gender archaeology.* Cambridge, England: Polity Press.

Sullivan, Deborah A. (2001). *Cosmetic surgery: The cutting edge of commercial medicine in America.* New Brunswick, NJ: Rutgers University Press.

Walters, Suzanna Danuta. (2001). *All the rage: The story of gay visibility in America.* Chicago: University of Chicago Press.

23

The Pink Dragon Is Female

Halloween Costumes and Gender Markers

BY ADIE NELSON

Adie Nelson's article offers a marvelously detailed analysis of one way in which the modern marketplace reinforces gender stereotypes—the gender coding of children's Halloween costumes. Nelson describes the research process she employed to label costumes as masculine, feminine, or neutral. She provides extensive information about how manufacturers and advertisers use gender markers to steer buyers, in this case parents, toward gender-appropriate costume choices for their children. Overall, Nelson's research indicates that gender-neutral costumes, whether they are ready-to-wear or sewing patterns, are a tiny minority of all the costumes on the market.

1. Many perceive Halloween costumes as encouraging children to engage in fantasy play. How does Nelson's research call this notion into question?
2. Describe some of the key strategies employed by manufacturers to "gender" children's costumes.
3. How do Halloween costumes help to reproduce an active-masculine/passive-feminine dichotomy?

★ ★ ★

[T]he celebration of Halloween has become, in contemporary times a socially orchestrated secular event that brings buyers and sellers into the marketplace for the sale and purchase of treats, ornaments, decorations, and fanciful costumes. Within this setting, the wearing of fancy dress costumes has such a prominent role that it is common, especially within large cities, for major department stores and large, specialty toy stores to begin displaying their selection of Halloween costumes by mid-August if not earlier. It is also evident that the range of masks and costumes available has broadened greatly beyond those identified by McNeill (1970), and that both children and adults may now select from a wide assortment of readymade costumes depicting, among other things, animals, objects, superheroes, villains, and celebrities. In addition, major suppliers of commercially available sewing patterns, such as Simplicity and McCall's, now routinely include an assortment of Halloween costumes in their fall catalogues. Within such catalogues, a variety of costumes designed for infants, toddlers, children, adults, and, not infrequently, pampered dogs are featured.

On the surface, the selection and purchase of Halloween costumes for use by children may simply appear to facilitate their participation in the world of fantasy play. At least in theory, asking children what they wish to wear or what they would like to be for Halloween may be seen to encourage them to use their imagination and to engage in the role-taking stage that Mead (1934) identified as play. Yet, it is clear that the commercial marketplace plays a major role in giving expression to children's imagination in their Halloween costuming. Moreover, although it might be facilely assumed that the occasion of Halloween provides a cultural "time out" in which women and men as well as girls and boys have tacit permission to transcend the gendered rules that mark the donning of apparel in everyday life, the androgyny of Halloween costumes may be more apparent than real. If, as our folk wisdom proclaims, "clothes make the man" (or woman), it would be presumptuous to suppose that commercially available children's Halloween costumes and sewing patterns do not reflect both the gendered nature of dress (Eicher & Roach-Higgens, 1992) and the symbolic world of heroes, villains, and fools (Klapp, 1962, 1964). Indeed, the donning of Halloween costumes may demonstrate a "gender display" (Goffman, 1966, p. 250) that is dependent on decisions made by brokering agents to the extent that it is the aftermath of a series of decisions made by commercial firms that market ready-made costumes and sewing patterns that, in turn, are purchased, rented, or sewn by parents or others. . . .

Building on Barnes and Eicher's (1992, p. 1) observation that "dress is one of the most significant markers of gender identity," an examination of children's Halloween costumes provides a unique opportunity to explore the extent to which gender markers are also evident within the fantasy costumes available for Halloween. To the best of my knowledge, no previous research has attempted to analyze these costumes nor to examine the ways in which the imaginary vistas explored in children's fantasy dress reproduce and reiterate more conventional messages about gender.

In undertaking this research, my expectations were based on certain assumptions about the perspectives of merchandisers of Halloween costumes for children. It was expected that commercially available costumes and costume patterns would reiterate and reinforce traditional gender stereotypes. Attempting to adopt the marketing perspective of merchandisers, it was anticipated that the target audience would be parents concerned with creating memorable childhood experiences for their children, envisioning them dressed up as archetypal fantasy characters. In the case of sewing patterns, it was expected that the target audience would be primarily mothers who possessed what manufacturers might imagine to be the sewing skills of the traditional homemaker. However, these assumptions about merchandisers are not the subject of the present inquiry. Rather, the present study offers an examination of the potential contribution of marketing to the maintenance of gender stereotypes. In this article, the focus is on the costumes available in the marketplace; elsewhere I examine the interactions between children and their parents in the selection, modification, and wearing of Halloween costumes (Nelson, 1999).

METHOD

The present research was based on a content analysis of 469 unique children's Halloween ready-made costumes and sewing patterns examined from August 1996 to November 1997 at craft stores, department stores, specialty toy stores, costume rental stores, and fabric stores containing catalogues of sewing patterns. Within retail stores, racks of children's Halloween costumes typically appeared in August and remained in evidence, albeit in dwindling numbers, until early November each year. In department stores, a subsection of the area generally devoted to toys featured such garments; in craft stores and/or toy stores, children's Halloween costumes were typically positioned on long racks in the center of a section devoted to the commercial paraphernalia now associated with the celebration of Halloween (e.g., cardboard witches, "Spook trees," plastic pumpkin containers). Costumes were not segregated by gender within the stores (i.e., there were no separate aisles or sections for boys' and girls' costumes); however, children's costumes were typically positioned separately from those designed for adults. . . .

All costumes were initially coded as (a) masculine, (b) feminine, or (c) neutral depending on whether boys, girls, or both were featured as the models on the packaging that accompanied a ready-to-wear costume or were used to illustrate the completed costume on the cover of a sewing pattern. . . . The pictures accompanying costumes may act as safekeeping devices, which discourage parents from buying "wrong"-sexed costumes.

The process of labeling costumes as masculine, feminine, or neutral was facilitated by the fact that these public pictures (Goffman, 1979) commonly employed recognizable genderisms. For example, a full-body costume of a box of crayons could be identified as feminine by the long curled hair of the model and the black patent leather pumps with ribbons she wore. In like fashion, a photograph depicting the finished version of a sewing pattern for a teapot featured the puckish styling of the model in a variant of what Goffman (1979, p. 45) termed "the bashful knee bend" and augmented this subtle cue by having the model wear white pantyhose and Mary-Jane shoes with rosettes at the base of the toes. Although the sex of the model could have been rendered invisible, such feminine gender markers as pointy-toed footwear, party shoes of white and black patent leather, frilly socks. make-up and nail polish, jewelry, and elaborately curled (and typically long and blonde) hair adorned with bows/barrettes/hairbands facilitated this initial stage of costume placement. By and large, female models used to illustrate Halloween costumes conformed to the ideal image of the "Little Miss" beauty pageant winner; they were almost overwhelmingly White, slim, delicate-boned blondes who did not wear glasses. Although male child models were also overwhelmingly White, they were more heterogeneous in height and weight and were more likely to wear glasses or to smile out from the photograph in a bucktooth grin. At the same time, however, masculine gender markers were apparent. Male models were almost uniformly shod in either well-worn running shoes or sturdy-looking brogues, while their hair showed little variation from the traditional little boy cut of short back and sides.

The use of gender-specific common and proper nouns to designate costumes (e.g., Medieval Maiden, Majorette, Prairie Girl) or gender-associated adjectives that formed part of the costume title (e.g., Tiny Tikes Beauty, Pretty Witch, Beautiful Babe, Pretty Pumpkin Pie) also served to identify feminine costumes. Similarly, the use of the terms "boy," "man," or "male" in the advertised name of the costume (e.g., Pirate Boy, Native American Boy, Dragon Boy) or the noted inclusion of advertising copy that announced "Cool dudes costumes are for boys in sizes" was used to identify masculine costumes.

Costumes designated as neutral were those in which both boys and girls were featured in the illustration or photograph that accompanied the costume or sewing pattern or in which it was impossible to detect the sex of the wearer. By and large, illustrations for gender-neutral ads featured boys and girls identically clad and depicted as a twinned couple or, alternatively, showed a single child wearing a full-length animal costume complete with head and "paws," which, in the style of spats, effectively covered the shoes of the model. In addition, gender-neutral costumes were identified by an absence of gender-specific nouns and stereotypically gendered colors.

Following this initial division into three categories, the contents of each were further coded into a modified version of Klapp's (1964) schema of heroes, villains, and fools. In his work, Klapp suggested that this schema represents three dimensions of human behavior. That is, heroes are praised and set up as role models, whereas villains and fools are negative models, with the former representing evil to be feared and/or hated and the latter representing figures of absurdity inviting ridicule. However, although Klapp's categories were based on people in real life, I applied them to the realm of make-believe. For the purposes of this study, the labels refer to types of personas that engender or invite the following emotional responses, in a light-hearted way from audiences: heros invite feelings of awe, admiration, and respect, whereas villains elicit feelings of fear and loathing, and fools evoke feelings of laughter and perceptions of cuteness. All of the feelings, however, are mock emotions based on feelings of amusement, which make my categories quite distinct from Klapp's. For example, although heroes invite awe, we do not truly expect somebody dressed as a hero to be held in awe. . . .

For the purposes of this secondary classification of costumes, the category of hero was broadened to include traditional male or female heroes (e.g., Cowboy, Robin Hood, Cinderella, Cleopatra), superheroes possessing supernatural powers (e.g., Superman, Robocop, Xena, the Warrior Princess) as well as characters with high occupational status (e.g., Emergency Room Doctor, Judge) and characters who are exemplars of prosocial conformity to traditional masculine and feminine roles (e.g., Team USA Cheerleader, Puritan Lady, Pioneer Boy). The category of villain was broadly defined to include symbolic representations of death (e.g., the Grim Reaper, Death, The Devil, Ghost), monsters (e.g., Wolfman, Frankenstein, The Mummy), and antiheroes (e.g., Convict, Pirate, The Wicked Witch of the West, Catwoman). Fool was a hybrid category, distinguished by costumes whose ostensible function was to amuse rather than to alarm. Within this category, two subcategories were distinguished. The first subcategory, figures of mirth, referred to costumes of clowns, court jesters, and harlequins. The second, nonhuman/inanimate objects, was composed of costumes representing foodstuffs (e.g., Peapod, Pepperoni Pizza, Chocolate Chip Cookie), animals and insects, and inanimate objects (e.g., Alarm Clock, Bar of Soap, Flower Pot). Where a costume appeared to straddle two categories, an attempt was made to assign it to a category based on the dominant emphasis of its pictorial representation. For example, a costume labeled Black Widow Spider could be classified as either an insect or a villain. If the accompanying illustration featured a broadly smiling child in a costume depicting a fuzzy body and multiple appendages, it was classified as an insect and included in the category of nonhuman/inanimate objects; if the costume featured an individual clad in a black gown, long black wig, ghoulish makeup, and a sinister mien, the costume was classified as a villain. Contents were subsequently reanalyzed in terms of their constituent parts and compared across masculine and feminine categories.

In all cases, costumes were coded into the two coding schemes on the basis of a detailed written description of each costume. . . .

RESULTS

The initial placement of the 469 children's Halloween costumes into masculine, feminine, or neutral categories yielded 195 masculine costumes, 233 feminine costumes, and 41 gender-neutral costumes. The scarcity of gender-neutral costumes was notable; costumes that featured both boys and girls in their ads or in which the gender of the anticipated wearer remained (deliberately or inadvertently) ambiguous accounted for only 8.7% of those examined. Gender-neutral costumes were more common in sewing patterns than in ready-to-wear costumes and were most common in costumes designed for newborns and very young infants. In this context, gender-neutral infant costumes largely featured a winsome assortment of baby animals (e.g., Li'l Bunny, Beanie the Pig) or foodstuffs (e.g., Littlest Peapod). By and large, few costumes for older children were presented as gender-neutral; the notable exceptions were costumes for scarecrows and emergency room doctors (with male/female models clad identically in olive-green "scrubs"), ready-made plastic costumes for Lost World/Jurassic Park hunters, a single costume labeled Halfman/Halfwoman, and novel sewing patterns depicting such inanimate objects as a sugar cube, laundry hamper, or treasure chest.

Beginning most obviously with costumes designed for toddlers, gender dichotomization was promoted by gender-distinctive marketing devices employed by the manufacturers of both commercially made costumes and sewing patterns. In relation to sewing patterns for children's Halloween costumes, structurally identical costumes featured alterations through the addition or deletion of decorative trim (e.g., a skirt on a costume for an elephant) or the use of specific colors or costume names, which served to distinguish masculine from feminine costumes. For example, although the number and specific pattern pieces required to construct a particular pattern would not vary, View A featured a girl-modeled Egg or Tomato, whereas View B presented a boy-modeled Baseball or Pincushion. Structurally identical costumes modeled by both boys and girls would be distinguished through the use of distinct colors or patterns of material. Thus, for the peanut M & M costumes, the illustration featured girls clad in red or green and boys clad in blue, brown, or yellow. Similarly, female clowns wore costumes of soft pastel colors and dainty polka dots, but male clowns were garbed in bold primary colors and in material featuring large polka dots or stripes. Illustrations for ready-to-wear costumes were also likely to signal the sex of the intended wearer through the advertising copy: models for feminine costumes, for example, had long curled hair, were made up, and wore patent leather shoes. Only in such costumes as Wrinkly Old Woman, Grandma Hag, Killer Granny, and Nun did identifiably male children model female apparel. . . .

[A]lthough hero costumes constituted a large percentage of both masculine and feminine costumes, masculine costumes contained a higher percentage of villain costumes, and feminine costumes included substantially more fool costumes, particularly those of nonhuman/inanimate objects. It may be imagined that the greater total number of feminine costumes would provide young girls with a broader range of costumes to select from than exists for young boys, but in fact the obverse is true. . . . [W]hen finer distinctions were made within the three generic categories, hero costumes for girls were clustered in a narrow range of roles that, although distinguished by specific names, were functionally equivalent in the image they portray. It would seem that, for girls, glory is concentrated in the narrow realm of beauty queens, princesses, brides, or other exemplars of traditionally passive femininity. The ornate, typically pink, ballgowned costume of the princess (with or without a synthetic jewelled tiara) was notable, whether the specific costume was labeled Colonial Belle, the Pumpkin Princess, Angel Beauty, Blushing Bride, Georgia Peach, Pretty Mermaid, or Beauty Contest Winner. In contrast, although hero costumes for boys emphasized the warrior theme of masculinity (Doyle, 1989; Rotundo, 1993), with costumes depicting characters associated with battling historical, contemporary, or supernatural Goliaths (e.g., Broncho Rider, Dick Tracy, Sir Lancelot, Hercules, Servo Samaurai, Robin the Boy Wonder), these costumes were less singular in the visual images they portrayed and were more likely to depict characters who possessed supernatural powers or skills.

Masculine costumes were also more likely than feminine costumes to depict a wide range of villainous characters (e.g., Captain Hook, Rasputin, Slash), monsters (e.g., Frankenstein, The Wolfman), and, in particular, agents or symbols of death (e.g., Dracula, Executioner, Devil boy, Grim Reaper). Moreover, costumes for male villains were more likely than those of female villains to be elaborate constructions that were visually repellant; to feature an assortment of scars, mutations, abrasions, and suggested amputations; and to present a wide array of ingenuous, macabre, or disturbing visual images. For example, the male-modeled, ready-to-wear Mad Scientist's Experiment costume consisted of a full-body costume of a monkey replete with a half-head mask featuring a gaping incision from which rubber brains dangled. Similarly, costumes for such characters as Jack the Ripper, Serial Killer, Freddy Krueger, or The Midnight Stalker were adorned with the suggestion of bloodstains and embellished with such paraphernalia as plastic knives or slip-on claws.

In marked contrast, the costumes of female villains alternated between relatively simple costumes of witches in pointy hats and capes modeled by young girls, costumes of the few female archvillains drawn from the pages of comic books, and, for older girls, costumes that were variants of the garb donned by the popular TV character Elvira, Mistress of the Dark (i.e., costumes that consisted of a long black wig and a long flowing black gown cut in an empire-style, which, when decorated with gold brocade or other trim at the top of the ribcage, served to create the suggestion of a bosom). The names of costumes for the female villains appeared to emphasize the erotic side of their villainy (e.g., Enchantra, Midnite Madness, Sexy Devil, Bewitched) or to neutralize the malignancy of the character by employing adjectives that emphasized their winsome rather than wicked qualities (e.g., Cute Cuddley Bewitched, Little Skull Girl, Pretty Little Witch).

Within the category of fools, feminine costumes were more likely than masculine costumes to depict nonhuman/inanimate objects (33.1% of feminine costumes vs. 17.4% of masculine costumes). Feminine costumes were more likely than masculine costumes to feature a wide variety of small animals and insects (e.g., Pretty Butterfly, Baby Cricket, Dalmatian Puppy), as well as flowers, foodstuffs (BLT Sandwich, Ice-Cream Cone, Lollipop), and dainty, fragile objects such as Tea Pot. For example, a costume for Vase of Flowers was illustrated with a picture of a young girl wearing a cardboard cylinder from her ribcage to her knees on which flowers were painted, while a profusion of pink, white, and yellow flowers emerged from the top of the vase to form a collar of blossoms around her face. Similarly, a costume for Pea Pod featured a young girl wearing a green cylinder to which four green balloons were attached; on the top of her head, the model wore a hat bedecked with green leaves and tendrils in a corkscrew shape. When costumed as animals, boys were likely to be shown modeling larger, more aggressive animals (e.g., Veliceraptor, Lion, T-Rex); masculine costumes were unlikely to be marketed with adjectives emphasizing their adorable, "li'l," cute, or cuddly qualities. In general, boys were rarely cast as objects, but when they were, they were overwhelmingly shown as items associated with masculine expertise. For example, a costume for Computer was modeled by a boy whose face was encased in the computer monitor and who wore, around his midtorso, a keyboard held up by suspenders. Another masculine costume depicted a young boy wearing a costume for Paint Can; the lid of the can was crafted in the style of a chef's hat, and across the cylindrical can worn from midchest to midknee was written "Brand X Paint" and, in smaller letters, "Sea Blue." Although rarely depicted as edibles or consumable products, three masculine costumes featured young boys as, variously, Root Beer Mug, Pepperoni Pizza, and Grandma's Pickle Jar.

DISCUSSION

Although the term "fantasy" implies a "play of the mind" or a "queer illusion" (Barnhart, 1967, p. 714), the marketing illustrations for children's Halloween costumes suggest a flight of imagination that remains largely anchored in traditional gender roles, images, and symbols. Indeed, the noninclusive language commonly found in the names of many children's Halloween costumes reverberates throughout many other dimensions of the gendered social life depicted in this fantastical world. For example, the importance of participation in the paid-work world and financial success for men and of physical attractiveness and marriage for women is reinforced through costume names that reference masculine costumes by occupational roles or titles but describe feminine costumes via appearance and/or relationships (e.g., "Policeman" vs. "Beautiful Bride"). Although no adjectives are deemed necessary to describe Policeman, the linguistic prompt contained in Beautiful Bride serves to remind observers that the major achievements for females are getting married and looking lovely. In addition to costume titles that employ such sexlinked common nouns as Flapper, Bobby Soxer, Ballerina, and Pirate Wench, sex-marked suffixes such as the *-ess* (e.g., Pretty Waitress, Stewardess, Gypsy Princess, Sorceress) and *-ette* (e.g., Majorette) also set apart male and female fantasy character costumes. Costumes for suffragettes or female-modeled police officers, astronauts, and fire fighters were conspicuous only by their absence.

Gender stereotyping in children's Halloween costumes also reiterates an active-masculine/passive-feminine dichotomization. The ornamental passivity of Beauty Queen stands in stark contrast to the reification of the masculine action figure, whether he is heroic or villainous. In relation to hero figures, the dearth of female superhero costumes in the sample would seem to reflect the comparative absence of such characters in comics books. Although male superheroes have sprung up almost "faster than a speeding bullet" since the 1933 introduction of *Superman,* the comic book life span of women superheroes has typically been abbreviated, "rarely lasting for more than three appearances" (Robbins, 1996, p. 2). Moreover, the applicability of the term "superhero" to describe these female characters seems at least somewhat dubious. Often their role has been that of the male hero's girlfriend or sidekick "whose purpose was to be rescued by the hero" (Robbins, 1996, p. 3).

In 1941 the creation of *Wonder Woman* (initially known as Amazon Princess Diana) represented a purposeful attempt by her creator, psychologist William Marston, to provide female readers with a same-sex superhero. . . .

Nevertheless, over half a decade later, women comic book superheroes remain rare and, when they do appear, are likely to be voluptuous and scantily clad. If, as Robbins (1996, p. 166) argued, the overwhelmingly male comic book audience "expect, in fact demand that any new superheroines exist only as pinup material for their entertainment," it would seem that comic books and their televised versions are unlikely to galvanize the provision of flat-chested female superhero Halloween costumes for prepubescent females in the immediate future.

The relative paucity of feminine villains would also seem to reinforce an active/passive dichotomization on the basis of gender. Although costumes depict male villains as engaged in the commission of a wide assortment of antisocial acts, those for female villains appear more nebulous and are concentrated within the realm of erotic transgressions. Moreover, the depiction of a female villain as a sexual temptress or erotic queen suggests a type of active passivity" (Salamon, 1983), whereby the act of commission is restricted to wielding her physical attractiveness over (presumably) weak-willed men. The veritable absence of feminine agents or symbols of death may reflect not only the stereotype of women (and girls) as life-giving and nurturing, but also the attendant assumption that femininity and lethal aggressiveness are mutually exclusive.

Building on the Sapir–Whorf hypothesis that the language we speak predisposes us to make particular interpretations of reality (Sapir, 1949; Whorf, 1956) and the assertion that language provides the basis for developing the gender schema identified by Bem (1983), the impact of language and other symbolic representations must be considered consequential. The symbolic representations of gender contained within Halloween costumes may, along with specific costume titles, refurbish stereotypical notions of what women/girls and men/boys are capable of doing even within the realm of their imaginations.

Nelson and Robinson (1995) noted that deprecatory terms in the English language often ally women with animals. Whether praised as a "chick," "fox," or "Mother Bear" or condemned as a "bitch," "sow," or an "old nag," the imagery is animal reductionist. They also noted that language likens women to food items (e.g., sugar, tomato, cupcake), with the attendant suggestion that they look "good enough to eat" and are "toothsome morsels." Complementing this, the present study suggests that feminine Halloween costumes also employ images that reduce females to commodities intended for amusement, consumption, and sustenance. A cherry pie, after all, has only a short shelf life before turning stale and unappealing. Although a computer may become obsolete, the image it conveys is that of rationality, of a repository of wisdom, and of scientifically minded wizardry.

In general, the relative absence of gender-neutral costumes is intriguing. Although it must remain speculative, it may be that the manufacturers of ready-to-wear and sewing pattern costumes subscribe to traditional ideas about gender and/or believe that costumes that depart from these ideas are unlikely to find widespread acceptance. Employing a supply–demand logic, it may be that marketing analysis of costume sales confirms their suspicions. Nevertheless, although commercial practices may reflect consumer preferences for gender-specific products rather than biases on the part of merchandisers themselves, packaging that clearly depicts boys or girls—but not both—effectively promotes gendered definitions of products beyond anything that might be culturally inherent in them. This study suggests that gender-aschematic Halloween costumes for children compose only a minority of both ready-to-wear costumes and sewing patterns. It is notable that, when male children were presented modeling female garments, the depicted character was effectively desexed by age (e.g., a wizened, hag-like "grandmother") or by calling (e.g., a nun).

The data for this study speak only to the gender practices of merchandisers marketing costumes and sewing patterns to parents who themselves may be responding to their children's wishes. Beyond this, the findings do not identify precisely whose tastes are represented when these costumes are purchased. It is always possible that, despite the gendered nature of Halloween costumes presented in the illustrations and advertising copy used to market them, parents and children themselves may engage in creative redefinitions of the boundary markers surrounding gender. A child or parent may express and act on a preference for dressing a male in a pink, ready-to-wear butterfly costume or a female as Fred Flinstone and, in so doing, actively defy the symbolic boundaries that gender the Halloween costume. Alternatively, as a strategy of symbolic negotiation, those parents who sew may creatively experiment with recognizable gender markers, deciding, for example, to construct a pink dragon costume for their daughter or a brown butterfly costume for their son. Such amalgams of gender-discordant images may, on the surface, allow both male and female children to experience a broader range of fantastical roles and images. However, like Persian carpets, deliberately flawed to forestall divine wrath, such unorthodox Halloween costumes, in their structure and design, may nevertheless incorporate fibers of traditional gendered images.

REFERENCES

Barnes, R., & Eicher, J. B. (1992). *Dress and gender: Making and meaning in cultural contexts.* New York: Berg.

Barnhart, C. L. (1967). *The world book dictionary: A–K.* Chicago: Field Enterprises Educational Corporation.

Bem, S. L. (1983). Gender schema theory and its implications for child development: Raising gender-aschematic children in a gender-schematic society. *Signs: Journal of Women in Culture and Society, 8,* 598–616.

Doyle, J. A. (1989). *The male experience.* Dubuque, IA: Wm. C. Brown.

Eicher, J. B., & Roach-Higgins, M. E. (1992). Definition and classification of dress: Implications for analysis of gender roles. In R. Barnes & J. B. Eicher (Eds.), *Dress and gender. Making and meaning in cultural contexts* (pp. 8–28). New York: Berg.

Goffman, E. (1966). Gender display. *Philosophical Transactions of the Royal Society of London, 279,* 250.

Goffman, E. (1979). *Gender advertisements.* London: Macmillan.

Klapp, O. (1962). *Heroes, villains and fools.* Englewood Cliffs, NJ: Prentice-Hall.

Klapp, O. (1964). *Symbolic leaders.* Chicago: Aldine.

McNeill, F. M. (1970). *Hallowe'en: Its origins, rites and ceremonies in the Scottish tradition.* Edinburgh: Albyn Press.

Mead, G. H. (1934). *Mind, self and society.* Chicago: University of Chicago Press.

Nelson, E. D. (1999). *Dressing for Halloween, doing gender.* Unpublished manuscript.

Nelson, E. D., & Robinson, B. W. (1995). *Gigolos & Madame's bountiful: Illusions of gender, power and intimacy.* Toronto: University of Toronto Press.

Peretti, P. O., & Sydney, T. M. (1985). Parental toy stereotyping and its effect on child toy preference. *Social Behavior and Personality, 12,* 213–216.

Rotundo, E. A. (1993). *American manhood: Transformations in masculinity from the revolution to the modern era.* New York: Basic Books.

Salamon, E. (1983). *Kept women: Mistress of the '80s.* London: Orbis.

Sapir, E. (1949). *Selected writings of Edward Sapir on language, culture and personality.* Berkeley: University of California Press.

Whorf, B. L. (1956). The relation of habitual thought and behavior to language. In J. B. Carroll (Ed.), *Language, thought, and reality* (pp. 134–159). Cambridge, MA: Technology Press of MIT.

24

The More You Subtract, the More You Add

Cutting Girls Down to Size

BY JEAN KILBOURNE

Jean Kilbourne is well-known for her films and lectures on images of women in advertising and the role of advertising in tobacco and alcohol consumption. In this chapter from her book, entitled Can't Buy My Love, *she focuses on the strategies advertisers use to sell products to adolescents. In particular, Kilbourne concerns herself with the negative, stereotypical messages sent to girls about who they should be and the impact of those messages on girls' self-esteem and body practices. She also explores the capacity of advertisers to turn adolescent rebellion and resistance into products.*

1. Why does Kilbourne argue that our culture can be toxic for girls' self-esteem?
2. What is the most important message sent to girls by advertisers, and why does Kilbourne view that message as detrimental to girls?
3. Why is this chapter entitled "The More You Subtract, the More You Add?"

Adolescents are new and inexperienced consumers—and such prime targets. They are in the process of learning their values and roles and developing their self-concepts. Most teenagers are sensitive to peer pressure and find it difficult to resist or even to question the dominant cultural messages perpetuated and reinforced by the media. Mass communication has made possible a kind of national peer pressure that erodes private and individual values and standards, as well as community values and standards. . . .

Advertisers are aware of their role and do not hesitate to take advantage of the insecurities and anxieties of young people, usually in the guise of offering solutions. A cigarette provides a symbol of independence. A pair of designer jeans or sneakers convey status. The right perfume or beer resolves doubts about femininity or masculinity. All young people are vulnerable to these messages and adolescence is a difficult time for most people, perhaps especially these days. According to the Carnegie Corporation, "Nearly half of all American adolescents are at high or moderate risk of seriously damaging their life chances." But there is a particular kind of suffering in our culture that afflicts girls. . . .

As Carol Gilligan, Mary Pipher and other social critics and psychologists have pointed out in recent years, adolescent girls in America are afflicted with a range of problems, including low self-esteem, eating disorders, binge drinking, date rape and other dating violence, teen pregnancy, and a rise in cigarette smoking. Teenage women today are engaging in far riskier health behavior in greater numbers than any prior generation.

The gap between boys and girls is closing, but this is not always for the best. According to a 1998 status report by a consortium of universities and research centers, girls have closed the gap with boys in math performance and are coming close in science. But they are also now smoking, drinking, and using drugs as often as boys their own age. And, although girls are not nearly as violent as boys, they are committing more crimes than ever before and are far more often physically attacking each other.

It is important to understand that these problems go way beyond individual psychological development and pathology. Even girls who are raised in loving homes by supportive parents grow up in a toxic cultural environment, at risk for self-mutilation, eating disorders, and addictions. The culture, both reflected and reinforced by advertising, urges girls to adopt a false self, to bury alive their real selves, to become "feminine," which means to be nice and kind and sweet, to compete with other girls for the attention of boys, and to value romantic relationships with boys above all else. Girls are put into a terrible double bind. They are supposed to repress their power, their anger, their exuberance and be simply "nice," although they also eventually must compete with men in the business world and be successful. They must be overtly sexy and attractive but essentially passive and virginal. It is not surprising that most girls experience this time as painful and confusing, especially if they are unconscious of these conflicting demands.

Of course, it is impossible to speak accurately of girls as a monolithic group. The socialization that emphasizes passivity and compliance does not apply to many African-American and Jewish girls, who are often encouraged to be assertive and outspoken, and working-class girls are usually not expected to be stars in the business world. Far from protecting these girls from eating disorders and other problems, these differences more often mean that the problems remain hidden or undiagnosed and the girls are even less likely to get help. Eating problems affect girls from African-American, Asian, Native American, Hispanic, and Latino families and from every socioeconomic background. The racism and classism that these girls experience exacerbate their problems. Sexism is by no means the only trauma they face. . . .

Of course, we must continue to pay attention to the problems of boys, as well. Two books published recently address these problems. In *Raising Cain: Protecting the Emotional Life of Boys,* Daniel Kindlon and Michael Thompson examine the "culture of cruelty" that boys live in and the "tyranny of toughness" that oppresses them. In *Real Boys: Rescuing Our Sons from the Myths of Boyhood,* psychologist William Pollock examines the ways that boys manifest their social and emotional disconnection through anger and violence. We've seen the tragic results of this in the school shootings, all by angry and alienated boys. . . .

Most of us understand that the cultural environment plays a powerful role in creating these problems. But we still have a lot to learn about the precise nature of this role—and what we can do about it. How can we resist these destructive messages and images? The first step, as always, is to become as conscious of them as possible, to deconstruct them. Although I am very sympathetic to the harm done to boys by our cultural environment, the focus of my work has always been on girls and women.

Girls try to make sense of the contradictory expectations of themselves in a culture dominated by advertising. Advertising is one of the most potent messengers in a culture that can be toxic for girls' self-esteem. Indeed, if we looked only at advertising images, this would be a bleak world for females. Girls are extremely desirable to advertisers because they are new consumers, are beginning to have significant disposable income, and are developing brand loyalty that might last a lifetime. . . .

Seventeen, a magazine aimed at girls about twelve to fifteen, sells these girls to advertisers in an ad that says, "She's the one you want. She's the one we've got." The copy continues, "She pursues beauty and fashion at every turn" and concludes with, "It's more than a magazine. It's her life." In another similar ad, *Seventeen* refers to itself as a girl's "Bible." Many girls read magazines like this and take the advice seriously. Regardless of the intent of the advertisers, what are the messages that girls are getting? What are they told?

Primarily girls are told by advertisers that what is most important about them is their perfume, their clothing, their bodies, their beauty. Their "essence" is their underwear. "He says the first thing he noticed about you is your great personality," says an ad featuring a very young woman in tight jeans. The copy continues, "He lies." "If this is your idea of a great catch," says an ad for a cosmetic kit from a teen magazine featuring a cute boy, "this is your tackle box." Even very little girls are offered makeup and toys like Special Night Barbie, which shows them how to dress up for a night out. Girls of all ages get the message that they must be flawlessly beautiful and, above all these days, they must be thin.

Even more destructively, they get the message that this is possible, that, with enough effort and self-sacrifice, they can achieve this ideal. Thus many girls spend enormous amounts of time and energy attempting to achieve something that is not only trivial but also completely unattainable. The glossy images of flawlessly beautiful and extremely thin women that surround us would not have the impact they do if we did not live in a culture that encourages us to believe we can and should remake our bodies into perfect commodities. These images play into the American belief of transformation and ever-new possibilities, no longer via hard work but via the purchase of the right products. As Anne Becker has pointed out, this belief is by no means universal. People in many other cultures may admire a particular body shape without seeking to emulate it. . . .

Women are especially vulnerable because our bodies have been objectified and commodified for so long. And young women are the most vulnerable, especially those who have experienced early deprivation, sexual abuse, family violence, or other trauma. Cultivating a thinner body offers some hope of control and success to a young woman with a poor self-image and overwhelming personal problems that have no easy solutions.

Although troubled young women are especially vulnerable, these messages affect all girls. A researcher at Brigham and Women's Hospital in Boston found that the more frequently girls read magazines, the more likely they were to diet and to feel that magazines influence their ideal body shape. Nearly half reported wanting to lose weight because of a magazine picture (but only 29 percent were actually overweight). Studies at Stanford University and the University of Massachusetts found that about 70 percent of college women say they feel worse about their own looks after reading women's magazines. Another study, this one of 350 young men and women, found that a preoccupation with one's appearance takes a toll on mental health. Women scored much higher than men on what the researchers called "self-objectification." This tendency to view one's body from the outside in—regarding physical attractiveness, sex appeal, measurements, and weight as more central to one's physical identity than health, strength, energy level, coordination, or fitness—has many harmful effects, including diminished mental performance, increased feelings of shame and anxiety, depression, sexual dysfunction, and the development of eating disorders.

These images of women seem to affect men most strikingly by influencing how they judge the real women in their lives. Male college students who viewed just one episode of *Charlie's Angels,* the hit television show of the 1970s that featured three beautiful women, were harsher in their evaluations of the attractiveness of potential dates than were males who had not seen the episode. In another study, male college students shown centerfolds from *Playboy* and *Penthouse* were more likely to find their own girlfriends less sexually attractive.

Adolescent girls are especially vulnerable to the obsession with thinness, for many reasons. One is the ominous peer pressure on young people. Adolescence is a time of such self-consciousness and terror of shame and humiliation. Boys are shamed for being too small, too "weak," too soft, too sensitive. And girls are shamed for being too sexual, too loud, too boisterous, too big (in any sense of the word), having too hearty an appetite. . . .

The situation is very different for men. The double standard is reflected in an ad for a low-fat pizza: "He eats a brownie . . . you eat a rice cake. He eats a juicy burger . . . you eat a low fat entree. He eats pizza . . . you eat pizza. Finally, life is fair." Although some men develop eating problems, the predominant cultural message remains that a hearty appetite and a large size is desirable in a man, but not so in a woman.

Indeed, a 1997 television campaign targets ravenous teenage boys by offering Taco Bell as the remedy for hunger (and also linking eating with sex via the slogan "Want some?"). One commercial features a fat guy who loses his composure when he realizes his refrigerator is empty. In another, two quite heavy guys have dozed off in front of a television set and are awakened by hunger pangs, which only Taco Bell can satisfy. It is impossible to imagine this campaign aimed at teenage girls.

Normal physiological changes during adolescence result in increased body fat for women. If these normal changes are considered undesirable by the culture (and by parents and peers), this can lead to chronic anxiety and concern about weight control in young women. A ten-year-old girl wrote to *New Moon,* a feminist magazine for girls, "I was at the beach and was in my bathing suit. I have kind of fat legs, and my uncle told me I had fat legs in front of all my cousins and my cousins' friends. I was so embarrassed, I went up to my room and shut the door. When I went downstairs again, everyone started teasing me." . . .

The obsession starts early. Some studies have found that from 40 to 80 percent of fourth-grade girls are dieting. Today at least one-third of twelve- to thirteen-year-old girls are actively trying to lose weight, by dieting, vomiting, using laxatives, or taking diet pills. One survey found that 63 percent of high-school girls were on diets, compared with only 16 percent of men. And a survey in Massachusetts found that the single largest group of high-school students considering or attempting suicide are girls who feel they are overweight. Imagine. Girls made to feel so terrible about themselves that they would rather be dead than fat. This wouldn't be happening, of course, if it weren't for our last "socially acceptable" prejudice—weightism. Fat children are ostracized and ridiculed from the moment they enter school, and fat adults, women in particular, are subjected to public contempt and scorn. This strikes terror into the hearts of all women, many of whom, unfortunately, identify with the oppressor and become vicious to themselves and each other.

No wonder it is hard to find a woman, especially a young woman, in America today who has a truly healthy attitude toward her body and toward food. Just as the disease of alcoholism is the extreme end of a continuum that includes a wide range of alcohol use and abuse, so are bulimia and anorexia the extreme results of an obsession with eating and weight control that grips many young women with serious and potentially very dangerous results. Although eating problems are often thought to result from vanity, the truth is that they, like other addictions and compulsive behavior, usually have deeper roots—not only genetic predisposition and biochemical vulnerabilities, but also childhood sexual abuse.

Advertising doesn't cause eating problems, of course, any more than it causes alcoholism. Anorexia in particular is a disease with a complicated etiology, and media images probably don't play a major role. However, these images certainly contribute to the body-hatred so many young women feel and to some of the resulting eating problems, which range from bulimia to compulsive overeating to simply being obsessed with controlling one's appetite. Advertising does promote abusive and abnormal attitudes about eating, drinking, and thinness. It thus provides fertile soil for these obsessions to take root in and creates a climate of denial in which these diseases flourish. . . .

Being obsessed about one's weight is made to seem normal and even appealing in ads for unrelated products, such as a scotch ad that features a very thin and pretty young woman looking in a mirror while her boyfriend observes her. The copy, addressed to him, says, "Listen, if you can handle 'Honey, do I look fat?' you can handle this." These two are so intimate that she can share her deepest fears with him—and he can respond by chuckling at her adorable vulnerability and knocking back another scotch. And everyone who sees the ad gets the message that it is perfectly normal for all young women, including thin and attractive ones, to worry about their weight.

"Put some weight on," says a British ad featuring an extremely thin young woman—but the ad is referring to her watch. She is so thin she can wear the watch on her upper arm—and this is supposed to be a good thing.

Not all of this is intentional on the part of the advertisers, of course. A great deal of it *is* based on research and *is* intended to arouse anxiety and affect women's self-esteem. But some of it reflects the unconscious attitudes and beliefs of the individual advertisers, as well as what Carl Jung referred to as the "collective unconscious." Advertisers are members of the culture too and have been as thoroughly conditioned as anyone else. The magazines and the ads deliberately *create* and intensify anxiety about weight because it is so profitable. On a deeper level, however, they *reflect* cultural concerns and conflicts about women's power. Real freedom for women would change the very basis of our male-dominated society. It is not surprising that many men (and women, to be sure) fear this.

"The more you subtract, the more you add," says an ad that ran in several women's and teen magazines in 1997. Surprisingly, it is an ad for clothing, not for a diet product. Overtly, it is a statement about minimalism in fashion. However, the fact that the girl in the ad is very young and very thin reinforces another message, a message that an adolescent girl constantly gets from advertising and throughout the popular culture, the message that she should diminish herself, she should be *less* than she is.

On the most obvious and familiar level, this refers to her body. However, the loss, the subtraction, the cutting down to size also refers to her sense of her self, her sexuality, her need for authentic connection, and her longing for power and freedom. I certainly don't think that the creators of this particular ad had all this in mind. They're simply selling expensive clothing in an unoriginal way, by using a very young and very thin woman—and an unfortunate tagline. It wouldn't be important at all were there not so many other ads that reinforce this message and did it not coincide with a cultural crisis taking place now for adolescent girls.

"We cut Judy down to size," says an ad for a health club. "Soon, you'll both be taking up less space," says an ad for a collapsible treadmill, referring both to the product and to the young woman exercising on it. *The obsession with thinness is most deeply about cutting girls and women down to size.* It is only a symbol, albeit a very powerful and destructive one, of tremendous fear of female power. Powerful women are seen by many people (women as well as men) as inherently destructive and dangerous. Some argue that it is men's awareness of just how powerful women can be that has created the attempts to keep women small. Indeed, thinness as an ideal has always accompanied periods of greater freedom for women—and as soon as we got the vote, boyish flapper bodies came into vogue. No wonder there is such pressure on young women today to be thin, to shrink, to be like little girls, not to take up too much space, literally or figuratively.

At the same time there is relentless pressure on women to be small, there is also pressure on us to succeed, to achieve, to "have it all." We can be successful as long as we stay "feminine" (i.e., powerless enough not to be truly threatening). One way to do this is to present an image of fragility, to look like a waif. This demonstrates that one is both in control and still very "feminine." One of the many double binds tormenting young women today is the need to be both sophisticated and accomplished, yet also delicate and childlike. Again, this applies mostly to middle- to upper-class white women.

The changing roles and greater opportunities for women promised by the women's movement are trivialized, reduced to the private search for the slimmest body. In one commercial, three skinny young women dance and sing about the "taste of freedom." They are feeling free because they can now eat bread, thanks to a low-calorie version. A commercial for a fast-food chain features a very slim young woman who announces, "I have a license to eat." The salad bar and lighter fare have given her freedom to eat (as if eating for women were a privilege rather than a need). "Free yourself," says ad after ad for diet products. . . .

Most of us know by now about the damage done to girls by the tyranny of the ideal image, weightism, and the obsession with thinness. But girls get other messages too that "cut them down to size" more subtly. In ad after ad girls are urged to be "barely there"—beautiful but silent. Of course, girls are not just influenced by images of other girls. They are even more powerfully attuned to images of women, because they learn from these images what is expected of them, what they are to become. And they see these images again and again in the magazines they read, even those magazines designed for teenagers, and in the commercials they watch.

"Make a statement without saying a word," says an ad for perfume. And indeed this is one of the primary messages of the culture to adolescent girls. "The silence of a look can reveal more than words," says another perfume ad, this one featuring a woman lying on her back. "More than words can say," says yet another perfume ad, and a clothing ad says, "Classic is speaking your mind (without saying a word)." An ad for lipstick says, "Watch your mouth, young lady," while one for nail polish says, "Let your fingers do the talking," and one for hairspray promises "hair that speaks volumes." In another ad, a young woman's turtleneck is pulled over her mouth. And an ad for a movie soundtrack features a chilling image of a young woman with her lips sewn together.

It is not only the girls themselves who see these images, of course. Their parents and teachers and doctors see them and they influence their sense of how girls should be. A 1999 study done at the University of Michigan found that, beginning in preschool, girls are told to be quiet much more often than boys. Although boys were much noisier than girls, the girls were told to speak softly or to use a "nicer" voice about three times more often. Girls were encouraged to be quiet, small, and physically constrained. The researcher concluded that one of the consequences of this socialization is that girls grow into women afraid to speak up for themselves or to use their voices to protect themselves from a variety of dangers. . . .

"Score high on nonverbal skills," says a clothing ad featuring a young African-American woman, while an ad for mascara tells young women to "make up your own language." And an Italian ad features a very thin young woman in an elegant coat sitting on a window seat. The copy says, "This woman is silent. This coat talks." Girls, seeing these images of women, are encouraged to be silent, mysterious, not to talk too much or too loudly. In many different ways, they are told "the more you subtract, the more you add." . . .

The January 1998 cover of *Seventeen* highlights an article, "Do you talk too much?" On the back cover is an ad for Express mascara, which promises "high voltage volume instantly!" As if the way that girls can express themselves and turn up the volume is via their mascara. Is this harmless wordplay, or is it a sophisticated and clever marketing ploy based on research about the silencing of girls, deliberately designed to attract them with the promise of at least some form of self-expression? Advertisers certainly spend a lot of money on psychological research and focus groups. I would expect these groups to reveal, among other things, that teenage girls are angry but reticent. Certainly the cumulative effect of these images and words urging girls to express themselves only through their bodies and through products is serious and harmful.

Many ads feature girls and young women in very passive poses, limp, dolllike, sometimes acting like little girls, playing with dolls and wearing bows in their hair. One ad uses a pacifier to sell lipstick and another the image of a baby to sell Baby Doll Blush Highlight. "Lolita seems to be a comeback kid," says a fashion layout featuring a woman wearing a ridiculous hairstyle and a baby-doll dress, standing with shoulders slumped and feet apart. In women's and teen magazines it is virtually impossible to tell the fashion layouts from the ads. Indeed, they exist to support each other.

As Erving Goffman pointed out in *Gender Advertisements,* we learn a great deal about the disparate power of males and females simply through the body language and poses of advertising. Women, especially young women, are generally subservient to men in ads, through both size and position. Sometimes it is as blatant as the woman serving as a footrest in the ad for Think Skateboards.

Other times, it is more subtle but quite striking (once one becomes aware of it). The double-paged spread for Calvin Klein's clothing for kids conveys a world of information about the relative power of boys and girls. One of the boys seems to be in the act of speaking, expressing himself, while the girl has her hand over her mouth. Boys are generally shown in ads as active, rambunctious, while girls are more often passive and focused on their appearance. The exception to the rule involves African-American children, male and female, who are often shown in advertising as passive observers of their white playmates.

That these stereotypes continue, in spite of all the recent focus on the harm done to girls by enforced passivity, is evident in the most casual glance at parents' magazines. In the ads in the March 1999 issues of *Child* and *Parents,* all of the boys are active and all of the girls are passive. In *Child,* a boy plays on the jungle gym in one ad, while in another, a girl stands quietly, looking down, holding some flowers. In *Parents,* a boy rides a bike, full of excitement, while a girl is happy about having put on lipstick. It's hard to believe that this is 1999 and not 1959. The more things change, the more they stay the same.

Girls are often shown as playful clowns in ads, perpetuating the attitude that girls and women are childish and cannot be taken seriously, whereas even very young men are generally portrayed as secure, powerful, and serious. People in control of their lives stand upright, alert, and ready to meet the world. In contrast, females often appear off-balance, insecure, and weak. Often our body parts are bent, conveying unpreparedness, submissiveness, and appeasement. We exhibit what Goffman terms "licensed withdrawal"—seeming to be psychologically removed, disoriented, defenseless, spaced out.

Females touch people and things delicately, we caress, whereas males grip, clench, and grasp. We cover our faces with our hair or our hands, conveying shame or embarrassment. And, no matter what happens, we keep on smiling. "Just smiling the bothers away," as one ad says. This ad is particularly disturbing because the model is a young African-American woman, a member of a group that has long been encouraged to just keep smiling, no matter what. She's even wearing a kerchief, like Aunt Jemima. The cultural fear of angry women is intensified dramatically when the women are African-American. . . .

Girls and young women are often presented as blank and fragile. Floating in space, adrift in a snowstorm. A Valentino clothing ad perhaps unwittingly illustrates the tragedy of adolescence for girls. It features a very young woman with her head seemingly enclosed in a glass bubble labeled "Love." Some ads and fashion layouts picture girls as mermaids or underwater as if they were drowning—or lying on the ground as if washed up to shore, such as the Versace makeup ad picturing a young girl caught up in fishing nets, rope, and seashells. An ad for vodka features a woman in the water and the copy, "In a past life I was a mermaid who fell in love with an ancient mariner. I pulled him into the sea to be my husband. I didn't know he couldn't breathe underwater." Of course, she can't breathe underwater either.

Breathe underwater. As girls come of age sexually, the culture gives them impossibly contradictory messages. As the *Seventeen* ad says, "She wants to be outrageous. And accepted." Advertising slogans such as "because innocence is sexier than you think," "Purity, yes. Innocence never," and "nothing so sensual was ever so innocent" place them in a double bind. "Only something so pure could inspire such unspeakable passion," declares an ad for Jovan musk that features a white flower. Somehow girls are supposed to be both innocent and seductive, virginal and experienced, all at the same time. As they quickly learn, this is tricky.

Females have long been divided into virgins and whores, of course. What is new is that girls are now supposed to embody both within themselves. This is symbolic of the central contradiction of the culture—we must work hard and produce and achieve success and yet, at the same time, we are encouraged to live impulsively, spend a lot of money, and be constantly and immediately gratified. This tension is reflected in our attitudes toward many things, including sex and eating. Girls are promised fulfillment both through being thin and through eating rich foods, just as they are promised fulfillment through being innocent and virginal and through wild and impulsive sex. . . .

Of course, some girls do resist and rebel. Some are encouraged (by someone—a loving parent, a supportive teacher) to see the cultural contradictions clearly and to break free in a healthy and positive way. Others rebel in ways that damage themselves. A young woman seems to have only two choices: She can bury her sexual self, be a "good girl," give in to what Carol Gilligan terms "the tyranny of nice and kind" (and numb the pain by overeating or starving or cutting herself or drinking heavily). Or she can become a rebel—flaunt her sexuality, seduce inappropriate partners, smoke, drink flamboyantly, use other drugs. Both of these responses are self-destructive, but they begin as an attempt to survive, not to self-destruct. . . .

There are few healthy alternatives for girls who want to truly rebel against restrictive gender roles and stereotypes. The recent emphasis on girl power has led to some real advances for girls and young women, especially in the arenas of music and sports. But it is as often co-opted and trivialized. The Indigo Girls are good and true, but it is the Spice Girls who rule. Magazines like *New Moon, Hues,* and *Teen Voices* offer a real alternative to the glitzy, boy-crazy, appearance-obsessed teen magazines on the newsstands, but they have to struggle for funds since they take no advertising. There are some good zines and Websites for girls on the Internet but there are also countless sites that degrade and endanger them. And Barbie continues to rake in two billion dollars a year and will soon have a postal stamp in her honor—while a doll called "Happy to be me," similar to Barbie but much more realistic and down to earth, was available for a couple of years in the mid-1990s (I bought one for my daughter) and then vanished from sight. Of course, Barbie's makers have succumbed to pressure somewhat and have remade her with a thicker waist, smaller breasts, and slimmer hips. As a result, according to Anthony Cortese, she has already lost her waitressing job at Hooter's and her boyfriend Ken has told her that he wants to start seeing other dolls.

Girls who want to escape the stereotypes are viewed with glee by advertisers, who rush to offer them, as always, power via products. The emphasis in the ads is always on their sexuality, which is exploited to sell them makeup and clothes and shoes. "Lil' Kim is wearing lunch box in black," says a shoe ad featuring a bikini-clad young woman in a platinum wig stepping over a group of nuns—the ultimate bad girl, I guess, but also the ultimate sex object. A demon woman sells a perfume called Hypnotic Poison. A trio of extremely thin African-American women brandish hair appliances and products as if they were weapons—and the brand is 911. A cosmetics company has a line of products called "Bad Gal." In one ad, eyeliner is shown in cartoon version as a girl, who is holding a dog saying "grrrr," surely a reference to "grrrrls," a symbol these days of "girl power" (as in cybergrrrl.com, the popular Website for girls and young women). Unfortunately, girl power doesn't mean much if girls don't have the tools to achieve it. Without reproductive freedom and freedom from violence, girl power is nothing but a marketing slogan.

So, for all the attention paid to girls in recent years, what girls are offered mostly by the popular culture is a superficial toughness, an "attitude," exemplified by smoking, drinking, and engaging in casual sex—all behaviors that harm themselves. In 1990 Virginia Slims offered girls a T-shirt that said, "Sugar and spice and everything nice? Get real." In 1997 Winston used the same theme in an ad featuring a tough young woman shooting pool and saying, "I'm not all sugar & spice. And neither are my smokes." As if the alternative to the feminine stereotype was sarcasm and toughness, and as if smoking was somehow an expression of one's authentic self ("get real").

Of course, the readers and viewers of these ads don't take them literally. But we do take them in—another grain of sand in a slowly accumulating and vast sandpile. If we entirely enter the world of ads, imagine them to be real for a moment, we find that the sandpile has completely closed us in, and there's only one escape route—buy something. "Get the power," says an ad featuring a woman showing off her biceps. "The power to clean anything," the ad continues. "Hey girls, you've got the power of control" says an ad for . . . hairspray. "The possibilities are endless" (clothing). "Never lose control" (hairspray again). "You never had this much control when you were on your own" (hair gel). "Exceptional character" (a watch). "An enlightening experience" (face powder). "Inner strength" (vitamins). "Only Victoria's Secret could make control so sensual" (girdles). "Stronger longer" (shampoo). Of course, the empowerment, the enlightenment, is as impossible to get through products as is anything else—love, security, romance, passion. On one level, we know this. On another, we keep buying and hoping—and buying.

Other ads go further and offer products as a way to rebel, to be a real individual. "Live outside the lines," says a clothing ad featuring a young woman walking out of a men's room. This kind of rebellion isn't going to rock the world. And, no surprise, the young woman is very thin and conventionally pretty. Another pretty young woman sells a brand of jeans called "Revolt." "Don't just change . . . revolt," says the copy, but the young woman is passive, slight, her eyes averted.

"Think for yourself," says yet another hollow-cheeked young woman, demonstrating her individuality via an expensive and fashionable sweater. "Be amazing" (cosmetics). "Inside every woman is a star" (clothing). "If you're going to create electricity, use it" (watches). "If you let your spirit out, where would it go" (perfume). These women are all perfect examples of conventional "femininity," as is the young woman in a Halston perfume ad that says, "And when she was bad she wore Halston." What kind of "bad" is this?

"Nude with attitude" features an African-American woman in a powerful pose, completely undercut by the brevity of her dress and the focus on her long legs. Her "attitude" is nothing to fear—she's just another sex object. Good thing, given the fear many people have of powerful African-American women.

The British ad "For girls with plenty of balls" is insulting in ways too numerous to count, beginning with the equation of strength and courage and fiery passion with testicles. What this ad offers girls is body lotion.

Some ads do feature women who seem really angry and rebellious, but the final message is always the same. "Today I indulge my dark side," says an ad featuring a fierce young woman tearing at what seems to be a net. "Got a problem with that?" The slogan is "be extraordinary not ordinary." The product that promises to free this girl from the net that imprisons her? Black nail polish.

Nail polish. Such a trivial solution to such an enormous dilemma. But such triviality and superficiality is common in advertising. How could it be otherwise? The solution to any problem always has to be a product. Change, transformation, is thus inevitably shallow and moronic, rather than meaningful and transcendent. These days, self-improvement seems to have more to do with calories than with character, with abdomens than with absolutes, with nail polish than with ethics.

It has not always been so. Joan Jacobs Brumberg describes this vividly in *The Body Project: An Intimate History of American Girls:*

> When girls in the nineteenth century thought about ways to improve themselves, they almost always focused on their internal character and how it was reflected in outward behavior. In 1892, the personal agenda of an adolescent diarist read: "Resolved, not to talk about myself or feelings. To think before speaking. To work seriously. . . . To be dignified. Interest myself more in others."
>
> A century later, in the 1990s, American girls think very differently. In a New Year's resolution written in 1982, a girl wrote: "I will try to make myself better in every way I possibly can with the help of my budget and baby-sitting money. I will lose weight, get new lenses, already got new haircut, good makeup, new clothes and accessories."

Not that girls didn't have plenty of problems in the nineteenth century. But by now we should have come much further. This relentless trivialization of a girl's hopes and dreams, her expectations for herself, cuts to the quick of her soul. Just as she is entering womanhood, eager to spread her wings, to become truly sexually *active,* empowered, independent—the culture moves in to *cut her down to size. . . .*

NOTES

234 *"According to the Carnegie Corporation":* Carnegie Corporation, 1995.
234 *"As Carol Gilligan, Mary Pipher":* Gilligan, 1982; Pipher, 1994; Sadker and Sadker, 1994.
235 *"Teenage women today":* Roan, 1993, 28.
235 *"a 1998 status report":* Vobejda and Perlstein, 1998, A3.
235 *"The socialization that emphasizes passivity":* Thompson, 1994.
235 *"Eating problems affect girls from African-American":* Steiner-Adair and Purcell, 1996, 294.
235 "*In* Raising Cain": Kindlon and Thompson, 1999.
235 "*In* Real Boys": W. Pollack, 1998.
236 *"A researcher at Brigham and Women's Hospital":* Field, Cheung, Wolf, Herzog, Gortmaker, and Colditz, 1999, 36.
236 *"Studies at Stanford University":* Then, 1992. Also Richins, 1991, 71.
236 *"this one of 350 young men and women":* Fredrickson, 1998, 5.
236 *"Male college students who viewed just one episode":* Strasburger, 1989, 757.
237 *"Taco Bell":* Garfield, 1997, 43.
237 *"A ten-year-old girl wrote to* New Moon*":* E-mail correspondence with Heather S. Henderson, editor-in-chief of HUES Magazine, New Moon Publishing, March 22, 1999.
237 *"from 40 to 80 percent of fourth-grade girls":* Stein, 1986, 1.
237 *"one-third of twelve- to thirteen-year-old girls":* Rodriquez, 1998, B9.
237 *"63 percent of high-school girls":* Rothblum, 1994, 55.
237 *"a survey in Massachusetts":* Overlan, 1996, 15.
237 *"our last 'socially acceptable' prejudice—weightism":* Steiner-Adair and Purcell, 1996, 294.
237 *"Although eating problems are often thought to result":* Smith, Fairburn, and Cowen, 1999, 171–76. Also Thompson, 1994. Also Krahn, 1991. Also Hsu, 1990. Also Jonas, 1989, 267–71.
238 *"Some argue that it is men's awareness":* Faludi, 1991. Also Kilbourne, 1986.
239 *"A 1999 study done at the University of Michigan":* Martin, 1998, 494–511.
240 *"Erving Goffman":* Goffman, 1978.
240 *"The exception to the rule involves African-American children":* Seiter, 1993.
241 *"Carol Gilligan terms the 'tyranny of nice and kind' ":* Brown and Gilligan, 1992, 53.
241 *"And Barbie continues to rake in":* Goldsmith, 1999, D3.
241 *"according to Anthony Cortese":* Cortese, 1999, 57.
243 *"Joan Jacobs Brumberg describes this difference vividly":* Brumberg, 1997, xxi.

REFERENCES

Becker, A. E. and Burwell, R. A. (1999). Acculturation and disordered eating in Fiji. Poster presented at the American Psychiatric Association Annual Meeting, Washington, DC, May 19, 1999.
Brown, J. D., Greenberg, B. S., and Buerkel-Rothfuss, N. L. (1993). Mass media, sex and sexuality. In Strasburger, V. C. and Comstock, G. A., eds. *Adolescent medicine: adolescents and the media.* Philadelphia, PA: Hanley & Belfus.
Brown, J. D., and Steele, J. R. (1995, June 21). *Sex and the mass media. A report prepared for the Henry J. Kaiser Family Foundation.*
Brown, L. M., and Gilligan, C. (1992). *Meeting at the crossroads: women's psychology and girls' development.* New York: Ballantine Books.

Brumberg, J. J. (1997). *The Body Project: An Intimate History of American Girls.* New York: Random House.

Carnegie Corporation (1995). *Great transitions: preparing adolescents for a new century.* New York: Carnegie Corporation.

Cortese, A. (1999). *Provocateur: women and minorities in advertising.* Lanham, MD: Rowman & Littlefield.

Faludi, S. (1991). *Backlash.* New York: Crown.

Field, A. E., Cheung, L., Wolf, A. M., Herzog, D. B., Gortmaker, S. L., and Colditz, G. A. (1999, March). Exposure to the mass media and weight concerns among girls. *Pediatrics,* vol. 103, no. 3, 36-41.

Fredrickson, B. L. (1998, Fall). *Journal of Personality and Social Psychology,* vol. 75, no. 1. Reported in *Media Report to Women,* 5.

Garfield, B. (1997, August 4). Taco Bell fills the bill for teens' tummies. *Advertising Age,* 43.

Gilligan, C. (1982). *In a different voice.* Cambridge, MA: Harvard University Press.

Goffman, E. (1978). *Gender advertisements.* Cambridge, MA: Harvard University Press.

Goldsmith, J. (1999, February 10). A $2 billion doll celebrates her 40th without a wrinkle. *Boston Globe,* D3.

Goodman, E. (1999, May 27). The culture of thin bites Fiji teens. *Boston Globe,* A23.

Grierson, B. (1998, Winter). Shock's next wave *Adbusters,* 21.

Hsu, L. K. (1990). *Eating disorders.* New York: Guilford Press.

Jonas, J. M. (1989). Eating disorders and alcohol and other drug abuse. Is there an association? *Alcohol Health & Research World,* vol. 13, no. 3, 267-71.

Kaiser Family Foundation, the Alan Guttmacher Institute, and the National Press Foundation (1996, June 24). *Emerging issues in reproductive health (fact sheet).*

Kilbourne, J. (1986). The child as sex object: images of children in the media. In Nelson, M. and Clark, K. (1986). *The educator's guide to preventing child sexual abuse.* Santa Cruz, CA: Network Publications.

Krahn, D. D. (1991). Relationship of eating disorders and substance abuse. *Journal of Substance Abuse,* vol. 3, no. 2, 239–53.

Martin, K. A. (1998, August). Becoming a gendered body: practices of preschools. *American Sociological Review,* vol. 63, no. 4, 494–511.

Overlan, L. (1996, July 2). 'Overweight' girls at risk. *Newton Tab,* 15.

Pipher, M. (1994). *Reviving Ophelia: saving the selves of adolescent girls.* New York: G. P. Putnam's Sons.

Pollack, W. (1998). *Real boys: Rescuing our sons from the myths of boyhood.* New York: Random House.

Reed, B. G. (1991). Linkages: battering, sexual assault, incest, child sexual abuse, teen pregnancy, dropping out of school and the alcohol and drug connection. In Roth, P., ed. *Alcohol and drugs are women's issues.* Metuchen, NJ: Scarecrow Press, 130-49.

Richins, M. L. (1991). Social comparison and idealized images of advertising. *Journal of Consumer Research,* 18, 71-83.

Roan, S. (1993, June 8). Painting a bleak picture for teen girls. *Los Angeles Times,* 28.

Rodriguez, C. (1998, November 27). Even in middle school, girls are thinking thin. *Boston Globe,* B1, B9.

Rothblum, E. D. (1994). "I'll die for the revolution but don't ask me not to diet": feminism and the continuing stigmatization of obesity. In Fallon, P., Katzman, M. A., and Wooley, S. C. (1994). *Feminist perspectives on eating disorders.* New York: The Guilford Press, 53-76.

Seiter, E. (1993). Different children, different dreams: racial representation in advertising. *Sold Separately: Aspects of Children's Consumer Culture.* Winchester, MA and London: Unwin Hyman.

Smith, K. A., Fairburn, C. G., and Cowen, P. J. (1999). Symptomatic relapse in bulimia nervosa following acute tryptophan depletion. *Journal of the American Medical Association,* vol. 56, 171-76.

Stein, J. (1986, October 29). Why girls as young as 9 fear fat and go on diets to lose weight. *Los Angeles Times,* 1, 10.

Steiner-Adair, C. (1996). Remarks at the 1996 Leadership Dinner of the Harvard Eating Disorders Center, October 29, 1996.

Steiner-Adair, C. and Purcell, A. (1996, Winter). Approaches to mainstreaming eating disorders prevention. *Eating Disorders,* vol. 4, no. 4, 294-309.

Strasburger, V. C. (1989, June). Adolescent sexuality and the media. *Pediatric Clinics of North America,* vol. 36, no. 3, 747-73.

Then, D. (1992, August). Women's magazines: Messages they convey about looks, men and careers. Paper presented at the annual convention of the American Psychological Association, Washington, DC.

Thompson, B. W. (1994). *A hunger so wide and so deep.* Minneapolis: University of Minnesota Press.

Vobejda, B. and Perlstein, L. (1998, June 17). Girls closing gap with boys, but not always for the best. *Boston Globe,* A3.

25

The Anthropometry of Barbie

Unsettling Ideals of the Feminine Body in Popular Culture

BY JACQUELINE URLA AND ALAN C. SWEDLUND

This reading by Jacqueline Urla and Alan Swedlund offers an interesting approach to understanding the relationship between the success of Barbie doll and the everyday body ideals and practices of girls and women in North America today. The authors apply the science of measuring bodies, or anthropometry, to Barbie doll and her "friends," comparing the extreme deviation of Barbie's body to the anthropometry of real women. Urla and Swedlund point out that Barbie exemplifies the commodification of gender in modern, consumer culture. They argue that the success of Barbie points to the strong desire of consumers for fantasy and for products that will transform them. Finally, the authors discuss the multiple meanings of Barbie for the girls and women who are her fans.

1. Why is Barbie a "perfect icon" of late capitalist constructions of femininity?
2. How has anthropometry, the science of measuring bodies, altered how we think and feel about gendered bodies?
3. How do Urla and Swedlund's findings underscore Kilbourne's arguments in the previous reading?

It is no secret that thousands of healthy women in the United States perceive their bodies as defective. The signs are everywhere: from potentially lethal cosmetic surgery and drugs to the more familiar routines of dieting, curling, crimping, and aerobicizing, women seek to take control over their unruly physical selves. Every year at least 150,000 women undergo breast implant surgery (Williams 1992), while Asian women have their noses rebuilt and their eyes widened to make themselves look "less dull" (Kaw 1993). Studies show that the obsession with body size and the sense of inadequacy start frighteningly early; as many as 80 percent of 9-year-old suburban girls are concerned about dieting and their weight (Bordo 1991: 125). Reports like these, together with the dramatic rise in eating disorders among young women, are just some of the more noticeable fallout from what Naomi Wolf calls "the beauty myth." Fueled by the hugely profitable cosmetic, weight-loss, and fashion industries, the beauty myth's glamorized notions of the ideal body reverberate back upon women as "a dark vein of self hatred, physical obsessions, terror of aging, and dread of lost control" (Wolf 1991: 10).

It is this conundrum of somatic femininity, that female bodies are never feminine enough, that they must be deliberately and oftentimes painfully remade to be what "nature" intended—a condition dramatically accentuated under consumer capitalism—that motivates us to focus our inquiry . . . on images of the feminine ideal. Neither universal nor changeless, idealized notions of both masculine and feminine bodies have a long history that shifts considerably across time, racial or ethnic group, class, and culture. Body ideals in twentieth-century North America are influenced and shaped by images from classical or "high" art, the discourses of science and medicine, and increasingly via a multitude of commercial interests, ranging from mundane life insurance standards to the more high-profile fashion, fitness, and entertainment industries.

Making her debut in 1959 as Mattel's new teenage fashion doll, Barbie rose quickly to become the top-selling toy in the United States. Thirty-four years and a woman's movement later, Barbie dolls remain Mattel's best-selling item, netting over one billion dollars in revenues worldwide (Adelson 1992), or roughly one Barbie sold every two seconds (Stevenson 1991). Mattel estimates that in the United States over 95 percent of girls between the ages of three and eleven own at least one Barbie, and that the average number of dolls per owner is seven (E. Shapiro 1992). Barbie is clearly a force to contend with, eliciting over the years a combination of critique, parody, and adoration. A legacy of the postwar era, she remains an incredibly resilient visual and tactile model of femininity for prepubescent girls headed straight for the twenty-first century.

It is not our intention to settle the debate over whether Barbie is a good or bad role model for little girls or whether her unrealistic body wrecks havoc on girls' self-esteem. . . . We want to suggest that Barbie dolls, in fact, offer a much more complex and contradictory set of possible meanings that take shape and mutate in a period marked by the growth of consumer society, intense debate over gender and racial relations, and changing notions of the body. . . . We want to explore not only how it is that this popular doll has been able to survive such dramatic social changes, but also how she takes on new significance in relation to these changing contexts.

We begin by tracing Barbie's origins and some of the image makeovers she has undergone since her creation. From there we turn to an experiment in the anthropometry of Barbie to understand how she compares to standards for the "average American woman" that were emerging in the postwar period.[1] Not surprisingly, our measurements show Barbie's body to be thin—very thin—far from anything approaching the norm. Inundated as our society is with conflicting and exaggerated images of the feminine body, statistical measures can help us to see that exaggeration more clearly. But we cannot stop there. First, as our brief foray into the history of anthropometry shows, the measurement and creation of body averages have their own politically inflected and culturally biased histories. Standards for the "average" American body, male or female, have always been imbricated in histories of nationalism and race purity. Secondly, to say that Barbie is unrealistic seems to beg the issue. Barbie *is* fantasy: a fantasy whose relationship to the hyperspace of consumerist society is multiplex. What of the pleasures of Barbie bodies? What alternative meanings of power and self-fashioning might her thin body hold for women/girls? Our aim is not, then, to offer another rant against Barbie, but to clear a space where the range of her contradictory meanings and ironic uses can be contemplated: in short, to approach her body as a meaning system in itself, which, in tandem with her mutable fashion image, serves to crystallize some of the predicaments of femininity and feminine bodies in late-twentieth-century North America.

A DOLL IS BORN

. . . Making sense of Barbie requires that we look to the larger sociopolitical and cultural milieu that made her genesis both possible and meaningful. Based on a German prototype, the "Lili" doll, Barbie was from "birth" implicated in the ideologies of the Cold War and the research and technology exchanges of the military-industrial complex. Her finely crafted durable plastic mold was, in fact, designed by Jack Ryan, well known for his work in designing the Hawk and Sparrow missiles for the Raytheon Company. Conceived at the hands of a military-weapons-designer-turned-toy-inventor, Barbie dolls came onto the market the same year that the infamous Nixon-Khrushchev "kitchen debate" took place at the American National Exhibition in Moscow. Here, in front of the cameras of the world, the leaders of the capitalist and socialist worlds faced off, not over missile counts, but over "the relative merits of American and Soviet washing machines, televisions, and electric ranges" (May 1988:16). As Elaine Tyler May has noted in her study of the Cold War, this much-celebrated media event signaled the transformation of American-made commodities and the model suburban home into key symbols and safeguards of democracy and freedom. It was thus with fears of nuclear annihilation and sexually charged fantasies of the perfect bomb shelter running rampant in the American imaginary, that Barbie and her torpedo-like breasts emerged into popular culture as an emblem of the aspirations of prosperity, domestic containment, and rigid gender roles that were to characterize the burgeoning postwar consumer economy and its image of the American Dream.

Marketed as the first "teenage" fashion doll, Barbie's rise in popularity also coincided with, and no doubt contributed to, the postwar creation of a distinctive teenage lifestyle.[2] Teens, their tastes, and their behaviors were becoming the object of both sociologists and criminologists as well as market survey researchers intent on capturing their discretionary dollars. While J. Edgar Hoover was pronouncing "the juvenile jungle" a menace to American society, retailers, the music industry, and moviemakers declared the thirteen to nineteen-year-old age bracket "the seven golden years" (Doherty 1988:51–52).

Barbie dolls seemed to cleverly reconcile both of these concerns by personifying the good girl who was sexy, but didn't have sex, and was willing to spend, spend, spend. . . .

Every former Barbie owner knows that to buy a Barbie is to lust after Barbie accessories. . . . As Paula Rabinowitz has noted, Barbie dolls, with their focus on frills and fashion, epitomize the way that teenage girls and girl culture in general have figured as accessories in the historiography of post-war culture; that is as both essential to the burgeoning commodity culture as consumers, but seemingly irrelevant to the central narrative defining cold war existence (Rabinowitz 1993). Over the years, Mattel has kept Barbie's love of shopping alive, creating a Suburban Shopper Outfit and her own personal Mall to shop in (Motz 1983:131). More recently, in an attempt to edge into the computer game market, we now have an electronic "Game Girl Barbie" in which (what else?) the object of the game is to take Barbie on a shopping spree. In "Game Girl Barbie," shopping takes skill, and Barbie plays to win.

Perhaps what makes Barbie such a perfect icon of late capitalist constructions of femininity is the way in which her persona pairs endless consumption with the achievement of femininity and the appearance of an appropriately gendered body. By buying for Barbie, girls practice how to be discriminating consumers knowledgeable about the cultural capital of different name brands, how to read packaging, and the overall importance of fashion and taste for social status (Motz 1987: 131–32). . . . In making this argument, we want to stress that we are drawing on more than just the doll. "Barbie" is also the packaging, spin-off products, cartoons, commercials, magazines, and fan club paraphernalia, all of which contribute to creating her persona. Clearly, as we will discuss below, children may engage more or less with those products, subverting or ignoring various aspects of Barbie's "official" presentation. However, to the extent that little girls *do* participate in the prepackaged world of Barbie, they come into contact with a number of beliefs central to femininity under consumer capitalism. Little girls learn, among other things, about the crucial importance of their appearance to their personal happiness and to their ability to gain favor with their friends. Barbie's social calendar is constantly full, and the stories in her fan magazines show her frequently engaged in preparation for the rituals of heterosexual teenage life: dates, proms, and weddings. . . .

Barbie exemplifies the way in which gender in the late twentieth century has become a commodity itself, "something we can buy into . . . the same way we buy into a style" (Willis 1991: 23). In her insightful analysis of the logics of consumer capitalism, cultural critic Susan Willis pays particular attention to the way in which children's toys like Barbie and the popular muscle-bound "HeMan" for boys link highly conservative and narrowed images of masculinity and femininity with commodity consumption (1991: 27). In the imaginary world of Barbie and teen advertising, observes Willis, being or becoming a teenager, having a "grown-up" body, is inextricably bound up with the acquisition of certain commodities, signaled by styles of clothing, cars, music, etc. . . .

BARBIE IS A SURVIVOR

. . . In the past three decades, this popular children's doll has undergone numerous changes in her fashion image and "occupations" and has acquired a panoply of ethnic "friends" and analogues that have allowed her to weather the dramatic social changes in gender and race relations that arose in the course of the sixties and seventies. . . .

[A] glance at Barbie's resumé, published in *Harper's* magazine in August 1990, while incomplete, shows Mattel's attempt to expand Barbie's career options beyond the original fashion model:

	Positions Held
1959–present	Fashion model
1961–present	Ballerina
1961–64	Stewardess (American Airlines)
1964	Candy striper
1965	Teacher
1965	Fashion editor
1966	Stewardess (Pan Am)
1973–75	Flight attendant (American Airlines)
1973–present	Medical doctor
1976	Olympic athlete
1984	Aerobics instructor
1985	TV news reporter
1985	Fashion designer
1985	Corporate executive
1988	Perfume designer
1989–present	Animal rights volunteer

It is only fitting, given her origin, to note that Barbie has also had a career in the military and aeronautics space industry: she has been an astronaut, a marine, and, during the Gulf War, a Desert Storm trooper. Going from pink to green, Barbie has also acquired a social conscience, taking up the causes of UNICEF, animal rights, and environmental protection. . . .

For anyone tracking Barbiana, it is abundantly clear that Mattel's marketing strategies are sensitive to a changing social climate. Just as Mattel has sought to present Barbie as a career woman with more than air in her vinyl head, they have also tried to diversify her otherwise lily-white suburban world. . . . With the expansion of sales worldwide, Barbie has acquired multiple national guises (Spanish Barbie, Jamaican Barbie, Malaysian Barbie, etc.).[3] In addition, her cohort of "friends" has become increasingly ethnically diversified, as has Barbie advertising, which now regularly features Asian, Hispanic, and African American little girls playing with Barbie. . . . This diversification has not spelled an end to reigning Anglo beauty norms and body image. Quite the reverse. When we line the dolls up together, they look virtually identical. Cultural difference is reduced to surface variations in skin tone and costumes that can be exchanged at will. . . .

"The icons of twentieth-century mass culture," writes Susan Willis, "are all deeply infused with the desire for change," and Barbie is no exception (1991: 37). In looking over the course of Barbie's career, it is clear that part of her resilience, appeal, and profitability stems from the fact that her identity is constructed primarily through fantasy and is consequently open to change and reinterpretation. As a fashion model, Barbie continually creates her identity anew with every costume change. In that sense, we might want to call Barbie the prototype of the "transformer dolls" that cultural critics have come to see as emblematic of the restless desire for change that permeates postmodern capitalist society (Wilson 1985: 63). Not only can she renew her image with a change of clothes, Barbie also is seemingly able to clone herself effortlessly into new identities—Malibu Barbie; Totally Hair Barbie; Teen Talk Barbie; even Afrocentric Barbie/Shani—without somehow suggesting a serious personality disorder. . . . The multiplication of Barbie and her friends translates the challenge of gender inequality and racial diversity into an ever-expanding array of costumes, a new "look" that can be easily accommodated into a harmonious and illusory pluralism that never ends up rocking the boat of WASP beauty

What is striking, then, is that, while Barbie's identity may be mutable—one day she might be an astronaut, another a cheerleader—*her hyper-slender, big-chested body has remained fundamentally unchanged over the years*—a remarkable fact in a society that fetishizes the new and improved. . . . We turn now from Barbie's "persona" to the conundrum of her body and to our class experiment in the anthropometry of feminine ideals. In so doing, our aim is deliberately subversive. We wish to use the tools of calibration and measurement—tools of normalization that have an unsavory history for women and racial or ethnic minorities—to destabilize the ideal. . . . We begin with a very brief historical overview of the anthropometry of women and the emergence of an "average" American female body in the postwar United States, before using our calipers on Barbie and her friends.

THE MEASURED BODY: NORMS AND IDEALS

★ ★ ★

As the science of measuring human bodies, anthropometry belongs to a long line of techniques of the eighteenth and nineteenth centuries concerned with measuring, comparing, and interpreting variability in different zones of the human body: craniometry, phrenology, physiognomy, and comparative anatomy. Early anthropometry shared with these an understanding and expectation that the body was a window into a host of moral, temperamental, racial, or gender characteristics. It sought to distinguish itself from its predecessors, however, by adhering to rigorously standardized methods and quantifiable results that would, it was hoped, lead to the "complete elimination of personal bias" that anthropometrists believed had tainted earlier measurement techniques (Hrdlicka 1939: 12).[4]

Under the aegis of Earnest Hooton, Ales Hrdlicka, and Franz Boas, located respectively at Harvard University, the Smithsonian, and Columbia University, anthropometric studies within U.S. physical anthropology were utilized mainly in the pursuit of three general areas of interest: identifying racial and or national types; the measurement of adaptation and "degeneracy"; and a comparison of the sexes. Anthropometry was, in other words, believed to be a useful technique in resolving three critical border disputes: the boundaries between races or ethnic groups; the normal and the degenerate; and the border between the sexes.

As is well documented by now, women and non-Europeans did not fare well in these emerging sciences of the body (see the work of Blakey 1987; Gould 1981; Schiebinger 1989, 1993; Fee 1979; Russett 1989; also Horn and Fausto-Sterling, this volume); measurements of women's bodies, their skulls in particular, tended to place them as inferior to or less intelligent than males. In the great chain of being, women as a class were believed to share certain atavistic characteristics with both children and so-called savages. Not everything about women was regarded negatively. In some cases it was argued that women possessed physical and moral qualities that were superior to those of males. Above all, woman's body was understood through the lens of her reproductive function; her physical characteristics, whether inferior or superior to those of males, were inexorably dictated by her capacity to bear children. . . .

With males as the unspoken prototype, women's bodies were frequently described (subtly or not) as deviations from the norm: as subjects, the measurement of their bodies was occasionally risky to the male scientists,[5] and as bodies they were variations from the generic or ideal type (their body fat "excessive," their pelvises maladaptive to a bipedal [i.e., more evolved] posture, their musculature weak.) Understood primarily in terms of their reproductive capacity, women's bodies, particularly their reproductive organs, genitalia, and secondary sex characteristics, were instead more carefully scrutinized and measured within "marital adjustment" studies and in the emerging science of gynecology, whose practitioners borrowed liberally from the techniques used by physical anthropologists. . . .

In the United States, an attempt to elaborate a scientifically sanctioned notion of a normative "American" female body, however, was taking place in the college studies of the late nineteenth and early twentieth centuries. By the 1860s, Harvard and other universities had begun to regularly collect anthropometric data on their male student populations, and in the 1890s comparable data began to be collected from the East Coast women's colleges as well. Conducted by departments of hygiene, physical education, and home economics, as well as physical anthropology, these large-scale studies gathered data on the elite, primarily WASP youth, in order to determine the dimensions of the "normal" American male and female. . . . Effectively excluded from these attempts to define the "normal" or average body, of course, were those "other" Americans—descendants of African slaves, North American Indians, and the many recent European immigrants from Ireland, southern Europe, and eastern Europe—whose bodies were the subject of racist, evolution-oriented studies concerned with "race crossing," degeneracy, and the effects of the "civilizing" process (see Blakey 1987). . . .

Between the two wars, nationalist interests had fueled eugenic interests and provoked a deepening concern about the physical fitness of the American people. Did Americans constitute a distinctive physical "type"; were they puny and weak as some Europeans had alleged, or were they physically bigger and stronger than their European ancestors? Could they defend themselves in time of war? And who did this category of "Americans" include? Questions such as these fed into an already long-standing preoccupation with defining a specifically American national character and, in 1945, led to the creation of one of the most celebrated and widely publicized anthropometric models of the century: Norm and Norma, the average American male and female. Based on the composite measurements of thousands of young people, described only as "native white Americans," across the United States, the statues of Norm and Norma were the product of a collaboration between obstetrician-gynecologist Robert Latou Dickinson, well known for his studies of human reproductive anatomy, and Abram Belskie, the prize student of Malvina Hoffman, who had sculpted the Races of Mankind series.[6] . . .

Described in the press as the "ideal" young woman, Norma was said to be everything an American woman should be in a time of war: she was fit, strong-bodied, and at the peak of her reproductive potential. Commentators waxed eloquent about the model character traits—maturity, modesty, and virtuousity—that this perfectly average body suggested. . . .

Norma and Norman were . . . more than statistical composites, they were ideals. It is striking how thoroughly racial and ethnic differences were erased from these scientific representations of the American male and female. Based on the measurements of white Americans, eighteen to twenty-five years old, Norm and Norma emerged carved out of white alabaster, with the facial features and appearance of Anglo-Saxon gods. Here, as in the college studies that preceded them, the "average American" of the postwar period was to be visualized only as a youthful white body.

However, they were not the only ideal. The health reformers, educators, and doctors who approved and promoted Norma as an ideal for American women were well aware that her sensible, strong, thick-waisted body differed significantly from the tall slim-hipped bodies of fashion models in vogue at the time.[7] . . . As the postwar period advanced, Norma would continue to be trotted out in home economics and health education classes. But in the iconography of desirable female bodies, she would be overshadowed by the array of images of fashion models and pinup girls put out by advertisers; the entertainment industry, and a burgeoning consumer culture. These idealized images were becoming, as we will see below, increasingly thin in the sixties and seventies while the "average" woman's body was in fact getting heavier. With the thinning of the American feminine ideal, Norma and subsequent representations of the statistically average woman would become increasingly aberrant, as slenderness and sex appeal—not physical fitness—became the premier concern of postwar femininity.

THE ANTHROPOMETRY OF BARBIE: TURNING THE TABLES

As the preceding discussion makes abundantly clear, the anthropometrically measured "normal" body has been anything but value-free. Formulated in the context of a race-, class-, and gender-stratified society, there is no doubt that quantitatively defined ideal types or standards have been both biased and oppressive. Incorporated into weight tables, put on display in museums and world's fairs, and reprinted in popular magazines, these scientifically endorsed standards produce what Foucault calls "normalizing effects," shaping, in not altogether healthy ways, how individuals understand themselves and their bodies. Nevertheless, in the contemporary cultural context, where an impossibly thin image of women's bodies has become the most popular children's toy ever sold, it strikes us that recourse to the "normal" body might just be the power tool we need for destabilizing a fashion fantasy spun out of control. It was with this in mind that we asked students in one of our social biology classes to measure Barbie to see how her body compared to the average measurements of young American women of the same period. Besides estimating Barbie's dimensions if she were life-sized, we see the experiment as an occasion to turn the anthropometric tables from disciplining the bodies of living women to measuring the ideals by which we have come to judge ourselves and others. We also see it as an opportunity for students who have grown up under the regimes of normalizing science—students who no doubt have been measured, weighed, and compared to standards since birth—to use those very tools to unsettle a highly popular cultural ideal. . . .

Since one objective of the course was to learn about human variation, our first task in understanding more about Barbie was to consider the fact that Barbie's friends and family do represent some variation, limited though it may be. Through colleagues and donations from students or (in one case) their children we assembled seventeen dolls for analysis. The sample included:

- 11 early '60s Barbie
- 4 mid-'70s-to-contemporary Barbies, including a Canadian Barbie
- 3 Kens
- 2 Skippers
- 1 Scooter
- Assorted Barbie's friends, including Christie, Barbie's "black" friend
- Assorted Ken's friends

To this sample we subsequently added the most current versions of Barbie and Ken (from the "Glitter Beach" collection) and also Jamal, Nichelle, and Shani, Barbie's more recent African American friends. As already noted, Mattel introduced these dolls (Shani, Asha, and Nichelle) as having a more authentic African American appearance, including a "rounder and more athletic" body. Noteworthy also are the skin color variations between the African American dolls, ranging from dark to light, whereas Barbie and her white friends tend to be uniformly pink or uniformly suntanned. . . .

Before beginning the actual measurements we discussed the kinds of data we thought would be most appropriate. Student interest centered on height and chest, waist, and hip circumference. Members of the class also pointed out the apparently small size of the feet and the general leanness of Barbie. As a result, we added a series of additional standardized measurements, including upper arm and thigh circumference, in order to obtain an estimate of body fat and general size. . . .

In scaling Barbie to be life-sized, the students decided to translate her measurements using two standards: (a) if Barbie were a fashion model (5' 10") and (b) if she were of average height for women in the United States (5' 4"). We also decided to measure Ken, using both an average male stature, which we designated as 5' 8" and the more "idealized" stature for men, 6'.

We took measurements of dolls in the current Glitter Beach and Shani collection that were not available for our original classroom experiment, and all measurements were retaken to confirm estimates. We report here only the highlights of the measurements taken on the newer Barbie and newer Ken, Jamal, and Shani, scaled at their ideal fashion-model height. For purposes of comparison, we include data on average body measurements from the standardized published tables of the 1988 Anthropometric Survey of Army Personnel. We have dubbed these composites for the female and male recruits Army "Norma" and Army "Norm," respectively.

Barbie and Shani's measurements reveal interesting similarities and subtle differences. First, considering that they are six inches taller than "Army Norma," . . . their measurements tend to be considerably less *at all points.* "Army Norma" is a composite of the fit woman soldier; Barbie and Shani, as high-fashion ideals, reflect the extreme thinness expected of the runway model. To dramatize this, had we scaled Barbie to 5' 4", her chest, waist, and hip measurements would have been 32"–17"–28", clinically anorectic to say the least. There are only subtle differences in size, which we presume intend to facilitate the exchange of costumes among the different dolls. We were curious to see the degree to which Mattel had physically changed the Barbie mold in making Shani. Most of the differences we could find appeared to be in the face. The nose of Shani is broader and her lips are ever so slightly larger. However, our measurements also showed that Barbie's hip circumference is actually larger than Shani's, and so is her hip breadth. If anything, Shani might have thinner legs than Barbie, but her back is arched in such a way that it tilts her buttocks up. This makes them appear to protrude more posteriorly, even though the hip depth measurements of both dolls are virtually the same (7.1"). Hence, the tilting of the lumbar dorsal region and the extension of the sacral pelvic area produce the visual illusion of a higher, rounder butt. . . . This is, we presume, what Mattel was referring to in claiming that Shani has a realistic, or ethnically correct, body (Jones 1991).

One of our interests in the male dolls was to ascertain whether they represent a form closer to average male values than Barbie does to average female values. Ken and Jamal provide interesting contrasts to "Army Norm," but certainly not to each other. Their postcranial bodies are identical in all respects. They, in turn, represent a somewhat slimmer, trimmer male than the so-called fit soldier of today. Visually, the newer Ken and Jamal appear very tight and muscular and "bulked out" in impressive ways. The U.S. Army males tend to carry slightly more fat, judging from the photographs and data presented in the 1988 study.[8]

Indeed, it would appear that Barbie and virtually all her friends characterize a somewhat extreme ideal of the human figure, but in Barbie and Shani, the female cases, the degree to which they vary from "normal" is much greater than in the male cases, bordering on the impossible. Barbie truly is the unobtainable representation of an imaginary femaleness. But she is certainly not unique in the realm of female ideals. Studies tracking the body measurements of *Playboy* magazine centerfolds and Miss America contestants show that between 1959 and 1978 the average weight and hip size for women in both of these groups have decreased steadily (Wiseman et al. 1992). Comparing their data to actuarial data for the same time period, researchers found that the thinning of feminine body ideals was occurring at the same time that the average weight of American women was actually increasing. A follow-up study for the years 1979–88 found this trend continuing into the eighties: approximately sixty-nine percent of *Playboy* centerfolds and sixty percent of Miss America contestants were weighing in at fifteen percent or more below their expected age and height category. In short, the majority of women presented to us in the media as having desirable feminine bodies were, like Barbie, well on their way to qualifying for anorexia nervosa.

OUR BARBIES, OUR SELVES

★ ★ ★

On the surface, at least, Barbie's strikingly thin body and the repression and self-discipline that it signifies would appear to contrast with her seemingly endless desire for consumption and self-transformation. And yet, as Susan Bordo has argued in regard to anorexia, these two phenomena—hyper-thin bodies and hyper-consumption—are very much linked in advanced capitalist economies that depend upon commodity excess. Regulating desire under such circumstances is a constant, ongoing problem that plays itself out on the body. As Bordo argues:

> [In a society where we are] conditioned to lose control at the very sight of desirable products, we can only master our desires through a rigid defense against them. The slender body codes the tantalizing ideal of a well-managed self in which all is "in order" despite the contradictions of consumer culture. (1990:97)

The imperative to manage the body and "be all that you can be"—in fact, the idea that you can *choose* the body that you want to have—is a pervasive feature of consumer culture. Keeping control of one's body, not getting too fat or flabby—in other words, conforming to gendered norms of fitness and weight—are signs of an individual's social and moral worth. But, as feminists Bordo, Sandra Bartky, and others have been quick to point out, not all bodies are subject to the same degree of scrutiny or the same repercussions if they fail. It women's bodies and desires in particular where the structural contradictions—the simultaneous incitement to consume and social condemnation for overindulgence—appear to be most acutely manifested in bodily regimes of intense self-monitoring and discipline. . . . Just as it is women's appearance that is subject to greater social scrutiny, so it is that women's desires, hungers, and appetites are seen as most threatening and in need of control in a patriarchal society.

This cultural context is relevant to making sense of Barbie and the meaning her body holds in late consumer capitalism. In dressing and undressing Barbie, combing her hair, bathing her, turning and twisting her limbs in imaginary scenarios, children acquire a very tactile and intimate sense of Barbie's body. Barbie is presented in packaging and advertising as a role model, a best friend or older sister to little girls. Television jingles use the refrain, "I want to be just like you," while look-alike clothes and look-alike contests make it possible for girls to live out the fantasy of being Barbie. . . . In short, there is no reason to believe that girls (or adult women) separate Barbie's body shape from her popularity and glamour.[9]

This is exactly what worries many feminists. As our measurements show, Barbie's body differs wildly from anything approximating "average" female body weight and proportions. Over the years her wasp-waisted body has evoked a steady stream of critique for having a negative impact on little girls' sense of self-esteem.[10] While her large breasts have always been a focus of commentary, it is interesting to note that, as eating disorders are on the rise, her weight has increasingly become the target of criticism. . . .

There is no doubt that Barbie's body contributes to what Kim Chernin (1981) has called "the tyranny of slenderness." But is repression all her hyperthin body conveys? Looking once again to Susan Bordo's work on anorexia, we find an alternative reading of the slender body—one that emerges from taking seriously the way anorectic women see themselves and make sense of their experience:

> For them, anorectics, [the slender ideal] may have a very different meaning; it may symbolize not so much the containment of female desire, as its liberation from a domestic, reproductive destiny. The fact that the slender female body can carry both these (seemingly contradictory) meanings is one reason, I would suggest, for its compelling attraction in periods of gender change. (Bordo 1990: 103)

. . . One could argue that, like the anorectic body she resembles, Barbie's body displays conformity to dominant cultural imperatives for a disciplined body and contained feminine desires. As a woman, however, her excessive slenderness also signifies a rebellious manifestation of willpower, a visual denial of the maternal ideal symbolized by pendulous breasts, rounded stomach and hips. Hers is a body of hard edges, distinct borders, self-control. It is literally impenetrable. Unlike the anorectic, whose self-denial renders her gradually more androgynous in appearance, in the realm of plastic fantasy Barbie is able to remain powerfully sexualized, with her large, gravity-defying breasts, even while she is distinctly nonreproductive. Like the "hard bodies" in fitness advertising, Barbie's body may signify for women the pleasure of control and mastery, both of which are highly valued traits in American society and predominantly associated with masculinity (Bordo 1990: 105). Putting these elements together with her apparent independent wealth can make for a very different reading of Barbie than the one we often find in the popular press. To paraphrase one Barbie-doll owner: she owns a Ferrari and doesn't have a husband—she must be doing something right![11] . . .

It is clear that a next step we would want to take in the cultural interpretation of Barbie is an ethnographic study of Barbie-doll owners.[12] In the meanwhile, we can know something about these alternative appropriations by looking to various forms of popular culture and the art world. Barbie has become a somewhat celebrated figure among avant-garde and pop artists, giving rise to a whole genre of Barbie satire, known as "Barbie Noire" (Kahn 1991). According to Peter Galassi, curator of *Pleasures and Terrors of Domestic Comfort,* an exhibit at the Museum of Modern Art, in New York "Barbie isn't just a doll. She suggests a type of behavior—something a lot of artists, especially women, have wanted to question" (quoted in Kahn 1991: 25). Perhaps the most notable sardonic use of Barbie dolls to date is the 1987 film *Superstar: The Karen Carpenter Story,* by Todd Haynes and Cynthia Schneider. In this deeply ironic exploration into the seventies, suburbia, and middle-class hypocrisy, Barbie and Ken dolls are used to tell the tragic story of Karen Carpenter's battle with anorexia and expose the perverse underbelly of the popular singing duo's candy-coated image of happy, apolitical teens. It is hard to imagine a better casting choice to tell this tale of femininity gone astray than the ever-thin, ever-plastic, ever-wholesome Barbie.

For Barbiana collectors it should come as no surprise that Barbie's excessive femininity also makes her a favorite persona of female impersonators, alongside Judy, Marilyn, Marlene, and Zsa Zsa. Appropriations of Barbie in gay camp culture have tended to favor the early, vampier Barbie look: with the arched eyebrows, heavy black eyeliner, and coy sideways look—the later superstar version of Barbie, according to BillyBoy, is just *too* pink. . . .

In the world of Barbie Noire, the hyper-rigid gender roles of the toy industry are targeted for inversion and subversion. While Barbie is transformed into a dominatrix drag queen, Ken, too, has had his share of spoofs and gender bending. Barbie's somewhat dull steady boyfriend has never been developed into much more than a reliable escort and proof of Barbie's appropriate sexual orientation and popularity. In contrast to that of Barbie, Ken's image has remained boringly constant over the years. He has had his "mod," "hippie" and Malibu-suntan days, and he has gotten significantly more muscular. But for the most part, his clothing line is less diversified, and he lacks an independent fan club or advertising campaign.[13] In a world where boys' toys are G.I. Joe-style action figures, bent on alternately saving or destroying the world, Ken is an anomaly. Few would doubt that his identity was primarily another one of Barbie's accessories. His secondary status vis-à-vis Barbie is translated into emasculation and/or a secret gay identity: cartoons and spoofs of Ken have him dressed in Barbie clothes, and rumors abound that Ken's seeming lack of sexual desire for Barbie is only a cover for his real love for his boyfriends, Alan, Steve, and Dave.

Inscrutable with her blank stare and unchanging smile, Barbie is thus available for any number of readings and appropriations. What we have done here is examine some of the ways she resonates with the complex and contradictory cultural meanings of femininity in postwar consumer society and a changing politics of the body. Barbie, as we, and many other critics, have observed, is an impossible ideal, but she is an ideal that has become curiously normalized. In a youth-obsessed society like our own, she is an ideal not just for young women, but for all women who feel that being beautiful means looking like a skinny, buxom, white twenty-year-old. It is this cultural imperative to remain ageless and lean that leads women to have skewed perceptions of their bodies, undergo painful surgeries, and punish themselves with outrageous diets. Barbie, in short, is an ideal that constructs women's bodies as hopelessly imperfect. It has been our intention to unsettle this ideal and, at the same time, to be sensitive to other possible readings, other ways in which this ideal body figures and reconfigures the female body. . . .

We have explored some of the battleground upon which the serious play of Barbie unfolds. If Barbie has taught us anything about gender, it is that femininity in consumer culture is a question of carefully performed display, of paradoxical fixity and malleability. One outfit, one occupation, one identity can be substituted for another, while Barbie's body has remained ageless, changeless untouched by the ravages of age or cellulite. She is always a perfect fit, always able to consume and be consumed. Mattel has skillfully managed to turn challenges of feminist protest, ethnic diversity, and a troubled multiculturalism to a new array of outfits and skin tones, annexing these to a singular anorectic body ideal. Cultural icon that she is, Barbie nevertheless cannot be permanently located in any singular cultural space. Her meaning is mobile as she is appropriated and relocated into different cultural contexts, some of which, as we have seen, make fun of many of the very notions of femininity and consumerism she personifies. As we consider Barbie's many meanings, we should remember that Barbie is not only a denizen of subcultures in the United States, she is also world traveler. A product of the global assembly line, Barbie dolls owe their existence to the internationalization of the labor market and global flows of capital and commodities that today characterize the toy industry, as well as other industries in the postwar era. Designed in Los Angeles, manufactured in Taiwan or Malaysia, distributed worldwide, Barbie™ is American-made in name only. Speeding her way into an expanding global market, Barbie brings with her some of the North American cultural subtext we have outlined in this analysis. How this teenage survivor then gets interpolated into the cultural landscapes of Mayan villages, Bombay high-rises, and Malagasy towns is a rich topic that begs to be explored.

NOTES

1. At the time of this writing, there was no definitive history of Barbie and the molds that have been created for her body. However, Barbie studies are booming and we expect new work in press, including M. G. Lord's *Forever Barbie: The Unauthorized Biography of a Real Doll* (1994), to provide greater insight into Barbie's history and the debates surrounding her body within Mattel and the press.
2. While the concept of adolescence as a distinct developmental stage between puberty and adulthood was not new to the fifties, Thomas Doherty (1988) notes that it wasn't until the end of World War II

that the term "teenager" gained standard usage in the American language.

3. Recent work by Ann duCille promises to offer an incisive cultural critique of the "ethnification" of Barbie and its relationship to controversies in the United States over multiculturalism and political correctness (duCille 1995). More work, however, needs to be done on how Barbie dolls are adapted to appeal to various markets outside the U.S. For example, Barbie dolls manufactured in Japan for Japanese consumption have noticeable larger, rounder eyes than those marketed in the United States (see BillyBoy 1987). For some suggestive thoughts on the cultural implications of the transnational flow of toys like Barbie dolls, TransFormers, and He-Man, see Carol Breckenridge's (1990) brief but intriguing editorial comment to *Public Culture.*
4. Closely aligned with the emergence of statistics, it was Hrdlicka's hope that the two would be joined, and that one day the state would be "enlightened" enough to incorporate regular measurements of the population with the various other tabulations of the periodic census, in order to "ascertain whether and how its human stock is progressing or regressing" (1939: 12).
5. In *Practical Anthropometry* Hrdlicka goes to some trouble to instruct field-workers (presumably male) working among "uncivilized groups" about the steps they need to take not to offend, and thereby put themselves at risk, when measuring women (1939: 57–59).
6. Norma is described in the press reports as being based on the measurements of 15,000 "real American girls." Although we cannot be sure, it is likely this data come from the Bureau of Home Economics, which conducted extensive measurements of students "to provide more accurate dimensions and proportions for sizing women's ready-made garments" (Shapiro 1945). For further information on the Dickinson collection and Dickinson's methods of observation, see Terry (1992.).
7. Historians have noted a long-standing conflict between the physical culture movement, eugenicists and health reformers, on the one hand, and the fashion industry, on the other, that gave rise in American society to competing ideals of the fit and the fashionably fragile, woman (e.g., Banner 1983; Cogan 1989).
8. One aspect of the current undertaking that is clearly missing is the possible variation that exists *within* individual groups of dolls that would result from mold variation and casting processes. Determining this variation would require a much larger doll collection at our disposal. We are considering a grant proposal, but not seriously.
9. This process of identification becomes mimesis, not only in Barbie look-alike contests, but also in the recent Barbie workout video. In her fascinating analysis of the semiotics of workout videos, Margaret Morse (1987) has shown how these videos structure the gaze in such a way as to establish identification between the exercise leader's body and the participant-viewer. Surrounded by mirrors, the viewer is asked to exactly model her movements on those of the leader, literally mimicking the gestures and posture of the "star" body she wishes to become. In Barbie's video, producers use animation to make it possible for Barbie to occasionally appear on the screen as the exercise leader/cheerleader—the star whose body the little girls mimic.
10. In response to this anxiety, Cathy Meredig, an enterprising computer software designer, created the "Happy to Be Me" doll. Described as a healthy alternative for little girls, "Happy to Be Me" has a shorter neck, shorter legs, wider waist, larger feet, and a lot fewer clothes—designed to make her look more like the average woman ("She's No Barbie, nor Does She Care to Be." *New York Times,* August 15, 1991, C-11).
11. "Dolls in Playland." 1992. Colleen Toomey, producer. BBC.
12. While not exactly ethnographic, Hohmann's 1985 study offers a sociopsychological view of how children experiment with social relations during play with Barbies.
13. Signs of a Ken makeover, however, have begun to appear. In 1991, a Ken with "real" hair that can be styled was introduced and, most dramatically, in 1993, he had his hair streaked and acquired an earring in his left ear. This was presented as a "big breakthrough" by Mattel and was received by the media as a sign of a broader trend in the toy industry to break down rigid gender stereotyping in children's toys (see Lawson 1993). It doesn't appear, however, that Ken is any closer to getting a "realistic" body than Barbie. Ruth Handler notes that when Mattel was planning the Ken doll, she had wanted him to have genitals—or at least a bump, and claims the men in the marketing group vetoed her suggestion. Ken did later acquire his bump (see "Dolls in Playland," Colleen Toomey, producer. BBC. 1993).

REFERENCES

Adelson, Andrea

1992 "And Now, Barbie Looks Like a Billion." *New York Times,* November 26, sec. D, p. 3.

Banner, Lois W.

1983 *American Beauty.* New York: Knopf.

Bartky, Sandra Lee
1990 "Foucault, Femininity, and the Modernization of Patriarchal Power." In *Femininity and Domination: Studies in the Phenomenology of Oppression,* pp. 63–82. New York: Routledge.
Blakey, Michael L.
1987 "Skull Doctors: Intrinsic Social and Political Bias in the History of American Physical Anthropology" *Critique of Anthropology* 7(2): 7–35.
Bordo, Susan R.
1990 "Reading the Slender Body." In *Body/Politics: Women and the Discourses of Science.* Ed. Mary Jacobus, Evelyn Fox Keller, and Sally Shuttleworth, pp. 83–112. New York: Routledge.
1991 "Material Girl: The Effacements of Post-modern Culture." In *The Female Body: Figures, Styles, Speculations.* Ed. Laurence Goldstein, pp. 106–30. Ann Arbor, Mich.: The University of Michigan Press.
Breckenridge, Carol A.
1990 "Editor's Comment: On Toying with Terror." *Public Culture* 2(2): i–iii.
Brumberg, Joan Jacobs
1988 *Fasting Girls: The History of Anorexia Nervosa.* Cambridge: Harvard University Press. Reprint, New York: New American Library.
Chernin, Kim
1981 *The Obsession: Reflections on the Tyranny of Slenderness.* New York: Harper and Row.
Cogan, Frances B.
1989 *All-American Girl: The Ideal of Real Woman-hood in Mid-Nineteenth-Century America.* Athens and London: University of Georgia Press.
Davis, Kathy
1991 "Remaking the She-Devil: A Critical Look at Feminist Approaches to Beauty" *Hypatia* 6(2): 21–43.
duCille, Ann
1995 "Toy Theory: Blackface Barbie and the Deep Play of Difference." In *The Skin Trade: Essays on Race, Gender, and the Merchandising of Difference.* Cambridge: Harvard University Press.
Fee, Elizabeth
1979 "Nineteenth-Century Craniology: The Study of the Female Skull." *Bulletin of the History of Medicine* 53: 415–33.
France, Kim
1992 "Tits 'R' Us." *Village Voice,* March 17, p. 22.
Goldin, Nan
1993 *The Other Side.* New York: Scalo.
Gould, Stephen Jay
1981 *The Mismeasure of Man.* New York: Norton.
Halberstam, Judith
1994 "F2M: The Making of Female Masculinity" In *The Lesbian Postmodern,* ed. Laura Doan, pp. 210–28. New York: Columbia University Press.
Hohmann, Delf Maria
1985 "Jennifer and Her Barbies: A Contextual Analysis of a Child Playing Barbie Dolls." *Canadian Folklore Canadien* 7(1–2). 111–20.
Hrdlicka, Ales
1925 "Relation of the Size of the Head and Skull to Capacity in the Two Sexes." *American Journal of Physical Anthropology* 8: 249–50.
1939 *Practical Anthropometry.* Philadelphia: Wistar Institute of Anatomy and Biology.
Jones, Lisa
1991 "Skin Trade: A Doll Is Born." *Village Voice,* March 26, p. 36.
Kahn, Alice
1991 "A Onetime Bimbo Becomes a Muse." *New York Times,* September 29.
Kaw, Eugenia
1993 "Medicalization of Racial Features: Asian American Women and Cosmetic Surgery" *Medical Anthropology Quarterly* 7(l): 74–89.
Lawson, Carol
1993 "Toys Will Be Toys: The Stereotypes Unravel." *New York Times.* February 11, sec. C, pp. 1, 8.
Lord, M. G.
1994 *Forever Barbie: The Unauthorized Biography of a Real Doll.* New York: William Morrow.
May, Elaine Tyler
1988 *Homeward Bound: American Families in the Cold War Era.* New York: Basic Books.
Morse, Margaret
1987 "Artemis Aging: Exercise and the Female Body on Video." *Discourse* 10 (1987/88): 20–53.
Motz, Marilyn Ferris
1983 "I Want to Be a Barbie Doll When I Grow Up: The Cultural Significance of the Barbie Doll." In *The Popular Culture Reader,* 3d ed. Ed Christopher D. Geist and Jack Nachbar, pp. 122–36. Bowling Green: Bowling Green University Popular Press.
Rabinowitz, Paula
1993 Accessorizing History: Girls and Popular Culture. Discussant Comments, Panel #150: Engendering Post-war Popular Culture in Britain and America. Ninth Berkshire Conference on the History of Women. Vassar College, June 11–13, 1993.
Russett, Cynthia Eagle
1989 *Sexual Science: The Victorian Construction of Womanhood.* Cambridge: Harvard University Press.

Schiebinger, Londa
1989 *The Mind Has No Sex?: Women in the Origins of Modern Science.* Cambridge: Harvard University Press.
1993 *Nature's Body: Gender in the Making of Modern Science.* Boston: Beacon.
Schwartz, Hillel
1986 *Never Satisfied: A Cultural History of Diets, Fantasies and Fat.* New York: Free Press.
Shapiro, Eben
1992 "'Totally Hot, Totally Cool.' Long-Haired Barbie Is a Hit." *New York Times.* June 22, sec. D, p. 9.
Shapiro, Harry L.
1945 *Americans: Yesterday, Today, Tomorrow.* Man and Nature Publications. Man and Nature Publications. (Science Guide No. 126). New York: The American Museum of Natural History.
Spencer, Frank
1992 "Some Notes on the Attempt to Apply Photography to Anthropometry during the Second Half of the Nineteenth Century" In *Anthropology and Photography,* 1860–1920. Ed. Elizabeth Edwards, pp. 99–107. New Haven: Yale University Press.
Sprague Zones, Jane
1989 "The Dangers of Breast Augmentation." *The Network News* (July/August), pp. 1, 4, 6, 8. Washington, D.C.: National Women's Health Network.
Stevenson, Richard
1991 "Mattel Thrives as Barbie Grows." *New York Times.* December 2.
Terry, Jennifer C.
1992 Siting Homosexuality: A History of Surveillance and the Production of Deviant Subjects (1935–1950). Ph.D. diss., University of California at Santa Cruz.
Williams, Lena
1992 "Woman's Image in a Mirror: Who Defines What She Sees?" *New York Times,* February 6, sec. A, p. 1, sec. B, p. 7.
Willis, Susan
1991 *A Primer for Daily Life.* London and New York: Routledge.
Wilson, Elizabeth
1985 *Adorned in Dreams: Fashion and Modernity.* London: Virago.
Wiseman, C., J. Gray, J. Mosimann, and A. Ahrens
1992 "Cultural Expectations of Thinness in Women: An Update." *International Journal of Eating Disorders* 11(1): 85–89.
Wolf, Naomi
1991 *The Beauty Myth: How Images of Beauty Are Used against Women.* New York: William Morrow.

26

"Where My girls At?"

Negotiating Black Womanhood in Music Videos

BY RANA A. EMERSON

In this reading, Rana Emerson explores the representations of African-American womanhood through close textual analysis of a large sample of music videos of African-American women performers. Her analysis focuses on visual images, the narrative and representations, and the accompanying musical tracks and lyrics of the music videos. Emerson argues that the videos contain contradictory and ambivalent themes. Many videos reflect how race, class, and gender constrain the autonomy and agency of African-American women. But not all do so. Some African-American women video artists resist controlling images and assert their control and worth.

1. Describe the stereotypes of African-American womanhood portrayed in many music videos.
2. How do the video images of some African-American women artists counter negative stereotypes?
3. Are there parallels between the array of representations of African-American women in music videos and the media images of women discussed in both the Kilbourne and the Urla and Swedlund articles?

Today's American youths of all racial and ethnic heritages are living in a cultural environment dominated by the idioms of Black youth and working-class culture that have been articulated since the late 1970s and early 1980s by hip-hop culture. Since its emergence in the mass media mainstream in the early 1990s, hip-hop culture has affected the arenas of film, fashion, television, art, literature, and journalism (Watkins 1998). In the mid- to late 1990s, African American youth emerged as an important segment of this teenage audience and consumer population (Watkins 1998). Recent ethnographic studies of Black youth in the 1990s have demonstrated the importance and impact that popular culture in general and hip-hop culture in particular have on the ways in which young African Americans make sense of their lives, social surroundings and the world around them (Arnett Ferguson 2000; Patillo-McCoy 1999). Therefore, it is important for those who wish to better understand the lives of young African Americans to investigate the attributes of the popular cultural products that inform their everyday lives and attempt to make sense of their participation with and within popular culture. Paying attention to the role of popular culture in the lives of these youths also contributes to sociological theory by further elucidating the significance of the mass media as a social institution and how ideologies of race, class, and gender are represented and reproduced within it.

While much has been written about the significance and impact of hip-hop culture on the lives of Black youth, young Black women, until very recently, have failed to be located as substantial producers, creators, and consumers of hip-hop and Black youth culture (George 1998; Perkins 1996; Rose 1994; Watkins 1998). Most of the contemporary research and criticism has focused on the experience of young men of African descent and, with rare exceptions, has implicitly and often explicitly identified Black popular culture, specifically hip-hop culture, with masculinity (George 1998; Perkins 1996; Rose 1991, 1994). . . .

The medium of the music video, the primary promotional vehicle for the recording industry today, is an especially rich space to explore the ways in which race, gender, class, and sexuality intersect in the construction and proliferation of ideologies of Black womanhood in the mass media and popular culture. This study explores Black women's representation in music video through the analysis of a sample of videos by African American women singers, rappers, and musicians produced and distributed at the end of the 1990s.

Most of the previous studies of Black women's representation in music videos have, on one hand, either focused on the hegemonic and stereotypical imagery and discourses of Black femininity or, on the other hand, exaggerated the degree of agency that Black female performers in music video have by emphasizing the resistant and counterhegemonic elements of the music video representations. Instead, this study demonstrates that in the cultural productions of Black women, music videos in this case, hegemonic and counterhegemonic themes often occur simultaneously and are interconnected, resulting in a complex, often contradictory and multifaceted representation of Black womanhood.

LITERATURE REVIEW

The vast majority of representations of Black women in popular culture are firmly grounded in the dominant ideologies surrounding Black womanhood in American society. Patricia Hill Collins (1991a) described these ideologies as controlling images that are rooted in the maintenance of hegemonic power and serve to justify and legitimize the continued marginalization of Black women. The media and popular culture are primary sites for the dissemination and the construction of commonsense notions of Black womanhood. Music videos, which have been criticized for their objectifying and exploitative depictions of women of all races and ethnicities (Aufderheide 1986; Dines and Humez 1995; Frith, Goodwin, and Grossberg 1993: Hurley 1994: Kaplan 1987; Stockbridge 1987; Vincent 1989; Vincent, Davis, and Boruszkowski 1987), often represent Black women according to the controlling images discussed by Hill Collins. The images that are seen most often are the hypersexualized "hot momma" or "Jezebel," the asexual "mammy," the emasculating "matriarch," and the "welfare recipient" or "baby-momma" (a colloquial term for young, unwed mothers).

Although Black female representation generally draws directly from the controlling images of Black womanhood, Black women's performances in popular culture often generate representations that counter the dominant ideological notions of Black womanhood. Consequently, the possibility that popular and expressive culture may exist as a site for resistance and revision of these stereotypical representations emerges. Hazel Carby (1986) and Angela Davis (1998) have shown that such phenomena occurred in the early part of this century. At the time, performers such as "Ma" Rainey, Bessie Smith, and Ethel Waters offered, in their music and onstage performances, a portrait of Black womanhood in which they asserted empowerment and sexual subjectivity. In both Carby's and Davis's views, this female blues culture was grounded in a Black feminist consciousness. Although some authors (Delano Brown and Campbell 1986; Kaplan 1987; Lewis 1990; Peterson-Lewis and Chennault 1986) have looked at race and gender representations in music videos, there have been few systematic studies of Black female representation within the medium (Goodall 1994; Roberts 1991, 1994;

Rose 1991, 1994). Even fewer studies have looked at the music videos by Black women performers themselves. . . .

[M]y study provides an empirical basis for identifying the ways in which Black women use the realm of culture and performance for social commentary and to respond to the controlling images of Black womanhood that were identified and discussed by Hill Collins, Carby, Davis, Rose, and other Black feminist theorists and critics. This study improves on the previous research on Black female music video performance because it problematizes the often-unexamined notion of resistance. Overall, this study furthers the inclusion of Black female youth in the conversation surrounding hip-hop culture by recognizing the active participation of Black female performers and audiences within it (McRobbie 1991, 1993, 1997; McRobbie and Nava 1984). In this way, it serves to question the identification of hip-hop culture with Black masculinity and Black male youth by demonstrating that music videos also serve as sites for expressing the lived experiences of Black female youth.

In this article, I will first identify how music videos exhibit and reproduce the stereotypical notions of Black womanhood faced by young African-American women. Next, I discuss the ways that Black woman performers use music videos for contesting hegemonic racist and sexist notions of Black femininity and asserting agency. Third, I demonstrate how contradictory themes in the music videos reflect a sense of ambivalence on the part of Black girls regarding the relationships between Blackness, womanhood, and sexuality. . . .

STEREOTYPES AND CONTROLLING IMAGES

The videos reflect how race, class, and gender continue to constrain and limit the autonomy and agency of Black women. Music videos contain imagery that reflects and reproduces the institutional context in which they are produced, and they are permeated by stereotypical controlling images of Black womanhood. Several stereotypes emerge in the ways Black women's videos are programmed, as well in the content of the videos themselves. First, the videos emphasize Black women's bodies. Second, they construct a one-dimensional Black womanhood. Finally, the presence of male sponsors in the videos and a focus on themes of conspicuous consumption and romance further exhibit the types of social constraints faced by young Black women.

The Body

The first way that patterns of social constraint emerge is in the emphasis on the body. It is clear that female rap and rhythm and blues (R&B) performers are required to live up to dominant notions of physical attractiveness and measure up to fairly rigid standards of beauty. The most striking example of this is the lack of variety in body size and weight. This was surprising, considering the conventional wisdom that the Black community possesses alternative beauty standards that allow for larger body types. Many authors have concluded that these standards contribute to a more positive body image among Black women (Cash and Henry 1995; Flynn and Fitzgibbon 1996; Harris 1994; Molloy and Hertzberger 1998). However, the majority of the videos I coded (30) featured artists who would be considered thin by most standards, while only 9 featured performers who would be considered overweight. The only women who are larger than the ideal are Missy "Misdemeanor" Elliott, Angie Stone, and a member of the group Xscape.

In those 30 videos, the thin, physically attractive performers are clearly constructed as objects of male desire. In Bobby Brown's video, *I've Got This Feelin',* featuring his real-life wife Whitney Houston, Whitney is broken up fetishistically into her body parts. The viewer is only allowed glimpses of her mouth and legs, her arms caressing Bobby's shoulder, and her hair. The implication is that the audience is not supposed to know who she is (although we do have our suspicions), until the shot widens to reveal her in her entirety, laughing knowingly and almost conspiringly with her husband. Cutting Whitney up into visual pieces undercuts her power. . . .

One-Dimensional Womanhood

For the most part, the portrait of Black womanhood that emerges from the video analysis is flat and one-dimensional. Black women are not represented in their full range of being. They are not multifaceted but are reduced to decorative eye candy. Black women performers are not allowed to be artists in their own right but must serve as objects of male desire. In the videos, only three of the featured artists were older than 30 (Janet Jackson, Whitney Houston, and Aretha Franklin). Indubitably, this reflects the youth-oriented nature of popular culture.

Pregnant women and mothers, as well as women older than 30, are not desirable as objects of the music video camera's gaze, reinforcing the sense that only women who are viewed as sexually available are acceptable in music videos. Only two of the videos depicted motherhood: Erykah Badu is visibly pregnant in *Tyrone,* and Joi is shown with her infant daughter in her video, *Ghetto Superstar.*

Sexual diversity is another element of Black womanhood that is conspicuously absent and also reflects the desirability of perceived sexual availability for men. None of the videos featured performers who were lesbian or bisexual, nor did they show even implicit homosexual or bisexual themes. This was interesting in light of the emergence of critically acclaimed and commercially popular bisexual and lesbian artists, most notably, Me'Shell Ndgeocello (whose most controversial video *Leviticus: Faggot* was censored by BET). As can be gleaned from the frequently homophobic rhetoric in hip-hop and R&B songs, sexual difference and nonconformity are still not legitimized in Black popular culture. As a result, it is not particularly surprising that bisexual and lesbian themes do not emerge in a sample of popular Black women's music videos.

"Man behind the Music": The Male Sponsor

The one-dimensional depiction of Black women as objects of male pleasure undermines their legitimacy and agency as artists. Because their role is primarily sexual, they are not taken seriously. Add to that mix the notion that legitimacy in hip-hop culture is identified with masculinity, and the result is that many Black women artists are presented to the public under the guidance of a male sponsor.

Although male sponsorship, defined as the prominence of a male producer, songwriter, or fellow artist, was only coded in four of the videos, when it does occur, it is fairly significant. In such videos, not only is the male sponsor (who is most often one of the so-called megaproducers such as Sean "Puffy" Combs or Jermaine "J. D." Dupri) prominent visually and narratively in the video, but he literally takes precedence over the artist herself, essentially becoming the true star of the video. . . .

Although Combs, Riley, and Dupri also appear in the videos for male artists that they produce, the impact that they have on the image of women artists appears to be greater. They occupy a primary position within the camera's gaze and on the musical track. . . .

Since Black women have little or no clout in the music industry and Black men dominate the hip-hop world, the presence of a male impresario undermines any sense of creative autonomy for woman artists. In fact, the producer in today's record industry wields an unprecedented amount of control over the musical product, often to the point of overriding an artist's creative decisions and input over the content of a song, and occasionally the video as well (George 1998). The producer and record company executives often choose the video director and contribute to the construction of the artist's image and presentation. These videos give the impression that women are unable to be successful without the assistance and creative genius of a male impresario.

BLACK WOMEN'S AGENCY: COUNTERING CONTROLLING IMAGES

Despite the continuing objectification and exploitation of Black women in music videos, I found evidence of contestation, resistance, and the assertion of Black women's agency in many others . . . as well. This agency emerged through the identification with signifiers of Blackness; an assertion of autonomy, vocality, and independence; and expressions of partnership. Collaboration, and sisterhood with other Black women and Black men.

Signifiers of Blackness: Black Aesthetic, Black Context

In these videos, Blackness does not carry a negative connotation. Instead, it is the basis for strength, power, and a positive self-identity. Darker skin is privileged among Black women artists, actresses, models, and dancers in the videos. Thirty of the videos featured women with darker complexions or a combination of lighter and darker skinned women. This was an especially interesting finding, after the controversies of the 1980s and 1990s about the frequent use of light-skinned women in music videos, which was criticized for valuing a white standard of beauty (Morgan 1999). In contrast, the videos examined in this study evinced a Black aesthetic in which standards of beauty, while problematic in themselves, were nevertheless based on a more African aesthetic.

The prevalence of a clear hip-hop sensibility supports the valuation of Black culture. Twenty of the videos were coded as being evocative of an urban hip-hop style. What emerges from these observations is the construction of a clear Black aesthetic. In fact, it becomes obvious that these videos exhibit an essentially Black universe. Although this was not specifically coded, white people appeared rarely if ever in the videos. When they do appear, they tend to be minor characters such as the gangsters in *Firm Biz* and Mariah's kidnappers in *Honey*. . . .

This construction of a Black universe leads to questioning the notion of Blackness as male or Black youth culture's association with masculinity. Instead, I found Black women firmly contextualized among signifiers and codes of Blackness. They explore themes of womanhood that directly associate them with Blackness and Black life, and they construct a significant and solid space (albeit limited by the fact that male artists continue to receive more representation than women in heavy video rotations) for girls and women in Black youth and hip-hop culture. By appropriating signs of Blackness, Black women artists are able to assert the particularity and forcefulness of Black femininity and agency through the music video.

Autonomy, Vocality, and Independence

Despite the predominance of traditional gender roles in the music videos, Black women performers are frequently depicted as active, vocal, and independent. This vocality is most frequent within the context of traditional relationships, where the performers express discontent with, and contest, the conditions faced by Black women in interpersonal relationships.

Instead of exhibiting representations of physical violence and aggression, sometimes found in men's videos, this sample of videos demonstrates the significance of verbal assertiveness. Speaking out and speaking one's mind are a constant theme. Through the songs and videos, Black women are able to achieve voice and a space for spoken expression of social and interpersonal commentary. . . .

Although they are not clearly and unequivocally rejecting the desirability and basic dynamics of heterosexual relationships, Black women in these videos assert their own interests and express dissatisfaction with the unequal state of Black men-women interpersonal relations. Black women also express their own agency and self-determination through direct action. What emerges is the ability of a Black woman to define her own identity and life outcomes.

Sisterhood, Partnership, and Collaboration

Although Black women assert independence, they do not accomplish their goals alone. In these videos, Black women look to each other for support, partnership, and sisterhood. Collaborations between women artists are a constant and recurring theme throughout the videos and suggest a sense of community and collectivity. This shows that women need each other for guidance and support to succeed and survive in the recording industry and the world at large. Within these collaborations, unlike the male sponsorships discussed above. the spotlight is shared, and the guest star does not overshadow the featured artist.

The most interesting video in which this occurs is *Sock It to Me,* in which Missy Elliott collaborates with the rappers Lil' Kim and Da Brat. It has an outer space, fantasy theme, and in the visual narrative, Missy and rapper Lil' Kim appear in red and white bubble space suits as explorers on a mission. As soon as they land on an unchartered planet, they are pursued by an army of monstrous robots under the control of the evil "mad scientist," portrayed by Missy's collaborator and producing partner Timbaland. They are chased throughout the rest of the video through space and on various barren planets. The chase scenes are interspersed with scenes of Missy, as she dances in the forefront of a troupe of dancers wearing futuristic attire. Missy also appears solo, seemingly suspended in space as she sings the track of the song. Just as Missy and Kim appear to be in danger of succumbing to Timbaland's goons, fellow rapper, Da Brat, during her rap sequence on the music track, come to the rescue on a jet ski-type spacecraft. They speed off through space, fighting off the mad scientist's crew, and arrive safely at Missy's mothership, prominently marked with the letter *M*. . . .

Overall, what emerges from this combination of agency, voice, partnership, and Black context is a sense of the construction of Black woman-centered video narratives. Within these narratives, the interests, desires, and goals of women are predominant and gain importance in contrast to those in which they are exploited and subsumed. Black women are quite firmly the subjects of these narratives and are able to clearly and unequivocally express their points of view.

AMBIVALENCE AND CONTRADICTION: NEGOTIATING BLACK WOMANHOOD

In this section, I discuss the ambivalent and contradictory relationship that young Black women appear to have with Black popular culture and how those contradictions are reflected in the music videos in this sample. In this regard, music videos exemplify a tension between the structural constraints of race and gender on one hand and women's resistance and self-affirmation on the other.

Every day, young Black women face conflicting messages about their sexuality and femininity, as well as their status both in the Black community and society at large. They must figure out how they should construct and assert their identity as Black women. Therefore, it is not surprising that within the cultural productions of young Black women, themes of contradiction and ambivalence would emerge.

While it sometimes appears that these artists are directly reflecting and capitulating to oppressive social forces, this seeming compromise can be interpreted more accurately as ambivalence regarding contradictory messages about Black female sexuality, namely, the coexistence of hypersexual images and the denigration and denial of the beauty of the Black female body. In response to these contradictory notions of Black womanhood, Black women performers frequently reappropriate often explicit images of Black female sexuality. This strategy of self-representation as sexual may, on one hand, be interpreted as a sort of false consciousness that reflects an acceptance of the controlling images of Black womanhood. However, I argue that instead, these sometimes explicit representations of Black women's sexuality actually exemplify a process of negotiating those contradictory and often conflicting notions and, more significantly, represent an attempt to use the space of the music video to achieve control over their own sexuality. The four themes that I located that indicate this process include collaboration between Black men and women, representation of a multidimensional sexuality, returning the gaze, and the indeterminate gaze.

"Together Again": Black Male-Female Collaboration

Black men and women are frequently seen in these videos as coworkers and collaborators. They are fellow group members, found in duets, and they appear as nonmusical guest stars in each other's videos. Fourteen of the videos portray Black men as fellow group members or platonic friends. This theme occurs nearly as often as when men appear as romantic interests (18 videos). Collaboration emerges as an important aspect of Black women's performance. Despite the fact that strides have been made in recent years, it remains difficult for young women to enter the music industry on their own. As suggested in the discussion of sponsorship above, entrée into the business can be easier if they are associated with a man who is already established.

As opposed to the sponsorship and/or impresario videos, Black men and women collaborated frequently in apparently equal working relationships. In this context of partnership, Black women performers wield a great deal of creative control as songwriters, producers, and video directors. What this suggests is that Black men and women can work together and provide each other with mutual support to achieve success in a competitive cultural field. This phenomenon also is embedded in the tradition of collectivity and collaboration as a theme in African and African American culture (Hill Collins 1991a, 1991b).

In these collaborative relationships, Black women performers gain an equal footing with their fellow male artists. For example, Missy "Misdemeanor" Elliott collaborates with her partner in crime, Timbaland, and Magoo in *Sock It to Me* and *Beep Me 911*. She also makes an appearance in the video for the male group Playa's party anthem, *Can't Stop the Music*. Lauryn Hill collaborates with Pras and Wyclef in the Fugees and is the director and costar in rapper Common Senses' reflection on fatherhood, *Retrospective for Life*. Not only do these women have the same level of creative control and autonomy as the men, they also are able to execute many actions previously assigned only to male performers. Most significantly, the Black women artists within these videos are able to construct themselves as textual subjects and wield their gaze in a similar manner to men. Woman performers are provided with space and opportunity to wield creative and artistic control and to construct their own narratives of Black womanhood that express their lived experience. In effect, these collaborative working relationships counter and overshadow the marginalizing and silencing that result from the sponsorship relationship. In fact, one could argue that Missy Elliott and Lauryn Hill have had more impact on American popular culture (and hip-hop culture) than their collaborators.

Multidimensional Sexuality: Reappropriating the Black Female Body

Most of the artists portray themselves with a highly stylized and glamorous image. Wearing designer gear, these women singers present themselves as sexy and provocative. In 21 of the videos, the artist was depicting a glamorous image, while in 17 they were coded as having a sexual image. This emphasis on appearance and physical attraction confirms the notion of the excessive sexuality of the Black woman. It supports the ideological controlling image of the hypersexual "sapphire" or "jezebel," effectively undermining Black womanhood and humanity.

However, in the videos analyzed, glamour and style are not the only salient attributes possessed by Black women artists. Instead, a sexualized image often occurs simultaneously with themes of independence, strength, a streetwise nature, toughness, and agency. Most of the time, the same artists express themselves in a single video as sexy and savvy, glamorous and autonomous. Fifteen of the videos depict artists having an independent image, and 13 are streetwise and tough. Many of these videos were also coded as glamorous and sexual.

What seems to emerge is a contradiction between the complex and often unconventional representations of Black women artists and the appearance of objectified and clearly one-dimensionally sexualized Black women dancers. Fifteen of the videos were coded as featuring female background dancers. For the most part, when these dancers appear on screen, they are scantily clad and move in a highly suggestive manner. Male dancers, in contrast, only appear in 8 of the 56 videos and are rarely explicitly sexualized. In Da Brat's *Give It to You,* which takes place at what appears to be a hip-hop industry party, Da Brat's tough and streetwise, even boyish, image contrasts sharply with the appearance of scantily clad female "groupies" who are mingling and dancing in the crowd. . . .

The fact remains that sex sells. In the entertainment industry, there is a call for bodies, namely, female bodies, to be on display to stimulate record sales. If it is not the artist herself, then models and dancers serve this purpose. Women remain the object of sexual desire, the selling point, and the bodies on display.

On the other hand, the juxtaposition and combination of sexuality, assertiveness, and independence in these videos can also be read as the reappropriation of the Black woman's body in response to its sexual regulation and exploitation. What emerges is an effort on the part of the Black female artist to assert her own sexuality, to gain her own sexual pleasure.

Whether this indicates compromise or capitulation to objectification and exploitation is not definitively clear. It is difficult to reach a conclusion on this solely from the data gathered from textual analysis. One would need to investigate the creative production decision-making process. However, the results of this analysis and interpretation indicate that trade-offs are made in the construction of an artist's image. Black womanhood, as expressed through Da Brat and Missy's performances, is the result of a process of negotiation in which objectification of the female body must be present in order for the performer to gain a level of autonomy, to gain exposure. While this seems on the surface like "selling out" to the dictates of patriarchy and the marketplace, I would argue that instead, it affirms the multidimensional nature of Black womanhood. . . .

Returning the Gaze: Sexuality on a Woman's Terms

An interesting manifestation of the phenomenon of contradiction and ambivalence is the pattern of a reversal and returning of the gaze. A critical mass of videos feature men as objects of women's desire, where men's bodies are the center of the camera's gaze. What also occurs in these videos is a reversing of traditional gender roles in which men are objectified. Simultaneously, women remain the object of the camera's gaze as well. In *Swing My Way,* KP and Envyi pursue a male love interest in a club. In *You're Making Me High,* Toni Braxton and actresses Erika Alexander, Vivica Fox, and Tisha Campbell rate male visitors on a numerical scale as they appear in an elevator, while Toni's *Unbreak My Heart* features Black male supermodel Tyson Beckford. TLC (group members T-Boz, Left Eye, and Chilli) are the only women players (and the only fully clothed individuals) in a sexy game of strip poker in *Red Light Special,* and Janet Jackson's *Love Will Never Do Without You* centers the well-chiseled Black body of Djamon Hounsou and the buffed white body of actor Antonio Sabato Jr. alongside her own washboard abs. What all of these videos have in common is the construction of the male body, and particularly the Black male body, as the object of Black female pleasure. The male body is not merely looked at; rather, it is actively pursued. These women clearly and unequivocally express what they want, how and when they want it, and that they frequently get it.

What results is a space where the erotic can become articulated on a woman's terms. When videos featuring themes of sexual desire and fulfillment were coded, signifiers of mutual sexual fulfillment predominated, and women's sexual fulfillment was more often portrayed than for men. Although women were usually visually constructed as the source of male pleasure, when issues of sexual pleasure were articulated either in the lyrical or visual text, or both, the importance of female sexual desire became key. This construction of a sphere of erotic agency does not simply symbolize the subjectivity of the individual Black woman but also results in the construction of agency at the social and cultural level. It results in a space for an articulation of themes of freedom and liberty.

A long-standing theme in Black popular culture and the African-American performance tradition has been the connection and interrelatedness of themes of sexuality to those of freedom (Davis 1998: Gilroy 1993). Angela Davis (1998) cites Audre Lorde's theory of "The Erotic as Power" in describing the ways in which the lyrics and performances of Black women artists included associations of sexuality as freedom and social commentary. In describing Billie Holliday's performance of "Some Other Spring," Davis elucidates how Holliday reappropriated the concept of love and sexual desire to symbolize liberty and autonomy:

> In a more complex racial and cultural context, she was able to carry on a tradition established by the blues women and blues men who were her predecessors: the tradition of representing love and sexuality as both concrete daily experience and as coded yearning for social liberation. (1998, 173)

Within the context of racial and sexual oppression and marginalization, love and sexuality have come to signify not only interpersonal relationships but also Black women's struggles for liberation and freedom at a broader level.

The Indeterminate Gaze

The address and gaze in these videos were frequently indeterminate. It was difficult to ascertain where the camera's gaze was intended to originate and to whom the video images and narrative were addressed. While clearly not ungendered, the gaze and address were frequently also neither male nor female. Both the male and female audience member or viewer appears to be constructed within these texts. The camera objectifies the Black female body in a traditional manner, while the lyrics of the song are addressed to a male subject. However, it becomes apparent that men are not the only intended audience. There appears to be a space constructed within the text that allows for Black women viewers to place themselves as subjects of the text, of the narrative.

A mélange of visual and aural strategies contribute to the construction of this indeterminate gaze. In these videos, the camera positioning, artist performance, and narrative structure are combined with visual omniscience. In addition, an indeterminate point of view and frequently non-gender-specific song lyrics contribute to the possibility of a multigendered and even ungendered gaze within music video texts. The Black female performers are not just looking at and talking to men but looking at and speaking with women as well. The unspecified and omniscient point of view constructed by camera positioning supports this by allowing both men and women to see themselves as subjects of the song and video.

The most compelling examples of this phenomenon occur in videos by the group En Vogue. In *Giving Him Something*, a remake of the Aretha Franklin R&B classic, En Vogue performs in a club for an all-male audience. They move seductively, gyrate their hips, and sing provocatively of "giving him something he can feel so he knows my love is real." The men in the audience are responding viscerally, biting their knuckles, and swooning. This scenario is interesting because while the group members are clearly objectified on stage and are explicitly sexualized, it is clear that they are gaining pleasure reciprocally along with a certain level of power over these men who are virtually losing control of their faculties as a result of their performance. Second, the men in the audience are extremely attractive themselves and are the objects of the camera's gaze. What is important here is that not only are men gaining pleasure from viewing the video, but women, as the viewers, are as well. This is not a mere role reversal but an example of an articulation of mutual pleasure and enjoyment. . . .

Informed by the context of the gender politics of Black male and female relationships, this construction of the unfixed, multiple gaze serves to level the sexual playing field. En Vogue, Toni Braxton, and TLC are not simply on display for men (although they surely are); their videos also place men on display for them and their fellow women viewers. In addition, and significantly, the simultaneous existence of their sexuality and independence contests inequality in man-woman relationships. As a result, instead of being the object of exploitation, the Black woman performer is able to construct a subject position for herself and her women viewers. While this is not articulated as a complete role reversal, which would ostensibly alienate male audiences, it is instead expressed as a mutual pursuit of sexual pleasure and satisfaction.

CONCLUSION

Despite the potentially limiting aspects of the frequently contradictory and stereotypical themes in music videos, I demonstrated that a more nuanced and complex depiction of Black womanhood emerges in the representations of Black woman performers.

My findings support and enhance the current literature in Black feminist theory. Whereas in *Black Feminist Thought* Hill Collins (1991a) demonstrated how the controlling images of Black womanhood are disseminated and legitimized through social institutions, my study extends her notion by showing how popular entertainment serves as a space for the proliferation of these controlling images. Hill Collins (1991a, 1991b) described the ways that Black women have countered these hegemonic notions of Black femininity through their culture, focusing on literature and performance in the Blues tradition. I show how Black women also are able to articulate other key themes of self-valuation, self-determination, and a critique of the interlocking nature of oppression. The themes of returning the erotic gaze and reappropriating the Black female body add an additional dimension to Black feminist theory by showing how Black women may use the sphere of culture to reclaim and revise the controlling images, specifically "the Jezebel," to express sexual subjectivity.

Of course, the conclusions drawn as a result of a textual content analysis of music videos are necessarily limited by the absence of inquiry into the production and reception of music videos and by the lack of a more comprehensive survey of the cultural landscape in which they exist. As a result, this study is not a complete analysis of the social context of Black female representation in music videos, and further investigation into Black women's reception and interpretation of music videos, as well as their role as cultural producers in the entertainment industry, is recommended.

REFERENCES

Arnett Ferguson, Ann. 2000. *Bad boys: Public schools in the making of masculinity.* Ann Arbor: University of Michigan Press.

Aufderheide, Pat. 1986. Music videos: The look of the sound. *Journal of Communication* 36:58–77.

Carby, Hazel V. 1986. It jus be's dat way sometime: The sexual politics of women's blues. *Radical America* 20:9–22.

Cash, Thomas F., and Patricia E. Henry. 1995. Women's body images: The results of a national survey in the U.S.A. *Sex Roles* 33:19–28.

Davis, Angela Yvonne. 1998. *Blues legacies and Black feminism: Gertrude "Ma" Rainey, Bessie Smith, and Billie Holiday.* New York: Pantheon Books.

Delano Brown, Jane, and Kenneth Campbell. 1986. Race and gender in music videos: The same beat but a different drummer. *Journal of Communication* 36:93–106.

Dines, Gail, and Jean Humez. 1995. *Gender race and class in media.* Thousand Oaks, CA: Sage.

Flynn, Kristin, and Marian Fitzgibbon. 1996. Body image ideals of low-income African American mothers and their preadolescent daughters. *Journal of Youth and Adolescence* 25:615–30.

Frith, Simon, Andrew Goodwin, and Lawrence Grossberg. 1993. *Sound and vision: The music video reader.* New York: Routledge.

George, Nelson. 1998. *Hip hop America.* Viking.

Gilroy, Paul. 1993. *The Black atlantic: Modernity and double consciousness.* Cambridge, MA: Harvard University Press.

Goodall, Nataki. 1994. Depend on myself: T.L.C. and the evolution of Black female rap. *Journal of Negro History* 79:85–94.

Harris, Shanette M. 1994. Racial differences in predictors of college women's body image attitudes. *Women & Health* 21:89–104.

Hill Collins, Patricia. 1991a. *Black-feminist thought: Knowledge, consciousness and the politics of empowerment.* New York: Routledge.

———. 1991b. Learning from the outsider within: The sociological significance of Black feminist thought. In *Beyond methodology: Feminist scholarship as lived research,* edited by Mary Margaret Fonow and Judith A. Cook. Bloomington and Indiana University Press.

hooks, bell. 1992. *Black looks: Race and representation.* Boston: South End.

Hurley, Jennifer M. 1994. Music video and the construction of gendered subjectivity (or how being a music video junkie turned me into a feminist). *Popular Music* 13:326–38.

Kaplan, E. Ann. 1987. *Rocking around the clock: Music television, postmodernism and consumer culture.* New York: Methuen.

Lewis, Lisa. 1990. *Gender politics and MTV.* Philadelphia: Temple University Press.

McRobbie, Angela. 1991. *Feminism and youth culture: From "Jackie" to "just seventeen."* Boston: Unwin Hyman.

———. 1993. Shut up and dance: Youth culture and changing modes of femininity. *Cultural Studies* 7 (3): 46.

———. 1997. *Back to reality? Social experience and cultural studies.* Manchester, UK, and New York: Manchester University Press.

McRobbie, Angela, and Mica Nava. 1984. *Gender and generation.* London: Macmillan.

Molloy, Beth L., and Sharon D. Hertzberger. 1998. Body image and self-esteem: A comparison of African-American and Caucasian women. *Sex Roles* 38:631–43.

Morgan. 1999. *When chickenheads come home to roost: My life as a hip-hop feminist.* New York: Simon & Schuster.

Patillo-McCoy, Mary. 1999. *Black picket fences: Privilege and peril among the Black middle class.* Chicago: University of Chicago Press.

Perkins, William Eric. 1996. *Droppin' science: Critical essays on rap music and hip hop culture.* Philadelphia: Temple University Press.

Peterson-Lewis, Sonja, and Shirley A. Chennault. 1986. Black artist's music videos: Three success strategies. *Journal of Communication* 1:107–14.

Roberts, Robin. 1991. Music videos, performance and resistance: Feminist rappers. *Journal of Popular Culture* 25:141–42.

———. 1994. Ladies first: Queen Latifah's Afrocentric feminist music video. *African American Review* 28:245–67.

Rose, Tricia. 1991. Never trust a big butt and a smile. *Camera Obscura* 23:9.

———. 1994. *Black noise: Rap music and Black culture in contemporary America.* Hanover, NH: Wesleyan University Press, University Press of New England.

Stockbridge, Sally. 1987. Music video: Questions of performance, pleasure and address. *Continuum: The Australian Journal of Media and Culture* 1(2). [cited 2 April 1997]. Available from www.mcc.murdoch.edu.au/readingroom/1.2/stockbridge.html.

Vincent, Richard C. 1989. Clio's consciousness raised? Portrayal of women in rock videos re-examined. *Journalism Quarterly* 66:155–60.

Vincent, Richard, Dennis K. Davis, and Lilly Ann Boruszkowski. 1987. Sexism on MTV: The portrayal of women in rock videos. *Journalism Quarterly* 64:750–55, 941.

Watkins, S. Craig. 1998. *Representing: Hip hop sulture and the production of Black cinema.* Chicago: University of Chicago Press.

27

Disco "Super-Culture"

Consuming Foreign Sex in the Chinese Disco

BY JAMES FARRER

James Farrer's study of the superculture of global disco in China is primarily based on participant observation in discos and dance halls in Shanghai and other towns and cities all over China from 1993 to 1996. He also collected data from focus group interviews on sexual relations in Shanghai. The disco in China is a public place in which young people can test their attractiveness as sexual and social commodities. Disco dance styles emphasize a gendered sexual image, one that differs markedly between women and men. Additionally, "foreigners" and "foreignness" play important roles in the Chinese disco.

1. Do the dance styles of men and women in Chinese discos resemble the gendered dance styles found in Western clubs and dance halls?
2. What does Farrer mean when he says that gendered sexuality is more show than talk or substance in the Chinese disco?
3. Why does Farrer describe global disco as a super-culture rather than a sub-culture?

COSMOPOLITAN DANCE CULTURE AND COSMOPOLITAN SEXUAL CULTURE

Sub-Culture or Super-Culture?

Global flows of media images, people and commodities have produced a globalization of sexual imagery, ideas and practices. Sociologists have begun to explore the sexual dimensions of globalization (e.g. Altman, 1997), but there are too few case studies of the particular social contexts in which transnational sexual cultures are locally practised and consumed. This article describes how a particular global culture—disco (*di-si-ke* in Chinese)—serves as a localized "cosmopolitan" context for novel forms of sexual expression by urban Chinese youth.

Dance is one of the most public and popular sites for the expression of sexual and gender identities (Hanna, 1988) and forms of popular dance are among the most globalized of sexual cultures. . . .

Yet, despite its prevalence and rich sexual content, social dance has been the "least theorized popular cultural activity and the least subject to the scrutiny of the social critic" (McRobbie, 1984: 132). Among China scholars, the popular disco has been dismissed as "second or third rate imitations of the West or Japan" by western proponents of traditional Chinese values and as "bourgeois decadence" or "slavish imitation of the West" by Chinese traditionalists (Schell, 1988: 356).

Such alarmist or dismissive views of cultural imports echo early sociological critiques of cultural "globalization" as the westernizing and homogenizing intrusion of the global capitalist system (Schiller, 1976; Wallerstein, 1990). This "cultural imperialism" thesis provoked responses —typically from ethnographers—that global culture is neither exclusively western nor homogenizing (Appadurai, 1996; Hannerz, 1992; Pieterse, 1994; Robertson, 1992). First, the origins of cultural objects do not have an obvious relationship to their uses; a popular cultural form that means one thing in its place of origin may mean something different in another local context (Liebes and Katz, 1990; Tobin, 1992). Second, imported cultural forms may be appropriated in local resistance to global influences (Sreberny-Mohammadi, 1991; Thompson, 1995).

These accounts of "appropriation", "localization" or "indigenization" have intellectual affinities with sub-cultural theories of youth cultures. Subcultural theorists emphasize how working-class youth create their own meanings out of the cultural products of a hegemonic capitalist order, express resistance, alterity and sub-group identification through music and dance (Hebdige, 1979; Martinez, 1997; Racz and Zetenyi, 1994) or, more pessimistically, use commercial leisure as an ineffective escape from deadend lives at school and home (Mungham, 1976; Willis, 1984). Despite critiques and reformulations of the sub-cultural paradigm, the questions still center on the processes of sub-group identity formation, the construction of "resistant" and "alternative" narratives, and the social uses of themes of resistance (Gore, 1997; Redhead et al., 1997; Thornton, 1996).

Disco, like some other globalized mass-commercial youth cultures, represents a problem for "sub-cultural" perspectives because disco from its inception made almost no claims to exclusive identities, to local cultural authenticity or even to musical authenticity (Frith, 1981). By replacing the live band with a DJ and a mixing board, the discotheque allowed the mixing and blending of musical styles without the average dancer even noticing. Although it had roots in diasporic black subcultures, by the late 1970s the discotheque was a global institution of borrowed sounds and symbols with no claim to a closed creative or folk community (Frith, 1981). African and Latin sounds dominated, though they were often popularized by performers from core countries (Chambers, 1985). Straights and gays mixed in a sexually fluid atmosphere (Dyer, 1990). . . .

Far from dead in the 1990s, as the culture of the discotheque circulates the globe, it has increasingly lost whatever exclusivity and specificity it once connoted. For instance, musical forms as diverse as Cantonese pop, reggae, rap and old disco standards arc played in turn in Shanghai's commercial discos, where dancers seldom express a strong identification with a particular genre of music. Associated in the US with a particular variety and era of dance music, the term *"di-si-ke"* in China covers all modern "partnerless" dances and any associated mixed or danceable music. Mass, global and commercial dance culture thus may require different sociological categories than those developed in studies of more exclusive dance "sub-cultures" or musically sophisticated and self-consciously "alternative" "club cultures"(Redhead et al., 1997; Thornton, 1996). The giant commercial discotheque—as it has been replicated in provincial centers throughout the world—is more a site for youth wishing to physically and symbolically *enter* global (mass/commercial/"modern") culture than for those trying to avoid it, resist it or distinguish themselves from it through "alternative" or "authentic" performances and identifications.

Global disco, in its mass-culture form, is perhaps more appropriately described as a *super-culture* rather than a *sub-culture.* Rather than spaces for identifying with a particular musical culture or sub-culture, large commercial discotheques are spaces where youth experience the larger society beyond their neighborhoods and their family and work lives (Willis, 1984: 38), sites for experiencing a glamorous modernity in which one does not distinguish oneself by class or locality. . . .

The "glamorous sophistication" which Walsh describes is now a global culture celebrating consumption, fashion and sexuality, in which youth on every continent participate, reinforcing an emergent global hegemony of consumer values. It would be wrong, however, to argue that national origins are irrelevant to transnational cultural forms such as disco. In the global cultural economy, where a thing comes from becomes an important aspect of its meaning for its cosmopolitan users. For instance, "tango" in Japan relies on its "authentic" Latin origins despite great differences in actual practices (Savigliano, 1992). Global cultural objects are thus nearly always "localized"—used in novel ways by local people—but their global origins are an important part of their meaning and can be a source of empowerment in local practices (Friedman, 1990). Nor is this merely a question of an accidental "mixing" of cultures ("creolization" in Hannerz, 1992; "hybridization" in Pieterse, 1994), but a purposeful participation in a cosmopolitan (or "foreign") space which is intentionally, kept separate from "local" practices. In particular, people may move daily between local and cosmopolitan identifications in order to enhance their own sense of personal autonomy (Hannerz, 1992). Thus the point of discussing Chinese discotheques is not to oppose an inauthentic invading global culture with authentic local cultural practices of appropriation and resistance, but to ask what participants make of their participation in this self-consciously cosmopolitan culture.

The Cosmopolitan Sexual Culture of the Disco

Substantively, I focus on the sexual culture of the disco. In early studies of sexual expression in dance culture, dance was usually described as a courtship activity (Rust, 1969), as a means of getting sex (Cressey, 1932), or as a male predatory activity (Mungham, 1976). "What is missing in views of this kind is the realization that dance is not just a means to sex (although of course it may well be such) but that it is or can be a form of sexual expression in itself" (Ward, 1993: 22). In particular, the dance hall can be a space for relatively "safe" and non "goal-oriented" sexual expression by young women (McRobbie, 1984; Peiss, 1986).

What is also missing in these views is an appreciation of the larger sexual culture that the dance hall spatially incorporates. Chinese discotheques are sites in a global circulation of sexual imagery and practices through commercial cultural practice—especially dance—which Savigliano (1992) describes as "the world economy of passion." . . .

The issue with which I am centrally concerned is the uses Chinese youth make of a deliberately constructed space of "foreign" sexuality. While, as I have already suggested, the disco is not a space for "resistance" to global cultural hegemony, nor is it a free space of cultural appropriation. The global sexual culture of disco provides particular opportunities and constraints for sexual expression. . . .

DISCO IN CHINA

The Social Background of Chinese Disco Culture

[D]isco dancing arrived in China in the mid-1980s, promulgated through American films like "Flashdance" and "Breakdance" and through the dance parties of foreign students at Chinese universities. The first specialized discotheques opened in the late 1980s in hotels restricted to foreigners. In the 1990s these restrictions were lifted and massive discotheques financed largely by Hong Kong entrepreneurs opened to a mostly local Chinese clientele. By the mid-1990s Shanghai boasted at least 10 multi-level "disco plazas", which could accommodate about 600 customers each and charged up to 100 yuan (US$12) for admission. The remaining 80 to 100 discos in Shanghai were small neighborhood youth clubs, which charged as little as 4 yuan (US$.50) for an afternoon admission but usually about 10–30 yuan. While 100 yuan represented about a tenth of the monthly income of a Shanghai factory worker, many youth spent much of their disposable income on entertainment. Working youth could afford tickets, and even unemployed youth were occasionally able to obtain free passes to expensive discos.

Young people described going to the disco to "get a little crazy" and "release tension"—reasons almost identical to those given by British youth in the 1960s (Rust, 1969: 171). Like these 1960s British youth, Shanghai youth still typically live at home with their parents and use dance clubs as escapes from domestic life (for arguments for the centrality of dance to contemporary UK youth culture see Thornton, 1996: 14–25).

Dance was the most popular commercial leisure activity for youth in Shanghai.[1] Moreover, in every county-level town I visited in several remote provinces of China, I was able to find some type of commercial dancing establishment. Fancy discotheques were most common in the flourishing coastal cities of China, but lesser imitations can be found in most inland cities, with names evocative of global entertainment centers ("L.A.", "Barcelona" and "New York"), Hollywood fantasy locales ("Broadway", "Casablanca"), sexual innuendoes ("Kiss", "Touch"), regionally famous discotheques (Taipei's "Kiss" or Shanghai's "JJ's") or globality itself ("Galaxy", "NASA"). In smaller locales disco and ballroom dancing were often combined by alternating the types of music or offering separate disco and ballroom dance sessions during different hours, catering to different age groups. In big cities like Shanghai, the market was more clearly divided between discotheques visited by youth (mostly aged 15–25 years) and social dance halls visited by middle-aged people (mostly 30–45 years). This article only discusses the discos, focusing on the large, metropolitan discos in Shanghai, which are the "models" for discos elsewhere in China.[2]

Consuming Foreign Sex/ Being Consumed

The commercial disco is described by Shanghaiese (by those who go there and those who do not) as "chaotic", where people are "unreliable" and just "playing around" or "fishing for big monies". Commercial discotheques are described as "in society" *(shehuishang)* as opposed to the non-commercial venues with controlled admissions and more decorous behavior. Students and youth are advised not to go to these commercial venues before they begin work and have themselves "entered society" (started work and started dating). Consequently, those who generally go to discos are those who have just "entered society" and have not been "in society" very long. In this urban Chinese "society", which is now a market society, youth must learn to imagine themselves as having a dual nature: that of both consumers and desirable commodities. The disco is a place to "enter society" and to test one's attractions as a sexual and social commodity in the chaotic marketplace which the word "society" now connotes. The disco is thus a simulacrum of the market society with its chaotic mix of temptations and pressures.

Unlike the earlier generation of British dancers surveyed by Rust (1969), Shanghai youth seldom described the discotheque as a place to make friends or find a mate. Most disco patrons agreed that people one meets in dance halls were "unreliable" or just "play friends". Both young men and young women dismissed those of the opposite sex in discos as "knowing how to play too well". Nor was a disco typically seen as a place to take a date. Men and women usually visited in groups of friends.

This does not mean that the Chinese discotheque is a sexless place. Far from it, the sexuality of the discotheque is especially visual, a chance to show oneself off in a sexualized fashion to an audience of glamorous strangers. Almost everyone puts on special clothing for their trips to the disco, including metallic miniskirts and stomach-revealing halter tops that first appeared in discos in the 1990s before they became acceptable on the streets. Some young men's clothing is intentionally "weird", as one trendy youth described it, including colorful shirts and stomach-revealing tops. Most young men, however, resemble the "teddy boys" of earlier decades in the West, their sober dress reflecting emotional cool and an attempt at social sophistication. In recent years local disco girls have moved away from traditional Shanghai standards of prettiness and taken to black lipstick, leather and metallic fabrics, also showing a new sense of urban cool. Young men and young women have different strategies of display, but for all showing off in the commercial leisure world of the discotheque is about displaying sexual self-confidence and poise in the anonymous self-reflecting gaze of the crowd.

Dance styles reflect both the globality of disco culture and its emphasis on the display of a gendered sexual image. Young men competed in "break-dance" (*zhanwu,* literally "cut dance") contests in which they hurled themselves on the dance floor and spun and flipped about in a style they picked up from US videos and films of the 1980s. These exercises were as much tests of physical and social daring as of dancing skill. In the smaller local discos young men and women gathered in a circle in the middle of the floor and watched these dancing exhibitions, prodding reluctant young men into the center of the ring. The display of masculine skill and daring was followed by the vigorous applause of the men and women surrounding them. Young women watched but seldom participated, most of them unwilling to engage in the ungainly spectacle of twirling about spreadlegged on the dirty floor.

Young women specialized in a different style of dance which some called "voluptuous dancing" *(saowu).* The dancers gyrated their hips and torsos to free-style dance steps derived from the provocative moves of back-up dancers on western music videos. This style of dancing became more popular in the mid-1990s when some clubs began hiring young women to dance on stages set up above the dance floor, where break-dance was impractical. Most dancers on the stages were not paid, however. Women customers, especially, frequently danced on the raised platforms with their hands above their heads and their eyes averted, welcoming the gaze of the dancers on the floor below. Young women also excelled at more complex Latin dances such as the cha-cha, samba and the lambada, generally with another female partner. Young women dancers usually described their sexual display and display of dancing skill as a rewarding form of self-expression as the following two quotes from two young women dancers illustrate:

> I like to see people watch me. I have a desire to show off myself. . . . People can think what they want when I dance.
>
> In dancing disco a girl would like to let herself go and show off her personality . . . if you can't let yourself go at a disco it's no fun.

As McRobbie suggests from her observations of British women, the sexual display in young women's dance suggests "a displaced, shared and nebulous eroticism rather than a straightforwardly romantic, heavily heterosexual 'goal-oriented' drive" (McRobbie, 1984: 134), a "whole body sexuality" as opposed to the "phallic sexuality" of "cock rock" (Dyer, 1990). Rather than expressing a goal-oriented sexual drive, looking good was the most constant theme of dance culture in the Shanghai disco. In its highly gendered forms, the projection of a sexual self-image through dance was used to explore one's own desirability and autonomy in the idiom of a safely "foreign" dance culture and was experienced subjectively as empowering and pleasurable.

The cosmopolitan definition of disco culture and this culture of sexual display are related on several levels. First, there is direct imitation of foreign styles of sexual display. Styles of dress and dance are studied from western videos. Second, there is the use of the "cosmopolitan" space of the disco for constructing an image of the sexualized self within a global market culture. Rather than creating a sense of local community, the cosmopolitan culture of the disco is a forum for exposing oneself—literally—to the anonymous gaze of a global image market, for reflecting upon the self as a modern consumer and also a prized sexual commodity.

The globality of the disco is deliberately engineered through imagery which incorporates "foreignness" into the ideal of the cosmopolitan metropolis. Shanghai's discos strive to capture the pleasures and dangers of metropolitan life, constructing the city as the center, rather than the periphery, of a global youth culture. DJs frequently identify the space of the discotheque with the cosmopolitan metropolis, shouting in English and Chinese: "Hello Shanghai!" The decor is Hollywood-style urban-industrial-chaos with exposed steel girders, old American cars and dayglow English graffiti. Videos are shown, including scenes of gyrating and scantily clad westerners at a beach dance club in Australia or sexy rap stars in US music videos.

A foreign (often Hong Kong or overseas Chinese) DJ is de rigueur at the biggest discos. International music dominates, though most clubs also play an occasional mix of Hong Kong and Taiwanese pop songs overlain with a disco beat. The lyrics of dance songs are typically in English and Shanghai favorites included the Village People's "YMCA" and Ace of Bass's "All That She Wants", which were played very regularly throughout my 3 years in Shanghai. One reason for the undying popularity of these tunes is that certain English words in the chorus are understood and repeated by the dancers, most of whom have studied English for at least 2 or 3 years. These familiar songs allow for a feeling of participation in the global culture of the disco. For instance, the "YMCA" dance is now a globalized dance ritual in which the dancers are encouraged to use their hands to make the shapes of the English letters, identifying themselves momentarily with a boundless global ecumene of sexy happy youth "at the YMCA". Similarly, Japanese, European, American and Chinese youth all were dancing the global fad-dance, the "Macerena", at Shanghai discos in 1996, shortly after it became popular in Europe and the US.

The sexual content of global disco culture was also accentuated by the music. As US hip-hop supplanted European techno in popularity in 1996, DJs took to playing several songs with sexually explicit lyrics in English. Although Chinese youth were largely unaware of the meaning of lyrics such as "I Like the Way your Pussy Tastes", and "I Don't Want a Short Dick Man", both foreign and Chinese DJs (who did understand the lyrics) played the songs in part to promote a more sexual atmosphere in the same way they used videos and paid dancers to encourage the sensual displays of the young patrons mounting the stages around the dance floor.

Actual "foreigners", especially white westerners, were an important prop in the construction of this cosmopolitan sexual culture, particularly in the metropolitan discos of cities like Beijing and Shanghai. Foreign students were often let in free. One Shanghai disco manager who was trying to push his second-rate club into first-rate status urged me to bring some foreign men. "The girls here are shy", he explained, "but they would like to make some foreign friends". This was not an unusual approach. Foreign men were perceived as a big attraction for local women. Young "foreigners" in general were part of the cosmopolitan sexual atmosphere consumed by Chinese in the biggest metropolitan discos. As one young woman said about her disco experiences, "The foreigners made it seem more modern".

Dancing with the energetic and exotic foreign women was a dare or challenge for Chinese men, who perceived such women as difficult to attain and as sexually and socially more sophisticated than local women. For Chinese women, on the other hand, dancing with the (often sexually aggressive) foreign men represented both sexual danger and a sudden increase in attention from other dancers. For both Chinese men and women, interactions with the foreigners thus represented excitement, a social challenge and an increase in attention. Yet these interactions were typically brief encounters, like most interactions in the disco, chances to play at being "cool", "desirable" and socially competent. Moreover, given the relatively smaller numbers of foreign visitors, most Chinese women and men never interacted in any depth with foreigners in discos (though conversely many foreign men pursued greater intimacies with local Chinese women). Like the videos, these foreigners were largely for visual consumption, but were even more useful as the "global audience" for the displays of local youth, male and female. To be watched (and desired) by the foreigners was to be directly affirmed as desirable in the cosmopolitan sexual market of the disco. Actual social interactions with foreigners were occasions for displaying poise and the capacity (both social and linguistic) to deal with the dangers and sexual temptations foreigners represented. Foreigners became the objects of sexual fantasy and occasional sexual adventures, but even more so were the mirrors for the construction of a cosmopolitan sexual self-image.

Disco dance in China and the overt sexual expression associated with it derive both legitimacy and persuasiveness from this larger global culture. Rather than being merely seen as the degenerate performances of juvenile delinquents influenced by foreign "spiritual pollution", such celebrations of sexual display can now be positively identified as the elite cosmopolitan culture of modern youth around the world. The specific forms and meanings this sex and gender display take, however, may be out of sync with global fashions even while they rely on their global origins for legitimacy. For instance, "voluptuous dancing" which reminded one American woman of "a striptease dance with the clothes left on" was interpreted by the Chinese woman performers quoted here as a form of pleasurable exhibitionism learned from foreign music videos. Music which would seem dated to London or New York youth was fresh and "western" for youth in Shanghai.

This appropriation of foreign sexual passion by Shanghai youth means that what Savigliano (1992) calls "the global economy of passion" does not always follow the pattern she describes of an over-rationalized industrial center appropriating the sexual culture of an economically backward "primitive other". Nor does it always take the form of the sexualization of a dominated "other" by the peoples of the center (Wallerstein, 1990: 44). Rather, in these Chinese discos, we find that dancers use dance stages, video images of sexy dancers and the anonymous crowds surrounding them to literally picture themselves in a transnational modernity of music, dance and sexual excitement. Within the disco, they become sexual "cosmopolitans" themselves with an enlarged repertoire of sexual strategies and styles.

The sexual repertoires of the disco also extend to more intimate forms of sexual expression. The separation of disco from everyday "real" life allows for forms of intimacy and behavior which would normally have been censored in other social spaces. The discotheque may even serve as a space for acting out "foreign" sexual ideologies. For instance, a 24-year-old professional woman I knew well once mentioned a "one night stand" to me, using the English expression. One day when we were alone I asked her about that experience:

> That was the only time I've done that ["one night stand"]! I went out with some friends, and I met him dancing. He was a really attractive man, and I really wanted to be with him. He took us out after the dancing and later I went home him. It was very "romantic and very wonderful". . . . We had been drinking. Maybe if I wasn't drinking then it wouldn't have happened so quickly. . . . The next morning we got up and went to the market to buy something for breakfast. It was really romantic.

Though her use of English is unusual, her English references to "one night stand" and "romance" in her narrative mark the accessibility of western sexual ideologies in the framing of her sexual adventure as part of a normal range of sexual behaviors originating in the heady atmosphere of the disco. For her, the disco provided a place where foreign sexual ideologies could be enacted in a "romantic", spontaneous and drunken encounter, a kind of behavior which would have been labeled as "hooliganism" a mere decade earlier (and still would be condemned by most Chinese adults).[3] In this case, though the relationship ended in a few days, she still was able to remember it fondly. Discotheques are thus spaces where youth—young women especially—can practice freedoms of sexual expression and interaction not usually allowed them in the larger society, using this cosmopolitan space for illicit sexual display and play without fear of social sanctions.

The "one night stand" is the extreme case of the "erotic play" of the disco but casual flirtations are more typical. Kissing with a new friend is not uncommon. It is actually because of its separation from everyday life that the disco becomes an important space for working on one's sexual self image and sense of desirability. In the more earnest environments of school and work casual liaisons and flirtations might be too costly. In the disco they can be dismissed as inconsequential. Inconsequentiality and anonymity are thus a cover for the important activities of sexual self-appraisal and self-appreciation in a kind of virtual sexual marketplace, cut off from the serious and consequential sexual marketplaces governed by family, community and friends. The separation of the disco from everyday life through its liminoid "foreignness" increases its utility for Chinese youth exploring alternative sexual images and behaviors.

In the culture of sexual display in the discotheque competing notions of sexual desire and desirability are simultaneously promoted by videos, dancers, DJs and advertising images. Canto-pop singers croon heroically about lost love, while hip-hop artists rap explicitly about sex acts in English. In the half-hour of slow dance in many discos, young people use the physical proximity of social dance to engage in tactile sexual intimacies. In the "voluptuous dance" young women—whose modesty had been encouraged in Chinese society—simulate sexual excitement with lithe pelvic motions. Most dancers are more modest with their hips, but not in their desire to look good. Being desirable is more important than modesty in the mutual visual consumption of the disco. But even here, there is no one standard. In their dress, traditional notions of feminine beauty and social status (pleated skirts, business suits and long hair) compete with images of urban cool (black t-shirts, halter tops and cropped hair). The cosmopolitan disco culture thus offers a varied palette of sexual themes and a place for sexual playfulness, its transnational melange of sounds and images marking it as a space apart from everyday life where such play is possible. Dancers use these transnational and local discourses tactically to enhance their own sense of pleasure, desirability and autonomous choice.

In summary, the disco is a deliberately engineered space of "foreign" sexual imagery, which Chinese youth appropriate to experiment with alternative sexual styles and sexual self-images. Actual foreigners are props in the exploration of sexual desirability and emotional poise in a global sexual and social marketplace. For youth faced with an increasingly free and competitive but still very earnest sexual marketplace in the "real world", the disco is a space apart from everyday life for proving their desirability through sexual display and casual and (usually) inconsequential flirtations with strangers. The "foreignness" of the disco abets the construction of the disco as a space apart in which sexual display and experimentation is possible. But in general the disco is still more a space for working on the *self*, rather than for working on serious relationships.

★ ★ ★

NOTES

1. According to the Shanghai Cultural Ministry there were more registered karaoke bars (about 2000) than dance halls (about 1200 including discos and social dance halls), but discos and dance halls tended to have far more daily visitors than karaoke bars. Beijing had similar numbers of discos and dance halls.
2. For a comparison of the forms of sociability in Shanghai's discos with Shanghai's social dance halls see Farrer (in press).
3. Unlike US or European dance clubs, there was very little drug use at these discotheques other than alcohol and tobacco. Women drank much less than men. Those who drank heavily tended to do so at dinner before coming to the disco, where drinks were comparatively expensive.

REFERENCES

Altman, Dennis (1997) "Global Gaze/Global Gays", *GLQ: A Journal of Lesbian and Gay Studies* 3: 417–36.

Barme, Geremie and Jaivin, Linda, eds (1992) *New Ghosts, Old Dreams: Chinese Rebel Voices*. New York: Times Books.

Chambers, Iain (1985) *Urban Rythms: Pop Music and Popular Culture.* New York: St Martin's Press.
Cressey, Paul G. (1932) *The Taxi-Dance Hall: A Sociological Study of Commercialized Recreation and City Life.* Chicago, IL: University of Chicago Press.
Dyer, Richard (1990) "In Defense of Disco", in Simon Frith and Andrew Goodwin (eds) *On Record: Rock, Pop, and the Written Word.* New York: Pantheon.
Farrer, James (2001) "Dancing Through the Market Transition: Disco and Dance Hall Sociability in Shanghai", in Deborah Davis (ed.) *The Consumer Revolution in Urban China.* Berkeley: University of California Press.
Friedman, Jonathan (1990) "Being in the World: Globalization and Localization", in Michael Featherstone (ed.) *Global Culture: Nationalism, Globalization and Modernity.* London: Sage.
Frith, Simon (1981) *Sound Effects: Youth, Leisure, and the Politics of Rock 'n' Roll.* New York: Pantheon.
Frith, Simon and McRobbie, Angela (1990) "Rock and Sexuality", in Simon Frith and Andrew Goodwin (eds) *On Record: Rock, Pop, and the Written Word.* New York: Pantheon.
Hanna, Judith Lynne (1988) *Dance, Sex and Gender: Signs of Identity, Defiance, and Desire.* Chicago, IL: University of Chicago Press.
Hannerz, Ulf (1992) *Cultural Complexity: Studies in the Social Organization of Meaning.* New York: Columbia University Press.
Hebdige, Dick (1979) *Subculture: The Meaning of Style.* London: Methuen.
Kunz, Jean Lock (1996) "From Maoism to ELLE: The Impact of Political Ideology on Fashion Trends in China", *International Sociology* 11(3): 317–35.
Liebes, Tamar and Katz, Elihu (1990) *The Export of Meaning: Cross-Cultural Readings of DALLAS.* New York: Oxford University Press.
McRobbie, Angela (1984) "Dance and Social Fantasy", in Angela McRobbie and Mica Nava (eds) *Gender and Generation.* London: Macmillan.
Mungham, Geoff (1976) "Youth in Pursuit of Itself", in Geoff Mungham and Geoff Peterson (eds) *Working Class Youth Culture.* London: Routledge.
Peiss, Kathy (1986) *Cheap Amusements: Working Women and Leisure in Turn-of-the-Century New York:* Philadelphia, PA: Temple University Press.
Pieterse, Jan Nederveen (1994) "Globalization as Hybridization", *International Sociology* 9(2): 161–84.
Racz, Josef G. and Zetenyi, Zoltan (1994) "Rock Concerts in Hungary in the 1980s", *International Sociology* 9(1): 43–53.
Robertson, Roland (1992) *Globalization: Social Theory and Global Culture.* Newbury Park, CA: Sage.
Savigliano, Marta E. (1992) "Tango in Japan and the World Economy of Passion", in Joseph J. Tobin (ed.) *Re-made in Japan: Everyday Life and Consumer Taste in a Changing Society,* pp. 235–51. New Haven, CT: Yale University Press.
Schell, Orville (1988) *Discos and Democracy: China in the Throes of Reform.* New York: Pantheon.
Schiller, Herbert (1976) *Communication and Cultural Domination.* White Plains, NY: International Arts and Science Press.
Sreberny-Mohammadi, Annabelle (1991) "The Global and the Local in International Communications", in James Curran and Michael Gerevitch (eds) *Mass Media and Society.* New York: Edward Arnold.
Thompson, John B. (1995) *The Media and Modernity: A Social Theory of the Media.* Stanford, CA: Stanford University Press.
Tobin, Joseph J., ed. (1992) *Re-made in Japan: Everyday Life and Consumer Taste in a Changing Society.* New Haven, CT: Yale University Press.
Wallerstein, Immanuel (1990) "Culture as the Ideological Battleground of the Modern World System", in Michael Featherstone (ed.) *Global Culture: Nationalism, Globalization and Modernity.* Newbury Park, CA: Sage.
Walsh, David (1993) "'Saturday Night Fever': An Ethnography of Disco Dancing", in Helen Thomas (ed.) *Dance, Gender and Culture.* New York: St Martin's Press.
Ward, Andrew H. (1993) "Dancing in the Dark: Rationalism and the Neglect of Social Dance", in Helen Thomas (ed.) *Dance, Gender and Culture.* New York: St Martin's Press.
Willis, Paul (1984) *Learning to Labor: How Working Class Kids Get Working-Class Jobs.* New York: Columbia University Press.

CHAPTER 6

Tracing Gender's Mark on Bodies, Sexualities, and Emotions

This chapter explores the ways in which gender patterns weave themselves into three of the most intimate aspects of the self: body, sexuality, and emotion. The readings we have selected make the general sociological argument that there is no body, sexuality, or emotional experience independent of culture. That is, all cultures sculpt bodies, shape sexualities, and produce emotions. One of the most powerful ways in which a gendered society creates and maintains gender difference and inequality is through its "direct grip" on these intimate domains of our lives (Schiebinger, 2000:2). Gender ideals and norms require work to be done in, and on, the body to make it appropriately feminine or masculine. The same ideals and norms regulate sexual desire and expression, and require different emotional skills and behaviors of women and men.

At first glance, it may seem odd to think about body, sexuality, and emotion as cultural and gendered products. But consider the following questions: Do you do gendered body work? Do you diet, lift weights, dehair your legs or face, or use makeup? If you were interviewing for a job, would you feel more comfortable sitting with legs splayed or legs crossed at the ankles? How do you feel and move your body through public spaces when you are alone at night?

Now think about the gendering of sexuality in the United States. The mark of gender on sexual desire and expression is clear and deep, tied to gendered body ideals and norms. Whose breasts are eroticized and why? Are women who have many sexual partners viewed in the same way as men who have many partners? Why are so many heterosexually identified men afraid of being perceived as homosexual? Why do women shoulder the major responsibility for contraceptive control?

Like sexuality, emotions are embodied modes of being. And like sexuality and body, emotions are socially regulated and constructed. They are deeply gendered.

Consider these questions: Do you associate emotionality with women or men? Is an angry woman taken as seriously as an angry man? Why do we expect women's body language (e.g., their touch and other gestures) to be more affectionate and gentle than men's? What is your reaction to these word pairs—tough woman/soft man?

The readings in this chapter explore the complex and contradictory ways in which bodies, sexualities, and emotions are brought into line with society's gender scheme. Two themes unite the readings. First, they demonstrate how the marking of bodily appearance, sexual desire and behavior, and emotional expression as masculine or feminine reinforces Western culture's insistence on an oppositional gender binary pattern. Second, the readings show how patterns of gender inequality become etched into bodies, sexualities, and emotions.

Gendered Bodies In their reading in this chapter, Paula Black and Ursula Sharma argue that all societies require body work of their members. But not all societies insist on the molding of men's and women's bodies into visibly oppositional and asymmetrical types; for example, strong male bodies and fragile female bodies. Only societies constructed around the belief and practice of gender dualism and hierarchy require the enactment of gender inequality in body work. To illustrate, consider the fact that the height ideal and norm for heterosexual couples in America consists of a man who is taller and more robust than his mate (Gieske, 2000). This is not a universal cultural imperative. As Sabine Gieske (2000) states, the tall man/short woman pattern was unimportant in eighteenth-century Europe. In fact, the ideal was created in the Victorian era under the influence of physicians and educators who defined men as naturally bigger and stronger than women. The expectation that men be taller than their female partners persists today, even though the average height gap between men and women is closing (Schiebinger, 2000). The differential height norm is so strong that many contemporary Americans react to the pairing of a short man and a tall woman with the kind of shock and disapproval typically directed at interracial couples (Schiebinger, 2000).

How are bodies made feminine and masculine as defined by our culture? It takes work, and lots of it. Well into the twentieth century, American gender ideology led men and women alike to perceive women as frail in body and mind. Boys and men were strongly exhorted to develop size, muscle power, physical skills, and the "courage" to beat each other on the playing field and the battlefield, while girls and women were deeply socialized into a world of distorted body image, dangerous dieting, and physical incompetence (Dowling, 2000). Throwing, running, and hitting "like a girl" was a common cultural theme that we now understand to be a consequence of the cultural taboo against girls developing athletic stature and skills. As Collette Dowling (2000) notes, "there is no inherent biological reason for girls not to throw as far, as fast, or as hard as boys do" (64). But there is a cultural reason: the embodiment of the belief in gender difference and inequality. We literally translate the "man is strong, woman is weak" dictum into our bodies. This dictum is so powerful that many people will practice distressed and unhealthy body routines and regimens to try to emulate the images of perfect male and female bodies. The mere threat of seeming masculine or mannish has kept a lot of girls and women from developing their own strength, and the specter of seeming effeminate or sissy has propelled many boys and men into worlds in which their bodies are both weapons and targets of violence (Dowling, 2000).

American men, especially young men, are expected to express stereotyped masculinity through their bodies by the way they move, sit, gesture, eat, and so forth. However, men's bodies are not held to equally severe ideals of bodily attractiveness and expectations of body work as are women's bodies. Gender inequality etches women's flesh in more debilitating and painful ways. In line with Kilbourne's article in Chapter 5, Black and Sharma argue in their reading that "women's bodies are regulated and commodified to an extent and in a form not experienced by men."

American cultural definitions of femininity equate attractiveness with a youthful, slim, fit body that, in its most ideal form, has no visible "flaws": no hair, pores, discolorations, perspiration marks, body odors, or trace of real body functions. In fact, for many women, the body they inhabit must be constantly monitored and managed—dehaired, deodorized, denied food—so that it doesn't offend. At the extreme, girls and women turn their own bodies into fetish objects to which they devote extraordinary amounts of time and money.

What are the models of feminine bodily perfection against which girls and women measure themselves and are evaluated by others? The images and representations are all around us: in magazine ads, on TV commercials and programs, in music videos, and in toy stores. The Barbie doll is an excellent example. As Urla and Swedlund argue in Chapter 5, Barbie is the "ever-thin, ever-plastic, ever-wholesome," excessively feminine American icon. Her body is, of course, impossible for real women to achieve.

The embodiment of women's subordination in gender stratified societies takes some extreme forms. For example, in highly restrictive patriarchies, women's bodies may be systematically deformed, decimated, and restricted. Foot-binding in traditional China is one of the most dramatic examples of an intentionally crippling gender practice. Although foot-binding may seem to be a practice that has no parallels in Western societies, in this chapter sociologist Fatema Mernissi challenges the ethnocentric tendency to dismiss practices such as foot-binding in China and the veiling of women in some nations today as alien or primitive. She does so by revealing the symbolic violence hammered directly on the Western female body by fashion codes and cosmetic industries that place Western women in a state of constant anxiety and insecurity. Mernissi's work suggests the usefulness of comparing practices such as foot-binding in China to cosmetic surgery, extreme dieting, and body sculpting among contemporary Western girls and women. Consider the following questions. What do these seemingly different forms of body work have in common? Do they serve similar functions? How do they replicate gender inequality?

The story of the gendering of bodies in the United States is not only about oppositional and assymetrical masculinity and femininity. Prisms of difference and inequality come into play. Let's consider one prism—race—to illustrate this point.

What's the answer to the question, "Mirror, mirror on the wall, who's the fairest of them all?" (Gillespie, 1998). You know what it is. The beauty standard is white, blonde, and blue-eyed. It is not Asian American, African American, Latin American, or Native American. In other words, there is a hierarchy of physical attractiveness, and the Marilyn Monroes and Madonnas of the world are at the top. Yes, there are African- and Asian-American models and celebrities. But, they almost always conform to white appearance norms. When they don't, they tend to be exoticized as the "Other." Just consider the fact that eyelid surgery, nasal implants, and nasal tip refinement procedures are the most common cosmetic surgery procedures undergone by Asian-American patients, largely women (Kaw, 1998). The facial features that Asian-American women seek to alter, including small, narrow eyes and a flat nose, are those that define them as racially different from white norms (Kaw, 1998). Racial and gender ideologies come together to reinforce an ethnocentric and racist beauty standard that devalues the "given" features of minority women. The medical system has cashed in on it by promoting a beauty standard that requires the surgical alteration of features that don't fit the ideal.

Gendered Sexualities Like the body, sexuality is shaped by culture. The sexualization process in a gendered society such as our own is tightly bound to cultural ideas of masculinity and femininity. In dominant Western culture, a real man and a real woman are assumed to be opposite human types, as expressed in the notion of the "opposite sex." In addition, both are assumed to be heterosexual as captured in the notion that "opposites attract." Conformity to this gendered sexual dichotomy is strictly enforced.

The term "compulsory heterosexuality" refers to the dominance of heterosexual values and the fact that the meanings and practices of non-hetero(sexuality) as well as hetero(sexuality) are shaped by the dominant heterosexual script. For example, "real sex" is generally conceived of as coitus or penile-vaginal intercourse. This "coital imperative," as Nicola Gavey, Kathryn McPhillips, and Marion Doherty call it in their reading, limits the control individuals have in determining what counts as sexual activity. The coital imperative frames women's sexuality as passive/receptive and men's as active/penetrative. That is, this imperative defines sex as something men "do" to women. Gavey and her coauthors also analyze the negative impact that privileging of men's sexual needs above women's has on the capacity for men and women to have truly reciprocal, safe sexual relations.

Interestingly, the Western obsession with the homosexual/heterosexual distinction is relatively new. Created in the nineteenth century, it became a mechanism by which masculinity and femininity could be further polarized and policed (Connell, 1999). Gay masculinity and lesbian femininity came to be defined as abnormal and threatening to the "natural gender order." Consequently, for many Westerners, the fear of being thought of as homosexual has a powerful impact on presentation of self. Men and boys routinely police each other's behavior and mete out punishment for any suggestion of "effeminacy" (Connell, 1999), while women and girls who engage in masculinized activities such as military service and elite sports (e.g., see article on women in pro basketball by Berlage in Chapter 7) risk being labeled lesbians.

Contemporary Western gender and sexuality beliefs have spawned stereotypes of lesbians as "manly" women and gay men as "effeminate." The reality is otherwise. For example, research shows that many gay men and lesbians are gender conformists in their expression of sexuality (Kimmel, 2000). Gay men's sexual behavior patterns tend to be masculine—oriented toward pleasure, orgasm, experimentation, and many partners—while lesbian sexuality is quite womanly in a Western sense, emphasizing sexual intimacy within romantic relationships. However, this account of non-heterosexuality is incomplete. Recent research suggests that compared to heterosexuals, non-heterosexuals have more opportunities to reflect upon and experiment with ways of being sexual. Although the meanings attached to sexuality in society-at-large shape their erotic encounters, they are also freer to challenge the dominant sexual scripts (Weeks, et al., 2001).

The imposition of gender difference and inequality on sexuality in societies such as the United States is also reflected in the sexual double standard. The double standard, which emphasizes and normalizes male pleasure and female restraint, is a widespread product and practice of patriarchy. Although Western sexual attitudes and behaviors have moved away from a strict double standard, the sexual lives of women remain more constrained than those of men. For example, girls and women are still under the control of the good girl/bad girl dichotomy (see the article by Tannenbaum in Chapter 4), a cultural distinction that serves to pit women against each other and to produce sexual relations between women and men that can be confusing and dissatisfying. Men grow up with expectations to embrace a sexual script by which they gain status from sexual experience (Kimmel, 2000). Women grow up with expectations to believe that to be sexually active is to compromise their value; but at the same time, to be flirtatious and sexy as well. Is this confusing or what? Imagine the relationship misunderstandings and disappointments that can emerge out of the meeting of these "opposite sex," sexual scripts.

The prisms of difference and inequality alter the experience and expression of gendered sexuality in many ways. Let's work with two prisms, age and ability/disability, to illustrate this point. In the United States age interacts with gendered patterns of sexuality in a complex fashion. Aging poses special problems for men and women who live in a culture in which sexual attractiveness equals youth. However, the problems of aging and sexual attractiveness are most compelling for women. Aging challenges cultural standards of femininity far more than standards of masculinity. With men, gray hair, wrinkles, and an expanding waistline do not diminish their value. In fact, aging may add value to people's perceptions of men. With women, however, gray hair, wrinkles, and an expanding waistline typically undermine their erotic worth.

The prism of ability/disability interacts with gendered sexuality in especially powerful ways. Pamela Block examines cultural assumptions about the sexuality and fertility of women with cognitive disabilities in her reading. She argues that disability remains a stigma in the United States, especially mental disability. The sexual stereotype of men with cognitive disabilities casts them as sexual predators who have to be carefully guarded and monitored, while controlling images of women with such disabilities depict them as sexually deviant (e.g., promiscuous) or perpetually childlike. Stereotyped as incapable of rational sexual decision-making, women with cognitive disabilities have been institutionalized and sterilized without their consent and routinely sexually disempowered.

Gendered Emotions Masculinity and femininity are emotionally opposed in Western culture (Bendelow and Williams, 1998). This opposition expresses itself in both obvious and subtle ways. The obvious is this: the emotions that can be expressed and the degree and forms of their expression are tied to gender. Boys and men must appear "hard" by hiding or shutting down feelings of vulnerability, such as fear, while girls and women are encouraged to be "soft"; that is, emotionally in touch, vulnerable, and expressive. Jennifer Lois's analysis of the emotional culture of a volunteer search-and-rescue team shows how the cultural definition of women and men as emotionally opposed plays itself out. Although the men and women in Lois's study were effective in both low- and high-risk crisis situations, women were more likely to be perceived as emotional deviants and inferiors compared to the men.

Consider the impact of learning and enacting different, even oppositional, emotional scripts on intimate relationships. If men are not supposed to be vulnerable, then how can they forge satisfying affectionate bonds with women and other men? Additionally, men who embrace the stereotype of hard masculinity may pay a price in well-being by concealing their own pain, either physical or psychological (Real, 2001). Also, it is important to recognize the negative consequences of gendered emotionality for girls and women. Although girls and women receive cultural encouragement to be "in touch with" themselves and others emotionally, this has strong associations with weakness and irrationality. When women express emotion according to cultural rules, they run the risk of being labeled hypersensitive, moody, temperamental, and irrational. The stereotype of the emotionally erratic and unstable woman has been widely used in efforts to undermine the advancement of women in politics, higher education, the professions, business, and other realms of public life. You know how it goes: we can't risk having a moody, irritable, irrational woman at the helm.

How do the prisms of difference and inequality interact with gendered emotions? Looking through a number of prisms simultaneously, we can see that the dominant definition of hard, stiff-upper-lip masculinity is white, heterosexual, and European (Seidler, 1998). The dominant definition of masculinity assigns rationality and reason to privileged adult men. All other people, including women, minorities, and children, are assumed to be more susceptible to influence by their bodies and emotions, and, as a consequence, less capable of mature, reasoned decision-making (Seidler, 1998).

We'd like to conclude this chapter by asking you to think about individual and collective strategies to reject conformity to patterns of gendered body work, sexual expression, and emotionality that demean, disempower, and prove dangerous to women and men. What would body work, emotional life, and sexual desire and experience be like if they were not embedded in and shaped by structures of inequality? How would personal growth, self-expression, and communication with others change if we were not under the sway of compulsory attractiveness, compulsory heterosexuality, the sexual double standard, and gendered emotional requirements? How would your life change? What can we do, as individuals and in concert, to resist and reject the pressure to bring our bodies, sexual experience, and emotional life into line with oppressive and dangerous ideals and norms?

REFERENCES

Bendelow, Gillian and Simon J. Williams. (1998). "Introduction: Emotions in social life." In *Emotions in social life,* G. Bendelow and S. Bendelow (Eds.), pp. xv-xxx. London: Routledge.

Connell, R. W. (1999). "Making gendered people." In *Revisioning gender,* M. Feree, J. Lorber, B. Hess (Eds.), pp. 449–471 Thousand Oaks, CA: Sage.

Dowling, Collette. (2000). *The frailty myth.* New York: Random House.

Gieske, Sabine. (2000). "The ideal couple: A question of size?" In *Feminism and the body,* L. Schiebinger (Ed.), pp. 375–394. New York: Oxford University Press.

Gillespie, Marcia Ann. (1998). "Mirror mirror." In *The politics of women's bodies.* R. Weitz (Ed.), pp. 184–188. New York: Oxford University Press.

Herdt, Gilbert. (1997). *Same sex, different cultures.* Boulder, CO: Westview.

Kaw, Eugenia. (1998). "Medicalization of racial features: Asian-American women and cosmetic surgery." In *The politics of women's bodies,* R. Weitz (Ed.), pp. 167–183. New York: Oxford University Press.

Kimmel, Michael. (2000). *The gendered society,* New York: Oxford.

Real, Terrence. (2001). "Men's hidden depression." In *Men and masculinity,* T. Cohen (Ed.), pp. 361–368. Belmont, CA: Wadsworth.

Schiebinger, Londa. (2000). "Introduction." In *Feminism and the body,* L. Schiebinger (Ed.), pp. 1–21. New York: Oxford.

Seidler, Victor Jeleniewski. (1998). "Masculinity, violence and emotional life." In *Emotions in social life,* G. Bendelow and S. Williams (Eds.), pp. 193–210. London: Routledge.

Weeks, Jeffrey, Brian Heaphy and Catherine Donovan. (2001). *Same sex intimacies: Families of choice and other life experiments.* New York: Routledge.

Weitz, Rose. (1998). "A history of women's bodies." In *The politics of women's bodies.* R. Weitz (Ed.), pp. 3–11. New York: Oxford University Press.

28

Men Are Real, Women Are "Made Up"[1]

Beauty Therapy and the Construction of Femininity

BY PAULA BLACK AND URSULA SHARMA

Pamela Black and Ursula Sharma analyze the relationship of beauty salons to the modern Western ideal of femininity. They based their study on interviews with a small number of women who teach beauty therapy and women who own or manage beauty salons. They also conducted observations in a college of beauty therapy and received beauty treatments themselves. In particular, Black and Sharma explore the tension between individual desire for a specific appearance and the generalized ideal of normative femininity. One of the key questions they pose is "How is the normative ideal of feminine beauty delivered to real women in a real situation?"

1. According to the authors, what are the positive and negative benefits of beauty therapy to women?
2. What is the relationship between the social class of beauty salon clientele and the ambience and services of salons and therapists?
3. How do beauty therapists define the services they provide?

All societies require work to be done to the body to change it from the "natural" to one that is specifically cultural (Falk, 1995). However, the ways in which feminine and masculine bodies become culturally acceptable differs greatly. In a late 20th century European and North American context, what might be called the ideal of femininity and the accompanying beauty industry are systems that regulate and commodify the bodies of women to an extent and in a form not experienced by men. In this context, male bodies require a very different form of maintenance in order to conform to hegemonic masculinity—men are "real" without this work. This is not to deny that in the acquisition of varying forms of masculinity intensive work is done to the body; body building is an example. Young men too are under increasing pressure to obtain and maintain a specific "look". However, on a routine day to day level men are not required to paint, moisturise, deodorise and de-hair their bodies in order to appear masculine. These activities, however, form part of the day to day routines of femininity (Holland *et al.* 1994). In this sense then femininity is a state to be constantly sought.

The beauty industry fuels this acquisition of femininity and even for those women who do not visit beauty parlours themselves, the beauty system is all pervasive. This does not necessarily mean that all women will equally achieve this ideal, or that all women will strive to attain it, but rather than as a feature of the everyday lives of women, femininity, and the discipline of the unruly body, form an inescapable backdrop (Holland *et al.*, 1998; Weekes, 1997; Skeggs. 1997).

In this article it is our aim to outline the work of beauty salons, and to investigate in some depth the claims surrounding what goes on there. The salon has been chosen as a site *par excellence*. where this attainment of femininity, and its definition and negotiation are being fought out. In its generally closed and intimate nature, the beauty salon is not only a feminised space, but also one in which the secret routines of femininity are commodified and exemplified.

* * *

THE NATURE AND SCALE OF THE BEAUTY INDUSTRY

A beauty salon has its own ambience. The uniforms of the staff, the decor, the layout reflect the aspirations of the owner. Some salons give the immediate impression of a clinic where staff dress in white, and where formality is emphasised. In other salons, the staff are required to dress in more flamboyant colours, often matched by the decor and the welcome received by the client. Walls are decorated with the qualification certificates of staff and membership certificates of professional associations. Salons generally contain a waiting area with comfortable seating and assorted magazines; a desk and a till close to the door where the client is taken after their treatment; a private area for staff; screened cubicles where treatments are carried out; and if the salon offers nail treatments, there is also a more communal treatment area where manicures are performed. The salon has its own routines and tracks along which staff walk in greeting clients, guiding them to treatment rooms, offering refreshments, and finally leading to the point where payment is made. Salons too have their own smell which is that of the equipment and chemicals used for treatments intermingled with the pleasant aromas of perfumed creams and lotions, cups of tea and coffee, and sometimes too the strong smell of nail products. In this atmosphere the intimate routines of body maintenance are carried out.

Beauty therapy is part of a vast multi-national industry. The value of the professional beauty industry in the UK in 1998 was £366 million, which represented a growth of almost 6% on the previous year. This figure includes beauty therapy treatments in a variety of sites including mobile, hair and beauty salons, health clubs as well as the conventional beauty salon. The growth in the customer base stood at over 13%. This means that 13% more people, the vast majority women, visited salons in 1998 compared to 1997 (The Beauty Industry Survey, Guild News, 1999).[2] . . .

The expansion of the beauty industry has accompanied the expansion of the leisure industry more generally and this was acknowledged by the therapists we spoke to:

> I think years ago there was almost not a sort of business ethic behind beauty therapy. It was a kind of luxurious service. And I think it was a lot of small salons working on quite outmoded and outdated lines and were seen very much as a wealthy woman's option. I think it has changed quite radically. . . . Now it is a lot slicker, more business oriented. Salons seem to plan a little better, analyse what they are doing to maximise their market. The bigger companies are a big influence. I think it's the growth of the leisure industry that has dragged us forward because, you know, the hotels are now big business, and gyms in hotels. So now we have a lot of salons in hotels alongside the leisure complex. . . . And I think it is far more accessible to people. With the advent of the fitness regime and people going to gyms and being more concerned and that, they are more exposed to beauty therapy and they are thinking well why not? I have worked out in the gym, why not treat myself to a massage? Why not have a facial?

However. it is important to note that one woman's leisure is another woman's work. The beauty industry itself, whilst providing the site for carefully packaged and segmented parcels of free time, is also the site of work involving physical labour, emotional work, long hours, low pay, and often poor work conditions for those employed within it.[3]

WHAT IS BEAUTY AND DO WOMEN WANT IT?

Perhaps it would come as no surprise if women did strive for beauty. Beauty is routinely associated with morality, sociability, kindness and other positive characteristics. . . . Of all of those who are subject to this valuing of beauty though, women are the group most routinely and consistently judged according to aesthetic ideals. Not all women wish to or are able to approximate to an idealise standard of beauty but all women are evaluated against this yard stick. Black women. for example, may reside outside of the idealised notions of the fair skinned beauty. In defining Black as not only a personal but a political identity, other aspects of appearance appear desirable and beautiful. For example, within the British context, Weekes (1997) describes how ambiguity arises from an internalisation of white standards of beauty by both Black men and women. In struggling to challenge this standard, a different Black standard of beauty is established, which is often itself essentialised. White people themselves internalise idealised standards of beauty. As the category of white is both internally differentiated and hierarchical, some groups of people may never feel that they are "white enough" (Dyer, 1997). In our study all of the interviewees except one were white, and the clientele of the salons visited whilst we were present was overwhelmingly white. Though we have analysed in detail the construction of femininity within beauty therapy, we have as yet paid less conceptual attention to the construction of whiteness, or other "racialised" categories. This is an avenue we intend to explore in future research.

Beauty may also be experienced in other contextually specific ways, for example, the symbols of beauty vary according to age and also class (Gimlin, 1991). It should also be noted that beauty is never a pure category, and femininity itself is always related to from a variety of subject positions. For example, Skeggs has shown in her study of white working class women that femininity was not an identity which the women unambiguously identified with (Skeggs, 1997). This ambivalence towards the trappings of femininity has been highlighted in our study by the refusal of therapists to acknowledge that their work is about beauty or the production of a highly feminised appearance.

The salon is also an explicitly heterosexual world. This does not mean that the clients and the therapists are all heterosexual, but rather that the body work performed there is set against the backdrop of an overt (female) heterosexuality. Lesbianism was not mentioned, and in the view of the therapists, the male visitors to beauty salons were likely to be gay, or "stressed" executives.

It seems then that discussion of beauty *per se* is inadequate. This inadequacy in terms of theorising is matched by the therapists' own mistrust of the term. There was a noted absence of any discourse of beauty in their interviews or wider discussions with us. Only in one sense did beauty arise and this was in the criticisms of the term in the title of their profession. This addition to "therapy" was seen to devalue the therapy and treatment side of their work, and to somehow trivialise their role. Beauty in this way contributed to the image of the beauty therapist as "bimbo" and was a much resented term. The following interviewee summarises the view she believes the general public holds of the profession:

> Beauty therapy is for the girls who could not hack it at school, they failed, they're stupid, all they do is paint finger nails and file and chew gum and that's it.

This kind of belief reflects the common-sense understanding of the "beauty/brains" split against which women are evaluated. In being associated with the beauty side of this dichotomy, beauty therapists felt that they were constantly battling to prove their intellectual and professional capabilities.

The treatments offered to clients in our study were carried out on a highly individualised basis, and the assessment of the best treatment as well as the most appropriate method of administering that treatment depended on the skills of the therapist. In fact this formed an important part of their training and professional rhetoric. Apart from a generalised sense of helping the client to look and feel good there was little evidence of a standardised form of beauty drawn upon in the work of the beauty salon. In fact, as the quote below illustrates. treatments provided without reference to the particular characteristics of the client were seen as inappropriate:

> Make-up is so subjective and what you think would look good on somebody is not their perception of what looks good on themselves. Students start out by putting their own make-up on everybody, which a sixty year old lady will not always suit! . . . They won't say "I hate it", but you know that they go into the toilet before they go home and take off all that the student has put on, because its a young person's make-up that has gone on to someone who really shouldn't be wearing so much make-up.

The professional standards of the industry and the working practices of each therapist were viewed as highly important but the ends towards which these carefully practised means were progressing were less carefully reflected upon. This does not mean that there are not generalised beliefs about the culturally acceptable forms of femininity, but rather that within this ideal there is no one way to be "beautiful", or even a desire to become so. In fact ideals of beauty were sometimes mentioned as specifically what the client was *not* aiming for:

> Not everyone wants to walk into a salon and see a blonde "bimbo", you know, with legs up to their armpits to make them feel intimidated. I mean people are ordinary. Ordinary people come through here. Elderly people come through here and they're pleased to see me because I'm past it (laughs).

Beauty therapists do not simply offer cosmetic services but see themselves as having some overlap with the medical profession. Their work on the body certainly strays in to the territory of both the medical profession and complementary therapies. Salons are not simply about nails and make-up!

> I would say that it does offer a very high therapeutic angle to it, it boosts people's confidence and self esteem. Some people, I think, misinterpret what vanity is. If somebody said years ago "I am going to a beauty therapist's", you instantly thought beauty and vanity, the two go together. But these days there's more to salons, there's more emphasis on stress related problems and that, massage is good to release stress and these other therapies have come forward, reflexology, aromatherapy, shiatsu.

The beauty ideal then must be widened and refined if it is to make any sense in the context of the beauty salon. The therapists themselves openly rejected the idea that they were producing a look for women as if on some sort of formalised conveyor belt. However, there is a problem here with why and how these individually experienced desires arise. An individual is never simply an individual. Selling individuality is a very different matter to demanding a product on the basis of purely individual desires.

IF NOT "BEAUTY" THEN WHAT DO BEAUTY THERAPISTS *DO*?

The beauty therapists then did not discuss beauty directly, instead their work was divided into "pampering", "treating", and "grooming". Pampering treatments are not seen as a necessity, but rather an indulgence for which the recipient often feels guilty. Pampering is associated with relaxation and implies that the need to relax arises from some stressful situation that the woman is escaping. Stress most often relates to work or family commitments. The cost of the beauty treatment, especially if seen as pampering rather than a necessity adds to the guilt associated with it, particularly if this is taken from family funds. One of the salon owners in our study had learnt to deal with this issue in her work:

> I mean we still get ladies in who say "oh I can't afford that". My answer is well, I ask them a question, "has he [husband] got a football season ticket", and if she says yes, "well spend an equal money on your face", and then they see it in a different light because the man is a bit old fashioned, you know, all that money on your face, you can't see what's happened".

The economic implications of "pampering" here are put in to perspective in terms of comparing the woman's time to herself and the use of household money to facilitate this, with the man's involvement in leisure activities. This struggle for women to create a space and a legitimacy for their own leisure activities has been well documented (Deem, 1986).

We were constantly told in the interviews that clients desire to look "normal", particularly in the case of treatments such as those for the removal of facial hair or acne, or that they simply want to "make the best of themselves". It is here that the work of the therapist becomes "treatment". The removal of facial hair is one area where therapies received are seen as being essential treatments for a legitimate, almost medicalised, problem:

> When you do a consultation there's two things a woman will always say. She will always say—jokingly—"I thought I was turning into a man". But it is a fundamental worry that they're losing their femininity . . . So they are convinced that they are on their own and they are convinced that they are no longer feminine. It's a awful thing for a woman to feel like that, very, very damaging. . . .

This is clearly an area where the facial characteristic of hair is seen as a signifier of masculinity, and in turn is a "terrible thing" to happen to a woman. These tiny hairs can question her social identity to such an extent that her appearance is kept under surveillance and regulated, sometimes with help from the beauty therapist, but always as a shameful secret from everyone else. This struggle for femininity questions essentialised notions of being born a woman, with a biologically female body. The artfulness of such an achievement illustrates the highly constructed nature of this supposed "natural" state.

Particularly where clients were busy working women, then the need to look adequately groomed was a further justificatory strategy for visiting the salon:

> Beauty therapist 1: A lot of women work more now so they feel they can take time for themselves, whereas before a lot of women just used to be housewives.
>
> Beauty therapist 2: Yeah, they hadn't got their own money.
>
> Interviewer: So a lot of clients you get here actually do work then?
>
> Beauty therapist 2: Yeah, yeah, this is why we've got a bit of a gap and then all of a sudden they're haring in for a quick half leg [wax] and an eye brow trim and dash back to work.

The work of beauty therapists then only obliquely, if at all, refers to the concept of beauty. The therapist contributes to the leisure time of the client: to the maintenance of an acceptable, feminine, but not overtly sexualized, appearance for the world of employment: and to the achievement of a narrowly defined standard of" normality". What this fragmentation of discourses around the work of the salon suggests is that femininity is produced in relation to several different external social institutions: or perhaps that different women in the salon invest in their femininity in different arenas. This multiplicity of roles has also created space for men to visit salons. We have anecdotal evidence of an increase in men's use of beauty salons in the UK. The type of treatments they were receiving appeared to fall in to the category of stress relief (eg. body treatments, massages etc.), or the area of grooming.

"ORDINARY PEOPLE COME THROUGH HERE"

The expansion of the leisure industry and the seemingly unhindered spread of rampant consumerism are not in themselves enough to explain the growth in the beauty industry. In order to do this we must look more widely at the role of the body in this consumer society, and more fundamentally at the role of women's bodies. In addition, by listening to beauty therapists themselves we can begin to evaluate some of the claims made in academic work.

Turner (1996) argues that patriarchy relied upon a comprehensive system of institutionalised discrimination against women in law, religion, employment, politics and so on. In Western democratic societies this systematic set of structures no longer explicitly discriminates against women and instead these societies may be characterised by what he terms "patrism". As a result of the shrinkage in institutionalised power of patriarchy, reaction shifts to other systems whereby male power is maintained. Could it be the case then that the "beauty myth" is a newly evolved system of oppression which has taken over where discrimination these other left off (Wolf, 1990)? Despite seeing the beauty system as all pervasive and damaging for women, Wolf does acknowledge the fact that often women desire to conform to this ideal. In this view then the role of the beauty industry is negative, even to the point of convincing women to become complicit in their own "torture". One of the therapists in our study commented on this pressure and her role within it:

> I wouldn't go out of the house without make-up. Which has got to be wrong. Image is very nice but it's not the be all and end all. It's just a vanity business isn't it?

The external pressures to conform to a certain ideal image have also been linked to the development of advertising and women's magazines which target women as consumers. Women are encouraged to create an *individual* look through consumption of *mass produced* products. The paradox in this situation is overcome through the woman's own labour to create her body or her home. Beauty is a tool which is used in the service of this push to consume (Lury, 1997; Winship, 1987). These pressures were to some extent acknowledged by the therapists themselves:

> Yes, there are always these magazines with gorgeous women staring back at you. You think that's the way men perceive you as, and that's the way you should look yourself.

The picture from the interviews, however, is not so straightforward. We would argue that the roots of this complicity, and the desire for a particular body image can be more usefully understood by linking the issues arising from our interviews to wider sociological themes.

One of the tensions within beauty therapy is that between the individualised treatment experience and the existence of a generalised feminine or beauty ideal. Individualisation and consumption go hand in hand. It has been claimed that the radical individualisation of late modernism has weakened communal ties and identifications, and that primary relationships now exist between individuals and the market. This has been documented in the case of leisure activities and lifestyle and even to the extent of intimate and love relations being restructured according to the demands of the market (Featherstone, 1982; Beck and Beck-Gernsheim, 1995). Whereas the focus for women in the past has been to achieve a certain look in order to gain love and win a husband, from the 1980s onwards love disappeared from the beauty books to be replaced by the goal of self-discovery and an individualised search for fulfilment (MacCannell and

MacCannell, 1987). In the beauty salon, the woman can perhaps convince herself that the product she is consuming actually is unique and tailored to her own very specific needs. At the same time beauty therapy as a social activity is important.

Studies of hairdressing salons and beauty parlours in the USA have found that a shared experience is central to the visit. Furman (1997) describes how for older Jewish women the visit to the salon is very much an opportunity to share details of everyday life events with friends and peers. Jacobs-Huey (1998) describes a similar process in a hairdressing salon with predominantly African-American clients. In both of these cases the layout of the salons and the type of treatments delivered facilitated a sharing of experiences between the women in a general open space. In the British context, at least in the salons we have visited, this is not the case. The client is led in to a private cubicle and interaction is facilitated only between the client and the therapist on a one to one basis. Privacy and individual attention to the client are part of the professional ethics of the beauty therapist. The role of therapist here is as unofficial counsellor rather than participant in a general conversation. Beauty therapists must carry out both body and emotion work. . . .

This counselling role was an ambiguous experience for the therapists. Though they were aware of the benefits to the client, and of the fact that this was one of the key motivating reasons for women to return for treatment, in terms of their own mental and emotional health, this role was a difficult and sometimes exhausting experience. At the same time they also derived immense personal satisfaction from having helped and listened to clients (Sharma and Black, 1999). However, in receiving manicures, and in the waiting areas, women do meet and spend some time sharing the social rituals mentioned in the studies from the USA. We would suggest that this may be one of the reasons for the growth in the popularity of nail treatments as women may attend salons partly in order to partake in this more social activity.

In another sense too, the woman enters the salon aiming to partake of thoroughly social activities. Not only the therapists but the clients themselves are fully aware that treatments must be *appropriate* to the category of woman that is being dealt with. This appropriateness is expressed implicitly in the therapists' discussion of age, or of tailoring treatments to clients' own desires. Women's knowledge of femininity and their own position in it is a collective achievement which is constructed not only through sources such as the media and the advertising industry, but also experienced through the milieu of friends, family, employment, partners and so on. Skeggs (1997) argues that women make differing levels of investment in femininity. In terms of the working class women in her study, relationship to femininity was ambiguous. Investment in its traditional trappings was higher when young, and decreased with age as time and money were invested elsewhere. The rituals of "dressing up" or shopping were social occasions, and the final appearance of the woman a combination of many factors, including her own social positioning, and those surrounding her. This is a regulated femininity, not wholly embraced by women themselves. The women here are working class and white. Class, "race" other than whiteness and sexuality will produce different relationships to femininity, and the women involved will be judged very differently. The individual does not enter the salon simply as an individual. Her decision to be there, her request for treatments, the desired outcome, the amount of money available to spend and a host of other factors enter the salon door with her.

Another recurring theme in our study has been the way in which therapists divide treatments into a series of dichotomies which we have described as, for example; aesthetics treatment; looking/feeling; and holism/bits. Although there is an overlap in the sense that treatments have effects for both appearance and feeling, generally this distinction holds:

> You're a bit like Jeckyll and Hyde when you're in the salon. You know, when you do somebody a make-up, talk about completely different things and you act in a completely

> different way to when you re giving them a massage or a facial. When you're doing their nails it's different again. . . . When you go out into industry you learn how to change conversation, how to change, erm, your persona.

One of the consequences of this distinction is the fact that one area lends more legitimacy to the work of beauty therapists than the other. It is the focus upon beauty, appearance and the face which has reinforced the image of therapists as "bimbos". Many of the therapists were keen to stress the level of treatment carried out in their salons, and the overlap with the medical profession.

This split also means that perhaps beauty can be separated out from treatment in any analysis of the role of beauty therapy in the lives of women. Both looking treatments (eg. facials, manicures, leg waxing) and feeling treatments (eg. reflexology, massage) can be evaluated in terms of their function in the production of a feminine body ideal. Body treatments which aim to relax the client and are focused upon dealing with stress could perhaps be compared to the executive game of squash which receives little negative comment from either work-mates, professionals or academics. Leaving aside the question of whether the reduction of stress simply perpetuates the stressful situation which the woman originally finds herself in, and the issue of whether beauty therapists are qualified to deal with "stress" in their pseudo-medical way, there has been less focus upon this area of the beauty industry in feminist critiques. In contrast, it is the beauty side of the work of salons which has attracted most negative comment.

Though tentative at this stage there is also some evidence in our interviews for clients to prefer different types of treatments according to factors such as social class. It appears that different salons and therapists will cater to different customer groups, guided by an implicit view of class:

> Maybe the smaller unregistered type salons that really couldn't give a damn, maybe it fulfils a place in the market place for those that don't want to pay more for better qualified, well turned out therapists who know what they're talking about. Maybe some people just want like the hairdressers that are on the corner of the street "oh I go there because it's handy", you know. OK your perm looks a total pig's ear and you didn't come out with the right colour on your hair, but it hasn't cost them much and it'll soon grow out. Perhaps it's the same sort of thing in beauty.

In general, salons which focus upon beauty treatments service a less middle class clientele. These salons do not strive to give the feel of a clinic and find difficulty in recruiting clients to holistic treatments. In contrast, those salons and therapists which cater to a more affluent and middle class market emphasise their therapeutic competencies and will emphasise body treatments at the expense of the more "bread and butter" techniques such as waxing and eyebrow shaping. This differing relationship to the body seems to confirm Bourdieu's claim that the working class have an instrumental relationship to their bodies, and that the middle class approach is characterised by seeing the body as an end in itself (Bourdieu, 1984). This differentiation in the types of treatments selected then does not appear to show that only a particular class of women visit salons, but that once in the salon the "habitus" of the woman leads to different activities.

DISCUSSION AND CONCLUDING COMMENTS: (JUST BECAUSE IT FEELS GOOD, DOESN'T MAKE IT RIGHT?)

One of the key issues we have had to face during this research has been the extent to which our natural sympathy with the subjects of research has had to be tempered with our reservations concerning the role of beauty in the oppression of women. As qualitative researchers we are aware of the professional discourse drawn upon by the beauty professionals in their conversations with us. The therapists were happy to talk to us and in many interviews we were the audience for their personal gripes and opinions concerning the industry. However critical of developments within the industry some of the interviewees were, none actually questioned the need for beauty therapy *per se*. They explained themselves and their role in terms of an occupational rhetoric (Fine, 1996). Part of the information we have gained during this research is undoubtedly a product of a professional discourse of beauty therapy. Though the testimonies undoubtedly carried an element of this rehearsed and professionalised rhetoric, the interviews and other conversations with therapists was clearly not simply limited to this level.

Despite this reservation. whilst conducting our research, we have been at times seduced by the professional rhetoric of the therapists, and also by the physically pleasurable sensations of beauty and body treatments. It would seem heavy handed and hypocritical to deny these sensations to other women, or to claim that they are suffering from some kind of false consciousness. At the same time, however, we have had to question whether the subjective experience of empowerment (rejuvenation, relaxation, time for oneself) is in fact what we could comfortably accept as empowerment (Holland *et al.,* 1998, 9).

A similar point is made by Davis in her exploration of cosmetic surgery (Davis, 1995; 1991). In her aim of moving away from women as "cultural dopes" at the mercy of the medical profession and commercialised discourses on beauty, she takes into account the narratives and experiences of women themselves. This presents her with a similar problem to ours, how to take the testimony of women seriously whilst remaining critical of the system into which they are buying. What she eventually concludes is that women are often aware of the contradictions between the oppressive nature of the beauty system, whilst at the same time trying to operate within that system to gain some advantage for themselves. This means that women are not ignorant cultural dopes, but rather knowledgeable and adept cultural actors.

Whilst we would agree that this is the case, it is important to continually bear in mind the commercial motivation of the beauty industry. As much as individual therapists are concerned with offering their clients the most effective and pleasurable treatments possible, underpinning all relationships between the (usually) women involved is the drive for profit. This is where beauty therapy as an industry differs from the social rituals surrounding the giving and receiving of what could be described as beauty treatments between friends or family.

As with the tension between generalised standards and individualised practices mentioned above, the question of the individual must be returned to here. The therapists argued that the service they were providing offered a chance for women to enjoy a female space and to take time for themselves away from the masculine world, as well as from the demands of work and family life . . .

This type of experience no doubt allows the woman to relax and to rejuvenate herself. The opportunity to relax in a female space and the chance to share problems with a sympathetic female listener are vital benefits of beauty treatments for the clients. These benefits are over and above the confidence drawn from the visual effects of treatments, the satisfaction gained from obtaining the desired "look", or from the pleasurable sensations derived from body treatments. For many of the women in the salons it was their own disposable income that they were choosing to spend in a way that gave them pleasure, and on treatments that focused upon their individual looks, bodies and emotional requirements. In this way beauty therapy can he seen as of positive benefit to women.

However, the discussion can not stop here. It seems logical that if the female body is something which must be worked on in order to produce a culturally recognisable product, then an industry will exist which serves this purpose, and in turn commodifies at least some of these necessary processes. However, this does not explain how this system works, or throw much light upon the different desires of the women involved in it. The tension between the individual desire for a specific look and a generalised view of what is acceptable is explained through the different positioning of the women who are clients. The generalised ideal is that of a normative femininity, which, as in the case of hegemonic masculinity, acts as an Ideal Type rather than an achievable way of being in the world. However, not all women have a similar relationship to this normative definition. In relation to body practices social class, "race", sexuality and age are vital factors which position women in very different relationships to any concept of "beauty" or "femininity". By choosing to spend leisure time and disposable income in the beauty salon, can we see women as asserting a right to their own space and independence, or as simply investing in femininity, in some cases to "put a floor under" the likely devaluation of their skills in other areas as Skeggs suggests (Skeggs, 1997)?

What remains interesting for us is this investigation of the delivery of a hegemonic or normative concept in a "real" situation to "real" people. How is "beauty" or "femininity" actually taken up (or not) by the women involved in its construction and maintenance? The issues we have raised here illustrate that beauty therapy provides an arena where conflicting discourses and practices concerning the body and femininity intersect. We have highlighted the fact that beauty therapy is an ambiguous industry providing as it does both positive and negative effects for individual women and the regulation of femininity more generally. It seems that in general women are not striving for beauty but rather desire to regulate their bodies in order to appear within the bounds of "normality". They also desire pleasure and relaxation through pampering and time for themselves in a predominantly female space. These are issues which undoubtedly require more work to be disentangled fully, and we are looking forward to examining such questions in future research.

NOTES

1. This phrase is taken from MacCannell and MacCannell. 1987: 212.
2. In this paper we refer to beauty therapy as a specific aspect of the wider beauty industry. We have studied beauty therapy as a profession and as a cultural practice. The beauty industry includes other arenas which we have paid less attention to, including fashion, cosmetics, advertising, dieting etc. In a similar way we refer to "beauty" as a specific aspect of the wider term of femininity. Beauty is a culturally constructed ideal. Femininity relates to the practices, identities and representations of what it means to be a "woman" in any given society.
3. On an international level it becomes clear that this interest in beauty, and its corollary of health and fitness, has grown in the past twenty years. From 1981 to 1990 there was a 70% increase in the number of cosmetic surgery operations performed in the USA, with around one and a quarter million reconstructive procedures being carried out in 1990 (Synnott, 1993: 75). At the same time measured levels of satisfaction that individuals feel with their bodies and their looks have decreased (Synnott, 1993).

REFERENCES

Beauty Industry Survey (1999). *The Guild News, Journal of the Guild of Professional Beauty Therapists,* Jan/Feb: 19–25.

Beck, U, and Beck-Gernsheim, E., (1995). *The Normal Chaos of Love,* Cambridge: Polity.

Bordo, S., (1993). *Unbearable Weight: Feminism, Western Culture and the Body,* Berkeley: University of California Press.

Bourdieu, P., (1999) [1984]. *Distinction: A Social Critique of the Judgement of Taste,* London: Routledge.

Butler, J., (1993) *Bodies That Matter: On the Discursive Limits of "Sex",* London: Routledge.

Chapkis, W., (1986). *Beauty secrets: Women and the Politics of Appearance,* Boston: South End Press.

Davis, K., (1995). *Reshaping the Female Body: The Dilemma of Cosmetic Surgery,* London: Routledge.

Davis, K., (1991). "Remaking the She-Devil: A Critical Look at Feminist Approaches to Beauty". *Hypatia,* 6(2): 21–43.

Deem, R., (1986). *All Work and No Play: The Sociology of Women and Leisure,* Milton Keynes: Open University Press.

Dyer, R., (1997). *White,* London: Routledge.

Faulk, P., (1995). "Written in the Flesh", *Body and Society,* 1(1): 95–105.

Featherstone, M., (1982). "The Body in Consumer Culture", *Theory, Culture and Society,* 1(2): 18–33.

Fine, G., (1996). "Justifying Work: Occupational Rhetorics in Restaurant Kitchens", *Administrative Science Quarterly,* 41: 90–112.

hooks, B., (1992). *Black Looks: Race and Representation,* Boston: South End Press.

Holland, J., Ramazanoglu, C., Sharpe, S. and Thomson, R., (1998). *The Male in the Head: Young People, Heterosexuality and Power,* London: Tufnell Press.

Holland, J., Ramazanoglu, C., Sharpe, S. and Thomson, R., (1994). "Power and Desire: the Embodiment of Female Sexuality", *Feminist Review,* 46: 21–38.

Lury, C., (1997). *Consumer Culture,* Cambridge: Polity.

MacCannell, D. and MacCannell, J.F., (1987). "The Beauty System", in Armstrong, N. and Tannenhouse, L., (eds). *The Ideology of Conduct: Essays on Literature and the History of Sexuality,* London: Methuen.

Sharma, U. and Black, P., (1999). *The Sociology of Pampering: Beauty Therapy as a Form of Work,* Centre for Social Research Working Paper. University of Derby.

Skeggs, B., (1997). *Formations of Class and Gender: Becoming Respectable,* London: Sage.

Synnott, A., (1993). *The Body Social: Symbolism, Self and Society,* London: Routledge.

Turner, B., (1996). *The Body and Society: Explorations in Social Theory* (2nd edition). London: Sage.

Weekes, D., (1997). "Shades of Blackness: Young Black Female Constructions of Beauty", in Mirza, H.S., (ed.). *Black British Feminism,* London: Routledge.

Winship. J., (1983). "Options—For the Way You Want to Live Now". *Theory, Culture and Society,* 1(3): 44–65.

Wolf, N., (1990). *The Beauty Myth,* London: Vintage.

29

Size 6

The Western Woman's Harem

BY FATEMA MERNISSI

Fatema Mernissi is a well-known Moroccan sociologist and Islamic scholar. She has done research in a number of Western nations, including the United States. This article is a chapter from her book entitled Scheherazade Goes West. *Mernissi challenges Westerners to think about the ways in which their feminine beauty images and practices can be as hurtful and humiliating to women as the enforced veiling of women in nations such as Iran and Saudi Arabia. In fact, she argues that the "size 6" ideal is a more violent restriction than the Muslim veil.*

1. How do definitions of feminine beauty differ in Morocco compared to the United States?
2. What does Mernissi mean when she states the Western man establishes male domination by manipulating time and light?
3. Read the final paragraph of this article carefully. Why does Mernissi end on this note?

It was during my unsuccessful attempt to buy a cotton skirt in an American department store that I was told my hips were too large to fit into a size 6. That distressing experience made me realize how the image of beauty in the West can hurt and humiliate a woman as much as the veil does when enforced by the state police in extremist nations such as Iran, Afghanistan, or Saudi Arabia. Yes, that day I stumbled onto one of the keys to the enigma of passive beauty in Western harem fantasies. The elegant saleslady in the American store looked at me without moving from her desk and said that she had no skirt my size. "In this whole big store, there is no skirt for me?" I said. "You are joking." I felt very suspicious and thought that she just might be too tired to help me. I could understand that. But then the saleswoman added a condescending judgment, which sounded to me like an Imam's *fatwa*. It left no room for discussion:

"You are too big!" she said.

"I am too big compared to what?" I asked, looking at her intently, because I realized that I was facing a critical cultural gap here.

"Compared to a size 6," came the saleslady's reply.

Her voice had a clear-cut edge to it that is typical of those who enforce religious laws. "Size 4 and 6 are the norm," she went on, encouraged by my bewildered look. "Deviant sizes such as the one you need can be bought in special stores."

That was the first time that I had ever heard such nonsense about my size. In the Moroccan streets, men's flattering comments regarding my particularly generous hips have for decades led me to believe that the entire planet shared their convictions. It is true that with advancing age, I have been hearing fewer and fewer flattering comments when walking in the medina, and sometimes the silence around me in the bazaars is deafening. But since my face has never met with the local beauty standards, and I have often had to defend myself against remarks such as *zirafa* (giraffe), because of my long neck, I learned long ago not to rely too much on the outside world for my sense of self-worth. In fact, paradoxically, as I discovered when I went to Rabat as a student, it was the self-reliance that I had developed to protect myself against "beauty blackmail" that made me attractive to others. My male fellow students could not believe that I did not give a damn about what they thought about my body. "You know, my dear," I would say in response to one of them, "all I need to survive is bread, olives, and sardines. That you think my neck is too long is your problem, not mine."

In any case, when it comes to beauty and compliments, nothing is too serious or definite in the medina, where everything can be negotiated. But things seemed to be different in that American department store. In fact, I have to confess that I lost my usual self-confidence in that New York environment. Not that I am always sure of myself, but I don't walk around the Moroccan streets or down the university corridors wondering what people are thinking about me. Of course, when I hear a compliment, my ego expands like a cheese soufflé, but on the whole, I don't expect to hear much from others. Some mornings, I feel ugly because I am sick or tired; others, I feel wonderful because it is sunny out or I have written a good paragraph. But suddenly, in that peaceful American store that I had entered so triumphantly, as a sovereign consumer ready to spend money, I felt savagely attacked. My hips, until then the sign of a relaxed and uninhibited maturity, were suddenly being condemned as a deformity.

"And who decides the norm?" I asked the saleslady, in an attempt to regain some self-confidence by challenging the established rules. I never let others evaluate me, if only because I remember my childhood too well. In ancient Fez, which valued round-faced plump adolescents, I was repeatedly told that I was too tall, too skinny, my cheekbones were too high, my eyes were too slanted. My mother often complained that I would never find a husband and urged me to study and learn all that I could, from storytelling to embroidery, in order to survive. But I often retorted that since "Allah had created me the way I am, how could he be so wrong, Mother?" That would silence the poor woman for a while, because if she contradicted me, she would be attacking God himself. And this tactic of glorifying my strange looks as a divine gift not only helped me to survive in my stuffy city, but also caused me to start believing the story myself. I became almost self-confident. I say almost, because I realized early on that self-confidence is not a tangible and stable thing like a silver bracelet that never changes over the years. Self-confidence is like a tiny fragile light, which goes off and on. You have to replenish it constantly.

"And who says that everyone must be a size 6?" I joked to the saleslady that day, deliberately neglecting to mention size 4, which is the size of my skinny twelve-year-old niece.

At that point, the saleslady suddenly gave me an anxious look. "The norm is everywhere, my dear," she said. "It's all over, in the magazines, on television, in the ads. You can't escape it. There is Calvin Klein, Ralph Lauren, Gianni Versace, Giorgio Armani, Mario Valentino, Salvatore Ferragamo, Christian Dior, Yves Saint-Laurent, Christian Lacroix, and Jean-Paul Gaultier. Big department stores go by the norm." She paused and then concluded, "If they sold size 14 or 16, which is probably what you need, they would go bankrupt."

She stopped for a minute and then stared at me, intrigued. "Where on earth do you come from? I am sorry I can't help you. Really, I am." And she looked it too. She seemed, all of a sudden, interested, and brushed off another woman who was seeking her attention with a cutting, "Get someone else to help you, I'm busy." Only then did I notice that she was probably my age, in her late fifties. But unlike me, she had the thin body of an adolescent girl. Her knee length, navy blue, Chanel dress had a white silk collar reminiscent of the subdued elegance of aristocratic French Catholic schoolgirls at the turn of the century. A pearl-studded belt emphasized the slimness of her waist. With her meticulously styled short hair and sophisticated makeup, she looked half my age at first glance.

"I come from a country where there is no size for women's clothes," I told her. "I buy my own material and the neighborhood seamstress or craftsman makes me the silk or leather skirt I want. They just take my measurements each time I see them. Neither the seamstress nor I know exactly what size my new skirt is. We discover it together in the making. No one cares about my size in Morocco as long as I pay taxes on time. Actually, I don't know what my size is, to tell you the truth."

The saleswoman laughed merrily and said that I should advertise my country as a paradise for stressed working women. "You mean you don't watch your weight?" she inquired, with a tinge of disbelief in her voice. And then, after a brief moment of silence, she added in a lower register, as if talking to herself: "Many women working in highly paid fashion-related jobs could lose their positions if they didn't keep to a strict diet."

Her words sounded so simple, but the threat they implied was so cruel that I realized for the first time that maybe "size 6" is a more violent restriction imposed on women than is the Muslim veil. Quickly I said good-bye so as not to make any more demands on the saleslady's time or involve her in any more unwelcome, confidential exchanges about age-discriminating salary cuts. A surveillance camera was probably watching us both.

Yes, I thought as I wandered off, I have finally found the answer to my harem enigma. Unlike the Muslim man, who uses space to establish male domination by excluding women from the public arena, the Western man manipulates time and light. He declares that in order to be beautiful, a woman must look fourteen years old. If she dares to look fifty, or worse, sixty, she is beyond the pale. By putting the spotlight on the female child and framing her as the ideal of beauty, he condemns the mature woman to invisibility. In fact, the modern Western man enforces Immanuel Kant's nineteenth-century theories: To be beautiful, women have to appear childish and brainless. When a woman looks mature and self-assertive, or allows her hips to expand, she is condemned as ugly. Thus, the wars of the European harem separate youthful beauty from ugly maturity.

These Western attitudes, I thought, are even more dangerous and cunning than the Muslim ones because the weapon used against women is time. Time is less visible, more fluid than space. The Western man uses images and spotlights to freeze female beauty within an idealized childhood, and forces women to perceive aging—that normal unfolding of the years—as a shameful devaluation. "Here I am, transformed into a dinosaur," I caught myself saying aloud as I went up and down the rows of skirts in the store, hoping to prove the saleslady wrong—to no avail. This Western time-defined veil is even crazier than the space-defined one enforced by the Ayatollahs.

The violence embodied in the Western harem is less visible than in the Eastern harem because aging is not attacked directly, but rather masked as an aesthetic choice. Yes, I suddenly felt not only very ugly but also quite useless in that store, where, if you had big hips, you were simply out of the picture. You drifted into the fringes of nothingness. By putting the spotlight on the prepubescent female, the Western man veils the older, more mature woman, wrapping her in shrouds of ugliness. This idea gives me the chills because it tattoos the invisible harem directly onto a woman's skin. Chinese foot-binding worked the same way: Men declared beautiful only those women who had small, childlike feet. Chinese men did not force women to bandage their feet to keep them from developing normally—all they did was to define the beauty ideal. In feudal China, a beautiful woman was the one who voluntarily sacrificed her right to unhindered physical movement by mutilating her own feet, and thereby proving that her main goal in life was to please men. Similarly, in the Western world, I was expected to shrink my hips into a size 6 if I wanted to find a decent skirt tailored for a beautiful woman. We Muslim women have only one month of fasting, Ramadan, but the poor Western woman who diets has to fast twelve months out of the year. *"Quelle horreur,"* I kept repeating to myself, while looking around at the American women shopping. All those my age looked like youthful teenagers.

According to the writer Naomi Wolf, the ideal size for American models decreased sharply in the 1990s. "A generation ago, the average model weighed 8 percent less than the average American woman, whereas today she weighs 23 percent less. . . . The weight of Miss America plummeted, and the average weight of Playboy Playmates dropped from 11 percent below the national average in 1970 to 17 percent below it in eight years."[1] The shrinking of the ideal size, according to Wolf, is one of the primary reasons for anorexia and other health-related problems: "Eating disorders rose exponentially, and a mass of neurosis was promoted that used food and weight to strip women of . . . a sense of control."[2]

Now, at last, the, mystery of my Western harem made sense. Framing youth as beauty and condemning maturity is the weapon used against women in the West just as limiting access to public space is the weapon used in the East. The objective remains identical in both cultures: to make women feel unwelcome, inadequate, and ugly.

The power of the Western man resides in dictating what women should wear and how they should look. He controls the whole fashion industry, from cosmetics to underwear. The West, I realized, was the only part of the world where women's fashion is a man s business. In places like Morocco, where you design your own clothes and discuss them with craftsmen and -women, fashion is your own business. Not so in the West. As Naomi Wolf explains in *The Beauty Myth,* men have engineered a prodigious amount of fetish-like, fashion-related paraphernalia: "Powerful industries—the $33-billion-a-year diet industry, the $20-billion cosmetic industry, the $300-million cosmetic surgery industry, and the $7-billion pornography industry—have arisen from the capital made out of unconscious anxieties, and are in turn able, through their influence on mass culture, to use, stimulate, and reinforce the hallucination in a rising economic spiral."[3]

But how does the system function? I wondered. Why do women accept it?

Of all the possible explanations, I like that of the French sociologist, Pierre Bourdieu, the best. In his latest book, *La Domination Masculine,* he proposes something he calls *"la violence symbolique":* "Symbolic violence is a form of power which is hammered directly on the body, and as if by magic, without any apparent physical constraint. But this magic operates only because it activates the codes pounded in the deepest layers of the body."[4] Reading Bourdieu, I had the impression that I finally understood Western man's psyche better. The cosmetic and fashion industries are only the tip of the iceberg, he states, which is why women are so ready to adhere to their dictates. Something else is going on on a far deeper level. Otherwise, why would women belittle themselves spontaneously? Why, argues Bourdieu, would women make their lives more difficult, for example, by preferring men who are taller or older than they are? "The majority of French women wish to have a husband who is older and also, which seems consistent, bigger as far as size is concerned," writes Bourdieu.[5] Caught in the enchanted submission characteristic of the symbolic violence inscribed in the mysterious layers of the flesh, women relinquish what he calls "les signes ordinaires de la hiérarchie sexuelle," the ordinary signs of sexual hierarchy, such as old age and a larger body. By so doing, explains Bourdieu, women spontaneously accept the subservient position. It is this spontaneity Bourdieu describes as magic enchantment.[6]

Once I understood how this magic submission worked, I became very happy that the conservative Ayatollahs do not know about it yet. If they did, they would readily switch to its sophisticated methods, because they are so much more effective. To deprive me of food is definitely the best way to paralyze my thinking capabilities.

Both Naomi Wolf and Pierre Bourdieu come to the conclusion that insidious "body codes" paralyze Western women's abilities to compete for power, even though access to education and professional opportunities seem wide open, because the rules of the game are so different according to gender. Women enter the power game with so much of their energy deflected to their physical appearance that one hesitates to say the playing field is level. "A cultural fixation on female thinness is not an obsession about female beauty," explains Wolf. It is "an obsession about female obedience. Dieting is the most potent political sedative in women's history; a quietly mad population is a tractable one."[7] Research, she contends, "confirmed what most women know too well—that concern with weight leads to a 'virtual collapse of self-esteem and sense of effectiveness' and that . . . 'prolonged and periodic caloric restriction' resulted in a distinctive personality whose traits are passivity, anxiety, and emotionality."[8] Similarly, Bourdieu, who focuses more on how this myth hammers its inscriptions onto the flesh itself, recognizes that constantly reminding women of their physical appearance destabilizes them emotionally because it reduces them to exhibited objects. "By confining women to the status of symbolical objects to be seen and perceived by the other, masculine domination . . . puts women in a state of constant physical insecurity. . . . They have to strive ceaselessly to be engaging, attractive, and available."[9] Being frozen into the passive position of an object whose very existence depends on the eye of its beholder turns the educated modern Western woman into a harem slave.

"I thank you, Allah, for sparing me the tyranny of the 'size 6 harem,'" I repeatedly said to myself while seated on the Paris-Casablanca flight, on my way back home at last. "I am so happy that the conservative male elite does not know about it. Imagine the fundamentalists switching from the veil to forcing women to fit size 6."

How can you stage a credible political demonstration and shout in the streets that your human rights have been violated when you cannot find the right skirt?

NOTES

1. Naomi Wolf, *The Beauty Myth: How Images of Beauty Are Used Against Women* (New York Anchor Books, Doubleday, 1992), p. 185.
2. Ibid., p. 11.
3. Ibid., p. 17.
4. Pierre Bourdieu: "La force symbolique est une forme de pouvoir qui s'exerce sur les corps, directement, et comme par magie, en dehors de toute contraine physique, mais cette magie n'opère qu'en s'appuyant sur des dispositions déposées, tel des ressorts, au plus profond des corps." In *La Domination Masculine* (Paris: Editions du Seuil, 1998), op. cit. p. 44.

   Here I would like to thank my French editor, Claire Delannoy, who kept me informed of the latest debates on women's issues in Paris by sending me Bourdieu's book and many others. Delannoy has been reading this manuscript since its inception in 1996 (a first version was published in Casablanca by Edition Le Fennec in 1998 as "Êtes-Vous Vacciné Contre le Harem").
5. *La Domination Masculine,* op. cit., p. 41.
6. Bourdieu, op. cit., p. 42.
7. Wolf, op. cit., p. 187.
8. Wolf, quoting research carried out by S. C. Woolly and O. W. Woolly, op. cit., pp. 187–188.
9. Bourdieu, *La Domination Masculine.* p. 73.

30

"If It's Not On, It's Not On"—Or Is It?

BY NICOLA GAVEY, KATHRYN MCPHILLIPS, AND MARION DOHERTY

Nicola Gavey, Kathryn McPhillips and Marion Doherty conducted in-depth interviews with 14 New Zealand women, ages 27 to 37, who came from diverse work experiences and educational backgrounds. The authors questioned the women about their past and current experiences with condoms, their heterosexual relationships and practices, their personal views of condoms, and how they thought others viewed condoms. The authors focus on the conflict between public health campaigns, which encourage women to demand condom use by male partners, and dominant heterosexual scripts that limit the ways in which women may control the course and outcomes of heterosexual encounters.

1. What are the two main "discursive arenas" or organizing principles of heterosexual sex that constrain women's sexual pleasure and activity?
2. Why do the authors believe that advice to women that encourages them to be assertive in sexual relationships with men is problematic?
3. Why did many of the women in this study have experiences of engaging in unwanted sex with men? How is this study's analysis of engaging in unwanted sex helpful to understanding the phenomenon of date rape?

"If it's not on, it's not on!" Slogans such as these exhort women not to have sexual intercourse with a man unless a condom is used. Health campaigns targeting heterosexuals with this approach imply that it is women who should act assertively to control the course of their sexual encounters to prevent the spread of HIV/AIDS and other sexually transmitted infections (STIs). Researchers and commentators, too, have sometimes explicitly concluded that it is women in particular who should be targeted for condom promotions on the basis of assumptions such as "the disadvantages of condom use are fewer for girls" (Barling and Moore 1990). . . .

Aside from the obvious question of whether women *should* be expected to take greater responsibility for sexual safety, this approach relies on various assumptions that deserve critical attention. For example, what constraints on women's abilities to unilaterally control condom use are overlooked in these messages? What assumptions about women's sexuality are embedded in the claim that the disadvantages of condoms are fewer for women? That is, is safer sex simply a matter of women deciding to use condoms at all times and assertively making this happen, or do the discursive parameters of heterosex work to subtly constrain and contravene this message? Moreover, are condoms as unproblematic for women's experiences of sex as the logic offered for targeting women implies? These questions demand further investigation given that research has repeatedly shown the reluctance of heterosexuals to consistently use condoms despite clear health messages about their importance.

There is now a strong body of feminist research suggesting that condom promotion in

Western societies must compete against cultural significations of condom-*less* sexual intercourse as associated with commitment, trust, and "true love" in relationships (e.g., Holland et al. 1991; Kippax et al. 1990; Willig 1995; Worth 1989; see also Hollway 1989). Here, we contribute to this body of work, which collectively highlights how women's condom use needs to be understood in relation to some of the complex gender dynamics that saturate heterosexual encounters. In particular, we critically examine (1) the concept of women's control over condom use that is tacitly assumed in campaigns designed to promote safer sex and condoms to women and (2) the foundational assumption of such campaigns that condoms are relatively unproblematic for women's sexual experiences. . . .

We suggest there are two main discursive arenas that need to be considered in critically evaluating strategies that assume women's ability to control condom use and the appropriateness of targeting women in particular for messages of (hetero) sexual responsibility. These are (1) a male sex drive discourse (Hollway 1984, 1989) and the corresponding scarcity of discourse around female desire (Fine 1988), and (2) the constraints of a "coital imperative" on how much control women (or men) have in determining what sexual activity counts as "real sex." We suggest that there are two important points where these kinds of discursive influences converge for women in ways that mediate the possibilities for condom use, that is, the sites of identity and pleasure.

* * *

ASSERTION, IDENTITY, AND CONTROL

* * *

Public sexual health education fervently promotes the rights of women to demand condom use by male partners. Against a backdrop of implied male aversion to condoms, both the Planned Parenthood Federation of America (1998–2001) and the Family Planning Association (2000) in New Zealand offer mock scripts that despite an air of gender neutrality suggest ways women can assert their right to insist on a condom being used for sexual intercourse. In this section, we consider what these kinds of hard-line "calls to assertiveness" might look like in practice. The excerpt below is taken from the interview with Rose, a young woman in her early 20s. At the very beginning of the interview, in response to a question about her "current situation, in terms of your relationships," she said, "My last experience was fairly unpleasant and so I thought I'll really try and um devote myself to singledom for a while before I rush into anything." In this one-night stand, approximately three weeks prior to the interview, Rose did manage to successfully insist that her partner use a condom. Her account of this experience is particularly interesting for the ways in which it graphically demonstrates the kinds of interactional barriers that a woman might have to overcome to be assertive about using a condom; moreover, it demonstrates how even embodiments of the male sex drive discourse that are not perceived to be coercive can act out levels of sexual urgency that provide a momentum that is difficult to stop. Although the following quote is unusually long, we prefer to present it intact to convey more about the flavor of the interaction she describes. We will refer back to it throughout this article and want to keep the story intact.

> NICOLA: So were condoms involved at all in that—
>
> ROSE: Yeah um, actually that's quite an interesting one because we were both very drunk, but I still had enough sense to make it a priority, you know, and I started to realize that things were getting to the point where he seemed to be going ahead with it, without a condom, and I was—had to really push him off at one point and—'cause I kept saying sort of under my breath, are you going to get a condom now, and—and he didn't seem to be taking much notice, and um—

NICOLA: So you were actually saying that?

ROSE: Yeah. I was—

NICOLA: In a way that was audible for him to—

ROSE: Yeah. (NICOLA: Yeah) And um—at least I think so. (NICOLA: Yeah) And—and it got to the point where [laughing] I had to push him off, and I think I actually called him an arsehole and—I just said, look fuck, you know, and—and so he did get one then but it seemed—

NICOLA: He had some of his own that he got?

ROSE: Yeah. (NICOLA: Yeah) Yeah it was at his flat and he had them. In fact that was something I'd never asked beforehand. I presumed he would have some. And um he did get one and then you know—and that was okay, but it was sort of like you know if I hadn't demanded it, he might've gone ahead without it. And I am fairly sure that he's pretty, um—you know he has had a pretty dubious past, and so that worried me a bit. I know he was very drunk and out of control and, um—otherwise the whole situation would never have occurred I'm sure, but still I had it together enough, thank God, to demand it.

NICOLA: And you actually had to push him off?

ROSE: Yeah, I think—I'm pretty sure that I did, you know it's—it's all quite in a bit of a haze, [laughing] (NICOLA: laughing) but um, it wasn't that pleasant, the whole experience. It was—he was pretty selfish about the whole thing. And um—yeah.

NICOLA: How much old—how—You said that he was a bit older than you.

ROSE: Ohh, he's only twenty-eight or nine, but that's quite a big difference for me. I usually see people that are very much in my own age-group. Mmm.

NICOLA: And um, you said that it was kind of disappointing sexually and otherwise and he was quite selfish, (ROSE: laughter) like at the point where you know you were saying, do you want to get—are you going to get a condom, at that point were you actually wanting to have sexual intercourse?

ROSE: Um, mmm, that's a good question. I can't remember the whole thing that clearly. And I seem to remember that I was getting um—I was just getting sick of it, or—[laughing] or—I might have been, but my general impression was that it was quite a sort of fumble bumbled thing, and—and it was quite—he just didn't—he didn't have it together. He wasn't—he possibly would've been better if he was less drunk, but he was just sort of all over the place and um—and I was just thinking, you know, I want to get this over and done with. Which is not the [laughing] best way to go into—into that sort of thing and um—yeah I—I remember at points—just at points getting into it and then at other points it just being a real mess. Like he couldn't—he wasn't being very stimulating, he was trying to be and bungling it 'cause he was drunk. And being too rough and just too brutish, just yeah. And um—

NICOLA: When you say he was trying to [laughing] be, what do you—

ROSE: Ohh, he was just like you know, trying to use his hands and stuff and just it—it was just like a big fumble in the dark (NICOLA: Right) type thing. I mean I'm sure h—I hope he's not usually that bad it was just like—I was in—sometimes—most of the time in fact I was just saying, look don't even bother. [laughing] You obviously haven't got it together to make it pleasurable. So in some ways—

NICOLA: You—you said you were thinking that or you said that?

ROSE: I just sort of—I did push his hand away and just say, look don't bother. Because—and I thought at that point that probably penetration would be, um, more pleasurable, yeah. But—and then—yeah. And I think I was actually wanting to just go to sleep and I kept—I think I um mentioned that to him as well and said, you know why don't we—we're we're not that capable at the moment, why don't we leave it. But he wasn't keen on [laughing] that idea.

NICOLA: What did he—is that from what he said, or just the fact that he didn't—

ROSE: I think he just said, ohh no, no, no we can't do that. It was kind of like we had to put on this big passionate spurt but um it just seemed quite farcical, considering the state we were both in. And um, so I—yeah I thought if—yeah I thought it would probably be the best idea to [laughing] just get into it and like as I—as I presumed—ohh, actually I don't know what I expected, but he didn't last very long at all, which was quite a relief and he was quite—he was sort of a bit apologetic and like ohh you know, I shouldn't have come so soon. And I was just thinking, oh, now I can go to sleep. [laughter] Yes, so I just um—since then I've just been thinking, um one-night stands ahh don't seem to be the way to go. They don't seem to be much fun and [laughing] I'm not that keen on the idea of a relationship either at the moment. So I don't know. Who—I—it's hard to predict what's going to happen but—mmmm.

This unflattering picture of male heterosexual practice painted by Rose's account classically illustrates two of the dominant organizing principles of heterosexual sex—a male sex drive discourse and a coital imperative. The male sex drive discourse (see Hollway 1984, 1989) holds that men are perpetually interested in sex and that once they are sexually stimulated, they need to be satisfied by orgasm. Within the terms of this discourse, it would thus not be right or fair for a woman to stop sex before male orgasm (normatively through intercourse). This discursive construction of male sexuality thus privileges men's sexual needs above women's; the absence of a corresponding discourse of female desire (see Fine 1988) or drive serves to indirectly reinforce these dominant perceptions of male sexuality.

The extent to which male behavior, as patterned by the male sex drive discourse, can constrain a woman's attempts to insist on condom use are graphically demonstrated in Rose's account. The man she was with behaved with such a sense of sexual urgency and unstoppability that although she was able to successfully ensure a condom was used, it was only as a result of particularly determined and persistent efforts. He was unresponsive to her verbal requests and did not stop proceeding with intercourse until Rose became more directly confrontational—calling him "an arsehole" and physically pushing him off. At this point, he eventually did agree to wear a condom and did not use his physical strength to resist and overcome her actions to retain or take control of the situation. We will come back to analysis of Rose's account later in the article. . . .

Pleasing a Man

The male sex drive discourse constructs masculinity in ways that directly affect women's heterosexual experiences as evidenced in the above description . . . of Rose['s] . . . experience. However, as we alluded to above, this discursive framework can also constitute women's sexual subjectivity in complex and more indirect ways. For example, for some women situated within this discursive framework, their ability to "please a man" may be a positive aspect of their identity. In this sense, a woman can be recruited into anticipating and meeting a man's "sexual needs" (as they are constituted in this discursive framework) as part of her ongoing construction of a particular kind of identity as a woman.

Sarah reported not liking condoms. She said she did not like the taste of them; she did not like the "hassle" of putting them on, taking them off, and disposing of them; and that they interfered with her sexual pleasure. However, it seemed that her reluctance to use them was also related to her sexual identity and her taken-for-granted assumptions about men's needs, desires, and expectations of her during sex. Sarah traced some of her attitudes toward sex to her upbringing and her mother's attitudes in which the male sex drive discourse was strongly ingrained.[1] She directly connected the fact that "when I've started, I never stop" to her difficulty in imagining asking a man to use a condom. As she said in an ironic tone,

> If a man gets a hard-on you've gotta take care of it because he gets sick. You know these are the mores I was brought up with. So you can't upset his little precious little ego by asking to use a condom, or telling if he's not a good lover.

Sarah explained her own ambivalence toward this male sex drive discourse by drawing on a psychoanalytic distinction between the conscious and unconscious mind:

> I mean I've hopefully done enough therapy to have moved away from that, but it's still in your bones. You know there's my conscious mind can say, that's a load of bullshit, but my unconscious mind is still powerful enough to drive me in some of these moments, I would imagine.

Thus, despite having a rational position from which she rejected the male sex drive discourse, Sarah found that when faced with a man who wanted to have sex with her, her embodied response would be to acquiesce irrespective of her own desire for sex. Her reference to this tendency being "in [her] bones" graphically illustrates how she regarded this as a fundamental influence. It is evocative of Judith Butler's suggestion "that discourses do actually live in bodies. They lodge in bodies, bodies in fact carry discourses as part of their own lifeblood" (Meijer and Prins 1998, 282). Sarah recalled an incident where her own desire for sex ceased immediately on seeing the man she was with undress, but she explained that she would not be prepared to stop things there:

> SARAH: He's the hairiest guy I ever came across. I mean—but no way I would stop. I mean, as soon as he took his shirt off I just kind of about puked [laughing], but I ain't gonna say—I mean having gone to all these convolutions to get this thing to happen there's no way I'm gonna back down at that stage.
>
> NICOLA: So when he takes his shirt off he's just about the most hairiest guy you've ever met, which you find really unappealing.
>
> SARAH: Terribly.
>
> NICOLA: Um, but you'd rather go through with it?
>
> SARAH: Well I wouldn't say rather, but I do it.

Sarah also described another occasion where she met a man at a party who said, "do you want to get together? and I said, well, you know, I just want to cuddle, and he said, well, that's fine, okay." She explained that it was very important to her to make her position clear before doing anything. However, they ended up having sexual intercourse, because as she said,

> SARAH: I was the one. I mean we cuddled, and then I was the one that carried it further.
>
> NICOLA: And what was the reason for that?
>
> SARAH: As I said, partly 'cause I want to and partly 'cause if ohh he's got a hard-on you have to.

For women like Sarah, it seemed that an important part of their identity involved being a "good lover." This required having sexual intercourse to please a man whenever he wanted it and, in Sarah's case, to the point of anticipating this desire on the basis of an erect penis. . . . Given these expectations, and her belief that most men do not like condoms, it is not surprising that she had developed an almost fatalistic attitude toward her own risk of contracting HIV, such that she could say, "There's a part of me that also says, as long as I'm clear and not passing it on, I'm not gonna worry. You know, and if I get it, hey, it was meant to be."

For understanding the actions of Sarah and women like her, the assertiveness model is not at all helpful. In these kinds of sexual encounters, Sarah's lack or possession of assertiveness skills is beyond the point. What stops her from acting assertively to avoid undesired sex are deeply inscribed features of her own identity—characteristics that are not related to fear of assertion so much as the production of a particular kind of self. Well before she gets to the point of acting or not acting assertively, she is motivated by other (not sexual) desires, about what kind of woman she wants to *be*.

In a similar way to Sarah, an important part of Sally's identity was having "integrity about sexuality." She explained her position in relation to not "leading somebody on" in terms of a desire for honesty:

> And so—and I—that really was clinched somehow that you didn't lead somebody on and so that's part of—that's one of the sort of ways in which I understand that contract notion, really. And so maybe that had something to do with how I see myself about being a person with a reasonable—with integrity about sexuality. I won't go into something with false promises kind of thing.

Prior to these comments, Sally talked about the origins of her beliefs about this kind of contractual notion where it was not possible to be physically intimate with someone unless you were prepared to have intercourse. She remembered feeling awful about touching a man's penis when she was younger:

> I had been unfair because of the—I suppose the feeling of sort of cultural value of—of the belief that men somehow you know it's tormenting to them to leave a cock unappeased (N:[laughter]) [laughing] basically or something like that.

Like Sarah, Sally too reflected on her experiences in a way that highlights the limitations of attempting to influence sexual behavior based on understandings of people as unitary rational actors. Sally discussed a six-month relationship with a past lover in which she had not used condoms. She said that she had made the decision not to use condoms because she already had an intrauterine device (IUD) and because they were seen to connote a more temporary rather than long-lasting relationship (a view ironically reinforced by the advice of a nurse at the Family Planning Clinic):

> SALLY: It's like condoms are about more casual kinds of encounters or I mean—I mean, I'm kind of—um they are kind of anti-intimacy at some level.
>
> NICOLA: And so if you'd used condoms with him, that would've meant—
>
> SALLY: Maybe it would have underscored its temporariness or its—yeah, its lack of permanence. I don't understand that. What I've just said really particularly. It doesn't [seem] very rational to me. [laughter]
>
> NICOLA: [laughter] No it very rational and I—
>
> SALLY: [indistinguishable] it seems to be coming out of you know, somewhere quite deeper about um—I think it goes back to that business about ideals stuff. And I think that's one of the things about not saying no, you know. And that the ideal woman and lover—the ideal woman is a good lover and doesn't say no. Something like that. And it is incredibly counterproductive [softly] at my present time in life. [sigh/laugh]

Sally's reference to the ideal woman who is a good lover (because) she does not say no implicitly recognizes the strength of male sex drive discourse and its effect on her sexual experiences. In a construction that is similar to Sarah's, Sally refers to this kind of influence as "deeper" than her "rational" views. She presents an appreciation of this kind of cultural ideal as internalized in some way that is capable of having some control over her behavior despite her assessment of this as "counterproductive."

Engaging in Unwanted Sex

Many of the women in this study recounted experiences of having sex with men when they did not really want to, for a variety of reasons. This now common finding (e.g., Gavey 1992) underscores the extent to which women's control of sex with men is limited by various discursive constraints in addition to direct male pressure, force, or violence. As Bronwyn said, it is part of "the job:"

> NICOLA: You said that you enjoy intercourse up to a point. Um, beyond that point, um, what are your reasons for continuing, given that you're not enjoying it?
>
> BRONWYN: Ohh I just think it's part of my function if you like [laughter] that sounds terribly cold-blooded, but it is, [laughing] you know, it's part of the job.
>
> NICOLA: The job of—
>
> BRONWYN: Being a wife. A partner or whatever.

Rose's account of her confrontational one-night stand, discussed previously, can be seen to be influenced in complex and subtle ways by the discursive construction of normative heterosexuality in which male sexuality and desire are supreme. She described the encounter as being quite unpleasant and disappointing, both sexually and in the way that he treated her—with respect to condoms and more generally throughout the experience. She described his actions during the encounter as clumsy and not the least bit sexually arousing ("He wasn't being very stimulating, he was trying to be and bungling it because he was drunk. And being too rough and just too brutish") and at one stage she suggested that they give up and go to sleep but he rejected this idea very strongly. Despite the extremely unsatisfactory nature of the sexual interaction and despite Rose's demonstrated skills of acting assertively, she did continue with the encounter until this man had had an orgasm through vaginal intercourse. It is difficult to understand why she would have done this without appreciating the power of the male sex drive discourse and the coital imperative in determining the nature of heterosexual encounters.

Although she did not define herself as a victim of the experience, Rose's account also makes clear that somehow she did not feel it was an option to end sex unless he gave the okay:

> It was sort of like—and I guess in that case he didn't have my utmost respect by that point. But it was sort of like, um, you know, he'd—he seemed to be just going for it, and I really really—I was drunk and sort of dishevelled and—and pretty resigned to having a bit of loose un—and unsatisfactory time which I didn't have a lot of control over.

She partially attributed this lack of control over the situation to the fact that the encounter took place at his flat, but this seems to be reinforced by underlying assumptions about sexuality and the primacy of his desires:

> Um, but it was kind of like he had his idea of what was going to happen and um I sort of realized after a while that he was so intent on it and the best thing to do was to just comply I guess, and make it as pleasurable as possible. Try and get into the same frame of mind that he was in. And—and yeah. Mmm, get it over with. It sounds really horrible in retrospect, it wasn't that bad it was just lousy, you know. It was just sort of a poor display of [laughing] everything. I suppose. Of intelligence and—and good manners. [laughter]

Part of the reason for the ambivalent nature of her account (swinging from describing what happened in very negative terms to playing down the experience as a poor display of manners) can be argued to originate from her positioning within a kind of liberal feminist discourse about sex. She had made a point of saying that she did not think things went in stages and defining herself in opposition to a model of female sexuality as fragile and in need of protection. The result of the connection between this liberal discourse of sexuality and that of the male sex drive is that she is left with no middle ground from which to negotiate within a situation of this kind. The absence of an alternative discourse of active female sexuality leaves her in a position of no return once she has consented to heterosexual relations and when certain minimal conditions are fulfilled (for her, this was the use of condoms). If heterosexuality were instead discursively constructed in such a way that women's sexual pleasure was central rather than optional, it makes sense that Rose may have felt able to call an end to this sex—which was, after all, so unpleasant that it led to her resolve to "devote [herself] to singledom for a while."

THE QUESTION OF PLEASURE

"I'm probably atypical from what I read of women in the fact that I personally don't like condoms." Sarah made this comment in the first minutes of the interview, and then much later she shared her assumption about how men regard using condoms: "Most men hate it. I've never asked them, but . . . that's the feeling I have from what I've read or heard." Sarah's generalized views about how women and men regard condoms echo dominant commonsense stories in Western culture. That is, most men do not like using condoms—a view shared by 80.5 percent of women in one large U.S. sample (Valdiserri et al. 1989), while women do not mind them. To explain her own dislike of condoms, Sarah was forced to regard herself as atypical. In the following section, we will discuss evidence that challenges the tacit assumption that condoms are relatively unproblematic for women.

Enforcers of the Coital Imperative: Condoms as Prescriptions for Penetration

Research on how men and women define what constitutes "real sex" has repeatedly found that a coital imperative exists that places penis-vagina intercourse at the center of (hetero) sex (Gavey, McPhillips, and Braun 1999; Holland et al. 1998; McPhillips, Braun, and Gavey 2001). The strength of this imperative was also reflected in the current research—as one women explained, "I don't think I worked out a model of being with someone like naked intimate touching which doesn't have sex at the end of it" (Sally defined sex as penis-vagina intercourse during the interview). Although this coital imperative could be viewed as forming part of the male sex drive discourse examined above, it is addressed separately here as it has particular consequences for safer sex possibilities.

Condoms seem to reinforce the coital imperative in two interconnected ways, both in terms of their symbolic reinforcement of the discursive construction of sex as *coitus* and through their material characteristics that contribute at a more practical level to rendering sex as finished after coitus. In the analyses that follow, we will be attending to the material characteristics of condoms as women describe them. As discussed earlier, we adopt a realist reading of women's accounts here. What the women told us about the ways in which condoms help to structure the material practice of heterosex casts a shadow over the assumption that condoms are unproblematic for women's sexual experience. These accounts illuminate how the male sexual drive discourse can shape not only the ways people speak about and experience heterosex, but also the ways in which a research lens is focused on heterosexual practice to produce particular ways of seeing that perpetuate commonsense priorities and silences (which, in this case, privilege men's pleasure above women's). That is, the ways in which condoms can interfere with a woman's sexual pleasure are relatively invisible in the literature, which tends, at least implicitly, to equate "loss of sexual pleasure" with reduced sensation in the penis, or disruption of desire and pleasure caused by the act of putting a condom on. As the following excerpts show, the material qualities of condoms have other particular effects on the course of sex. These effects are especially relevant both to the question of a woman's pleasure and to the way in which the coital imperative remains unchallenged as the definitive aspect of heterosex.

While many of the women said they like (sometimes or always) and/or expect sexual intercourse (i.e., penis-vagina penetration) when having sex with a man, many of the women noted how condoms operated to enforce intercourse as the finale of sex. Several women found this to be a disadvantage of condoms, in that they tend to limit what is possible sexually, making sex more predictable, less spontaneous, playful, and varied. That is, once the condom is on, it is there for a reason and one reason only—penile penetration. It signals the beginning of "the end." Women who identified this disadvantage tended to be using condoms for contraceptive purposes and so were comparing sex that involved condoms unfavorably to intercourse with some less obtrusive form of contraception such as the pill or an IUD, or with no contraception during a "safe" time of the month.

For example, Julie found that condoms prescribed penetration at a point where she could be more flexible if no condom was involved:

> That's what I mean about the condom thing. It's like this is *the act* you know, and you have to go through the whole thing. Whereas if you don't use condoms, you know, like he could put it in me and then we could stop and then put it in again, you know, you can just be a bit more flexible about the whole thing.

The interconnection between the material characteristics of condoms (its semen-containing properties require "proper use") and the discursive construction of the encounter (coitus is spoken about as "*the act*") has the effect of constraining a woman's sexual choices and leaving this generally unspoken coital imperative unchallenged. The discursive centrality of coitus within heterosex is materially reinforced by the practical difficulties associated with condoms, as Deborah said,

> Once you've put on the condom . . . that limits you. Once you've got to the stage in sex that you put a condom on, you then—it's not that you can't change plans, but it's a hassle if you then decided that you might like to move to do um—you might like to introduce oral sex at this stage, as opposed to that stage, then you have to take it off, or you don't, or—

⋆ ⋆ ⋆

Thus, condoms not only signalled when penetration would take place, but their use served to reinforce the taken-for-granted axiom of heterosexual practice, that coitus is the main sexual act. Furthermore, one woman (Sally) described how the need for a man to withdraw his penis soon after ejaculation when using a condom disrupted the postcoital "close feeling" she enjoyed. These excerpts can be seen to represent a form of resistance to the teleological assumptions of the coital imperative; women's accounts of desiring different forms of sexual pleasure (including, but by no means limited to, emotional pleasures) may provide rich ground for exploring safer sex options. This potentially productive area has yet to be fully exploited by traditional health campaigns, which perhaps reflects the lack of acknowledgment given to discourses of female sexual desire and pleasure in Western culture in general (see also Fine 1988).

Some women also talked about the effect using condoms had on their sexual pleasure by using the language of interruption and "passion killing" more commonly associated with men. Sarah, who rarely used condoms, said that "it breaks the flow":

> I have a lot of trouble reaching an orgasm anyway, and it's probably one of the reasons I don't like something that's interrupting, because I do go off the boil very quickly. Um, once it's on and it's sort of decided that penetration tends to be what happens. I don't suppose it's a gold rule, but it seems to be the way it is. So you know there's no more warm-up.

Unlike health campaigns directed at the gay community, which have emphasized the range of possible sex acts carrying far less risk of HIV infection than penile penetration, campaigns aimed at heterosexuals have done little to challenge the dominant coital imperative.[2] Health campaigns that promote condoms as the only route to safer sex implicitly reinforce this constitution of heterosexuality and the dominance of the male sex drive discourse. As the responses of the women in this study demonstrate, this reluctance to explore other safer sex possibilities may be a missed opportunity for increasing erotic possibilities for women at the same time as increasing opportunities for safer sex. The fact that all of these women spent some time talking about the ways in which condoms can operate to enforce the coital imperative or reduce their desire indicates the importance of taking women's pleasure into account when designing effective safer sex programs. That is, it may simply not be valid to assume that "the disadvantages of condom use are fewer for girls" (Barling and Moore 1990) if we expect women's sexual desires and pleasures to be taken as seriously as men's. Special effort may be required to ask different questions to understand women's experiences in a way that doesn't uncritically accept a vision of heterosex as inherently constrained through the lens of the coital imperative and male sex drive discourses.

DISCUSSION

* * *

Holland et al. (1998) have argued that heterosexual relations as they stand are premised on a construction of femininity that endangers women. Evidence for this position can be drawn from the current study as the interaction of discourses determining normative heterosexuality produces situations in which women are unable to always ensure their safety during sexual encounters. Holland et al. (1998) have argued that a refiguring of femininity is needed to ensure that women have a greater chance of safer heterosexual encounters. One of the prerequisites for change of this kind would be acknowledgment of the discourses of active female desire, which have traditionally been repressed (Fine 1988). Indeed, our research here suggests that the claim that condoms are "relatively unproblematic" for women is based on a continued relegation of the importance of women's sexual pleasure, relative to men's. Without challenging the gendered nature of dominant representations of desire, and more critically examining the coital imperative, condom promotion to women is likely to remain a double-edged practice. As both a manifestation and a reinforcement of normative forms of heterosex, it may be of limited efficacy in promoting safer heterosex.

NOTES

1. It should be emphasized that Sarah's mother's attitudes would have been in line with contemporary thought at the time. Take, for example, the advice of "A Famous Doctor's Frank, New, Step-by-Step Guide to Sexual Joy and Fulfulment for Married Couples" (on front cover of Eichenlaub 1961, 36), published when Sarah was nearly an adolescent:

   Availability: If you want good sex adjustment as a couple, you must have sexual relations approximately as often as the man requires. This does not mean that you have to jump into bed if he gets the urge in the middle of supper or when you are dressing for a big party. But it does mean that a woman should never turn down her husband on appropriate occasions simply because she has no yearning of her own for sex or because she is tired or sleepy, or indeed for any reason short of a genuine disability. (Eichenlaub 1961, 36)
2. While some sexuality education directed at teenagers might be more likely to encourage alternatives to coital sex, and hence broader definitions of safer sex (e.g., Family Planning Association 1998; see Burns 2000), this is still less evident in material designed for the "mature sexuality" of heterosexual adults.

REFERENCES

Barling, N. R., and S. A. Moore. 1990. Adolescents' attitudes towards AIDS precautions and intention to use condoms. *Psychological Reports* 67:883–90.

Burns, M. 2000. "What's the word?" A feminist, post-structuralist reading of the NZ Family Planning

Association's sexuality education booklet. *Women's Studies* 16:115–41.

Crawford, M. 1995. *Talking difference: On gender and language.* London: Sage.

Eichenlaub, J. E. 1961. *The marriage art.* London: Mayflower.

Family Planning Association. 1998. *The word—On sex, life & relationships* [Booklet]. Auckland, New Zealand: Family Planning Association. [ISBN 0-9583304-8-4]

———. 2000. *Condoms* [Pamphlet]. Aukland, New Zealand: Family Planning Association. (Written and produced in 1999. Updated 2000)

Fine, M. 1988. Sexuality, schooling, and adolescent families: The missing discourse of desire. *Harvard Educational Review* 58:29–53.

Gavey, N. 1992. Technologies and effects of heterosexual coercion. *Feminism & Psychology* 2:325–51.

Gavey, N., K. McPhillips, and V. Braun. 1999. Interruptus coitus: Heterosexuals accounting for intercourse. *Sexualities* 2:37–71.

Holland, J., C. Ramazonaglu, S. Scott, S. Sharpe, and R. Thomson. 1991. Between embarrassment and trust: Young women and the diversity of condom use. In *AIDS: Responses, interventions, and care,* edited by P. Aggleton, G. Hart, and P. Davies. London: Falmer.

Holland, J., C. Ramazonaglu, S. Sharpe, and R. Thomson. 1998. *The male in the head: Young people, heterosexuality and power.* London: The Tufnell.

Hollway, W. 1984. Gender difference and the production of subjectivity. In *Changing the subject: Psychology, social regulation and subjectivity,* edited by J. Henriques, W. Hollway, C. Urwin, and V. Walkerdine. London: Methuen.

———. 1989. *Subjectivity and method in psychology: Gender, meaning, and science.* London: Sage.

Kippax, S., J. Crawford, C. Waldby, and P. Benton. 1990. Women negotiating heterosex: Implications for AIDS prevention. *Women's Studies International Forum* 13:533–42.

McPhillips, K., V. Braun, and N. Gavey. 2001. Defining heterosex: How imperative is the "coital imperative"? *Women's Studies International Forum* 24:229–40.

Meijer, I. C., and B. Prins. 1998. How bodies come to matter: An interview with Judith Butler. *Signs: Journal of Women in Culture and Society* 23:275–86.

Planned Parenthood Federation of America, Inc. 1998–2001. *Condoms.* Retrieved 18 June 2001 from the World Wide Web: http://www.plannedparenthood.org

Statistics New Zealand. 2001. *Statistics and information about New Zealand.* Retrieved 4 July 2001 from the World Wide Web:http://www.stats.govt.nz/default.htm

Valdiserri, R. O., V. C. Arena, D. Proctor, and F. A. Bonati. 1989. The relationship between women's attitudes about condoms and their use: Implications for condom promotion programs. *American Journal of Public Health* 79:499–501.

Willig, C. 1995. "I wouldn't have married the guy if I'd have to do that. "Heterosexual adults' accounts of condom use and their implications for sexual practice. *Journal of Community and Applied Social Psychology* 5:75–87.

Worth, D. 1989. Sexual decision-making and AIDS: Why condom promotion among vulnerable women is likely to fail. *Studies in Family Planning* 20:297–307.

31

Sexuality, Fertility, and Danger

Twentieth-Century Images of Women with Cognitive Disabilities

BY PAMELA BLOCK

This reading examines the intersections of gender, sexuality, and cognitive disabilities. The author, Pamela Block, primarily focuses on cultural images of disabled women, although she does offer a brief analysis of images of the sexuality of men with cognitive disabilities. Block argues that despite greater understanding of disability in U.S. culture today, mental disability remains a stigma that manifests itself in discrimination and abuse. One form that this takes is the severe regulation of the sexuality and fertility of women with cognitive disabilities.

1. What are the controlling images of the sexuality of women with cognitive disabilities?
2. How has the sexuality of men with cognitive disabilities been portrayed?
3. What does Block mean when she says cognitive disability is a gendered cultural construct?

The sexuality of women with cognitive disabilities has been a subject of concern to social service professionals and policy makers in the United States for over a century. Historically, in professional treatises, newspaper accounts, freak shows, literature, and film, women with cognitive disabilities have been portrayed in contradictory ways as both sexually vulnerable and socially threatening, needing professional management and control. Discussions of the "sexual nature" of this group are still present in medical, legal, and popular cultural discourse. This presentation considers five examples (four nonfiction and one fictional) where the sexual identities of women with cognitive disabilities received national scrutiny. Two of the cases are historical: the life-long institutionalization of "Deborah Kallikak" at the end of the 19th century and the 1927 sterilization of Carrie Buck. Three are contemporary: the sterilization of Cindy Wasiek in 1994, and the 1989 Glen Ridge sexual assault of a young woman with a cognitive disability, and the fictional 1998 movie "The Other Sister." These examples reveal that implicit cultural assumptions (modern fairy tales) about the sexuality and fertility of women with cognitive disabilities are embedded in United States history, and still very much in evidence today.

Disability, when applied as medical or psychological diagnoses, takes the culturally, socially, and historically derived identity of an individual and subsumes it beneath a designation of pathology. When an individual enters the biomedical and psychosocial service-systems as disabled every other personal characteristic becomes secondary; the person becomes defined by their disability.

Whether the disability is physical, mental, or imaginary, labeling a person in this way attaches stigma and results in social exclusion (1,2,3). Disability studies theorists stress the importance of separating the *disability* (physiological condition) from the *impairment* (the social ramifications of the condition) (4,5). For example, having no legs is a physiological condition, but it is the inaccessibility of buildings that creates a barrier and results in exclusion.

Following feminist and other critiques of science (6,7,8,9,10,11,12,13,14), I would call into question the notion of psychiatry or biomedicine as representative of pure empirical science. An illusory shelter of scientific rationalism obscures the fundamentally ideological nature of the "treatment" of individuals with cognitive disabilities and psychiatric illness. In *The Science Question in Feminism* (1986), Sandra Harding states:

> Will not the selection and definition of problems always bear the social fingerprints of the dominant groups in a culture? With these questions we glimpse the fundamental value-ladenness of knowledge-seeking . . . (8, p. 22)

In addition, the particular influence of professional theories and practices on social relations and cultural representations may vary according to the context in which they persist (15).

Early twentieth century theories of mental development (and deficiency) in the United States were usually linked to social status. Although "mental deficiency" was considered a medical diagnosis, the decision to label an individual "mentally deficient" was closely tied to structures of power, i.e., ideologies of race, class, and sexuality, theories of modernization and racial degeneracy, and cultural perceptions of urbanization, immigration, masculinity and femininity. The late 1800s and early 1900s was a period of excelerated social change due to immigration and rapid urbanization. Former slaves, rural people, and immigrants (from places other than Western Europe) crowded into the cities (16,17). Elite groups feared that large influxes of people they considered to be of poor mental and physical quality would result in "degeneration" on a national scale. The need to control marginal populations resulted in new prisons and mental institutions, and the development of new professions, such as social work (18). It also resulted in the growth of the eugenics movement.

Throughout this century people with cognitive disabilities in the United States have been constructed in two ways: as social threats that must be segregated in order to protect the social order, or as socially vulnerable. Without the skills to survive in a dangerous and rejecting society. The main barrier faced by people with cognitive disabilities has been social exclusion. In the early decades of this century, institutionalization of people with cognitive disabilities was not uncommon. Even today, people with cognitive disabilities have difficulty finding independence outside of segregated programs, not because of their personal limitations, but because they are not wanted (19). Expressions of rejection range from banishment to freak shows, refusals to implement inclusive educational and employment policies, to crimes of violent hatred (5,3).

Although this paper focuses primarily on cultural images of women, it is important to note that powerful images concerning the sexuality of men with cognitive disabilities also exist. Many people believe that men with cognitive disabilities are sexual predators (20). Historically, men with cognitive disabilities were institutionalized for fear of their supposed potential for physical or sexual aggression. There are many recent cases where men with cognitive disabilities have been arrested and convicted on charges of physical and/or sexual assault, and even murder, with no evidence other than a personal confession. Confessions of people with cognitive disabilities are questionable because they are often eager to please, and easily intimidated. They may attempt to tell questioners what they want to hear and/or simply fail to understand the implications of what they are saying (21). Robert Perske (1991) describes cases where men with cognitive disabilities were convicted and even sentenced to death despite the absence of any corroborating evidence. This was true even in cases where evidence pointing to other suspects existed.

Cultural beliefs concerning the sexual danger presented by men with cognitive disabilities are pervasive. It is common that when a new group home is established for neighbors to voice concerns for the safety of their children. Such fears have been around longer than the story of Frankenstein's monster. Despite popular assumptions that men with cognitive disabilities are likely to be child molesters, the obverse comes closer to the truth. Although girls and women with cognitive disabilities are at higher risk, men and boys of this group are far more likely to experience sexual aggression than boys without disabilities. As is the case for men without disabilities, men and boys with cognitive disabilities who have experienced sexual aggression may, in turn, begin to hurt others.

The production of cognitive disability as a gendered cultural construct is a complex process involving the interplay of biomedical and psychological theory, social policy and practice and symbolic representations of people with disabilities as freaks and medical oddities or as dangerous threats within the popular culture. Older theories, policies, practices and representations intermingle, producing fractured and contradictory bits of cultural data, which, in turn, are reformulated into new theories, policies, practices, and representations. This results in a layering process where old images persist alongside of newer representations, women with cognitive disabilities are transformed into figures of folklore, archetypal characters representing wider cultural messages about the role of women in twentieth-century United States society and the penalties for failure to comply with this role. A closer look at some of these "fairy tales" of women with cognitive disabilities may shed some light on the historical dimension of current practices and beliefs concerning this group, specifically barriers to social change.

The stories presented below have a fairy tale quality: They depict demonic succubae, imprisoned women, controlling mothers, extreme poverty, stolen children, demonic changelings, cruel foster parents, evil wizards (doctors, psychologists, and psychiatrists), and human wolves. The final (and only fictional) story depicts an irascible woman who, like Cinderella, fights powerful forces and overcomes all barriers to win true love and happiness with her prince. How sad that only the fictional story has a happy ending.

GODDARD AND THE "KALLIKAKS"

Psychologist Henry H. Goddard was an early-twentieth century eugenics theorist specializing in the detection and treatment of feeble-mindedness. He viewed low intelligence as the rout of all other types of degeneracy, including prostitution, criminality, poverty and alcoholism. His most famous work was the 1913 eugenics family study entitled *The Kallikak Family: A Study in the Heredity of Feeble-Mindedness.* Critiques of Goddard's research revealed that his methods were shoddy at best, and fraudulent at worst (22,23). Although Goddard's conclusions were proved false, the legacy of his work is still visible in the United States and abroad. His persuasive arguments for the large-scale segregation of people considered "feeble-minded," and the custodial training-school model he advocated persisted virtually intact in the United States until the 1980s and still exists in places. In addition, Goddard's theories and his institutional model influenced modern popular-cultural conceptions of people with cognitive disabilities in the United States.

Goddard researched the family history of a 23-year-old woman he called "Deborah Kallikak," a resident of the Training School for Feeble-minded Boys and Girls in Vineland, New Jersey. Goddard coined the term Kallikak from the Greek words *kallos* (beauty) and *kakos* (bad) (23). By tracing Deborah's family history,

Goddard claimed to have found a recessive gene for feeblemindedness passed down by her family for five generations. Goddard classified Deborah as a moron, a term he coined from the Greek word for "foolish." Goddard believed that morons were particularly dangerous to society because there was no physical manifestation of disability. According to historian David J. Smith:

> The label moron came to be widely applied to people who were considered to be "high grade defectives"—those who were not retarded seriously enough to be obvious to the casual observer and who had not been brain-damaged by disease or injury. Morons were characterized as being intellectually dull, socially inadequate, and morally deficient. (23, p. 12)

Morons could be lovely, (as shown in Goddard's book by pictures of Deborah in beautiful dresses and neat hair), but sinister, because they could easily "pass" for normal. "Moronic" traits were intangible: the inability to understand complex emotional or social situations resulting in "regressed behaviors," poor judgement, poor insight, and poor decision making abilities, and an "increased vulnerability to life events,"[1] According to Goddard, the only way to protect society from degeneration was to segregate feeble-minded individuals and prevent them from marrying and passing their recessive traits to their descendents.

Although Goddard described Deborah as "valuable to the institution" (24, p. 2), he did not hesitate to form conclusions about what her fate might have been if she were allowed to leave Vineland:

> Today if this young woman were to leave the institution, she would at once become prey to the designs of evil men or evil women and would lead a life that would be vicious, immoral, and criminal, though because of her mentality she herself would not be responsible. There is nothing that she might not be led into, because she has no power of control, and all her instincts and appetites are in the direction that would lead to vice. (24, p. 12)

Goddard warned that there were families like the Kallikaks everywhere, multiplying at twice the rate of the rest of the population. He described one such family living in urban misery:

> In one arm she held a frightful looking baby, while she had another by the hand. Vermin were visible all over her. In a room with few chairs and a bed, the latter without any washable covering and filthy beyond description. There was no fire and both mother and baby were thinly clad. They did not shiver, however, nor seem to mind. The oldest girl, a vulgar, repulsive creature of fifteen came into the room and stood looking at the stranger. She had somehow managed to live. All the rest of the children, except the two that the mother was carrying, had died in infancy. (24, pp. 73–4)

According to Goddard, this life could have been Deborah's fate, had she been safely kept in the custody of the training school.

Eventually, Goddard's research was criticized for his use of lay field-workers who made diagnoses of feeble-mindedness based on observation, interviews, or even decades-old stories told by relatives or neighbors (23). It was not until more than sixty years after Goddard published his study that Stephen J. Gould and his colleague Steven Selden noticed that the photographs of the supposedly "feeble-minded" branch of the Kallikak family were retouched to make the subjects appear stupid and ominous. In the photos, Deborah's family members were shown in rural settings in front of rough shacks. . . .

Smith was able to find and interview surviving family members who were characterized as feeble-minded in Goddard's book. He found no indication of cognitive disability in any of them. Many graduated from high school or college and worked in professions such as teaching and mechanics. Deborah entered the Vineland training school in 1889 when she was 8 years old. She remained institutionalized until her death at 89 years of age. By today's standards she would never have been institutionalized at all. Despite Goddard's diagnosis of moron, Smith found her academic challenges more indicative of what today would be called a learning disability (23). Goddard used his diagnosis of Deborah in order to promote his theories of "feeble-mindedness" and his institutional model of service provision. Through this model, implemented nationally and abroad, Goddard influenced the lives of thousands of men and women who received this diagnosis and were forced to live in training schools and mental institutions. Additionally, Goddard's representation of the "Kallikak" family as a threatening source of social and genetic degeneration caught the popular imagination. Images of sexually deviant "feeble-minded" families who, for generation after generation have lived in rural or urban degradation are recurrent figures in literature, film, and television. Goddard's theories provided professional legitimacy to cultural assertions that cognitive disability was shameful and must be hidden. He introduced the notion of "feeble-mindedness" as a pathology that must be extracted from society like a cancer. Through images of degradation and defect, women considered feebleminded were depicted as less than human, yet with an insidious power to corrupt and transform society if they were not removed from it. This position would later be advocated prominently by the United States eugenics movement.

EUGENICS AND THE STERILIZATION OF CARRIE BUCK

Eugenics, the science of the genetic improvement of the human race, was influenced by Darwin's theory of the importance of heredity in the evolutionary process, and Mendel's research on the transmission of genetic traits over generations. Meaning to "harness the force of heredity for the improvement of man," eugenics was used to establish race and class distinctions as "natural" and incontrovertible (25,26). The eugenics movement was not homogeneous; it included a variety of different political and scientific views. However, it was the more extreme theories that had the greatest influence on the development of United States national policy.

Deborah Kallikak's case is illustrative of the special treatment women diagnosed as feebleminded received at the hands of professionals. As early as the 1870s "feeble-minded" women were identified as a population in need of control and stewardship (17,27). The government, medical establishment, and society-at-large became wary of the assumed "obsessive sexual nature" of this group (28,29,30,31,32). Distorted sexual stereotypes are typical of many marginalized social groups (33,34), but unlike people stigmatized for their gender, economic status, or race, tens of thousands of "feeble-minded" US citizens were forcibly institutionalized, segregated by sex, and denied the right to have sexual relations and raise families (35,36,37,38). African Americans with disabilities were sometimes turned away from institutions and sent to prison instead (39).

Goddard believed segregation was the ultimate solution to feeble-mindedness and mentioned sterilization merely as a "makeshift" measure until enough facilities could be built (24, p. 117). However, by the 1920s it became apparent that it was too expensive to maintain such a large population in segregated institutions, even when the higher-functioning residents performed tasks to defray the cost of their upkeep (19). Sterilization and parole came to be seen as a more economically viable solution.

Harry H. Laughlin of the Eugenics Record Office drafted a model law, which included a list of ten "socially inadequate" groups targeted for sterilization:

> 1) feeble-minded; 2) insane (including psychopathic); 3) criminalistic (including the delinquent and wayward); 4) epileptic; 5) inebriate (including drug habituees):

> 6) diseased (including the tuberculous, the syphilitic, the leprous, and others with chronic, infectious, and legally segregable diseases); 7) blind (including those with seriously impaired vision; 8) deaf (including those with seriously impaired hearing: 9) deformed (including the crippled); and 10) dependent (including orphans. ne'er-do-wells, the homeless, tramps and paupers) (40).

In the wording of this law, "the state eugenics agent was empowered to investigate a person's heredity, to make arrests, and to cause the offender to be sterilized" (38, p. 35). Between 1907 and 1931 eugenics laws permitting the involuntary sterilization of criminals, degenerates, and imbeciles were passed in 30 states. Sterilization might entail tubal ligation or even full removal of the uterus, a much more complicated and expensive procedure (38, p. 36).

The sterilization of Carrie Buck in 1927 was the precedent for the large-scale movement throughout the United States to sterilize people diagnosed as mentally deficient. At the time of the court case, Buck, her mother, and her sister were residents of the State Colony for Epileptics and Feeble-minded in Lynchburg, Virginia. It was later determined that, as with Deborah Kallikak, none of them would be defined today as having cognitive disabilities. Between the ages of 3 and 17, Buck was the "foster-child" (i.e., unpaid servant) of the Dobbs family of Charlottesville, Virginia. According to Buck, she became pregnant in 1923 after being raped by the Dobbs' nephew. The Dobbs family then arranged for Buck, after cursory testing, to be certified as feebleminded and institutionalized—just as her mother had been three years previously.

In 1924, a eugenic sterilization law was passed in Virginia. Alfred Priddy superintendent of the State Colony, attorney Aubrey Strode, working closely with the Eugenic Records Office in New York, decided to use Carrie Buck as a test case to determine the constitutionality of the law. Through the court case Buck vs. Bell, the law allowing eugenic sterilization was upheld by the state of Virginia in 1925 and the United States Supreme Court in 1927 (27). Supreme Court Justice Oliver Wendell Holmes, Jr. wrote:

> She may be sexually sterilized without detriment to her general health and that her welfare and that of society will be promoted by her sterilization. We have seen more than once that the public welfare may call upon the best citizens for their lives. It [the state] would be strange if it could not call upon those who already sap the strength of the State for these lesser sacrifices, often not felt to be such by those concerned, in order to prevent our being swamped with incompetence. It is better for all the world, if instead of waiting to execute degenerate offspring for a crime, or let them starve for their imbecility, society can prevent those who are manifestly unfit from breeding their kind. . . . Three generations of imbeciles are enough. (41)

Carrie Buck was sterilized in 1927 and released into the community as a domestic servant. Her sister Doris, aged 16, was also sterilized and paroled (27). Doris Buck had been told the operation was an appendectomy and was unaware she was sterilized until 1979 (27, p. 216). She was later part of a successful lawsuit undertaken by the American Civil Liberties Union (ACLU) in 1980 on behalf of the 8,300 people sterilized in Virginia institutions between 1927 and 1974 (27, p. 251). Winifred Kempton and Emily Kahn (1991) reported that between 1907 and 1957 roughly 60,000 individuals, a conservative estimate, were involuntarily sterilized in the United States, many without being told (42, p. 96).

Although the eugenics movement was discredited after World War II because of the extreme measures taken by Nazi Germany, sterilization was still advocated and used (both legally and illegally) in the United States through the 1970s. Although policy-makers dropped this strategy for social control, certain doctors and social workers continued to advocate for the practice on an individual level, sometimes with the support and encouragement of the woman's parents. After the 1980s, due to several high profile lawsuits such as the ACLU suit in which Doris Buck participated, doctors became less willing to perform the procedure without a clear legal mandate. Yet sterilization was still considered an option by many families wishing to "protect" their daughters.

THE STERILIZATION OF CINDY WASIEK

The 1994 sterilization of Philadelphia resident Cindy Wasiek, following a seven-year court battle, shows that the nonconsensual sterilization of women with cognitive disabilities is still advocated and practiced, although such practice is sometimes contested (43,44). Cindy Wasiek was described in newspapers as having a mental age of 5, and being "severely retarded." Her mother, Dorothy Wasiek, feared her daughter might be raped and become pregnant. Because of antiseizure medications, she could not place her daughter on contraceptive medication. She decided that sterilization would be the best way to protect her daughter. The central theme in this story was not Cindy Wasiek's safety, but rather how to allay her mother's fears. Cindy Wasiek's entire life was structured on her mother's fear of her being raped and becoming pregnant. She was even placed in a group home where all residents and staff were women. Most parents do not have the power to make decisions about the sexuality and fertility of their children. However, parental and public perceptions considered Cindy Wasiek's disability to be so severe that she was prevented from being an active participant in the life decisions that concerned her. Cultural perceptions that people with cognitive disabilities are perpetual children allow parents to influence or control all aspects of their adult lives.

People with disabilities are more likely to experience sexual abuse than the general population, but the chances of Cindy Wasiek being raped and becoming pregnant were statistically remote (45). In any case, sterilization is not a protection from rape or sexually transmitted diseases. Based on the argument that a mother should have the authority to decide what is best for her daughter, Dorothy Wasiek had her daughter sterilized after Supreme Court Justice Souter refused to grant what is ordinarily a routine stay until the court had heard the case (43). Although this was not eugenic sterilization, the opportunity was available because of the continued existence of a law (passed by virtue of the eugenics movement) allowing nonconsensual sterilization. Because of the legacy of the eugenics movement, involuntary sterilization of people with cognitive disabilities remains culturally acceptable in the United States, although it would be unacceptable for use on other marginalized groups. Individuals with cognitive disabilities continue to be denied the rights granted to other United States citizens. Decisions regarding their bodies and their lives continue to be made without their consent.

GLEN RIDGE SEXUAL ASSAULT

On March 1, 1989, exactly one hundred years after Deborah Kallikak was institutionalized at Vineland, New Jersey, a young woman with a cognitive disability was playing ball in a neighborhood park in Glen Ridge, New Jersey. A group of young men from her high school, many of whom had known her for over a decade, came up and promised her a date with a popular high-school athlete if she would accompany them to a nearby house. When they arrived at the house, where two of the young men lived, the woman was told to undress and perform various sexual acts on herself and several of the young men. Eventually, some of them took turns inserting a fungo bat, a broom handle, and a stick into her vagina (46).

In the winter of 1992–1993, when three of these young men were tried for sexual assault, a defending attorney criticized the young woman's mother because "she took no measures to protect young men from her daughter" (47). The defense was attempting to prove that the young woman "craved sex" (48) and was "aggressive in her attitude and approach toward boys" (49). This is the modern legacy of the eugenics scholars who wrote about the "immoral" and "uncontrollable sexual nature" of women with cognitive disabilities. Defense lawyers in sexual-assault cases will sometimes assert that a woman is sexually promiscuous, but how often is the victim presented as a social threat? The defense's statements evoke images not of mere promiscuity, but of a sexually dangerous woman from which young men must be protected. This image was affirmed in the courtroom by a psychiatrist, and outside the courtroom by neighbors who, according to one journalist, "spoke of her as some kind succubus, with unknowable needs and unfathomable wants" (50).

The young woman grew up with the young men who assaulted her. Her sister stated that, as a child, she was pinched and called "piggy," "dummy," and "retarded" by neighborhood children. Once, she was tricked into eating dog feces by a group of children that included two of the young men on trial (51). Journalist Anna Quindlen wrote:

> They behaved as though she were an inflatable doll, an inanimate object. Subtract the stereotypes about loose girls and uncontrollable male urges, and you come up with a clear picture of what went on in that basement: young men doing a cruel and reprehensible thing to a woman they chose specifically because they knew her limitations and tractability. This case isn't about boys being boys. It's about boys being predators. I guess it wasn't much of a leap, from the dog feces to the broomstick. (52)

During the trial, both the prosecution and the defense attempted to use stereotypes about people with cognitive disabilities to their advantage. Instead of focusing on the character and history of the perpetrators, discourse revolved almost exclusively on the young woman's sexuality or vulnerability to abuse. Researcher Bernard Lefkowitz (46) found a pattern where many nondisabled young women in Glen Ridge were abused and harassed by this same group of young male athletes, but these events were never mentioned during the trial. Instead, the defense resurrected images of the disabled woman's obsessive sexuality that, although discredited decades ago, still have a powerful presence in our culture. The imagery used by prosecution was evocative of Goddard's description of Deborah Kallikak. They described the young woman from Glen Ridge as innocent, but yearning for social acceptance and so eager to please that she was incapable of saying "no."

In order to prove she was "mentally defective," the prosecution agreed to forego protection afforded by the rape shield law. Unlike most sexual assault trials, where information on sexual history is barred by law, the young woman's experiences were discussed in minute detail. The defense argued that "the case's complexities forced them to explore the woman's sexual past to prove that she knew what she was doing and wanted it" (53). The prosecutors made no objection, asserting that the woman's sexual history supported their contention that she was "mentally defective," as defined by New Jersey's sexual assault laws, and thus incapable of understanding her right to refuse sexual activity (54).

The jury was persuaded by this argument. In 1993, the three young men were found guilty of sexual assault, and sentenced to remain for an "indeterminate" minimum term of imprisonment in what was described as a "campuslike complex for young offenders" (55). They were immediately free on bail and remained so for the duration of the appeals process. They began serving prison time in 1997.

Both the woman's lawyers and the journalists covering the case continually referred to her pliability, low self-esteem, and passivity. It was repeatedly asserted that her "mental age" was 6 or 8 and that her I.Q. was 64 at most. They were more interested in what she was than in who she was. Without the issue of "mental defect," however, the case would have been difficult, if not impossible, to win.

THE OTHER SISTER

In complete contrast to nonfiction examples above, the 1998 movie *The Other Sister* presented non-disabled actress Juliette Lewis as Carla Tate, a beautiful, determined, and resourceful woman with a cognitive disability who was institutionalized as a child by her wealthy parents. At the beginning of the movie she left this elite facility to return home to her family. The story centered around Carla's efforts to develop professionally and personally. She passed a college level course. She achieved independence by getting her own apartment. She fell in love, found a life partner, and married him. Ultimately, she succeeded at everything she wanted to achieve. The only significant barrier to Carla's success was her overprotective and controlling mother, who resisted all of Carla's efforts to achieve independence.

Needless to say, most people with cognitive disabilities living in the real world are not able to sail so effortlessly past the financial, interpersonal, social, educational, and sexual barriers placed in their path. Realistically, success in a single one of these areas constitutes a major victory. Most people with or without disabilities do not conform to Hollywood standards of physical beauty. Most people with disabilities neither have the money that Carla had at her disposal nor the many opportunities that resulted from her access to money. Most do not have such extensive social support networks of loving and understanding family and friends as did Carla and the man who became her husband. Carla's easy victories trivialize the real-life struggles faced by people with disabilities. Other than a little teasing on the college campus, no loneliness, rejection, prejudice, or hatred mars the movie's perfect fantasy. Carla presents such a contrast to the other women portrayed here that one must wonder about the significance of her invention. The movie represents people with disabilities as wanting and achieving the same things to which all United States citizens are supposed to aspire, and most never fully attain. When Carla succeeds, she represents not just people with disabilities but all people who strive and dream despite all barriers; she symbolizes the universal potential for success. How unfortunate that this easy success is an illusion.

Michel Taussig (1987) wrote about a culture of terror in which the line between fiction and reality is blurred saying "the unstable interplay of truth and illusion becomes a phantasmic social force" (56, p. 121). Although Taussig was speaking of societies whose foundations were formed through atrocities of violent colonialism, I suggest that the concept of a culture of terror also applies to the atrocities experienced by women with cognitive disabilities living in the United States during the twentieth century. Just as the colonial native was vilified through images of violent savagery and idealized through saintly images of shamanic healing, so do women with cognitive disabilities in the United States symbolize both demonic succubae and heroic Cinderellas. Through the lens of *The Other Sister,* the wrongs of the past and the barriers in the present are symbolically neutralized. In an illusory process of symbolic healing, the lived experiences of real women such as "Deborah Kallikak," Carrie Buck, Cindy Wasiek, and the young woman from Glen Ridge are justified, trivialized, and forgotten.

CONCLUSION

It is clear from these stories that despite advances in recent decades, disability remains a stigma in the United States, especially mental disability. Although *The Other Sister* remains largely a fairytale, it is true that higher education, independent living, marriage and child rearing are realistic goals for many people with cognitive disabilities. However the lives of people with cognitive disabilities are still consistently judged to have less worth than the lives of others'. . . .

By considering the lives of people with disabilities "intolerable" and the people themselves "better off dead," by infantilizing people with disabilities, or treating them like sexual monsters, a cycle of discrimination and abuse is perpetuated. . . .

Despite the many barriers to overcome, the future is hopeful. Women and men with cognitive disabilities in national and international self advocacy movements such as Self Advocates Becoming Empowered are working together to achieve for inclusion into US communities and cultures and equal protection under the law. They are asserting their rights to care, counseling and educational services, and empowering people with cognitive disabilities to assert, protect, and heal themselves.

REFERENCES

1. Goffman, E: Stigma: Notes on the Management of Spoiled Identity. New Jersey, Prentice Hall, 1963.
2. Mercer, J: Labelling the Mentally Retarded: Clinical and Social System Perspectives on Mental Retardation. Berkeley, University of California Press, 1973.
3. Waxman, B. F: Hatred: The Unacknowledged Dimension in Violence Against Disabled People, Sexuality and Disability 9(3):185–99, 1991.
4. Asch, A., Fine, M: "Introduction: Beyond Pedestals." In Women with Disabilities: Essays in Psychology, Culture and Politics. Philadelphia, Temple University Press, 1988. pp. 1–37.
5. Thompson, R. G: Extraordinary Bodies: Figuring Physical Disability in American Culture and Literature. New York, Columbia University Press, 1997.
6. Fausto-Sterling, A: Life in the XY Corral. Women's Studies International Forum 12(3): 319–331, 1989.
7. Fausto-Sterling, A: Myths of Gender: Biological Theories About Women and Men. New York, Basic Books. 1985.
8. Harding, S: The Science Question in Feminism. Ithaca, Cornell University Press, 1986.
9. Haraway, D: A Manifesto for Cyborgs: Science Technology and Socialist Feminism in the 1980s. Socialist Review 90:65–105, 1985.
10. Haraway, D: Situated Knowledges: The Science Question in Feminism and the Privilege of Partial Perspective. Feminist Studies, 14(3):575–596, 1988.
11. Hubbard, R: The Politics of Women's Biology. New Brunswick, Rutgers University Press, 1990.
12. Kuhn, T. S: The Structure of Scientific Revolutions (2nd ed.). Chicago, University of Chicago Press, 1970.
13. Martin, E: The Egg and the Sperm: How Science has Constructed a Romance Based on Stereotypical Male-Female Roles. Signs 16(3): 485–501, 1991.
14. Mies, M: Sexist and Racist Implications of New Reproductive Technologies. Alternatives. 12:323–42, 1987.
15. Strathern, M: Reproducing the Future: Essays on Anthropology, Kinship and New Reproductive Technologies. New York, Routledge, 1992.
16. Block, P: Biology, Culture, and Cognitive Disability: Twentieth Century Professional Discourse in Brazil and the United States. Duke University, Unpublished Doctoral Dissertation, 1997.
17. Rafter, N. H: Claims Making and Socio-Cultural Context in the First US Eugenics Campaign. Social Problems, 39(1): 17–34, 1992.
18. Rafter, N. H.: White Trash: Eugenics as Social Ideology. Society 26(1):43–49, 1988.
19. Trent Jr., J. W: Inventing the Feeble Mind: A History of Mental Retardation in the United States. Berkeley, University of California Press, 1994.
20. Schilling. R. F., Schinke, S. P: Mentally Retarded Sex Offenders: Fact, Fiction and Treatment. Journal of Social Work and Human Sexuality 7(2):33–48, 1989.
21. Perske, R: Unequal Justice: What Can Happen When Persons with Retardation or Other Developmental Disabilities Encounter the Criminal Justice System. Nashville, Abingdon Press, 1991.
22. Gould, S. J: The Mismeasure of Man, New York; W. W. Norton and Co., 1981.
23. Smith, D. J: Minds Made Feeble: The Myth and Legacy of the Kallikaks, Maryland, Aspen Systems Corp., 1985.
24. Goddard, H. H: The Kallikak Family: A Study in the Heredity of Feeble-Mindedness. New York, The Macmillan Company, 1913.
25. Fairchild, H. P: The Melting Pot Mistake. Boston; Little Brown and Company, 1926.
26. Osborn, F: Preface to Eugenics. New York, Harper & Brothers, 1940.
27. Smith, D. J. and Nelson, K. R: The Sterilization of Carrie Buck. New Jersey, New Horizon Press, 1989.
28. Abramson Paul R. et al: Sexual Expression of Mentally Retarded People: Educational and Legal Implications. American Journal of Mental Retardation 93(3): 328–34, 1999.
29. Edmonson, B. et al: What Retarded Adults Believe about Sex. American Journal of Mental Deficiency 84(1):11–18, 1979.
30. Heshusius, L: Sexual Intimacy, and Persons We Label Mentally Retarded: What They Think—What We Think. Mental Retardation 20(4): 164–8, 1982.
31. Kempton, W: The Sexual Adolescent Who is Mentally Retarded. Journal of Pediatric Psychology 2(3): 104–7, 1977.

32. Sinason, V: Uncovering and Responding to Sexual Abuse in Psychotherapeutic Settings. In Thinking the Unthinkable: Papers on Sexual Abuse and People with Learning Difficulties. H. Brown and A. Craft (eds.). London, FPA Education Unit, 39–49, 1988.
33. hooks, bell: Selling Hot Pussy: Representations of Black Female Sexuality in the Cultural Marketplace, Black Looks: Race and Representation. Boston, South End Press, 1992.
34. Parker, R. G: Bodies, Pleasures, and Passions: Sexual Culture in Contemporary Brazil. Boston, Beacon Press, 1991.
35. Blank, R. H: Fertility Control: New Techniques, New Policy Issues. New York, Greenwood Press, 1991.
36. Gordon, L: Women's Body, Women's Right. New York, Penguin, 1976.
37. Reilly, P. R: The Surgical Solution: The History of Involuntary Sterilization in the United States. Baltimore, Johns Hopkins University Press, 1991.
38. Shapiro, T. M: Population Control Politics: Women, Sterilization, and Reproductive Choice. Philadelphia, Temple University Press, 1985.
39. Noll, S: Feeble-Minded in Our Midst: Institutions for the Mentally Retarded in the South 1900–1940. Chapel Hill. The University of North Carolina Press, 1995.
40. Lauglin, H: The Legal Status of Eugenical Sterilization, Washington: Eugenics Record Office, 1993. Cited in Rodriguez-Trias, H., Sterilization Abuse. In Biological Woman: The Convenient Myth. R. Hubbard. M. S. Henifin and B. Fried (eds.). Cambridge, Schenkman Publishing Company, Inc. pp. 147–160, 1982.
41. Buck vs. Bell, quoted in 38, p. 3, 1927.
42. Kempton, W., Kahn E: Sexuality and People with Intellectual Disabilities: A Historical Perspective. Sexuality and Disability 9(2):93–111, 1991.
43. Bowden, M: A Fight Over Sterilization May Finally Be Finished. Philadelphia Inquirer, A1, A14, Sat. November 12, 1994.
44. Goldberg, D: Woman's sterilization may end long Pennsylvania legal fight. Washington Post, A:2 col. 5, January 7, 1995.
45. Sobsey, D: Violence and Abuse in the Lives of People with Disabilities: The End of Silent Acceptance. Baltimore, Paul H. Brookes Publishing Co., 1994.
46. Lefkowitz, B: Our Guys: The Glen Ridge Rape and the Secret Life of the Perfect Suburb. Berkeley, University of California Press, 1997.
47. Hanely, R: Prosecutor Mocks Defense in Trial on Sexual Assault. New York Times B6: Col. 5. February 18, 1993b.
48. Hanley, R: Sex-Assault Trial Stresses Woman's Past. New York Times A26: Col. 4, October 31 1992a.
49. Hanley, R: Young Woman Was Aggressive Toward Boys, Psychiatrist Says. New York Times, B5: Col. 5, January 1, 1993a.
50. Junod, T: Ordinary People: Were the Suburban Youths Who Raped a Retarded Girl "Star Athletes" or "Just Typical Kids?" Sports Illustrated 78(12):68, March 23, 1993.
51. Hanley, R: Sister Calls Woman in Assault Case Pliable. New York Times B6: Col. 5, November 11, 1992c.
52. Quindlen, A: 21 Going on 6. New York Times, Sec. 4:17, December 13, 1992.
53. Manegold, C. S: A Rape Case Worries Advocates for the Retarded. New York Times Sec. 4:3, March 14, 1993.
54. Hanley, R: Accuser's Past at Issue in Assault Case. New York Times, late edition B26, November 5, 1992b.
55. Nieves, E: Sentences in Sexual Assault Divide Glen Ridge Jurors. New York Times Sec 1:48, April 25, 1993.
56. Taussig, M: Shamanism, Colonialism, and the Wild Man. Chicago: University of Chicago Press, 1987.

NOTES

1. The traits still form an important part of modern definitions for mental retardation. For a more detailed exploration of the evolution of definitions for mental retardation, see the first chapter of Block 1997.

32

Peaks and Valleys

The Gendered Emotional Culture of Edgework

BY JENNIFER LOIS

Jennifer Lois explores the "gendered emotional culture of high-risk takers," specifically a volunteer search-and-rescue group made up of women and men, with whom she did more than five years of ethnographic fieldwork. She describes the intense emotions and emotion work of rescuers before, during, and after the most dangerous rescues. Lois identifies four stages of edgework or voluntary risk-taking and the events and emotions that marks each stage, presenting a distinct model of gendered emotional culture.

1. What were the major differences in the emotional experiences and management techniques of the women and men rescuers? Similarities?
2. How does "doing gender" shape the experience of adrenaline rushes?
3. How and why was the feminine "emotion line" downgraded?

* * *

Action in emergency situations calls for rational thinking in the face of potentially overwhelming emotions. This article is about how male and female members of "Peak," a volunteer search and rescue group, "managed" their emotions (Hochschild 1983) before, during, and after their most dangerous, stressful, or gruesome rescues. At times their emotions and corresponding management techniques were consistent with broader gender stereotypes; at other times they were not. In this article, I explore these similarities and differences, showing how rescuers negotiated the (sometimes) conflicting demands of both gender and emotion norms in high-risk crisis situations. I also reveal how these negotiated courses of action resulted in power differences between the women and men in Peak. . . .

Lyng (1990) has noted that men tend to take more physical risk than women and that they are more likely to engage in "edgework" (a term he borrows from Hunter S. Thompson [1971]). Lyng details how risk takers, such as skydivers and firefighters, negotiate the boundary, or "edge," between safety and danger, defining "the archetypical edgework experience [as] one in which the individual's failure to meet the challenge at hand will result in death or, at the very least, debilitating injury" (1990, 857). He further contends that the edgework concept encapsulates a wider array of activities in which individuals also need to negotiate the edge or boundary line between two physical or mental states: "life versus death, consciousness versus unconsciousness, sanity versus insanity, an ordered sense of self and environment versus a disordered self and environment" (1990, 857). Thus, although the quintessential edgework experience is life threatening, the concept also has a broader application that extends beyond pure physical danger.

Edgework gives individuals a feeling of control over their lives and environment while they push themselves to their physical and mental limits. At a psychological level, surviving the edge leads them to experience intense highs. It is this sensory experience on the edge that compels edgeworkers to pursue it repeatedly, each time pushing their physical and mental limits further to control the seemingly uncontrollable (Lyng 1990; see also Palmer 1983). Although it appears that emotions play an important role in edgeworkers' motivations to continue pursuing such high-risk activity, little consideration has been given to what seems to be the prominent emotional culture of edgework.

In this article, I examine the gendered emotional culture of edgeworkers. Through their search and rescue activity in the mountains of the western United States, the members of Peak Volunteer Search and Rescue experienced many physically and emotionally threatening situations, such as searching for missing skiers in avalanche-prone terrain and extracting mutilated bodies from planes that crashed in the wilderness. These extreme conditions—the most crucial life-and-death circumstances—called for members to engage in edgework. They had to be able to complete their task under intense stress, and members who could remain in control during risky situations—those whose edge was farther out—were more often sent on challenging rescues because they were considered better suited to handle them. Although it was important for them to have the skills to accomplish the mission, it was more important that they were able to regulate the intense feelings that arose from such dangerous or gruesome tasks; uncontrolled feelings rendered them useless. . . .

SETTING AND METHOD

These data are drawn from a five-and-one-half-year ethnographic study of Peak, a volunteer search and rescue group in a Rocky Mountain resort town. Peak County consisted of 1,700 square miles, 1,300 of which were undeveloped national forest or wilderness area lands. Local residents and tourists alike used this "backcountry" land year-round for various recreational purposes such as hiking, camping, rock climbing, white-water rafting/kayaking, snowmobiling, and backcountry skiing. Occasionally, recreational enthusiasts became lost or injured in these vehicle-inaccessible regions. Because the county sheriff's deputies did not have the skills or resources to venture into these remote areas, the sheriff commissioned Peak, a volunteer group of local citizens, to act as the public safety agent in the backcountry. . . .

Peak's members had to have many specialized rescue skills to reach and help victims who were incapacitated doing a wide variety of recreational activities. For example, some members were adept at riding snowmobiles and were frequently sent to search for lost snowmobilers or backcountry skiers. Others possessed extensive white-water skills, and as such, their expertise was used for rafting or kayaking accidents. Many members, however, had only basic skill levels in several areas, for example, operating the rope and pulley systems used to maneuver victims and rescuers over cliffs, surviving for several days in the wilderness, and searching avalanche debris with radio signal receiving devices. . . . Of the 30 or so members in Peak, approximately 20 were men, and 10 were women.[1] All members were white, and most were middle class to upper-middle class. Their ages ranged from 22 to 55, and their education levels ranged from high school to the MD degree. . . .

I became interested in Peak after reading several local newspaper accounts of their rescues. With no specialized backcountry skill or experience, I joined the group in 1994 to study it sociologically. I began attending the biweekly business meetings, weekly training sessions, posttraining social hours at the local bar, and a few missions. Through these initial interactions, I became intrigued by how members defined their participation in rescue work and how these definitions affected their lives (Blumer 1969). . . .

PEAKS AND VALLEYS

The levels of difficulty, danger, and stress varied greatly among Peak's missions. At times, members were asked to perform only slightly demanding, low-urgency tasks such as hiking a short distance up a trail to carry a hiker with a twisted ankle out to the parking lot. Other times, they were asked to perform very difficult, dangerous, or gruesome tasks such as entering a potential avalanche zone to search for a missing skier, negotiating the rapids of a rushing river to reach a stranded rafter, being lowered down a cliff face to rescue an injured rock climber, or recovering a body from a violent accident. It was these physically and emotionally demanding situations that most threatened rescuers' sense of control, requiring them to engage in edgework—to negotiate the boundary between order and chaos—not only during the missions but before and after as well.

There were four stages of edgework that members experienced in Peak's missions: preparing for the edge, performing on the edge, going over the edge, and extending the edge. These stages were distinctly marked not only by the flow of rescue events but also by members' feelings in each stage. Yet, despite passing through the same stages of edgework, women and men experienced edgework differently, interpreting and managing feelings in gender-specific ways before, during, and after the missions.

Preparing for the Edge: Anticipating the Unknown

Missions were variable events, and members were often required to use whatever resources they had to accomplish their task. Generally, the men in the group found it exciting not to know what to expect from a rescue, and they felt challenged by the prospect of relying on their cognitive and technical skills to quickly solve any puzzle that suddenly presented itself. Peak's female rescuers, however, tended to view the missions' unpredictability as stressful, and they worried in anticipation about performing under certain conditions.

Women commonly worried that they might be physically unable to perform a task either because they would not be strong enough or because they would not know what to do. . . .

Not only did Elena, a member of four years, worry about her preparedness to help victims, she also saw this apprehension as problematic; importantly, she felt that her lack of confidence in her ability was the source of the problem, not her ability level itself.

Women also tended to worry about their ability to maintain emotional control, realizing that they could encounter a particularly upsetting scene on a mission. For example, Maddie, a 10-year member, told me that one situation she dreaded was encountering a dead victim whom she knew. She expected that this situation would be one that most threatened her emotional control. . . .

Worrying about what could arise on a future mission compelled many women to make a plan of action ahead of time, speculating about their potential reactions to stressful events. Preparing for edgework by imagining numerous different scenarios gave them some sense of control over the unpredictable future, and through such planning, they were able to manage their uncomfortable anticipatory feelings about the unknown, a dynamic found in other research on high-risk takers (Holyfield 1997; Lyng 1990).

Maddie's statement also typified another technique many female rescuers used in conjunction with planning and rehearsing future scenarios: They set low expectations for themselves. Part of their planning process was to prepare for the most demanding possible situations, the ones in which they were most likely to fail. This emotion management strategy served two functions. First, it made women acutely aware of their progress toward the edge on missions. Maddie said that on "gruesome" missions, she remained highly cognizant of her emotional state, always prepared to hand off her task to someone else. The second function of women's low expectations was that they would probably perform beyond them, which allowed them to remain within their limits while feeling good about surpassing their expectations.

Anticipating a poor performance was not very common among the men in the group, however. Most of the men in Peak used the opposite technique—sheer confidence—to prepare for emergency action. Brooke, a four-year member in her late 20s, told me that two seven-year members, 28-year-old Gary and 32-year-old Nick, were able to perform at very high levels because of their high expectations for themselves:

> I think that both of those guys see themselves as Superman. Which is not necessarily a good thing. They sort of see themselves as being invincible [and] I think that they might test their physical limits more than I would. They might go into a situation that I would stand back and say, "I don't think that's safe." But they're convinced that nothing's going to happen to them. . . .

[M]en were extremely confident too, even when they were accused by others of overestimating their own ability. Roger, a 27-year-old member of six years, was highly experienced. He described his ability to assess avalanche danger, a highly unpredictable phenomenon, as better than most other members'. On one occasion when the team was practicing avalanche skills, Roger walked out to the edge of a cornice (a windblown pile of snow overhanging a steep hill or cliff, which can break off and cause an avalanche). Another highly experienced rescuer, Shorty, questioned Roger's judgment because if he broke the cornice and caused an avalanche, he could easily have been swept up in it, carried down the mountain, and buried under several feet of heavy snow. Roger was angered because he was very confident in his ability to assess how far out he could walk on any cornice without breaking it, reasoning that his past experience gave him this knowledge. . . .

Roger's confidence helped him prepare for edgework in the event of a real mission. By stating where he thought cornices would break, he was quite literally reaffirming his ability to assess the edge—the boundary between safety and danger—which allowed him to feel in control of the situation. . . .

Many of Peak's men used these confidence displays to assert that they could outperform each other, which created a highly competitive environment. I witnessed many of these bravado sessions, mostly during social hours at the bars, where men discussed their own strengths as they anxiously awaited the opportunity to prove themselves on future missions. Several gender scholars have suggested that masculinity, but not femininity, must constantly be proven—that men are "only as masculine as [their] last demonstration of masculinity" (Beneke 1997, 43; see also Connell 1987; Kimmel 1996; Messner 1992). Thus, perhaps Peak's men were not only anticipating a chance to prove themselves as rescuers but also as men.

Wanting to prove oneself and having confidence was much less tolerated for women in Peak, however. For example, when Robin, a six-year member in her early 30s, did display confidence about her abilities and experiences, other members doubted and criticized her. Although she held high rescue certifications and had a great deal of experience in many of the activities required for search and rescue, such as white-water rafting and searching, other rescuers found her to be strange and weird, basing these judgments on her displays of confidence. Elena described what she thought bothered her (and others) about Robin:

> She brags a lot. About herself. And she shows off a lot. And I think that people perceive that as not necessarily a good thing. . . . Sometimes I wonder if she doesn't feel like she just needs to always prove herself.

Clearly, Robin was being held to different standards than male rescuers, as she was negatively sanctioned for performing what I observed to be "masculine" behavior. . . .

There are several explanations for this gendered difference in preparation strategies. First, in general, men were more experienced than women. Through their own recreation as well as through group-related activities, men's exposure to risk was both more frequent and more hazardous than women's. Yet, this gendered confidence pattern was not totally explained by differential risk exposure. For example, when I talked to equally experienced men and women, apprehension still dominated women's anticipatory feelings (except for Robin's), and confidence dominated men's. Furthermore, even when women performed well on missions, it did not seem to boost their confidence for future situations, while conversely, men's poor performance did not erode theirs.[2]

A second factor in explaining this pattern was that the masculine nature of rescue work made men feel more at ease in the setting, and thus they tended to display unwavering certainty that they could handle any situation that might arise. Women felt disadvantaged by this masculine environment and, taking this into account, set low expectations for themselves. In one way, their feelings were based in reality: They were aware that, on the whole, the men in the group were physically stronger and thus able to perform harder tasks than they. . . .

In another way, though, women's insecurities were due to cultural and group stereotypes about men's superior rescue ability. For example, the belief that men are emotionally stronger than women made women question whether they would be able to perform edgework in potentially upsetting situations, while the same stereotype enabled men to have confidence that they would maintain control in those situations. Yet, my observations (discussed later in this article) yielded no gendered pattern of emotional control. Another stereotype that made women worry about their rescue ability was the belief that men were more technically inclined than they were. This stereotype came into play during trainings and missions when the group used any kind of mechanization, such as rope and pulley systems, helicopters, or snowmobiles. Cyndi, a three-year member in her late 20s, told me she felt "hugely" intimidated in her first year by the technical training, yet she later became quite adept in setting up and operating rope and pulley systems. Elena had a similar experience. She told me that during her first training, she looked around at "all the guys" and thought, "What am I doing here? I'm not even qualified for any of this." Cyndi's and Elena's feelings of inferiority acted as "place markers" where "the emotion conveys information about the state of the social ranking system" (Clark 1990, 308). In this case, they felt inferior because performing gender-appropriate behavior, or "doing femininity" (West and Zimmerman 1987), was inconsistent with "doing" edgework, which they perceived as a much more "masculine" endeavor.

Because women felt this gendered tension, they may have devised these distinct ways to moderate it. By remaining trepidacious and maintaining low expectations that they would often exceed, the women reaffirmed their place in the group as useful. . . .

Performing on the Edge: Suppressing Feelings

During Peak's urgent missions, clear thinking and rational action, core features of edgework, were seen as especially crucial. However, in such demanding situations, members' capacity for emotional and physical control was seen as more tenuous: Emotions threatened to push them over the edge, preventing them from physically performing at all. Rescuers who were easily scared, excited, or upset by a mission's events were considered undependable. Members employed several strategies to control these feelings during the missions, allowing them to perform under pressure.

Rescuers were particularly wary of the onset of adrenaline rushes because such potent physiological reactions threatened their composure; they felt that the emotions they experienced could "get in the way" of their performance. Yet, adrenaline was not totally undesirable; in fact, at lower levels, both male and female rescuers welcomed it because it helped them focus and heightened their awareness. Mostly, though, Peak's emotional culture cast adrenaline rushes—the involuntary physiological response that causes increased heart rate and breathing—as an important situational cue, one that rescuers should heed as a warning that they were at risk of losing control. . . .

After exploring what rescuers meant by the term *adrenaline rush,* I discovered they were actually referring to two distinct (and potentially problematic) emotional states associated with adrenaline: fear and urgency.

Excessive fear was dangerous because it could paralyze rescuers, rendering them ineffective and thus increasing risk for both their teammates and the victim. . . .

Loss of control due to fear, however, was almost always associated with women's reactions to adrenaline, while becoming too "excitable" or feeling excessive urgency was only associated with men's reactions to adrenaline. Like too much fear, members considered excessive urgency to be a detrimental emotion during missions because it could cause rescuers to act irrationally and thus, unsafely. On one occasion, the mission coordinator selected two experienced members to enter a dangerous avalanche gully and evacuate a snowboarder who had a broken leg. While they were tending to the victim, two newer members, Patrick and Mitch, skied down into the gully without getting authorization from the mission coordinator, which endangered the two experienced members already in the gully by creating the possibility of triggering an avalanche above them. Furthermore, Patrick was unable to ski the difficult terrain, causing him to spend extraordinarily more time in the gully, thus exponentially increasing the risk to himself and others. Group members accused Patrick and Mitch of letting their adrenaline override rational, controlled action: They skied into the gully because it was exciting and risky, not because they were needed on the scene. . . .

One reason men and women might have experienced adrenaline rushes differently was because they were simultaneously "doing gender" (West and Zimmerman 1987). Men tended to be confident at the prospect of undertaking risk, which may have caused them to interpret their adrenaline during the mission as pleasurable and exciting. Since women, on the other hand, reported more cautious mind-sets in preparing for missions, worrying about their ability to exercise emotional and physical control in risky situations, perhaps they were more likely to define their adrenaline rushes as fear. It is important to note, however, that while too much adrenaline, whether it be feminine fear or masculine urgency, was considered detrimental to the safety of the missions, urgency was less stigmatized than fear, perhaps because it was seen as easier to conquer or because it was a more pleasurable sensation than fear. As a result, women had a harder time gaining status from their successful mission performances because they (and others) defined their adrenaline as fear.

Nevertheless, men and women managed their feelings of urgency and fear similarly: They suppressed them. For example, the most critical mission I experienced had four casualties. A van had driven off the side of a dirt road and tumbled to the bottom of a 400-foot ravine. Search and rescue was called because the accident was inaccessible to the paramedics, who needed ropes to get down to the victims and a hauling system to get them out. Cyndi told me that while on that mission, she was in control of her emotions, successfully suppressing them, because she was working the rope systems up on the road, unable to see beyond the drop-off down to the accident site. She felt differently, however, when one of the accident victims reached the top of the hill in a panicked state. The victim, who had a broken arm, had managed to climb up the 400-foot embankment in an effort to catch up to the rescue team who was evacuating her critically injured mother. Cyndi was thrown off kilter by this sight:

> Because I was up at the top, it wasn't real. You know, I could sort of disassociate, it's like, "Okay, let's just get the job done and not think about it." But then you're meeting this person [climbing out of the accident scene] who is just out of it. I mean, she was panicked and [she had] adrenaline [rushing], and I was just kind of like, "Okay, there

> really are real people down there, but I'm not gonna get panicked. I need to calm this person down, because she's not gonna help rushing up to the scene, and getting in the way of the paramedics [while] trying to get to her mother."

Cyndi's emotional control was threatened when the victim emerged from the trauma scene. The sight forced her to the edge, where her ordered, controlled action was threatened by her feeling of chaotic, uncontrolled panic. She quickly narrowed her focus further, successfully managing her own impending panic by monitoring the victim's behavior. In this way, she was able to keep her feelings at bay while she continued working. High-risk takers frequently report similar reactions: They narrow their focus so dramatically that they lose awareness of everything extraneous to the risk activity itself (Holyfield 1997; Lyng 1990).

Another way emotions interfered with performance was when members were disturbed by the graphic sight of the accidents they encountered. Recovering the body of a dead victim, for example, held great potential for upset feelings, especially if the death was violent or gruesome, leaving the body in pieces, excessively bloody, or positioned unnaturally (such as having the legs bent backward or a limb missing). Such situations could cause extreme reactions in rescuers, possibly preventing them from doing the job they were assigned. On the whole, men were assumed to be better suited for these graphic jobs because they were perceived to be emotionally stronger than women. . . .

[M]embers stated these gendered expectations more blatantly. Maddie told me she had noticed a common pattern in the 10 years she had been in the group:

> I think there's an emotional consideration [to being in this group], because our society says men need to hold their emotions in check more so than women. It's expected. It's an expectation from our society. So in any kind of situation where emotions could come into play, you know, something that's really gruesome, [the mission coordinators] aren't gonna ask [the women], they'll ask the guys first.

Maddie's statement highlights how emotional stoicism was not only a critical feature of doing edgework but also of "doing masculinity." These two concepts, edgework and masculinity, were often so confounded that the gender order in Peak was implicitly justified, a point I return to later. . . .

Yet, men were not immune to the potentially disturbing effects of gruesome rescues. In fact, Meg, a member for 10 years, told me that despite stereotypes of masculine emotional strength, she had seen experienced men who had trouble dealing with dead bodies, even though they were willing to assist in the recovery task:

> I've seen people that are very, very macho and strong and opinionated become very sheepish in those situations. . . . [They] march right in and as soon as they get a visual on [the body], they're off doing something else. [They] walk away. Can't look, can't touch. . . . And for me, a body recovery is just like recovering a living person. You know, it's just a body of who was there, and the "who" part is gone. . . . So body recoveries are not so difficult for me, but for some people it's a real struggle.

Thus, emotional upset even threatened some men who were expected to be emotionally tough.

In these gruesome situations, both male and female rescuers reported using one primary emotion management strategy to combat their upset feelings: depersonalizing the victims. Meg alluded to this by saying a body is not a person because the "who" part is gone. Such detachment is a common way people maintain instrumental control in emotionally threatening situations (DeCoster 1997; Jones 1997; Smith and Kleinman 1989). . . .

Fear, urgency, and emotional upset were some of the powerful feelings that threatened rescuers' control during missions in very gendered ways.

Male and female rescuers, however, reported very similar ways of dealing with these threatening emotions to maintain "affective neutrality" (Parsons 1951) during missions: They suppressed their feelings by closely focusing on their task and depersonalizing the victims. This group norm of displaying affective neutrality signified a cool-headedness that was considered safe and effective, and the group considered those who could achieve it in the most critical of circumstances—those who could push the edge the farthest—their most valuable members. At times, these evaluations were based on individuals' past performances; at other times, they were based on gendered stereotypes of emotional capabilities.

Going over the Edge: Releasing Feelings

Immediately after missions, members' suppressed feelings began to surface. Both women and men viewed the sensations they got from successful mission outcomes, like reuniting victims with their family, as the ultimate reward, and I often witnessed them expressing these positive feelings upon hearing the news of a saved victim. They instantly discarded their objective demeanor and became jovial, slaphappy, and chatty. They released the pent-up stress that had been tightly managed throughout the missions by shouting, high-fiving each other, making jokes, and talking about what they had been thinking and feeling throughout the mission. Generally, they felt energized, which they regarded as a positive feeling of control and competence. . . .

Not surprisingly, after missions with negative outcomes, both male and female rescuers reported highly unpleasant emotions, which usually hit them once they got home and were alone. At times, these feelings rushed forward uncontrollably, taking rescuers over the edge into emotional disorder and chaos. One source of upset feelings was recurring visual memories of emotionally disturbing scenes, which are common when people, such as medical students (Smith and Kleinman 1989), first see dead bodies. . . .

There were two ways in which members released these feelings, and these management techniques appeared to be highly related to gender. Women tended to cry. I talked to Elena, the four-year member who initially felt unqualified for search and rescue, shortly after her first (and only) dead-body recovery. She told me that she thought she was "okay" until she got home and was in the shower, where she started to cry. She felt that this initial release was enough to reduce the backlog of feelings that had piled up while she was suppressing them during the mission. It allowed her to regain her composure, reducing her stress and anxiety to manageable levels. . . . Men, however, never reported crying as a means of dealing with the emotional turmoil of missions. Although it is possible that men and women did cry with equal frequency and masculinity norms prevented men from reporting it, it was more likely that women saw this as a more acceptable emotion management technique and coped in this way more often than men (see Gove, Geerken, and Hughes 1979; King et al. 1996; Mirowsky and Ross 1995; Roehling, Koelbel, and Rutgers 1996; Thoits 1995).

After the most traumatic missions, such as one occasion when members extricated the charred remains of several forest firefighters caught in a "fire storm" (an extremely hot, quick-moving, and dangerous type of forest fire), the group provided a professionally run "critical incident debriefing" session where they could talk about their feelings after the mission. While these sessions encouraged men (who were the ones most often involved in such intense missions) to express their feelings, there were only two of these sessions offered in my five and one-half years with Peak. As a general rule, Peak's culture did not encourage men to express their feelings after emotionally taxing rescues. For example, neither female nor male rescuers asked men how they felt after such events. Women, however, were accorded much more displayed concern after disturbing missions. I regularly witnessed both women and men asking women how they felt and touching them supportively, for example, by sympathetically rubbing their back or hugging them. These observations do not suggest that rescuers were not concerned about the men, however; in fact, they were often so concerned that they made sure to phone male rescuers a few days after a disturbing event and ask them "how things were going," an indirect way of checking up on them and reaching out to them emotionally.

That Peak's culture did not encourage men to display their emotions is not surprising, given that toughness and emotional stoicism are central features of many cultural conceptions of masculinity (Connell 1987; Kimmel 1996; Messner 1992). Instead, Peak's men coped with their anxiety and unpleasant feelings by drinking alcohol after negative-outcome missions, such as dead-body recoveries. After Tyler, Nick, and Shorty recovered the trapped kayaker's body in another county, Tyler told me that they bought a 12-pack of beer for Nick and Shorty to drink while Tyler drove them home. . . .

In this way, members used "bodily deep acting," manipulating their physiological state to change their emotional state (Hochschild 1990), by relaxing themselves with alcohol in an effort to dampen the chaos of their surfacing feelings.

Men coped in this way more than women. Although women went drinking with the men after intense and upsetting missions, they drank substantially less alcohol (averaging one to two drinks) than the men (who frequently drank five or more) and left the bar much earlier to go home. . . .

Extending the Edge: Redefining Feelings

In the long term, positive-outcome missions allowed rescuers to extend the edge; members' success served as evidence that they could push their limits further next time. Negative outcomes, however, threatened to compress the edge, leaving rescuers wondering if they were capable and unsure of the risk they were willing to assume in the future. The fourth stage of edgework, then, was marked by members' ability to regain control of these negative feelings by cognitively processing and redefining their experiences, a process Kitsuse (1962) termed "retrospective interpretation." In this way, rescuers employed another type of "deep acting" where they "visualiz[ed] a substantial portion of reality in a different way" (Hochschild 1990, 121), which transformed their feelings about it. This helped them maximize their future edgework ability.

Women and men shared the same emotions and management techniques in this stage of edgework. For example, guilt was a stressful emotion for both male and female rescuers in the wake of unsuccessful missions because they could feel personally responsible for the outcome, for example, if they failed to save a victim. . . .

One way in which members neutralized their guilt was by redefining their part in missions. One technique was "denying responsibility" (Sykes and Matza 1957) for the victim's fate, which could take the form of blaming the victims themselves. Cyndi told me how she reconciled her conflicting feelings about a dead victim she helped recover. He had taken his brand-new pickup truck up a narrow, steep hiking trail to see how well the four-wheel drive worked and was killed when he rolled it off the trail into a ravine. . . .

Another technique both female and male rescuers used to counter the stress of emotionally taxing missions in the long term was to weight the successes more than the failures. . . . Personally accepting credit for successes protected rescuers by increasing their confidence and making them feel that they had control over risky conditions, a phenomenon found among other edgeworkers (Lyng 1990). . . .

Many female and male rescuers used this technique: They defined their overall participation in search and rescue as valuable and thus were able to extend the edge—risk more emotionally and physically—because the rewards outweighed the costs. The gender similarities in these data suggest that in this stage of edgework, how rescuers accomplished gender did not conflict with how they redefined their feelings.

CONCLUSION

★ ★ ★

Peak's members constructed an emotional culture that prioritized suppressing all emotions during missions and releasing them only after the crisis ended (see Irvine [1999] and Stearns [1994] for analyses of similar emotional patterns).

Peak's women and men shared this emotional culture, agreeing on the potentially disruptive nature of emotions as well as the corresponding need to suppress them and remain cool during crises. They also agreed on the need to release these pent-up feelings after crisis situations. But Peak's men and women differed in the steps they took to bring themselves in line with these cultural beliefs. Thus, the abstract assumptions about emotions were shared, but the norms instituted to achieve them differed along gender lines.

These two ways of accomplishing edgework constitute two distinct "emotion lines," which Hochschild (1990, 123) has called a "series of emotional reactions [resulting from] . . . a series of instigating events." For example, women and men in Peak tended to interpret missions' "instigating events" differently, which set off a chain of feelings and management techniques unique to each gendered emotion line. The masculine emotion line was constructed around the interpretation of edgework as exciting. The men in the group tended to be confident in their abilities even before they knew what a mission might require of them and often held the belief that the more demanding the mission, the better. In general, they looked forward to being challenged by very difficult situations, and their vocabulary reflected this as they referred affectionately to these situations as on the verge of "going to shit" and to themselves as being "put in the hot seat." They thrived on excitement during the missions, interpreting their heightened arousal as urgency, and continued to expect that they would succeed. When missions ended unfavorably, they did not release the built-up tension all at once but let it leak out slowly, referring to it with telling metaphors such as "unwinding." Later, they neutralized their failure with emotional "justifying ideologies" (Cancian and Gordon 1988) that helped them maintain a positive self-image. Thus, it appears that the men tended to approach and engage in edgework with positive feelings (perhaps already suppressing negative feelings) and in the event of failure released these pent-up emotions slowly: They followed an "excitement/slow leak" emotion line of failed edgework.

The feminine emotion line was based on the idea of edgework as anxiety producing. Peak's women tended to be unsure of their ability to engage in edgework and were often anxious in anticipation of many physically and emotionally challenging situations. Many women openly questioned their potential for physical competence and emotional self-control. During critical missions, they generally remained anxious, often interpreting their heightened arousal as fear, and constantly worrying that they might fail. When missions ended unfavorably, they released their emotions abruptly by bursting into tears. They later used emotional justifying ideologies, like the men did, to reconceptualize their actions, which neutralized potentially damaging definitions of the self. Thus, it appears that the women in the group tended to enact an "anxiety/outburst" emotion line of failed edgework.

To negotiate the often conflicting demands of edgework and gender, Peak's men and women devised these distinct but gender-appropriate emotion lines. These two emotion lines, however, were not equally respected ways of enacting edgework. In fact, it was the distinction between the two that stratified the group members, creating a hierarchy of emotional competence for edgework, with men at the top. When members evaluated the gendered ways of preparing for and enacting edgework, both men and women recognized the superiority of masculine "excitement" over feminine "anxiety." Although most women reported managing their anxiety in a relatively effective way (i.e., they performed edgework competently), they viewed themselves as "emotional deviants" (Thoits 1990) when it came to the first two stages: preparing for and acting in crisis. (Recall how they tended to consider their lack of confidence as problematic, to decline tasks they thought might overwhelm them emotionally, and to interpret their adrenaline as fear.)

Thoits (1990) has hypothesized that people who are marginalized in a subculture may recognize their own emotional deviance more frequently than nonmarginal members because their own emotions often conflict with those dominant in the subculture. By virtue of their fringe status, marginal subcultural members might, for example, feel pulled between two different emotional subcultures: the one in which they are marginal, and another with different norms and values, in which they better fit. Most women in Peak accepted their status as emotional deviants: They rarely challenged the low expectations others had for them and often held low expectations for themselves, generally believing that their feelings and management techniques were inferior to men's. . . . Clark (1990, 314) has suggested that when it comes to emotions and status, "[h]aving no place, or feeling 'out of place,' can be more painful even than having an inferior place." Thus, Peak's women tended to validate their membership by volunteering to do less challenging tasks. In this way, they used inferior "place claims" (Clark 1990) to accept and reinforce their emotional place—subordinate though it was—in Peak's missions.

In the period after the missions, however, Peak's women generally did not feel that their norm of "outburst" was inferior to men's slow-leak method of ventilating emotions. In fact, they viewed their method as superior to men's and disparaged the slow-leak norm, because they believed that it caused negative emotions to become trapped and to fester. . . . Yet, women's superior emotional place claims went unacknowledged. Most of Peak's men did not accept an inferior status when it came to their slow-leak method of releasing emotions. (Recall they were "too cool" to attend counseling.) . . . Thus, the third phase of edgework, releasing emotions, was contested gender terrain; both women and men vied for the right to define normative ventilation methods.

It is possible that many women in the group perceived their position as inferior when it came to accomplishing edgework because they often felt as though it was a masculine domain, as much high-risk taking tends to be (Harrell 1986). After the danger had passed, however, when it came to dealing directly with emotions, women may have considered themselves the "emotional specialists." In her landmark study of a male-dominated corporation, Kanter (1977) identified this common stereotype of women, noting that both genders assumed women to be better equipped "naturally" to deal with emotional issues. More recently, other scholars have found evidence of this pervasive stereotype in more contemporary male-dominated settings as well (see Hochschild 1983; Pierce 1995), which suggests that specializing in emotional issues is still a core feature of "doing femininity" (West and Zimmerman 1987). Given this powerful belief, on one hand, it is easy to see why Peak's women felt justified in asserting the superiority of their "outburst" norm; by the same token, however, it is puzzling that women's place claims in the emotional realm were given little credibility by Peak's men.

One explanation for this phenomenon might be that norms of masculinity, including the norm of masculine emotional stoicism (Connell 1987; Kimmel 1996; Messner 1992), were so strongly entrenched and intricately connected to the edgework subculture that it gave men the "means of emotional culture production" (Cancian and Gordon 1988): They controlled the standards by which edgeworkers were judged. Furthermore, if emotions were the main avenue through which men distinguished themselves from women before, during, and after edgework, they may have felt that their appropriate gender performance—their very masculinity—would be threatened if they were to display emotions associated with a feminized edgework performance. This interpretation resonates with Connell's (1987) conception of "hegemonic masculinity," which is sustained because it dominates over other gendered forms, such as alternative masculinities held by gay men or nurturing fathers, and any kind of femininity.

These data show that there can be contested emotional terrain within one emotional culture. . . . My research . . . shows how Peak's women evaluated themselves in terms of . . . discrepant cultural messages and how these self-interpretations guided their future action. In some cases, they acted to resist their subordinate position, drawing from the larger emotional culture to bolster their claims to a more respected place in the group. In other cases, they drew on gendered emotion norms to reinforce their subordinate status. Similarly, the men in the group accepted women's place claims in some cases yet denied them in others. By examining how women and men reacted to gendered cultural messages about emotions, these data reveal how gender may be constructed selectively by relying on culturally specific (and occasionally contradictory) emotion norms.

NOTES

1. Because membership fluctuated greatly in the course of five and one-half years, I am only able to estimate the number of members who possessed certain attributes at any given time.
2. There were very few exceptions to this pattern, and they were extreme cases. For example, when Patrick overestimated his skiing ability and created a dangerous situation (discussed later), he became more humble, but only after the board of directors formally disciplined him and group members overtly and harshly criticized him. His error was so egregious that members teased him for several years following the event. Perhaps Patrick was humbled by this situation because of the extraordinary pressure the group put on him; in almost all other cases, however, male rescuers maintained their confidence even after performing poorly.

REFERENCES

Beneke, Timothy. 1997. *Proving manhood.* Berkeley: University of California Press.

Blumer, Herbert. 1969. *Symbolic interactionism.* Englewood Cliffs, NJ: Prentice Hall.

Cancian, Francesca M., and Steven L. Gordon. 1988. Changing emotion norms in marriage: Love and anger in U.S. women's magazines since 1900. *Gender & Society* 2:308–42.

Clark, Candace. 1990. Emotions and micropolitics in everyday life: some patterns and paradoxes of "place." In *Research agendas in the sociology of emotions,* edited by Theodore D. Kemper. Albany: State University of New York Press.

Connell, R. W. 1987. *Gender and power.* Stanford, CA: Stanford University Press.

DeCoster, Vaughn A. 1997. Physician treatment of patient emotions: An application of the sociology of emotion. In *Social perspectives on emotion,* Vol. 4, edited by Rebecca J. Erickson and Beverley Cuthbertson-Johnson. Greenwich, CT: JAI.

Eisenberg, Nancy. 1986. *Altruistic emotion, cognition, and behavior.* Hillsdale, NJ: Lawrence Erlbaum.

Gordon, Steven L. 1989. Institutional and impulsive orientations in selectively appropriating emotions to self. In *The sociology of emotions: Original essays and research papers,* edited by David D. Franks and E. Doyle McCarthy. Greenwich, CT: JAI.

Gove, Walter R., Michael Geerken, and Michael Hughes. 1979. Drug use and mental health among a representative national sample of young adults. *Social Forces* 58:572–90.

Hochschild, Arlie R. 1983. *The managed heart.* Berkeley: University of California Press.

———. 1990. Ideology and emotion management: A perspective and path for future research. In *Research agendas in the sociology of emotions,* edited by Theodore D. Kemper. Albany: State University of New York Press.

Holyfield, Lori. 1997. Generating excitement: Experienced emotion in commercial leisure. In *Social perspectives on emotion,* Vol. 4. edited by Rebecca J. Erickson and Beverley Cuthbertson-Johnson. Greenwich, CT: JAI.

Irvine, Leslie. 1999. *Codependent forevermore.* Chicago: University of Chicago Press.

Jones, Lynn Cerys. 1997. Both friend and stranger: How crisis volunteers build and manage unpersonal relationships with clients. In *Social perspectives on emotion,* Vol.4, edited by Rebecca J. Erickson and Beverley Cuthbertson-Johnson. Greenwich, CT: JAI.

Kanter, Rosabeth Moss. 1977. *Men and women of the corporation.* New York: Basic Books.

Kimmel, Michael. 1996. *Manhood in America: A cultural history.* New York Free Press.

King, Gary, Steven R. Delaronde, Raymond Dinoi, and Ann Forsberg. 1996. Substance use, coping, and safer sex practices among adolescents with hemophilia and human immunodeficiency virus. *Journal of Adolescent Health* 18:435–41.

Kitsuse, John. 1962. Societal reactions to deviant behavior: Problems of theory and method. *Social Problems* 9:247–56.

Lyng, Stephen. 1990. Edgework: A social psychological analysis of voluntary risk taking. *American Journal of Sociology* 95:851–86.

Messner, Michael A. 1992. *Power at play.* Boston: Beacon.

Palmer, C. Eddie. 1983. "Trauma junkies" and street work: Occupational behavior of paramedics and emergency medical technicians. *Urban Life* 12:162–83.

Parsons, Talcott. 1951. *The social system*. New York: Free Press.

Smith, Allen C. III, and Sherryl Kleinman. 1989. Managing emotions in medical school: Students' contacts with the living and dead. *Social Psychology Quarterly* 52:56–69.

Steams, Peter N. 1994. *American cool*. New York: New York University Press.

Sykes, Gresham, and David Matza. 1957. Techniques of neutralization: A theory of delinquency. *American Sociological Review* 22:664–70.

Thoits, Peggy A. 1990. Emotional deviance: Research agendas. In *Research agendas in the sociology of emotions*, edited by Theodore D. Kemper. Albany: State University of New York Press.

———. 1995. Identity-relevant events and psychological symptoms: A cautionary tale. *Journal of Health and Social Behavior* 36:72–82.

Thompson, Hunter S. 1971. *Fear and loathing in Las Vegas*. New York: Warner.

West, Candace, and Don H. Zimmerman. 1987. Doing gender. *Gender & Society* 1:125–51.

Whalen, Jack, and Don H. Zimmerman. 1998. Observations on the display and management of emotion in naturally occurring activities: The case of "hysteria" in calls to 9-1-1. *Social Psychology Quarterly* 61:141–59.

CHAPTER 7

Gender at Work and Leisure

Throughout this book we emphasize the social construction of gender, a dominant prism in people's lives. This chapter explores some of the ways that the social and economic structures within capitalist societies create gendered opportunities and experiences at work, play, and leisure. The gendered patterns of work and leisure that emerge in capitalist systems are complex, like those of a kaleidoscope. These patterns reflect the interaction of gender with other social prisms such as race, sexuality, and class. Readings in this chapter support points made throughout the book. First, women's presence, interests, orientations, and needs tend to be diminished or marginalized within work and leisure. Second, one can use several of the concepts we have been studying to understand the relationships of men and women at work and leisure, including hegemonic masculinity, "doing gender," the commodification of gender, and the idea of separate spaces for men and women.

In this chapter, we explore the construction and maintenance of gender within both paid and unpaid work in the United States, including an article that looks at gender in the global workplace. We also examine what happens when we are not at work; that is, we look at some gendered patterns at play and leisure. We use the terms *play* and *leisure* to describe two related patterns. Leisure is defined in juxtaposition with paid work; however, not all time off from paid work is spent at play (Roberts, 1999). The time not spent at paid work is also spent in household chores or in self-care activities, such as bathing and sleeping. Because women spend more time doing household chores, the amount of time they have for leisure is typically less than is the case for men (Robinson and Godbey, 1997).

WORK

We begin with a discussion of work. The history of gender discrimination in the paid labor market is a long one (Reskin and Padavic, 1999). Social science research has documented gendered practices in workplace organizations. The first reading, by Amy Wharton, describes research on topics including studies of individual actions and behaviors (i.e., individual hiring decisions and distribution of the labor force) to institutional practices within organizations (Acker, 1999). Wharton's description of the research on gender and work reveals the multiple ways gender is woven into both paid and unpaid work. Throughout this chapter we explore four different but overlapping areas of research on gender and work: gender segregation in the workforce, pay inequality, sexual harassment, and the organizational practices that maintain gender inequality at work.

The workforce in the United States is gendered. Think about different jobs (e.g., nurse, engineer, teacher, mechanic, domestic worker) and ask yourself if you consider them to be "male" or "female" jobs. Now, take a look at Table 7.1, which lists job categories used by the Bureau of Labor Statistics (2002). You will note that jobs tend to be gender typed; that is, men and women are segregated into particular jobs. The consequences for men and women workers of this continuing occupational gender segregation are significant in the maintenance of gendered identities. We included in Table 7.1 jobs predominantly held by men (engineers, judges and lawyers, and construction trades), predominantly held by women (health assessment and treating occupations, including nursing; teachers; and health service workers), and equally distributed across gender (executive, administrative, managerial, and sales) to illustrate the gender segregation of the labor force.

Gender segregation of jobs is linked with pay inequity in the labor force. As you look through Table 7.1, locate those jobs that are the highest paid and determine whether they employ more men or more women. Also, compare women's to men's salaries across occupational categories. Wharton refers to the difference between men's and women's salaries as the "gender wage gap" in her reading. This gap in salaries is evident in Table 7.1. Even in those job categories that are predominantly filled by women, men earn more than women.

The pattern you see does not deny that *some* women are CEOs of corporations, such as Xerox and Hewlett Packard, and that today we see women workers everywhere including on construction crews. However, although a few women crack what is often called "the glass ceiling," getting into the top executive or hypermasculine jobs is not easy for women and minority group members. The glass ceiling refers to the point at which women and others, including racial minorities, are "blocked from any further upward movement" in organizations (Baxter and Wright, 2000:275). Informal networks generally maintain the impermeability of glass ceilings, with executive women often isolated and left out of "old boys networks," finding themselves "outsiders on the inside" (Davies-Netzley, 1998:347). Similar internal mechanisms within union and trade-related organizations also keep women out because "knowing" someone often helps to get a job in the higher paid, blue-collar occupations.

TABLE 7.1 Year 2000 Median Weekly Salary and Percentages of men and women in Selected Occupational Categories by Gender and Race[1]

Occupational Category	% White Female (Median Salary) In Category	% White Male (Median Salary) In Category	% Black Female (Median Salary) In Category	% Black Male (Median Salary) In Category	% Hispanic Female (Median Salary) In Category	% Hispanic Male (Median Salary) In Category
Total work force	34.9 ($500)	47.6 ($669)	6.6 ($429)	6.0 ($503)	4.5 ($364)	7.3 ($414)
Executive, administrative, managerial	39.8 ($693)	47.4 ($1,038)	5.0 ($623)	3.4 ($747)	2.9 ($630)	2.8 ($792)
Engineers	8.2 ($950)	75.4 ($1,128)	1.0 ($982)	4.5 ($896)	0.4 ($800)	3.3 ($988)
Health diagnosing occupations	23.3 ($934)	53.1 ($1,536)	2.4 ($722)	3.4 ($1,907)	1.3 ($791)	2.2 ($1,046)
Health assessment and treating occupations	67.1 ($780)	14.6 ($944)	9.6 ($735)	1.1 ($848)	2.5 ($773)	1.0 ($864)
Teachers (except college and university)	63.7 ($681)	22.6 ($834)	9.0 ($630)	2.3 ($764)	4.1 ($614)	1.3 ($780)
Lawyers and judges	23.6 ($1,079)	66.3 ($1,451)	3.6 ($1,060)	3.3 ($1,055)	1.6 ($1,030)	3.3 ($1,242)
Sales representatives, commodities (except retail)	22.9 ($679)	70.6 ($839)	1.1 ($462)	1.7 ($596)	6.2 ($542)	4.3 ($735)
Sales, retail, and personal	43.2 ($308)	37.2 ($478)	9.2 ($275)	4.6 ($413)	7.3 ($275)	5.1 ($342)
Health service	52.8 ($349)	6.9 ($367)	31.1 ($320)	4.4 ($381)	8.3 ($318)	1.1 ($337)
Construction trades	1.7 ($491)	87.9 ($605)	0.3 ($386)	7.2 ($ 527)	0.2 ($465)	19.5 ($467)
Machine operators, assemblers and inspectors	26.8 ($357)	51.6 ($506)	6.4 ($350)	9.1 ($449)	7.8 ($329)	11.9 ($381)

1. Taken from unpublished data from the Current Population Survey, Bureau of Labor Statistics, 2000 annual average data. Usual weekly earnings of employed full-time wage and salary workers.

GENDER, RACE, AND CLASS AT WORK

When we incorporate the prism of race, segregation in the work force and pay inequality becomes more complex. Another look at Table 7.1 indicates that Hispanic and African-American men and women earn less than white men and women, with a couple exceptions (African-American women engineers and African-American men in health diagnosing occupations). In addition, Hispanic and African-American women and men are much less likely to be found in the job categories with higher salaries than their percentages in the labor force would suggest. The continuing discrimination against African Americans, Hispanics, and other ethnic minority groups (as indicated in Table 7.1) shows patterns similar to the discrimination against women, both in the segregation of certain job categories as well as in the wage gap that exists within the same job category. These processes operate to keep individuals—particularly women—from African-American, Hispanic and other marginalized groups "contained" in the labor force.

The inequities of the workplace carry over into retirement, as Toni M. Calasanti and Kathleen F. Slevin describe in their reading. Women and other marginalized groups are at a disadvantage as they leave the workforce and retire. Calasanti and Slevin give detailed descriptions of the social inequalities in retirement income, which indicate that the inequalities in the labor force have a long-term effect for women. They argue that only a small group of privileged white men are able to enjoy their "golden years" and the reasons for this situation are monetary.

Efforts to change inequality in the workplace by combatting wage and job discrimination through legislation have included both gender and race. In 1963, Congress passed the Equal Pay Act, prohibiting employment discrimination by sex, but not by race. Men and women in the same job, with similar credentials and seniority, could no longer receive different salaries. Although this legislation was an important step, Blankenship (1993) cites two weaknesses in it. First, by focusing solely on pay equity, this legislation did not address gender segregation or gender discrimination in the workplace. Thus, it was illegal to discriminate by paying a woman less than a man who held the same job, but gender segregation of the work force and differential pay across comparable jobs was legal. As Blankenship (1993) notes, this legislation saved "men's jobs from women" (220) because employers could continue to segregate their labor force into jobs that were held by men and those held by women and pay jobs held by men at a higher rate. Second, this legislation did little to help minority women, as a considerable majority of employed women of color were in occupations such as domestic workers in private households or employees of hotels/motels or restaurants that were not covered by the act (Blankenship, 1993).

In 1964, Congress passed Title VII of the Civil Rights Act. They drafted this Act to address racial discrimination in the labor force. This act prohibited discrimination in "hiring, firing, compensation, classification, promotion, and other conditions of employment on the basis of race, sex, color, religion, or national origin" (Blankenship, 1993:204). Sex-based discrimination was not originally part of this legislation, but was added at the last minute, an addition that some argue was to assure the bill would not pass Congress. The Civil Rights Act did pass Congress and women were protected along with the other groups. However, the enforcement of gender discrimination was much less enthusiastic than that of race discrimination (Blankenship, 1993).

Blankenship (1993) argues that the end result of these two pieces of legislation to overcome gender and race discrimination was to "protect white men's interests and power in the family" (221) with little concern about practices that kept women and men of color out of higher-paying jobs. Unfortunately, these attempts seem to have had little impact on race and gender discrimination (Strum and Guinier, 1996). Take another look at Table 7.1 and think about the ways the different allocations of jobs and wages affects women's and men's lives across race and social class—their ability to be partners in relationships and their ability to provide for themselves and their families.

As you think about the differences that remain in wage inequality, consider what still needs to be accomplished. Pay equity may seem like a simple task to accomplish. After all, now we have laws that should be enforced. However, the process by which most companies determine salaries is quite complex. They rank individual job categories based upon the degree of skill needed to complete job-related tasks. Ronnie Steinberg (1990), a sociologist who has studied comparable worth of jobs for over thirty years, portrays a three-part process for determining wages for individual jobs. First, jobs are evaluated based upon certain job characteristics such as "skill, effort, responsibility, and working conditions" (457). Second, job complexity is determined by applying a "value to different levels of job complexity" (457). Finally, the values determined in the second step help to set wage rates for the job.

On the surface, this system of determining salaries seems consistent and "compatible with meritocratic values," where each person receives pay based upon the value of what he or she actually do on their jobs (Steinberg, 1987:467). What is recognized as "skill," however, is a matter of debate and is typically decided by organizational leaders who are predominately white, upper-class men. The gender and racial bias in the system of determining skills is shocking. Steinberg gives an example from the State of Washington in 1972 in which two job categories, legal secretary and heavy equipment operator, were evaluated as "equivalent in job complexity," but the heavy equipment operator was paid $400 more per month than the legal secretary (Steinberg, 1990:456). Although it appears that all wages are determined in the same way based upon the types of tasks they do at work, Steinberg (1987, 1990, 1992) and others (including Wharton in this chapter) argue that the processes used to set salaries are highly politicized and biased.

GENDER DISCRIMINATION AT WORK

One way of interpreting why these gendered differences continue in the workforce is to examine workplaces as gendered institutions, as discussed in the Introduction to this book. Researchers have examined work as a gendered institution (Acker, 1999; Reskin, McBrier, and Kmec, 1999). For example, Martin (1996) studied managerial styles and evaluations of men and women in two different organizations: universities and a multinational corporation. She found that when promotions are at stake, male managers mobilized hegemonic masculinity to benefit themselves, thus excluding women. Understanding the processes and patterns by which hegemonic masculinity is considered "normal" within organizations is one avenue to understanding how organizations work to maintain sex segregation and pay inequity.

Peter Levin describes some of these subtle mechanisms of discrimination in the reading in this chapter. Men keep women out of particular jobs through displays of masculinity that make women uncomfortable in work spaces. They also employ stereotypes of women that disadvantage women, as Levin describes for women traders on the floor of the American Commodities Exchange. As you read the piece by Levin, consider those mechanisms and others that gender segregate workplaces and keep women from advancing into particular jobs.

Sexual Harassment

Some of the situations that Levin describes can be considered sexual harassment. The definition of sexual harassment includes two different types of behaviors (Welsh, 1999). Quid pro quo harassment is that which involves the use of sexual threats or bribery to make employment decisions. Supervisors who threaten to withhold a promotion to someone they supervise, unless that person has sex with them, engage in quid pro quo harassment. The second form of sexual harassment is termed "hostile environment" harassment, which the EEOC defines as consisting of behavior that creates an "intimidating, hostile, or offensive working environment" (Welsh, 1999:170). A hostile environment occurs when a company allows situations that interfere with an individual's job performance. As you read Levin, identify examples of hostile environment harassment such as the use of sexual jokes or unwanted touching that might make an individual uncomfortable on the job.

Unfortunately, we do not know exactly how prevalent sexual harassment is. Reports of the incidence of sexual harassment are not consistent, primarily due to the various ways sexual harassment is measured across studies (Welsh, 1999). Although various studies report that anywhere from 16 to 90 percent of women experience sexual harassment at their workplace, an attempt to summarize 18 different surveys estimated the prevalence rate at 44 percent (Welsh, 1999). Any incidence of sexual harassment, however, creates an uncomfortable workplace for the target and perpetuates gender inequality.

Gender discrimination at work is much more than an outcome of cultural or socialization differences in women's and men's behaviors in the workplace. Corporations have vested interests in exploiting gender labor. The exploitation of labor is a key element in the global as well as the U.S. economy, particularly as companies seek to reduce labor costs. Women, in particular, are likely targets for large corporations. In developing nations, companies exploit poor women's desires for freedom for themselves and responsibilities to their families. Ching Kwan Lee illustrates these points in her reading describing the lives of two women workers in factories in two different parts of Asia. Their lives help us to better understand how the structures and patterns of work can reinforce and reproduce gender inequality both in and out of the workplace. As you read Lee's description of how corporations drew the women workers into the plants and their work experiences, ask how yourself how other structures and patterns within the workplace reinforce gender difference and inequality in men's and women's identities, relationships, and opportunities.

THE EFFECT OF WORK ON OUR LIVES AND LEISURE

As illustrated in the Lee reading, work connects us to our identities and affects people's expectations for themselves and others (Kohn, Slomczynski, and Schoenbach, 1986) and their emotions (as indicated in the reading on emergency medical technicians by Lois in Chapter 6). It is not just paid work that affects our orientations toward self and others, but also work done in the home (Spade, 1991). In Western societies, work also defines leisure, with leisure related to modernization and the definition of work being "done at specific times, at workplaces, and under work-specific authority" (Roberts, 1999:2). Although the separation of leisure from work is much more likely to be found in developed societies, work is not always detached from leisure, as evidenced by the professionals who carry home a briefcase at the end of the day, or the beepers that summon individuals to call their workplaces.

WOMEN'S WORK

Care work is one gendered pattern that restricts women's leisure more so than men's leisure. Women's leisure is often less an escape from work and more a transition to another form of work, domestic work. In an international study using time budgets collected from almost 47,000 people in 10 industrialized countries, Bittman and Wajcman (2000) found that men and women have a similar amount of free time; however, women's free time tends to be more fragmented by demands of housework and caregiving. Another study using time budgets found that women spent 30.9 hours on average in various different family care tasks such as cooking, cleaning, repairs, yard work, and shopping, while men spent 15.9 hours per week performing those tasks (Robinson and Godbey, 1997:101). Women reported more stress in the Bittman and Wajcman (2000) study, which the authors attributed to the fact that "[f]ragmented leisure, snatched between work and self-care activities, is less relaxing than unbroken leisure" (185).

Domestic work, while almost invisible and generally devalued, cannot be left out of a discussion of work and leisure (Gerstel, 2000). Wharton describes feminist research on "women's work" in her reading. She argues that women carry an unequal burden of caregiving because women's labor in the market place is underpaid and their jobs are less likely to be seen as important. Gerstel (2000), refers to the contribution of women to care work as "the third shift." As a result, domestic labor and caregiving, being unpaid, are done by people least valued in the paid market. The undervaluation of caregiving carries over to the paid market as well. Look again at Table 7.1 and identify those job categories that encompass caregiving such as health care workers. Now, compare salaries and percentages of men and women in these caregiving jobs. As you start to consider these issues, ask why we undervalue caregiving—the unpaid caregiving in the home as well as caregiving in the workplace? Why are men encouraged not to participate in care work and, when that strategy is successful, why do women become the default caregivers? How is it that the work of the home is undervalued and how is it related to the workplace and leisure?

OVERLAP OF WORK AND PLAY

Leisure and play often overlap with work, particularly for women, as noted in many of the readings in this chapter. Alexis Walker describes how women often do household and carework while at leisure, such as folding clothes, paying bills, or changing diapers while watching television. Or as Gerstel (2000) describes, a leisurely family picnic typically involves the woman preparing, serving, and cleaning up the meal. Play overlaps with work in the workplace as well. In his reading, Levin defines the commodities trading floor as one in which work ebbs and flows, with the down times as occasions for play. Levin finds that commodities exchange workers continually respond to and reinforce traditional gender behaviors and expectations across their workdays, both while working and, particularly, when playing.

Gender and Sports Work and play also overlap in organized sports. For amateurs, considerable work goes into preparing for play, but for professional athletes, such as those described by Gai Berlage in this chapter, sporting activities are a job. The structure and patterns of sporting activities resemble work organizations in many ways. Sporting activities, including recreational sports, are much like other social institutions with fixed sets of rules and relationships that lead to predictable patterns of behavior. These fixed patterns in sporting activities construct and maintain gender (Fine, 1987). For example, Schacht's (1996) study of collegiate men's rugby teams, found that gender is constructed and maintained on teams composed entirely of men. The dynamics of the sport—embedded in fixed rules—include both a competitive element ("survival of the fittest") as well as the reinforcement of the definition of males as strong in both body and spirit, as exemplified by the ideology of "no pain, no gain." The practices that maintain hegemonic masculinity are similar to that at top levels of workplaces (Martin, 1996). As Schacht describes, the informal codes surrounding rugby encourage disdain for femininity and all things feminine by men who play it. Rugby does so by demanding of players that they repeatedly put women down and denigrate men who don't do "rugby masculinity."

Men's patterns at work and play can both be described as what Bird (1996) calls homosociality, or relationships based upon same-gender, nonsexual attractions. Bird argues that hegemonic masculinity is maintained by homosociality. She describes homosocial relationships as characterized by emotional detachment, competitiveness, and sexual objectification of women. The practice of separating women and men in organizations reinforces homosociality and maintains gender difference and inequality.

Separating men and women on sports teams also reinforces homosociality and maintains gender inequality. Women also participate in sports such as rugby, basketball, and baseball, but find themselves doing the same athletic activities with less financial or emotional encouragement than men, unless their team happens to be a national champion. The passage of Title IX in 1972 tried to equalize financial support for girls and women sports teams in schools and colleges. However, there is still a tendency to treat women's athletic abilities and their gender as anomalous and to give them less support than their male peers. For example, gender makes a difference in what sports appear on television. The Amateur Athletic Foundation of Los Angeles (2000) reported that in 1999 women's sports on local news programs received only 8.7 percent of the coverage, only a slight improvement from 5 percent in 1989. Organized, for-profit sports are even more likely to incorporate gendered patterns because they are part of the consumer culture described in Chapter 5.

Berlage shows how gender is used to entice us to pay to watch women's sporting activities. In her reading on women's professional basketball, you will find that sports reinforce traditional gender identities for women. The patterns she describes are similar to the dilemmas women athletes in other sports face. Even if female athletes are physically strong and capable on the playing field, they are still expected to maintain a feminine identity. Today, women professional and nonprofessional basketball players are playing a "male" sport, and playing it well. Thus, on the court these women may be perceived to be more masculine than feminine. Merging quality athletic abilities with feminine appearances is something that women professional basketball players and those who promote their teams struggle with. Although some things changed, many contradictions remain as women and men move in new directions but are still plagued by traditional definitions of gender. Furthermore, when we discourage women from developing their athletic abilities by encouraging women athletes to be "feminine," we may be disadvantaging girls and women in other areas. For example, Hanson and Kraus (1998) find that girls' participation in sports (with the exception of cheerleading) relates to their success in science.

LEISURE

A variety of activities, including sports, fill leisure time. Television takes up the most hours (15 hours on average per week) followed by socializing (6.7 hours), with reading, hobbies, recreation/sports/outdoor activities, and adult education each averaging between 2–3 hours per week. (Robinson and Godbey, 1997:125). Thus, many of us spend a significant portion of our leisure time at home in front of the television. According to Robinson (1990), "television viewing is increasingly the primary leisure activity of most people, taking up 40% of the average person's free time as a primary activity" (24). The last reading in this chapter by Walker provides an intimate glance at how this rather inert activity (some refer to avid television viewers as "couch potatoes"), or more specifically the use of the remote control, actively supports gendered patterns in relationships.

We can only illustrate a few patterns of work, play, and leisure in this chapter. The rest you can explore on your own as you take the examples from the readings and apply them to your own life. As you read through the articles in this chapter, consider the effects of maintaining gendered patterns at work and play. While you are at it, consider why these patterns still exist and what the patterns of inequality in work and play look like in your life.

REFERENCES

Acker, Joan. (1999). Gender and organizations. In *The handbook of the sociology of gender,* Janet Saltzman Chafetz (Ed.), pp. 171–194. New York: Kluwer Academic/Plenum.

Amateur Athletic Foundation of Los Angeles. (2000). *Gender in televised sports.* Los Angeles: Author.

Baxter, Janeen and Erik Olin Wright. (2000). The glass ceiling hypothesis: A comparative study of the United States, Sweden, and Australia. *Gender & Society* 14(2): 275–294.

Bird, Sharon R. (1996). Welcome to the men's club: Homosociality and the maintenance of hegemonic masculinity. *Gender & Society* 10(2): 120–132.

Bittman, Michael and Judy Wajcman. (2000). The rush hour: The character of leisure time and gender equity. *Social Forces* 79(1): 165–189.

Blankenship, Kim M. (1993). Bringing gender and race in: U.S. employment discrimination policy. *Gender & Society* 7(2): 204–226.

Bureau of Labor Statistics (2002). Current Population Survey, Bureau of Labor Statistics, 2000 annual average data. Unpublished data received 10/16/01.

Davies-Netzley, Sally Ann. (1998). Women above the glass ceiling: Perceptions on corporate mobility and strategies for success. *Gender & Society* 12(3): 339–355.

Fine, Gary Alan. (1987). *With the boys: Little League baseball and preadolescent culture.* Chicago: University of Chicago Press.

Gerstel, Naomi. (2000). The third shift: Gender and care work outside the home. *Qualitative Sociology* 23(4): 467–483.

Hanson, Sandra L. and Rebecca S. Kraus. (1998). Women, sports, and science: Do female athletes have an advantage? *Sociology of Education* 71 (April): 93–110.

Kohn, Melvin L., K. M. Slomczynski, and Carrie Schoenbach. (1986). Social stratification and the transmission of values in the family: A cross-national assessment. *Sociological Forum* 1: 73–102.

Martin, Patricia Yancey (1996). Gendering and evaluating dynamics: Men, masculinities, and managements. In *Men as managers, managers as men,* Dave Collinson & Jeff Hearn (eds.), pp. 186–209. Thousand Oaks, CA: Sage.

Reskin, Barbara F., Debra R. McBrier, and Julie A. Kmec. (1999). The determinants and consequences of workplace sex and race composition. *Annual Review of Sociology* 23: 335–361.

Reskin, Barbara F. and Irene Padavic. (1999). Sex, race, and ethnic inequality in United States workplaces. In *Handbook of the sociology of gender,* Janet Saltzman Chafetz (ed.), pp. 343–374. New York: Kluwer Academic/Plenum Publishers.

Roberts, Kenneth. (1999). *Leisure in contemporary society.* Oxon, England: CABI Publishing.

Robinson, John P. and Geoffrey Godbey. (1997). *Time for life: The surprising ways Americans use their time.* University Park: Pennsylvania State University Press.

Robinson, John P. (1990). I love my T.V. *American Demographics* 12(9): 24–27.

Schact, Steven P. (1996). Misogyny on and off the "pitch": The gendered world of male rugby players. *Gender & Society* 10(5): 550–565.

Spade, Joan Z. (1991). Occupational structure and men's and women's parental values. *Journal of Family Issues* 12(3): 343–360.

Steinberg, Ronnie J. (1992). Gendered instructions: Cultural lag and gender bias in the Hay System of job evaluation. *Work and Occupations* 19(4): 387–423.

———. (1990). Social construction of skill: Gender, power, and comparable worth. *Work and Occupations* 17(4): 449–482.

———. (1987). Radical changes in a liberal world: The mixed success of comparable worth. *Gender & Society* 1(4): 446–475.

Sturm, Susan and Lani Guinier. (1996). Race-based remedies: Rethinking the process of classification and evaluation: The future of Affirmative Action. California Law Review.

Welsh, Sandy. (1999). Gender and sexual harassment. *Annual Review of Sociology* 25: 169–190.

33

Feminism at Work

BY AMY S. WHARTON

Amy Wharton is a sociologist whose research interests include work and gender. In this reading she reviews feminist contributions to research on work, looking at the historical development of this research in three areas: the unpaid labor of women, wage inequality and the stratification of the labor force, and the gendering of work organizations. Much of the early research on gender and work was in response to the plight of women in the labor force, attempting to improve the conditions and compensation for women's work. This article provides an important overview of the research on work and gender, while also describing the connections between theory and action in the research she reviews.

1. What is housework labor and should it be compensated?
2. How does her description of gender segregation and wage inequality relate to the data in Table 7.1?
3. Find two examples of a gendered institution from this reading. How is this different from "doing gender"?

This article examines the contributions of feminist scholarship to the study of work, occupations, and organizations. I focus primarily but not exclusively on sociology and use a broad definition of what counts as feminist scholarship. For the purposes of this article, it includes theory and research aimed at uncovering and reducing gender inequalities in the workplace. Feminist scholarship on work has evolved considerably over the past few decades. This evolution corresponds to some extent to developments in feminism as a social movement. Feminist activists have shaped the direction of feminist scholarship on work, and this scholarship in turn has influenced feminist activists' strategies and orientations.

Although feminist scholarship on work is vast, certain themes have received particular attention over the years. Each theme can be linked to a distinct set of feminist political goals. In this article, I discuss three key themes in feminist scholarship and activism, and I trace the evolution of each theme over time.

* * *

[1] WOMEN'S WORK, DOMESTIC LABOR, AND THE WORK OF CAREGIVING

One of the most important expressions of the feminist movement in colleges and universities in the 1960s and 1970s was its critique of academic disciplines, like sociology and other social sciences, for ignoring women. Research on work, occupations, and organizations was singled out as being especially inattentive to women. Documenting women's experiences in the paid labor force was seen as one way to compensate for their previous neglect and an emphasis on men. This research on women's work focused on a range of occupations. Women in female-dominated clerical and service jobs, such as beautician, waitress, and office worker, for example, were the subject of Louise Kapp Howe's *Pink Collar Workers* (1977). During the 1970s, the vast majority of employed women worked in these types of jobs. Several edited collections, such as one by Stromberg and Harkness (1978), took a somewhat broader view, focusing on issues for women employed in a range of settings, from clerical and service to professional. The occupation-specific focus of these studies was consistent with a tradition of older sociological work on occupations.

The literature on women's work did not entirely ignore theoretical questions and issues of work structure and organization, but, like the sociology of occupations more generally, these were not central concerns. This growing literature on women's work had other omissions as well. There was relatively little discussion of women of color, while professional women received attention disproportionate to their share of the labor force. In addition, as noted previously, while various writers described problems faced by women working in particular settings, there was little systematic focus on work structure and organization or on gender inequality as a pervasive feature of work.

Because most sociological research on work focused on paid work, unpaid work done by women in the home received virtually no attention. Indeed, sociologists rarely treated households as places of work, viewing them instead as realms of "expressive activity" in the tradition of sociologist Talcott Parsons. Feminist scholars attempted to show that housework could be understood in the same terms as paid work. However, while work done in the home provided useful goods and services, it was unrecognized and undervalued.

Oakley's book *The Sociology of Housework* (1974) was among the first products of this new scholarship. She concluded that housework could be understood in the same terms as paid work: "Women define housework as labour, akin to that demanded by any job situation. Their observations tie in closely with many findings of the sociology of work; the aspects of housework that are cited as satisfying or dissatisfying have their parallels in the factory or office world" (41). The homemakers Oakley interviewed, however, did not see themselves as an oppressed group and were unsympathetic to feminism. Consistent with the consciousness-raising agenda of the feminist movement at this time, Oakley explored ways to increase homemakers' awareness of their oppression. "The 'deconditioned' housewife," she observed, "is thus a potential revolutionary" (197).

Marxist feminists also attended to these sorts of questions. As Secombe (1974) noted, mainstream sociologists and economists were not the only ones who had ignored the work of housewives. In what became known as the "domestic labor debates," Marxist feminist scholars sought to increase the recognition given homemakers by Marxist theorists by showing how homemakers were—like their working-class husbands—exploited by capital (Secombe 1974; Beechey 1976). This debate had both a scholarly and political agenda; Marxist feminists challenged male-dominated notions of the working class and sought to uncover the potential political power of housewives.

Feminist scholarship on housework and homemakers called attention to the productive, but unpaid, work performed in the home, and it also highlighted the precarious economic position of full-time housewives. Although wages for housework did not become a major feminist demand, feminist efforts to calculate the economic value of homemaking showed that women in this occupation worked long hours and performed services that would command a substantial wage in the paid labor market (Bergmann 1986). Because homemakers are unpaid, however, their contributions to the household are unrecognized. This becomes particularly problematic when divorce occurs or when the husband dies or becomes disabled. Feminist economists like Bergmann (1986) and Sawhill (1983) showed how current policies designed to protect wage earners, such as Social Security, offer little protection to displaced full-time homemakers.

Caregiving as a Feminist Issue

Women continue to perform most household work, and feminist activists and scholars continue to explore this issue. The traditional feminist approach to housework—treating it as an occupation—has given way to new orientations, however. Compare this quote from DeVault's book, *Feeding the Family* (1991), to the earlier quote from Oakley:

> The women I talked with referred to their activity as something other than "work" in any conventional sense, as activity embedded in family relations. . . . Though they recognize that they work at feeding, and that the work includes many repetitive, mechanical tasks, their language reveals an unlabeled dimension of caring as well: some speak of their efforts as "love," while others talk about caring for children as not quite a job, but as "something different." (10)

DeVault (1991) is one of a growing number of feminist scholars throughout the social sciences interested in understanding caregiving. In contrast to Parsons, who separated the instrumental from the affective, and in contrast to early feminists, who focused only on the instrumental aspects of household work and caring, current feminist scholarship unites these two strands: caregiving is both an instrumental and an expressive activity. This approach can also be seen in studies of emotional labor and other forms of caring work done for pay (Hochschild 1983; Steinberg and Figart 1999).

[2] WAGE INEQUALITY AND GENDER SEGREGATION

Simpson (1989) describes the 1970s and 1980s as a time when sociologists interested in work abandoned more traditional sociological concerns and embraced an economic approach to work-related issues. In Simpson's view, this shift was marked by a move away from "the behavior of flesh-and-blood workers at the level of the face-to-face work group" toward a more macrostructural, economic approach (564). Simpson's description of this change could also apply to feminist scholarship on work during the same era. Feminist scholarship began to more systematically attend to structural and economic issues, and attention shifted from women's work to issues of gender and gender inequality. Feminist interest in these issues was also motivated by changes occurring in women's lives. Rising divorce rates meant that more women were working for pay to support themselves and their children. Economic independence proved difficult, however, as most women worked in low-paying, predominantly female clerical and service jobs.

Explaining the Gender Wage Gap

Economists have devoted considerable attention to understanding what determines the worth of jobs and why some jobs pay more than others. For the most part, these scholars have emphasized the role of human capital in wage determination. Human capital refers to the portfolio of skills that workers acquire through various kinds of investments in education, training, or experience. Applied to the gender wage gap, human capital theory implies that women earn less than men because of differences in the kind and amount of human capital each has accumulated.

Feminist scholars were skeptical that this provided a complete understanding of the gender wage gap, and they critiqued economic arguments for overlooking the many ways in which social factors shape wage setting and wage inequality (for example, England and Farkas 1986; Steinberg and Haignere 1987). In addition, feminist researchers have suggested that "institutional" or structural factors, as well as individual-level factors, contribute to the gender wage gap (Roos and Gatta 1999). By focusing on these issues, feminist scholars helped broaden the study of gender inequality and the gender wage gap from a narrow focus on individuals' human capital to a concern with social structural and institutional factors. On a more general level, feminists' interest in understanding women's earnings and work experiences has contributed to a more accurate view of labor markets. Researchers who focus only on men—especially white men—tend to "exaggerate the extent to which labor market processes are meritocratic" (Reskin and Charles 1999, 384).

During the last few decades, feminist research on the wage gap has proliferated. Women's lower earnings relative to men's has become, as Roos and Gatta (1999) observe, "a social truism" (95). Moreover, researchers' understanding of how the gender-based wage inequality occurs, how it has changed over time, and its variations between women and between men has become increasingly complex and nuanced. Two major themes in this work illustrate recent developments.

First, research on the gender wage gap has become increasingly attentive to issues of race and ethnicity (Browne 1999; Kilbourne, England, and Beron 1994; Tomaskovic-Devey 1993). As Reskin and Charles (1999) note, the historical tendency has been a "balkanization of research on ascriptive bases of inequality" (380). Studies of the gender wage gap, in particular, have often ignored racial and ethnic variations among women and men. Not only does this produce potentially inaccurate results, but it also has hindered efforts to understand the forces generating and maintaining wage inequality. A second, emerging theme in recent studies of the gender wage gap are attempts to relate gender-based wage inequality to wage inequality more generally. Wage inequality in the United States has increased in recent decades—a pattern that reflects industrial and occupational restructuring, changing labor force demographics, globalization, and political trends (Morris and Western 1999). This widening inequality reflects not only earnings differences between women and men but also differences between women and between men (Mishel, Bernstein, and Schmitt 1999). As researchers now realize, a narrow focus on the gender wage gap misses these broader patterns of inequality.

The Gender Segregation of Occupations and Jobs

Gender segregation is an entrenched and pervasive feature of the industrial workplace. Women make up almost half of the paid labor force, but women and men are employed in different occupations, firms, and jobs. Although occupational segregation has declined over the twentieth century (with most of the decline occurring in the 1970s), its persistence in the face of so many other changes both inside and outside the workplace has inspired a tremendous amount of feminist research.

One line of research on segregation involves large-scale, quantitative studies of occupational segregation in the labor force as a whole. These studies document trends in segregation over time and identify factors associated with changes in segregation levels. Recent findings suggest that the declines in occupational segregation that took place in the 1970s appeared to level off in the 1990s (Jacobs 1999). Other researchers have examined segregation cross-nationally, looking for clues as to the economic, political, and cultural factors that produce gender segregation and inequality.

This literature on occupational segregation has been supplemented in recent years by research at other levels of analysis. Bielby and Baron (1984), for example, moved segregation research from the occupation to the job and firm levels of analysis; they showed that job segregation within firms was considerably higher than segregation at the occupational level. Tomaskovic-Devey (1993) also found extremely high levels of job segregation by gender in his study of North Carolina firms. Even if occupations appear gender integrated, these studies demonstrate that women and men rarely work together, holding the same job in the same firm. A second extension of segregation research moves it in a more macro direction by focusing on labor markets—rather than jobs or occupations. Cotter et al. (1997) show that occupational integration at the local labor market level improves earnings for all women in that labor market, regardless of the gender composition of a woman's own occupation.

Links to Feminist Activism: Affirmative Action and Pay Equity

Feminist research on gender segregation and the gender wage gap has always been more than an academic pursuit; it has also been used to advance efforts to reduce gender inequality in earnings and to improve women's position in the workforce. These two goals are closely connected; women's concentration in lower-paying predominantly female jobs and their exclusion from higher-paying jobs held mostly by men both help explain the gender wage gap. Improvement in women's labor market situation, then, requires that women have greater access to higher-paying male jobs or that the wages for predominantly female jobs increase. These are not mutually exclusive objectives, and feminist research provides evidence supportive of both strategies.

Affirmative action, which grew out of Title VII of the 1964 Civil Rights Act and was later elaborated through numerous executive orders, focuses on improving women's and minorities' access to jobs from which they have been excluded. This policy thus aims to break down gender and racial segregation by moving women and minorities into jobs traditionally held by men and whites. Recent studies evaluating the effectiveness of affirmative action policies and programs suggest that they have been relatively successful (Reskin 1998). Although these programs have become politically unpopular, affirmative action does seem to improve women's and minorities' access to higher-paying jobs held disproportionately by men and whites.

Despite this success, most feminist scholars believe that affirmative action alone is insufficient for eliminating discrimination and increasing women's wages (England 1992). The vast majority of women work in predominantly female jobs, and affirmative action is unlikely to ever completely remedy this situation. As a result, feminist scholars and women workers have attempted to improve the pay of predominantly female jobs by compensating these jobs at wages equivalent to comparable jobs held by men. The pay equity (or comparable worth) movement uses job evaluation methods to determine the relative worth of jobs. Job evaluation is a strategy for determining how pay is assigned to jobs and thereby to justify (or critique) relative pay rates. Employers use job evaluation as a tool to decide how to compensate different jobs, and feminists use it to demonstrate gender bias in wage setting.

Job evaluation techniques have been used in several state and local settings, including Washington, Oregon, New York State, and the city of San Jose, California (Acker 1989; Steinberg and Haignere 1987; Blum 1991). In all these cases, this technique was proposed as a way to correct perceived gender biases in the ways in which wages were attached to jobs. Most notably, job evaluation showed that jobs evaluated as comparable in terms of their skill requirements, working conditions, and the like were often compensated at different levels depending upon their sex composition. Predominantly female jobs tended to be devalued relative to jobs of comparable skill filled by men. These results called into question the notion that wages were set according to gender-neutral processes and instead revealed an important source of gender bias. In fact, as England (1992) notes, "if a single job evaluation plan is used to set pay throughout a firm or government, *it nearly always gives women's jobs higher wages relative to men's than most employers pay*" (205; emphasis in original). Feminist scholars and labor organizations often cooperated in their efforts to use job evaluation as a means of identifying and correcting gender bias in wage setting.

[3] GENDERED JOBS, ORGANIZATIONS, AND WORK ACTIVITIES

Rosabeth Moss Kanter's 1977 classic, *Men and Women of the Corporation,* reshaped feminist scholarship on work and gender. Kanter argued that many differences between women's and men's work-related behaviors and attitudes that previous researchers had attributed to gender could be better understood as being due to women's and men's different structural positions in organizations. Although some of Kanter's claims have been widely criticized, her contention that much gender-related behavior on the job stems from how work is organized rather than the characteristics of workers has gained wide acceptance among feminists.

Kanter's claims about the effects of group composition are among her most important, though intensely debated, contributions to feminist understandings of work. These claims rest on the proposition that, "as proportions shift, so do social experiences" (Kanter 1977, 207). Kanter was particularly interested in "skewed groups," in which one social type is numerically dominant and the other constitutes a very small numerical minority (for example, 15 percent or less). Her focus on this type of group stemmed from the fact that this is likely to be the situation experienced by newcomers to a social setting. Members of the numerical minority in skewed groups are called tokens. Kanter argued that relations between tokens and dominants in skewed groups worked to the tokens' disadvantage, and she attributed these negative dynamics to social structural factors. The relative proportions of different social types—not the particular gender, race, ethnic background, and so forth, of group members—were to blame for the effects of tokenism.

For two decades, feminists have been debating Kanter's claims. Most now agree with Williams (1998) that, while Kanter's search for a structural explanation for gender-related behavior was an important contribution, she was "wrong about tokenism" (141). As Williams's research (1989) shows, gender does matter when considering the consequences of tokenism: women who enter predominantly male jobs and work settings typically encounter much more resistance and hostility from their male coworkers than do men who enter predominantly female jobs and work settings.

Konrad, Winter, and Gutek (1992) refer to Kanter's perspective (1977) as a "generic" approach and contrast it with an "institutional" orientation. Though not specifically directed to feminist scholarship, this distinction captures an important difference between feminist approaches to the workplace: Kanter argued that gender differences in the workplace were real manifestations of something else—the effects of social structure. Institutionalists, by contrast, suggest that gender is embodied in social structures and other forms of social organization. Hence, aspects of the workplace that are conventionally treated as genderless or gender neutral in fact are expressions of gender. This latter view represents a gendered institutions perspective.

Gendered Institutions

Institutions are those features of social life that seem so regular, so ongoing, and so permanent that they are often accepted as just "the way things are." Because highly institutionalized social arrangements require relatively little effort to sustain them, it is much more difficult to alter something that is highly institutionalized than it is to perpetuate it. The concept of a gendered institution is consistent with these ideas. In her 1992 article on this theme, Acker contends that many of the institutions that constitute the "rules of the game" in American society—and, indeed, most societies—embody aspects of gender. As Acker defines it, to say that an institution is gendered means

> that gender is present in the processes, practices, images, and ideologies, and distributions of power in the various sectors of social life. Taken as more or less functioning wholes, the institutional structures of the United States and other societies are organized along the lines of gender. . . . [These institutions] have been historically developed by men, currently dominated by men, and symbolically interpreted from the standpoint of men in leading positions, both in the present and historically. (567)

This way of thinking about gender directs us to the organization, structure, and practices of social institutions, such as the workplace, and it calls attention to the ways in which these entrenched, powerful, and relatively taken-for-granted aspects of the social order produce and reproduce gender distinctions and inequality.

Drawing on the notion of gendered institutions, feminists suggest that the structures and practices of work organizations are gendered at all levels. As Steinberg (1992) notes, "Masculine values are at the foundation of informal and formal organizational structures. . . . Images of masculinity and assumptions about the gendered division of labor organize institutional practices and expectations about work performance" (576). As an example of this line of argument, some research has focused on how cultural beliefs about gender infuse people's understandings of jobs, occupations, and particular work activities. By establishing certain work roles, jobs, and occupations as appropriate for one gender and off-limits to another, these cultural beliefs establish the "way things are" or a set of commonsense understandings of who should engage in what type of work.

That jobs dominated by a particular gender come to be seen as most appropriate for that gender may seem unproblematic and inevitable, but this association is produced through a complex process of social construction. As Reskin and Roos (1990) note, virtually any occupation can be understood as being more appropriate for one sex or another "because most jobs contain both stereotypical male and stereotypical female elements" (51). Nursing, for example, requires workers to be skilled in the use of complex medical technologies. Emphasizing the caring aspects of this occupation, however, allows it to be cast as an occupation particularly appropriate for women. Most jobs and occupations contain enough different kinds of characteristics that they can be construed as appropriate for either women or men.

The gendering of work can also be seen in gender-integrated positions or in jobs that contain a minority of the other gender, and it applies to jobs held by men as well as those that are predominantly female. As a growing literature on work and masculinity has shown, many predominantly male jobs implicitly and explicitly require incumbents to display traditionally masculine behaviors, such as aggressiveness. Maier (1999) argues that managerial practices and organizational cultures—not merely specific jobs—embody a "corporate masculinity" that privileges individualism, competitiveness, and technical rationality (71).

In some situations, the work tasks may be gendered as feminine, but the worker performing them is male. Hall's study (1993) of table servers offers a useful example of these arrangements. Styles of table service are laden with gender meanings. A familial style of service, which Hall labels "waitressing," has been historically associated with women working in coffee shops and family restaurants. By contrast, "waitering" is a more formal style, usually associated with male servers in high-prestige restaurants. Hall suggests that even in sex-integrated restaurants, "work roles, job tasks, and service styles" continue to be gendered, such that "waitering"—whether performed by women or men—is more highly valued by employers and customers than is "waitressing" (343).

The gendering of work also shapes the relations between jobs, especially hierarchical relations. In his writings on bureaucracy, the classical sociologist Max Weber provided one of the definitive sociological understandings of work hierarchies. For Weber, bureaucratic work arrangements were necessarily hierarchical and involved specialization, a fixed division of labor, and meritocratic rules and regulations (Weber 1946). Feminists have inserted gender into these arguments in two important respects. First, some argue that gender is an aspect of bureaucracy itself; that is, gender is embedded in this formal system of organization (Ferguson 1984). In *The Feminist Case Against Bureaucracy,* Ferguson suggests that, as a hierarchical system, bureaucracy perpetuates the interests of the powerful over the powerless. In this respect, it is a metaphor for the dominance of men over women. By rewarding character traits such as "impression management, need to please, conformity, identification with the organization, dependency and so forth," bureaucracies "feminize" those at the bottom of the hierarchy while concentrating power at the top (116).

While Ferguson (1984) suggests that bureaucracy itself is gendered, more recent work suggests that gendering is more central to the informal social relations of workplace hierarchies. For example, Pierce (1995) explored the relations between lawyers (mostly male) and paralegals, predominantly female. Although the lawyers and paralegals she studied engaged in some of the same kinds of tasks (for example, legal research and writing) and were very interdependent in many respects, the relations between these positions were highly gendered. As Pierce states, "Structurally, paralegal positions are specifically designed for women to support high-status men, and the content of paralegal work is consistent with our cultural conceptions of appropriate behavior for traditional wives and mothers" (86). Paralegals thus are expected to defer to and serve lawyers, who in turn rely on paralegals to perform this caretaking labor.

Links to Feminist Activism: Gender on the Job

Research deriving from a structural or institutional perspective on gender has been especially useful in feminists' efforts to improve women's and men's lives inside work organizations. Kanter's recognition (1977) that the relative power and opportunity of individuals and groups at work can be altered by redesigning organizational structures has inspired some efforts in this direction. In addition, recent years have seen an infusion of feminist scholarship into organizational research. Some of this research has its roots in Kanter's work, as it explores how the demographic composition of work groups shapes interaction and behavior (Chemers, Oskamp, and Costanzo 1995; Ruderman, Hughes-James, and Jackson 1996). An important finding to emerge from this research is that differences between people—such as those deriving from gender or race—are not always salient in the workplace. While sex category is probably more salient in more situations than many other attributes of a person, organizational research suggests that it is not always an important factor in workplace social relations. Understanding when and why gender (or race or some other characteristic) matters and for what outcomes are important tasks facing feminist scholars studying organizations.

Research informed by a gendered institutions perspective has helped feminists identify and combat the "glass ceiling" (U.S. Department of Labor 1991). The metaphor of the glass ceiling speaks to the powerful, yet invisible, forces that prevent women and minorities from reaching the highest levels of organizations. Research addressing how gender is built into job requirements, work activities and hierarchies, and cultural beliefs about job worth has identified the more subtle factors that maintain gender and racial inequality in organizations. This research also contributes to efforts to combat sexual harassment at work and to understand how sexuality as well as gender is embedded in work organizations.

Feminist scholarship in the tradition of a gendered institutions perspective has also addressed issues of gender inequality and the pay gap. For example, while early pay equity research relied on job evaluation to identify gender bias, recent studies suggest that job evaluation methods themselves contain their own sources of bias (Acker 1989; England 1992). Bias in job evaluation occurs when predominantly female, "nurturant" jobs are given fewer points than they merit, while predominantly male jobs are given a boost in ranking. An example cited by England (1992) illustrates the point: "Attendants at dog pounds and parking lots (usually men) were rated more highly than nursery school teachers, and zookeepers more highly than day care workers" (199). Researchers question whether it is possible to objectively measure the worth of jobs.

★ ★ ★

REFERENCES

Acker, Joan. 1989. *Doing Comparable Worth: Gender, Class, and Pay Equity.* Philadelphia: Temple University Press.

———. 1992. Gendered Institutions. *Contemporary Sociology* 21:565–69.

Beechey, Veronica. 1976. Some Notes on Female Wage Labour in Capitalist Production. *Capital and Class* 3:43–66.

Bergmann, Barbara. 1986. *The Economic Emergence of Women.* New York: Basic Books.

Bielby, William T. and James N. Baron. 1984. A Woman's Place Is with Other Women: Sex Segregation Within Organizations. In *Sex Segregation in the Workplace,* ed. B. F. Reskin. Washington, DC: National Academy Press.

Blum, Linda M. 1991. *Between Feminism and Labor: The Significance of the Comparable Worth Movement.* Berkeley: University of California Press.

Browne, Irene, ed. 1999. *Latinas and African-American Women at Work.* New York: Russell Sage Foundation.

Chemers, Martin M., Stuart Oskamp, and Mark A. Costanzo. 1995. *Diversity in Organizations.* Thousand Oaks, CA: Sage.

Collins, Sharon M. 1997. *Black Corporate Executives: The Making and Breaking of a Black Middle Class.* Philadelphia: Temple University Press.

Cotter, David A., JoAnn DeFiore, Joan M. Hermsen, Brenda Marsteller Kowalewski, and Reeve Vanneman. 1997. All Women Benefit: The Macro-Level Effect of Occupational Integration. *American Sociological Review* 62:714–34.

DeVault, Marjorie L. 1991. *Feeding the Family: The Social Organization of Caring as Gendered Work.* Chicago: University of Chicago Press.

England, Paula. 1992. *Comparable Worth: Theories and Evidence.* New York: Aldine de Gruyter.

England, Paula and George Farkas. 1986. *Households, Employment, and Gender: A social, Economic and Demographic View.* New York: Adine de Gruyter.

Ferguson, Kathy E. 1984. *The Feminist Case Against Bureaucracy.* Philadelphia: Temple University Press.

Hall, Elaine J. 1993. Waitering/Waitressing: Engendering the Work of Table Servers. *Gender & Society* 7:329–46.

Hochschild, Arlie Russell. 1983. *The Managed Heart: The Commercialization of Human Feeling.* Berkeley: University of California Press.

Howe, Louise Kapp. 1977. *Pink Collar Workers.* New York: Avon Books.

Jacobs, Jerry A. 1999. The Sex Segregation of Occupations: Prospects for the 21st Century. In *Handbook of Gender and Work,* ed. G. N. Powell. Thousand Oaks, CA: Sage.

Kanter, Rosabeth Moss. 1977. *Men and Women of the Corporation.* New York: Basic Books.

Kilbourne, Barbara S., Paula England, and Kurt Beron. 1994. Effects of Individual, Occupational, and Industrial Characteristics on Earnings: Intersections of Race and Gender. *Social Forces* 72:1149–76.

Konrad, Alison M., Susan Winter, and Barbara A. Gutek. 1992. Diversity in Work Group Sex Composition. *Research in the Sociology of Organizations* 10:115–40.

Maier, Mark. 1999. On the Gendered Substructure of Organization: Dimensions and Dilemmas of Corporate Masculinity. In *Handbook of Gender and Work,* ed. G. N. Powell. Thousand Oaks, CA: Sage.

Mishel, Lawrence, Jared Bernstein, and John Schmitt. 1999. *The State of Working America,* 1998–99. Ithaca, NY: ILR Press.

Morris, Martina and Bruce Western. 1999. Inequality in Earnings at the Close of the Twentieth Century. *Annual Review of Sociology* 25:623–57.

Oakley, Ann. 1974. *The Sociology of Housework.* New York: Pantheon Books.

Pierce, Jennifer. 1995. *Gender Trials: Emotional Lives in Contemporary Law Firms.* Berkeley: University of California Press.

Reskin, Barbara F. 1998. *The Realities of Affirmative Action in Employment.* Washington, DC: American Sociological Association.

Reskin, Barbara F. and Patricia A. Roos. 1990. *Job Queues, Gender Queues: Explaining Women's Inroads into Male Occupations.* Philadelphia: Temple University Press.

Roos, Patricia A. and Mary Lizabeth Gatta. 1999. The Gender Gap in Earnings: Trends, Explanations, and Prospects. In *Handbook of Gender and Work,* ed. G. N. Powell. Thousand Oaks, CA: Sage.

Ruderman, Marian N., Martha W. Hughes-James, and Susan E. Jackson, eds. 1996. *Selected Research on Work Team Diversity.* Washington, DC: American Psychological Association; Greensboro, NC: Center for Creative Leadership.

Sawhill, Isabel V. 1983. Developing Normative Standards for Child Support Payments. In *The Parental Child Support Obligation,* ed. J. Cassetty. Lexington, MA: Lexington Books.

Secombe, Wally. 1974. The Housewife and Her Labour Under Capitalism. *New Left Review* 83:3–24.

Simpson, Ida Harper. 1989. The Sociology of Work: Where Have the Workers Gone? *Social Forces* 67:563–81.

Steinberg, Ronnie J. 1992. Gender on the Agenda: Male Advantage in Organizations. *Contemporary Sociology* 21:576–81.

Steinberg, Ronnie J. and Deborah M. Figart. 1999. Emotional Labor Since *The Managed Heart.* The Annals of the American Academy of Political and Social Science 561:8–26.

Steinberg, Ronnie J. and Lois Haignere. 1987. Equitable Compensation: Methodological Criteria for Comparable Worth. In *Ingredients for Women's Employment Policy,* ed. C. Bose and G. Spitze. Albany: State University of New York Press.

Stromberg, Ann H. and Shirley Harkness, eds. 1978. *Women Working: Theories and Facts in Perspective.* Palo Alto, CA: Mayfield Press.

Tomaskovic-Devey, Donald. 1993. *Gender and Racial Inequality at Work*. Ithaca, NY: ILR Press.
US. Department of Labor. 1991. *A Report on the Glass Ceiling Initiative*. Washington, DC: Department of Labor.
Weber, Max. 1946. Bureaucracy. In *From Max Weber: Essays in Sociology*, ed. H. H. Gerth and C. Wright Mills. New York: Oxford University Press.
Williams, Christine L. 1989. *Gender Differences at Work*. Berkeley: University of California Press.
———. 1998. What's Gender Got to Do with It? In *Required Reading: Sociology's Most Influential Books*, ed. D. Clawson. Amherst: University of Massachusetts Press.

34

Gendering the Market

Temporality, Work, and Gender on a National Futures Exchange

BY PETER LEVIN

In this reading, Peter Levin provides a detailed description of how gender gets constructed throughout the work day on a exchange market floor, both during peak work times and times when things are slower. Levin's description illustrates how men and women relate to each other within the structure of work during busy times, and "play" during quieter times. He argues that gender is not simply something that is "done" by men and women, but something embedded in the structure of work. This reading provides a day-to-day examination of what Wharton describes as gendered organizations.

1. What does Levin mean by hegemonic masculinity?
2. How does gender get defined during the times when the market is active and when the market is less active?
3. What does Levin mean when he says that competency is constructed in masculine and sexual terms?

On the often-chaotic trading floor of the American Commodities Exchange (ACE),[1] time matters. In the world of organized commodity futures trading, both gender and the markets themselves are organized temporally. Scholars have long argued that gender is not a quality inherent to individuals but rather it consists of a set of socially produced, hierarchically organized relations between men and women (Connell, 1987,1995; Stacey & Thorne, 1985; West & Zimmerman, 1987). Men and women face differential conditions in the context of organizations and in labor markets more broadly (Epstein, 1970; Hartmann, 1976; Kanter, 1977). Furthermore, a number of scholars have documented how bureaucracies as well as the microlevel processes of workplaces are also "gendered" (Acker, 1990; Britton, 1997; Ferguson, 1984; Hall, 1993; Leidner, 1991, 1994; Pringle, 1989; Salzinger, 1997). Increasingly detailed evidence shows how gender changes across factory settings and management strategies so that gendered meanings "take place within the framework of local, managerial subjectivities and strategies" (Salzinger, 1997, p. 550). With variations in factory-level labor processes, gender takes on different subjective meanings and varies in distributional effects.

Perhaps less noted and less studied are the effects of temporal variations on gender. For many occupations and in many organizations, the pacing of work significantly affects both the subjective understanding and the structural arrangement of gender. Temporality complements studies of gender variation across spatial locales. The ACE is not a continuous or homogeneous work environment. Rather, the temporal shifts in market activity shape the ways men and women understand and constitute gender.

The ACE floor operates within two distinct gender repertoires, one of *competence* and the other of *sexualized difference.* In the modality of work, gender is constituted within a language of competence, which constructs the trading floor as gender neutral even as it privileges a particular form of dominant masculinity. During slower times of play, gender reemerges in a more overtly sexualized form. These two repertoires can shift abruptly; men on the floor refer to women's bodies as suitable for sexual objectification in one moment and unsuitable for handling the physicality of the market in the next. In this article, I show the ways in which gender vacillates between these two repertoires.

* * *

AT WORK AND PLAY

My work confirms Baker's (1984) analysis of trading patterns at a U.S. stock options market, identifying the market as diurnally curvilinear, with heightened activities in the morning and afternoon and a slow, nonactive midday period. Investor orders build up overnight and are released as the market opens, providing a spurt of trading in the early part of the morning. Furthermore, economic data is often released within the first hour of trading, providing additional impetus for speculative market activity. This was readily apparent to me while I was working as a clerk during the market's opening. As the clock ticks toward the opening bell, there emerges an almost palpable sense of expectation and excitement. Cordiality and muted greetings to colleagues and competitors turn to intermittent shouts about "the call," the anticipated opening range based upon overnight market activity. In the moments before the bell rings to announce the opening of trading, the volume can climb to a loud, continuous shout. It is what many respondents called "electric" . . .

Similarly, activity picks up again as traders make their final trades of the day. There are categories of trades made explicitly during the market's closing range. Furthermore, the Exchange sounds a bell for the last minute, last 10 seconds, and the end of trading, creating a heightened sense of urgency.

Participants anticipate other busy times as well—the release of economic data, meetings of the federal reserve board, and planned political speeches. During these spells, lasting anywhere from moments to hours, traders and employees adopt a triage mentality. When the boards posting market prices go "fast market," indicating an inability of the electronic boards to keep up with the pace of changes in trading prices, the most important thing to do is minimize errors, stay controlled, and remain alert for unanticipated market movements and customer responses. Breaks for lunch or coffee are shortened or cancelled. Because it is impossible to predict what will happen when the market moves, all participants remain in a state of high readiness. . . .

The market does slow down, particularly in the middle of the day. When trading diminishes, a different set of rules applies. The pace and intensity of actual working time ranges from day to day. As one respondent said, "You could be for 20 minutes under an unbearable amount of stress, and then once that 20 minutes is over, you could have 2 hours of doing nothing" (24-year-old male pit clerk). During slow periods, traders would leave the pits or stand around waiting for something to happen. People on the floor amused themselves by telling jokes, goofing around, doing crossword puzzles, reading newspapers, or standing around engaging in idle conversation.

Fieldwork observations distinguishing the ACE as two distinct modalities depending on the amount of market activity were confirmed in focused interviews. Respondents routinely distinguished the exchange's fast and slow periods. A 28-year-old female trade checker described it in the following manner, typical of my respondents:

> When it's fast, it's all about business, it's like you're in an accounting firm during tax time, and it is all business, nobody smiling . . . you might not be pleasant, some of them are fighting; as you know, they get upset. And then when it's slow, it's laid back, people are smuggling in food, they're doing the stupid little sharking,[2] which I think is so, it's hilarious . . . and then, the wrestling between the boys. . . . It's like kindergarten, when it's slow, and you're looking for something to do. . . .

At Work: Gender as Competence

These fast and slow periods provide the temporal context for distinguishing the two gender repertoires operating on the ACE floor. Although men and women did not physically change during work and play periods, the "constituting narratives" (Salzinger, 1997) of gender as a set of social relations between men and women varied widely within these two contexts. These temporal shifts and the different gender repertoires operating during work and play reveal the localized content of gender on the trading floor. During work, men on the floor rarely noted women as women. Instead, men gauged women's success and failure in the pits with seemingly gender-neutral criteria. Although women were somewhat more likely to see the different application of these criteria to men and women (e.g., in the different meanings of aggressiveness assigned to men and women), they too were as likely as men to accept competence as gender neutral.

Thus, men and women insisted that gender itself was not the cause of success or failure on the trading floor. Instead of pointing to gender, participants would talk about competence as being able to "get the job done," particularly under the heavy stress of a fast market. Gender, however, did not disappear during work times. Competence, though imbued with a gender-neutral veneer, smuggled in a distinctively "gendered logic" (Acker, 1990; Smith, 1987). Rather than being subsumed or displaced by competence, gender operated through it. Within this logic, competence and masculinity coincided considerably. Enacting a hegemonic masculinity coincided with proving oneself to be an aggressive, assertive participant in the market. Because the components of competence were interpreted as masculine, women were often put in positions where they were forced to compromise between being competent and distancing themselves from conventional ideals of femininity. The operation of this gender repertoire is illustrated in three key facets of competence on the trading floor: handling stress, being aggressive, and being physical.

One element of competence at the ACE is the ability to handle stress under fast market conditions. The fact that the markets potentially change very quickly creates specific challenges for working on the floor. For example, at the height of trading in 1994, contracts traded hands at a rate averaging 50 per second. Incremental changes in the price of a contract, known as a tick, range between $12.50 and $32 per tick. For a 100-contract order, a single tick is often worth more than $2,000; a 1,000-contract order can be worth $25,000.

Although participants talked about handling stress as a universal activity, interviews with participants showed how they linked stress management to distinctly masculine attributes. Often, clerks and traders coped with stressful, mistake-laden days by trying to deliberately forget about errors made under pressure:

> It's like when you make an error when you're playing baseball. I played sports my whole life, and whenever you made an error, you sat there and moped about it, chances were the next time . . . the ball was hit to you, you're going to make an error again. So you, it's like, so key to just forget about it. And how weird is that, that you say to yourself, O.K., forget about the fact that I just lost $12,000 for that guy, let's go back to work. That's challenging. (24-year-old male pit clerk)

Sports metaphors are a significant and often pervasive component of accounts of the market, and, in fact, a number of former athletes actually work on the floor. Linking competence to sports allowed men to interpret their experiences in the context of a competitive masculinity unavailable to women (Hearn & Parkin, 1987; Messner, 1992, pp. 17–19). Metaphors such as sports and battle were often used to describe the pressures of a fast-moving market. Women do participate in both athletics and armed service, but for men, the metaphor of sports denoted manliness as much as masculinity. Without explicit reference to gender, these narratives nevertheless tied together masculinity and the ability to handle pressure.

Because handling pressure acts explicitly as a gender-neutral concept but was implicitly constituted as a masculine ideal, women who do excel under pressure are in a position where their success must be explained. Under these circumstances, men grudgingly acknowledged successful women as competent but not also as women. That is, women traders could be respected as a trader or treated as a woman but rarely both. A handful of female brokers, traders, and clerks on the floor were identified as competent in this manner, as this 54-year-old male trader made clear:

> Take Susan, that's a perfect example. Now there's a person. There's a player, there's a market maker, and so for her, you have to respect her. Forget about the fact that she's a woman. You have to respect her as a person, because she was in there, was constantly in the market. I think you just know that there's a person that I can go to with a 50 [contract order] and know that it's going to clear the next day at that price.

In this trader's account, the example of a successful woman who could handle the pressure served to downplay gender as a constituent element of competence, focusing instead on the importance of being reliable under pressure. But later in the same interview, this same trader remarked that Susan's ability to compete "maybe makes her less of a woman." Despite the rhetoric of gender neutrality, high-status women often compromised their femininity. Successful male traders, by contrast, were held in higher esteem through their abilities to "step up" in fast markets.

The second element of competence as a gender repertoire is the ability to be aggressive. In an active market, clerks compete with each other to get orders from customers, and traders compete with each other to buy and sell contracts at the best possible prices. Almost 80% of my respondents (15 of 19) explicitly listed aggressiveness as an important element of competence. The imagery that respondents used to describe aggressiveness was vivid, often sexual or violent, and revealed its gendered character: "You have to want to cut someone's balls off" (54-year-old male); "From 7:20 to 2:00, I turn it on" (35-year old male); "I'm trying to buy 1,000 at a better price, we're not going to sit there and discuss it over a cup of tea" (25-year old female); "It's survival of the fittest. It's a war" (26-year old male).

Although both women and men considered themselves aggressive traders, gender again operated through, rather than instead of, competence. Most women on the floor described themselves as aggressive, a quality heightened by the work environment. A 25-year-old female pit clerk, proud of her ability to "hold her own" in a male-dominated environment, conceptualized aggressiveness as universally practiced and at the same time highly defeminizing:

> Everything about the pit goes against what would be, I think, considered feminine. You're yelling, you're screaming, you're spitting, the guys fart and burp all day long, the place smells, it's sweaty, the language is foul, it's aggressive, you're competing aggressively for business. Whether you're a clerk or if you're in the pit as a broker, you're competing aggressively to get your order filled ahead of the next guy. I don't think it's a very feminine environment.

With respect to the actual practices of being aggressive, women were virtually indistinguishable from their male counterparts in their ability and willingness to get in the face of a recalcitrant clerk or trader who was not allowing them to get their orders filled.

Nevertheless, aggressiveness continued to be coded as an eminently masculine characteristic. Women on the floor were often considered "bitchy," a term applied widely to women in men's worlds (see Kanter, 1977; Williams, 1989). For instance, this same woman, when I asked if people called her Deborah or Debbie, replied that "they mostly just call me 'bitch.'" By contrast, men were often criticized for not being aggressive enough. Nonaggressive men were considered ineffective: they did not command enough attention, fill their orders, or get good trades.

Being physical is the third element of competency on the floor. In one respect, the job is in fact physically demanding. The exchange requires all floor participants to remain standing while on the floor, which often means that people stand on their feet for hours at a time. Yelling is an integral part of the labor process, and some traders go to voice therapists to strengthen their voices to be heard. Finally, particularly during busier times on the floor, there is quite a bit of physical jostling as traders struggle to execute their orders and clerks attempt to get the attention of both traders and their customers.

Working on the ACE floor highlights the importance of physical size and space. Depending on the day, there were anywhere between 400 and 1,000 people standing in the trading area where I worked. Clerks often observed that the amount of space that was "theirs" during the day is roughly the space of their body: "The space of my body, pretty much. I mean, I stand, I just stand there, and that's my office. You know, I just stand in a little spot, like the area of a floor tile, all day long" (24-year-old male pit clerk). The press of bodies on the floor emphasizes physical size and floor presence, especially height. Large physical size, being both big and tall, is an advantage in this environment.

Being physical is the component of competence that most closely dovetails with connotations of gender as a reference to male and female bodies. Not surprisingly, physical differences became a locus for discussion about women on the floor. Respondents assumed that most women were at a disadvantage due to the physical nature of the floor. Women's voices are "not as heavy" and seemingly not able to carry as well as men's. Women are also seen as less physically able to hold their own. In addition, many of the respondents pointed out that even for the women who could "hack it," the floor would not be a desirable work environment:

> With girls, and being in the pits, you're like this [claps his hands together]. You're pancaked, man. Some women don't feel comfortable with that, probably. You know, having a guy pressed right up, you know, you're pressed up against a guy, and having a guy pressed up against you from behind. All day long. I would think that'd be uncomfortable for a woman. (39-year-old male clerk)

Invoking women's perceived inability to be as physical as necessary to hold their own in the pits cast gender language of competence yet attributed this difficulty to the characteristics of women's bodies.

The definition of competency that emerges from the discussion of work is a gender repertoire that constructs in masculine, and often sexual, terms while at the same time maintaining a veneer of gender neutrality. Language, even when sexualized, was likely to be directed at the market itself rather than at women as women. Traders often spoke in quite coarse manners about getting "fucked" by the market or accidentally "screwing" a customer, but these constituting discourses were captured in competence rather than in sexualized difference. In periods of work, gender is not made less salient, instead, masculinity is codified in ways that give shape to ACE work activities.

At Play: Gender as Sexualized Difference

If during the period of work, gender is interpreted as a form of competence, during play periods, gender becomes much more directly tied to sex and heterosexual imagery. Here women's bodies became objects for heterosexual masculinity. I focus on joking and getting along as important mechanisms through which the informal social structure of the floor is maintained in gender-dichotomous ways. When the market is less active, a dominant part of the exchange's atmosphere or cultural context consists largely of risqué storytelling, practical-joke playing, and joke telling. Masculinity becomes more explicitly sexualized and women more fully excluded from the men's world of trading.

In addition to acting as a stress reliever, as in classical accounts of humor at work (Haas, 1972; Radcliffe-Brown, 1965; Wilson, 1979), joking acts as a primary language through which group solidarity is formed and maintained (Hughes, 1958, p. 109; Lyman, 1987; Norman, 1994). As such, jokes act as a key element of the constituting discourse of gender. Kanter's (1977) study of a large, male-dominated corporation treated joking as a part of corporate culture where men would use off-color or sexual jokes to emphasize women's differences from their male counterparts (pp. 225–226). In my setting, many women did attempt to participate in the joking culture. They spoke about "playing the game" or being able to joke without being offended by the men's apparently juvenile and sexualized behavior. Many women on the floor stress their thick skins and their aptitude for taking a joke. Despite these seemingly "honorary men"—women who could be expected to laugh at jokes and listen to ongoing banter —the repertoire of sexual joking during times of play highlighted rather than minimized the differences between men and women and created visible in-groups and outgroups.

The sexual content of the trading floor corresponds to other accounts of merchant bankers (McDowell & Court, 1994) and bond traders (Lewis, 1989). In my observations, joking often had very explicit sexual connotations. In one typical example, after an altercation between two male clerks, one said to the other, "You weigh 100 pounds more than me, you could probably beat up my sister too." The second clerk's response, both to the clerk and laughing onlookers, was "Yeah, I could, but I'd fuck her first. Up the ass!" Violent and sexually aggressive jokes in particular facilitate the identification of the ACE as a man's world. These jokes are ubiquitous. They include reworking comic strips in sexually suggestive ways, alluding to the sexual practices of coworkers (and their relatives), putting sexual spins on current events, and making jokes about individual women on the floor.

The *hetero*-sexualization of jokes in this male bonding precludes women from being able to be, as in McDowell's (1995) characterization of British merchant bankers, "honorary big swinging dicks" during these periods of play (see also Acker, 1990, regarding "honorary men" in organizations). Although many of the women on the floor did swear and occasionally act sexually coarse with men, their status as women made it difficult for them to be sexual subjects; men continued to see women, as a group, as sexual objects. For instance, although some women considered themselves one of the guys, this did not preclude them from being sexually objectified. Women on the floor who told dirty jokes were seen as having their femininity eroded, and participants spoke of this erosion as "not very delicate" or "unladylike," and such women were said to "talk like a truck driver." The contradiction lies in the fact that for a woman to be one of guys, she has to stop being feminine. For men on the floor, discursively constructing even unladylike women as potential sexual objects maintains their ability to assert themselves as masculine men. . . .

With some exceptions—there was one female phone clerk who was jokingly sharing pictures from *Playgirl* to a disgusted audience of men—there is little opportunity for women to joke as sexual aggressors.

Men used joking and sexual banter about women as a way to reinforce a highly gendered group solidarity. This took place particularly during play periods. Talk about sexual exploits over the weekends was pervasive, graphically describing receiving oral sex from a date, picking up women and taking them home from a bar, or paying for prostitutes to come to a party. Comments and joking stories told throughout the less active moments of the trading day provided a way for the men on the floor to communicate their manliness to each other. . . .

CONCLUSION

The primary claim I make is that attention to temporality on the ACE floor highlights different gender repertoires that serve as the constituting discourses of gender, that gender actually operates differently according to the temporal rhythms of the market. When the market is active, gender is articulated through the language of competence. The components of competence—handling stress, being aggressive, being physical—are understood as gender neutral on their face but at the same time obscure highly gendered logics of action. This explains why both men and women perceived women as having to fit in a man's world by getting in peoples' faces, shouting, pushing, and shoving. This construction of competence is hegemonic: It postures as gender neutral but actually tilts the playing field in favor of men.

When the market is less active, the more overtly sexualized repertoire of joking and getting along emerges. Men and women use jokes to pass time, fit in, and relieve tension, but a direct result of men's sexual banter is to facilitate group solidarity among men to the exclusion of women. Strong heterosexual joking is predicated on men being the sexual agents of jokes and women being the objects. Although a few long-tenured women were able to joke with the men, for most, this was not the case. Women could not easily participate in these jokes precisely because the concept of women as agents disrupts the normal pattern of female objectification. If both men and women were able to be subjects of sexual banter, who would be left to be the objects?

My second, more general claim is that temporal rhythms are a key to understanding variations in gendered work practices. Salzinger (1997), for example, makes a convincing argument that the meaning of gender can vary greatly at the shop-floor level depending on such local conditions as management attitudes and labor processes. My argument is that even at the local level, gender changes dramatically depending on the pace of work. Particularly in workplaces characterized by lots of temporal variation in the workday—hospital emergency rooms, police departments, restaurants—time matters a great deal with regard to how gender is articulated. In most workplaces, there are lunch hours, coffee breaks, speedups, or slowdowns, all of which have important consequences for the study of gender. . . .

NOTES

1. The ACE, and all names of participants, are pseudonyms.
2. "Shark fins" are trading cards ripped into the shape of a fin. These fins are then surreptitiously attached to an unaware person's jacket collar, and the "sharked" individual is then often sent on a bogus errand. As he or she passes along the lines of clerks and runners, people will scream out "Shark!" until the individual, often turning red in embarrassment or anger, notices he or she has been tagged.

REFERENCES

Acker, J. (1990). Hierarchies, jobs, bodies: A theory of gendered organization. *Gender and Society, 4,* 139–158.

Baker, W. E.(1984). The social structure of a national securities market. *American Journal of Sociology, 89,* 775–811.

Britton, D. M. (1997). Gendered organizational logic: Policy and practice in men's and women's prisons. *Gender and Society, 11,* 796–818.

Connell, R. W. (1987). *Gender and power.* Cambridge, MA: Polity Press.

Connell, R. W. (1995). *Masculinities.* Berkeley: University of California Press.

Epstein, C. F. (1970). *Woman's place: Options and limits in professional careers.* Berkeley: University of California Press.

Ferguson, K. (1984). *The feminist case against bureaucracy.* Philadelphia: Temple University Press.

Haas, J. (1972). Binging: Educational control among high steel ironworkers. *American Behavioral Scientist 16,* 27–34.

Hall, E. J. (1993). Waitering/waitressing: Engendering the work of table servers. *Gender and Society, 7,* 329–346.

Hartmann, H. (1976). Capitalism patriarchy, and job segregation by sex. In M. Blaxall & B. Reagan (Eds.), *Women and the workplace: The implications of occupational segregation* (pp. 137–169). Chicago: University of Chicago Press.

Hearn, J., & Parkin, P. W. (1987). Gender and organizations: A selective review and critique of a neglected area. *Organization Studies,* 4, 219–242.

Hughes, E. (1958). *Men and their work.* Glencoe, IL: Free Press.

Kanter, R. M. (1977). *Men and women of the corporation.* New York: Basic Books.

Leidner, R. (1991). Serving hamburgers and selling insurance: Gender, work, and identity in interactive service jobs. *Gender and Society, 5,* 154–177.

Leidner, R. (1994). *Fast food, fast talk: Service work and the routinization of everyday life.* Berkeley: University of California Press.

Lewis, M. (1989). *Liar's poker: Two cities, true greed.* New York: Norton.

Lyman, P. (1987). The fraternal bond as a joking relationship: A case study of the role of sexist jokes in male group bonding. In M. Kimmel (Ed.), *Changing men: New directions in research on men and masculinity* (pp. 148–163). Newbury Park, CA: Sage.

McDowell. L. (1995). Body work: Heterosexual gender performances in city workplaces. In D. Bell & G. Valentine (Eds.), *Mapping desire: Geographies of sexualities* (pp. 75–95). New York: Routledge.

McDowell, L., & Court, G. (1994). Missing subjects: Gender, power, and sexuality in merchant banking. *Economic Geography, 70,* 229–251.

Messner, M. A. (1992). *Power at play: Sports and the problem of masculinity.* Boston, MA: Beacon.

Norman, K. (1994). The ironic body: Obscene joking among Swedish working-class women. *Ethnos, 59,* 187–211.

Pringle, R. (1989). *Secretaries talk: Sexuality, power, and work.* London: Verso.

Radcliffe-Brown, A. R. (1965). *Structure and function in primitive society.* New York: Free Press.

Salzinger, L. (1997). From high heels to swathed bodies: Gendered meanings under production in Mexico's export-processing industry. *Feminist Studies, 28,* 549–574.

Smith, D. E. (1987). *The everyday world as problematic: A feminist sociology.* Boston: Northeastern University Press.

Stacey, J., & Thorne, B. (1985). The missing feminist revolution in sociology. *Social Problems. 32,* 301–316.

West, C., & Zimmerman, D. (1987). Doing gender, *Gender and Society, 1,* 125–151.

Williams, C. L. 1989. *Gender differences at work: Women and men in nontraditional occupations.* Berkeley: University of California Press.

Wilson, C. (1979). *Jokes: Form, content, use and function.* New York: Academic Press.

35

Gender and the South China Miracle

Two Worlds of Factory Women

BY CHING KWAN LEE

Ching Kwan Lee, a sociologist whose specialization is the globalization of women's work, presents a portrait of two factory workers in Hong Kong and South China. These excerpts from her book of the same title provides a broad view of the ways large corporations exploit gender and familial relations to maintain a cheap and dependable labor force. Large corporations often move from one area to another country to maximize profits, depending upon changes in the market and the lure of other localities with more favorable laws. The lives of these women represent two scenarios in that pattern of profit seeking.

1. How is gender manipulated differently in the two plants?
2. What is the relationship of women to their families in these two plants?
3. Why are the women willing to work under the restrictive conditions of localistic despotism in the Shenzhen plant?

Since the mid-1980s, China has become the world's new "global factory," with the southern province of Guangdong (including Hong Kong) as its powerhouse. Millions of women workers are toiling in sweat shops and modern factories, churning out Mickey Mouse toys, Barbie dolls, Nike sports shoes, Apple jeans, watches, radios, televisions, and computers for worldwide consumption. These mass-produced commodities may be highly standardized, but the factory regimes that produce them, which spring up along the trail of mobile international capital, are not. The stories of two Chinese women, Yuk-ling and Chi-ying, highlight both the differences and the similarities between the worlds of labor where the south China economic miracle is manufactured, and where labor politics and women's identities are made and remade.

YUK-LING: WORKING MOTHER AMIDST ECONOMIC RESTRUCTURING

On a brightly lit, air-conditioned shop floor in a modern factory building in Hong Kong, about a hundred women workers sat along both sides of three conveyor belts, assembling mini hi-fi products. They worked for Liton, an electronics factory producing household audio equipment. Starting with printed circuit boards, then cassette decks, CD players, tuners, and remote controls, these women assembled hi-fi products to be sold under international brand names like Schneider, Mitsubishi, Packard Bell, and Techwood to German, Japanese, Mexican, and American households. No one cared to pronounce these names correctly. Instead, women workers deliberately blended Cantonese accents into, for instance, the German "Schneider," to result in the playful but meaningless sounds of "Si-nai-daa." This was one of the ways to bring some collective authenticity to bear on the nine-hour-fifteen-minute work day. For many of these women, days like this had filled more than twenty years. "I have spent half my life in this factory," they said with pride and occasional sighs.

Yuk-ling, age forty-three, the line leader of Line HK1, was a short, slim, spirited woman who would look much younger than her age if not for the bulging bags under her eyes. Like other young women of her generation in the 1960s, Yuk-ling quit school after sixth grade and started working full-time when she was thirteen. At the age of nineteen, after several factory jobs, Yuk-ling came to work at Liton in the early 1970s, which at that time was still named Mo's, a subsidiary of an American electronics corporation. Attracted to an industry that provided factory work then considered more modern, clean, and feminine than alternatives like garment-making and wig-making, she came to try it out. "And then, one day you counted and it was already some twenty years," she said. She started as a line girl, assembling printed circuit boards and transistor radios, and was later promoted to tester, material handler, and finally line leader. Yuk-ling met her husband through her coworkers, got married at age twenty-nine, and had two little girls by the time we met in 1992. Her husband was the leader of a group of construction workers and was responsible for getting project contracts for the group. Although he earned more than Yuk-ling when he worked, his contribution to the family income fluctuated. Yuk-ling's monthly income, around HK $5,000 (US $600), was critical for the entire family, especially since they started paying the mortgage on their apartment under a government subsidized home-ownership program several years ago.

Like many of her coworkers, Yuk-ling had a tightly packed daily work and family schedule, and her physical mobility was confined to the neighborhood where she worked and lived. Each morning at 7:30, Yuk-ling prepared breakfast for her eldest daughter and got her dressed for school. At 7:45, she took both daughters on a ten-minute bus ride to deposit her older daughter at kindergarten, repeating her routine motherly advice of "no fighting with other kids, no sweets, listen to your teacher, and work hard." She then took another bus to a nearby public housing estate where her baby-sitter lived, and left her younger daughter with the woman, who would prepare breakfast and lunch for the girl. By then, Yuk-ling had exactly seven minutes to walk to Liton, where work began at 8:15. If she was late, other women workers knew it was because her daughters were sick and she had to take them to the doctor before coming to work. When that happened, the line leader from another line would pitch in for her until Yuk-ling showed up. On an average day, however, she was seldom late, but she had to hide behind the pantry door to eat her breakfast—fried noodles or freshly baked bread that Lan, a woman coworker, bought for her. Everyone, including her foreman and the production manager, knew that she was sneaking away to eat breakfast, but no one found it problematic. They knew that she had to do this, and that when she came back from behind the door, she would be a brisk, responsible, and indispensable line leader as she had always been for the past two decades.

Yuk-ling's work involved everything required to keep production on schedule. In the past few years, after Liton extended its production lines into Shenzhen, China, the work pace in this Hong Kong plant had slowed down a bit. Instead of 400 hi-fi units, average daily output was scaled down to 300. This was partly because the orders for the Hong Kong plant tended to be for small volume, but involved more design changes than the orders filled by the Shenzhen plant. Moreover, this, plant now concentrated on pilot production of models that would then be mass-produced by the Shenzhen plant. Both these trends meant that Yuk-ling had to rearrange the production lines more frequently and that her "line girls" had to change their line seats in response to different assembling procedures for different models. "Line girls," once an apt description, had become an anachronistic reminder of the length of time these women had spent working on the lines. Although women workers at Liton were around forty years old on average and were married with children, they enjoyed exploiting the absurdity arising from the gap between their actual age and the youthful "line girls" label to have some fun. From time to time, they yelled loudly, "Mother, Mother, help, I've messed up!" to get Yuk-ling's assistance when they had problems with electronics components that had become smaller and smaller over the years.

The few men on the shop floor were repair workers, foremen, or production managers. All these middle-aged men were also longtime employees of Liton, but unlike the women, who stayed on the line, they had moved up the plant hierarchy from positions of apprentice and quality control operator. Women workers understood the reason for the men's promotion: when men had families, they needed and wanted promotions, whereas for women, having a family meant that they could not be managers or be given similar opportunities. On the shop floor, women were not shy about teasing their foremen whenever the latter made production mistakes, or embarrassing them with sexual innuendoes. In this factory, labor control seemed invisible, unnecessary, and above all hardly felt by women workers.

About half of these women were local Cantonese, while half came from Fujian, a province neighboring Guangdong. Most of these Fujianese women had moved to Hong Kong with their families more than ten or twenty years earlier, and most of them spoke Cantonese. The two groups of women got along well at work, although Cantonese women tended to make fun of Fujianese frugality and dietary habits. In vivid exchanges of family news or purchases of discount items for each other, these women knew no local boundaries. As "line girls," these women earned about HK $3,000 (US $400) per month, with some individual adjustments of a few hundred dollars depending on the length of service. The lack of promotion prospects and the meager income might have led to self-teasing remarks, but not to utter frustration.

What seemed to have anchored them so permanently in this factory was that this employer allowed them to integrate their dual responsibilities as mothers and workers. Yuk-ling and other women with children found that the fixed working hours and the five-day work week that Liton offered more than compensated for its lower wage rate. "When the kids have their school holidays, we also have our day off," they said, justifying their acceptance of low wages. Moreover, when women had emergencies to take care of, such as when their children's school teachers wanted to talk to them or when their children were sick, Liton's management turned a blind eye to their absence if it was restricted to an hour or so during the work day. Women at Liton, therefore, found themselves in a low-level equilibrium—they managed to balance family and work, their lives were stable, and everything was within the neighborhood. In the meantime, on the shop floor, years of repeating similar work procedures had made work bearable and routines a source of relative comfort. The work day was punctuated by women's talk, which, at times playful, at times sour, was satisfying enough to make the day feel shorter.

"Work life is hard. Whatever I do, I do it for my kids, so that they will have a better life in the future," Yuk-ling remarked. She found her husband dependable, "as long as he supports the family and does not gamble or smoke." She preferred her role as a mother in a network of kin bounded by familial interdependence and mutual obligations to my independence and freedom as a single, professional woman. "In the end, women need to have families," she advised me.

Recently, Yuk-ling and other women at Liton were concerned about losing the stability of the integrated family and work life they had managed to maintain for so long. The general trend of plant relocation to mainland China might push them into the service sector, where work hours were not compatible with family hours, upsetting the tightly coupled daily schedule they had cherished all these years. That would threaten not only the amount of money they brought home, but also their deeply cherished beliefs about proper motherhood.

CHI-YING: PEASANT DAUGHTER IN THE BORDERLAND

Liton operated another electronics plant just across the northern border of Hong Kong. Traveling from Liton's Hong Kong plant to this Shenzhen plant would take an hour and a half by bus. The same range of hi-fi products was made on production lines arranged in exactly the same way as those in Hong Kong. Every step of the production process was specified by "work procedure sheets," xeroxed copies of those used in Hong Kong, which were hung above every work station in this Shenzhen plant. Two senior production managers and several foremen commuted between Hong Kong and Shenzhen every three days to oversee production on both sides of the border. Other managers, who were stationed in Shenzhen six days per week, had worked in the Hong Kong plant for a long time before they were assigned to Shenzhen. Yet the world of labor here could not be more different from that in Hong Kong.

Chi-ying was a twenty-two-year-old peasant girl from a rural village in the northern Chinese province of Hubei. She came to Shenzhen two years ago and, through an introduction by Hubei locals who worked at Liton, she was recruited as a material handler. All her coworkers were young women, usually in their late teens or early twenties. Several Hubei locals worked on the line, and Chi-Ying would talk to them in their village dialect. With women from other provinces, such as Jiangxi, Hunan, and Sichuan, she would speak in Mandarin, the national language. Her line leader was a Guangdong woman and her supervisor a Guangdong man, so she had picked up a few words in Cantonese.

All workers wore blue uniforms with shoulder stripes of different colors to distinguish their roles and ranks. Control at work was very explicit. A clerk from the personnel office appeared intermittently to check on operators' fingernails. Anyone who had long nails was fined two *renminbi* (RMB) and had a misdemeanor record put in her personnel file. Every visit to the bathroom required a permit from the line leader. A normal work day lasted eleven hours, with a one-hour lunch break around noon. Whenever Chi-ying was late to work, the time clock would print her card with red ink and her supervisor would warn her in rude Cantonese. That was also why many northern workers learned foul language in Cantonese well before they could use the dialect in everyday life. Because absenteeism was heavily penalized and fined, Chi-ying came to work even when she was sick. Many times, she had seen line girls suffering from fever or menstrual pain clinging to the line, sobbing or cursing. Overtime shifts were frequent and mandatory in busy seasons, work lasted until eleven at night. If workers refused to do overtime work, they would first be fined and later dismissed if they repeatedly refused.

Like many *buk-mui* (literally, maidens from the north), as women workers from outside of Guangdong were pejoratively called in Shenzhen, Chi-ying believed that her supervisor only promoted his own locals to be line leaders. Easier positions on the line were also reserved for Guangdong women. Position on the line made a difference in how hard they had to work, but all workers were paid the same fixed daily wages. Women workers especially disliked soldering because of the smell of melting iron and the smoke they inhaled while doing that job. Everyone noticed that only women from the north who had no locals in the managerial ranks were assigned to do soldering. Because some of Chi-ying's locals were line leaders, she knew she was marginally better off than those from Jiangxi or Hunan "with no one up there." Yet, she also realized that she was in no way comparable to women workers from Longchuan, her supervisor's county. These Longchuan women were all testers, line leaders, or senior line leaders, the best positions available to women at Liton. Although all senior managerial positions were occupied by Hong Kong people, shop-floor management was monopolized by the kin group from Longchuan, headed by four young men who were cousins.

Lacking the ambition to get promoted in any particular factory, Chi-ying was satisfied knowing that if she wanted, she could switch to another factory that would want her for her factory experience. Factory jobs were plentiful in Shenzhen. Nevertheless, while at Liton she learned to make good use of her locals in eluding management's strict control. In trying to get permission to take a two-week home-visit leave, she carefully orchestrated an emergency telegram from home and asked her locals at Liton to spread the news that her mother was deadly sick so that her supervisor would not doubt the authenticity of the telegram. She also asked one of her male locals who was a technician and a roommate of her supervisor not to deduct the RMB 100 for her leave. Deduction of wages was a normal practice when workers took home-visit leave, although exceptions were allowed for "good" workers with legitimate grounds to take leave.

Chi-ying's closest friends were all from Hubei. Because it was company policy to disperse workers from the same village or county into different production lines, Chi-ying and her locals got together mainly in the canteen and the dormitories, where they exchanged gossip, complaints, and news from home. Her aunt and her cousins all lived in the same dormitory room, and Chi-ying would inform them of her whereabouts every time she went out. That was her pseudo-family away from home. Although resentful of the despotic management, long hours of closely monitored work, and poor food in the canteen, Chi-ying wanted to work in Shenzhen. From what she gathered from other locals, Liton might not be the best factory in terms of pay and work conditions, but neither was it the worst. Her monthly paycheck amounted to about RMB 300, about one-third more than that of an assembly worker. She could make as much as RMB 400, given more overtime shifts. Back home, her peasant father earned on average RMB 700 a year.

Having a cash income to herself epitomized a totally new way of life that would have been beyond her means had she stayed in her Hubei village, and above all, her factory job in Shenzhen allowed her to decide on her own marriage. Several years ago, when she was twenty, Chi-ying's parents found her a fiancé through a matchmaker. "The guy had a residence in Wuhan [the capital city of Hubei] and they thought I'd have a better life in the city," Chi-ying recalled. It was the usual practice in the village to wait for several years before the couple formally married each other. Chi-ying did not resist the arrangement although she hardly knew the young man. Then, one Chinese New Year, when some of her cousins and uncles went back to their village from Shenzhen, she decided to try out something new while she was still young. "The name Shenzhen had an aura of excitement to us village kids. I had never seen a high-rise or paved road that people talked about," Chi-ying said, nostalgic for her past innocence.

Sometime later, the arranged marriage dissolved when Chi-ying declared her intention of working in Shenzhen for a few more years. She sent back part of her wages to compensate for the presents and money the young man had sent her parents when they were still engaged to each other. She kept half of the money herself, for future use, and she sent the rest to her parents. In Shenzhen, Chi-ying met a Hubei local and they decided to get married in a year's time. On one of her home visits, Chi-ying brought him home to meet her parents, and they agreed to her plan.

Despite the hardships inside the factory and the daily discrimination against out-of-province workers, Shenzhen offered young peasant women like Chi-ying an expanded horizon of modernity. Interestingly, for Chi-ying and her friends, hardship rather than idleness was what a modern way of life entailed. A pair of cheap earrings bought with her own money, a visit to the barber shop, a trip to the shopping mall, going to the movies, and simply strolling along the main street seeing other young people all brought her the satisfaction of feeling "I have been there." The realization that she had to go back home eventually only reinforced her attachment to her life as a Shenzhen sojourner: modern, free, and young.

Watching Hong Kong television broadcasts from across the border, Chi-ying was aware of a supposedly more modern pattern of womanhood than what she followed in Shenzhen. From time to time, she expressed her polite admiration for Hong Kong women's opportunities and their glamorous, comfortable lives as portrayed in television series. But then, occasionally, she would ponder aloud whether women could really be happy in a city as competitive and stressful as Hong Kong. Most of the time, though, when Chi-ying contemplated her life, she compared herself with her grandmother and her mother at home. "They have never left the village. They have not had their own jobs," she remarked with quiet complacency.

★ ★ ★

I have found that in south China, where capitalism and socialism meet, different modes of control over women workers are used by mobile capital, creating different lifeworlds of production. My ethnographic study focuses on two factory regimes, localistic despotism and familial hegemony, which are formed in the two plants owned and managed by the same enterprise, producing the same range of electronics products by using exactly the same technology. Whereas in Shenzhen, young migrant women workers are subjected to overt, punishment-oriented control mediated by patriarchal localistic networks, Hong Kong women workers enjoy more autonomy under a regime of covert control mediated by a set of familialistic practices. Whereas the constitution of "maiden workers" is central to the exercise and contest of class power in the Shenzhen plant, the construction of "matron workers" is the linchpin of the regime in the Hong Kong plant. But why such differences, and with what consequences?

Systematic comparison across the two factories and their institutional embeddedness point to the central importance of gender and labor markets. More specifically, because the colonial state in Hong Kong and the client state in Shenzhen both adopt a permissive stance toward labor relations, managerial autonomy is maintained and management responds less to the state-level political apparatus than to the conditions of the local labor market. Yet the structure and the processes of labor market supply and demand are not merely economic forces but are institutions undergirded by the social organization of gender. Thus, in Shenzhen, for instance, the massive supply of single, young rural daughters is predicated on women's marginalized position in their families of origin, women's intent to flee from patriarchal demands on their labor and from arranged marriages, and their subscription to a cultural notion equating factory work with appropriate femininity. Moreover, the localistic networks that channel them from the fields to the factories embody traces of patriarchal authority: male locals and relatives become guardians of women away from home. No less than the supply, the demand for female labor is gendered in that foreign capitalists adopt and reproduce the gendered notion of women as more docile, more dexterous, and cheaper laborers for labor-intensive work than men. Capitalists respond to all these conditions of the labor market by incorporating localistic networks into the factories, as doing so can lower the cost of labor reproduction and facilitate control.

In Hong Kong, gender is at work in the organization of the labor market, but in different ways. In the dwindling, sunset manufacturing labor market there, I have found a supply of middle-aged working mothers whose relatively advanced age, low qualifications, and gendered family responsibilities combine to lock them into a declining sector of the Hong Kong economy. Stuck with the same employer for many years, they have unwittingly acquired firm-specific experience that their employers treasure as long as they want to keep part of the production going in Hong Kong. Working mothers' gender role in the family profoundly shapes the conditions of their employment, so much so that management has an interest in incorporating familialism as a control strategy. By facilitating women's fulfillment of motherly responsibilities through shop-floor practices, management succeeds in stabilizing the supply of experienced laborers and keeping wages low, as part of the cost of labor reproduction is transferred to the women's families. . . .

36

Gender, Social Inequalities, and Retirement Income

BY TONI M. CALASANTI AND KATHLEEN F. SELVIN

This reading by Toni Calasanti and Kathleen Selvin helps us to understand some of the consequences of workplace inequalities for men and women. They provide information about Social Security that helps us to understand how gendered policies and practices in both the workplace and government can disadvantage women. An understanding of how these policies and practices evolve can help us to plan for a better future in the United States, both for ourselves individually, and for society collectively. As you read this article, remember that the policies and the practices described are social constructions within a particular society.

1. What are the consequences of seeing retirement as a male experience in terms of income, health care, and "free time"?
2. What workplace practices and gender-based assumptions about work disadvantage women in Social Security retirement income?
3. What impact does Social Security policies and workplace practices have on the retirement incomes of individuals from racial and ethnic groups other than whites, individuals across social classes, and individuals who are not legally married (including same-sex partners)?

Most people believe that retirement equals freedom: freedom from a forty-hour work week, from the same routine, from supervision or, conversely, from feeling the weight of too much responsibility. This attitude comes across in questions asked of retirees such as, "So what do you do with all your free time now?" Or by fears expressed by some contemplating labor force withdrawal, "What will I do with my time?" Similarly, investigations of the retirement transition presume such freedom and measure satisfaction in terms of personal responses to this new freedom, the "lack of structure." . . .

In this chapter, our primary focus is on retirement income. We chose this emphasis partly because finances can have an enormous impact on later life. Monetary difficulties can lead to a situation where "retirement does not mean freedom but restriction, and, in the extreme, imprisonment" (Braithwaite and Gibson 1987, 11). In addition, the obvious relationship between paid and unpaid work on one hand and retirement income on the other makes the latter an ideal choice for examining the gendered nature of retirement. In this way we show that, for instance, women's greater likelihood of poverty in old age than men's is not, strictly speaking, either a natural result of demographic trends, such as greater life expectancy (itself partly influenced by social factors) nor is it random: It is a result of patterned social processes, such as the ways in which old people get access to income, among other things. Further, these social processes are racialized such that White women may in fact have higher retirement incomes than men of color due, in part, to their marriages to White men. It is women of color who are particularly disadvantaged.

To clarify the arguments we put forward . . . we need to lay out the broad context undergirding our analyses. Putting women's lives at the center of analysis reveals the wide range of productive activities which men and women perform that have economic value, including paid labor, unpaid labor, and services provided to others (Calasanti and Bonanno 1992; Herzog et al. 1989). When we define work this way, we see that, at all ages, women perform more productive activities than do men (Herzog et al. 1989) and that retirement does not mean leisure or freedom from labor. . . . Gender relations structure the productive activities in which men and women engage: which ones they do, the rewards for these, and so on. Gender relations play out over the life course and through retirement such that men's and women's experiences of this time differ in critical ways: in terms of income and the meaning of retirement. When combined with sensitivity to race and class, we find that the "golden years" await only a select group of predominantly White and privileged men. Many women, men of color, and members of the working class must continue a range of productive activities in retirement, in both the formal and informal economy, unless poor health precludes this. In addition, women continue their unpaid domestic labor. By contrast, more privileged men have choices in this regard: They can choose to engage in paid work, and they can choose to be involved in domestic labor. The voluntary nature of these activities underscores the power differences based on gender, as well as race and class. Who has the freedom to choose to perform labor, and what types, in their retirement years?

Women's continued work in retirement is not "freely" chosen when it is predicated on financial need or power relations within the family. Instead, it is built into workplace and state policies, and also results from "normal" behavior within families—the expectation (and mandate) that women will have primary responsibility for domestic labor. Simply put, family obligations of employed women constitute a "second job" which men generally do not undertake. Women continue this domestic labor (typically regardless of class or race) over the course of their lives while men retire relatively free of it. Although most men perform some domestic labor, they generally do not take primary responsibility nor do as much as most women. At the same time, class and race play a role. On one hand, women of higher class, most of whom are White, may maintain responsibility for domestic labor while paying other women, working-class and often women of color, to perform it (Glenn 1992). On the other hand, many men are also not living this "dream": White, middle-class retired men's unearned advantage in the labor market, and their resultant ability to secure stable, "career" jobs, is based on the disadvantages experienced by women, working-class White men, and people of color.

GENDER, THE STATE, AND RETIREMENT INCOME: SOCIAL SECURITY

One of the most important sources of income in retirement in the United States is the public pension, Social Security (Atchley 1997). However, gender inequities in both paid work and family have been embedded in this program since its inception in 1935. Social policies that assume the existence of a traditional nuclear family tend to reinforce the gender inequities embedded in that family form (Estes 1991). Social Security is but one example of this process, with critical consequences for men's and women's retirement benefits.

Social Security originally covered only those who contributed to the program—specifically, retired workers. In so doing, it also assumed a particular form of unequal gender relations: that women always depend upon men in heterosexual (marital) relationships, and that women would be homemakers and men would be breadwinners. Just as the "family wage" assumed a patriarchal family head (May1987) who would provide for other members, Social Security assumed the same in retirement. Thus, men's presumed labor force history—a long-term, stable career with ever-increasing rewards—formed the basis of benefit eligibility and calculations. Never mind that some minority and working-class men were virtually excluded by this formulation as they were shut out of family wages and "careers."

At the same time that reproductive labor was seen to be a woman's "job" and her implicit basis for economic support, it was not valued as highly as men's paid labor. As Harrington Meyer (1996) notes, when wives and widows were added as beneficiaries in 1939, their eligibility was non-contributory. That is, their benefits were not based on their own contributions as workers but instead by virtue of their marital status alone—solely as (former) wives of eligible workers. Further, a spouse "dependent"—originally referred to as "the 'wife allowance'" (Harrington Meyer 1996, 458)—is entitled to only half of the main benefit amount. Importantly, while domestic labor may have been assumed by virtue of marital status, this was not the basis for the benefit; actual performance of such tasks was never a condition of benefit receipt (Harrington Meyer 1996). Social Security further reinforced women's subordination by distinguishing between deserving and undeserving women. Widows could collect Social Security based on their spouse's work histories, but divorced women could not (Rodeheaver 1987). Regardless of the reasons for the divorce, women were felt to be at fault for somehow calling upon themselves their abuse or abandonment. Despite changes in Social Security over time, such as the ability of divorced women to collect benefits if they were married for at least ten years, the reduced spousal benefit remains in effect.

In addition, Social Security ignored the reality of female breadwinners—despite the fact, for instance, that in 1940, 40 percent of Black women held jobs (compared to 15 percent of White women) (Amott and Matthaei 1996). Indeed, by assuming only one breadwinner and tying benefit levels to earnings, Social Security pays some dual-earner couples lower benefits than it pays to a traditional couple in which the man earns that amount alone (Harrington Meyer 1996). Thus, Social Security legislation concerned itself neither with women's lower wages nor with family obligations that might interfere with continuous labor force participation. Again, it devalued women's reproductive labor by not counting years engaged in reproductive labor in their benefit levels.

We can see the cumulative impact of gender relations within the family when we look at how Social Security benefits are calculated. First, benefit levels are tied to earnings: The more one earns, the greater the likelihood that one will receive the maximum benefit, which was $1,373.10 per month in 1999 (Social Security Administration 1999). Women tend to be clustered in a relatively small array of low-paid jobs, a factor which in itself deflates Social Security benefits. But in addition, benefits are also based on the earnings of the best 35 years of work. Due to family obligations, women are far more likely than men to have worked fewer than 35 years, and thus have years of zero earnings included in the calculation. Thus, women who leave the labor market usually receive less pay upon their return and lower Social Security benefits later on. Men's ability to have, on average, only one zero year out of 35, compared to women's average of 12 zero years (Harrington Meyer 1996) is firmly rooted in the gender division of family labor.

Thus, gender relations in family and work influence the retirement experiences of both men and women through the formation of pensions. Public pensions, such as Social Security, and private pension schemes are fashioned on the basis of men's experiences of work and production, as well as traditional, heterosexist notions about the domestic sphere. The emphasis on traditional couples excludes those who are never married, including for reasons of sexual preference. Further, while both paid and unpaid activities make important economic contributions (Calasanti and Bonanno 1992; Herzog et al. 1989), most pensions are based on White, middle-class men's work history, and therefore ignore reproductive labor (Quadagno and Harrington Meyer 1990; Scott 1991). As a result, pension plans treat men's labor as more valuable, and reward it more highly than women's in retirement, despite the fact that men's ability to engage in more highly paid labor likely relied upon women's reproductive work—that is, the primarily domestic work involved in maintaining people. Conversely, the assumption that individuals spend their lives as members of traditional nuclear families translates into policies that reinforce women's dependence on men for financial security in old age (Harrington Meyer 1996; Rodeheaver 1987), such as Social Security's spousal benefit.

To illustrate, in their study of a national sample of men and women, DeViney and Solomon (1995) found that, even after controlling for type of industry and occupation, gender was still a significant predictor of retirement income for two reasons. First, in terms of age, the older a man is, the higher his retirement income. The same is not true for women. Second, being continuously married to the same person is much more important for a woman's retirement income than for a man's. Only for women, even among those presently married, does discontinuous marital history matter; having been divorced or widowed earlier in their lives has an important impact on their retirement income. Women who had been continuously married to the same person had a monthly average retirement income of only $83.52 less than similar men. By contrast, women whose marital history was interrupted received an average of $356.35 less than men with interrupted marital histories. Thus, it appears that women still accrue a substantial penalty for "[d]eviation from 'traditional' marital careers" (DeViney and Solomon 1995, 98). Similarly, research in Great Britain uncovered a link between marital status and pension receipt for women but not for men (Ginn and Arber 1999).

A racial bias embedded in the original Social Security legislation excluded occupations typically held by people of color, particularly agricultural labor and domestic labor. These exclusions interest us because they reveal the intersections of gender with race and class. As we have noted, domestic labor is devalued as it is seen to be part of the private sphere and remains unpaid. When such devalued labor is in fact paid—that is, when others are hired to perform such tasks as housecleaning and caregiving for the old or young—it is relegated to those groups with the least amount of power: women, but especially women of color (Glenn 2000). The exclusion of domestic labor, then, particularly disadvantaged working-class women of color, who find both their paid and unpaid labor devalued. Figures from the 1940s, important years in the earnings history of present retirees, reveal that retired women of color were often employed as domestic laborers. Depending upon the group one is examining, between one-fifth and one-half of women of color were employed as domestic laborers, compared to only 12 percent of European American women (Amott and Matthaei 1996; King 1992). Despite legislative changes that have broadened coverage to almost all workers, still only 83 percent of Blacks aged 65+ (men, 81 percent and women, 84 percent) and 74 percent of non-White Hispanics (men, 77 percent and women, 72 percent) received Social Security in 1996, compared to over 90 percent of White men and women (Social Security Administration 1998a, table 1.9).

Lack of Social Security coverage also jeopardizes health in later life. In addition to the loss of pension benefits, Medicare benefits automatically accrue only to those who receive Social Security. If one is ineligible for Social Security, one also does not receive Medicare.

The intersection of race, class, and gender is also evident when we examine other ways in which marital status is rewarded. Social Security recipients are considered "dually entitled" if they qualify for benefits both as retired workers and as a present/former spouse, but receive the larger, spouse's benefit. Again, the larger benefit is based on marital status—having been married continuously for at least 10 years. However, not only is the proportion of ever-married women decreasing, but Black women are far less likely to have been married at least ten years. As a result, they are less likely to qualify as dually entitled than are White women (Harrington Meyer 1996). At the same time, Black men as a group receive much lower wages than do White men. As a result, those Black women who would be able to opt for a spouse's benefit are also less likely to find that it is appreciably higher than their own retired worker benefit. Thus, their class position also prevents them from having equal access to a higher, spousal benefit.

Finally, as implied above, Social Security benefits also maintain class privilege. First, working-class members enjoy less job stability than do middle-class workers. As a result, they receive lower benefits due to the impact of number of years of continuous work on payment levels. Second, benefit levels are tied to past earnings through a progressive formula, which means that while people with low lifetime earnings receive a higher replacement rate (their benefits represent a larger percentage of their previous earnings), people with high lifetime earnings receive higher absolute benefit amounts. Overall, then, tying benefits to past earnings advantages high-income workers.

We now turn briefly to some of the ways in which gender relations within the family and workplace also influence retirement income.

GENDER RELATIONS IN THE FAMILY AND THE WORKPLACE

Though most women work for pay, they still bear primary responsibility for household tasks (Twiggs, McQuillan, and Ferree 1999; Press and Townsley 1998; Coverman and Sheley 1986). Decades of research have shown that women take on the vast majority of housework, in part because no one else will do it and in part because everyone, including women, expects them to do so, as if to prove that they're real women. Men, on the other hand, tend to avoid such labor as "women's work" (Szinovacz 2000, 78). The power relations and advantages that accrue to men in this domestic division of labor have long appeared in household research. As Hartmann (1981) noted decades ago, the addition of a husband to the household increased a wife's domestic burden. Examining men and women over several years, Gupta (1999) finds that, while men decrease their domestic labor when they take on female partners, and increase it only when single, women do the opposite. They put in a lot more housework to take care of male partners than they do when they're alone. Women consistently do more across a wide array of living arrangements and marital statuses, including cohabitation. But the biggest gender discrepancy is among those who are married (South and Spitze 1994). Men add work to women's lives, whereas women toil to make men's lives easier.

This domestic division of labor influences retirement in terms of both the jobs that people take and their upward mobility. Among today's retirees, women often entered the labor force later than men, had to work particular shifts, turn down promotions or enter particular types of jobs in order to maintain their domestic labor roles. Domestic labor time also has an impact on other labor market outcomes, such as earnings (Coverman 1983). Importantly, this does not imply that women expend less energy at work; in fact, evidence indicates that they work harder than men (Bielby and Bielby 1988).

Men are privileged by this division of labor in ways that go beyond pay levels. White middle-class men are able to take jobs, promotions, and geographic moves to maximize their economic security. Indeed, their ability to even have "careers" or engage in paid work is based upon their not having to be concerned with household work (Acker 1990). They can take advantage of particular job opportunities, for example, a move into a more time-consuming position or to a different city, well supported by their wives' unpaid work at home. Race and class shape this situation as well. For instance, the opportunity to even consider mobility for a promotion is not equally available to different minority or class groups.

In addition to household tasks, women predominate in a number of care work activities throughout their lives. Although care work may bring with it a number of benefits, it may also have a negative impact on such things as retirement income. Many women appear to partake in what has been called serial caregiving. That is, they do not commonly care for children and elder adults at the same time, but instead tend to follow one with another: spouses, grandchildren, or others (AARP 1995b). While we return to the topic of caring for others later, at this point we emphasize the impact of this activity for women's retirement incomes. Among older employed women, it is important to note that acting as unpaid caregivers for frail elderly can impede their labor force activity (Stoller 1993; Harrington Meyer 1996). Although women, particularly African American women (Hatch and Thompson 1992), typically do not drop out of the labor force when involved in caring for adults (Moen, Robeson, and Fields 1994), they may reduce hours or productivity levels, or change jobs to accommodate their care work demands. These factors typically translate into lower retirement incomes.

The cumulative impact of gender relations within the family on retirement income is evident in many ways. The assumption of a male breadwinner has served to justify women's lower wages and mobility, which in turn further justifies women's predominance in care work and other domestic obligations. From this vantage point, we can see that both women's and men's retirement income is firmly rooted in the gender division of family labor.

The highest incidence of poverty among old women occurs among the never married. In part, this is because they lack spousal benefits—benefits denied to those who do not form traditional couples—from Social Security as well as possible private pensions. At the same time, never-married old men—who do have lower retirement incomes than married men (Mitchell, Levine, and Phillips 1999) —are not as likely to be poor as a result. Thus, gender relations in the family tell only part of the story. Women receive low wages over the life course, whether working continuously, as is likely to be the case among never-married women, or intermittently.

Gender relations in the workplace result in occupational segregation: Women and men tend to work in different jobs. In itself, this might not be a problem; but compared to men's jobs, women's jobs—primarily in the service sector—pay substantially less (Mitchell et al. 1999). These jobs also offer less mobility and fewer benefits, including pensions (Farkas and O'Rand 1998). Even women who work in traditionally male occupations earn less than their male counterparts. Before the 1970s, women retirees earned, on average, 50 to 60 percent of men's wages; today, women still only make 73 cents to the male dollar (U.S. Bureau of the Census 1999a). This labor force discrimination has a cumulative effect; women are less able to save for the future while employed, and they receive less Social Security income later. . . .

REFERENCES

AARP. 1995b. "The Sandwich Generation: Does It Really Exist?" *Horizons* 5(1): 5.

Acker, Joan. 1990. "Hierarchies, Jobs, Bodies: A Theory of Gendered Organizations." *Gender & Society,* 4(2): 139–58.

Amott, Teresa, and Julie Matthaei. 1996. *Race, Gender and Work: A Multicultural History of Women in the United States,* revised edition. Boston: South End Press.

Atchley, Robert C. 1997. "Retirement Income Security: Past, Present, and Future." *Generations* 21(2): 9–12.

Bielby, Denise D., and William T. Bielby. [1988] "She Works Hard for the Money: Household Responsibilities and the Allocation of Work Effort." *American Journal of Sociology* 93: 1031–59.

Braithwaite, Virginia A., and Diane Gibson. 1987. "Adjustment to Retirement: What We Know and What We Need to Know." *Ageing and Society,* 7(1): 1–18.

Calasanti, Toni M., and Alessandro Bonanno. 1992. "Working 'Over-time': Economic Restructuring and Class Retirement." *The Sociological Quarterly* 33(1): 135–52.

Coverman, Shelly. 1983. "Gender, Domestic Labor Time, and Wage Inequality." *American Sociological Review* 48(6): 623–37.

Coverman, Shelly, and Joseph Sheley. 1986. "Change in Men's Housework and Child-Care Time, 1965–1975." *Journal of Marriage and the Family* 48: 413–22.

DeViney, Stanley, and Jennifer Crew Solomon. 1995. "Gender Difference in Retirement Income: A Comparison of Theoretical Explanations." *Journal of Women & Aging* 7(4): 83–100.

Estes, Carroll L. 1991. "The New Political Economy of Aging: Introduction and Critique." Pp. 19–36 in *Critical Perspectives on Aging: The Political and Moral Economy of Growing Old,* ed. Meredith Minkler and Carroll Estes. New York: Baywood Publishing Co.

Farka, Janice I., and Angela M. O'Rand. 1998. "The Pension Mix for Women in Middle and Late Life: The Changing Employment Relationship." *Social Forces* 76(3): 1007–32.

Ginn, Jay, and Sara Arber. 1999. "Changing Patterns of Pension Inequality: The Shift from State to Private Sources." *Ageing and Society* 19(3): 319–42.

Glenn, Evelyn Nakano. 1992. "From Servitude to Service Work: Historical Continuities in the Racial Division of Paid Reproductive Labor." *Signs* 18:1–43.

———. 2000. "Creating a Caring Society." *Contemporary Sociology* 29(1): 84–94.

Gupta, Sanjiv. 1999. "The Effects of Transition in Marital Status on Men's Performance of Housework." *Journal of Marriage and the Family* 61(3): 700–11.

Harrington Meyer, Madonna. 1996. "Family Status and Poverty among Older Women: The Gendered Distribution of Retirement Income in the U.S." Pp. 464–79 in *Aging for the Twenty-First Century,* ed. Jill Quadagno and Debra Street. New York: St. Martin's Press.

Hartmann, Heidi. 1981. "The Family as the Locus of Gender, Class, and Political Struggle: The Example of Housework." *Signs* 6(3): 366–94.

Hatch, Laurie Russell, and Aaron Thompson. 1992. "Family Responsibilities and Women's Retirement." Pp. 99–113 in *Families and Retirement,* ed. Maximilliane Szinovacz, David J. Ekerdt, and Barbara Vinick. Thousand Oaks, Calif.: Sage Publications.

Herzog, A. Regula, R. L. Kahn, John N. Morgan, James S. Jackson, and Toni Antonucci. 1989. "Age Differences in Productive Activities." *Journal of Gerontology* 44(4): S129–S138.

King, Mary C. 1992. "Occupational Segregation by Race and Sex, 1940–1988." *Monthly Labor Review* 114(7): 30–36.

May, Martha. 1987. "The Historical Problem of the Family Wage: The Ford Motor Company and the Five Dollar Day." Pp. 111–31 in *Families and Work,* ed. Naomi Gerstel and Harriet Engel Gross. Philadelphia: Temple University Press.

Mitchell, Olivia S., Phillip B. Levine, and John W. Phillips. 1999. "The Impact of Pay Inequality, Occupational Segregation, and Lifetime Work Experience on the Retirement Income of Women and Minorities." Public Policy Institute AARP: Washington D.C.

Moen, Phyllis, Julie Robeson, and Vivian Fields. 1994. "Women's Work and Caregiving Roles: A Life Course Approach." *Journal Of Gerontology: Social Sciences* 49(4): S176–S186.

Press, Julie E., and Eleanor Townsley. 1998. "Wives' and Husbands' Housework Reporting: Gender, Class, and Social Desirability." *Gender & Society* 12(2): 188–218.

Quadagno, Jill S., and Madonna Harrington Meyer. 1990. "Gender and Public Policy." *Generations* 14(2): 64–66.

Rodeheaver, Dean. 1987. "When Old Age Became a Social Problem, Women Were Left Behind." *The Gerontologist* 27(6): 741–46.

Scott, Charles G. 1991. "Aged SSI Recipients: Income, Work History, and Social Security Benefits." *Social Security Bulletin* 54(8): 2–11.

Social Security Administration. 1998a. Annual Statistical Supplement, *Social Security Bulletin,* Washington, D.C.: U.S. Government Printing Office.

Social Security Administration. 1999. Annual Statistical Supplement, *Social Security Bulletin,* Washington, D.C.: U.S. Government Printing Office.

South, Scott J., and Glenna Spitze. 1994. "Housework in Marital and Nonmarital Households." *American Sociological Review* 59(3): 327–47.

Stoller, Eleanor Palo. 1993. "Gender and the Organization of Lay Health Care: A Socialist-Feminist Perspective." *Journal of Aging Studies* 7(2): 151–70.

Szinovacz, Maximilliane E. 2000. "Changes in Housework After Retirement: A Panel Analysis." *Journal of Marriage and the Family* 62(1): 78–92.

Twiggs, Joan E., Julia McQuillan, and Myra Marx Ferree. 1999. "Meaning and Measurement: Reconceptualizing Measures of the Division of Household Labor." *Journal of Marriage and the Family* 61(3): 712–24.

U.S. Bureau of the Census. 1999a. "Household Income at Record High; Poverty Declines in 1998, Census Bureau Reports." Economics and Statistics Division. www.census.gov/Press-Release/www/1999/cb99-188.html. Accessed March 14, 1999.

37

Marketing and the Publicity Images of Women's Professional Basketball Players from 1977 to 2001

BY GAI INGHAM BERLAGE

Sociologist Gai Ingham Berlage examines the way women professional basketball teams market their players. In doing so, she provides a brief history of women's professional basketball as well as a discussion of women's participation in sports in general. Women basketball players must balance their athletic ability and prowess against demands by team owners and promoters that they be "feminine." Gender, sexuality, and sexual orientation are once again linked, with excellent athletes often thought to be lesbians and lesbians closeting themselves to avoid being shunned by fans.

1. Why do owners market and pressure the professional women basketball players to present themselves as "feminine"?
2. How much has changed in marketing the women professional basketball players?
3. Some women players engage in what is called "role splitting." What is role splitting and why do some women fall victim to it?

The question of how to publicize and market women's sport has been fraught with problems for sport promoters. This is due to American cultural contradictions that have existed between traditional definitions of femininity and of athleticism. Traditionally sports were defined as a male domain. Strength, aggression, and competitiveness, the qualities necessary for becoming a great athlete, were culturally defined as masculine characteristics. Therefore, women who dared to compete often found themselves in the conflicting position of having to prove on the one hand that women were capable of athletic feats and on the other that they were really women. Stereotypes of the female athlete as masculine or lesbian were prevalent.

Even as late as 1980, prominent sport sociologists Don Sabo and Ross Runfola, stated: "sport and masculinity are virtually synonymous in American culture" and "a primary function of sport is the dissemination and reinforcement of such tradition American values as male superiority, competition, work, and success."[1]

Traditionally, masculinity and femininity were thought to be biologically preordained. An American cultural ideology of male superiority and the cultural conditioning of males and females into proscribed gender roles reinforced the paternalistic, hegemonic order. Social acceptance for men and women was dependent upon displaying culturally defined masculine or feminine characteristics. Women, the weaker sex, were believed to be frail, passive, emotional, and dependent upon men. Their major function was to marry and to have children. Defined primarily by their sexuality and beauty, appropriate roles were that of girlfriend, wife, and mother. To be attractive to the opposite sex meant a woman had to be beautiful and sexually alluring, but not promiscuous. There were "good" and "bad" girls. Good girls were sexually attractive, but chaste. Bad girls were "loose women" or prostitutes.

GENDER AND SPORT

Sports such as figure skating, gymnastics and tennis that emphasized form, beauty and grace were the appropriate sports for women. Team sports such as football, baseball and basketball were the quintessence of masculinity. Competitiveness, aggressiveness, physical strength and dominance over one's opponent created winning teams and were the very characteristics that defined traditional masculinity. Team sports served as a male rite of passage. Women's participation in professional basketball challenges traditional hegemonic definitions of "masculinity" and "femininity" and makes sports a contested terrain. For many men sports provide the last bastion for proof of male superiority.

The female athlete has often found herself in a dilemma. How can she be athletic and still be perceived of as feminine by the public and not be considered a freak or anomaly. In 1974 Jan Felshin described "the apologetic," three ways that female athletes have managed this identity. One way is that female athletes de-emphasize the importance of sport in their lives and deny the importance of their accomplishments. In other words, they apologize for being athletic. Secondly, they can separate their athletic self from their public self. They do this by being competitive on the athletic field, but off the field they stress their femininity. They do this by appearing as feminine as possible in demeanor and appearance, always wearing skirts or dresses, make-up and jewelry when in public. When asked by the press or someone else about their sports career, they emphasize that their main goal in life is to find a husband and get married, sport is secondary. And a third way that some resolve the conflict is by participating only in more acceptable female sports such as tennis or golf.[2]

Even women who played so-called women's sports felt conflicted as athletes and felt they needed to appear feminine. Proper female tennis and golf attire has traditionally been golf skirts and tennis skirts or dresses. But dress was not enough, with many tennis and golf stars making sure they had a feminine hairdo and wore make-up and lipstick when they played. To downplay their sporting persona, they often told reporters that their primary goal in life was to have a family. Chris Evert is a good example of this.

Prior to the 1960s the white heterosexual female athlete was the dominant cultural ideal. This ideal reflected America's racist and sexist prejudices. African American women usually had to compete separately and apart from their white counterparts. For example, until the mid-1950s African American basketball teams were barred from participating in Amateur Athletic Union (AAU) national basketball tournaments. The All-American Redheads a professional women's basketball team that existed from 1936 to 1986 never had a black player. The All-American Girls' Baseball League of 1943–1954 was also all white.

After the Civil Rights Act of 1964, race became less of an issue. When the first women's basketball leagues were founded in the late 1970s, race was not a significant issue in recruitment. Today it is a non-issue. The deciding factor in recruitment is how well a woman can play and how well known she is as a player. African-American players such as Cynthia Cooper, Lisa Leslie and Sheryl Swoopes are WNBA stars. Both black and white fans proudly wear their WNBA name shirts.

Sexual preference, however, continues to be an issue in American society. In the public's mind, femininity is equated with heterosexuality, thus "masculine" women are assumed to be lesbian. Because of this many lesbian athletes remained in the closet and have projected an image of femininity in appearance and demeanor off the athletic field, fearful that the public and sponsors would not accept them.

Traditionally, promoters of women's sports have portrayed women athletes as "feminine" in order to gain public acceptance of women's sports. "Female-athletes" were seen as playing feminine versions of men's sports. This created an illusion that female athleticism was very different from male athleticism. In this way, women's athleticism was seen as neither challenging cultural perceptions of male athletic superiority nor the image of sport as a male domain.

EARLY HISTORY OF WOMEN'S BASKETBALL

The history of women's basketball reflects this role conflict that existed between femininity and athleticism. In 1892, when Senda Berenson, a physical educator at Smith College, first introduced women to basketball she carefully feminized the rules to create a less strenuous women's game. Until 1971 most girls and women played basketball with girls basketball rules. It wasn't until the 1971–1972 season that women's college basketball rules officially adopted the men's five-player, full-court game with a 30-second shot clock. Now women were playing "real basketball," the men's game.

Early promoters of women's professional basketball stressed that the women's game was different from the men's, emphasizing that the players were "real" women and thus very feminine. The All-American Redheads, a women's professional basketball team that existed from 1936-1986, found its marketing niche by barnstorming across the country playing by men's rules against men's teams in the "battle of the sexes." The sensationalism of having the "weaker sex" take on men was a good draw. To make sure that the women never challenged the dominant ideology of sports as a male domain and that men were physically superior, the owners billed the games as entertaining basketball shows. By calling the games "shows," the contests were placed outside the realm of "true" professional sport and into the real of entertainment. Showmanship then became as important as athleticism. The femininity of the players' was also stressed. The team trademark was women's flaming red hair. These natural or dyed red hair beauties always wore make-up and lipstick on the court. Publicity emphasized their comedy routines, their half-time show of basketball wizardry, and their attractiveness.

Women basketball players traditionally have been caught in a "Catch 22" situation. If sports exemplified male characteristics and demonstrated male athletic superiority, then if a woman didn't play like a man she was considered an inferior athlete. However, if she played like a man, then she wasn't a woman. She either suffered from a "masculinity complex" or was a lesbian. Either way she was a freak. It was a no win situation.

MARKETING WOMEN'S PROFESSIONAL BASKETBALL

Promoters of women's professional basketball from the first leagues in the 1970s to those of today have found themselves in a cultural quandary as to the best way to publicize the sport and the women players. The question of how to market women's professional basketball has revolved around three issues.

One, how do you set women's basketball apart from men's without stressing that the players are female and falling victim to the cultural femininity trap? It is easy for a promoter to get trapped into saying either that women play a uniquely female game or that women are "feminine" athletes or both.

Two, for marketing purposes, historic milestones or "firsts" sell. Promoters wanted to market the "first" professional women's basketball league and capitalize on the historic event. But, how do you market it as a new event in sport, when women are playing a game that men have played professionally for years? How do you prevent the public and the press from making invidious comparisons between the men's and women's game?

Three, how do you gain public acceptance for women's basketball and women players when cultural stereotypes of highly athletic women are often negative? Stereotypes abound of women athletes as masculine, Amazons, butches and freaks. To gain acceptance do you stress their femininity? If so, do you undermine the athleticism of the women?

Promoters of women's team sports have found themselves on a slippery slope, because of the cultural conflict between traditional roles of female and athlete. This cultural ambivalence toward women athletes has led many promoters of women's sports as well as women players to stress that on the court women are tough, rough and competitive like male athletes, but off the court they are feminine. Basically they have employed what Felshin has called "the apologetic." One way the Amateur Athletic Union resolved this conflict was that, for years, they featured a free-throw contest and a beauty contest to name the queen of women's basketball tournament as part of their national women's championship basketball tournament. Press coverage and social acceptance have often necessitated this type of compromise and kept the woman athlete in the conflicting position of being a hyphenated female-athlete.

CHANGES SINCE THE 1960S

Since the 1960s historic changes in public perceptions of gender roles have occurred. The catalyst for these changes in perception came from the Women's Liberation Movement of the 1960s. The Movement led to greater acceptance of the equality of men's and women's roles and provided for greater opportunities for women to participate in all aspects of American life from business to sports.

Title IX in 1972 made it illegal for any school receiving Federal funding to discriminate on the basis of sex. To comply with the law, high schools and colleges across the country began to provide girls and women with interscholastic and intercollegiate sports. Girls' high school and women's college basketball teams sprang up across the country.

In 1972, the Association for Intercollegiate Athletics for Women (AIAW) established the first national collegiate women's basketball championships. In 1976, women's college basketball had grown so popular, that Mel Greenberg, a writer for the *Philadelphia Inquirer,* created the first top twenty women's college basketball teams poll. Many people began to think that there should be opportunities for women college graduates to play professional basketball in the United States. At the time they had to go overseas to Europe, South America or Japan if they wanted to play pro ball.

In 1974, the U.S. Supreme Court ruled that Little League Baseball had to admit girls. Since then more and more parents have encouraged their daughters to participate in soccer, basketball, softball and baseball leagues. As more girls have participated, public attitudes have changed. Today girls' athleticism is expected and accepted. The term "tomboy" that was once used to refer to young girls who enjoyed sports has become an anachronism. Today large numbers of girls are participating in community and school sports.

In high schools girls' basketball is the most popular sport for girls. In 1976, women's basketball became an Olympic sport for the first time and the American women's basketball team took home a silver medal at the Montreal Olympics. These events showed that American women could play world class basketball.

FIRST WOMEN'S PROFESSIONAL BASKETBALL LEAGUES

With the growing popularity of girls' and women's basketball in the United States and with women's basketball becoming an Olympic sport for the first time in 1976, it was only natural that some business entrepreneurs would consider forming a professional women's basketball league in the United States. Since 1977, eleven or more professional women's basketball leagues have been formed. Only two lasted having three seasons or more. The Women's Basketball League, the WBL lasted three seasons from 1978-1981. The Women's National Basketball Association, the WNBA, will start its sixth season in 2002. The National Basketball Association, the NBA, formed the WNBA in 1997. Today it is the only women's league in operation.

The first professional women's basketball league was the brainchild of Jason Frankfurt, New York restaurant owner and former stockbroker. In 1976 he decided to create the Women's Basketball Association (WBA). The league was to play a six-month, sixty-two game schedule beginning in October of 1977. To make sure the advertising for the league appealed to women as well as men he hired Lois Geraci Ernest, president of Advertising to Women, Inc., to be the commissioner of the WBA and to be in charge of promotion.

Ernest quickly announced that this was a women's league not only in terms of players, but also in terms of having women in front office positions. With the hiring of Ernest, it seemed that publicity for the league was headed in the right direction. It was assumed that publicity would stress the athletic skills of the women and avoid any hints of sexism. But the January press conference to announce the league was a sexist promotional embarrassment and disaster. At the press conference Frankfurt was introduced as "chairman of the broads." Cocktail waitresses in spike heels and satin shorts served the assembled reporters.[3] A reference was also made to transsexual professional men's tennis player Richard Raskin, who had undergone a sex change operation and was now playing on the women's tennis circuit as Renee Richards. Ernest stated, "If there is a Renee Richards who can play basketball, she is welcome in our league."[4] With all the missteps the league was doomed before it played a single game.

The idea to form a league didn't end there. Sports entrepreneur, Bill Byrne decided to undertake the challenge of re-launching the league as the Women's Basketball League (WBL) in 1978. Byrne had some experience with startup leagues, including the American Football League, the World Football League, and the American Professional Slow-Pitch Softball League. Although all had folded after relatively short periods, Byrne had learned a number of lessons from these failures. He felt he now had the expertise to market a women's professional basketball league.

Byrne's WBL lasted three seasons and was the most successful of the women's leagues prior to the American Basketball League (1996–1998) and the WNBA of today. Byrne to his credit made a concerted effort from the very beginning to stress the athleticism of the women players and to avoid any hints of sexism. However, some of the names of the teams were gender specific such as the Milwaukee Does, the Minnesota Fillies and the Dayton Rockettes.

WBL management tried to stress the professionalism of the women's play. At the first League game between the Milwaukee Does and the

Chicago Hustle, Chuck Bekos, the Milwaukee Does publicity director, discussed the problem of sexist misconceptions with Jill Lieber, a Milwaukee sports columnist. He said, "A guy came into my office when we were first starting out and tried to sell me a sleazy, bikini type uniform. And I told him to get out, that this wasn't some kind of roller derby production."[5]

Lieber was very impressed by the professionalism of the League and wrote, "Too often women athletes have depended on sex appeal to sell themselves and their sport."[6] Unfortunately, as the season progressed teams found that their games got very little press coverage. Often a local reporter would write one feature article about "feminine" basketball players and then ignore the team. Without press coverage it was difficult to attract fans and to make more people aware of the league. When a fan wrote to the *Chicago Tribune* to complain about Chicago Hustle coverage, a sports columnist sarcastically replied, "There hasn't been much coverage of them because news space is needed to report on the hangnails of male athletes."[7]

When the press did write about the games they often stressed the femininity of the players as much as their athleticism. For example, Rita Easterling, a Chicago Hustle star who regularly scored 20 points or more was referred to by [the Chicago press] as "the point guard the Bulls of the NBA are looking for," the "Queen of the Floor Burns" and as the "All-American" girl. The press made sure to portray her as a "feminine athlete" and Easterling also accepted the duality of her role. She said, "A lot of people say I'm crazy for taking charges, but I'd do anything to get back the ball. Still, on and off the court, I'd like to be looked at as a lady."[8]

Janie Fincher another Hustle player who often scored 15 points or more a game was better known for her looks than her basketball skills. In a celebrity exhibition game, sportscaster Al Lerner had patted her on the fanny. The press played up the incident and she became known as "Fincher-Pincher." She became a "sex object" and some Chicago fans began wearing T-shirts emblazoned with "Fincher-Pincher." Fincher disliked the stereotype, but was philosophical about it saying, "there's always going to be a stereotype from certain people. . . . I made up my mind when I hear them to take it as a compliment. I think, 'If they notice you, fine, they must have noticed the team.'"[9] Although she didn't complain, it must have distressed her to hear rude male fans making distasteful remarks. Some male fans would yell "breathe deeply," a reference to her big breasts, every time she made a foul shot.[10]

A *Sports Illustrated* writer seemed more interested in writing about the good looks of Iowa Cornet star, Molly Bolin, than her athletic skill. He wrote, "Off the court? Well, suffice it to say that if beauty were a stat, Molly Bolin would be in the Hall of Fame."[11] Bolin in a mistaken belief that her sex appeal might help to attract male fans to Cornet games, at her own expense had a 18"x24' "pin-up" type poster made of herself. She said, "People always warned me about exploitation like it was a dirty word. But it's all about putting people in the seats isn't it? You don't have to look like a man, act like one or play like one in this game. And I just wanted to show that women aren't trying to be men. If you really want to make it when you're new, you've got to grab everything you've got and go with it."[12]

Even well-known basketball celebrities such as Anne Meyers believed that public acceptance was dependent on the players appearing feminine. She said, "In order for people to accept the WBL, they want to know that we are women. It's not fair, but it's a fact of life."[13] Advertisers reinforced the idea that sex appeal sells. They selected the most attractive players to be spokeswomen for their products. Identical twins, Kay and Faye Young, who played for the New York Stars were better known for their Dannon Yogurt TV commercials then their basketball skills. Other players on the team often complained that the Dannon Yogurt twins got more publicity than the star players.

The owners of the California Dreamers franchise in an attempt to increase press coverage sent their players to five weeks of charm school. This gimmick may have worked for Philip K. Wrigley in 1943 when he established the All-American Girls' Baseball League (AAGBL), but publicity wise it wasn't effective in 1979. It didn't attract a great deal of press nor fans to the games.[14]

The worst example of sexist marketing of a professional women's basketball league was the Liberty League. Jim Drucker, who formed the league in 1991, believed that the game should be miniaturized so women could play a game that was faster and more similar to the men's game. He shortened the length of the court, lowered the basketball rim, and made the ball smaller. To entice men to the game, he had the women wear flashy tight fitting spandex uniforms.[15] Many players were upset with the sexually exploitative uniforms. Cary McGehee said, "I want to be known as a ballplayer, not as a ballplayer with a nice butt and hips."[16] Nikata Lowry and Tonya Edwards said, "Hey, if the LBA resembles the NBA, why humiliate the players? No one asked Charles Barkley to wear form-fitting uniforms."[17]

As might be expected there were male sportswriters who thought the idea was a good one. Merrell Noden, editor of *Sport's Illustrated's* "Scorecard" page endorsed the idea in a sexist way. He said, "historically women's professional basketball leagues have lasted about as long as overcoats in Madonna's videos. But if adaptation counts for anything, The Liberty Basketball Association . . . might have a chance."[18]

Miniaturized sexist NBA basketball was not the panacea for women's pro basketball that Drucker thought. The LBA folded after the All-Star Exhibition game. The LBA marked a new low in women's search for athletic acceptance.

Today with the public's greater acceptance of female athleticism, women athletes should be judged purely on the basis of their athletic skills without reference to their femininity and promotion should emphasis their athleticism. But, is this the case?

The WNBA

When the WNBA was formed in 1997, NBA Properties President Dick Wilts stated, "You're not going to see anything that's focused on gender."[19] True to his word, there are no gendered team names or uniforms as there were in earlier leagues. However, no attempt was made to change the name of the men's division from NBA to MNBA and the women's game was pushed to a summer schedule so as not to compete with NBA games or the use of NBA arenas. The WNBA is treated as the little sister of the NBA.

The focus of WNBA marketing was supposed to be the sport. Yet, the opening WNBA press release photo that appeared in *USA Today* was not an action shot of women basketball players, but a very carefully posed picture of three attractive and feminine looking marquee players, Sheryl Swoopes, Rebecca Lobo, and Lisa Leslie. Although they were pictured in WNBA uniforms, the players poses were those of fashion models not basketball players. Not a hair is out of place, and not a drop of sweat was to be seen. The emphasis was on movie star status rather than athleticism. Images of supermodels not basketball stars is not the non-gendered image management promised.

The two stars, Lisa Leslie and Rebecca Lobo, have also been co-opted into accepting the idea that as basketball players they must project an image of femininity and heterosexuality to be accepted by the public, the media and the powers that be in sport. Both are victims of the "femininity trap" or what has been called "the apologetic."[20] In an interview, Leslie states: "When I'm playing, I'll sweat and talk trash. However, off the court, I'm lipstick, heels and short skirts. I'm very feminine, mild-mannered and sensitive."[21] Leslie is separating her social self from her athletic self or, to use Curry and Jiobu's term, role splitting.[22] On the court the female is aggressive and competitive, essentially masculine traits, but off the court she is "femininely" soft and passive.

According to sport sociologist Pat Griffin, women athletes' stress their femininity in order to avoid being stigmatized or labeled lesbian. The reason for this behavior is that in American society there exists an intense blend of homophobia and sexist standards of feminine attractiveness. It is these standards that exert pressure on women athletes to monitor their behavior and appearance at all times, so that they project an image of femininity.[23]

These standards no doubt also influenced WNBA management decisions as to how to market and promote their female athletes. For example, first year marketing for the WNBA focused on making Rebecca Lobo, Lisa Leslie, and Sheryl Swoopes WNBA superstars. Each was a superior player. However, publicity carefully created an image of each of the athletes as personifying a socially acceptable traditional female role. Rebecca Lobo was the clean cut, attractive, All-American girl type, Lisa Leslie, the glamorous model, and Sheryl Swoopes, the devoted mother. This was no doubt a conscious effort by WNBA management to create an image of a heterosexual "feminine" league. While espousing athleticism, they carefully adhered to the traditional image of "feminine athlete" or hyphenated female-athlete.

The cover page of the inaugural issue of *Sports Illustrated Women Sport* magazine showed a very pregnant Sheryl Swoopes. She was in her basketball uniform with her shirt tightly drawn over her very round stomach. To maximize the impact of her pregnant appearance, the shot is only from her head to her very round stomach. The roundness of her stomach is further accented by the roundness of the basketball she is holding. The caption reads, "A Star Is Born: Sheryl Swoopes and the WNBA Are Both Due in June." If there was any doubt about women's biological destiny, the picture is a quick reminder that women's primary role is that of mother.

On the cover, directly across from Swoopes pregnant stomach are the listings for two other articles. The titles are: "The Coach as Sexual Predator" and "Why I Fell for Grant Hill."[24] The first article reinforces the idea that women are especially vulnerable and in need of protection. The second reinforces the notion that women's thoughts revolve around catching a man. Sexuality overshadows athleticism even in a magazine for women devoted solely to women's sports.

The WNBA Program also combines sexuality with athleticism. Although the cover creates an image of the WNBA as making women's basketball history, the table of contents shows a very sexual picture of Lisa Leslie. Her lips are pursed in kiss like fashion. The feature article shows Swoopes, Lobo and Leslie at the opening press conference. On the following page is an action shot of Swoopes with a basketball. The caption reads, "As a new mother, Swoopes, will provide Houston a boost when she returns to action." Rebecca Lobo is also pictured modeling the new Champion WNBA shirt that is available to fans. The caption reads, "Lobo looks to strike a championship pose with the Liberty this season." The program also contains an article picturing some of the clothing items designed by Champion for purchase by WNBA fans. The article is titled, "Make a Fashion Statement." Examples of sexuality and fashion appear to dominate over athleticism.

This emphasis on WNBA athletes as fashionable women was not just an early promotional gimmick. Even after three successful seasons, the WNBA was still creating an image of feminine athletes. The 2000 New York Liberty Yearbook contains a four-page spread of popular WNBA athletes in fashion poses. Accompanying each photo is a statement by the athlete as to how she likes to look feminine and ladylike off the court such as that by Tina Thompson, "Just because I'm an athlete and play tough on the court doesn't mean that I don't enjoy looking feminine off the court."[25]

The female apologetic is very much a part of many WNBA players' persona. Cynthia Cooper, four-time WNBA MVP and star of the four time WNBA championship Houston Comets in her autobiography, *Cynthia Cooper She Got Game,* relates how she's as comfortable in cocktail dresses and spike heels as in Nike sneakers and sweats. She also says, "its important for WNBA players to project the image of athleticism on the court and femininity off the court. Some girls are leery of playing competitive sports because they're afraid of being labeled too masculine."[26] Sponsors also seem to believe that for women basketball players to be accepted by the public they must be attractive. A recent ESPN promotional campaign for the WNBA is called "Basketball is Beautiful."[27]

Strategically the WNBA has marketed the league as family entertainment. And it has done a good job of attracting parents and their children. Low ticket prices have made it affordable family entertainment. For example, the average price of a WNBA ticket in is $15.50 compared to $51.01 for a NBA ticket.[28]

WNBA management has made a conscientious effort to give the league a heterosexual image. An example of this is an August 1998 press release about nine players who were combining a WNBA career with motherhood. A feature article appeared in *USA Today* titled, "In Focus: Mothers in the WNBA." A smaller accompanying article was titled, "Husband at Home Provides Big Assist."[29]

Females make up seventy-five percent of those who attend WNBA games. This female audience includes single women, married women, and mothers with daughters. Although the majority of the women are heterosexual, a significant proportion are lesbian and the WNBA management is cognizant of the fact some of its most loyal fans are lesbians. The league wants the continued support of the gay community, but finds itself in a difficult position. How to market to the lesbian community while still maintaining an image of the WNBA as family entertainment. In the American public's mind, family entertainment usually conveys an image of heterosexuality.

As attendance has continued to decline from the high set in the second season of play, management has searched for new ways to attract more people to the games. Many franchises have reached out to the lesbian community. Today it is estimated that at least nine of the sixteen franchises do some direct marketing to gay fans. The Los Angeles Sparks as a result of seeing their attendance decline staged a pre-season prep rally on May 4, 2001 at West Hollywood's Girl Bar, the largest lesbian club in the United States with approximately 12,000 members. The Miami Sol and the Houston Comets have also reached out to the gay community by making official appearances at lesbian bars. Last year the Seattle Storm hosted a Gay Pride Night. And this year the Sacramento Monarchs hosted their "First Annual Gay Pride Event."[30]

Reaching out to lesbians is a bold step for the WNBA. Although the American public is more acceptant of homosexuality today than in past, issues of sexual preference are still hotly debated. While the public is aware that many of the WNBA fans, players and coaches are lesbian, they may feel more comfortable with lesbians remaining silent as to their sexual preference.

The WNBA offers same-sex partner benefits, but only two individuals are openly acknowledged as gay. Liberty player, Sue Wicks and Liberty president and general manager, Carol Blazejowski.[31] Fear of being stigmatized by the public and shunned by sponsors have kept lesbian WNBA players and coaches in the closet.

Hopefully the WNBA will be a catalyst for greater recognition of and acceptance of differences in sexual preference. And women athletes will be accepted purely on the basis of their athleticism and not on the basis of their femininity or heterosexuality.

NOTES

1. Don Sabo, Jr. and Ross Runfola, *Jock: Sports & Male Identity,* Englewood Cliffs, NJ: Prentice-Hall, A Spectrum B, 1980; x–xi.
2. Patricia Del Rey, "The Apologetic and Women in Sport," pp. 107–108, in *Women in Sport from Myth to Reality* by Carol Oglesby, Philadelphia: Lea & Febiger, 1978.
3. Grace Lichtenstein, "Women's Pro Basketball League: The New Million-Dollar Baby," *MS* magazine, March 1980, p. 70.
4. Margaret Roach, "The 12-Team Pro League Fulfills Basketball Dreams," *The New York Times,* 23 January 1977, Sec. 5 p. 5.
5. Jill Lieber, "Does Selling the Real Thing," Milwaukee newspaper clipping, circa 8 Dec. 1978, NBHF library file.
6. Lieber.
7. Quoted in Mary Jo Festle, *Playing Nice: Politics and Apologies in Women's Sports,* New York: Columbia University Press, 1996, p. 254.
8. *WBL Women's Pro Basketball League Official Guide 1979–1980,* p. 44; Bill Jauss, "Hustle's Rita Easterling—the 'All-American' girl," *Chicago Tribune,* 17 April 1979.

9. Bill Jauss, "Fincher Fights Her Sexy Stereotype with Hustle," *Chicago Tribune,* 16 November 1979, Sec. 6 p. 2.
10. Festle, p. 255.
11. Roy S. Johnson, "The Lady is a Hot Shot," *Sports Illustrated,* 6 April 1981, p. 32.
12. Johnson, p. 34.
13. Susan Silton, "How to Be Charming While Dribbling," *Ms. Magazine,* October 1980, p. 25.
14. Silton, p. 25.
15. Michael Hiestand, "Sports Business: Women's Basketball League Plans to Target Male Viewers," *USA Today,* 18 Dec. 1990, 3C.
16. Donna Carter, "A New E.R.A. Professional Basketball League Is Hoping to Take Women's Game to Greater Heights," *Denver Post,* 15 April 1991, p. 8D.
17. Mike Conklin, "Sports," *The Chicago Tribune,* 25 January 1991, p. 9.
18. Merrell Noden editor, "Scorecard," *Sports Illustrated,* 24 December 1990, p. 11.
19. Michael Hiestand, "NBA Puts Clout to Test: The Sport, Not Gender, at Core of Marketing New League," *USA Today,* 5 Feb. 1997, p. 3C.
20. Janet Felshin, "The Triple Option . . . For Women in Sport," *Quest,* 21, 1974, 36–40.
21. Robyn Marks, "Supermodels: The WNBA Is Banking on a Tall, Talented, Attractive Trio to Take Women's Hoops to Another Level," *Sport,* July 1997, p. 47.
22. Timothy Curry and Robert Jiobu, *Sports: A Social Perspective,* Englewood Cliffs, NJ: Prentice-Hall, p. 168.
23. Pat Griffin, "Changing the Game: Homophobia, Sexism, and Lesbians in Sport," in D. Stanley Eitzen, *Sport in Contemporary Society: An Anthology,* 5th Edition, NY: St. Martin's Press, 1996, p. 396.
24. *Sport Illustrated Women Sport,* Spring 1977, cover.
25. *New York Liberty 2000 Yearbook,* pp. 24–27.
26. Cynthia Cooper with Russ Pate, *She Got Game,* New York: Warner Books, Inc., 1999, p. 218.
27. Anthony Schoettle, "Sports Business: WNBA Launches 'Phase 2,' 4 June 2001, About.com.
28. Kelli Anderson, "Scorecard: WNBA Finals: Finishing Up Strong," *Sports Illustrated,* Sept. 7, 1998, p. 22.
29. Valerie Lister, "In Focus Mothers In The WNBA," *USA Today,* August 4, 1998, 3C.
30. Therese Jansen, "WBA Sparks Court Lesbians," 18 May 2001, about.com; Debbie Arlington, "WNBA Alters Strategy, Courts Lesbians," *The Sacramento Bee,* 4 July 2001, Tim Weir, "WNBA Explores Lesbian Fan Base," *USA Today,* 24 July 2001, 2C.
31. Kristie Ackert, "Progress Report, Gays Scoring Points in Fight for Rights," *New York Daily News,* 11 August 2002, Sports Sec., 56.

38

Couples Watching Television

Gender, Power, and the Remote Control

BY ALEXIS J. WALKER

Alexis Walker's research provides a careful analysis of the ways couples "do gender" while viewing television, a form of leisure that predominates in many people's lives. She collected the data for this study in an upper-division undergraduate class using a semistructured interview. The sample was a convenience sample, with students selecting those who participated. The sample of 36 couples (72 individuals) was diverse, with length of relationship varying across couples from 1 year to 15 or more years together. Most couples were white (77%) with the remainder African American, Hispanic or mixed race and most were married (24 couples), with 15 percent gay or lesbian (5 couples). This analysis provides an unusual opportunity to examine gendered patterns in a leisure activity that others do not typically observe.

1. What does Walker mean when she says the couples are "doing gender"?
2. What does she mean by "latent power" and how does it affect the couples' interactions around the remote control?
3. How does the relationship of the lesbian couple, Becky and Mary, differ from the other couples in this study?

Over the past 20 years, feminist scholars have shown that ordinary, routine, run-of-the-mill activities that take place inside homes every day bear an uncanny resemblance to the social structure. For example, the distribution of household labor and of child care is gendered in the same way that paid work is gendered: The more boring and less desirable tasks are disproportionately performed by women, and status has a way of reducing men's, but not women's, participation in these tasks. (See Thompson & Walker, 1989, for a review.) Examining television-watching behavior is a way to extend the feminist analysis to couples' leisure.

* * *

[I]ndividuals watched television quite often—on average almost daily for nearly 3 hours per day ($M = 2.77$, $SD = 1.48$). During the week prior to the interview, they had, on average, watched television together on 4.87 days ($SD = 2.09$). Nearly all, 94% ($n = 29$), of the women and 87% ($n = 27$) of the men reported that, regarding watching television with their partners, they were happy with the way things are. Yet two thirds of the women and three fifths of the men reported that there were things about their joint television watching that were frustrating to them. The interview transcripts were revealing about these frustrations. Women complained about their partners' grazing behavior, both during a show and when they first turned on the television set. One woman in a 3-year cohabiting relationship said:

> I would say that the only thing that's frustrating for me is when we first turn on the TV and he just flips through the channels. It drives me crazy because you can't tell what's on, because he just goes through and goes through and goes through.

Another woman, in the 17th year of a first marriage, reported, "[I get frustrated] only if I get hooked into one show and then he flips it to another one. As soon as I get hooked into something else, he flips it to something else." Such reports from women were common. A married man spontaneously agreed: "We don't watch TV a lot together; I would rather do other activities with my wife. Channel switching wasn't a problem until . . . the remote control." Indeed, many men indicated that their women partners were bothered by this behavior.

In contrast, men reported being frustrated with the quality of the programming or the circumstances of watching, rather than with RCD [Remote Control Device] activity. For example, one husband said, "I wish we had a VCR. . . . I wish we had one of those TVs where you could watch two things on the screen at once." Another said, "It's sort of frustrating when I want to watch something she doesn't, and she goes into the other room and gets sort of pouty about it." A third reported, "No, [nothing is frustrating], but she does talk a little."

I looked specifically at the RCD; for example, where is the RCD usually located? Men were more likely than women to say that they usually hold the RCD or have it near them . . . and they were less likely than women to say that their partner usually holds it or has it near them. . . . The transcriptions support this general pattern of RCD location, as well. A husband reported, "I usually use the remote because I know how to use it, and it usually sits right in front of me while I am on the couch." A young married woman said, "I had the baby [the RCD] this time. This was a rare occasion." Roger (all names are pseudonyms), a married man reported:

> I frequently have the remote at my side. I won't change the channel until we are ready to look for something else. If there is someone who wants to change the channel at a commercial, it will be Sally [his wife]. I will hand the remote to her, and she will change it to another favorite show, and then back. And that is very typical.

Sally agreed. The last time they watched television together, the RCD was in "Roger's pocket! Either in his shirt pocket or bathrobe pocket." A young, married man reported:

> I don't hold [the RCD], but I pretty much have control of it, and if I don't care what's on, then I let her have it. Sometimes we fight over it. Not like fight, but, I mean, it's like, "You always have the remote control."

Women were significantly more likely than men to say that RCD use was frustrating to them. . . . Only 10% (n = 3) of the men, but 42% (n = 13) of the women evidenced such frustration. Furthermore, women . . . reported that significantly more RCD behaviors were frustrating to them than men reported. . . . Yet 30% (n = 9) of the women in the sample and 16% (n = 5) of the frustrated men reported that they would like to change how the RCD is used during their joint television watching. This difference was not significant.

What was frustrating about RCD use? Respondents reported being frustrated by the amount of grazing, the speed of grazing, heavy use of the RCD, and the partner taking too long to go back to a channel after switching from it during a commercial. A few respondents actually indicated concern about their own frequent RCD use. Women and men, however, reported similar percentages of other television-watching behaviors that were frustrating (e.g., too much time watching television; bothersome behaviors of the partner, such as making fun of a program); 58% (n = 18) of the women and 48% (n = 15) of the men were frustrated by these other behaviors.

Thus far, I have shown that men control the RCD more than women and that women are more frustrated by RCD behaviors than men are. I also asked about the other activities engaged in while watching television. Two types of activities were mentioned: family work (e.g., child care, cooking, laundry) and pleasurable activities, such as doing nothing (i.e., relaxing), eating, drinking, playing computer games, and so on. When activities within each type were summed, the findings were revealing. When asked about their most recent joint television-watching episode, men . . . responded that they were significantly more likely than women . . . to engage in pleasurable behaviors while watching television . . . Women were not more likely than men . . . to do family work while they watched television, although the data suggested a trend in this regard . . . . The small proportion of households with children (32%, n = 10) may have contributed to this finding. At least 80% of both women and men described this most recent experience as typical of their joint television watching and of their RCD use. Interestingly, women . . . tended more than men . . . to describe the particular show they watched as their partner's preference rather than their own.

Recall that 30% of the women said they would change the use of the RCD during their joint television watching if they could. Only half as many men would make such a change. The open-ended data support these results. For example, a young married woman described her technique of standing in front of the television to interrupt the signal from the RCD. Another young married woman said that her partner used the RCD to watch more than one program at a time. "I should get him one of those TVs with all the little windows so he can watch them all," she said sarcastically. A middle-aged married woman said that she would like to change their television watching so that she would have "control of the remote for half of our viewing time." Of those who would like to make any changes in their television watching, one in five women expected that they would *not* be successful.

Men typically admitted their heavier RCD use. For example, a middle-aged married man said that he switched channels to avoid commercials. I'm the guilty party," he said. "My [family members] would leave it there and watch the commercial. I just change it because I'd rather not be insulted by commercials."

One of the most provocative questions asked of respondents was "How do you get your partner to watch a show that you want to watch?" The results were enlightening. A cohabiting woman said, "I tell him that would be a good one to watch, and he says, 'No,' and keeps changing [channels]. I whine, and then usually I don't get [my way]." A middle-aged married woman said:

> Let me think here, when does that occur? [Laughing.] If I really want to watch, I'll say, "I want to watch this one." . . . I'll say, "Come in here and watch this" if he's not in the room, but pretty much we watch the same things a lot, whether or not that's because I let him. He, a lot of the time, turns it on, and I'll come in and join him. But, if it's something I really want to watch, I'll say, "Don't flip the channel; I want to see this."

In contrast, a young married man said that he gets his partner to watch a show he wants to watch in this way:

> I just sneak the remote away from her if she has it, or, if I'm there first then. . . . I mean, if there's sports on, that's usually what we watch unless there's something else on. I mean, usually if there's . . . some kind of sports game on, we usually watch that, but unless there's . . . another show on that, you know, she can talk me into, deter my interest, or something. . . .

When asked how his partner gets him to watch something that she wants to watch, he reported:

> Oh, I guess, if there's not anything that I'm . . . real big on watching then I'll let her choose, or if she, you know, she's interested in something. . . . A bunch of times, we watch TV, and it's like, well, we'll go back, and, well, that's kind interesting, we go back and forth.

His wife agreed:

> I usually don't have to beg him. I don't know. [Laughing.] I tie him down, and say,

> "You're watching this." I don't know. He usually just comes over, and if it's not what he wants, then he'll take the remote and try to find sports.

In other words, this couple watches sports when it is available. If there is no sports program on television and if the husband does not have something else he really wants to watch, then the wife may choose a show, but her husband will be looking for a sports program while her show is on, or at least he will go back and forth between her show and others.

A woman who has been married for 18 years was deliberate in her efforts to watch a particular show:

> I usually start a couple of days ahead of time when I see them advertised, and it is something that I am going to want to watch. I tell him to "get prepared!" I have to be relatively adamant about it. When the time comes up, I have to remind him ahead of time that I told him earlier that I want to watch the program.

When her husband wants to watch a program, however, she said, "He just watches what he wants. He doesn't ask." Finally, a married man reported, "I just say I want to watch something, and if she wants to watch something really bad, I will let her watch what she wants to watch." Ultimately, the authority is his.

The data are much the same when people report on changes they would like to see in the way they watch television together. One man who has been cohabiting with his partner for one year said, "I should probably let her 'drive' sometimes, but [it] would bug me too much not to be able to do it." A woman who has been married for 37 years painted a brighter picture. When asked, "How do you feel about watching TV with your husband? Are you happy with the way things are?" she responded:

> Yes, right now. But see, without the VCR we'd be in trouble because I just tape anything I want to see. Without that, there'd be more conflict. . . . Buying a second TV has changed the way we watch TV. It's made it easier—less stress, less conflict.

A young married man also was more positive. When asked, "Have you changed the way you watch television together?" he replied, "We take turns watching our programs, and I let her hold the remote during her programs."

Earlier. I mentioned that 14% (n = 5) of the couples in this study were gay or lesbian. In these couples, too, one partner usually is more likely than the other to use the RCD. In a gay male couple, one nearly always used the RCD, and the other almost never used it. When the RCD user was asked why they have this typical pattern, he responded: "Why? I don't know. I just like using the remote. I think I'm better at it than he is." In answer to a question regarding whether he used the RCD at all, his partner indicated: "He doesn't let me." In a second gay male couple, one partner again was far more likely to use the remote than the other partner, but both reported using controlling strategies to get their partner to watch a show they wanted to watch. Greg said, "I just tell him I want to watch it and we do." Rob said, "I just turn it on, and that is what we watch." When asked, "How does your partner get you to watch a program he wants to watch?" Greg replied: "I usually don't watch programs I don't want to watch. If he asked me to watch it with him for a purpose, I would."

In contrast, one partner in a lesbian pair reported, "If we are both here, we try to make sure it's something that we both like." In fact, this couple limited their television viewing to avoid potential problems resulting from their different styles of RCD use. They also made it a practice to talk to each other during the commercials, in part, so that the one partner who tended to do so, would not graze. A second lesbian pair reported similar behavior. When asked, "Think back to the beginning of your relationship with [your partner]. Have your expectations about watching television with her changed over time?" she responded:

> In the beginning. . . . TV watching was something we could do when we didn't know each other very well yet. You know, it was kind of like a sort of a neutral or a little bit less personal activity that we could sit and watch TV together as a shared activity. And it's still a shared activity. . . . We don't use it to tune each other out, and if someone wants to talk, we just click the mute button or turn it off.

Becky's partner, Mary, used the RCD much more often than Becky did. According to Becky, however, when Mary grazes, "she's perfectly willing, if I say, 'This looks really good,' she'll stop. She doesn't dominate that way." In fact, when Mary grazes, "she'll just say, 'Is this bothering you?'" Mary agreed that she was the one who usually held the RCD, but that they shared, too. "If Becky has a show she really likes, then I give her the remote so I'm sure I don't play with the TV while she's watching her show." Mary does not "let" her partner hold the RCD; she asks her to hold it to keep her own behavior in check. Indeed, Mary's frustration with their joint television watching comes from her own behavior: "Well, I feel self-conscious about how much I change the channels because I know that she doesn't like to change as often or as fast as I do." . . .

DISCUSSION

These data confirm that for women in heterosexual pairs leisure is a source of conflict—conflict between their own enjoyment and the enjoyment of their partners (Shank, 1986; Shaw, 1985, 1991). The data expose the contradictions between the goal and the reality of leisure for women. Support also comes front the findings that men, more than women, combine other pleasurable pursuits with television watching. Others (Coverman & Sheley, 1986; Firestone & Shelton, 1994; Hochschild, 1989; Mederer, 1993) have shown that women more than men dovetail family labor with their leisure activity.

The data also support previous work suggesting that, when heterosexual couples watch television together, men dominate in program selection and in the use of the RCD (Copeland & Schweitzer, 1993; Cornwell et al., 1993; Eastman & Newton, 1995; Heeter & Greenberg, 1985; Krendl et al., 1993; Lindlof et al., 1988; Morley, 1988; Perse & Ferguson, 1993). Indeed, unnegotiated channel switching by male partners was a frequent occurrence in this sample. Men use the RCD to avoid commercials, to watch more than one show at a time, and to check what else is on (Perse & Ferguson, 1993). And they do so even when their partners are frustrated by these behaviors.

The data reveal that men have power over women in heterosexual relationships (Komter, 1989). Men are more likely than women to watch what they want on television and to do so without considering their partner's wishes. Men control the RCD, which gives them the means to watch what they want, when they want, in the way that they want. Men also persist in RCD use that is frustrating to their women partners. These are examples of manifest power. Men make overt attempts to get their way and are successful at doing so. Men's power is evident, as well, in the lesser power of their women partners. For example, women struggle to get their male partners to watch a program they want to watch and are less able than men to do so. Furthermore. women watch a preferred show on a different television set or videotape it so that they can watch it later. These options do not prevent a husband or male partner from watching a show that he wants to watch when he wants to watch it.

Men's latent power over women is evident as well. Even though women rarely control the RCD, fewer than half report that RCD use is frustrating to them, and only 30% say they would like to change their partner's RCD behavior. According to Komter (1989), resignation to the way things are is evidence of latent power. Another illustration of the effect of latent power is anticipation of a negative reaction. Only four women feel they would be successful if they attempted to change the way their partners use the RCD. In the heterosexual sample, women seem less able than men to raise issues of concern to them, they anticipate the struggles they will encounter when and if their own preferences are made known, and they predict a negative reaction to their wishes from their male partners.

Confirmation of men's latent power over their women partners also was demonstrated by a series of auxiliary analyses. I was unable to explain the dependent variables of respondent's relationship happiness or respondent's enjoyment of the time the couple spends together with independent variables such as frustration with remote control use, dominance of the remote control, or desire to change frustrating remote control use. As Komter (1989) suggested, both lesser power and resignation on the part of women contribute to the appearance of balance in these pairs.

Joint television watching in heterosexual couples is hardly an egalitarian experience. . . . [S]ome women use a second television or a videocassette recorder to level the playing field (i.e., so that they are able to watch the shows they want to watch). A second television set, however, reduces joint leisure time among those couples who can afford it, and a VCR means that a woman may have to wait to watch her show. Even these solutions to conflict around joint television watching demonstrate that couples watching television are not simply passive couch potatoes. They are doing gender, that is, acting in ways consistent with social structures and helping to create and maintain them at the same time.

Everyday couple interaction is hardly mundane and run-of-the-mill. It is a systematic recreation and reinforcement of social patterns. Couples' leisure behavior is gendered in the same way that household labor is gendered: Social status enhances men's leisure activity relative to women's. Thus, leisure activity has gendered meanings (Ferree, 1990). Through it, women and men are creating and affirming themselves and each other as separate and unequal (Ferree, 1990; Thompson, 1993). In other words, leisure activity is both an occasion for relaxation and an occasion for doing gender (Fenstermaker, West, & Zimmerman, 1991: Shaw, 1991).

As Osmond and Thorne (1993) point out, "Gender relations are basically power relations" (p. 593). Because the power of men in families is legitimate—that is, backed by structural and cultural supports—it constrains the less powerful to act to maintain the social order and the stability of their relationship (Farrington & Chertok, 1993). Few women make demands of their heterosexual partners so that their patterns of television watching change. Instead, they say they are "happy" with their joint television watching. This same pattern is evident when we examine family labor. Most women describe the objectively uneven distribution of household work as fair (Thompson, 1991).

Hochschild (1989) argued that women give up leisure as an indirect strategy to bolster a myth of equality. Rather than resenting her male partner's leisure time, a woman uses the time when he is pursuing his own leisure or interests to engage in what she describes as her interests: housework and child care. Overall, she defines their level of involvement at home as equal, a view that can be sustained only if she ignores her own lack of leisure time, as well as the amount of leisure time her partner has. Hochschild also suggested that a woman sees her male partner's leisure time as more valuable than her own because she feels that more of his identity and time than hers are committed to paid work. She concludes, therefore, that he deserves extra relaxation. In Hochschild's view, and in Komter's (1989), women cannot afford to feel resentment in their close relationships.

In a review of the literature, Szinovacz (1987) wrote that there were few studies of how people in families exercise power. She argued that such studies are needed, as are studies on strategies of resisting power. The data reported here suggest that the exercise of power around couples' television watching behavior can be overt and relentless. Men's strategies to control the content and style of viewing are ways in which they do gender. Women's resistance strategies (e.g., getting a second television set, using the VCR) are also ways of doing gender. They do little to upset the intracouple power dynamics. Indeed, most women whose male partners are excessive grazers do not describe resistance strategies at all. Instead, they maintain the status quo (Komter, 1989).

Of interest is that in lesbian and gay couples one person was more likely to be the heavier RCD user, as well. Yet these couples had some unique patterns. The behavior of the lesbian couples, in particular, is suggestive for those of us wishing to establish and maintain egalitarian partnerships. One lesbian woman demonstrated a solution to the conflict between partners when one is distressed by the other's RCD behavior. Asked, "Is there anything else you'd like to tell us?" she responded:

> Well, I think that the most important thing for me is to remember to be sensitive to the fact that she doesn't have the same tastes as me, and I try to think about that. And, if she mentions that she likes something, then I ask her before I change the channel if she's done watching it, or if she's not interested, if I could change the channel.

In this act, she elevates the importance of her partner's wishes to the level of her own. She demonstrates the consideration that her partner desires and deserves (Hochschild, 1989; Thompson, 1991). When asked if it would be important for them to make changes in the way they watch television together, her partner expressed insight into her own behavior. Mary, the RCD user, likes to "veg out" and watch TV, but Becky likes to:

> pretend I'm not going to watch, I'm going to get a magazine or the newspaper, . . . or I'll bring some desk paper work over to do. . . . I think well, I'll just sit there in the living room while Mary watches TV. I'll work on our bills or something. . . . Then, what happens is I'll look around and think that looks kind of interesting. Although usually by the time I've looked around, she's changed the channel. . . . I think what happens is that she's more up front about saying, "Hey, I'm going to veg out here and watch some TV," and I pretend I'm going to do more worthwhile things, and I end up just watching TV anyway.

Perhaps these two women, with their honesty to themselves, their sensitivity to each other, and their concern about the ways in which their own behavior is or could be a problem in their relationship, are doing gender, too. They are concerned with the relationship, rather than with getting their own way. This is how women are said to make connection and to demonstrate care, to give what Hochschild (1989) described as a gift of gratitude. Using these strategies, they maximize joint enjoyment of leisure and minimize power imbalances. Rather then reproducing structural hierarchies, they create a bond of equality and provide a different course for the resolution of inherent conflict within couples.

The results from this study are hardly definitive. They are based on a small, volunteer sample, albeit one sufficiently diverse to include different types of close, romantic relationships. Additionally, the very small number of lesbian and gay couples suggests a need to exercise caution when generalizing from these findings. Further study with larger, representative samples will be required to extend these findings beyond the couples interviewed here.

Nevertheless, the patterns I identified are similar to those found in other studies of television watching in families and of the intersection of gender and power in close relationships. Mundane activities are important for understanding the intersection of gender and power in close relationships. Indeed, as Lull (1988) noted:

> Television is not only a technological medium that transmits bits of information from impersonal institutions to anonymous audiences, [but] it is a social medium, too—a means by which audience members communicate and construct strategies to achieve a wide range of personal and social objectives. (p. 258)

Others (Morley, 1988; Spigel, 1992) have pointed out that the way men engage with television programming and women watch more distractedly are illustrations of cultural power. Daytime television programming in the 1950s, for example, was designed to be repetitive and fragmented to facilitate joint housework and television watching for women (Spigel, 1992), thus helping to create and reinforce the view that leisure at home is problematic for women. The availability of the RCD does not change the fact that women's leisure is fragmented. . . .

REFERENCES

Copeland, G. A., & Schweitzer, K. (1993). Domination of the remote control during family viewing. In J. R. Walker & R. V. Bellamy, Jr. (Eds.), *The remote control in the new age of television* (pp. 155–168). Westport, CT: Praeger.

Cornwell, N. C., Everett, S., Everett, S. E., Moriarty, S., Russomanno, J. A., Tracey, M., & Trager, R. (1993). Measuring RCD use: Method matters. In J. R. Walker & R. V. Bellamy, Jr. (Eds.), *The remote control in the new age of television* (pp. 43–55). Westport, CT: Praeger.

Coverman, S., & Sheley, J. F. (1986). Change in men's housework and child-care time, 1965–1975. *Journal of Marriage and the Family,* 48, 413–422.

Eastman, S. T., & Newton, G. D. (1995). Delineating grazing: Observations of remote control use. *Journal of Communication,* 45, 77–95.

Farrington, K., & Chertok, E. (1993). Social conflict theories of the family. In P. G. Boss, W. J. Doherty, R. LaRossa, W. R. Schumm, & S. K. Steinmetz (Eds.) *Sourcebook of family theories and methods: A contextual approach* (pp. 357–381). New York: Plenum.

Fenstermaker, S., West, C., & Zimmerman, D. H. (1991). Gender inequality: New conceptual terrain. In R. L. Blumberg (Ed.), *Gender, family, and economy: The triple overlap* (pp. 289–307). Newbury Park, CA: Sage.

Ferree, M. M. (1990). Beyond separate spheres: Feminism and family research. *Journal of Marriage and the Family,* 52, 866–884.

Firestone, J., & Shelton, B. A. (1994). A comparison of women's and men's leisure time: Subtle effects of the double day. *Leisure Sciences, 16,* 45–60.

Heeter, C., & Greenberg, B. S. (1985). Profiling the zappers. *Journal of Advertising research, 25,* 15–19.

Hochschild, A. (with Machung, A.). (1989). *The second shift: Working parents and the revolution at home.* New York: Viking.

Komter, A. (1989). Hidden power in marriage. *Gender and Society, 3,* 187–216.

Krendl, K. A., Troiano, C., Dawson, R., & Clark, G. (1993). "OK, where's the remote?" Children, families, and remote control devices. In J. R. Walker & R. V. Bellamy, Jr. (Ed.), *The remote control in the new age of television* (pp. 137–153). Westport, CT: Praeger.

Lindlof, T. R., Shatzer, M. J., & Wilkinson, D. (1988). Accommodation of video and television in the American family. In J. Lull (Ed.), *World families watch television* (pp. 158–192). Newbury Park, CA: Sage.

Lull, J. (1988). Constructing rituals of extension through family television viewing. In J. Lull (Ed.), *World families watch television* (pp. 237–259). Newbury Park, CA: Sage.

Mederer, H. (1993). Division of labor in two-earner homes: Task accomplishment versus household management as critical variables in perceptions about family work. *Journal of Marriage and the Family, 55,* 133–145.

Morley, D. (1988). Domestic relations: The framework of family viewing in Great Britain. In J. Lull (Ed.), *World families watch television* (pp. 22–48). Newbury Park, CA: Sage.

Osmond, M. W., & Thorne, B. (1993). Feminist theories: The social construction of gender in families and society. In P. G. Boss, W. J. Doherty, R. LaRossa, W. R. Schumm, & S. K. Steinmetz (Eds.), *Sourcebook of family theories and methods: A contextual approach* (pp. 491–623). New York: Plenum.

Perse, E. M., & Ferguson, D. A. (1993). Gender differences in remote control use. In J. R. Walker & R. V. Bellamy, Jr. (Eds.), *The remote control in the new age of television* (pp. 169–186), Westport, CT: Praeger.

Shank, J. W. (1986). An exploration of leisure in the lives of dual career women. *Journal of Leisure Research, 18,* 300–319.

Shaw, S. M. (1985). Gender and leisure: Inequality in the distribution of leisure time. *Journal of Leisure Research, 17,* 266–282.

Shaw, S. M. (1991). Gender, leisure, and constraint: Towards a framework for the analysis of women's leisure. *Journal of Leisure Research 26,* 8–22.

Spigel, L. (1992). *Make room for TV: television and the family idea in postwar America.* Chicago: University of Chicago.

Szinovacz, M. E. (1987). Family power. In M. B. Sussman & S. K. Steinmetz (Eds.), *Handbook of marriage and the family* (pp. 651–693). New York: Plenum.

Thompson, L. (1991). Family work: Women's sense of fairness. *Journal of Family Issues,* 12, 181–196.

Thompson, L. (1993). Conceptualizing gender in marriage: The case of marital care. *Journal of Marriage and the Family, 55,* 557–569.

Thompson, L., & Walker, A. J. (1989). Gender in families: Women and men in marriage, work, and parenthood. *Journal of Marriage and the Family, 51,* 845–871.

CHAPTER 8

Gender in Intimate Relationships

Although institutions and organized activities such as work and play provide frameworks for our lives, it is the relationships within these activities that hold our lives together. What surprises many people is that everyday relationships are patterned and institutionalized. We don't mean "the daily routine" kinds of patterns, we mean patterns across relationships related to gender. For example, sociologists consider the family to be more than just a personal relationship; they view it as a social institution. That is, the family, as an institution, has relatively fixed roles and responsibilities and meets some basic needs in society such as caring for dependent members and providing emotional support for its members. As you read through this chapter, you will come to realize how social norms influence all gendered relationships, including intimate relationships. This introduction and the readings in this chapter illustrate the following points. First, gendered intimate relationships always evolve, often in response to social changes unrelated to the relationships themselves. Second, gender embeds itself in an idealized version of intimacy, the traditional family, that is not the reality in the United States and most parts of the world today, as illustrated in readings in Chapters 1 and 3.

Let's stop for a moment and look at relationships in general. The word "relationship" takes on many different meanings in our lives. We can have a relationship with the teller at our bank because he or she is usually there when we deposit our paycheck. We have relationships with our friends; some we may have known most of our lives, others we have met more recently. And we have relationships with our family and with people who are like family. Then, we have relationships with; those whose relationships surround us with love, economic support, intimacy, and almost constant engagement. All of these relationships are shaped by gender. You've already read about relationships at work and play in Chapter 7; in this chapter you will read about how gender shapes more intimate relationships from friendships to partnering to parenting.

GENDERED FRIENDSHIPS

Consider the impact of gender on friendships. We can have many friends, all of whom we love or have affection for. In the past, researchers often argued that friendships varied by gender in predictable and somewhat stereotypical ways. That is, they described women's friendships as more intimate, or focused on sharing feelings and private matters, while describing men's friendships as more instrumental or focused around doing things. To illustrate, Cancian (1990) argued that men were more instrumental or task-oriented in their love relationships, whereas women expected emotional ties. For example, when asked how he expressed his love to his wife in Cancian's study, one man told the interviewer that he washed her car.

Karen Walker reviews the arguments about gender differences in friendships in the first reading in this section. In her research, Walker finds that men and women hold stereotypical views of men's and women's friendships. However, she finds unexpected patterns as the participants in her study talked about their friendships—men described intimate and emotional discussions with their friends and women told of how they shared time with friends in common activities that were less intimate than researchers had previously assumed. These exceptions to stereotypical views of gendered friendships, however, also are patterned; that is, they appear to vary by social class and the constraints of differing lifestyles. If men and women are more similar in their friendship relationships than we had assumed, then men may not be from Mars and women may not be from Venus, as one best-selling author suggests (Gray, 1992).

GENDER AND CHANGING LIVING ARRANGEMENTS

Regardless of what planet you are from, society tells us that men and women are supposed to fall in love and marry. As we have emphasized throughout this book, American culture assumes idealized intimate relationships to be heterosexual, accompanied by appropriate gendered behaviors, and, of course, based in nuclear families. You will notice that we did not include the word "family" in the title of this chapter. We did this because the traditional, idealized family—mom, dad, two kids and a dog all living in a house behind a white picket fence—is only a small percentage of households today and it was not the predominant form of relationships in times past. Instead, intimate relationships of all types shape and define our self-perceptions, genders, and daily activities. As we explore those relationships in this chapter, it is important to begin by examining the reality of how households actually are patterned. In Table 8.1, we list the various household configurations and the percentage in each category in the United States today. Do these data surprise you? Relationships in the United States are changing as indicated by the diversity of households as noted in Table 8.1. Almost 31 percent of the households in the United States in 1998 were occupied by single persons and other non-family combinations. The increase in single and non-family households reflects a

Table 8.1 Household composition in the United States in 1998[1]

Household type	Percentage of all households
Married Couple with own children under age 18 (only 21.2% of these families have single-income households with a male wage earner)	24.6%
Married Couple with no own children under age 18	28.3%
Single parent with children under age 18	11.7%
Other family	4.5%
Single person	25.7%
Other nonfamily	5.2%

1. Data for this table were taken from the 1999 Population Profile of the United States, p. 21.

trend toward postponing marriage and an increase in the divorce rates. According to the U.S. Census Bureau (1999:22), of those adults who were unmarried in 1998, 64 percent never married, 21 percent were divorced, and 15 percent were widowed, with women being 81 percent of all widowed adults. The median age at first marriage has risen for both men and women. In 1980, the median age for women to marry was 22 and for men 25, whereas in 1998 that increased to 25 for women and 27 for men.

Although 56 percent of all Americans of age 15 or older were married in 1998 (U.S. Census Bureau, 1999:22), 28 percent of households were married couples without children, a percentage higher than that of married couples with children under age 18 in the home at that time. The percentage increases to 30.3 percent of all households when you include children over the age of 18 while 20.4 percent of married-couple families included a member 65 or older (U.S. Census Bureau, 1999:22). The latter figures should not be surprising, given the later median age at marriage and the increase in longevity in the United States. Life expectancy for children born in 1998 is 79.4 years for women and 73.9 years for men (Calasanti & Slevin, 2001). Thus, a family today is more likely to include a grandmother and adult children than little children and a puppy.

FEMINIZATION AND JUVENILIZATION OF POVERTY

Another deviation from the idealized, traditional family in Table 8.1 is that 11.7 percent of households were single parents with children under age 18. This change in household composition challenges the traditional gendered relationships expected in families, but also relates to social class inequalities tied to gender. The increase in the number of households headed by poor women raising children has been called the feminization or juvenilization of poverty (Bianchi, 1999; McLanahan and Kelly, 1999). Diana Pearce coined the term, "the feminization of poverty" in 1978 (Bianchi, 1999:308), at a time when the number of poor women-headed families rapidly increased. The rate of women's poverty relative to men's fluctuates over time and is about 50 to 60 percent higher than the poverty rate of men (Bianchi, 1999:311).

The juvenilization of poverty refers to an increase in the poverty rates for children that began in the early 1980s, whether in single or two-parent families (Bianchi, 1999). The Urban Institute (2001) published 1997 data indicating that only 47 percent of children from low-income households live with two parents, whereas 75 percent of children from families with higher incomes do so. Children's poverty rates in the 1980s and 1990s have stayed above 20 percent; during the 1980s and 1990s, children were 20 to 23 percent of those officially categorized as poor in the United States (Bianchi, 1999:314–315). The feminization and juvenilization of poverty are serious problems and are both gender as well as race and social class issues.

SAME-GENDER COUPLES

A change not reflected in Table 8.1 is the growing number of same-gender couples. It is difficult to know how many same-gender households there actually are; the category listed as "non-family" in Table 8.1 includes boarders or roommates (U.S. Census Bureau, 1999:21). One report indicated that 51 percent of same-gender couples identified themselves as married in a U.S. Census Bureau test of its questionnaires (Freiberg, 1999). As of 2001, however, same-gender couples could not marry legally in the United States, although some states allowed civil unions. In other countries, such as the Netherlands and Germany, same-gender couples could marry legally. Even if they cannot do so legally in the United States, many lesbian and gay couples choose to profess their love and commitment formally in ceremonies that are both similar and different from traditional heterosexual weddings. Same-gender couples also create families and redefine gender, parenthood, and family as Gillian Dunne describes in her article in Chapter 10.

THE IDEALIZED FAMILY

The growing diversity of household and family configurations in the United States has reshaped the ways we enter, confirm, maintain, and envision long-term intimate relationships as well as challenging how we think of gender. Rigid gender roles that solidified the image of an idealized family are being challenged. Some new patterns arise from choice, including an increased tolerance for lifestyle diversity, such as single person households. Others, such as single parents (typically mothers) who raise children alone, often arise out of divorce and poverty. Whatever the reasons, the idealized, traditional family with its traditional gendered relationships never was the norm in American households.

To illustrate how rare that idealized family is, only one-fourth of all households in Table 8.1 have children under age 18 in the home. However, most of these households do not resemble the culturally idealized pattern of the father at work and the mother and the children at home (Coontz, 2000). That idealized, traditional family type comprises only about 5 percent of all households in the United States. In 1999, of the 34,340,000 families with children under age 18, just over one-fifth of these families (21.2 percent) were households in which only the father was employed (U.S. Census Bureau, 2000). Thus, both partners are employed in almost 80 percent of the 24.6 percent of married couple households with children under age 18 in Table 8.1. The fastest growing segment of the workforce is mothers of preschool children, increasing from 41.9 percent of mothers employed in 1980 to 62.2 percent in 1988 (U.S. Bureau of Labor Statistics, 2002). Like gender, "the family" is a culturally constructed concept that often bears little resemblance to reality. The idealized, traditional family with separate and distinct gender roles does not exist in most people's lives.

Historically, families have changed considerably in terms of how they are formed and how they function. While enduring relationships typically involve affection, economic support, and concern for others, marriage vows of commitment are constructed around love in the United States today. In previous generations, most marriage vows promised commitment and love "until death do us part." Marriages in the 1800s, even when rooted in love, often were based on economic realities. These nineteenth century marriages were likely to evolve into fixed roles for men and women linked to the economy of the time, roles that reinforced gender difference but not necessarily gender inequality. For example, farm families developed patterns that included different, but not always unequal roles for men and women (Smith, 1987). In the latter half of the nineteenth century and into the early 1900s, families, particularly immigrant families, worked together to earn enough for survival, with women often taking in work in the home, such as laundry and boarders, and children working in the factories (Bose, 2001; Smith, 1987).

What changed and how did the current idealized roles of men and women within the stereotypical traditional family come to be? Martha May (1982) argues that the father as primary wage earner was a product of early industrialization in the United States. She notes that unions introduced the idea of a family wage in the 1830s to try to give men enough wages so their wives and children were not forced to work. In the early 1900s, Henry Ford expanded this idea and developed a plan to pay his male workers $5.00 per day *if* their wives did not work for money (May, 1982). The Ford Motor Company even hired sociologists to go into the homes of his workers to check to make sure that the wives were not working for pay either outside or inside the home (i.e., taking boarders or doing laundry), before paying this family wage to male workers (May, 1982). In fact, very few men actually were paid the higher wage (May, 1982). You might ask why Ford Motor Company was so interested in supporting the family with an adequate wage. At that time, factories faced high turnover because work on these first assembly lines was unpleasant, paid very little, and the job was much more rigid and unpleasant than the farm work that most workers were accustomed to. Ford enacted this policy to reduce turnover and lessen the threat of unionization. Thus, according to May, one reason behind the social construction of the "ideal family" was capitalist motivation to tie men to their jobs for increased profits, not individual men wanting to control their families.

GOVERNMENT POLICIES AND FAMILY RELATIONSHIPS

Governmental policies also play a role in shaping families in the form of tax laws, health and safety rules, and other legislation. Early health and safety laws increased the age of employment for factory workers (Bose, 2001) and are still in effect in terms of what age a child can begin to work for pay. These laws relate to the creation of adolescence, with children remaining in school throughout high school. The dependent deduction on tax returns was intended to encourage families to have more children at a time when politicians worried about the declining birth rate in this country. At the same time and for the same reason, Canada instituted a policy that is still in place in which women are given a payment each month for every child under the age of 18. The government makes this payment directly to the mother. While only a nominal sum, it is distributed across social classes as an incentive to bear and raise children.

Policies and laws that affect all relationships are often written in such a way that idealizes women's role in two-parent families and reinforced hegemonic masculinity and emphasized femininity, while ignoring other choices and life situations, such as same gender couples. The impact of policies and practices on relationships in poor families is also heavy, making it difficult to "do" idealized gender because living, in and of itself, is challenging, as illustrated in the reading by Karen Pyke.

CHANGING RELATIONSHIPS

Heterosexual, marital, and parenting relationships have changed a lot over time. In addition to government regulations, the feminization of the workforce has affected men's and women's roles in marriages (Blackwelder, 1997). The need for families to have two workers is intricately connected to the economic structure of the country and has changed considerably relationships in the home. As Kathleen Gerson discusses in her reading, couples have started to modify expectations for marriage and marital responsibilities in response to changes in educational attainment and workforce participation of women. Gerson describes the dilemmas faced and strategies tried by young people as they attempt to balance commitment and autonomy in their lives.

Gender continues to permeate and frame ever-evolving relationships. Mothers continue to worry about child care arrangements while they work and construct various interpretations of how these arrangements affect their children, all based upon traditional views of women's role in the family (Uttal, 1996). While men and women struggle to share household labor, women continue to do at least twice as much housework as men (Coltrane, 2000; Shelton and John, 1996). Thus, the idealized, traditional gendered responsibility for women in the United States to care for children and the household remains, even though most women work outside of the home. Jacobs and Gerson (2001) argue that the changes in family composition and gender relations have created situations in which members of families, particularly women and most particularly single women, are overworked with little free time left for themselves or their families (as discussed in Chapter 7). Jacobs and Gerson (2001) describe the situation as particularly acute for those couples whose work weeks are 100 hours or more and who tend to be highly educated men and women with prestigious jobs. However, as described by Pyke, relationships across social class face different constraints. Young men and women face many choices in terms of how they will form relationships, including the moral dilemma between commitment and autonomy that Gerson discusses in this chapter.

CULTURE AND GENDERED RELATIONSHIPS

The changes in coupling have affected parenting relationships, but these relationships are culturally specific, depending on the construction of gender in a particular society. The article by Gracia Clark on mothering and work in Asante culture describes a different way of looking at how changes related to modernization affect gender identification, as well as the relationships of women and men to their children and each other. In Asante culture, parenting has traditionally been defined very differently from that in the West—women are good mothers if they provide for their children. Being able to provide for one's children is particularly difficult in a shaky economy, as is the Asante urban culture in Ghana that Clark describes. The result of the economic uncertainty and cultural values relating a mother's role to providing for her children is that a good mother goes back to work soon after the birth of her child. It is interesting that in Asante urban culture, providing food is linked to mothering and men who provide for their children are called "good mothers." What makes a "good mother" in Western societies? Who defines or specifies these characteristics? And how do we learn what is expected of mothers?

Expectations about what makes a good father in U.S. society have changed, too, at least in terms of what magazines say. In their reading, Maxine Atkinson and Stephen Blackwelder provide an interesting overview of articles from 1900 to 1989 on fathering. They analyze these magazine articles to track the way our definitions of motherhood and fatherhood evolved in that time period. As you read through this chapter, try to predict what fatherhood and motherhood will look like in your generation and your children's generation.

The readings in this chapter provide a fuller understanding of how our most intimate relationships are socially constructed around gender. As friends, lovers, parents, and siblings, we are defined in many ways by our gender. Compare the dilemmas young people face in Gerson's reading, the realities of the couples in Pyke's reading, the relationships in Asante families, and those of same-gender couples described in the reading in Chapter 10. Ask yourself what choices you have made or wish to make as you consider how gender influences what you expect in your intimate relationships.

REFERENCES

Bianchi, Suzanne M. (1999). Feminization and juvenilization of poverty: Trends, relative risks, causes, and consequences. *Annual Review of Sociology* 25:307–333.

Blackwelder, Julia Kirk. (1997). *Now hiring: The feminization of work in the United States, 1900–1995.* College Station: Texas A&M University Press.

Bose, Christine E. (2001). *Women in 1900: Gateway to the political economy of the 20th century.* Philadelphia: Temple University Press.

Calasanti, Toni M. and Kathleen F. Selvin. (2001). *Gender, social inequalities, and aging.* Walnut Creek, CA: AltaMira Press.

Cancian, Francesca M. (1990). The feminization of love. In *Perspectives on the family: History, class and feminism,* Christopher Carlson (Ed.), pp. 171–185. Belmont, CA: Wadsworth.

Coltrane, Scott. (2000). Research on household labor: Modeling and measuring the social embeddedness of routine family work. *Journal of Marriage and the Family* 62: 1208–1234.

Coontz, Stephanie. (2000). *The way we never were: American families and the nostalgia trap.* New York: Basic Books.

Freiberg, Peter. (1999). Gay couples identify as married. June 21, 1999. http://www.gfn.com/archives/story/phtml?sid=1259. Downloaded June 6, 2002.

Gray, John. (1992). *Men are from Mars, women are from Venus: A practical guide for improving communication and getting what you want in your relationships.* New York: HarperCollins.

Jacobs, Jerry A. and Kathleen Gerson. (2001). Overworked individuals or overworked families? Explaining trends in work, leisure, and family time. *Work and Occupations* 28(1): 40–63.

May, Martha. (1982). The historical problem of the family wage: The Ford Motor Company and the five dollar day. *Feminist Studies* 8(2): 399–424.

McLanahan, Sara S. and Erin L. Kelly. (1999). The feminization of poverty: Past and future. In *Handbook of the Sociology of Gender,* Janet Saltzman Chafetz (Ed.), pp. 127–145. New York: Kluwer Academic/Plenum Publishers.

Shelton, Beth Anne and Daphne John. (1996). The division of household labor. *Annual Review of Sociology* 22: 299–322.

Smith, Dorothy. (1987). Women's inequality and the family. In *Families and Work,* Naomi Gerstel and Harriet Engel Gross (Eds.), pp. 23–54. Philadelphia: Temple University Press.

Urban Institute. (2001). *National survey of American families II 1997 results.* http://newfederalism.urban.org.nsaf/snapshots_index.html. Downloaded 2/12/03.

U.S. Bureau of Labor Statistics. (2002). http://www.bls.gov/opub/traw/pdf/table06-07.pdf. Downloaded 2/12/03.

U.S. Census Bureau. (2000). *Statistical abstracts of the United States: 2000.* Washington, DC: U.S. Government Printing Office.

———. (1999). *Population profile of the United States: America at the close of the 20th century.* Washington, DC: U.S. Government Printing Office.

Uttal, Lynet. (1996). Custodial care, surrogate care, and coordinated care: Employed mothers and the meaning of child care. *Gender & Society* 10(3): 291–311.

39

Men, Women, and Friendship

What They Say, What They Do

BY KAREN WALKER

Karen Walker is a social scientist who conducted in-depth interviews with 19 men and 33 women between the ages of 24 and 48 to examine theories of how men and women relate in intimate friendships. She used what is called a "snowball sample" to find the participants in her study. That is, she asked the individuals she interviewed to refer her to someone else who was either a friend or someone they knew. Most of her sample was white, but they represented many different ethnic groups. In addition, one respondent was African American, one was Puerto Rican, and one was Arab American. The group represented both middle- and working-class individuals. Her interviews provide a unique opportunity for us to see how this group of individuals viewed gender and friendships, as well as what they did in their own friendships.

1. What did the men and women in her sample expect in same-gender friendships?
2. How did the men and women in this study actually describe their same-gender intimate friendships?
3. How did social class influence the types of same-gender relationships these men and women had?

Stereotypes about friendship represent women's friendships as intimate relationships in which sharing feelings and talk are the most prevalent activities. Men's friendships are represented as ones in which sharing activities such as sports dominate interaction. In this article, however, I argue that the notions that women share infinite feelings whereas men share activities in their friendships are more accurately viewed as cultural ideologies than as observable gender differences in behavior. Using data from in-depth interviews with working- and middle-class men and women, I show that men and women use these ideologies to depict their friendships and to orient their behavior. Responses to global questions about friendship indicate that men and women think about their friendships in culturally specific ways that agree with stereotypes about men's and women's friendships.

Responses to questions about specific friends, however, reveal more variation in same-sex friendships than the stereotypes or the social scientific literature lead one to expect. When men and women discuss friendship they emphasize the behavior that corresponds to their cultural notions of what men and women are like. Men focus on shared activities, and women focus on shared feelings. When specific friendships are examined, however, it becomes clear that men share feelings more than the literature indicates, whereas women share feelings less than the literature indicates; furthermore, the extent to which they do so varies by class. Employed middle-class women indicate that they are sometimes averse to sharing feelings with friends. Working-class men, on the other hand, report regularly sharing feelings and discussing personal problems.

This approach differs significantly from much of the social scientific literature on friendship by closely examining the link between behavior and ideology and seeing gender as an ongoing construction of social life. Many friendship studies over the past decade have emphasized the extent to which men and women have different kinds of friendships; the conclusions concur with the most prevalent stereotypes. Lillian Rubin (1985) has argued that men bond through shared activities, whereas women share intimate feelings through talk. She ascribes these differences to two phenomena. First, socialization of children encourages attention to relationships for girls and competition among boys and men. Second, the psychic development of girls leads girls and women to develop permeable ego boundaries and relational, nurturing capacities that encourage them to seek intimacy within friendship with other females. Boys and men, on the other hand, are threatened by having close, intimate friendships. Intimacy threatens their sense of masculinity because it touches that feminine part of their psyche that they were forced to repress in early childhood. According to Rubin, shared activities and competition are compensatory structures for men that prevent them from becoming too intimate.

Other authors differ somewhat on the causes and evaluation of these differences. Scott Swain (1989), for example, argues that perceiving men as deficient in intimate capacities as measured by verbal communication ignores nonverbal intimacy. In a study of intimate friendships among antebellum men, Karen Hansen (1992) has argued that arguments resting on psychic development are essentialist and neglect historical changes in intimate behavior. Other authors attribute observed gender differences in friendships solely to socialization rather than psychic development (Allan 1989; Swain 1989). There is, however, little debate that contemporary men characteristically engage in shared activities, whereas women engage in verbal sharing of feelings with their friends (Bell 1981; Caldwell and Peplau 1982; Eichenbaum and Orbach 1989; Oliker 1989; Sherrod 1987; Swain 1989).

In contrast to the most commonly developed explanations in friendship studies, some sociologists have argued in recent years that gender is constructed on an ongoing basis in social life (Connell 1987; Leidner 1991; West and Zimmerman 1987). These approaches account not only for the variation within same-sex friendships that I observed but also for the strength of the ideologies about friendship. The process of representing themselves as adhering to cultural norms when they discuss friendship is one way men and women create coherent understandings about themselves as gendered humans; furthermore, because social processes of gender construction emphasize the naturalness of gender differences, individuals rarely question the extent to which those differences exist. Leidner observes that even "the considerable flexibility of notions of proper gender enactment does not undermine the appearance of inevitability and naturalness that continues to support the division of labor by gender" (1991, 158). To this I would add that even when individuals understand gender differences as socially rather than biologically caused, as several of my respondents did, they see those social causes as having shaped their personalities in such a way that change is difficult, if not impossible, to achieve.

In addition to exploring how women and men construct gender-specific understandings of their friendships, I also explore how social class influences men's and women's capacities for conforming to gendered behaviors. Women's or men's material circumstances affect their abilities to conform to gendered norms about friendship. The friendships of women who did not work in the paid labor force or women whose family took priority over their labor market participation generally conformed to a model of intimate friendship. Middle-class men and women who were geographically or occupationally mobile tended to report a lack of intimacy in specific friendships. Working-class women and men who participated in dense social networks and whose resources limited the extent of their social activities often spent time talking to their friends and sharing feelings about events in their lives, thereby creating intimate friendships; thus social class shapes the experiences men and women have with friendship, and it may do so in ways that contradict stereotypes about men's and women's behaviors.

* * *

Perceived Differences in Men's and Women's Friendships

In response to questions about whether men's and women's friendships were similar or different, respondents' answers reflected cultural gender ideology. In a conversation immediately preceding our interview, one professional woman, Anna,[1] asked me what my dissertation was about. Told that it was a study of men's and women's friendships, she said that was very interesting: Her friendships were much different than her husband's friendships. Her husband had a friend named Jim whom Tom (Anna's husband) referred to as his "buddy." But even though Jim and Tom had a lot of fun together, Tom could not establish a more intimate friendship with Jim. Jim did not express his feelings. Anna thought this was because Jim was afraid his feelings would overwhelm him.

Although Anna was unique in bringing up the subject of men's and women's differences herself, about 40 percent of the respondents reported that men were not as open as women.[2] When asked "Do you think that men's friendships are different than women's friendships?", one man responded this way:

> Well, we know that they are often. But whether that's a result of social construction or genetics I have no idea. [How are they different?] Well, there's a lot more sense of competition between a lot of men. A lot less openness about personal matters. It's just not as personal as being friends, uh, as your women friends. (middle-class man)

Other responses sounded remarkably similar in focusing on openness, the degree of closeness, and intimacy:

> Men keep more to themselves. They don't open up the way women do. Some women will spill their guts at the drop of the hat. (working-class man) . . .

Although a focus on intimacy and closeness was the most common response to questions about gender differences in friendships, another 25 percent of the respondents suggested that men and women do different things with their friends. Men play sports and share activities; women talk. Women discuss children and families; men discuss cars, work, and politics. Thus, about 65 percent of the respondents expressed and engaged in the construction of dominant cultural ideology about friendship that women are more intimate and talk in their friendships, whereas men engage in activities and do not share feelings.

Responses of the remaining one-third who gave other answers about men's and women's friendships were fairly evenly divided between those who said that women's friendships showed greater conflict, those who did not know if there were differences, and those who said men's and women's friendships were essentially similar in that loyalty and trust were requisites of all friendships.

Not only did men and women give gendered accounts of friendship, they also assumed that men and women were different. In doing so, they did not necessarily know how men and women differed, but they believed that they did. Individuals sometimes understood the response of someone from the opposite gender as characteristic of that gender if it differed from their own responses. Men and women frequently told me how their spouses were different from them. . . .

EXPERIENCES IN SPECIFIC FRIENDSHIPS

Many respondents gave general accounts of their friendships that agreed with their understandings of what appropriate gendered behavior is. When I asked respondents detailed questions about activities they did with friends and things they talk about, however, I discovered that there was frequently a disparity between general representations that indicated cultural beliefs and specific information.

For instance, [one man] who thought that men friends shared activities such as hunting and fishing never went hunting or fishing with friends. His economic resources were severely limited, he worked two jobs, and had little time to socialize except in the evenings. His wife worked part-time in the evenings and on weekends, and while she worked he took care of their two children. Sometimes friends came over, and they watched TV and drank together. Most frequently, however, he socialized with friends at work in a high school where he was a janitor. He and his work friends discussed retirement, their wives, their children, as well as shared interests in sports.

When asked what he had talked about with his closest friend at work that day, he said they had discussed their wives' preferences for marital courtship. One wife liked to have the scene set before sex. She and her husband went out to eat together, they went shopping together, and there was a period in which they both knew what would happen later in the evening. My respondent told his friend that he and his wife liked spontaneity. Preparing for sex through a series of rituals seemed superficial to them. This is the kind of detailed talk about a significant relationship in which women are reputed to engage. It is a far cry from the stereotype of men friends who discuss sports and who do not have intimate discussions. Nor does it fit with my respondent's own perception of what friendships are like.

There were many similar examples. Seventy-five percent of the men reported engaging in nongendered behavior with friends—all of whom reported that they spoke intimately about spouses, other family members, and their feelings. Furthermore, one-third of those men reported that they engaged in other nongendered behavior as well as intimate talk. A lawyer who said that he did not discuss marital issues or personal matters with his friends later reported that he and his friends had talked with some regularity about fertility problems they were having with their wives. They discussed the fact that their wives were much more concerned about their difficulties in conceiving than they were.

Some men reported the impact of divorce on friendship interaction. One professional man told me of at least two men whom he had supported through their divorces, listening to them talk about their feelings. He told me of friends who had helped him through his divorce, men who allowed him to "ventilate." This man repeatedly told me that he was very emotive and expressive and that he complained a lot to friends. That these conversations are disregarded in how men shape their understandings about being men is indicated by what happened later in the interview. He told me that he thought women were more open than men were. He said he had gone to hear Robert Bly talk, and Robert Bly had said that women had an ability to get to the heart of the matter, to really articulate things, whereas men were unable to do so. My respondent said he really thought Bly was right, and he regretted his own inability to be more open. He also said that he never discussed marital problems or issues with his friends. He reported that he and his friends usually discussed sports, even though my respondent did not much care for sports. . . .

In addition to talking about things that men are reputed not to talk about—feelings and relationships—sometimes men did things that did not fit with their ideas of what men typically did. For instance, both men and women thought of shopping as something women, but not men, do together. Most men, in fact, said they did not shop with friends. The response of Joe, a working-class man, illuminates the meaning of shopping as a gendered activity. He reported that sometimes he went food shopping with a married friend: "Anita gives him the list and we go to the supermarket like two old ladies and we pick out the things, 'Well, this one's cheaper than that one so let's get this.'" Joe seemed a little embarrassed by this activity. He laughed softly as he reported these shopping trips. His married friend, whom I also interviewed, denied that he shopped with friends.

Gene, a middle-class man, said that he was concerned about moisturizer, but he did not discuss it with friends. I asked why not, and he said it was "faggotty" to do so, blushed, and laughed with embarrassment. Then he talked about going to visit a gay friend in another city. The morning after he got there he and his friend went out early to buy moisturizer and vegetable pâté at Neiman-Marcus—activities he thought of as stereotypically gay and not things most heterosexual men do (his friend was in the entertainment industry in Hollywood, and his behavior could just as easily have been ascribed to the demands of a lifestyle valuing appearance and style). Gene's use of the term "faggotty" indicated that there are certain things that he did not do as he constructed his masculinity in everyday interaction. This accords with Carrigan, Connell, and Lee's (1987) discussion about the importance of hegemonic heterosexual masculinity in constructing power relations among men. Men who construct alternative forms of masculinity, such as being gay, are perceived as not masculine. Like the working-class man who did not feel like a man when he shopped for food with his friend, Gene did not feel like a man when he discussed moisturizer, and so he did not do it with straight friends.

Given that gay men are perceived as not masculine, it is interesting to note that Gene's friendship with Al, his gay friend, did not appear to threaten Gene's masculinity. Gene's account of how Al came out to him emphasized friendship over sexual orientation. Longtime college friends, Al told Gene he was gay several years after college, and he added that he knew his coming out might end their friendship. Gene reported being "very insulted" that Al thought he might stop being his friend. Whether or not being friends with Al threatened his masculinity was unimportant compared to Gene's public reaffirmation of the norm of loyalty to friends; furthermore, Gene emphasized joking behavior as important to friendship and connected joking to masculinity. Al, it turned out, had an acerbic and imaginative sense of humor that Gene appreciated.

Even in a situation where behavior contradicts gendered norms, some aspect of the interaction often constructs gender in ways consonant with ideology. Whereas the action of one person may contradict ideology, the response of the other may construct gender in ways conforming to dominant ideology. This pattern of emphasizing friendship activities that fit with their ideologies of masculinity is analogous to what Leidner (1991) observed in her study of men and women interactive service workers who emphasized those aspects of their jobs that they thought of as consonant with their gender identities. For instance, Gene and a friend were sitting together one night:

> And he says, "So, how are things going?"
> And I, I had been through a real bad period.
> And I said, "Fuck it, I'm going to tell him."
> So I told him how everything was wrong in my life, and toward the end of it he said, "Oh, that's the last time I ask you how you're feeling." . . . I mean, he was actually offensively unsupportive.

In this instance, Gene violated gender ideology dictating that he not discuss his problems with his friend. His friend responded by letting Gene know that he did not want to hear Gene's problems. By belittling Gene's disclosures, his friend also constructed masculinity in ways conforming to dominant norms. Gene, who tended to reject the legitimacy of the dominant norms and who self-consciously violated those norms, interpreted his friend's behavior as typically male. Gene used this event to describe masculine behavior to me, even though it also included Gene's own transgressions of gender ideology.

The construction of gender, therefore, is a highly complex process that occurs through behavior, through ideology, and through accounts of both. One person may act in ways that contradict gender ideology, and the response may be disapproval, which reinforces the ideology. In addition, there are many events and forms of interaction that occur simultaneously in one situation, and flexibility in the construction of one aspect of gender may not threaten the overall construction of gender if other interactions, such as joking, reinforce gender ideology.

Nongendered Behavior among Women

Similar processes of understanding their own behavior in terms of cultural ideologies and norms occurred among women. Women reported that they could tell their friends anything. They readily volunteered that they talked about their husbands and lovers. Few women volunteered that shared activities were essential to friendship; the middle-class women were more likely to do so than the working-class women—they had more resources with which to engage in shared activities. Few women volunteered that they go to spectator sports with friends. Few women volunteered that they engaged in athletic activities with friends.

Just as there were disparities between men's statements about what their friendships were generally like and what their specific experiences were, there were also disparities between what women said their friendships were like generally and what specific friendships were like. About 65 percent of all women reported engaging in behavior that did not conform to gender ideology. Some women, like some men, occasionally went to spectator sporting events with friends, but, with one exception, I only heard about these activities when I asked directly if they went to sporting events or did athletics with friends. Unlike men, women did not volunteer that they did so. About 25 percent of all women worked out together, went to aerobics classes together, or belonged to local sports teams, but they primarily defined these occasions as other times to see each other and talk; the activities themselves were defined as relatively unimportant.

Another 15 percent of the women respondents said that many friendship interactions occurred while they were doing things with friends. Several middle-class women belonged to musical groups where they met friends. A few working-class women went out to clubs together, leaving their spouses at home. Sharing activities, therefore, sometimes provided the basis for women's friendships just as it provided the basis for some men's friendships.

Finally, 25 percent of the women respondents reported that they considered certain information private and would not discuss it with friends. Ilana, the professional woman quoted above who said that friendship entailed showing openness and caring, reported that there were many things about which she would not talk to friends. For instance, she did not discuss her relationships with men with other women.

Ilana indicated that her reticence was unusual, but it did not force her to reconsider her femininity. Instead, she both acknowledged and denied the masculine stereotype of sharing activities without talk. On the one hand, she said, "There are a lot of male friendships based on doing certain male things together without saying much." On the other hand, she said, "There are also men that surprise me in that it seems to me that they like to gossip and talk about women even more than women may want to talk about men." . . .

Two academic women reported doing very little intimate sharing with friends. One woman reported that she had no friends whatsoever, although she was on friendly relations with many colleagues. She did not discuss her relationship with her husband or her family, or many intimacies with anyone other than her husband. She also did not talk to people about worries or problems because she said she did not like to burden people. The woman who referred me to her told me that she, too, rarely discussed intimate matters with friends. Like the first woman, she said she did not do so because she did not want to impose her problems on others; she did not think they would be interested.

One group of four working-class women defined openness as an essential component to friendship, and three of the women regarded the fourth woman, Susan, as not open enough, a failing they urged her to overcome. Susan concurred with her friends' opinions that she was not open enough, but she chose silence and reserve as a form of rebellion because she was angry with the woman in the group to whom she was closest. Although Susan's behavior did not conform to the ideal of feminine openness and intimacy, the talk of her friends constructed femininity even as they censured Susan for her secrecy.

Several professional women regretted that they did not have close intimate friendships, or they accepted what they took to be the fact that they did not make good friends. They felt like failures for not having friends with whom they could "say anything." On the one hand, they gave me reasons why they did not have those friendships. On the other hand, they felt as if they were unusual and "bad" friends. This belief in their own failure did not emerge in interviews with men. Whereas some men regretted not being able to open up, no men said they were bad for not doing so. For men, it was characteristically masculine to lack the ability to share intimate feelings with friends.

CLASS AND FRIENDSHIP

Sociologists have noted that social class as well as gender influences friendship patterns (Allan 1989; Willmott 1987). In this study, one way social class influenced behavior was that behavior conflicted with popular ideology, particularly for middle-class women. Middle-class women in traditionally male-dominated occupations such as university professors, lawyers, and doctors most frequently reported a lack of intimacy in friendships. Many had been very mobile. They moved from city to city while they pursued their educations. Some reported having intimate friendships when they were in college or graduate school, but indicated that recent friendships were less likely to be so. Whereas some women regularly corresponded with old friends with whom they were intimate, some did not and therefore reported a lack of intimate friendships. One doctor reported:

> Well, pretty much all my life, although I find this sort of dwindling down now . . . but I think for a long time I always had a couple really close friends, like people I'd see almost every day, talk to a lot, and um, would tell virtually anything and everything to. . . .

In addition to being occupationally mobile, this woman, like many of the middle-class respondents, had been geographically mobile. She had infrequent contact with friends who lived in other cities, and she experienced attrition in groups of older friendships from college and medical school; furthermore, she did not replace friends as quickly as she once did. In addition, drawing on the expectation that friends could be drawn from peers with whom she was relatively equal, she found there were fewer equals in her pool of potential friends. Finally, competition at work dictated against the formation of intimate friendships.

Helen Gouldner and Mary Symons Strong (1987) noted similar experiences among executive women who lacked the time to form friendships off the job and hesitated to form friendships with people at work. Time was a major factor in limiting intimate friendships for the employed middle-class women I interviewed. For many, the development of intimacy required some minimal amount of time. Professional women often lacked the time to give to friends, particularly if they were married and had children. Most of their socializing with friends was done with work colleagues with whom they limited the amount of personal information they exchanged or with their husbands and other couples. Socializing with couples was not very intimate, but were occasions when men and women alike reported that they discussed politics, children, and work.

Professional men had similar friendships in these respects. They, too, had been mobile. Their groups of college friends had shrunk over the years. Friends had not been replaced as quickly as they had been lost. Time was limited for many men, particularly married men with children. In addition, much of their socializing was also done in couples, when intimate sharing was at a minimum. These constraints on professional men's and women's abilities to form long-lasting friendships with extensive interaction produced friendships resembling the masculine model of friendship more than the feminine model, especially when compared to unemployed women or working-class men and women.

When middle-class respondents had intimate friendships, they tended to have them with long-time friends and levels of intimacy varied over time. Periods of crisis, such as divorce, were most likely to be times when men and women had intimate discussions, but those times tended to be limited, and after the troubles subsided, they returned to their old patterns of interaction.

The middle-class women who had friendships conforming to gender stereotypes of intimacy tended to be those women who stayed at home with small children and had neighborhood friends. They had more opportunities to see friends alone, which increased their opportunities for intimate discussions. In addition, collegial norms against sharing intimate details did not exist for these women.

Working-class respondents' lives were structured differently from those of professional men and women, and their friendships were correspondingly different.[3] Only one working-class man was from another state (whereas three-fourths of the middle-class respondents grew up in other states). Working-class men and women tended to know their friends for much longer periods of time—they met in school or in their neighborhoods. They had fewer friends, but they saw them more often. In addition, they frequently saw their friends informally at home or in the neighborhood; as a result, they knew about each other's troubles in a way that professional men and women who socialized occasionally with friends did not. Their social lives were more highly gender segregated, which provided greater opportunities for intimate discussion. Men, as well as women, reported that they talked to their friends about intimate things on a regular basis. Thus nongendered behavior among working-class respondents tended to occur among men who reported frequent intimate discussions ("We're worse than a bunch of girls when it comes to that").

The working-class respondents also had more problems—financial, substance abuse, family, health—and having these problems meant they tended to be topics of conversations with their friends. When professional men and women faced troubles, they also talked to their friends, but their lives were more stable in many respects. As a group the middle-class respondents were affluent, and they reported having few financial difficulties. Middle-class respondents reported that heavy periods of drug and alcohol use tended to have been in their late adolescence and early twenties, times when they were not married. The working-class respondents married young, and drug and alcohol use tended to be a problem because it interfered with marriage and family. Financial problems, drug problems, and marital problems often seemed to occur together among the working-class respondents with each problem exacerbating the others, whereas middle-class respondents tended to deal with single problems at a time and to emerge from periods of crisis more quickly. To the extent that discussing problems is a mark of intimacy for many individuals, working-class men appeared to have more intimate friendships than professional men and women.

CONCLUSION

I have conceptualized gender as an ongoing social creation rather than as a role individuals learn or a personality type they develop that causes differences in behavior. This approach to gender accounts for many of the differences between how men and women represent their friendships in general and the specific patterns their friendships have. In interviews, men and women gave me general accounts of their friendships based on specific cultural norms of masculinity and femininity. These accounts often came early in the interview when I asked respondents to give me their definition of friendship and to tell me why friendship was important to them, if it was. They also came up toward the end of the interview when I asked them if they thought men's friendships differed from women's friendships. And they often contradicted specific information gathered during the friendship history.

It is not, however, sufficient to understand these processes as the ongoing production of gender. The process of gender construction is universal, but the forms and content of the construction vary. One must, therefore, ask why these particular ideologies about gender differences in friendship dominate rather than others. Earlier ideologies of friendship represented women as incapable of loyalty and true friendship and men as noble friends. Why has a reversal occurred? Asking this question, Barry Wellman (1992) argues that the domestication of the community through suburbanization has decreased the importance of men's public and semipublic social ties, at the same time increasing the importance of private domestic ties, at which women have traditionally excelled; thus men's friendships are beginning to resemble women's friendships in that they are carried out in the home and as "relations of emotional support, companionship, and domestic services" (Wellman 1992, 101). Wellman is effectively addressing a perceived change over time in friendship behavior among men, not the question of why women's friendships are privileged in contemporary culture.

The answer to that lies in the emergence of the women's movement and the self-conscious attempts to valorize women—and the response to that valorization by men. Much of the academic literature about women's friendship patterns is written by feminist social scientists who respond to older cultural ideologies that men were capable of true friendship and women were not (Rubin 1985; Smith-Rosenberg 1975). Rubin's argument, which claims that men are psychically incapable of the kind of intimacy characterizing women's friendships, draws explicitly on Carol Gilligan's and Nancy Chodorow's feminist work. Although it is often difficult to determine the source of cultural ideology, the efflorescence of articles and books about women's friendships and relationships occurs after the beginning of second wave feminism, and it is only then that men begin to respond systematically to those works.

In everyday interaction, these cultural ideologies are reproduced through discussions with friends, through the kind of generalizing people come to regarding gender that I mentioned earlier, through socialization, and through exposure to ideas about gender differences by the media. If men and women accept cultural definitions of reality, and many of them do, they interpret their behavior in light of such depictions and feel either deficient or validated depending on whether they fit or not. Many also orient their behavior toward these prescriptions for behavior. The fact that so much of their behavior does not match the cultural prescriptions, however, indicates the limits of those prescriptions for determining behavior. It also calls into question the adequacy of explanations that conclude that women and men are essentially different. Men and women respond to the demands of their lives in a variety of ways; some of those responses may disaffirm those same men's and women's gender ideologies.

In addition to asking why ideologies about women's openness and men's activities dominate other ideologies, one may also ask why men and women accept the stereotypes when their own behavior differs significantly from them. Several reasons account for this. First, friends sometimes frown on men's and women's gender-inappropriate behaviors. Such disapproval reinforces gender norms, and is interpreted by the one who receives disapproval as an indication that the behavior of the one who disapproves conforms to gender ideologies; therefore, the person whose behavior does not conform comes to see his or her behavior as anomalous. Second, women and men sometimes do not see the disparity between behavior and ideology because they do not reflect on their behavior. Given the enormous number of interactions in which people participate in daily life, it is not surprising that they neglect to reflect on behavior that they do not think meaningful in the construction of their identities.

Finally, friendship interactions are very complex, and men and women do gender in a variety of ways in single interactions. Often their behaviors do not conform to gender stereotypes, but other elements of the interaction construct gender along normative lines. The usual outcome of doing gender, buttressed by life's experiences, is that the actors have no doubts about their gender identities as women or men. In the case of friendship and gender stereotypes, men's and women's flexibility in behavior does not threaten their identities as men or women because so many other aspects of interaction reinforce their identities, which they reify. They are not, therefore, normally faced with a problem of gender identity that they must reflect on and solve either through modifying their behaviors or through modifying their ideologies to better reflect behavior. As I noted earlier, the change in ideology from one valorizing men's friendships to one valorizing women's friendships was initiated by feminists who had identified problems and self-consciously acted to solve them. Their success may rest on the fact that, when asked to reflect on their own experiences, others also recognized the problems in the way men's and women's friendships were perceived and the way they were experienced; but it was first necessary to bring the disparities to people's attention. Even this change did not fundamentally alter the notion that women's friendships are emotionally expressive, whereas men's friendships are not; rather, it altered the *value* of men's and women's friendships.

One must be wary of arguments that posit the early development of gendered personalities that cause differences in friendship; however, provocative psychoanalytic theories of masculine or feminine personalities cannot account for the variation one sees in men's or women's behaviors. Nor does it account for the disparity between people's general accounts of friendship and their specific experiences, a disparity that makes sense if one understands those general representations as attempts to provide coherence, and understanding, and to find norms by which to live as gendered humans.

NOTES

1. All the names of the respondents have been changed.
2. Results about perceived differences in men's and women's friendships are based on responses of 46 individuals. At the beginning of the study, four women and two men were not asked to compare men's and women's friendships.
3. Clyde Franklin makes similar observations about Black men: "While working class black men's same-sex friendships are warm, intimate, and holistic, upwardly mobile black men's same-sex friendships are cool, non-intimate, and segmented (1992, 212).

REFERENCES

Allan, G. 1989. *Friendship: Developing a sociological perspective.* Boulder, CO: Westview. Bell, R. 1981. *Worlds of friendship.* Beverly Hills, CA: Sage.

Carrigan, T., B. Connell, and J. Lee. 1987. Hard and heavy: Toward a new sociology of masculinity. In *Beyond patriarchy: Essays by men on pleasure, power, and change,* edited by M. Kaufman. Toronto: Oxford University Press.

Connell, R. W. 1987. *Gender and power.* Stanford, CA: Stanford University Press.

Eichenbaum, L., and S. Orbach. 1989. *Between women: Love, envy, and competition in women's friendships.* New York: Penguin.

Franklin, C. W., II. 1992. Friendship among Black men. In *Men's friendships,* edited by P. M. Nardi. Newbury Park, CA: Sage.

Gouldner, H., and M. S. Strong. 1987. *Speaking of friendship: Middle class women and their friends.* New York: Greenwood.

Hansen, K. V. 1992. Our eyes behold each other: Masculinity and intimate friendship in antebellum New England. In *Men's friendships,* edited by P. M. Nardi. Newbury Park, CA: Sage.

Leidner, R. 1991. Serving hamburgers and selling insurance: Gender, work, and identity in interactive service jobs. *Gender & Society* 5:154–77.

Oliker, S. 1989. *Best friends and marriage.* Berkeley: University of California Press.

Rubin, L. 1985. *Just friends: The role of friendship in our lives.* New York: Harper & Row.

Sherrod, S. 1987. The bonds of men: Problems and possibilities in close male relationships. In *The making of masculinities: The new men's studies,* edited by H. Brod. Boston: Allen & Unwin.

Smith-Rosenberg, C. 1975. The female world of love and ritual. *Signs* 1:1–29.

Swain, S. 1989. Covert intimacy: Closeness in men's friendships. In *Gender in intimate relationships:*

A microstructural approach, edited by B. Risman and P. Schwartz. Belmont, CA: Wadsworth.

Wellman, B. 1992. Men in networks: Private communities, domestic friendships. In *Men's friendships,* edited by P. M. Nardi. Newbury Park, CA: Sage.

West, C., and D. Zimmerman. 1987. Doing gender. *Gender & Society* 1: 125–51.

Willmott, P. 1987. *Friendship networks and social support.* London: Policy Studies Institute.

40

Moral Dilemmas, Moral Strategies, and the Transformation of Gender

Lessons from Two Generations of Work and Family Change

BY KATHLEEN GERSON

Sociologist Kathleen Gerson has studied work and family patterns for some time. Her first book examining women's and men's roles in families, published in 1985, examined the decisions women make about work, career, and motherhood. This book was aptly titled Hard Choices. *In this reading, she considers what she has learned from studying men's and women's choices about work and family, exploring the moral dilemmas that she has observed and considering those that young people face as they negotiate familial relationships and work responsibilities.*

1. What does Gerson mean by "the tension between autonomy and commitment?"
2. What changes does she observe in heterosexual relationships?
3. What is neotraditionalism and how does it affect men and women in heterosexual relationships?

Choosing between self-interest and caring for others is one of the most fundamental dilemmas facing all of us. To reconcile this dilemma, modern societies in general—and American society in particular—have tried to divide women and men into different moral categories. Since the rise of industrialism, the social organization of moral responsibility has expected women to seek personal development by caring for others and men to care for others by sharing the rewards of independent achievement.

Although labeled "traditional," this gendered division of moral labor represents a social form and cultural mandate that rose to prominence in the mid-twentieth century but reached an impasse as the postindustrial era opened new avenues for work and family life. . . . At the outset of the twenty-first century, women and men face rising conflicts over how to resolve the basic tensions between family and work, public and private, autonomy and commitment. They are searching for new strategies for reconciling an "independent self" with commitment to others.

While the long-term trajectory of change remains unclear, new social conditions have severely undermined the link between gender and moral obligation. The young women and men who have come of age amid this changing social landscape face risks and dangers, but they also inherit an unprecedented opportunity to forge new, more egalitarian ways to balance self-development with commitment to others. To enable them to do so, however, we must reshape work and family institutions in ways that overcome beliefs and practices that presume gender differences in moral responsibility. . . .

While it is important to assert that it is just as valuable to pursue emotional connection and provide care as it is to create an independent self or provide economically for a family, it is also critical to question the premise that women and men can be separated into distinct, opposed, or unchanging moral categories. As Epstein argues, any vision of dichotomous gender distinctions is not only inaccurate; it is also an ideological construct that justifies and reinforces inequality. Connell points out that "masculinities" and "femininities" vary across historical time and space. Lorber and Risman, among others, question the concept of gender itself, pointing to the social paradoxes and cultural contradictions to which all human actors must respond in constructing their public and private selves. These theorists recognize that gender is a social institution, not an inherent trait, and that it shapes organizations and opportunity structures as well as personal experiences (Connell 1987, 1995; Epstein 1988; Lorber 1994; Risman 1998).

There are good analytic and empirical reasons to reject the use of gender to resolve the knotty moral conflicts between public and private, work and family, self and other. It is difficult to avoid the conclusion that using gender in this way is more prescriptive than descriptive. Such approaches may depict how women and men *should* behave, but they do not provide an accurate description or explanation of how women and men actually *do* behave or how they *would* behave if alternative options were available. Certainly, the proportion who have conformed to gendered injunctions about appropriate moral choices has varied substantially across societies, subcultures, and historical periods. Countless women and men have been labeled "deviant" for their reluctance or inability to uphold idealized conceptions of gender. A framework of gendered moralities helps justify inequalities and stigmatize those who do not conform.

★ ★ ★

FROM MORAL DIFFERENCE TO MORAL DILEMMAS

The erosion of social supports for traditional conceptions of moral obligation has prompted the search for new ways to balance family and work, but the contradictory nature of this change has also produced enduring political debates and rising social dilemmas. Cultural conservatives decry the rise of nontraditional families and the expansion of public opportunities for women, claiming that these changes represent a decline of morality in America. Social progressives, in contrast, applaud the expansion of opportunity, personal choice, and tolerance for diversity but are concerned that the spheres of family and work are colliding. Workplaces seem more demanding, local communities seem to be losing civic engagement, and families seem starved for time and resources (Hochschild 1997; Jacobs and Gerson forthcoming; Putnam 2000; Schor 1992). As it has become increasingly difficult to carve out equal space for the unpaid caring work that women have historically performed, women and men alike are facing apparently irreconcilable choices between caring and self-sufficiency. The classic tension between individualism and commitment now assumes new forms and is being experienced in intensified ways.

In a context where broad, multilayered social changes clash with continuing inequalities, it is more fruitful to focus on moral dilemmas than on moral differences. A dilemma is a difficult, perplexing, or ambiguous choice between equally undesirable (or desirable) alternatives, while *moral* involves a concern with the rules of right conduct (*Webster's dictionary* 1992). Of course, we experience many moral dilemmas in deeply personal ways. They become social, however, when institutional and cultural contexts make it difficult or impossible for individuals to make a socially sanctioned choice—when, for example, all options invite disapproval but action is nevertheless required.[1]

The current period poses many such situations. In forming adult relationships, how do women and men weigh the need and desire for achieving autonomy with the hope of establishing an enduring commitment? In caring for a new generation, how do parents balance the need to spend time with their children and the need to earn enough to support them? In crafting a personal identity, how do individuals choose between attaining independence and building connections to others? These situations are *dilemmas* because they pose choices that have no institutionalized or unambiguously "correct" resolutions. They are *social* because they arise from the way that social change structures available options and creates conflicts, ambiguities, and inconsistencies. They are *moral* because others judge these choices, forcing new generations either to change or to reproduce prevailing moral codes. Socially structured moral dilemmas force us to move beyond habits and routines to develop and justify new actions and beliefs.[2]

Studying the creation of, and strategic responses to, socially structured moral dilemmas helps to illuminate the processes of gender change. It allows us to see how the definitions and practices of gender unfold as individuals develop responses to contradictory social options. Focusing on dilemmas allows us to view gender as an "incomplete" institution. Gendered responses do not reflect inherent gender differences but are instead strategies developed by differently situated social actors coping with ambiguous circumstances. Change is possible in this framework but never guaranteed. It becomes more likely when social arrangements create conflicts that require innovative responses. To expand on West and Zimmerman's (1987) notion of "doing gender," such circumstances create conditions in which it becomes possible to "redo" gender and, potentially, to "undo" gender, that is, to either recreate or change the daily experiences and practices of gender. Contemporary young women and men confront just such circumstances. Analyzing their coping strategies provides a lens through which to view the future contours of gender, work, and family.

STUDYING GENDER CHANGE: FINDINGS FROM A NEW GENERATION

During the last several decades, I have studied two pivotal generations. My earlier research examined how the women and men who came of age in the 1970s and 1980s helped forge changes in gender, work, and family life as they reacted to new structural and cultural conditions (Gerson 1985, 1993). My current research focuses on how the generation who grew up in these changing households and are now entering adulthood are responding to a world where nontraditional family forms predominate and gender inequality has been seriously questioned. In significant ways, the older group can be viewed as the "parents of the gender revolution" and the generation now coming of age as the "children of the revolution" (Gerson 2001). They have watched their parents cope with the erosion of the breadwinner-homemaker ethos, and they must now devise their own strategies in the face of continuing work and family change.

To discover how new generations are experiencing and responding to these vast social changes, I conducted in-depth, life history interviews with 120 young women and men between the ages of 18 and 32. They were randomly selected from a range of economic and social contexts, including inner-city, outer-city, and suburban neighborhoods throughout the New York metropolitan area. They are evenly divided between women and men, with an average age of 24, and are economically and racially diverse, with 54 percent non-Hispanic whites, 21 percent African Americans, 18 percent Hispanics, and 8 percent Asians.

Most lived in families that underwent changes that cannot be captured in the static categories of household types. That said, a large majority lived in some form of nontraditional arrangement before reaching 18. About 40 percent lived in a single-parent home at some point in their childhood, and 7 percent saw their parents break up after they left home.[3] About one-third grew up in homes where both parents held full-time jobs of relatively equal importance, at least at some point during their childhoods.[4] The remaining 27 percent described growing up in homes that were generally traditional in the sense that mothers worked intermittently, secondarily, or not at all, although most of these households underwent some form of change as mothers went to work or marriages faced crises. As a whole, the group experienced the full range of changes now emerging in family, work, and gender arrangements. . . .

NEW DILEMMAS, AMBIGUOUS STRATEGIES

How does this generation view its moral choices? As adult partnerships have become more fluid and voluntary, they are grappling with how to form relationships that balance commitment with autonomy and self-sufficiency. As their mothers have become essential and often sole breadwinners for their households, they are searching for new ways to define care that do not force them to choose between spending time with their children and earning an income. And in the face of rising work-family conflicts, they are looking for definitions of personal identity that do not pit their own development against creating committed ties to others. As young women and men wrestle with these dilemmas, they are questioning a division of moral responsibility that poses a conflict between personal development and caring for others.

Seeking Autonomy, Establishing Commitment

The decline of permanent marriage has raised new and perplexing questions about how to weigh the need and desire for self-sufficiency against the hope of creating an enduring partnership. In wrestling with this quandary, young women and men draw on lessons learned in their families and personal relationships. Yet, they also recognize that past experiences and encounters can provide, at best, a partial and uncertain blueprint for the future.

Few of the women and men who were interviewed reacted in a rigidly moralistic way to their parents' choices. Among those whose parents chose to divorce (or never marry), about 45 percent viewed the breakup as a prelude to growing difficulty, but the other 55 percent supported the separation and felt relief in its aftermath. Danisha, a 21-year-old African American, concluded that conflict would have emerged had her parents stayed together. . . .

And at 26, Erica, who grew up in a white middle-class suburb, supported her parents' decision to separate and received more support from each of them in its aftermath:

> I knew my parents were going to get divorced, because I could tell they weren't getting along. They were acting out roles rather than being involved. They were really drifting apart, so it was something perfectly

> natural to me. In the new situation, I spent more valuable time with my parents as individuals. So time with my father and mother was more meaningful to me and more productive.

Among those whose parents stayed together, almost 60 percent were pleased and, indeed, inspired by, their parents' lifelong commitment, but about 40 percent concluded that a breakup would have been better than the persistently unhappy, conflict-ridden relationship they watched unfold. Amy, a 24-year-old Asian American, explains:

> I always felt my parents would have divorced if they didn't have kids and didn't feel it was so morally wrong. They didn't really stick together because they were in love. I know all couples go through fights and stuff, but growing up, it seemed like they fought a lot, and each of them has made passing comments—like. "Oh, I would have divorced your mom by now" or "I would have left your dad a thousand times." [So] I wouldn't have broken down or been emotionally stressed if my parents divorced. I didn't want to hear the shouting, and I didn't want to see my mom cry anymore. And I was also afraid of my dad, because he would never lay a hand on my mom, but he's scary. He could be violent.

Whether their parents stayed together or parted, most concluded that neither steadfast commitment nor choosing to leave has moral meaning in the abstract. The value of enduring commitment depends on the quality of the relationship it embodies.

When considering their own aspirations, almost everyone hopes to establish a committed, lasting relationship with one partner. Yet, they also hold high standards for what a relationship should provide and anticipate risks in sustaining such a commitment. Across the divides of gender, race, and class, most agree that a satisfying and worthwhile relationship should offer a balance between autonomy and sharing, sacrifice and support. . . .

Amy imagines a partnership that is equal and fluid, capable of adapting to circumstances without relinquishing equity:

> I want a fifty-fifty relationship, where we both have the potential of doing everything. Both of us working, and in dealing with kids, it would be a matter of who has more flexibility with regard to their career. And if neither does, then one of us will have to sacrifice for one period, and the other for another.

Most acknowledge, however, that finding a lasting and satisfying relationship represents an ideal that is hard to reach. If it proves unattainable; they agree that being alone is better than remaining in an unhappy or destructive union. Building a full life thus means developing the self in multiple ways. . . .

Across the range of personal family experiences, most also agree that children suffer more from an unhappy home than from separated parents.[5] Miranda, whose parents parted when her father returned to Mexico in her teens, looks back from the vantage point of 27 and concludes,

> For people to stay together in spite of themselves, just for the child, they're damaging the child. It's almost like a false assumption that you can do something for the sake of the child while you're being drained. Because the life is getting sucked out of you. How can you give life when it's sucked out of you?

Women and men both wonder if it is possible to establish relationships that strike a good balance between self-affirmation and commitment, providing and receiving support. Having observed their parents and others struggle with varying degrees of success against the strictures of traditional gender categories, they are hopeful but guarded about the possibilities for resolving the tension between autonomy and commitment in their own lives.[6] . . .

Care as Time, Care as Money

If the rise of fluid adult partnerships has heightened the strains between commitment and autonomy, then the rise of employed mothers and the decline of sole male breadwinners have made the meaning of care ambiguous. Now that most children—whether living in single-parent or two-parent households—depend on the earnings of their mothers, parents face conflicts in balancing the need to provide economic support with the need to devote time and attention.

Rigid notions of gendered caring do not fit well with most family experiences, and the majority express support for parents who transgressed traditional gender categories. Among those who grew up in two-earner households, four out of five support such an arrangement, most with enthusiasm. Across race, class, and gender groups, they believe that two incomes provided the family with increased economic resources, more flexibility against the buffeting of economic winds, and greater financial security. . . .

Of course. this means they see a mother's employment as largely beneficial. Whether in a two-parent or single-parent home, women and men agree that an independent base enhanced a mother's sense of self, contributed to greater parental equality, and provided an uplifting model. Rachel, 24 and from a white, working-class background, explains,

> I don't think that I missed out on anything. I think it served as a more realistic model. I've heard all that stuff about how children need a parent at home, but I don't think that having her stay home with me, particularly considering her temper, would have been anything other than counterproductive. The reality is that I'm going to have to work, and a lot of women in her generation chose not to work and did or didn't have the option. She had a choice, and she did what she wanted, and I think that's really great.

Kevin, 25 and from a middle-class, white family, agrees:

> For quite a while, my mom was the main breadwinner. She was the one who was the driving force in earning money. My mother's persona was really hard working, and that's something I've strived to be with and to emulate. I didn't think it was wrong in any way. I actually feel it's a very positive thing. Whatever my relationships, I always want and appreciate people who work, and I'm talking about female involvement. It's part of who I am, and it makes me very optimistic knowing that hard work can get you somewhere.

They also deemed highly involved fathers, whether in two-earner or single-parent households, as worthy examples. Daniel, now 23, describes his Irish father's atypical working hours and parental involvement:

> My father was always around. He's a fire fighter, so he had a lot of free time. When he was home, he was usually coaching me and my brother or cooking dinner or taking us wherever we wanted to go. He was the only cook up until me and my brother started doing it. So I want to make sure that, if I get married and have kids, I'm there for my kids.

In contrast, those who grew up in a largely traditional household expressed more ambivalence. Although half felt fortunate to have had a mother devoted primarily to their care, the other half would have preferred for their mothers to pursue a more independent life. At 21, Justin, who grew up in a white, largely middle-class suburb, looks back on his mother's domestic focus with a strong conviction that it took its toll on the whole household:

> She was very involved [and] always around. And I appreciated it, but I felt guilty that maybe I was taking too much. It's just that she wasn't happy. And she didn't give us any responsibilities at all. I guess that made her feel good to have someone rely on her. She felt needed more. And in the long run, obviously that's not something good. . . .

Breadwinning fathers may also elicit mixed reactions. Their economic contributions are appreciated but not necessarily deemed sufficient. A good father, most concluded, takes time and offers emotional support as well. At 29, Nick, who grew up in a white working-class neighborhood and remembers feeling frustrated by his own father's distance, is seeking joint custody of his own young daughter:

> I have seen a lot of guys who have kids and have never changed a diaper, have never done anything for this child. Don't call yourself daddy. Even when she was saying, "Oh, she might not be yours," it didn't matter to me. This child is counting on me.

In this context, care becomes a slippery concept. Across family circumstances, these young adults judge an ideal parent—whether mother or father—to be one who supports her or his children both economically and emotionally. At 21, Antonio, who grew up in a three-generational Hispanic household and whose father died of alcoholism, has concluded that fathers should give their children the time and emotional support typically expected of a mother:

> [An ideal father] is a strong, balanced man. He's a daddy but he has the understanding of a mommy. He can care for you and protect you and guide you. . . . That's what I want to do with my kids. I want to make sure that I have time. I don't want to leave them in front of a TV set all day, because what they're learning is not coming from me. So I want to be there or, if not, I want to be in a position where I can take you with me.

If fathers should resemble traditional conceptions of mothers, then mothers should resemble fathers when it comes to work outside the home. Gabriel, a white 25-year-old who was raised by his father after his parents divorced when he was in grade school, explains,

> In terms of splitting parental stuff, it should be even. Kids need a mother and a father. And I'm really not high on the woman giving up her job. I have never wanted to have a wife who didn't make a salary. But not for the sake of leeching off of her, but so that she was independent. . . .

If such an ideal proves beyond reach, as many expect it will be, women and men agree that families should apportion moral labor however best fits their circumstances—whether or not this means conforming to classic notions of gender difference. Mothers can and often do demonstrate care through paid work and fathers through involvement. Now 26 and raising a child on her own, Crystal, an African American, rejects a natural basis for mothering:

> I don't really believe in the mother instinct. I don't believe that's natural. Some people really connect with their children, and some people just don't. I think it should be whoever is really going to be able to be there for that child.

In the end, the material and emotional support a child receives matters more than the type of household arrangement in which it is provided. Michelle, a 24-year-old of Asian descent who watched her parents struggle in an unhappy marriage and then separate after she and her brother left home, focuses on emotional support rather than family composition:

> As long as the child feels supported and loved, that's the most important thing. Whether it's a two-parent home, a single-parent home, the mother is working, or anything, it's just really important for the child to have a good, strong foundation.

Identity through Love, Identity through Work

In a world where partnerships are fragile and domesticity is devalued, young women and men are confronting basic questions about identity and self-interest. Do they base their personal well-being and sense of self on public pursuits or private attachments? What balance can or should be struck between them?

In pondering their parents' lives, most could find no simple way to define or measure self-interest. While a minority uphold traditional gendered identities, most do not find such resolutions viable. Women are especially likely to conclude that it is perilous to look to the home as the sole source of satisfaction or survival. Reflecting on the many examples of mothers and other women who languished at home, who were bereft when marriages broke up, or who found esteem in the world of paid work, 9 out of 10 express the hope that their lives will include strong ties to the workplace and public pursuits. Sarah, now a psychologist with a long-term lesbian partner who works "constantly," has high hopes but also nagging worries:

> I have a lot of conflicts now—work versus home and all of that stuff. But I would feel successful if I had a life with a lot of balance and that I'd made time for people who were important to me and made a real commitment to the people that I care about. And also, to work—I would be dedicated to work. And work and home would be connected. It would all be integrated, and it would be an outgrowth of my general way of being.

On the other side of the gender divide, many men have also become skeptical of work-centered definitions of masculine identity. As traditional jobs have given way to unpredictable shifts in work prospects, they are generally guarded about the prospect of achieving stable work careers. Having observed fathers and friends who found work either dissatisfying or too demanding, two-thirds of the men concluded that, while important, work alone could not provide their lives with meaning. These young men hope to balance paid work and personal attachments without having to sacrifice the self for a job or paycheck. Traditional views persist, but they increasingly compete with perspectives that define identity in more fluid ways. Widely shared by those who grew up in different types of families, these outlooks also transcend class and race differences. They cast doubt on some post-feminist assertions that a "new traditionalism" predominates among young women and men (Crittenden 1999). When asked how he would like to divide caretaking and breadwinning, Kevin considers the possibilities:

> Whoever can do it and whoever's capable of doing it, but it should be divided evenly. If there's something I can't do, just that I don't have the talent to do it, I would hope the other person would be able to. And the same goes the other way. My parents were like that. It was a matter of who was able to do what. There were hundreds of times when my dad made our lunches. And my sister claims that his were better than my mom's.

Yet, beyond the apparent similarities, a gender divide emerges. With one-third of men—but far fewer women—preferring traditional arrangements over all others, women are more likely to uphold flexible views of gender for themselves and their partners. More important, women and men both distinguish between their ideals and their chances of achieving them. If most hope to integrate family and work—and to find partners with whom to share the rewards and burdens of both—far fewer believe they can achieve this lofty aspiration. It is difficult to imagine integrating private with public obligations when most workplaces continue to make it difficult to balance family and job. And it is risky to build a life dependent on another adult when relationships are unpredictable. In this context, both women and men acknowledge that their actual options may fall substantially short of their ideals. For women, finding the right job and the right partner may seem too much to expect. Maria laments,

> Sometimes I ask myself if it's unrealistic to want everything. I think a lot of people would settle for something that is not what they wished, and, to me, that feels worse. It's a Catch 22, because you could wait so long, you never get anything, or you could settle for something and then be cut off from something else.

And men agree, although they are more likely to focus on the constraints of the workplace, as Peter, 27 and white, implies: "I want as even a split as possible. But with my hours, I don't think it could be very even."

AN EMERGING GENDER DIVIDE: AUTONOMY AND NEOTRADITIONALISM AS FALLBACK POSITIONS

The ideal of a balanced self continues to collide with an intransigent social world. New generations must thus develop contingent strategies for less than ideal circumstances. If egalitarian aspirations cannot be reached, what options remain? Here, women and men tend to diverge. Indeed, even as they are developing similar ideals, they are preparing for different outcomes. If an egalitarian commitment proves unworkable, most men would prefer a form of "modified traditionalism" in which they remain the primary if not sole family breadwinner and look to a partner to provide the lion's share of domestic care. Women, in contrast, tend to look toward autonomy as preferable to any form of traditionalism that would leave them and their children economically dependent on someone else.

As young women and men consider the difficulties of building balanced, integrated lives, they move from ideals to consider the fallback positions that would help them avert worst-case scenarios. Here, as we see below, the gender gap widens. Women, in hoping to avoid economic and social dependence, look toward autonomy, while men, in hoping to retain some traditional privileges, look toward modified forms of traditional arrangements. Yet, both groups hope to resolve these conflicts as they construct their lives over time.

Women and Autonomy

Among the women, 9 out of 10 hope to share family and work in a committed, mutually supportive, and egalitarian way. Yet, most are skeptical that they can find a partner or a work situation that will allow them to achieve this ideal. Integrating caretaking with committed work remains an uphill struggle, and it seems risky to count on a partner to sustain a shared vision in the long run. Even a modified version of traditionalism appears fraught with danger, for it creates economic vulnerability and constricted options in the event that a relationship sours or a partner decides to leave. Four out of five women thus prefer autonomy to a traditional marriage, concluding that going it alone is better than being trapped in an unhappy relationship or being abandoned by an unreliable partner. . . .

Autonomy for women means, at its core, economic self-sufficiency. A life that is firmly rooted in the world of paid work provides the best safeguard against being stuck in a destructive relationship or being left without the means to support a family. Healthy relationships, they reason, are based on a form of economic individualism in which they do not place their economic fate in the hands of another. Rachel declares,

> I'm not afraid of being alone, but I am afraid of being with somebody who's a jerk. I can spend the rest of my life alone, and as long as I have my sisters and my friends, I'm okay. I want to get married and have children, but I'm not willing to just do it. It has to be under the right circumstances with the right person.

Men and Neotraditionalism

Young men express more ambivalence about the choice between autonomy and traditionalism. If a committed, egalitarian ideal proves out of reach, about 40 percent would opt for independence, preferring to stress the autonomous self so long associated with manhood and now increasingly affirmed by women as well. But six out of 10 men would prefer a modified traditionalism in which two earners need not mean complete equality. This split among men reflects the mix of options they confront. Work remains central to constructing a masculine identity, but it is difficult to find work that offers either economic security or good opportunities for family involvement.

Without these supports, men are torn between avoiding family commitments and trying to retain some core advantages provided by traditional arrangements.

From men's perspective, opting for the autonomy conferred by remaining unmarried, unattached, or childless relieves them of the economic burden of earning a family wage in an uncertain economy, but it also risks cutting them off from close, committed, and lasting intimate connections. A neotraditional arrangement, in contrast, offers the chance to create a family built around shared breadwinning but less than equal caretaking. In this scenario, men may envision a dual-earner arrangement but still expect their partner to place family first and weave work around it. Josh, a white 27-year-old who was raised by his father after his mother was diagnosed with severe mental illness, asserts,

> All things being equal, it should be shared. It may sound sexist, but if somebody's gonna be the breadwinner, it's going to be me. First of all, I make a better salary. If she made a much better salary, then I would stay home, but I always feel the need to work, even if it's in the evenings or something. And I just think the child really needs the mother more than the father at a young age.

Modified traditionalism provides a way for men to cope with economic uncertainties and women's shifting status without surrendering some valued privileges. It collides, however, with women's growing desire for equality and rising need for economic self-sufficiency.

Resolving Moral Dilemmas over Time

In the absence of institutional supports, postponing ultimate decisions becomes a key strategy for resolving the conflicts between commitment and self-development. For women as much as men, the general refrain is, "You can't take care of others if you don't take care of yourself." Michael wants to be certain his girlfriend has created a base for herself at the workplace before they marry, hoping to increase the chances the marriage will succeed and to create a safety net if it fails:

> There are a lot of problems when two people are not compatible socially, economically. When Kim gets these goals under her belt, and I have my goals established, it'll be a great marriage. You have to nurture the kind of marriage you want. You have to draw it out before you can go into it.

For Jennifer, 19 and white, autonomy also comes first. Commitment may follow, but only when she knows there is an escape route if the relationship deteriorates:

> I will have to have a job and some kind of stability before considering marriage. Too many of my mother's friends went for that—let him provide everything—and they're stuck in a relationship they're not happy with because they can't provide for themselves or the children they now have. The man is not providing for them the way they need, or he's just not a good person. Most of them have husbands who make a lot more money, or they don't even work at all, and they're very unhappy, but they can't leave. So it's either welfare or putting up with somebody else's crap.

Establishing an independent base becomes an essential step on the road to other goals, and autonomy becomes a prerequisite for commitment. This developmental view rejects the idea that individualism and commitment are in conflict by defining the search for independence as a necessary part of the process of becoming able to care for others. To do that, women as well as men tend to look to work, and its promise of autonomy, to complete the self. For those with children as well as those who are childless, lifelong commitments can be established when "you feel good enough about yourself to create a good relationship." Shauna, a 30-year-old African American who was raised by her mother and stepfather, explains,

> If you're not happy with yourself, then you can't be happy with someone else. I'm not

> looking for someone to fill a void. I think that's what a lot of people do when they look for relationships, and that's not what it's about. It's about sharing yourself with the other person, and when you're content and happy with who you are, then you can give more of yourself to someone else, and that's the type of person that I want to be with.

These strategies are deeply felt and intensely private responses to social and personal conflicts that seem intractable. More fundamental solutions await the creation of systematic supports for balancing work and family and for providing women and men with equal opportunities at the workplace and in the home. Without these supports, new generations must cope as best they can, remaining both flexible and guarded. Andrew, a white 27-year-old, has concluded that rigid positions are not helpful in an unpredictable world:

> I would like to have an equal relationship, but I don't have a set definition for what that would be like. I would be fine if both of us were working and we were doing the same thing, but it would depend on what she wants, too. If she thought, "Well, at this point in my life, I don't want to work," or if I felt that way, then it would be fine for one person to do more work in some respects. But I would like it to be equal—just from what I was exposed to and what attracts me. . . .

CONCLUSION: TOWARD A NEW MORAL ORDER?

Deeply rooted social and cultural changes have created new moral dilemmas while undermining a traditional gendered division of moral labor. The widespread and interconnected nature of these changes suggests that a fundamental, irreversible realignment is under way. Less clear is whether it will produce a more gender-equal moral order or will, instead, create new forms of inequality. The long-term implications are necessarily cloudy, but this ambiguity has created some new opportunities along with new risks.

While large-scale social forces are propelling change in a general direction, the specific forms it takes will depend on how women and men respond, individually and collectively, to the dilemmas they face. Those who have come of age during this period are adopting a growing diversity of moral orientations that defies dichotomous gender categories. Their experiences point to a growing desire for a social order in which women and men alike are afforded the opportunity to integrate the essential life tasks of achieving autonomy and caring for others.

Yet, persistent inequalities continue to pose dilemmas, especially for those who aspire to integrate home and work in a balanced, egalitarian way. To understand these processes, we need to focus on the social conditions that create such dilemmas and can transform, and potentially dissolve, the link between gender and moral responsibility. Of course, eradicating this link might only mean that women are allowed to adopt the moral strategies once reserved for men. We also need to discover how to enable everyone, regardless of gender, class, or family situation, to balance care of others with care of self.

The possibilities have never been greater for creating humanistic, rather than gendered, conceptions of moral obligation. New moral dilemmas have prompted women and men to develop innovative strategies, but the long-term resolution of these dilemmas depends on reorganizing our social institutions to foster gender equality and a better balance between family and work. Freud once commented that a healthy person is able "to love and to work." Achieving this vision depends on creating a healthy society, where all citizens are able to combine love and work in the ways they deem best.

NOTES

1. A vivid example of a "damned if you do and damned if you don't" situation can be found in Hays's (1996) analysis of the "cultural contradictions of motherhood," in which women are expected to practice

intensive mothering even as they seek a life outside the home.

2. In Giddens's (1979) language, they cause people to move from "practical consciousness" to "discursive consciousness." Behavior becomes action because it has new social meaning.
3. Of this group, more than 27 percent lived largely with a single parent, including 7 percent whose parents shared joint custody and 5 percent who lived with single, custodial fathers. The rest saw one or both of their parents remarry and form a new, two-parent household.
4. A larger proportion of households were dual-earning, but they varied in the degree of equality between parents' jobs and did not necessarily include both biological parents.
5. Amato and Booth (1997) confirmed this viewpoint. Respondents also argue that both parents should sustain strong ties to their children whether or not they remain together.
6. Cancian (1987) provided an in-depth analysis of innovative attempts among couples to create interdependent relationships, in which both women and men are responsible for love.

REFERENCES

Connell, R. W. 1987. *Gender and power.* Stanford, CA: Stanford University Press.

———. 1995. *Masculinities.* Berkeley and Los Angeles: University of California Press,

Crittenden, Danielle. 1999. *What our mothers didn't tell us: Why happiness eludes the modern woman.* New York: Simon & Schuster.

Epstein, Cynthia F. 1988. *Deceptive distinctions: Sex, gender and the social order.* New Haven, CT: Yale University Press.

Gerson, Kathleen. 1985. *Hard choices: How women decide about work, career, and motherhood.* Berkeley and Los Angeles: University of California Press.

———. 1993. *No man's land: Men's changing commitments to family and work.* New York: Basic Books.

———. 2001. Children of the gender revolution: Some theoretical questions and findings from the field. In *Restructuring work and the life course,* edited by Victor W. Marshall, Walter R. Heinz, Helga Krueger, and Anil Verma. Toronto, Canada: University of Toronto Press.

Hochschild, Arlie R. 1997. *The time bind: When work becomes home and home becomes work.* New York: Metropolitan Books.

Jacobs, Jerry A., and Kathleen Gerson. Forthcoming. *The time divide: Work, family, and social policy in a hurried era.* Cambridge. MA: Harvard University Press.

Lorber, Judith. 1994. *Paradoxes of gender.* New Haven CT: Yale University Press.

Putnam, Robert. 2000. *Bowling alone: The collapse and revival of American community.* New York: Simon & Schuster.

Risman, Barbara J. 1998. *Gender vertigo: American families in transition.* Hew Haven, CT: Yale University Press.

Schor, Juliet. 1992. *The overworked American: The unexpected decline of leisure.* New York: Basic Books.

Webster's dictionary of the English language. 1992. Chicago: J. G. Ferguson.

West, Candace, and Don H. Zimmerman. 1987. Doing gender. *Gender & Society* 1:125–51.

41

Mothering, Work, and Gender in Urban Asante Ideology and Practice

BY GRACIA CLARK

This description of Ghanian culture by anthropologist Gracia Clark provides an insightful look an different ways in which motherhood, fatherhood, and parenting can be defined. Clark describes these relationships in urban Asante communities, a matrilineal society. In matrilineal societies the lineage or lines of descent follow the mother's family. She describes how this group maintained basic cultural ideals about family relationships within the framework of Western influences. We can learn a lot about our own culture as we explore this system of relationships, values, and norms.

1. How does the Asante definition of working mother differ from that in Western societies?
2. What is the father's relationship to the child and how do his financial responsibilities differ from that of the mother's?
3. How do men's and women's responsibilities in marriage and to their families differ from that in Western cultures?

★ ★ ★

THE WORKING MOTHER IDEAL

Asante market traders work at least as long hours as the double day faced by U.S. working mothers, but without the same supposed guilt. The good, self-sacrificing Asante mother does not stay home with her children, but goes out working hard for them. Asante women explain their financial obligations to feed their children as the dominant bond of motherhood in everyday life. Women in the archetypal women's occupations of rural food farming and urban trading can feed their children directly with their produce, but those in other jobs nonetheless say they work to look after their children. Several traders mentioned that they did not need to settle down and work seriously until they began having children to look after. The subsistence or commercial farming of Asante village women is likewise considered work, not homemaking.

Biological motherhood remains a key responsibility, but one that logically or naturally mandates income-generating work rather than personal responsibility for childcare. Women who were describing their trading careers to me in 1994 and 1995 used the phrase "and that's how I had Kwame, Abenaa, and Kofi" about specific kinds of work they did. At first I thought that these women were simply remembering associated pregnancies or births as a convenient way of dating specific periods in their own lives. However, the most frequent way of specifying times in the past used a slightly different phrase.

"At that time, I carried Kwame on my back, *saa bere no, na medi Kwame hye me kyi.*" A narrator trying to explain exactly when she moved to a new town and tried a new occupation, in that example, recalled which child she had carried on her back then. The first statement actually uses the same word *wo* used for birth and conception, *"Saa na mewo Kwame."* This sentence merges, without exactly conflating, the physical and economic processes of creating and raising children.

Asante ideals and practices underline the importance of economic support in enacting motherhood, as a continuation of childbirth itself. Pregnancy and childbirth are essential parts of motherhood, painful and risky ones that make birth mothers irreplaceable and entitled to unconditional support. But these relatively brief initial biological stages are fruitless, in the final analysis, without the capacity to complete the process by raising the baby to adulthood. Successful reproduction means producing someone "to replace yourself," as a functioning Asante adult with children. Both daily survival and social sponsorship require financial investments that many women must struggle hard to provide, and whose outcome is as uncertain as pregnancy. Providing adequate economic support is in fact more difficult and problematic for most Asante women, given their life circumstances, than providing childcare and a good moral example.

An Asante mother does accept the responsibility to make sure that her children are well cared for, but she does not feel that a mother would care for her children better than another attentive, capable relative of reasonable age might. One mother explained that "no one will sit and listen to a child cry," so anyone around will see that they have water, food and shelter, but "no one will work for them like I do." Her devotion to her children, far from making her feel ambivalent about working, drove her to work harder and longer. Taking time off work to raise children seems almost a contradiction in terms.

In several interviews, I asked about a possible positive choice to stay at home. These questions elicited responses about mothers who might be deaf, retarded, or otherwise mentally or physically handicapped. Indeed, I met one deaf woman who was taking care of the combined children of several of her sisters in the family home, supported out of her sisters' trading incomes. They recognized her as intelligent and capable, and she had successfully married and had several children of her own. I also encountered several older women without handicaps who had likewise stayed home to supervise their own and several sisters' children, supported by the cash income of the other sisters.

The concept that paid work is an integral part of motherhood was also expressed by the African American women hospital workers interviewed by Karen Sacks (1988). They considered their jobs an extension of their maternal responsibilities, along with their unpaid housework at home. Wage work demonstrated, rather than contradicted, their willingness to take care of their children by any means necessary. They stressed this continuity more than the cleaning and organizing skills they consciously employed in both locations. Of course, there was no lineage organization available to provide an ideological framework for them, but their responsibilities were acknowledged and respected within the African American community. Still, disrespect for these roles within the dominant white community helped keep their work classified as part-time or casual, with less security and upward mobility.

MOTHERING CHILDREN

Asante ideology endorses economic responsibility as the emotional heart of motherhood as well as its practical bottom line. In the absolute bond between mother and child, the conditionality and renegotiability so conspicuous in the fluid relations between Asante kin and spouses are supposedly absent. Individual autonomy and negotiated mutual aid dominate Asante ideals of both lineal and marital relations, instead of the pooling of income, labor, or assets, for central decision making so central to many universalist models of the household (Hammel 1984; Laslett 1972)[1] Children's loyalty to their mothers is explicitly rooted in the economic merging or fusion expected and experienced in this relationship alone. Since this economic support is the hardest to delegate or replace, it becomes the most definitive of maternal devotion and a greater focus of maternal identity and child bonding.

The popular song "Sweet Mother" expresses this sentiment by incorporating the common proverb: "When I don't eat, she doesn't eat." This phrase resonates on at least two levels. One is literal, that if she had food you would also be eating it. Sharing food shortages is far from a trivial issue for the many mothers facing chronic financial crisis. As a metaphor, the phrase also evokes complete emotional identification, sharing your worries and sorrows. For Asante, persons grieving at a close relative's funeral or facing some other upsetting situation show strong feelings by not wanting to eat, while others refrain out of respect for those feelings.

Such tight mother/child bonding is naturalized by Asante culture into an inevitable biological response that should last through life, so that deviations are considered unnatural. The bond between mother and child is seen as so strong that virtually nothing can dissolve it. The biological event of childbirth establishes a culturally undeniable debt to the mother for her pain and blood and the invaluable gift of life. This debt can never be repaid or canceled, but must be honored by passing it on to one's own children. In this matrilineal society, the structural importance of maternal ancestry makes it an ineradicable part of personal identity, even when the biological mother dies in childbirth or abandons the child. This fact of birth need not be reconfirmed by childhood experiences, and long fostering never erases it, but it also cannot grow through reciprocation or negotiation like other kin links.

The maternal bond normally does strengthen throughout childhood, maintained as a live relationship through continual interaction. Domestic work by growing children seems to intensify Asante bonding more than domestic work by mothers. One mother explained her painful and continuing grief for the death of a daughter at age seven or eight by describing how the girl had been so helpful around the house, compared to babies she had lost at younger ages. Mothers' hard work supporting the family is the dominant image in the powerful child attachment that remains into adulthood. Most children experience the physical intimacy of childcare with siblings and other relatives, and delegating primary childcare is not seen as a threat to maternal-child bonding. By contrast, delegating cooking, even to a husband's own daughters, is seen as easily weakening and endangering the marital relationship (Clark 1989a).

The cases showing the fullest substitution of maternal relationships were when childless women had adopted one of their sister's infant daughters or when motherless orphans had joined an aunt's family at an early age. In these cases, either the living birth mother or the living birth children were absent. These older women relatives would have been called "mother" by those children in any case, but now took emotional priority over the birth mother. Mother and child did not erase or conceal knowledge of the birth relationship, as is standard in U.S. adoptions, but they rarely mentioned it.

The weak relation between physical childcare and maternal child bonding is the more remarkable because childcare is considered the basis for warm emotional ties to other relatives. It is recognized as creating special bonds between a baby and its primary caretakers, most often older sisters. Men and women both recognize a special debt to "the sister who carried me on her back," even compared to other siblings. The physical intimacy of childcare establishes warm loyalty to men as well as to women, reflecting Asante boys' and mens' noticeable willingness to carry around and play with infants and toddlers, as well as older children.

Children are also fostered out to various relatives for a number of reasons, especially to grandparents, aunts, and cousins.[2] Strengthening the child's relationships with these relatives is seen as a useful side effect of fostering, and sometimes the primary reason for it. Children also go live with relatives to receive or provide childcare, cook, run errands for elders, and to receive vocational training or education. In these cases, it is the children's work (domestic or income-generating) that generates the bonding, rather than the foster parent's childcare.

MALE PARENTS

The stereotype of economic and emotional merging between mother and child is invoked so powerfully for mothers that, by implication, fathers, uncles, or siblings may not show the same degree of concern. Their support is more conditional and variable, like relations between other relatives and spouses. Financial support in childhood or young adulthood is important in bonding children to fathers and male kin, as with mothers. In order to claim paternity publicly, the biological father should participate in the baby's naming ceremony by presenting a cloth and other gifts that at least symbolically support the mother during her resting period after birth. Children often solicit school fees and other important support in person, so that fathers who consistently refuse or disown their children face old age with little reciprocation (van der Geest 1997).

Fatherhood does have a more absolute spiritual aspect, from the precolonial *ntoro* affiliation passed on from father to child. Just as the blood of childbirth shows the shared blood, or *mogya,* of the matrilineage, or abusua, so a person's bones (white like semen) serve as the biological referent for the paternal ntoro spirit. Without a close relationship with the father, a child lacks the protection of the father's ntoro spirit. This is said to be especially important for the health of very young children whose own ntoro spirit has not yet grown strong. That spiritual affiliation is not transferable to other male sponsors, although a father's paternal brothers would have the same one. While Asantes usually confirmed this metaphysical association when asked in 1995, they spontaneously referred to the bodily intimacy of infancy as the epitome of fatherhood. The formulaic reason for supporting fathers in old age is "I peed on his knee." Of course, this does presume the father was present. Christian beliefs also lend strong spiritual reinforcement to the father as head of the family, by analogy to God the Father.

Ideally the father's brothers (his matrilineal successors) take over full paternal financial responsibility if the father dies or becomes incapable. I did know several successors in the 1970s who took such claims seriously, but the standard of support given is expected to be markedly less than for their biological children. One man in his fifties praised his own uncle as a near saint, because he had educated him and his siblings exactly like his own children, after his father died.

Maternal and paternal uncles represent valued alternative sources of financial support and moral guidance, even when the father is still alive. According to matrilineal principles, the mother's brother is as much her reproductive partner as her husband (Mitchell 1966). In a structural sense this uncle is the male mother, like the father's sister is the female father. A woman raises her children for the lineage her brother shares, as her husband will remind him (Maxwell Owusu, personal communication, 1993). Male financial responsibility for children can thus be spread over several individuals, forming a range of potential substitute or supplementary fathers (Douglas 1969).

PRAISES AND INSULTS ACROSS THE GENDER FRONTIER

The contrast between fatherhood and motherhood is rigid enough to remain consistent even when used across gender lines. When fathers show the devotion associated with motherhood they are praised as good mothers, not fathers. One young woman described her father as a very good one and, in the next breath, its a real *obaatan,* or nursing mother. He proved his unusual level of love and concern, not by staying home like a U.S. "househusband," but by his eagerness to provide for all of his children's needs as soon as he heard of them. His children by both wives enjoyed the same willing financial support; none of the usual nagging for school fees here. This economic motherliness made him more, not less, of a father. Among Jamaicans of African descent, mothers who took this economic role were conversely respected and praised for being like fathers (Clarke 1966).

Asante apparently saw nothing odd about using the word *obaatan* for a man, if he showed the appropriate qualities. I even heard the Christian God called a "real nursing mother," or *obaatan paa,* in public prayer without raising an eyebrow in a fairly conservative audience. It was not a frequent usage, but familiar to those I asked about it. Only a few elderly informants knew an archaic male version of the term *obarimatan,* but they agreed that it was never used now and was previously very rarely used.

There are few ethnographic examples of such a central positive gender concept from one gender being applied to a member of the other without ambivalence. One close parallel is found in a discussion of chiefship ideals with a male chief who was serving under a female paramount in Sierra Leone (MacCormack 1972). He said that a chief is like a mother, "because she settles our quarrels and tells us what we should do." Strathern (1980), concerned with establishing the situational character of gender in a New Guinea case, argues that this rather detaches it from bodily sex markers like genitalia. Male traits remain positive and female traits negative, regardless of who takes them on. Asante attribute positive as well as negative female traits to men. However, the transfer to women of core male traits is more frequent than the reverse, and this does spark some ambivalence (Clark 1994).

In Asante, the male gender ideal most often transferred from men to women is the positive capacity for economic self-aggrandizement. This mirrors the transfer of economic motherhood from mother to father and likewise leaves behind the specifically sexual or biological content of the original term. The phrase *obaa barima,* which could be translated as "manly or brave woman," uses the word *obarima* that refers to young male bravery and strength, but is also the most polite positive way of referring to sexual virility. The sexual conduct of an obaa barima was not suspect, although her subservience to her husband might be reduced. In the same way, the nursing-mother father was no less male in the positive senses of sexual and business competence.

The market women I heard called obaa barima in 1979 and 1994 were those who had achieved the level of financial success and economic independence considered essential for men, not those showing unusual physical strength, bravery, or sexual prowess. The Asante image of womanly beauty already includes more strength than in Western beauty ideals, consonant with historic female responsibilities for farming, so physically strong women did not attract particular comment. One market women was known as unusually loud and aggressive in personal interactions, even by market standards, but people called her "tough," not obaa barima.

The separation kept between sexuality and economic activity in parenthood contradicts their close linkage in the relations of marriage. If praises often define gender in terms of the economic sphere, negatively gendered insults also indicate strong connections between sexual and economic power. Ama Ata Aidoo reveals the linkage succinctly, if sarcastically, with this categorical statement:

> In Ghanaian society, women themselves believe that only two types of their species suffer—the sterile, that is those incapable of bearing children—and the foolish. And by the foolish, they refer to the woman who depends solely on her husband for sustenance. [Aidoo 1970:x]

The force of this judgment depends partly on understanding the implied sexual insult behind the Twi word translated as *fool (kwasea).* It is a very common insult, but a strong one, especially when aimed at a man, as it often is. It carries a strong connotation of sexual impotence, one of the most damaging accusations for a man and the most visible form of sterility. In this one sentence, therefore, Aidoo associates and equates voluntary female economic impotence with both female reproductive impotence and male sexual impotence, both presumably involuntary.

Even stronger male insults continue this conflation of economic and sexual inadequacy, often translated for me in conversations as the "useless" husband. He literally may be a person who does nothing, *oyehweeni,* or even worse, an empty person, *onipa hunu.* I was told that this last phrase was so terrible that a woman would never use it in her husband's hearing without expecting both violence and divorce to result. These insults carry sexual and biological connotations not explicit in the words themselves.

Conversely, the term used in praise of economic motherhood contains biological and domestic references to childbirth, childcare, and cooking. Both the literal Twi meaning of *obaatan* and its usual translation into English lack any mention of paid work. Literally, it refers to the period of seclusion after childbirth, ideally 40 days. The woman, or *obaa,* stays inside the compound near the hearth, or *otan,* keeping herself and the new baby warm and protected while she recovers from birth, until the baby's outdooring ceremony. The word *obaatan* thus addresses the pain and danger she has just survived, rather than her long-term devotion.

The term makes no verbal reference to nursing, although the mother would certainly be nursing her baby then and for about a year to come. She also is more likely to take full physical care of the baby during this period, since she has already suspended her usual work schedule to rest. Nonetheless, speakers knowing English invariably used the English term *nursing mother* to translate it, and those who did not speak English pantomimed holding a baby to their breast, for instance when I asked, "Did you just say your father was a real nursing mother?" The association of obaatan behavior with work, rather than with physical childbirth or nursing, facilitates its application to men.

NURSING-MOTHER WORK AND PROPERTY

The most common use of the word *obaatan,* on the other hand, refers not to that seclusion or rest period but to what follows, when the mother has returned to work with renewed motivation. People usually spoke of new mothers at home with the phrase *wawo foforo,* literally "she has newly given birth." The hearth and kitchen, while not strictly private places in the usual shared or compound house, are considered decidedly less public than the street or the market. Several proverbs that apply to both men and women contrast the home with the street. The stereotypical obaatan was hard at work, precisely back in the street, the market, or even on the road.

She was off doing *obaatan adjuma,* or nursing-mother work, just as she should be. This was defined not by its location, at home, or its occupational content, light or otherwise feminine, but by its purpose, single-minded devotion to the child. Because of this purpose, nursing-mother work absolutely had to provide a steady, reliable daily income to feed the children, the more urgently after she stopped actually nursing. The priority placed on low risk meant such work was also likely to employ a low level of capital and give a relatively low rate of return, even on that small investment.

It is capital accumulation, not income-generating work, that is the gendered category for Asante. A real man or *obarima paa,* (by which most Asantes imply an Asante man) should aspire beyond nursing-mother work to a more lucrative occupation that enables him to accumulate. A real woman aspires beyond this also, but there is less expectation that she will succeed. Some women traders did manage to build houses or accumulate other property, but many others bemoaned their inability to leave such *gyapaadee,* or inheritance, for their children.

Asante values encourage and celebrate the achievement of wealth by women as well as men, but they assume that men, on average, will have more. This affects lineage relations as well as marital relations. A popular proverb says "men bring money to the lineage, and women bring children." When lineages assess contributions per capita for funerals, court costs or other joint expenses, each woman owes half of what a man owes. People justify this diffential burden by explaining how children limit women's earning capacity. They mention the constant drain of spending money on the children, not the time constraint of caring for them, as the factor that restricts women's business growth and limits their income. The ideal division of financial responsibilities between mothers and fathers makes mothers responsible for daily subsistence. Fathers pay for major items such as school fees or new houses, which come later in the children's lives and can be done without if necessary.

The nursing-mother work a woman does helps her husband to accumulate faster because the woman uses her own income or farm produce to feed his children and himself. The husband can then use his income and any possible inheritance for reinvestment in property that remains his, not subject to routine subsistence demands. This dynamic is sharply visible in cocoa farming areas, where women's farms are fewer and smaller, but applies equally to urban commercial enterprises (Mikell 1989; Okali 1983). Women's access to this kind of food subsidy is rarer. They can establish a stable accumulative dynamic if they begin early enough in life, often before marriage, and can still rely on their mothers for free food. These fortunate and energetic young women may later have enough capital that its derived income can continue accumulating after their children enter school and begin to require serious investment in school fees.

The ambivalent relationship between motherhood and work lies in this competition for income. Motherhood demands work but also constrains it, not through the demands of childcare or other domestic work, but through a financial demand for daily income that can preclude capital accumulation. At the same time, the most devoted mother needs to carefully balance the immediate needs of her children against their long-term interests, since she also knows that her own accumulation is the best way to provide long-term security for them and higher levels of schooling in the future. An avowed motive of taking care of their children provides a cloak of respectability for women's wealth and even greed. Higher aspirations for their education, reaching to university education abroad, can justify almost infinite accumulation. Children's financial security is an extension of their mothers', whereas men's accumulation is more likely to compete with child support both in the short term, as personal aggrandizement, and in the long term, through matrilineal inheritance by a brother or nephew.

MARRIAGE AND SEXUALITY

Asante women are severely criticized for putting the interests of marriage and husband before those of their children, which would be both immoral and stupid. Individual men as husbands cannot and should not be relied upon, because of circumstances beyond their control as well as possibly unsuspected moral failings. Added to the legitimate prior claims of their matrilineal kin are the ever-present dangers of death, illness, divorce, or polygyny, any one of which might abruptly end their financial support of wife and children. A woman who ignored this common knowledge was considered either below normal intelligence or an irresponsible mother, because she was not taking the most ordinary precautions for her children's future.

The practical and moral issues here merge in the popular proverb "you can get a new husband, but not a new brother." I often heard it repeated, in the male or female version, to explain the priority Asante give to matrikin. Women expect their brothers to remain loyal, since lineage continuity depends on the sister's children. When one elderly woman wanted to express her close relationship with a sweetheart of her youth, she said "we were like twins," that is, brother and sisters. Armah puts it more poetically, but just as forcefully:

> A father is only a husband, and husbands come and go; they are passing winds bearing seed. They change, they disappear entirely, and they are replaced. An uncle remains. [Armah 1970:139]

After remarrying in middle age, men can and often do beget a new set of children who may displace the first wife and her children emotionally and financially. This specter also weakens children's confidence in a father's absolute commitment to them. A mother's excessive attachment to her husband implies a self-indulgence in sexual gratification or vanity almost as obvious as in adultery.

The opposition between motherhood and sexuality here seems at first to resonate with Western ideals of the asexually pure (if not virgin) Mother. But Asante sexuality and its relational counterpart, marriage, maintain some separation from motherhood, with its lineage context, despite the acknowledged sequence of sex and conception. This configuration of roles does not require a diminuation or devaluation of women's sexuality per se, or of wives' sexuality in relation to their husbands.

Far from repressing sexuality, Asante culture may go even farther than U.S. culture in considering sex as necessary to health for both men and women. This was brought home to me during my first fieldwork trip, when several of my middle-aged aunties visited me during a repeated mild malaria attack that was causing visible weight loss. "You know why you are getting ill," they began delicately, but they had to explain more baldly to this obtuse foreigner. The long separation from my husband, who had visited me in the field, was the obvious problem. "We know that you people are very faithful to your husbands, and we respect that, but you shouldn't take it so far as to make yourself sick. There are lots of nice men here." On another occasion, equally respectable matrons remarked jokingly that they could not imagine going without sex for more than two or three months—for instance, if their husbands migrated to Nigeria, as many Asante men then did.

Despite strict historical and contemporary sanctions against adultery, backed by Christian and indigenous religion, there was little support shown for long-term fidelity to marriage without regular sexual satisfaction. Neither older nor younger traders expected a woman to remain in a marriage where age, chronic illness, or neglect meant no prospect of sex. For one thing, she would be wasting years of her reproductive potential, to broach a subject much more unmentionable than sexual activity. Pronatalist values entitle older women to regular sex right up until menopause, since they might well want additional children.

Nor does this attitude mean that Asante women have a purely practical, procreative attitude to sex, with no ideas of romantic love. There is a lively tradition of adolescent courtship, placed within the classic stereotype of the farm village, where moonlit nights become the "white nights" of first love or early marriage, because the pair get so little sleep. But this exciting period should come to an end with full adulthood, at least by age thirty or so. For one thing, adults should have settled down with an adequate steady partner by now and be getting regular sex without so much fuss. By this age, both men and women should have more important things to think about, such as business, property, and school fees for their children. Those who self-indulgently continue to pursue sexual adventures suffer from a kind of arrested development, and I heard specific insults for this, for both men and women.

Since celibacy is considered unhealthy for both sexes, the self-sacrifice involved in postpartum abstinence is even more significant. The only circumstance in which a woman expects to voluntarily deny herself sex for a considerable length of time is, significantly enough, when she is nursing each baby. Powerful values support birth spacing of about two years, since early pregnancy amounted to a death sentence for the first child from premature weaning, and for the second from its likely small size. The high price of infant formula and its sanitary dangers, fortunately well known to Kumasi mothers, made them respect the protection proper birth spacing provides for breastfeeding.

When pregnancies came closer together than two years, both the mother and father drew criticism as weak or undisciplined people and selfish, neglectful parents. Censorious neighbors compared one young couple who had four children aged four and under (including twins) to rabbits, although they were devout Christians who had deliberately planned their family that way. A lurid poster in the Kumasi Central Post Office in 1994 depicted a selfish modern woman who refused to breastfeed her baby in order to look young for and resume sex with her husband, who was shown sucking at her breast. The cautionary results were diarrhea for the baby and breast cancer for the mother. Many Kumasi women said they distrusted medical contraceptive methods for their side effects and unreliability. Some attributed their increasing popularity to the pressures of Christian monogamy, which left self-indulgent husbands with no accepted alternative sexual outlet.

The connection between nursing mothers and sexual abstinence also had some positive repercussions for women. Women traders sometimes mentioned that they could travel more freely when nursing a baby, since they had fewer marital duties and greater immunity from sexual harassment away from home. Even if a stranger did not respect her person, he would not be cruel enough to threaten her child so directly. In fact, some women referred to the nursing mother as a sexy image, a kind of forbidden fruit, who was evidently sexually experienced and presumably missing it since the child was born.

* * *

CONCLUSION

Tracing the thread of gender through these various arenas of Asante life, each with its own dynamic of contestation, reveals an intricate web of contradiction and confirmation at each intersection of arenas. The balance of competing demands from marriage and motherhood, or from investment in children or business, continually shifts at the individual and societal levels in response to internal dynamics as well as external economic and ideological pressures. Some of the relations, between trading and childcare or trading and capital accumulation, for example, show historical trends that have little to do with gender directly, but whose ramifications affect gender relations in disparate arenas such as marriage and school. The constant renegotiation of all these gendered ideals and practices, far from indicating their weakness or imminent disappearance, signals that they remain valuable sociocultural assets that give powerful leverage still needed to meet new challenges in contemporary life. Among Asante, this renegotiation process need not even be much disguised, since cultural norms endorse it as an inevitable and desirable fact of life. Personally remembered change does not prevent Asante from naturalizing as biological mandates the current status of these various historical processes, but it does seem to help them accept further changes with some equanimity.

The preeminence of economic issues for Asante in many of these contestations reminds analysts that sexuality is not always the most salient aspect of gender practices or ideas. Anthropologists have frequently pointed out instances where economic, political, or ritual relations can be firmly gendered without necessary implications for sexual practice or identity (Amadiume 1987; Strathern 1980). As in Asante, this detachment from sexuality facilitates the transfer of gendered attributes or statuses from men to women and vice versa. Third World feminists have also repeatedly criticized the reliance on sexuality to indicate femininity and liberation alike, linking it to hypersexualization of contemporary Western media and commercial life, starting as early as 1972 (Davis 1989).

These issues have arisen most often in contexts where specific gender relations are being discussed as deviant, unusual, or exceptional. Amadiume condemns sexualized interpretations of the Igbo female husband (Amadiume 1987). For Angela Davis's Egyptian interlocutors, the issues were veiling, seclusion, and female circumcision (Davis 1989). Studies of "third genders" or transgendered persons in Native American cultures also suggest that occupation, dress, and other criteria were more central than sexual orientation or sexual practice to establishing the gender minority status of these individuals (Blackwood 1984; Whitehead 1981). By contrast, for Asante the importance of economic factors in gender applies to parenting and trading relations that express and conform to expected majority gender roles, as well as to deviance (positive or negative). Without devaluing sexuality for women or men, the economic performance of gender appears more central and more notable.

In analyses of parenting in particular, the biologically rooted relationship with birth mothers and fathers is tacitly accepted as the primary model or norm. Insightful discussions of the nonsexual aspects of gender roles come from the literature on child fostering. The absence of biological relationship with the foster parent makes the emphasis on other aspects of the important gender-linked roles of mother and father both justified and somewhat deviant. Pioneering studies of fostering by Goody spelled out the separability of biological and economic aspects of parenting (Goody 1978). Weismantel describes nurturing and financial support as creating virtual biological parenthood in the Bolivian Andes, where a man feeding and caring for an unrelated boy said this would make him his father after several years, even though the birth father was still alive back in the village (Weismantel 1995). Discussions of "othermothering" in African American communities also mention highly valued economic as well as emotional nurturing from women who are not the birth mothers (Collins 1990; James 1993).

In the case of urban Asante, economic mothering also takes precedence in the daily lives of ordinary biological mothers, whether single or married. The economic implications of mothering become a central aspect of gender, whether in the lineage, marriage, or workplace, for better or for worse. Competing loyalties in both matrilineal and marital relations find their most consistent expression in decisions about financial support. Hostility and distrust between men and women focus on economic relations more openly than on sexual relations, so that financial tensions bear a displaced sexual charge (Clark 1989a). Praises and insults about gender conformity and deviance often feature economic performance, although they are far from silent on sexual performance.

The Asante experience also suggests that cultural endorsement of economic mothering and consensus about the structure of negotiation among competing claims from motherhood, wifehood, and trading do not eliminate significant material and ideological conflict. Instead, gender issues still play a key role in structuring other arenas of conflict. In multicultural societies like the United States, subcultures based on ethnicity, race, and gender show different norms and practices from the hegemonic ideology or "master narrative" of family and are stigmatized and penalized because of that. Despite the hegemony of Akan family patterns within Ghana, and the flexibility and negotiability those norms endorse, bitter public and private conflict is all too frequent over how to fulfill the three central, conventional gender roles of wife, mother, and trader. Internal structural tensions, in this case between wifehood and motherhood, sharpen under pressure from local and international economic crisis. In return, tensions over family loyalties intensify national conflicts over commercial policy and wealth differences (Clark 1988). Teasing out the gendered connections between arenas of conflict distinct from sexuality, marriage, and child rearing, the intimate core of much gender analysis, turns out to be an essential step for understanding either the overall configuration of gender or the course of political and economic conflicts.

NOTES

1. These issues of comparative household structure are addressed at more length in Clark (1989b).
2. The variety of arrangements for fostering and sibling childcare is discussed in an extensive and sophisticated literature. Conveying even a general sense of their emotional, economic, and labor implications is beyond the scope of this paper. The wide range of fruitful approaches can be seen in several edited volumes, including Bledsoe and Cohen (1993); Goody (1982); and Scheper-Hughes and Sargent (1998).

REFERENCES CITED

Aidoo, Ama Ata
1970 No Sweetness Here. London: Longman.

Amadiume, Ifi
1987 Male Daughters, Female Husbands. London: Zed Press.

Armah, Ayi Kwei
1970 Fragments. Boston: Houghton Mifflin.

Blackwood, Evelyn
1984 Sexuality and Gender in Certain Native American Tribes: The Case of Cross-Gender Females. Signs: Journal of Women in Culture and Society 10:27–42.

Bledsoe, Caroline, and Barney Cohen, eds.
1993 Social Dynamics of Adolescent Fertility in Sub-Saharan Africa. Washington, DC: National Academy Press.

Clark, Gracia
1989a Money, Sex and Cooking: Manipulation of the Paid/Unpaid Boundary by Asante Market Women. *In* The Social Economy of Consumption. B. Orlove and H. Rutz, eds, Pp. 323–348. Monographs in Economic Anthropology No.6, Society for Economic Anthropology. Lanham, MD: University Press of America.
1989b Pooling and Sharing Inside and Across Household Boundaries. *In* The Household Economy: Reconsidering the Domestic Mode of Production. Richard Wilk, ed. Pp. 191–218. Boulder, CO: Westview.
1994 Onions Are My Husband: Survival and Accumulation By West African Market Women. Chicago: University of Chicago Press.

Clark, Gracia, ed.
1988 Traders vs. the State. Boulder, CO: Westview.

Clarke, Edith
1966 My Mother Who Fathered Me. London: Allen and Unwin.

Collins, Patricia Hill
1990 Black Feminist Thought. Boston: Unwin Hyman.

Davis, Angela
1989 Women, Culture and Politics. New York: Random House.

Douglas, Mary
1969 Is Matriliny Doomed? *In* Man in Africa. Mary Douglas and Phyllis Kaberry, eds. Pp. 121–136. London: Tavistock.

Goody, Esther
1978 Some Theoretical and Empirical Aspects of Parenthood in West Africa. *In* Marriage, Fertility and Parenthood in West Africa. Christine Oppong, G. Adaba, M. Bekombo-Priso, and J. Modey, eds. Pp. 1227–1272. Changing African Family Monograph 4. Canberra: Australian National University, Dept. of Demography.

Goody, Esther, ed.
1982 Parenthood and Social Reproduction: Fostering and Occupational Roles in West Africa. Cambridge: University Press.

Hammel, E. A.
1984 On the *** of Studying Household Form and Function. *In* Households: Comparative and Historical Studies of the Domestic Group. Robert Netting, R. Wilk, and E. Arnould, eds. Pp. 129–143. Berkeley: University of California Press.

James, Stanlie
1993 Mothering: A Possible Black Feminist Link to Social Transformation? *In* Theorizing Black Feminisms: The Visionary Pragmatism of Black Women. Stanlie James and Abena Busia, eds. Pp. 144–154. New York: Routledge.

Laslett, Peter, ed.
1972 Household and Family in Past Time. Cambridge: Cambridge University Press.

MacCormack, Carol (Hoffer)
1972 Mende and Sherbro Women in High Office. Canadian Journal of African Studies 6:151–164.

Mikell, Gwendolyn
1989 Cocoa and Chaos in Ghana. New York: Paragon House.

Mitchell, J. Clyde
1966 The Yao Village. Manchester: Manchester University Press.

Okali, Christine
1983 Cocoa and Kinship in Ghana. London: Kegan Paul.

Sacks, Karen
1988 Caring by the Hour: Women, Work and Organizing at Duke Medical Center. Urbana: University of Illinois Press.

Scheper-Hughes, Nancy, and Carolyn Sargent, eds.
1998 Small Wars: The Cultural Politics of Childhood. Berkeley: University of California Press.
Strathern, Marilyn
1980 No Nature, No Culture: The Hagen Case. *In* Nature, Culture and Gender. Carol MacCormack and Marilyn Strathern, eds. Pp. 174–221. Cambridge University Press.
van der Geest, Sjaak
1997 Elderly People in Ghana. Ghana Studies Council Newsletter 10.
Weismantel, Mary
1995 Making Kin: Kinship Theory and Zumbagua Adoptions. American Ethnologist 22:685–704.
Whitehead, Harriet
1981 The Bow and the Burden Strap: A New Look at Institutionalized Homosexuality in Native North America. *In* Sexual Meanings. Sherry Ortner and Harriet Whitehead, eds. Pp. 80–115. Cambridge: Cambridge University Press.

42

Fathering in the 20th Century

BY MAXINE P. ATKINSON AND STEPHEN P. BLACKWELDER

Sociologists Maxine Atkinson and Stephen Blackwelder conducted a study of popular magazines published from 1900 to 1989 to examine the ways the definition of fatherhood has changed during this time period. Systematically sampling all magazines during this time period using The Reader's Guide to Periodical Literature, *they approached their analysis of the magazines in two ways. First they counted all articles relating to parenting, coding them as relating to mothers, fathers, parents, or parent-child. They then coded the articles that related to single mothers, single fathers, divorced fathers, divorced mothers, desertion, the "mommy track," the "daddy track," single parents, and stepparents. They discuss their surprising findings here.*

1. How has the depiction of fatherhood changed over the time period they studied?
2. What is the relationship of increased labor force participation to interest in fathering?
3. How does the depiction of fathering in this study differ from that of the Asante in Gracia Clark's article in this chapter?

"Fatherhood has a very long history, but virtually no historians" (Demos, 1982, p. 425). While scholars (e.g., Lewis & Pleck, 1979; Rossi, 1984) assert that interest in fatherhood has increased during the past century, we have virtually no evidence upon which to base such a claim. While it is clear that social scientists are interested in fathering, including cultural definitions and behavior, it is no more than assertion that the lay public shares this interest.

* * *

In [our] study, we questioned whether or not the lay interest in fatherhood has increased over time, an assumption widely held in the professional literature. A content analysis of popular magazines published from 1900 through the 1980s indicated that popular interest in fatherhood has not increased but rather fluctuated over the last century.

Addressing the question of lay interest in fatherhood led to an unexpected but important finding in the social history of the culture of the family. Using interest in mothers and parents in general as comparison categories for interest in fathers, we found that the popular conceptualization of parenthood has changed remarkably over the past century. Interest in gender-nonspecific parenting overtook interest in mothering or fathering by the twenties and increased in emphasis in later decades. We suggest that, during the twenties, the conceptualization of parenting began to change from a very gendered view, with mothers and parents regarded as equivalent, to a less gendered perspective. While it would be tempting to use these data to infer that interest in fathering has therefore increased, to do so risks committing an ecological fallacy. It is just as likely that people began to use the term *parenting* when they were actually referring to the behaviors of mothers and simply found that parenting was a more acceptable term. The data on division of labor in child care supports this interpretation. Even so, this change in language indicates a less gendered view of the world. Perhaps most likely, the increased use of the term *parenting* may indicate that most people think that fathers and mothers should be engaging in the same behavior regardless of what they actually do. Whether the public has simply relabeled mothering, or has changed its norms about the extent to which both parents should be actively engaged in childrearing, an important cultural change has occurred.

We also examined trends in the changing definition of fatherhood to see if fathers were more likely to be defined as nurturers than providers in later decades. While it is true that, especially during the forties, seventies, and eighties, fathers were much more likely to be seen as nurturers than providers, the century is characterized more by fluctuation than by any sustained trend. The fluctuation in interest in fathers and in the definition of fathering is consistent with the variance in the perception of fathers' competence, which has been documented for the earlier part of the century (LaRossa et al., 1991).

Both Easterlin (1972, 1980) and LaRossa (1988) suggest possible behavioral explanations for the change in the culture of fatherhood. Inasmuch as men's child-care behavior is relatively unchanged in recent decades, LaRossa suggests women's behavior as the predictor of any change in the culture of fatherhood. We thus examined the influence of married women's labor force participation and fertility rates on interest in and definitions of fathering.

We found that neither married women's labor force participation nor fertility rates are substantially related to interest in fathering; however, both trends are related to interest in gender-nonspecific parenting. As married women's labor force participation increases and fertility decreases popular interest in parenting increases.

We also examined the relationships between labor force participation, fertility rates, and the definition of fathers as providers versus nurturers. While there appears to be a weak relationship between married women's labor force participation and an emphasis on fathers as nurturers, there is a much stronger relationship between fertility rates and the emphasis on fathers in their provider roles. Plotted against popularity of the provider definition, fertility rates neatly coincide with the public's changing emphasis on fathers as providers. When fertility rates are high, fathers are more likely to be defined as providers than as nurturers. However, we suggest that the relationship between fertility and the definition of fathering is closely tied to economic conditions. Easterlin's argument that, during "good" economic times, fertility rates increase and that, during "bad" economic times, fertility rates decrease informs these findings. Under economic conditions when fathers are able to meet the provider ideal, fertility increases and the father's provider role is emphasized. When economic conditions do not allow the achievement of the provider role, fertility decreases and an alternative definition of fathers as nurturers is adopted. Fertility is probably an intervening variable between economic conditions and the definition of fathering. We might be well-advised to include economic indicators in future research on fathering. . . .

One of the limitations of this study is that popular magazine articles are more likely to emphasize positive, rather than negative, aspects of parenting. The use of other historical data, such as cartoons (LaRossa et al., 1991), that are less likely to be biased in a positive direction, is important for a more complete view of historical reality.

A second limitation of this research is the class bias that results from the analysis of popular magazines. Other popular media that appeal to a wider range of socioeconomic levels, or publications more likely to be read by other than middle-class white Americans (cf. Gecas, 1972), need to be examined to provide us with a richer and more accurate view of the culture of fatherhood.

Although there may be a race bias in this research, net of the obvious class bias, the extent to which it exists is not clear. For example, of the 283 articles that we analyzed for fathering definition, 5% are from magazines (*Ebony, Jet,* and *Essence*) targeted at an African American audience. At the same time, magazines such *Look, Newsweek, Reader's Digest,* and *Sports Illustrated* certainly are not read solely by people of one race or ethnicity. Research focusing on magazines targeted at black readers may very well provide different findings than those resulting from this sample. In addition, other racial/ethnic minorities, such as Latinos and Asian Americans, who may prefer their recreational reading in a language other than English are not represented in the current study.

This research points to the need for a historical view of the family with increased generalizability beyond what can be accomplished with case studies or other anecdotal evidence. A systematic analysis of popular magazines is one means by which increased generalizability can be accomplished. Studies of the family may be one of the few areas in which we know more about behavior than culture. Increased understanding of the historical culture of the family can place our knowledge of behavior in a sociohistorical context and thereby inform our theoretical models.

REFERENCES

Demos, J. (1982). The changing faces of fatherhood: A new exploration in American family history. In S. H. Cath, A. R. Gurwitt, & J. M. Ross (Eds.), *Father and child: Developmental and clinical perspectives* (pp. 425–445). Boston: Little, Brown.

Easterlin, R. A. (1972). Relative economic status and the American fertility swing. In E. B. Sheldon (Ed.), *Family economic behavior: Problems and prospects* (pp. 170–223). Philadelphia: J. B. Lippincott.

Easterlin, R. A. (1980). *Birth and fortune: The impact of numbers on personal welfare.* New York: Basic Books.

LaRossa, R. (1988). Fatherhood and social change. *Family Relations, 37,* 451–457.

LaRossa, R., Gordon, B. A., Wilson, R. J., Bairan, A., & Jaret, C. (1991). The fluctuating image of the 20th century American father. *Journal of Marriage and the Family, 53,* 987–997.

Pleck, E. H., & Pleck, J. H. (1980). *The American man.* Englewood Cliffs, NJ: Prentice-Hall.

Rossi, A. S. (1984). Gender and parenthood. *American Sociological Review, 49,* 1–19.

43

Class-Based Masculinities

The Interdependence of Gender, Class, and Interpersonal Power

BY KAREN D. PYKE

This reading helps us to understand how the prism of social class modifies the privilege and power of males in interpersonal relationships. Karen Pyke uses data from a survey of 215 individuals, along with excerpts from interviews with 70 divorced and remarried men and women who participated in the larger survey. The experiences of the men and women she interviewed illustrate how class privilege combines with hegemonic masculinity to shape the interpersonal power dynamics of men and women, and disadvantages lower-class men as well as women.

1. Is hegemonic masculinity a characteristic of individual men or the system within which gender is shaped?
2. What are the effects of men's class position on wives of upper- and lower-class men?
3. In what ways does class undermine the privilege of hegemonic masculinity for lower-class men?

* * *

MEN'S JOBS AND CONJUGAL PRIVILEGES

Leisure and Autonomy

An important indicator of marital power is the ease with which husbands can free themselves from the boundaries of family life to pursue other interests. Roughly half of the 69 husbands in first marriages did not spend an inordinate amount of time away from their families beyond what was required of their work day. Among those who did, however, some interesting class differences emerge. The ideological supremacy of the male career provided a means by which higher-class husbands could absent themselves in the evenings and on weekends. These absences often were due to legitimate business trips, though not always as necessary as portrayed to their wives and sometimes lengthened for the pursuit of leisure. Some higher-class husbands used work as a smoke-screen for leisure time with friends or extramarital affairs. For example, one husband, the owner of a textile firm and father of two, extended his foreign business trips to add some pleasure, which included sexual affairs.

In fact, middle-class wives were more often shocked than were working-class wives to learn of their husbands' sexual affairs, which had been easily obscured by the broad cloak of the male career. When her first husband, a salesperson who spent a few nights a month out of town, confessed that he had been with 10 different women, one respondent said, "'When? When?' I couldn't believe that he even had the time to do that. . . . It was a real shock." Another respondent recalled how she "would go out of [her] way to make sure that [her husband] was ready to go" when he had his weekend business trips. She did so for years before she learned they weren't business trips at all and that he had been having an affair with one woman in particular for the previous two years. "And you'd think that if I was bright enough or something I would've noticed it."

For lower-class men there is less blurring of the line between work and leisure, often delineated by the punch of a time card. Men's time with male friends, often drinking or "tinkering" with cars (sometimes resold for a profit), was viewed by their wives as self-centered leisure (see also Halle 1984, 58). In reality, however, many higher-class husbands also were drinking with other men, but it was associated with "working" and "getting ahead." These varying meanings shaped wives' resentment and acceptance. Working-class wives viewed husbands who spent a lot of time with their male friends as "lazy," "not ambitious," "self-centered," "carefree," "immature," and "irresponsible." But higher-class men who spent a lot of time away from home pursuing leisure that was at least nominally associated with work were more often viewed as "ambitious" and doing so out of necessity, even if their wives wished they could cut back on their hours.

When working-class men, such as self-employed contractors, had jobs that could have provided a smokescreen for leisure, they relied on them as a cover less often. Instead they tended to be more blatant in their pursuit of leisure away from their families. They also were more careless in hiding their extramarital affairs and, consequently, were more likely to get caught. For example, with his wife in the hospital after having given birth, one husband, a truck driver, brought another woman home without concealing her from his wife's visiting brother, who later reported the infidelity to his sister.

Half of all working-class men (20 of 38) did not engage in rebellious behavior or stray outside of the boundaries of "good husbands." Those who did typically moved in a social milieu of like men, often coworkers, who encouraged such behavior (see Connell 1991; Halle, 1984; LeMasters 1975; Rubin 1976; Willis 1977). For example, Nick, who had several jobs in construction and other trades as well as periods of unemployment in his first marriage, drank heavily, often with coworkers. He said, "The people I worked with, that's just what you did, especially on the weekends, on Friday nights, you'd get hammered."

This interplay between social milieu and the construction of a defiant masculinity is evident in Ted and Debbie's 9-year-long marriage. At the age of 15, Debbie married Ted, a self-employed plumber 9 years her senior. She finished high school and, later, stayed home and raised their son. Ted was often away from home drinking with male friends and "running around" with women. Although he sometimes used work as an excuse, he wasn't covering it up very well. For example, as Debbie explained, he would say, "'I'm going out to buy a pack of cigarettes' and wouldn't come back until the next day."

Debbie regards Ted's affairs as having been "quick thrills." She said, "He wasn't emotionally involved. It was more part of the recreation of being drunk, being high, being part of that group of people." She referred to that group as "low lifes," . . .

Lower-class husbands who ostentatiously pursued drugs, alcohol, and sexual carousing are constructing a compensatory form of masculinity. Such behavior was worn like a badge of masculinity in the work and social environments they inhabited. By drinking with other working-class men at the bar and openly engaging in extramarital relationships, they appear to be defying existing power structures, displaying their independence from the control of their wives and "the establishment" (i.e., higher-status men). This exaggerated masculinity compensates for their subordinated status in the hierarchy of their everyday work worlds. It gave them a sense of autonomy and self-gratification, entitlements that higher-status men acquire more easily and with greater impunity, thereby creating the illusion of ascendant masculinity. Although this behavior is characteristic of some and not all working-class men, it reinforces a stereotype of subordinated heterosexual masculinity that higher-class men call on as evidence of their own civility and gender equity, thereby further obscuring their power and privilege and reaffirming their ascendant masculinity.

In sum, lower-class men do not enjoy the same ideological legitimations for personal autonomy and leisure in their marriages that higher-class men acquire as part of their career package. Instead, some working-class husbands engage in defiant behavior and construct a compensatory masculinity (see Collinson 1992; Connell 1991, 1995; Willis 1977). In the next section, I describe this overt form of power in more detail.

Overt Domination of Wives

Because working-class men's jobs do not provide a shortcut to marital power, they must either concede power to wives or maintain dominance by some other means. They were, overall, both more egalitarian (especially in sharing housework and child care) and more *explicitly* domineering in their marriages than were higher-class husbands. Domineering lower-class husbands draw more directly and overtly on personal masculine privilege as their essential right as a means of bolstering their conjugal power (see Collinson 1992; Rubin 1976). The following case provides an example of such overt power and illustrates its link to the denigrated status of working-class men.

Nick, age 38, remarried his first wife, Nina, following 4 years of divorce and a tumultuous 10-year marriage that was marked by his drinking, violence, and chronic depression. In his first marriage, his dissatisfaction with several jobs, mostly in construction, led to his current position as a splicer for a utility company. Nick's description of his transition to a splicer reflects the centrality of his work in affecting his low self-esteem. . . .

Nick's self-esteem as a man also plummeted when Nina returned to work. He recalled, "That probably hit me really hard. . . . I wanted to be the provider. When she went to work, it took that away, it took away my status as the man of the house, I thought."

Nick's heavy drinking in his first marriage often was accompanied by violent attacks on his wife and their house, usually prompted by violations of the traditional and submissive role Nick wanted Nina to fulfill. "Small things would trigger it," Nick explained, such as his wife's "lousy" housekeeping. "Plus I was a real jealous person, and whenever [her] old flames would appear I just couldn't handle that, even though I'm sure she was pretty dedicated." Nick's violent rage also was triggered when Nina challenged his domination. . . .

Nick's low self-esteem, alcoholism, and violence eventually gave way to chronic depression and thoughts of suicide that landed him in the hospital. "All I was trying to do was provide for the family and be with the kids, but I was sinking the whole way. . . . Even though I was trying harder, I was still getting violent, and things were getting worse and worse," he explained.

Nick was not exceptional in his abuse. Among the 36 women interviewed, 52 percent married the first time to lower-class men said their first husbands had hit them, compared to 20 percent of those married to higher-class men. The greater incidence of wife abuse (based mostly on self-reports) committed by lower-status, underemployed, and unemployed husbands who cannot fulfill the provider role has been documented elsewhere (Dibble and Straus 1980; Gelles and Cornell 1985; Levinger 1966; O'Brien 1971; Straus, Gelles, and Steinmetz 1980). Other research links economic disadvantage with the husband's increased hostility toward his wife (Conger, Ge, and Lorenz 1994; Liker and Elder 1983).

Working-class husbands' subordinated class status in relation to other men—and women—in the labor force seems to exacerbate their need to use their marriage as a place where they can be superior (Ferraro 1988; O'Brien 1971; Pyke 1994; Rubin 1976). With their power base on shakier ground, they are more likely to resort to explicit and relentless tactics, such as violence, as well as criticism and constant surveillance of their wives.

Some lower-class husbands, particularly those who were violent and adulterous, greatly feared their wives' infidelity. They expressed this fear in baseless accusations, demands for their wives' constant attention, restrictions on their wives' movement and employment, or spying (see also Ferraro 1988). Seemingly irrational, this fear appears to be a reflection of their sense that they offered too little compared to other men to hold onto their wives. These feelings of powerlessness led some to use terrorizing tactics to bolster their control over their wives; however, it had devastating consequences on the marriage, typically pushing wives to leave. . . .

Husbands' Entitlement to Housework

The hidden power of higher-class men and the explicit power of lower-class husbands are also evident by the mechanisms that excuse them, to varying degrees, from performing domestic work. Male careers provide a rationale for higher-class husbands' freedom from family work, whereas working-class men are more likely to rely on blatant and increasingly contested patriarchal ideologies for similar entitlements. In both higher- and lower-class marriages, husbands' absences precluded their doing household chores. However, because the absences of higher-class men were more likely due, ostensibly at least, to evening or out-of-town business obligations or night classes, wives viewed this division as fair. In addition, any work higher-class husbands brought home from their jobs also excused them from family tasks. For example, a female accounting manager remarried to a systems analyst said, "I do 75 percent of the housework because he also works at home. . . . He's always on the computer so I don't know if it's work or play."

Higher-class husbands also derived from their careers greater entitlement to a stay-at-home wife. They were more likely than working-class husbands to veto, discourage, or limit their wives' labor force participation, especially in first marriages (Pyke 1994). For example, Jane was married the first time to a prominent attorney who earned $300,000 annually. His position entitled him to limit her teaching to part-time. She said,

> But he made sure that we both knew that his job came first and if I was working too many hours, he made it clear that I should cut back on my hours. . . . He had a very prominent job, a lot of public recognition that came with his job. . . . So I supported him in that and I was sort of content to be in his shadow. . . . His career came before my career and was much more important to me at the time.

Another attorney, who was childless prior to remarriage, gave a description of the kind of wife he sought after his divorce that emphasized her supporting role to his career. He said, "I made good money as a lawyer, so my wife didn't need to work to support the household. I needed someone to take care of my children and my house when I am not there."

This need for a wife to serve as a maid and nanny propelled many higher-status men to look in the same places for a wife as they do for paid domestic help: among the lower class. An interclass mate selection occurred in remarriage between men employed in high-status occupations and women who were unhappy with dead-end, low-skilled jobs and who had worked out of necessity in their first marriages. Among 102 women whom I surveyed whose first and second marriages were identified as being either working or middle to upper class, 25 percent moved from a working-class first husband to a middle-class second husband (70 percent remarried husbands of the same social class as their first husbands, and 5 percent moved from middle to working class). Similarly, Gerson (1985, 1993) found that men are more likely to seek domestically oriented partners as their breadwinning ability increases, and women are more likely to veer toward domesticity when faced with blocked job opportunities and married to men who earn enough to be the sole provider.

Even though both higher- and lower-class husbands tended to avoid household labor, they enjoyed varying levels of legitimacy for doing so. Higher-class husbands were more likely to be excused by the priorities granted to their career and provider role (see also Gerson 1993; Hochschild with Machung 1989), which also served as the places they most prominently produced ascendant masculinity. Lower-class husbands, on the other hand, whose jobs and lower-earnings provided them with little justification for not sharing chores—especially when wives worked for pay also—more *directly* relied on rigid gender divisions of labor in the home as a means of producing masculinity (see Game and Pringle 1983). However, explicit traditional ideologies about the proper roles of husbands and wives were likely to be challenged and resented by wives. Hence, they were a less reliable basis of men's freedom from housework than was the ideology of the husband's career. This again suggests that the power of lower-class husbands that rests on notions of masculine privilege is likely to be undermined, especially in long-term, stable marriages.

What about the minority of husbands actively embracing egalitarian divisions of labor? Higher-class husbands more often were constrained by their job demands from doing so, even when they professed a sense of obligation or desire. In contrast, there is a greater structural incentive for working-class families to adopt egalitarian practices. Lower-class husbands often cared for children while wives worked a different shift from their own. And, for some, such as Nick in his remarriage, greater involvement with children and family life provided a sense of self-worth and meaning that compensated for the degradation endured on the job (Connell 1991; Gerson 1993; Pyke 1994). These men do not appear to put stock in ascendant or exaggerated masculinities and instead produce an egalitarian masculinity involving expressiveness and high levels of family involvement (referred to as the "New Man"; Messner 1993).

The pressures exerted by the structural conditions of working-class life may lead some men to juggle a Dr. Jekyll and Mr. Hyde existence in which they produce hypermasculinity in male cliques and on the job and an egalitarian masculinity in their family relations. For example, working-class men might use talk of masculine superiority, privilege, and authority as a means of producing hypermasculinity while nonetheless sharing power and family work with their wives on a day-to-day basis. That is, when the situational context changes, the form of masculinity produced, even within categories of social class, can change as well. Furthermore, the consistent construction and maintenance of hypermasculinity across all arenas of social life, including family relations, are so costly as to become less desirable and untenable to individual men. In fact, as they approach midlife, working-class men tend to drop out of the male cliques in which hypermasculinity is, collectively produced (Rubin 1976).

Egalitarian masculinity may not be appreciated by some wives who view it as a threat to their feminine identity. In the next section, I discuss some surprising insights about how the supremacy of the male career leads some middle-class wives to negatively evaluate egalitarian husbands out of preference for male dominance.

Middle-Class Men and Egalitarian Masculinity

Ideological hegemonies present elite interests as everyone's interests. Thus, they lead subordinates, as well as elites, to sanction those who fail to reproduce the dominant group's power advantage. For example, higher-class egalitarian men who violate notions of ascendant masculinity often attract hostility, even from wives who would appear to derive benefits from their husbands' defection but are convinced otherwise. This was evident in a few middle-class first and second marriages in which husbands were unable to live up to the ethic of masculine ambition and high earnings to which their wives felt entitled. It was not that these husbands were poor providers or husbands; on the contrary, they tended to be very family-involved men with moderate earnings. However, their wives, who expected them to have *greater* ambition and earnings, were less likely to view them as "real" men and reacted to their "inadequacy" with disdain. Disappointment in the "failure" of their husbands to live up to the ideal of ascendant masculinity bolstered wives' marital power. Because explicit patriarchal ideologies are not prominent in the discourse of the higher-class cultural milieu, middle-class husbands less often call them up as a means of boosting their power.

The resultant power sharing is not always appreciated by wives. The ideological supremacy of the male career and the "doing" of essential femininity fosters the desire among some wives to be dependent and to "look up" to their husbands. For example, Jean resented her shared power as a further indication of her first husband's failure as a man. Her disappointment in him for not being less "submissive" and more successful like the people she worked with, even though he earned what she considered to be a good income, resulted in the break up of their otherwise happy marriage. They married when she was 19 and he was 24. She finished her education and became a successful accountant. It bothered her that her husband, Phil, an insurance title representative, did not want to finish his bachelor's degree. . . .

Such unhappiness endured despite Phil's instrumental and emotional support of Jean and her career. For example, when she had to work until late at night or on Saturdays, "it was no problem. . . . He would get dinner started." After she had a car accident, he drove her to and from work. "He was very good, generous," she said. The marriage faltered despite all the positive qualities of her husband, which enabled her to advance in her career. Indeed, it was those very qualities that led her to lose respect for him as a man; he was not as "egotistical" as other men. Instead, she said, "He was generally pretty accommodating. I was really the one who forced my decision. I think that's one of the things that bothered me. I had a very strong personality and he would back down sometimes."

The very traits she desired in Phil, such as egotism, ambition, and dominance, would have undermined her marital power, obligating her to more instrumental and expressive support of his career at the potential expense of her own. This illustrates the pressure on men to accomplish ascendant masculinity via a successful career, its connection to male dominance and feminine identity, and the negative effects endured by some higher-class men who are not hyperambitious.

One middle-class man described the strain in both his first and second marriages as relating to his failure to achieve a higher level of success. He said,

> I think from [ex-wife's] perspective, and my current wife is still kind of the same way, it's almost like . . . they both felt that when you get married it's . . . you hit a button and all of a sudden your husband is out there making $75,000 a year. And they [wives] work as a lark.

His second wife pressures him to leave advertising, where his earnings have suffered, and go into commercial real estate like one of her relatives who earns $100,000 a year. Their individual highest annual earnings during their marriage are equal at $40,000. Her resentment, which he gingerly tiptoes around, is a dominant theme in their marriage. He described her as "contentious" and a "battle ax". . . .

Although her husband performs 60 percent of their weekly total household labor hours, her resentment for his failure to live up to the middle-class standard of success overshadows any gratitude she might have for his greater household work.

The previous two examples illustrate how some women's acceptance of the dominant ideology about what it means to be a man reflects and contributes to the hegemony of the male career. These examples suggest that when middle-class men are not very successful in their careers, power can shift to a resentful wife. Because methods of accomplishing femininity often rest on women's subordination to men, some women may resent their power. They may use it, as exemplified here, to steer men toward the production of masculinity in ways that emphasize male power and female subordination. These examples underscore the ways cultural notions of what constitute a "real" man and a "real" woman elicit women's participation in the project of male dominance and female subordination.

SUMMARY AND CONCLUSION

In contrast to oversimplified, gender-neutral or gender-static approaches, the theoretical framework presented here integrates interpersonal power with broader structures of class and gender inequality. I used empirical examples of conjugal power to illustrate how interpersonal powering processes and gender and class relations can be considered components of an interacting system. Specifically, structures of inequality are expressed in ideological hegemonies, which construct gender in ways that reemphasize and normalize the domination of men over women and that of privileged men over lower-class men. Furthermore, the relational constructions of ascendant and subordinated masculinities have different implications for interpersonal power dynamics.

For example, the different conjugal power processes available across social class further feed into the cultural legitimations of higher-class men's superior position. In the absence of legitimated hierarchical advantages, lower-class husbands are more likely to produce hypermasculinity by relying on blatant, brutal, and relentless power strategies in their marriages, including spousal abuse. In so doing, they compensate for their demeaned status, pump up their sense of self-worth and control, and simulate the uncontested privileges of higher-class men. The production of an exaggerated masculinity in some working-class subcultures also serves the interests of higher-class men by deflecting attention from their covert mechanisms of power and enabling them to appear egalitarian by contrast.

The coercive power strategies of lower-class men appear unmatched in degree. This is misleading, however. It is precisely their demeaned status and weak power base that have propelled many working-class men to rely on extreme methods of control as a kind of last resort in asserting power and producing masculinity. Thus, those who study power relations must be careful not to equate brutality of power with quantity of power and examine the ways that power inequalities may be obscured in other seemingly egalitarian relationships by hegemonic cultural ideologies. In addition, brutal styles of masculinity, such as displayed in the workplace or tavern, should not be assumed to be automatically linked to brutal power strategies in marriage. Some men may balance more egalitarian practices in their personal life with more public displays of hypermasculinity and claims to male dominance. Similarly, higher-class men who are mild mannered and civil in the workplace may nonetheless exercise brutal forms of power in their family life.

The omnirelevance of gender to social life and the ways it is taken for granted as essential and inevitable makes it an especially effective ideology in normalizing and mystifying gendered power relations. In doing gender, men and women engage in practices that promote male dominance and female subordination in most social contexts. Specifically, I have shown how some women pursue marital arrangements that contribute to male dominance as a method of accomplishing their gender; that is, they do so to affirm their "essential nature" (West and Fenstermaker 1993). It is not necessarily that they consciously desire male dominance, but the methods they employ in "doing gender" produce conditions that foster such power differentials. It is thus important that researchers studying interpersonal power consider how it is a symbolic artifact of the routine production of gender as well as the structural conditions of men's and women's lives.

Although I have used the case of marital power and the hegemony of the male career, other hegemonic ideologies similarly affect power dynamics in marriage as well as other social relationships and reinforce "essential" gender differences. For example, a white heterosexual masculine ethic pervades capitalist, managerial ideologies that stress rationality, success orientation, impersonality, emotional flatness, and a disregard for family concerns. Because these traits are associated with "essential" masculinity and are antithetical to notions of "essential" femininity, this ethic would appear to exclude women from management positions and undermine the power of those who have successfully acceded to such ranks while (re)constructing "essential" femininity and ascendant masculinity. Similarly, the masculine ethic of management associated with higher-class men embodies traits that reflect and perpetuate the negative evaluation of lower-class men, men of color, young men, and homosexual men. This, in turn, reinforces the construction of compensatory masculinities, such as the "cool pose" associated with African American men (Majors and Billson 1992) and the hypermasculinity described here among lower-class white men and also common among male youths (Messerschmidt 1993). Thus, ideological hegemonies have a different impact on men and women across race, sexuality, age, and social class in ways that reflect and (re)construct relational conceptions of masculinities and femininities with different implications for interpersonal power. The ensuing practices of interpersonal power, in turn, reinforce structures of inequality and their ideological legitimations.

Femininity is also cross-cut into diverse forms by the structural and cultural conditions of race, social class, sexuality, and age. For the sake of clarity, however, I have centered this discussion almost exclusively on white, heterosexual, class-based masculinities. The construction of femininities can perhaps best be understood in relation to men. As Connell observed (1987, 186–87), all forms of femininity are constructed in the context of the overall subordination of women to men. The interplay of diverse femininities does not reemphasize a hierarchy among women as much as intermale hierarchies of dominance as well as gender hierarchies.

The degree to which women are accommodating to men provides a useful basis for conceptualizing femininities. "Emphasized femininity" (Connell 1987, 187) is produced among women who view their role as naturally subservient to men. Noncompliant femininity, on the other hand, emphasizes women's independence and desired equality with men. It is displayed by the woman who can do it all: maintain a good job, a clean house, well-behaved children, and a loving marriage. Noncompliant femininity obscures women's subordination to men by associating their paid labor with equality and downplays how their employment benefits elite males who purchase women's discounted labor.

How the construction of femininities reflects and (re)constructs (or resists) the gender order and intermale hierarchies needs to be further explored. An examination of hidden power dynamics might reveal that the key difference in the ways these two forms of femininities are played out has less to do with quantity of male dominance than with quality. For example, women who display greater egalitarianism in some arenas of their marriage or job may feel pressed to accomplish their gender with greater submission in other arenas (see Hochschild with Machung 1989; Pyke 1994). On the other hand, women who emphasize their femininity may be able to wield considerable power from behind a smokescreen of female subservience. By examining the underlying cultural ideologies at play and the actual practices, we can learn how the construction of these and other forms of femininity shapes interpersonal power, plays into the construction of masculinities, and obscures while (re)producing inequality. . . .

REFERENCES

Collinson, David L. 1992. "Engineering humour": Masculinity, joking, and conflict in shop-floor relations. In *Men's lives,* edited by Michael S. Kimmel and Michael A. Messner, 232–46. New York: MacMillan.

Conger, Rand D., Xiao-Jia Ge, and Frederick O. Lorenz. 1994. Economic stress and marital relations. In *Families in troubled times: Adapting to change in rural America,* edited by Rand D. Conger and Glen H. Elder, Jr., 187–203. New York: Aldine de Gruyter.

Connell, R. W. 1987. *Gender and power.* Stanford, CA: Stanford University Press.

———. 1991. Live fast and die young: The construction of masculinity among young working-class men on the margin of the labour market. *The Australian and New Zealand Journal of Sociology* 27:141–71.

———. 1995. *Masculinities.* Los Angeles: University of California Press.

Dibble, Ursula, and Murray S. Straus. 1980. Some social structure determinants of inconsistency between attitudes and behavior: The case of family violence. *Journal of Marriage and the Family* 42:71–80.

Ferraro, Kathleen J. 1988. An existential approach to battering. In *Family abuse and its consequences,* edited by Gerald T Hotaling, David Finkelhor, John T. Kirkpatrick, and Murray A. Straus, 126–38. Newbury Park, CA: Sage.

Game, Ann, and Rosemary Pringle. 1983. *Gender at work.* Boston: Allen & Unwin.

Gelles, Richard J., and Claire Pedrich Cornell. 1985. *Intimate violence in families.* Beverly Hills, CA: Sage.

Gerson, Kathleen. 1985. *Hard choices: How women decide about work, career, and motherhood.* Los Angeles: University of California Press.

———. 1993. *No man's land: Men's changing commitments to family and work.* New York: Basic Books.

Halle, David. 1984. *America's working man: Work, home, and politics among blue-collar property owners.* Chicago: University of Chicago Press.

Hochschild, Arlie, with Anne Machung. 1989. *The second shift: Working parents and the revolution at home.* New York: Viking.

LeMasters, E. E. 1975. *Blue-collar aristocrats: Life-styles at a working class tavern.* Madison: University of Wisconsin Press.

Levinger, George. 1966. Sources of marital dissatisfaction among applicants for divorce. *American Journal of Orthopsychiatry* 36:803–7.

Liker, Jeffrey K., and Glen H. Elder, Jr. 1983. Economic hardship and marital relations in the 1930s. *American Sociological Review 48*:343–59.

Majors, Richard, and Janet Mancini Billison. 1992. *Cool pose: The dilemmas of black manhood in America.* Lexington, MA: Lexington Books.

Messerschmidt, James W. 1993. *Masculinities and crime: Critique and reconceptualization of theory.* Lanham, MD: Rowman & Littlefield.

Messner, Michael. 1993. "Changing men" and feminist politics in the United States. *Theory and Society* 22:723–37.

O'Brien, John E. 1971. Violence in divorce-prone families. *Journal of Marriage and the Family* 33:692–98.

Pyke, Karen. 1994. Women's employment as a gift or burden? Marital power across marriage, divorce, and remarriage. *Gender & Society* 8:73–91.

Rubin, Lillian. 1976. *Worlds of pain: Life in the working-class family.* New York: Basic Books.

Straus, Murray A., Richard J. Gelles, and Suzanne K. Steinmetz. 1980. *Behind closed doors: Violence in the American family.* New York: Doubleday.

West, Candace, and Sarah Fenstermaker. 1993. Power, inequality, and the accomplishment of gender: An ethnomethodological view. In *Theory on gender/feminism on theory.* edited by Paula England, 151–74. New York: Aldine de Gruyter.

Willis, Paul. 1977. *Learning to labor.* New York: Columbia University Press.

CHAPTER 9

Enforcing Gender

Throughout Part Two, we have discussed patterns, including those of learning; selling; and doing gender at work, play, and in intimate relationships. In this chapter, we look at the patterns surrounding the enforcement of gender. Enforcing gender is about more than just *doing* gender; it is about assault, coercion, and explicit constraints on people's behaviors, as well as more subtle and tacit constraints on identities to enforce gender conformity. Enforcing gender involves people using a range of social control strategies, such as physical abuse and rape, harassment, gossip and name calling, as well as laws and rules created by governments, organizations, and religions to coerce people to conform to gender norms they might not otherwise wish to obey. Many readings throughout this book are about the enforcement of gender, particularly those in Chapter 4. This chapter explicitly focuses on the different forms of social control used to enforce gender.

The enforcement of gender can have profound effects on women's and men's choices, self-esteem, relationships, and abilities to care for themselves. We argue two main points in this chapter. First, doing gender is, by and large, not something that we freely choose; rather, there are many times that we are forced to do gender, whether we would wish to do it or not. Second, there are many occasions whereby the very act of maintaining a gendered identity hurts ourselves and others, either physically, emotionally, or both.

SOCIAL CONTROL

Enforcing gender is about the physical and emotional control of everyone. Berger (1963) describes the processes by which we learn to conform to the norms of society as "a set of concentric rings, each representing a system of social control" (73). At the middle of the concentric rings, Berger (1963) places the individual.

Table 9.1 Statistics on the Incidences of Sexual Violence in the United States

Type of Assault	Men		Women	
Rape/sexual assault[1]	White:	.2/1,000	White:	2.7/1,000
	Black:	.5/1,000	Black:	3.3/1,000
	Hispanic:	0/1,000	Hispanic:	1.7/1,000
Intimate partner homicide[2]	White:	.49/100,000	White:	1.11/100,000
	Black:	4.11/100,000	Black:	3.55/100,000
	Asian:	.19/100,000	Asian:	.92/100,000
	AIAN[3]:	1.20/100,000	AIAN[3]:	2.26/100,000

1. U.S. Dept. of Justice, 2000. Rate per year, age 12 and older for year 1998.

2. Paulozzi, Saltzman, Thompson, and Holmgreen, 2001. Estimated age-adjusted annual rates, United States, 1981–1998. *Note:* Hispanic rates were not specified in this report.

3. American Indian/Alaska native

Social control mechanisms including family and friends are in the next ring, and the legal and political system of a society are in the outer ring. He argues that most social control of behavior occurs in the inner rings which he describes as "broad coercive systems that every individual shares with a vast number of fellow controlees" (Berger, 1963:75). As such, most gendered behavior is enforced in those center rings, with the forms of social control differing slightly by gender. For example, homophobia is a control mechanism that is often used to enforce gender patterns for men. Friends or peers who call a boy a "fag" can make him uncomfortable and force him to display hegemonic masculinity. The commodification of gender most often enforces emphasized femininity, with young girls comparing themselves to the idealized images of the media, as described in Chapter 5, particularly the reading by Kilbourne.

GENDER VIOLENCE

However, although most of the enforcement of gender happens as part of normal interaction, some forms of social control are more coercive, including physical assault, rape, sexual harassment, and even murder. We will begin by discussing the harm done when violence is used to enforce gender. We hope you have never experienced physical violence, but many individuals have. The statistics in Table 9.1 illustrate how often such violence is reported to authorities. Clearly, women are more likely than men to be victims of intimate violence. However, as the first reading by Julia Hall indicates, much of the violence against women is not reported to the authorities, making the incidence of violence for women higher than the statistics in Table 9.1 suggest.

Violence, particularly gender violence, turns women's lives upside down. It is estimated that approximately one quarter of all women have experienced some form of intimate violence in their lives (Mahoney, Williams, and West, 2001). Intimate violence covers a wide range of incidents, including physical assault; slapping; sexual assault; threats against the woman, her children, or loved ones; control over finances and daily activities; and emotional assaults and abuse. Abusers exert "coercive control" over their victims, making victims fearful for their lives and the lives of loved ones (Mahoney, Williams, and West, 2001). As such, victims often become psychologically battered and become emotionally dependent on their perpetrators (Mahoney, Williams, and West, 2001). Domestic violence not only lowers the victim's self-esteem, but also her ability to leave an abusive situation is difficult because abusers often control women's freedom of movement and finances (even for women who work for pay outside the home). Abusers also prevent women from getting the psychological support they need to leave the abusive situation (Mahoney, Williams, and West, 2001). In her reading in this chapter, Hall describes the situation of poor women returning to the men that beat and abuse them, a pattern found across social class relationships. Hall's article illustrates that the effects of violent acts extend beyond the abuser and abused, such as upon the young girls she interviewed who witnessed domestic violence. These poor, white middle-school girls constructed their lives and planned for their futures around gendered violence; they distrusted men and expressed the wish to never marry and have children because of the violence they witnessed.

Date Rape Date rape is another form of intimate abuse that almost entirely affects women. Studies find that one in four college women have been raped or the victims of attempted rape (Bachar and Koss, 2001). The situation of date rape evolves in a rape culture that encourages and justifies rape (Boswell and Spade, 1996). The literature on the prevalence of rape has been consistent over the last 15 years; however, there is little work on prevention of intimate violence. Attempts to work with mixed-gender audiences in preventing rape are not as successful as those directed at helping women to understand that they do not cause rape or trigger intimate violence (Bachar and Koss, 2001).

We argue that the underlying reasons for intimate violence and sexual abuse are complicated and relate to the maintenance of hegemonic masculinity. Hegemonic masculinity depends upon the degradation of women, sexually, physically, and emotionally. Inherent in "becoming" a man is learning to disdain women. As Scott Straus, Steven Ortiz, and Beth Quinn suggest in their readings in this chapter, many men are expected to disrespect women and other men congratulate them when they do so. The same patterns and practices we have been describing throughout this book also explain how men learn to do and justify the physical and emotional abuse of women and what allows them to get away with it. Although there have been some changes in the way police and courts handle intimate violence, there is a long-standing attitude that the "victims ask for it" (Mahoney, Williams, and West, 2001).

INSTITUTIONALIZED ENFORCEMENT OF GENDER

Although the enforcement of gender mainly occurs in daily activities, as Berger (1963) notes, institutionalized settings also sometimes enforce gender. Organizations either participate in or simply ignore more subtle forms of gender enforcement. These patterns can be found in schools (e.g., fraternities), the workplace, and men's sports teams (Schact, 1996). In patriarchal societies worldwide, it is men who make the rules (and laws) and write the informal scripts that prescribe behaviors for themselves, other men, and women. These rules become institutionalized in daily patterns of life.

Religion is one institution that enforces gender difference and inequality. Many religions have or have had rules that exclude or segregate women within the practice of that religion. For example, women are excluded from the priesthood in the Catholic Church and marginalized in many other Christian churches (Nesbitt, Baust, and Bailey, 2001). Some Jewish congregations segregate women physically and do not allow women to study the Torah with the same seriousness as men (Rose, 2001). You will read about how Islamic women negotiate the rules of gender imposed by their religion in the reading by Shahin Gerami and Melodye Lehnerer in Chapter 10.

Many rules subtly maintain the domination of women—including the sexual violation of women—by men. The article by Andrea Marie Bertone describes sex trafficking in the international economy. Politicians and traffickers, who are almost always men, construct and maintain the rules supporting sexual trafficking. These rules allow poor girls and women to enter countries under dubious documentation and fail to provide the help needed to escape lives of prostitution and abuse (Kempadoo and Doezema, 1998). The women sold into virtual sexual slavery, and the men who take advantage of their domination, are part of a brutal pattern of gender enforcement. The sexual trafficking of women helps to maintain the sexual domination of men over women, just as fraternities help to maintain the sexual domination of individual brothers in a fraternity (Straus reading) over women.

Another, more subtle form of gender enforcement maintained because those in control of the situation and the victims themselves don't always "notice" it happening is sexual harassment as discussed in Chapter 7. Sexual harassment begins early, among schoolchildren (American Association of University Women, 2001; Hand and Sanchez, 2000, Sandler in Chapter 4). Although the gender difference in the experience of harassment is only 9 percent greater for girls (76% of boys and 85% of girls experience sexual harassment), girls endure the more demeaning forms of harassment and at a higher rate than do boys (Hand and Sanchez, 2000). Unfortunately, sexual harassment continues long beyond high school, as noted by Quinn in this chapter and discussed in Chapter 7.

Although it is subtler than sexual assault, sexual harassment has serious consequences for women. In her reading, Quinn helps us to understand how sexual harassment enforces gender for men and women by making women objects of men's masculinity. The "girl watching" that Quinn describes creates a wall between men and women that makes it impossible for them to relate on an equal basis.

Unfortunately, there are too many walls created through the enforcement of gender for women and men. The reading by Steven Ortiz describes the "code of conduct" for major league baseball wives who travel with their husbands during the season. This reading describes another form of gender enforcement institutionalized within a specific organization, one which maintains a code among men wherein women are kept "in their place." The baseball clubs and players would just as soon have their wives stay at home; however, it is not easy for those women to do so because they are alone for most of the year, solely in charge of the household (Gmelch and San Antonio, 2001). Although traveling seems like a better option to many baseball wives, those who do travel have explicit instructions on where to go and what to say when traveling with their husbands' teams.

The enforcement of gender integrates many prisms, not all of which we can cover in this chapter. We can learn a lot about the enforcement of gender by considering the situation for transgendered individuals or persons who don't conform to typical gendered forms, such as Betsy Lucal in Chapter 1. We all feel pressure to fit into a binary system of gender; however, these individuals in particular face considerable pressure to be forced into either masculine or feminine identities (Gagné and Tweksbury, 1998).

Sometimes we only see gender enforcement when we look at someone else's life circumstances. Often it is only by looking at someone else's circumstances, such as the Asante women who faced social pressure when they did not provide for their children in Chapter 8, or Straus' description of his pledging experiences in this chapter, that we become more aware of those systems of enforcement that shape our behaviors. Think about your life as you read this chapter and consider the ways your gender is enforced.

REFERENCES

American Association of University Women. (2001). *Hostile hallways: Bullying, teasing, and sexual harassment in school.* Washington, DC: American Association of University Women Educational Foundation.

Bachar, Karen and Mary P. Koss. (2001). From prevalence to prevention: Closing the gap between what we know about rape and what we do. In *Sourcebook on violence against women,* Claire M. Renzetti, Jeffrey L. Edleson, and Raquel Kennedy Bergen (Eds.), pp. 117–142. Thousand Oaks, CA: Sage.

Berger, Peter L. (1963). *An invitation to sociology: A humanistic perspective.* Garden City, NY: Anchor Books.

Boswell, Ayres and Joan Z. Spade. (1996). Fraternities and rape culture: Why are some fraternities more dangerous places for women? *Gender & Society* 10(2): 133–147.

Gagné, Patricia and Richard Tweksbury. (1998). Conformity pressures and gender resistance among transgendered individuals. *Social Problems* 45(1): 81–101.

Gmelch, George and Patricia Mary San Antonio. (2001). Baseball wives: Gender and the work of baseball. *Journal of Contemporary Ethnography* 30(3): 335–356.

Guiffre, Patti A. and Christine L. Williams. (1994). Boundary lines: Labeling sexual harassment in restaurants. *Gender & Society* 8(3): 378–401.

Hand, Jeanne Z. and Laura Sanchez. (2000). Badgering or bantering? Gender differences in experience of, and reactions to, sexual harassment among U.S. high school students. *Gender & Society* 14(6): 718–746.

Hochschild, Arlie. (1983). *The managed heart: Commercialization of human feeling.* Berkeley: University of California Press.

Kempadoo, Kamala and Jo Doezema. (1998). *Global sex workers: Rights, resistance, and redefinition.* New York: Routledge.

Mahoney, Patricia, Linda M. Williams, and Carolyn M. West. (2001). Violence against women by intimate relationship partners. In *Sourcebook on violence against women,* Claire M. Renzetti, Jeffrey L. Edleson, and Raquel Kennedy Bergen (Eds.), pp. 143–178. Thousand Oaks, CA: Sage.

Nesbitt, Paula, Jeanette Baust, and Emma Bailey. (2001). Women's status in the Christian church. In *Gender mosaics: Societal perspectives,* Dana Vannoy (Ed.), pp. 386–396. Los Angeles: Roxbury Publishing Company.

Paulozzi, Leonard J., Linda E. Saltzman, Martie P. Thompson, and Patricia Holmgreen. (2001). *Surveillance for homicide among intimate partners—United States, 1981–1998.* Washington, DC: Government Printing Office.

Rose, Dawn Robinson. (2001). Gender and Judaism. In *Gender mosaics: Societal perspectives,* Dana Vannoy (Ed.), pp. 415–424. Los Angeles: Roxbury Publishing Company.

Schact, Steven P. (1996). Misogyny on and off the "pitch": The gendered world of male rugby players. *Gender & Society* 10(5): 550–565.

U.S. Department of Justice. (2000). *Criminal victimization in United States, 1998 Statistical Tables.* Washington, DC: Government Printing Office.

44

It Hurts to Be a Girl

Growing Up Poor, White, and Female

BY JULIA HALL

Julia Hall uses her sociological training to examine the effects of shifting economic tides on domestic violence. She studies the effects of domestic violence on the female children of those whose lives have been turned upside down by the closing of factories and loss of jobs. This extensive case study of nine poor, white girls—age 11 to 13, living in a Northeast urban area—describes the effects of domestic violence on future generations. These girls interpreted their lives based upon the anger, abuse, and violence they witnessed.

1. In what ways are incidents of violence normalized in these girls' lives and their communities?
2. What strategies have the girls developed to escape the violence they live with on a daily basis?
3. How have schools failed to deal with the violence these children face?

In this investigation, I contend that a group of poor white middle school young women in the postindustrial urban Northeast are living among high concentrations of domestic violence. I refer to this group as "Canal Town" girls.[1] These young women are envisioning lives in which, by charting a course of secondary education, they hope to procure jobs and self-sufficiency. As their narrations indicate, such plans are fueled by the hope that they will live independent lives as single career women and, therefore, will bypass the domestic violence that currently rips through their own and their mothers' lives. . . .

While there are many analyses that focus on the ways in which institutions and the formation of female youth cultures contribute to inequitable futures (Finders 1996; Holland and Eisenhart 1990; McRobbie 1991; Raissiguier 1994; Smith 1988; Valli 1988; Weis 1990), none of this work picks up on the issue of violence. Fine and Weis (1998) examine this theme as it boldly emerges in data on the lives of poor and working-class adult white women and the production of identity. They found that white working-class women experience more abuse, as compared with working-class women from other cultural backgrounds, and are more apt to treat their abuse as a carefully guarded secret (Weis et al. 1997; Weis, Marusza-Hall, and Fine 1998).[2]

Informed by such work, I turn this critical lens on middle school girls. What I found is that their lives are also saturated with domestic abuse. They are not talking about it, not reporting it, and covering their bruises with clothes. They are also hiding it from others and themselves to such an extent that it is not openly dealt with at a critical level at all. Nowhere in their narrations is there any sense that males are accountable for their violent behavior.

This research is contextualized in a postindustrial economy characterized by the systematic dismantling of the basic productive capacity of a nation, a trend sharply experienced in the United States during the 1970s and 1980s. During these decades, the U.S. steel industry had already begun a process of shifting to foreign, less expensive, less regulated markets, as did other areas of manufacturing and production. As a result, smaller businesses that were dependent on industry also closed (Bluestone and Harrison 1982). No longer able economically to support its own populace, the city in this analysis currently relies on shrinking state resources. Left in the wake of global restructuring are empty factories, gutted warehouses, and people who can no longer make a decent living. Canal Town is an urban neighborhood reflecting these changes.

No longer able to find the wage-earning jobs they once enjoyed, today residents are often unemployed (Perry 1996). Many rely on food stamps and Aid to Families with Dependent Children (AFDC). The demography of Canal Town has also shifted from white to racially diverse. This change is reflected in the neighborhood school, which was transformed into a Spanish-English bilingual magnet in the late 1970s. The community center, which has traditionally been staffed by white adults, however, is almost exclusively visited by local white youth. . . .

Although still socially and economically privileged by their whiteness, among most white former workers, a family wage has disappeared. The cushion of wealth that white laborers were often able to amass for their families across generations is quickly eroding. Still, there may be pockets of accumulated resources that are shared among white families in Canal Town, for example, in the form of home ownership or a pension (Fine and Weis 1998; Oliver and Shapiro 1995). . . .

The Canal Town girls' families have historically been working-class, most having fathers and grandfathers who worked in industry while their mothers and grandmothers stayed home. The subordination of women to men within white working-class families has been heavily investigated (Smith 1987). Others explore this subordination through the notion of the *family wage*. The family wage appeared advantageous to all family members, but in reality it supported the notion that women should receive lower wages than men or stay home (May 1987; Woodcock Tentler 1979). Although the young women in this investigation contend that the adult men around them are no longer employed in full-time labor jobs, present-day gender arrangements in the Canal Town community are linked to the ideology prevalent during the days of heavy industry. . . .

CANAL TOWN GIRLS

To obtain some sense of the nine poor white sixth-, seventh-, and eighth-grade girls who participated in this research, I share information from the individual interviews pertaining to their home life. Out of the nine girls—Anne, 11; Rosie, 11; Sally, 11; Jamie, 12; Elizabeth, 12; Lisa, 12; Katie, 13; Christina, 13; and Lisette, 13—only Jamie says she lives with both parents. Elizabeth, Anne, and Katie maintain they live with their mothers, siblings, and their mothers' steady boyfriends. The rest of the girls—Christina, Lisette, Lisa, Rosie, and Sally—report they live with their mothers, siblings, and on occasion, their mothers' different boyfriends. Only Jamie and Christina said they were in contact with their biological fathers, while the remaining girls contend they have no knowledge of their fathers' whereabouts. In terms of employment, seven of the girls state their mothers are not presently working, nor have they been in the past. Only Elizabeth says that her mother used to work as a secretary before she was born. Of the adult men who contribute to household expenses, Jamie's father holds a part-time job in the trucking industry, while Katie is the only girl to claim her mother's boyfriend earns money for their family. As Katie explains, he collects items on trash day that he sells to pawn shops. Jamie says her family receives food stamps, while all of the other girls say their families rely on food stamps and AFDC. Jamie lives in a house that her parents inherited from her grandparents, while the rest of the girls say they live in apartments.

DREAM JOBS

As the Canal Town girls begin to talk about what they want their lives to look like after high school, they stress going to college and/or obtaining a good job, and only mention marriage or family after being asked. Since the girls are only in middle school, their plans for the future may not yet be specific or thought out, but the positioning of a job or career as central to the production of identity is worth noting.

> Christina: I want to be a doctor . . . I'll have to go to college for a long time . . . I don't know where I'll go [to college], hopefully around here . . . I'm not sure what type of doctor, but I'm thinking of the kind that delivers babies.

> Lisette: I want to be a leader and not a follower . . . I want to be a teacher in [the neighborhood] because I never want to leave here . . . I want to go to [the local] community college, like my sister, learn about teaching little kids . . . I definitely want to be a teacher. . . .

All of these young girls envision further education in their future, but most do not yet have a clear sense of what school they hope to attend or how long they plan to go. . . . Any of these girls may switch ideas about careers a number of times, yet when asked about the future, all of them focus their energies on the single pursuit of furthering their schooling and landing a job. Christina says she worries about being homeless, which is likely a chronic fear among poor youth.

These white young adolescents are the daughters of presently poor adults. None of their parents continued their education beyond high school and a few did not graduate from grade 12. College, they say, is not an option that is really discussed much at home. Perhaps Lisette has the clearest idea of where she would like to go to school because she is the only white girl who I worked with who has an older sibling enrolled in an institution of higher education. Lisette's sister attends a nearby community college and studies early childhood education, a circumstance that likely influenced her little sister's plans. Interestingly, three of these white girls indicate that although they want to break out of cycles of dependence and have careers, they do not want to leave their neighborhood—whether for school or work.

The importance of a job or career is emerging within the identities of these girls, but it is too soon to say whether they will follow through on their plans for further education or training. The outlook is not promising, as all but a few of their older siblings are negotiating lives riddled with substance abuse and early pregnancies. My conversations with the principal of the area high school reveal that very few local teenagers are enrolling in any form of advanced studies.

Even though the Canal Town girls view education as important in obtaining their goals, they both accept and reject academic culture and knowledge. I observed that on a daily basis while in class, these girls copy homework, pass notes, read magazines and/or books, or, in other words, participate in the form rather than the content of schooling. Time spent in school involves passively skipping across the surface of learning.

In only a few instances did the girls actually narrate resentment toward school authority. As part of a tradition of working-class women whose personal choices have been mediated by structural constraints, the Canal Town girls, by virtue of gender, are not part of this legacy of expressed resentment. Animosity toward institutional authority is typically male and is linked to the historical contestation between capital and workers (Everhart 1983; Weis 1990). Since white women generally labored in the private sphere or as marginalized wage workers, they did not directly engage in such struggles.

FAMILY PLANNING

★ ★ ★

[M]ost of the Canal Town girls, with the exception of Lisa, contend they do not wish to have husbands, homes, and families at all. Rather, the Canal Town girls claim they are looking to the life of a single career woman as a way to circumvent the abuse that they see inscribed in future families or relationships with men. It quickly becomes clear that seeking refuge from domestic violence plays a big role in constructing identities. For many of these girls, the future includes avoiding marriage and family altogether and getting a job so they can rely on themselves.

> Christina: I don't want to be married because if I was married, my husband would want a kid. I don't want to have a kid because its father may not treat us right . . . hitting and stuff. . . . There's not enough for everybody and the kid shouldn't have to suffer . . . I want to always stay in [this neighborhood] . . . live alone. . . . At least I know trouble here when I see it.

> Lisette: I don't want to get married and be told to stay at home . . . and be someone's punching bag . . . I'II get a one-bedroom apartment and live alone and just try to be the best teacher I can be. . . .

The girls are not devising career-oriented plans simply to escape a patriarchal-dominant home. Rather, they specifically say they view a job as the ticket to a life free of abuse. By concentrating energies on the world of work instead of family, some of the Canal Town girls feel they can spare bringing children into the world, whom they feel often bear the brunt of adult problems. . . .

While the girls say they want to live as independent women in the public sphere, they are developing such identities in response to violent men. Nowhere, in more than one year of observations and interviews, did I hear these girls hold men and boys accountable for their abusive behavior. While it may be the case that they hold such a critique, the absence of any such discourse in the data is glaring, especially given the frequency and detail in which abuse was mentioned in the private space of an interview or in hushed conversations with friends.

IT HURTS TO BE A GIRL

When these young girls are asked to describe their neighborhood, they soon begin to tell stories of women being abused at the hands of men. The women in their narrations seemingly work to conceal their abuse from authorities and ultimately end up "going back."

> Jamie: It's a pretty good place to live. . . . There's lots of auto crashes, drunk people. Lots of people go to the bars on Friday and Saturday and get blasted. They're always messing with people. Some guy is always getting kicked out of the bar for fighting. Guys are mostly fighting with their girlfriends and are getting kicked out for punching so they continue to fight in the street; I see it from my bedroom window; only the girl mostly gets beat up really bad . . . but later she was saying it was her fault.

> Christina: There's lots of violence in this neighborhood. Like there's this couple that's always fighting. When the guy gets mad, he hits her. It happens upstairs in their house. She's thrown the coffee pot at him and the toaster, they [the coffeepot and toaster] landed in the street . . . I saw it while walking by. . . . The guy would show off all the time in front of his friends. One day when he was hitting her, she just punched him back and told him she wasn't going to live with him anymore. He used to hit her hard. She used to cry but she would still go out with him. She said she loved him too much to dump him. A lot of people go back. . . .

Even though the community is seen as "a pretty good place to live" and "overall a nice neighborhood," the girls' descriptions of residency quickly devolve into stories of violence—mostly violence directed toward women by the hands of men in both public and private spaces. . . .

Although these girls may look at abuse differently, they all are quick to recognize violence as a defining feature of their community. Many mention that women "go back" to their abuser as if it were acceptable or normal for men to abuse women, and that it is the women's duty to negotiate their way around this violence. Again, missing in these arguments is the recognition that men are responsible for their abusive behavior. The only critique articulated is raised by Lisette, but it is directed toward neighborhood mothers whom she feels are not adequately putting their children's needs first.

Despite their young ages, the Canal Town girls have heard of abuse in a variety of different contexts and forms and, according to their narrations, women in this neighborhood do not always endure their violence in isolation. Rosie, for example, reveals that a neighbor sought refuge with her mother after a severe beating. Abuse, however, is seemingly concealed within the community—that is, complaints rarely reach a more public forum.

For girls, abuse does not just exist in public places and in the private dwellings of others. Violence also occurs in their own homes. In talking about personal experiences with abuse, they typically contextualize violence as part of the past, as "things are better now." The younger girls, though, are not consistent in packaging such events in history. For instance, Elizabeth and Sally shift from present to past in describing the abuse in their homes. Many women recall chilling vignettes of unbridled rage that pattern their upbringing. . . .

> Sally: I used to think of myself as a zero, like I was nothing. I was stupid; I couldn't do anything . . . I don't anymore because we're all done with the violence in my house . . . I've tried to keep it out since I was a kid. . . . My mom and John [her mother's boyfriend] will argue over the littlest things. My mom is someone who is a violent person too. Sometimes she hits us, or he does. Then she would take a shower and we would get all dressed up, and we would all go out somewhere. After something bad would happen, she would try to make it better. She's a real fun person. . . . We're really close. We make cookies together and breakfast together.

> Anne: My mom and her boyfriend constantly fight because they drink. When I was little, I remember being in my bed. I was sleeping, only my other sisters came and woke me up because my mom and her boyfriend were fighting. We [Anne and her sisters] started crying. I was screaming. My sisters were trying to calm me down. Our door was above the staircase and you could see the front door. I just had visions of me running out the door to get help because I was so scared. My oldest sister was like nine or ten and I had to go the bathroom and we only have one and it was downstairs. She sneaked me downstairs and into the kitchen and there was glasses smashed all over, there were plants underwater, the phone cord was underwater in the kitchen sink. It was just a wreck everywhere. But most of all, there were streams of blood mixing in with the water, on the floor, on the walls. . . .

As these narrations indicate, domestic violence patterns the lives of these girls. . . . Mom can offer little salvation as she is often drunk, violent herself, or powerless as the man in her life is on an abusive rampage. . . . [T]hese females have little recourse from the extreme and terrifying conditions that govern their lives. In Sally's case, her mother is also violent yet is thought of as making up for that abuse by involving her daughter in family-style activities. Sally learns, therefore, not to see or feel pain. Given these accounts, it is easy for me to conclude that the effects of domestic violence are not something that can be contained at home, and the Canal Town girls indicate that exposure to abuse profoundly shapes their behavior in other places, such as school.

> Elizabeth: About twice a month they [her mother and mother's boyfriend] fight. But not that far apart. Last time he [the boyfriend] smacked me, I had a red hand on my face. I walked around with a red hand on my face, only I wouldn't let anybody see it . . . I skipped school and the [community] center for, like, three days so no one would ask me about it . . . I hid in my closet until you could barely see it. Then when I went back to school, I stayed real quiet because I didn't want people to look at me, notice the hand on my face.

> Christina: When I had a boyfriend, he [her father] got so mad at me. He told me I wasn't allowed to have a boyfriend. I didn't know that because he never told me. He said that if he ever saw him again, I would get my ass kicked. So one day he heard that Robbie [her boyfriend] walked me to school. Well, he [her father] came over that night and pulled down my pants and whipped me with his belt. I was bloody and the next day full of bruises. But I hurt more from being embarrassed to have my pants pulled down at my age. It hurt to sit all day long at school; that's all I could concentrate on. I couldn't go to the nurse because then she would find out. Nobody knew how I hurt under my clothes. I couldn't go to gym because people would find out, so I skipped. I hid in the bathroom but got picked up by the hall monitor who accused me of skipping gym to smoke. I just got so mad when I heard this, I pushed her [the hall monitor] away from me and yelled. I was out of control with anger when they were dragging me down to the principal's. I got suspended for a week and had to talk to a school psychologist for two weeks about how bad smoking is for your health. . . .

The glimpses into these lives suggest children from violent homes are learning at very young ages how to negotiate lives that are enmeshed inside a web of overwhelming circumstances. Elizabeth talks about how her mother's boyfriend blames her and her mother and sister for all that is wrong. . . . As they devise ways to conceal their bruises, they each face their pain alone. Elizabeth skips school and seeks shelter from the world in the same closet in which she is punished by her mother's boyfriend. Christina is choked on her anger and pain and separates herself from school activity only to become embroiled in another set of problems. . . .

The narrations of the poor white girls in this study reflect findings in much of the existing research (Weis, Marusza-Hall, and Fine 1998). As these girls indicate, abuse at home makes it difficult to concentrate in school, and the hurt, anger, and fear that they harbor inside often render them silent, which also corroborates these studies (Elkind 1984; Jaffe, Wolfe, and Wilson 1990). According to Afulayan (1993), some children blame themselves for the abuse and skip school to protect a parent from the abuser, while other children become ill from worry. Depression, sleep disturbances, suicidal tendencies, and low self-esteem are other symptoms exhibited among children living in violent homes (Hughes 1988; Reid, Kavanaugh, and Baldwin 1987).

All of these girls reveal they spend incredible energy on keeping their abuse a secret while in school. This is likely in response to a number of fears, including fear of public embarrassment, fear of further angering an abuser, or fear that families will be torn apart by authorities. While observing the girls at school, I noticed that some of them sustained bruises that could not be so easily hidden under long sleeves or turtlenecks. One day, for example, Christina came to school wearing an excessive amount of eye makeup, which was noticeable, considering she usually did not wear any. While talking to her outside after school, I realized this was probably an attempt to conceal a black eye, which could clearly be seen in the harsh light of day.

Interestingly, I did not hear any talk of domestic violence at school—critical or otherwise. This finding parallels the poor and working-class white women in the study of Weis et al. (1997) who also were silent about the abuse in their lives, which was similarly not interrupted by schools, the legal system, and so forth. It did not seem to me that any of the girls sought help at school from their white female peers, teachers, or anyone else in coping with abuse. Instead, in the space of the school, a code of silence surrounding domestic violence prevailed, even though the girls articulate an awareness of others' abuse throughout the community. Not once did I hear students or teachers query others about violence, nor was abuse even mentioned as a social problem in classes in which human behavior was discussed. Even on the day that Christina came to school attempting to camouflage a bruised eye, I did not observe a teacher pull her aside to talk, nor did I hear her friends ask her if she was all right. Dragged by their families from one violent situation to the next, it is remarkable that these girls are, for the most part, able to get through the school day, go home, and come back again tomorrow.

CONCLUSION

The Canal Town girls are from families that had been working-class for generations. Born into the snares of a postindustrial economy, today these girls are growing up in poverty. As their narrations on work and family indicate, gender arrangements in their lives echo that of the working class in which women are subordinate to men. Embedded in this subordination is a silencing of domestic violence.

Domestic violence runs painfully deep in the lives of the Canal Town girls. These girls are socialized at an early age to conceal abuse from those outside the community who might take action. As a method of coping, they have learned to work around abuse to such an extent that by envisioning their future lives as financially independent, they hope to sidestep violence. The sting of abuse provides much of the scaffolding for how these girls wish to construct their lives, and men are seemingly not taken to task. In this tight-knit community, it often hurts to be a girl.

During an entire year of fieldwork, I never saw or heard a teacher approach a student concerning domestic violence. I also never witnessed a teacher initiate a discussion on the topic of abuse in class. Throughout the year, I had the opportunity to ask all of the teachers if they had knowledge of the extent of violence in the Canal Town community or the possibility of abuse in the lives of their students. The teachers had little to say on this topic, many indicating they had not thought much about domestic violence, although they "wouldn't be surprised."

By not responding to violence in the home, institutions that structure the lives of these girls, such as schools, arguably contribute to its concealment. The guidelines already in place in some schools, the counselors, and child abuse training for teachers do not typically address the needs of battered youth. Due to shame or punishment that awaits at home, youth do not always visit a counselor. It is often the case that teachers also are afraid to report abuse—afraid of upsetting the students, parents, and school administrators. Perhaps educators feel "unsure" about their suspicions and "wait" to see more evidence. Indeed, alerting Child Protective Services many times leads to further abuse by an angered parent. Likewise, police investigations and court appearances often prove unproductive and humiliating for women and children (Weis, Marusza-Hall, and Fine 1998).

Educators must come to the conclusion that at least some students in their classes go home to abusive situations. Teachers and policy makers, therefore, are confronted with the task of formulating more tangible responses. In English and History classes, boys and girls can often be led in critical discussions about domestic violence as it relates to classroom material and to daily life. Through these lessons, abuse must be positioned as abnormal behavior, with social and historical roots that can be unraveled. Older kids can also be encouraged to enter internships at domestic violence shelters and hot lines, so youth can learn that abuse is wrong, it is not a personal problem, and there is some recourse (Weis, Marusza-Hall, and Fine 1998).

Educators and social scientists must additionally seek out other safe spaces in students' lives where critical conversations can take place (Fine and Weis 1998; Weis, Marusza-Hall, and Fine 1998). The Canal Town girls, for example, are regular visitors to a neighborhood community center. Places such as community centers, arts programs, and youth groups offer a location that is unbounded by state guidelines where such talk can happen. By conducting workshops in these sites by those who run domestic violence shelters, youth can be led to think critically about abuse and can come realize they are not alone and that there is a possibility for a different way of life.

NOTES

1. This name is based on the fact that during the early 1800s, this area of the city was selected as the last stop on a major canal that was constructed across the state. This opened up the city, transforming it into a formidable site for the production and transport of steel. During the past few decades, however, most of this industry has left the area.
2. Boys living in violent homes may experience more abuse than girls. This has been found to be the case because when angry, boys typically act out more than girls. Because this acting out is more apt to enrage a violent adult, boys often end up as a more primary target (Jouriles and Norwood 1995).

REFERENCES

Afulayan, J. 1993.Consequences of domestic violence on elementary school education. *Child and Family Therapy* 15:55–58.

Bluestone, B., and B. Harrison. 1982. *The deindustrialization of America: Plant closings, community abandonment, and the dismantling of basic industry.* New York: Basic Books.

Elkind, P. 1984. *All grown up and no place to go.* Reading, MA: Addison-Wesley.

Everhart, R. 1983. *Reading, writing and resistance: Adolescence and labor in a junior high school.* Boston: Routledge and Kegan Paul.

Finders, M. 1996. *Just girls: Hidden literacies and life in junior high.* New York: Teachers College Press.

Fine, M., and L. Weis. 1998. *The unknown city: The lives of poor and working-class young adults.* New York: Beacon.

Holland, D., and M. Eisenhart. 1990. *Educated in romance: Women, achievement, and college culture.* Chicago: University of Chicago Press.

Hughes, H. 1988. Psychological and behavioral correlates of family violence in child witnesses and victims. *American Journal of Orthopsychiatry,* 58:77–90.

Jaffe, P., S. Wolfe and S. Wilson.1990. *Children of battered women.* Newbury Park, CA: Sage.

Jouriles, E., and W. Norwood. 1995. Physical aggression toward boys and girls in families characterized by the battering of women. *Journal of Family Psychology* 9:69–78.

May, M. 1987. The historical problem of the family wage: The Ford Motor Company and the five dollar day. In *Families and work,* edited by N. Gerstel and H. E. Gross. Philadelphia, PA: Temple University Press.

McRobbie, A. 1991. *Feminism and youth culture: From Jackie to just seventeen.* Boston: Unwin Hyman.

Oliver, M., and T, Shapiro. 1995. *Black wealth, white wealth: A new perspective on racial inequality.* New York: Routledge.

Perry, D. 1996. *Governance in Erie County: A foundation for understanding and action.* Buffalo: State University of New York Press.

Raissiguier, C. 1994. *Becoming women, becoming workers: Identity formation in a French vocational school.* Albany: State University of New York Press.

Reid, J., T. Kavanaugh, and J. Baldwin. 1987. Abusive parents' perception of child problem behavior: An example of paternal violence. *Journal of Abnormal Child Psychology* 15:451–66.

Smith, D. 1987. *The everyday world as problematic: A feminist sociology.* Boston: Northeastern University Press.

———. 1988. Femininity as discourse. In *Becoming feminine: The politics of popular culture,* edited by L. Roman, L. Christian-Smith, and E. Ellsworth. London: Falmer.

Valli, L. 1988. Gender identity and the technology of office education. In *Class, race, and gender in American education,* edited by L. Weis. Albany: State University of New York Press.

Weis, L. 1990. *Working class without work.* New York: Routledge.

Weis, L., M. Fine, A. Proweller, C. Bertram, and J. Marusza-Hall. 1997. I've slept in clothes long enough: Excavating the sounds of domestic violence among women in the white working class. *Urban Review* 30:43–62.

Weis, L., J. Marusza-Hall, and M. Fine. 1998. Out of the cupboard: Kids, domestic violence, and schools. *British Journal of Sociology of Education* 19:53–73.

Woodcock Tentler, L. W. 1979. *Wage earning women: Industrial work and family life in the US, 1900–1930.* New York: Oxford University Press.

45

Escape from Animal House

Frat Boy Tells All

BY SCOTT STRAUS

In this brief reflection on his fraternity days, journalist Scott Straus describes the ways in which fraternities use misogyny—the hatred of women—to create an in-group bonding that isolates frat brothers from women. The fraternity he describes uses homophobia and misogyny to promote sexual violence and sex as a way to control brothers' feelings and loyalties. He calls these practices a "foolproof recipe for raping women." Straus gives us a look inside a subculture of American society that systematically has promoted misogyny and sexual violation of women as a basis for membership in an elite group.

1. What are the practices of this fraternity that promote misogyny?
2. What practices of this fraternity systematically define appropriate sexuality for brothers?
3. Why don't frat brothers leave, as did Straus?

Six years ago I was an 18-year-old prep-school kid, a readymade frat boy lured to Dartmouth College by its reputation for drinking and ruggedness. I pledged Alpha Delta, which, was notoriously raucous; *National Lampoon's Animal House* was based on a screenwriter's experience there. But shortly after my initiation I dropped out, disgusted and angry, and now I tell secrets that I swore not to.

Typical frat boys wore T-shirts identifying Dartmouth as, the place for COLD BEER, COLD WEATHER, COLD WOMEN, and an AD rival's shirt declared membership in the HE-MAN WOMAN HATERS CLUB. Frat graffiti was also rife with homophobia—I found the words ALL FAGS MUST DIE carved into a library table.

A favorite fraternity ditty, written after Dartmouth went coed in 1972 and sung to the tune of "This Old Man" goes like this:

> Our cohogs, they play one/they're all here to spoil our fun/with a knick-knack, paddy-whack/send the bitches home/our cohogs go to bed alone.

The word *cohogs* refers to coeds and to a thick-shelled New England clam, hence female genitalia.

In rural Hanover, New Hampshire, Greek houses dominated the social circuit, offering free beer and music. Over half the student body belonged, most of my freshman-class friends pledged, but I had some reservations. I had just stopped drinking for fear of alcoholism (in prep school I used to drink as much as I could, to make a name for myself). But I was persuaded that AD was changing for the better, and the brothers assured me I could abstain.

The civilities lasted all of one week. The brothers made clear that for the next two months, the pledges were to sit on the floor, fetch brothers beer, tote lunch boxes, and endure insults. Each week two of us were forced to guzzle the "Rack of Gnarl," a 100-ounce concoction that invariably induced vomiting. (Oysters with mayonnaise and dog food was a favorite, as was a mix of Listerine, Diet Coke, and blue cheese.)

The brothers' goal was to humiliate us and make us feel excluded—like women and gay men. In a ceremony in the so-called Sex Room, pledges were forced to perform fellatio on something phallic. As an AD bully explained: "We try to kind of let you experience what a girl goes through when she sucks your dick."

Once pledges were branded *the other,* we had to fight our way into *the brother:* We had to become men, proof of which lay in drinking endless cups of beer (I was given prune juice) and in demonstrating heterosexual prowess, hatred for women, and loyalty to the fraternity. This was a foolproof recipe for raping women, though I did not know that then.

KEEPING SCORE

The frat-house basement floor was rimmed with a gutter for vomit, piss, and stale beer. Called the gorf, it was often filled on weekends. Whenever a woman entered the cellar, ADs ritually dropped their pants and chanted lewdly about Iowa cornfields.

Each week, who had sex and with whom was official "house business," and the house scribe read off the list of the lucky. Brother X scored with sister Y, he said, and we broke into song. Here's to brother X, we chanted, then drank. Here's to sister Y, and we drank again.

Misogyny was our pass, to show each other we were cool. And so was the sex we had. Caring for the "girl" was just a way to loosen her pants; this was sex to "score" and then tell the brothers about. Sex for any one brother was for the communal consumption of the brotherhood.

We were goaded to visit the strip joints of Montreal. A designated driver, I solemnly followed the posse of pledges from bar to bar during, their preclimax drinking binge. At the performance I winced, wishing I was somewhere else. Another pledge noticed my reticence and slipped a stripper money to wag her ass in my face. I didn't move, struck dumb. *Did I like it? Was I a brother?* they seemed to be taunting. I felt scrutinized, embarrassed, and I wondered how the woman felt.

The final test was called Hell Night. Rooms in the frat house were converted into stations we had to pass; at the end was the infamous Sex Room. Blindfolded, pledges were brought in one by one. There before the brothers we each had to tell our first or most exciting sex experience. I lied; I made up a story about fucking a woman who shaved between her legs. I got kudos for it; the brothers cheered.

PLAYING THE GAME

Masculinity never came easily to me; it was always my bane. In high school at Andover, in our all-male dorm rooms, I used to chime in late at night when we called women "pelts" and complained that so and so's girlfriend was taking him away from us. Awed as a child by men's and older boys' tales about the great mystery of sex—which, once had, would make me into a man—I "did the nasty" the first chance I got, then told my friends. I didn't really like the sex, I had no idea what I was doing, and when I boasted I got laid, my pals made fun of me. My score was low—the girl wasn't pretty enough. Still, I felt relief: I had shown them I could screw, and that proved I was an authentic guy.

Almost every male I know has been seduced by such qualifying rounds for becoming one of the guys. Even though AD's rigamarole was more formalized and rigorous than any I'd ever faced, I knew the pattern well. I had tried to prove that I'm a man on those terms for most of my life.

Although I played the game, I never really felt comfortable with what men required one another to be. I didn't understand why men made one another feel stupid all the time, whether for having sex with someone ugly or for being thoughtful. Why were we "pussy-whipped" if we wouldn't join in citing women as our common problem? And why were we "abandoning tradition" if we refused to be cruel to one another? I harbored this dissent even as a pledge, but I had no framework then to think about gender critically, and no way to talk about my feelings.

GETTING SCREWED

During most frat meetings, I sat in the back row on the floor. When I finally passed Hell Night and attended my first meeting as a brother, I was excited. My lunch box was gone; I claimed a brothers-only seat and even moved closer to the front of the hall. Maybe the bravado would subside, I thought; maybe now I could speak my mind and try to change the system from within. So when the fraternity president announced that he had some heartbreaking news for AD dichards, I leaned forward.

The Dartmouth administration, he said, was forcing all fraternities on campus to admit women. The older brothers had withheld this information from the pledges, he continued, but now that we were brothers they wanted to hear our thoughts on the idea. The first speakers railed against it booing the administration. Others said that "cracks" would destroy the fraternity and all its traditions. Finally I stood up and declared my solid support for the idea at some length, saying that AD men had a lot to learn from women. After I sat down I noticed that no one met my eyes, but neither was anyone rebutting my points. Soon the debate fizzled out, and a few brothers began laughing.

My heart sank. In fact, no such directive from the administration existed. I had fallen for a trap. The fuckers had beat me, finally fished out of me that I no longer thought of women as "pelts." What did I get for ending the misogynistic masquerade? Derision. The fuckers made me endure two months of abuse as a pledge only to trick me, to make it humiliating ever to think that our club could be anything but all-male. This was brotherhood and bonding between men based on betrayal, I thought then, and I basically never went back.

TAKING ACTION

I officially dropped out of AD during my next term, moved off campus, and began to try to find my way. I flirted with New Age spiritualism, embraced a liberal environmental agenda, and, after the first bombs were dropped on Baghdad, turned to Marx.

But I reacted to feminism with the most emotion. It happened during a poetry class taught by Ivy Schweitzer, who spun a dazzling feminist analysis into her readings of Eliot and Pound. When she finished, I stewed in my seat: She had misrepresented men, I thought; so the next afternoon I rushed to her office to protest. Not all men were bad, I proffered; some men were even stifled by the expectations of masculinity. Listening patiently to my defensiveness, Professor Schweitzer suggested she and I were saying much the same thing, and she encouraged me to read feminist books written by men.

My initial response to reading male feminism was to work toward a "redefinition" of masculinity. I wanted men to change, to talk about their feelings, and to end the misogyny. But it was reading Marx that led me to think about how institutions had to change. If I was to challenge sexism, I realized, I had to strike at the structures that perpetuated it. That's when I made the connection to fraternities. They were a logical target for me; I had firsthand knowledge of what went on behind their closed doors.

Two years after I pledged AD, I set out with Professor Schweitzer's spouse, Tom Luxon, an English professor and women s studies instructor, on an independent study of fraternities. I wrote a 50-page paper exposing as many secrets as I could; showing how fraternities promoted misogyny, homophobia, and anti-intellectualism; and arguing that frat culture promoted sexual assault.

Professor Luxon and I circulated the paper, spun out my analysis in a cafeteria backroom, and showed a leaked videotape of a Hell Night off campus. Many faculty were supportive, and a few administrators were as well. But an appointment with the dean of students proved pointless, lawyers for the college thwarted our showing the video on campus, and I learned secondhand that my former brothers in AD had marked an X through my face in the house composite photo.

Then during the spring term of my senior year, a string of sexual assaults caught the school's attention. Several survivors had been raped in fraternities or by pledges. I published an abbreviated version of my independent study in the school's lone lefty student newspaper. Several well-attended debates ensued. From where I sat, it looked like we were winning the intellectual arguments. Many students, almost all women, quietly voiced support for our side. A small anti-fraternity group was formed. A journalist, sent to Hanover to cover the controversy for *Rolling Stone* magazine, observed, "These could be scenes from an *Animal House* sequel with a no-fun ending."

LOOKING BACK

Our contribution is hard to measure. Fraternities at Dartmouth were not abolished, and they don't seem to be on the way out anytime soon. Nostalgia centers for alumni, they are vital for fundraising—and still the place to go to hear dated songs that reminisce about when the school was all-male. Three years ago, when pledge rush was shifted from freshman to sophomore year, Greek membership dipped slightly, from just over half to just under half of each entering class.

To this day I find it hard to believe that everyone else in my 20-strong pledge class stuck it out. Did they really feel that bonding in a strip joint, or through bragging in the Sex Room, was meaningful and harmless? Was I the only one to see the absurdity because I was the only one who was sober?

Though I occasionally wonder if I betrayed my former brothers, I don't regret my activism. My initial exposure to feminism pitched me into self-flagellation and guilt—I didn't like myself for the man that I had been. But in aiming at the frats, I picked a battle with my name on it—I was fighting over my body, my socialization; and my personal fight against sexism and homophobia lay in challenging the violence between men.

I don't consider myself "above" sexism or homophobia, because I am still a man in this culture. My sexuality remains indeterminate as I seek to unpack its social influences, and I have a lot to work through. The difference now is that I believe I can make the right judgments. On that night I stood and spoke up only to be made a fool of, I flunked my brothers' test for me—but I passed my own. My anti-frat activism, more than anything I had done before to prove myself—to test and trust my convictions—helped me know who I am. My fight against traditional masculinity marked a moment of courage, a personal beginning, and a place from which I am becoming the man that I want to be.

46

Sexual Trafficking in Women

International Political Economy and the Politics of Sex

BY ANDREA MARIE BERTONE

Political scientist Andrea Bertone describes how the international sex trade exploits women from formerly undeveloped areas in a system of sexual services and sexual tourism. Although there have been attempts to halt the sexual trafficking of women, most have not been successful because the control over the legal, criminal, and political systems both in the sending and receiving countries gives little legitimacy to the rights of the exploited women. Women taken from their homeland to be used in sexual trade in a developed country have few rights and are essentially at the mercy of their exploiters. This sexual exploitation of women enforces gender through the enslavement of women for the sexual satisfaction of men.

1. Bertone argues that the changing role of women in industrialized countries relates to an increase in sexual trafficking of women from poorer countries. Why might this be so?
2. What countries are destinations for sexual trafficking?
3. What are the consequences of sexual trafficking for the women involved?

★ ★ ★

INTERNATIONAL TRAFFICKING IN GIRLS AND WOMEN

A recent manifestation of the North/South, East/West political-economic divide is the international sex trade in women. Sex tourism, mail order brides, prostitution in brothels, pornography, and militarized sexual services are examples of this market. Trafficking in women is a large subset of the business in which women are coerced, enslaved, kidnapped, tortured, or raped in order to sexually service men for the profit of others (Raghu, 1997, p. 145). Trafficking in girls and women is one form of migrant trafficking, but one that has special characteristics. Trafficking in women for purposes of sexual employment can involve situations in which the woman is aware of the circumstances before she travels. However, it also involves situations in which a girl and/or woman is kidnapped for purposes of trafficking, or sold into prostitution or forced marriage, and therefore it is not considered voluntary.

The definition of trafficking and the exploitation and prostitution of others is set out in articles 1 and 2 of the 1949 Convention for the Suppression on the Traffic in Persons and of the Exploitation of the Prostitution of Others. The Convention refers to actions at both the national and international levels. Since 1949, the concept of trafficking has been extended to include trafficking for the purpose of other forms of exploitation of women. The wider view of trafficking is reflected in the Beijing Declaration and Platform for Action, which also includes forced marriages and forced labor (www.unifem.undp.org). The International Organization for Migration (IOM), a Geneva-based Inter-governmental organization, states that trafficking occurs when: "a migrant is illicitly engaged (recruited, kidnapped, sold, etc.) and/or moved, either within national or across international borders; [or when] intermediaries (traffickers) during any part of this process obtain economic or other profit by means of deception, coercion and/or other forms of exploitation under conditions that violate the fundamental human rights of migrants" (IOM, 1999). It also includes those cases where the woman is aware of the nature of the work at the point of leaving but on arrival finds herself in a situation where her fundamental human rights and freedoms are violated (Qweb Sweden, 1999).

Another compelling definition of the international traffic in women "includes any situation where women or girls cannot change the immediate conditions of their existence; where, regardless of how they got into those conditions; they cannot get out (without grave consequence); and where they are subject to sexual violence and exploitation. The prime conditions for trafficking arise when developing nations commence the transformation of their economies" (Raghu, 1997, p. 145). Rapid economic industrialization in formerly undeveloped countries and regions, coupled with historical structures at the societal level which support the demoralization and subordination of women, are the primary facilitators of the sex trade in Southeast Asia, especially Thailand and Malaysia.

AN INTERNATIONAL POLITICAL ECONOMY OF SEX

The study of international political economy attempts to clarify the complexity of relations throughout the world. Industrial capitalism has emerged as the most favored economic system, though the ways in which it is manifested through political bodies, such as the nation-state, differs greatly throughout the world. Any theory of politics or economics separately cannot hope to explain or understand the interstices of human interaction. The trafficking in women is a perversion of the interaction of politics and economics and it proves globalization to be a process by which humans may be commodified in the most base and demoralizing ways. It is widely agreed that the contemporary, international sex trade has its roots in the international political economy of the capitalist, world market (Enloe, 1989; Pettman, 1996; Raghu, 1997; Skrobanek et al., 1997). The international political economy of sex not only includes the supply side—the women of the third world, the poor states, or exotic Asian women—but it cannot maintain itself without the demand from the organizers of the trade—the men from industrialized and developing countries. The patriarchal world system hungers for and sustains the international subculture of docile women from underdeveloped nations. The women themselves, who are forced or lured into the trade, believe that providing international sexual services and sex tourism outfits is the acceptable order of things. The men accept this world order as well, regardless of their background.

THE POLITICS AND CONSEQUENCES OF TRAFFICKING IN WOMEN

Networks

Although trafficking in women for the purposes of sexual services is not a new global phenomenon, the networks of international prostitution rings have proliferated enormously in the last ten years. The Geneva-based International Organization for Migration (IOM) has stated that as many as 500,000 women are annually trafficked into Western Europe alone. The primary reasons for this are the international spread of capitalism, the growing gap between rich and poor countries, the increasing demand for it in industrialized countries, and the changing roles of the woman as independent from her family and/or as the primary, family breadwinner. Sex trafficking is a large scale, highly organized and profitable international business venture transcending state borders and nationalities of women who supply the commodity of sex and of men who demand it. Through networks of traffickers, women are transported from Asia, Southeast Asia, the former Soviet Union, Latin America, and Africa to Western Europe, North America, Australia, and Japan.

There are three different kinds of networks of trafficking in women which Mr. Marco Gramegna, an official of the IOM, has identified. The large-scale network has political and economic international contacts in both the countries of origin and destination. Women are recruited in a variety of seemingly legal ways as au pairs or language students. The medium scale networks specialize in trafficking women from one specific country. The small-scale network traffics one or two women at a time, whenever a club or brothel owner places an order through contacts. The contacts recruit the woman, accompany her to the country of destination, and deliver her to the club owner. It is apparent that the route and mode of transport used will depend on the location of the sending country (IOM, 1998).

Most often, little information gets back to the small villages from where these women have come. Either they come back and do not talk about their experiences because they are afraid of being ostracized in their family, or they do not return because they are kept in situations of bondage. Therefore, those who are committed to taking concrete action to put an end to trafficking cannot reliably depend upon women to warn other women. The whole allure of traveling abroad will often win out to any rumors that women may have heard about other women overseas. Sometimes women will simply hear that there is much money to be made overseas.

Not only has international labor migration expanded dramatically in recent decades, but there are clear patterns of migration and avenues where people from certain states or regions send most of their migrants to other states or areas. It is a commonly recognized phenomenon in the study of international migration. There are many implications for this kind of migration. Unless we are referring to traditional countries of immigration (United States, Australia, Canada), many times when a pattern of labor migration grows between two places, the country of destination will operate under the assumption that the workers will return home after a certain amount of time. This political and social position is naive on the part of host countries. National policies have not reflected the growing number of migrant laborers who may decide to stay in the host country, thereby possibly encouraging the delayed rights of certain peoples. This is most notable in countries of Western Europe where the "guest worker" programs of the 1960s and 1970s resulted in the permanent settlement of millions of men and women from North Africa and the Middle East, much to the chagrin of the political and social establishments.

Because of the illegal nature of sex trafficking, girls and women may or may not be sending home remittances to family. If they are not, it is difficult for them to return home because of the social stigma that prostitutes have. If they have contracted HIV or other sexually transmitted diseases, they will no longer be able to work due to the illness and may be sterile from their disease. Nevertheless, certain regions are more dependent than others on those remittances.

Prostitution has existed for thousands of years in many different societies. However, South and Southeast Asia are one of the original areas of the world where sexualized work and sex trafficking developed. For example, Thailand's sex tourism can be traced back through local forms of prostitution and concubinage, and colonial sex trading. Its scope and numbers dramatically changed in the face of another international process, militarization, linked especially to the use of Bangkok for rest and relaxation during the Vietnam War (Pettman, 1996a, p. 200; Lim, 1998).

Militarized prostitution developed around the huge foreign military bases like those which were until recently in the Philippines. Militarized prostitution is seen as providing for the sexual needs of the soldier, rationalized in different ways as "boys will be boys," maintaining morale and rewarding long, overseas service. There has been a well-documented history of international politics surrounding military prostitution, with colonial authorities and more recently commanders from the foreign military negotiating with local governments to procure sex for the soldiers while at the same time attempting to cause the least political impact and disruption (Enloe, 1993). Managing a military base is a foreign policy issue, a community relations issue, and a law and order issue. It is also a public health issue (Pettman, 1996a, p. 201). Asian governments have done little in the way of formalized actions to discourage the trafficking of its women. Historically they have looked the other way as more and more girls and women are being lured to the cities and abroad for sexualized employment.

There is also a growing traffic in "exotic" women from sex destination states to rich states, mainly Western but including Japan. The term "entertainers" is often a transit category or euphemism for prostitution. Some 286,000 Filipinas and some 50,000 Thai women entered Japan as entertainers between 1988 and 1992. They are particularly vulnerable as young women, in jobs that are sex related, who frequently become overstayers, in a country that is both largely unknown to them and where they are subjected to gendered, racialized stereotyping and treatment (David, 1992). This in turn affects labor migrants in more respectable jobs. For example, Filipino maids avoid the company of "entertainers" (David 1992 as quoted in Pettman, 1996b, p. 207).

According to Pettman, there are an estimated 200,000 Thai women in Western European brothels, and many more in other states. Often they are on temporary or tourist visas or they are overstayers, caught in the familiar binds of debt, poverty, violence, and control. These forms of traffic make the international/internal state distinction even more problematic.

After Asia, Eastern Europe and the former Soviet Union are extremely fast growing markets for young women. In fact, trafficking of women out of this area began so recently that almost no academic research has been conducted on it. Stories which surface about unfortunate women in uncompromising situations show up in newspaper articles in Western Europe and Canada and reports placed on the Internet by the JOM. Little is known about exactly how many women from Eastern Europe and the former Soviet Union are involved in this trade. Since the fall of Communism, criminal networks have flourished both in the trading of illegal drugs and weapons, and also in the trafficking of women to the Middle East and Western Europe. Economic desperation has superseded hope that conditions will improve because capitalism has been introduced.

Centered in Moscow and the Ukrainian capital, Kiev, the networks trafficking women run east to Japan and Thailand, where it is estimated that thousands of young Slavic women now work against their will as prostitutes, and west to the Adriatic coast and beyond. The routes are controlled by Russian crime gangs based in Moscow. Even when they do not specifically move the women overseas, they provide security, logistical support, liaison with brothel owners in many countries, and false documents. Women often start their journey by choice. Seeking a better life, they are lured by local advertisements for good jobs in foreign countries at wages they could never imagine at home. The travel networks are often complicated. Women will respond to ads in the newspapers and are lured to Italy, Germany, Turkey, and Israel where promises are made of good jobs which simply do not exist. A study done by the Washington-based nonprofit group Global Survival Network found that police officials in many sending countries simply disregard evidence that the trafficking is taking place (Specter, 1998).

Israel is a common destination for the hundreds of thousands of women who leave Ukraine. Prostitution is not illegal and there is a great demand for sexual services. In Milan, Italy, the police broke up a ring that was holding auctions in which women abducted from the countries of the former Soviet Union were put on blocks, partially naked, and sold at an average price of just under $1,000. Michael Platzer, head of operations for the UN's Center for Crime Prevention, Vienna, Austria, stated, "The Mafia is not stupid. There is less law enforcement since the Soviet Union fell apart and more freedom of movement. The earnings are incredible. The overhead is low—you do not have to buy cars or guns. Drugs you sell once and they are gone. Women can earn money for a long time. Laws help the gangsters." Prostitution is semi-legal in many places and that makes enforcement tricky. In most cases punishment is very light (Specter, 1998).

Consequences

What are the consequences of trafficking in women? There is no doubt that the consequences of trafficking are grave for the women and countries involved. Mr. Gramegna of the IOM identified a number of consequences including a threat to orderly, legal migration and a growth in clandestine immigration. These can have serious implications for national sovereignty and relations between states, as well as internal political and economic consequences. Socially, trafficking can feed popular fear of uncontrolled borders and xenophobic sentiments. Security is put at risk by the growth in criminality that trafficking in women involves. Powerful networks are controlling the trade as the activity becomes ever more lucrative.

The gravity of the consequences for the individual cannot be overstated. These women face sexual abuse, with all the dangers of injury and severe health risks it entails. The women may be deprived of their documents and forced into a situation of severe dependence, comparable to being a hostage (or a slave). They are often subject to violence by traffickers and clients alike, deprived of basic human rights and forced to live in unendurable conditions. Some women die as a direct result of abuse and exploitation by traffickers. The mental and emotional consequences for the victims can be as severe as and longer lasting than physical scars. For many, it is difficult to talk about the ordeal and impossible to return to normal life. In some countries, a woman may be ostracized from the community if it becomes known she has worked as a prostitute. Few trafficked women receive any counseling or rehabilitation assistance (www.iom.ch/doc/trafficking).

The lethal combination of poverty, powerlessness, and poor health is evident in the figures. Many prostitutes know little or nothing of AIDS, but even if they did they would be in no position to demand that their clients use condoms. The clients' fears of AIDS have had the apparent effect of sending them in search of younger and younger boys and girls. AIDS itself is very much a part of the international political economy of sex, demonstrating how permeable state borders and people's bodies are to certain kinds of international traffic. A different reading of AIDS as a threat to national security is made by Filipina feminist members of GABRIELA, who see American military men and foreign sex tourists as infecting the Philippines body politic, and invading national sovereignty. The impact of a politics of unequal trade and debt, World Bank conditionality, restructuring, and the government's search for hard currency is linked with a feminist analysis of patriarchy and the eroticization of women's bodies (Enloe, 1990a, p. 38; Pettman, 1996a, pp. 200–202).

* * *

The world is undeniably shaped by a patriarchal structure. The combination of persistent patriarchy and rapid economic expansion places women in great disadvantage to their male counterparts in endeavors of labor migration. If industrialized countries have a problem with inequality, the problem is magnified in underdeveloped countries where poverty is the norm, education levels are low, and people are driven to find ways to better their economic situation. Allowing women equal access to education will reduce their tendency to choose gendered employment because as skilled laborers, they can contribute to a stronger national economy and improve the ability of the country to compete on a more equal footing with industrialized countries. Even if women, once educated, decide to migrate for economic reasons, their personal security will increase and they will contribute to bettering the perception of women, in general, and the country from which they came, in particular. However, all of these positive changes are contingent on changing the patriarchal structure in underdeveloped countries. This does not necessarily mean that the structure should or will become matriarchal, but that the men will understand the import of women's education and will not encourage women to be farmed into dangerous work situations where they will be at risk of their life and well-being.

REFERENCES

Enloe, Cynthia. 1989. *Bananas, Beaches and Bases: Making Feminist Sense of International Politics.* London: Pandora Publishers.

———. 1993. *The Morning After: Sexual Politics at the End of the Cold War.* Berkeley, CA: University of California Press.

International Organization for Migration (IOM). 1998. "Information Campaign Against Trafficking in Women from Ukraine." www.IOM ch/ defaultmigration web.asp.

Lim, Lin Lean. (Ed.) 1998. *The Sex Sector.* Geneva: International Labour Office.

Pettman, Jan Jindy. 1996. "An International Political Economy of Sex?" in Eleanore Kofman and Gillian Young (eds.), *Global Theory and Practice.* New York: Pinter Publishing.

———. 1996. *Worlding Women.* London: Routledge Press.

QWeb Sweden. 1999. "Crossing Borders Against Trafficking in Women and Girls." www.qweb.kvinnoforum.se/trafficking/index.html.

Raghu, Maya. 1997. "Sex Trafficking of Thai Women and the United States Asylum Law Response." *Georgetown Immigration Law Journal,* Vol. 12, No. 1 (Fall) pp. 145–186.

Skrobanek, Siriporn, Nataya Boonpakdee, and Chutima Jantateero. 1997. *The Traffic in Women.* London: Zed Books Ltd.

Specter, Michael. 1998 "Traffickers' New Cargo: Naive Slavic Women." *New York Times,* May 18.

47

Sexual Harassment and Masculinity

The Power and Meaning of "Girl Watching"

BY BETH A. QUINN

Beth Quinn's research explores the accounts men and women give for what on the surface is a relatively benign form of sexual harassment, watching and rating women at work. She conducted 43 interviews with men and women, lasting from 1 to 3 hours, asking about relationships at work. She randomly sampled over half of her sample (25) from one organization, and recruited the remaining individuals from a nearby community college and university. The accounts men give for sexual harassment indicate that they realize the effects of their girl watching on the women they work with and would not like it if they were women. The interviews reveal that sexual harassment is instrumental in bonding men together and separating men from women in the workplace.

1. What is the role of objectification and (dis)empathy in maintaining masculinity?
2. How does sexual harassment work to enforce gender for men in the workplace and what are the effects of such behavior for women and men?
3. What would have to change if "girl watching" were to disappear from the workplace?

Confronted with complaints about sexual harassment or accounts in the media, some men claim that women are too sensitive or that they too often misinterpret men's intentions (Bernstein 1994; Buckwald 1993). In contrast, some women note with frustration that men just "don't get it" and lament the seeming inadequacy of sexual harassment policies (Conley 1991; Guccione 1992). Indeed, this ambiguity in defining acts of sexual harassment might be, as Cleveland and Kerst (1993) suggested, the most robust finding in sexual harassment research. . . .

This article focuses on the subjectivities of the perpetrators of a disputable form of sexual harassment, "girl watching." The term refers to the act of men's sexually evaluating women, often in the company of other men. It may take the form of a verbal or gestural message of "check it out," boasts of sexual prowess, or explicit comments about a woman's body or imagined sexual acts. The target may be an individual woman or group of women or simply a photograph or other representation. The woman may be a stranger, coworker, supervisor, employee, or client. For the present analysis, girl watching within the workplace is centered.

The analysis is grounded in the work of masculinity scholars such as Connell (1987, 1995) in that it attempts to explain the subject positions of the interviewed men—not the abstract and genderless subjects of patriarchy but the gendered and privileged subjects embedded in this system.

Since I am attempting to delineate the gendered worldviews of the interviewed men, I employ the term "girl watching," a phrase that reflects their language ("they watch girls").

I have chosen to center the analysis on girl watching within the workplace for two reasons. First, it appears to be fairly prevalent. For example, a survey of federal civil employees (U.S. Merit Systems Protection Board 1988) found that in the previous 24 months, 28 percent of the women surveyed had experienced "unwanted sexual looks or gestures," and 35 percent had experienced "unwanted sexual teasing, jokes, remarks, or questions." Second, girl watching is still often normalized and trivialized as only play, or "boys will be boys." A man watching girls—even in his workplace—is frequently accepted as a natural and commonplace activity, especially if he is in the presence of other men.[1] Indeed, it may be required (Hearn 1985). Thus, girl watching sits on the blurry edge between fun and harm, joking and harassment. An understanding of the process of identifying behavior as sexual harassment, or of rejecting this label, may be built on this ambiguity.

Girl watching has various forms and functions, depending on the context and the men involved. For example, it may be used by men as a directed act of power against a particular woman or women. In this, girl watching—at least in the workplace—is most clearly identified as harassing by both men and women. I am most interested, however, in the form where it is characterized as only play. This type is more obliquely motivated and, as I will argue, functions as a game men play to build shared masculine identities and social relations.

Multiple and contradictory subject positions are also evidenced in girl watching, most notably that between the gazing man and the woman he watches. Drawing on Michael Schwalbe's (1992) analysis of empathy and the formation of masculine identities, I argue that girl watching is premised on the obfuscation of this multiplicity through the objectification of the woman watched and a suppression of empathy for her. In conclusion, the ways these elements operate to produce gender differences in interpreting sexual harassment and the implications for developing effective policies are discussed.

PREVIOUS RESEARCH

The question of how behavior is or is not labeled as sexual harassment has been primarily through experimental vignettes and surveys.[2] In both methods, participants evaluate either hypothetical scenarios or lists of behaviors, considering whether, for example, the behavior constitutes sexual harassment, which party is most at fault, and what consequences the act might engender. Researchers manipulate factors such as the level of "welcomeness" the target exhibits and the relationship of the actors (supervisor-employee, coworker-coworker).

Both methods consistently show that women are willing to define more acts as sexual harassment (Gutek, Morasch, and Cohen 1983; Padgitt and Padgitt 1986; Powell 1986; York 1989; but see Stockdale and Vaux 1993) and are more likely to see situations as coercive (Garcia, Milano, and Quijano 1989). When asked who is more to blame in a particular scenario, men are more likely to blame, and less likely to empathize with, the victim (Jensen and Gutek 1982; Kenig and Ryan 1986). In terms of actual behaviors like girl watching, the U.S. Merit Systems Protection Board (1988) survey found that 81 percent of the women surveyed considered "uninvited sexually suggestive looks or gestures" from a supervisor to be sexual harassment. While the majority of men (68 percent) also defined it as such, significantly more men were willing to dismiss such behavior. Similarly, while 40 percent of the men would not consider the same behavior from a coworker to be harassing, more than three-quarters of the women would.

The most common explanation offered for these differences is gender role socialization. This conclusion is supported by the consistent finding that the more men and women adhere to traditional gender roles, the more likely they are to deny the harm in sexual harassment and to consider the behavior acceptable or at least normal (Gutek and Koss 1993; Malovich and Stake 1990; Murrell and Dietz-Uhler 1993; Popovich et al. 1992; Pryor 1987; Tagri and Hayes 1997). Men who hold predatory ideas about sexuality, who are more likely to believe rape myths, and who are more likely to self-report that they would rape under certain circumstances are less likely to see behaviors as harassing (Murrell and Dietz-Uhler 1993; Pryor 1987; Reilly et al. 1992).

These findings do not, however, adequately address the between-group differences. The more one is socialized into traditional notions of [gender], the more likely it is for both men and women to view the behaviors as acceptable or at least unchangeable. The processes by which gender roles operate to produce these differences remain underexamined.

Some theorists argue that men are more likely to discount the harassing aspects of their behavior because of a culturally conditioned tendency to misperceive women's intentions. For example, Stockdale (1993, 96) argued that "patriarchal norms create a sexually aggressive belief system in some people more than others, and this belief system can lead to the propensity to misperceive." Gender differences in interpreting sexual harassment, then, may be the outcome of the acceptance of normative ideas about women's inscrutability and indirectness and men's role as sexual aggressors. Men see harmless flirtation or sexual interest rather than harassment because they misperceive women's intent and responses.

Stockdale's (1993) theory is promising but limited. First, while it may apply to actions such as repeatedly asking for dates and quid pro quo harassment,[3] it does not effectively explain motivations for more indirect actions, such as displaying pornography and girl watching. Second, it does not explain why some men are more likely to operate from these discourses of sexual aggression contributing to a propensity to misperceive.

★ ★ ★

FINDINGS: GIRL WATCHING AS "HOMMO-SEXUALITY"

> [They] had a button on the computer that you pushed if there was a girl who came to the front counter. . . . It was a code and it said "BAFC"—Babe at Front Counter. . . . If the guy in the back looked up and saw a cute girl come in the station, he would hit this button for the other dispatcher to [come] see the cute girl.
>
> —PAULA, POLICE OFFICER

In its most serious form, girl watching operates as a targeted tactic of power. The men seem to want everyone—the targeted woman as well as coworkers, clients, and superiors—to know they are looking. The gaze demonstrates their right, as men, to sexually evaluate women. Through the gaze, the targeted woman is reduced to a sexual object, contradicting her other identities, such as that of competent worker or leader. This employment of the discourse of asymmetrical heterosexuality (i.e., the double standard) may trump a woman's formal organizational power, claims to professionalism, and organizational discourses of rationality (Collinson and Collinson 1989; Gardner 1995; Yount 1991).[4] As research on rape has demonstrated (Estrich 1987), calling attention to a woman's gendered sexuality can function to exclude recognition of her competence, rationality, trustworthiness, and even humanity. In contrast, the overt recognition of a man's (hetero)sexuality is normally compatible with other aspects of his identity; indeed, it is often required (Connell 1995; Hearn 1985). Thus, the power of sexuality is asymmetrical, in part, because being seen as sexual has different consequences for women and men.

But when they ogle, gawk, whistle and point, are men always so directly motivated to disempower their women colleagues? Is the target of the gaze also the intended audience? Consider, for example, this account told by Ed, a white, 29-year-old instrument technician.

> When a group of guys goes to a bar or a nightclub and they try to be manly. . . . A few of us always found [it] funny [when] a woman would walk by and a guy would be like, "I can have her." [pause] "Yeah, OK, we want to see it!" [laugh]

In his account—a fairly common one in men's discussions—the passing woman is simply a visual cue for their play. It seems clear that it is a game played by men for men; the woman's participation and awareness of her role seem fairly unimportant.

As Thorne (1993) reminded us, we should not be too quick to dismiss games as "only play." In her study of gender relations in elementary schools, Thorne found play to be a powerful form of gendered social action. One of its "clusters of meaning" most relevant here is that of "dramatic performance." In this, play functions as both a source of fun and a mechanism by which gendered identities, group boundaries, and power relations are (re)produced.

The metaphor of play was strong in Karl's comments. Karl, a white man in his early thirties who worked in a technical support role in the Acme engineering department, hoped to earn a degree in engineering. His frustration with his slow progress—which he attributed to the burdens of marriage and fatherhood—was evident throughout the interview. Karl saw himself as an undeserved outsider in his department and he seemed to delight in telling on the engineers.

Girl watching came up as Karl considered the gender reversal question. Like many of the men I interviewed, his first reaction was to muse about premenstrual syndrome and clothes. When I inquired about the potential social effects of the transformation (by asking him, Would it "be easier dealing with the engineers or would it be harder?") he haltingly introduced the engineers' "game."

> Karl: Some of the engineers here are very [pause] they're not very, how shall we say? [pause] What's the way I want to put this? They're not very, uh [pause] what's the word? Um. It escapes me.
>
> Researcher: Give me a hint?
>
> Karl: They watch women but they're not very careful about getting caught.
>
> Researcher: Oh! Like they ogle?
>
> Karl: Ogle or gaze or [pause] stare even, or [pause] generate a commotion of an unusual nature.

His initial discomfort in discussing the issue (with me, I presume) is evident in his excruciatingly formal and hesitant language. The aspect of play, however, came through clearly when I pushed him to describe what generating a commotion looked like: "'Oh! There goes so-and-so. Come and take a look! She's wearing this great outfit today!' Just like a schoolboy. They'll rush out of their offices and [cranes his neck] and check things out." That this is as a form of play was evident in Karl's boisterous tone and in his reference to schoolboys. This is not a case of an aggressive sexual appraising of a woman coworker but a commotion created for the benefit of other men. . . .

Producing Masculinity

I suggest that girl watching in this form functions simultaneously as a form of play and as a potentially powerful site of gendered social action. Its social significance lies in its power to form identities and relationships based on these common practices for, as Cockburn (1983, 123) has noted, "patriarchy is as much about relations between man and man as it is about relations between men and women." Girl watching works similarly to the sexual joking that Johnson (1988) suggested is a common way for heterosexual men to establish intimacy among themselves.

In particular, girl watching works as a dramatic performance played to other men, a means by which a certain type of masculinity is produced and heterosexual desire displayed. It is a means by which men assert a masculine identity to other men, in an ironic "hommo-sexual" practice of heterosexuality (Butler 1990).[5] As Connell (1995) and others (Butler 1990; West and Zimmerman 1987) have aptly noted, masculinity is not a static identity but rather one that must constantly be reclaimed. The content of any performance—and there are multiple forms—is influenced by a hegemonic notion of masculinity. When asked what "being a man entailed, many of the men and women I interviewed triangulated toward notions of strength (if not in muscle, then in character and job performance), dominance, and a marked sexuality, overflowing and uncontrollable to some degree and natural to the male "species." Heterosexuality is required, for just as the label "girl" questions a man's claim to masculine power, so does the label "fag" (Hopkins 1992; Pronger 1992). I asked Karl, for example, if he would consider his sons "good men" if they were gay. His response was laced with ambivalence; he noted only that the question was "a tough one."

The practice of girl watching is just that—a practice—one rehearsed and performed in everyday settings. This aspect of rehearsal was evident in my interview with Mike, a self-employed house painter who used to work construction. In locating himself as a born-again Christian, Mike recounted the girl watching of his fellow construction workers with contempt. Mike was particularly disturbed by a man who brought his young son to the job site one day. The boy was explicitly taught to catcall, a practice that included identifying the proper targets: women and effeminate men.

Girl watching, however, can be somewhat tenuous as a masculine practice. In their acknowledgment (to other men) of their supposed desire lies the possibility that in being too interested in women the players will be seen as mere schoolboys giggling in the playground. Taken too far, the practice undermines rather than supports a masculine performance. . . . A man must be interested in women, but not too interested; they must show their (hetero)sexual interest, but not overly so, for this would be to admit that women have power over them.

The Role of Objectification and (Dis)Empathy

As a performance of heterosexuality among men, the targeted woman is primarily an object onto which men's homosocial sexuality is projected. The presence of a woman in any form—embodied, pictorial, or as an image conjured from words—is required, but her subjectivity and active participation is not. To be sure, given the ways the discourse of asymmetrical sexuality works, men's actions may result in similarly negative effects on the targeted woman as that of a more direct form of sexualization. The crucial difference is that the men's understanding of their actions differs. This difference is one key to understanding the ambiguity around interpreting harassing behavior.

When asked about the engineers' practice of neck craning, Robert grinned, saying nothing at first. After some initial discussion, I started to ask him if he thought women were aware of their game ("Do you think that the women who are walking by . . . ?")

Robert did not want to admit that women might not enjoy it ("that didn't come out right") but acknowledged that their feelings were irrelevant. Only subjects, not objects, take pleasure or are annoyed. If a woman did complain, Robert thought the guys wouldn't know what to say." In her analysis of street harassment, Gardner (1995, 187) found a similar absence, in that "men's interpretations seldom mentioned a woman's reaction, either guessed at or observed."

The centrality of objectification was also apparent in comments made by José, a Hispanic man in his late 40s who worked in manufacturing. For José, the issue came up when he considered the topic of compliments. He initially claimed that women enjoy compliments more than men do. In reconsidering, he remembered girl watching and the importance of intent.

> There is [pause] a point where [pause] a woman can be admired by [pause] a pair of

> eyes, but we're talking about "that look." Where, you know, you're admiring her because she's dressed nice, she's got a nice figure, she's got nice legs. But then you also have the other side. You have an animal who just seems to undress you with his eyes and he's just [pause], there's those kind of people out there too.

What is most interesting about this statement is that in making the distinction between merely admiring and an animal look that ravages, José switched subject position. He spoke in the second person when describing both forms of looking, but his consistency in grammar belies a switch in subjectivity: you (as a man) admire, and you (as a woman) are undressed with his eyes. When considering an appropriate, complimentary gaze, José described it from a man's point of view; the subject who experiences the inappropriate, violating look, however, is a woman. Thus, as in Robert's account, José acknowledged that there are potentially different meanings in the act for men and women. In particular, to be admired in a certain way is potentially demeaning for a woman through its objectification. . . .

When asked to envision himself as a woman in his workplace, like many of the individuals I interviewed, Karl believed that he did not "know how to be a woman." Nonetheless, he produced an account that mirrored the stories of some of the women I interviewed. He knew the experience of girl watching could be quite different—in fact, threatening and potentially disempowering—for the woman who is its object. As such, the game was something to be avoided. In imagining themselves as women, the men remembered the practice of girl watching. None, however, were able to comfortably describe the game of girl watching from the perspective of a woman and maintain its (masculine) meaning as play.

In attempting to take up the subject position of a woman, these men are necessarily drawing on knowledge they already hold. If men simply "don't get it"—truly failing to see the harm in girl watching or other more serious acts of sexual harassment—then they should not be able to see this harm when envisioning themselves as women. What the interviews reveal is that many men—most of whom failed to see the harm of many acts that would constitute the hostile work environment form of sexual harassment—did in fact understand the harm of these acts when forced to consider the position of the targeted woman.

I suggest that the gender reversal scenario produced, in some men at least, a moment of empathy. Empathy, Schwalbe (1992) argued, requires two things. First, one must have some knowledge of the other's situation and feelings. Second, one must be motivated to take the position of the other. What the present research suggests is that gender differences in interpreting sexual harassment stem not so much from men's not getting it (a failure of the first element) but from a studied, often compulsory, lack of motivation to identify with women's experiences.

In his analysis of masculinity and empathy, Schwalbe (1992) argued that the requirements of masculinity necessitate a "narrowing of the moral self." Men learn that to effectively perform masculinity and to protect a masculine identity, they must, in many instances, ignore a woman's pain and obscure her viewpoint. Men fail to exhibit empathy with women because masculinity precludes them from taking the position of the feminine other, and men's moral stance vis-à-vis women is attenuated by this lack of empathy.

As a case study, Schwalbe (1992) considered the Thomas-Hill hearings, concluding that the examining senators maintained a masculinist stance that precluded them from giving serious consideration to Professor Hill's claims. A consequence of this masculine moral narrowing is that "charges of sexual harassment . . . are often seen as exaggerated or as fabricated out of misunderstanding or spite" (Schwalbe 1992, 46). Thus, gender differences in interpreting sexually harassing behaviors may stem more from acts of ignoring than states of ignorance.

The Problem with Getting Caught

But are women really the untroubled objects that girl watching—viewed through the eyes of men—suggests? Obviously not; the game may be premised on a denial of a woman's subjectivity, but an actual erasure is beyond men's power! It is in this multiplicity of subjectivities, as Butler (1990, ix) noted, where "trouble" lurks, provoked by "the unanticipated agency of a female 'object' who inexplicably returns the glance, reverses the gaze, and contests the place and authority of the masculine position." To face a returned gaze is to get caught, an act that has the power to undermine the logic of girl watching as simply a game among men. Karl, for example, noted that when caught, men are often flustered, a reaction suggesting that the boundaries of usual play have been disturbed.[6]

When a woman looks back, when she asks, "What are you looking at?" she speaks as a subject, and her status as mere object is disturbed. When the game is played as a form of hommosexuality, the confronted man may be baffled by her response. When she catches them looking, when she complains, the targeted woman speaks as a subject. The men, however, understand her primarily as an object, and objects do not object.

The radical potential of sexual harassment law is that it centers women's subjectivity, an aspect prompting Catharine MacKinnon's (1979) unusual hope for the law's potential as a remedy. For men engaged in girl watching, however, this subjectivity may be inconceivable. From their viewpoint, acts such as girl watching are simply games played with objects: women's bodies. Similar to Schwalbe's (1992) insight into the senators' reaction to Professor Hill, the harm of sexual harassment may seem more the result of a woman's complaint (and law's "illegitimate" encroachment into the everyday work world) than men's acts of objectification. For example, in reflecting on the impact of sexual harassment policies in the workplace, José lamented that "back in the '70s, [it was] all peace and love then. Now as things turn around, men can't get away with as much as what they used to." Just whose peace and love are we talking about?

★ ★ ★

CONCLUSIONS

In this analysis, I have sought to unravel the social logic of girl watching and its relationship to the question of gender differences in the interpretation of sexual harassment. In the form analyzed here, girl watching functions simultaneously as only play and as a potent site where power is played. Through the objectification on which it is premised and in the nonempathetic masculinity it supports, this form of girl watching simultaneously produces both the harassment and the barriers to men's acknowledgment of its potential harm.

The implications these findings have for anti–sexual harassment training are profound. If we understand harassment to be the result of a simple lack of knowledge (of ignorance), then straightforward informational sexual harassment training may be effective. The present analysis suggests, however, that the etiology of some harassment lies elsewhere. While they might have quarreled with it, most of the men I interviewed had fairly good abstract understandings of the behaviors their companies' sexual harassment policies prohibited. At the same time, in relating stories of social relations in their workplaces, most failed to identify specific behaviors as sexual harassment when they matched the abstract definition. . . . [T]he source of this contradiction lies not so much in ignorance but in acts of ignoring. Traditional sexual harassment training programs address the former rather than the later. As such, their effectiveness against sexually harassing behaviors born out of social practices of masculinity like girl watching is questionable.

Ultimately, the project of challenging sexual harassment will be frustrated and our understanding distorted unless we interrogate hegemonic, patriarchal forms of masculinity and the practices by which they are (re)produced. We must continue to research the processes by which sexual harassment is produced and the gendered identities and subjectivities on which it poaches (Wood 1998). My study provides a first step toward a more process-oriented understanding of sexual harassment, the ways the social meanings of harassment are constructed, and ultimately, the potential success of antiharassment training programs.

NOTES

1. For Example, Maria, an administrative assistant I interviewed, simultaneously echoed and critiqued this understanding when she complained about her boss's girl watching in her presence: "If he wants to do that in front of other men . . . you know, that's what men do."
2. Recently, more researchers have turned to qualitative studies as a means to understand the process of labeling behavior as harassment. Of note are Collinson and Collinson (1996), Giuffre and Williams (1994), Quinn (2000), and Rogers and Henson (1997).
3. Quid pro quo ("this for that") sexual harassment occurs when a person with organizational power attempts to coerce an individual into sexual behavior by threatening adverse job actions.
4. I prefer the term "asymmetrical heterosexuality" over "double standard" because it directly references the dominance of heterosexuality and more accurately reflects the interconnected but different forms of acceptable sexuality for men and women. As Estrich (1987) argued, it is not simply that we hold men and women to different standards of sexuality but that these standards are (re)productive of women's disempowerment.
5. "Hommo" is a play on the French word for man, *homme*.
6. Men are not always concerned with getting caught, as the behavior of catcalling construction workers amply illustrates; that a woman hears is part of the thrill (Gardner 1995). The difference between the workplace and the street is the level of anonymity the men have vis-à-vis the woman and the complexity of social rules and the diversity of power sources an individual has at his or her disposal.

REFERENCES

Bernstein, R. 1994. Guilty if charged. *New York Review of Books,* 13 January.

Buckwald, A. 1993. Compliment a woman, go to court. *Los Angeles Times.* 28 October.

Bumiller, K. 1988. *The civil rights society: The social construction of victims.* Baltimore: Johns Hopkins University Press.

Butler, J. 1990. *Gender trouble: Feminism and the subversion of identity.* New York: Routledge.

Cleveland, J. N., and M. E. Kerst. 1993. Sexual harassment and perceptions of power: An under-articulated relationship. *Journal of Vocational Behavior* 42 (1): 49–67.

Cockburn, C. 1983. *Brothers: Male dominance and technological change.* London: Pluto Press.

Collinson, D. L., and M. Collinson. 1989. Sexuality in the workplace: The domination of men's sexuality. In *The sexuality of organizations,* edited by J. Hearn and D. L. Sheppard. Newbury Park, CA: Sage.

———. 1996. "It's only Dick": The sexual harassment of women managers in insurance sales. *Work, Employment & Society* 10 (1): 29–56.

Conley, F. K. 1991. Why I'm leaving Stanford: I wanted my dignity back. *Los Angeles Times,* 9 June.

Conley, J., and W. O'Barr. 1998. *Just words.* Chicago: University of Chicago Press.

Connell, R. W. 1987. *Gender and power.* Stanford, CA: Stanford University Press.

———. 1995. *Masculinities.* Berkeley: University of California Press.

Estrich, S. 1987. *Real Rape.* Cambridge, MA: Harvard University Press.

Garcia, L., L. Milano, and A. Quijano. 1989. Perceptions of coercive sexual behavior by males and females. *Sex Roles* 21 (9/10): 569–77.

Gardner, C. B. 1995. *Passing by: Gender and public harassment.* Berkeley: University of California Press.

Giuffre, P., and C. Williams. 1994. Boundary lines: Labeling sexual harassment in restaurants. *Gender & Society* 8:378–401.

Guccione, J. 1992. Women judges still fighting harassment. *Daily Journal,* 13 October, 1.

Gutek, B. A., and M. P. Koss. 1993. Changed women and changed organizations: Consequences of and coping with sexual harassment. *Journal of Vocational Behavior* 42 (1): 28–48.

Gutek, B. A., B. Morasch, and A. G. Cohen. 1983. Interpreting social-sexual behavior in a work setting. *Journal of Vocational Behavior* 22 (1): 30–48.

Hearn, J. 1985. Men's sexuality at work. In *The sexuality of men,* edited by A. Metcalf and M. Humphries. London: Pluto Press.

Hopkins, P. 1992. Gender treachery: Homophobia, masculinity, and threatened identities. In *Rethinking masculinity: Philosophical explorations in light of feminism,* edited by L. May and R. Strikwerda. Lanham, MD: Littlefield, Adams.

Jensen, I. W., and B. A. Gutek. 1982. Attributions and assignment of responsibility in sexual harassment. *Journal of Social Issues* 38 (4): 121–36.

Johnson, M. 1988. *Strong mothers, weak wives.* Berkeley: University of California Press.

Kenig, S., and J. Ryan. 1986. Sex differences in levels of tolerance and attribution of blame for sexual harassment on a university campus. *Sex Roles* 15 (9/10): 535–49.

MacKinnon, C. A. 1979. *The sexual harassment of working women.* New Haven, CT: Yale University Press.

Malovich, N. J., and J. E. Stake. 1990. Sexual harassment on campus: Individual differences in attitudes and beliefs. *Psychology of Women Quarterly* 14 (1): 63–81.

Murrell, A. J., and B. L. Dietz-Uhler. 1993. Gender identity and adversarial sexual beliefs as predictors of attitudes toward sexual harassment. *Psychology of Women Quarterly* 17 (2): 169–75.

Padgitt, S. C., and J. S. Padgitt. 1986. Cognitive structure of sexual harassment: Implications for university policy. *Journal of College Student Personnel* 27:34–39.

Popovich, P. M., D. N. Gehlauf, J. A. Jolton, J. M. Somers, and R. M. Godinho. 1992. Perceptions of sexual harassment as a function of sex of rater and incident form and consequent. *Sex Roles* 27 (11/12): 609–25.

Powell, G. N. 1986. Effects of sex-role identity and sex on definitions of sexual harassment. *Sex Roles* 14: 9–19.

Pronger, B. 1992. Gay jocks: A phenomenology of gay men in athletics. In *Rethinking masculinity: Philosophical explorations in light of feminism,* edited by L. May and R. Strikwerda. Lanham, MD: Littlefield Adams.

Pryor, J. B. 1987. Sexual harassment proclivities in men. *Sex Roles.* 17 (5/6): 269–90.

Quinn, B. A. 2000. The paradox of complaining: Law, humor, and harassment in the everyday work world. *Law and Social Inquiry* 25 (4): 1151–83.

Reilly, M. E., B. Lott, D. Caldwell, and L. DeLuca. 1992. Tolerance for sexual harassment related to self-reported sexual victimization. *Gender & Society* 6:122–38.

Rogers, J. K., and K. D. Henson. 1997. "Hey, why don't you wear a shorter skirt?" Structural vulnerability and the organization of sexual harassment in temporary clerical employment. *Gender & Society* 11:215–38.

Schwalbe, M. 1992. Male supremacy and the narrowing of the moral self. *Berkeley Journal of Sociology* 37:29–54.

Smith, D. 1990. *The conceptual practices of power: A feminist sociology of knowledge.* Boston: Northeastern University Press.

Stockdale, M. S. 1993. The role of sexual misperceptions of women's friendliness in an emerging theory of sexual harassment. *Journal of Vocational Behavior* 42 (1): 84–101.

Stockdale, M. S., and A. Vaux. 1993. What sexual harassment experiences lead respondents to acknowledge being sexually harassed? A secondary analysis of a university survey. *Journal of Vocational Behavior* 43 (2): 221–34.

Tagri, S., and S. M. Hayes. 1997. Theories of sexual harassment. In *Sexual harassment: Theory, research and treatment,* edited by W. O'Donohue. New York: Allyn & Bacon.

Thorne, B. 1993. *Gender Play: Girls and boys in school.* Buckingham, UK: Open University Press.

U.S. Merit Systems Protection Board. 1988. *Sexual harassment in the federal government: An update.* Washington, DC: Government Printing Office.

Welsh, S. 1999. Gender and sexual harassment. *Annual Review of Sociology* 1999:169–90.

West, C., and D. H. Zimmerman. 1987. Doing gender. *Gender & Society* 1: 125–51.

Wood, J. T. 1998. Saying makes it so: The discursive construction of sexual harassment. In *Conceptualizing sexual harassment as discursive practice,* edited by S. G. Bingham. Westport, CT: Praeger.

York, K. M. 1989. Defining sexual harassment in workplaces: A policy-capturing approach. *Academy of Management Journal* 32:830–50.

Yount, K. R. 1991. Ladies, flirts, tomboys: Strategies for managing sexual harassment in an underground coal mine. *Journal of Contemporary Ethnography* 19:396–422.

48

Traveling with the Ball Club

A Code of Conduct for Wives Only

BY STEVEN M. ORTIZ

This excerpt from an article by Steven Ortiz provides us with a detailed description of some of the more subtle methods of gender difference and inequality in his examination of the treatment of players' wives by the team management during the baseball season. Women's and men's identities directly relate to the desires of the teams' managers. The team management gains by selling the players to the public as dominant, sexy men. On the other hand they encourage the wives to be feminine, docile, and silent. The "code of conduct" has considerable implications for both women and men and relationships both on the road and at home.

1. How are the women and men who travel with the ball club supposed to act?
2. Why is it so important to the managers that men and women act this way when traveling on the road?
3. In what other settings, both in and out of the workplace, have you seen similar "codes of conduct"?

★ ★ ★

Most wives of major league baseball players, like the two-person career wives of other professional athletes,[1] stay at home when their husbands are traveling during the season, to meet the many responsibilities associated with parenting and family life. Those who do travel rarely do so for the entire season, for various reasons: because they have children, or because the team would require them to pay their own expenses, or, in some cases, because they have their own jobs or careers. Therefore, wives who travel throughout the season usually do not have children, do not have their expenses paid by the team organization, and are not employed. As a result, it is more common for wives to limit their traveling to an occasional "family trip," which means bringing the children along.[2] Moreover, not all ball clubs allow wives to travel with the team, and many restrict or discourage it. Indeed, given the ultramasculine nature of the world of professional sport, and the fact that most team organizations are not "family friendly" (e.g., Lutz 1993; *San Francisco Chronicle* 1993), it is understandable that wives are not encouraged to travel with their husbands, and are accorded a subordinate status when they do. Though team management may see wives as a source of stability for their married ballplayers, which may make them more productive, they may also believe that a wife's place is at home and not on the road.

As female outsiders in a highly masculine arena of life, wives who travel on road trips are essentially "strangers" with second-class status,[3] and, in certain specific road-trip situations, they actually have nonperson status (Goffman 1959; 1963, p. 84; 1967, pp. 67–68). In this context, traveling on the road is part of the socialization process of being the wife of a major league ballplayer; as a temporary participant in her husband's occupational world, she is an outsider who must learn how to "play the game." Gradually, traveling wives become aware of an unwritten "code of conduct" they must observe.

This code of conduct is shaped by shared definitions of appropriate male or teammate interaction, as well as expectations, and by an acknowledgment of the sexual dalliances of married teammates. It is socially constructed and enforced by the husbands or their teammates, and consists of various strategies and rituals for keeping the wives under control.

* * *

STAYING OUT OF HOTEL BARS: MARITAL INEQUITY AND THE CODE OF CONDUCT

> First, they tell us when we can go with them on the road, and then they have all these ground rules. We were not allowed to go in hotel bars. You know, "You can't go in there." Oh, that used to drive me nuts! I hated that.
>
> —ROBYN

In her travels with her husband and his team, the wife . . . discovers that she must deal with the code and its application to hotel bars and similar public settings. The code of conduct is essentially a reflection of male dominance. Whether it reflects gender inequity in the husband's occupational world, or marital inequity in the sport marriage, the result is the same: it is used by the men on the team to keep the wives under control.

Learning how the code applies to hotel bars, and comparable public settings, is a complex socialization process for the wife. She is often put in the position of juggling what is and what is not permitted in learning how the code applies to the hotel bars and other similar public places: blending in with the immediate environment, avoiding suspicious definitions of innocent situations, not engaging in loose talk, and socially constructing a "cool" reputation.

The hotel bar, located in the hotel where the visiting ball club stays, is not only a home territory for the teammates (e.g., Cavan 1966, pp. 205–233), but it is defined by the teammates as another male region for privacy. However, as a team refuge for male-bonding interaction where they often unwind from the pressures of the game, the hotel bar is also a very popular gathering place for female groupies. It is common knowledge among teammates and wives that female groupies congregate in the hotel bar for the purpose of meeting major league ballplayers. But the hotel bar also serves another purpose for certain teammates: it provides a setting for married teammates to rendezvous with girlfriends, or to meet female groupies. Thus, the primary reason the hotel bar is off-limits to the wives is because it provides married teammates with a place to meet other women. Marsha explains:

> We're not supposed to go in certain hotel bars because that's where the guys that are married, that don't have their wives with them, are going to be with the groupies. It's like a groupie hangout. So you don't have the wives in there.

Essentially, the teammates are territorial, and, according to the code, the wives must stay out of the hotel bar. Among some couples, this is a crucial dimension of the code. As an attempt to define a boundary of exclusion, this aspect of the code is based on the premise that if a wife sees a married teammate with another woman, she will inform others, or perhaps tell the married teammate's wife what she has seen.

Although a wife is usually accepted as an insider in various road situations, her outsider status is again clearly illustrated when she is told to stay out of the hotel bar. In criticizing this part of the code, Stacy angrily maintained:

> It's the most ridiculous rule in the world! Why are these guys doing it in the hotel when they could pick any bar anywhere? They would be so much more secluded. If they're going to screw around, you'd think they'd want to go somewhere else to be separate from the other guys. I don't know why they do that, except for the ones that want to show off. And that way, by being in the hotel bar, the other guys will see them there. And a lot of the guys probably think it's like another notch on their belt. They're proud of it. That's the only reason I would think that they would want to be in the hotel bar.

A wife may feel a potential solidarity with the wives of adulterous teammates, and uphold the principle of family, by forming a united front among wives. This is her response to the united front among the teammates who want to show off their exploits in front of each other, as an audience to extramarital dalliances, and bond through their displays of masculinity. Thus, there is an emergence of two competing "interest groups" and two sets of "allies." As an inner sanctum, the hotel bar, like the back of the airplane, serves as an exclusively male region where the wives are concerned.

In following this dimension of the code, many of the wives will often walk straight from the elevator, or somewhere else in the lobby, to the street without lingering near the entrance of the bar. Shelia said, in her insistence that some wives will not look in the bar as they walk past it, "You don't even look in there." Stacy also admitted, "You just look straight ahead. You don't look in."

This aspect of the code varies from team to team, and, on some ball clubs, it may not exist at all. On other ball clubs it is not only strictly enforced, but the wives are not even allowed in the hotel lobby. In fact, on some ball clubs, the entire hotel is off-limits to the wives. Furthermore, this part of the code, which prohibits the wives from staying with the team at the hotel, is not limited to the world of major league baseball. For example, in the world of professional football, the hotel may also be off-limits to the wives on away games. Gwen, whose husband is a basketball player, argued that this requirement also exists in the world of professional basketball: "It's prevalent in the NBA. They don't let the wives travel with the players. They couldn't even stay in the same hotel."

While most wives do not agree with the dimension of the code which excludes them from the hotel bar, and although they question it, they still follow it. Another reason why this aspect of the code exists is because many of the unfaithful husbands are quite indiscreet about their sexual activities with other women on the road. When some ballplayers leave the field and enter the backstage recesses of their occupational world, or participate in various frontstage activities within it, they claim a certain amount of sexual freedom. However, because many of them have a self-absorbed perception of life and often conspire to preserve male privileges, which their celebrity status contributes to (in what can be viewed as the "spoiled athlete syndrome" [Ortiz 1994a]), philandering husbands are not very discreet with their girlfriends, or in concealing their affairs with other women.

It is also a common practice for the unfaithful husband to meet his girlfriend away from the hotel bar, or to pick up other women in the popular watering holes he frequents, in the different cities scattered along road-trip circuits. Therefore, the wives are told that they must also stay out of the various bars where the teammates hang out because they also constitute male sanctuaries for their extramarital liaisons. This brings us to another part of the code: certain bars, restaurants, dance clubs, or nightspots are off-limits to the wives. In fact, it not only applies to these public places on road trips, but to the public places where the teammates hang out during spring training.

When a wife happens to be alone, what does she do if she accidently sees a married teammate with another woman and he notices that she sees him? Because this usually occurs in the general vicinity of the hotel where the team is staying, more than one wife has found herself in this predicament. Depending on the wife and her husband's shared understanding of this dimension of the code, and depending on whether she is with her husband when she finds herself in this situation, if she happens to see a married teammate with another woman, she often will not acknowledge him. This may not be a self-defined requirement for most wives, but it is frequently a wife's initial response. In this awkward situation, she shapes the line of action that enables her to blend in and tries to emphasize her nonperson status. For example, if she sees him before he sees her, she will look away before he catches her looking at him, so he will think she did not see him, or if she sees that he sees her first, she will quickly look in the opposite direction or pretend she did not see him. However, such nonperson strategies are often pointless when she is recognized by a married teammate because she is defined as a potential threat—even if she maintains her silence.

Trying to manage social invisibility in these uncomfortable encounters is fairly common for most wives. As Robyn admitted, since relying on nonperson strategies is a reflection of feeling intimidated and trapped, "I wouldn't say anything because my husband would make me leave. So I wouldn't say anything. I'd pretend I didn't see him." In recalling a particularly unsettling incident Shelia revealed:

> There was also a guy in Borderville. He was just there for one year. He had this girl who traveled with him, and he had a wife and four kids back in Borderville. And I saw them together on the elevator in Fernville. I got on the elevator with them, and I actually tried to act like I didn't see him. I turned around, and I looked at the wall, and he said, "Hello." Well, later on, he asked Frank if I would say anything. And Frank said, "Well, no, I don't think she would. It's none of her business."

A wife's fear of getting caught witnessing "forbidden behavior" is another important basis for her need to be socially invisible. Ironically, although it is the unfaithful husband who should be afraid of getting caught with another woman, it is often the wife who is afraid of getting caught watching his forbidden behavior. . . .

Initially, when a wife sees a married teammate with another woman, she often tries to be cautious in trying to define the situation as carefully as possible, and in trying not to leap to the wrong conclusion. After all, he may be completely innocent of any wrongdoing because he is only entertaining a visiting female relative, female family friend, or sister-in-law. However, there are the "blabbers." These are the wives who, the other wives say, are quick to jump to the wrong conclusion, to think the worst of the married teammate, and to announce to others what they have seen. This angers many of the wives because they contend that one should know all of the facts before saying anything, if anything is discussed at all, or keep it to oneself because it should not be one's concern.

There are reasons why certain wives are quick to jump to the wrong conclusion, and to spread their suspicious definitions of what could be innocent situations. For example, they may be projecting their own fears or compensating for the fact that their husbands are having affairs. Spreading their suspicions about the extramarital dalliances of certain married teammates raises the issue of revealing what wives see and implicates that part of the code which holds that they are not to disclose to others what they have seen. As Robyn told me, in discussing why this is so important to the teammates, "If someone was in the hotel bar, just talking to another girl, then they don't want the wife to run back and tell his wife that she saw him with a girl." Some of the wives speculate that the wives who gossip or spread rumors are insecure about their own marriages, or they have difficulty in solving their own marital problems. Nevertheless, the consequences of their actions are potentially damaging for them as well as those they gossip about. It is also possible that some wives sanction each other for gossiping because they do not want to know that their husbands were unfaithful. In fact, some of the wives who stay home may resent the wives who travel with the team because they are afraid of what they will learn about their husbands. . . .

Not revealing to anyone, except her husband, what she has seen on the road is another important aspect of the code because it influences how the teammates define the traveling wife. They are not likely to trust her if she does not maintain her silence. . . . Not jumping to the wrong conclusion, not spreading gossip, and not revealing what she has seen, all play a part in the social construction of the wife's reputation on the ball club, and often throughout the league. In their travels with the wives, the teammates soon learn which wives they can trust. The longer her husband is in the world of major league baseball, the more it shrinks, and the more likely she is to acquire some kind of a reputation on the team or in the league. Her reputation, whether it is positive or negative, precedes her among the teammates and wives on new ball clubs.

Establishing a reputation for being "cool" plays an important part in becoming "one of the guys." However, establishing a reputation as a wife who can be trusted can have serious drawbacks. Because some of the unfaithful husbands on the team know they can depend on her silence, they may often try to take advantage of her silence, or take her silence for granted, as they engage in their extramarital sexual activities. Thus, in maintaining her silence, she often becomes a co-conspirator in the sexual escapades of these men—or other backstage secrets—despite the fact that she strongly disapproves of their extramarital relationships. Indeed, under these circumstances, the fact that she can be trusted is very ironic.

★ ★ ★

NOTES

1. See Papanek (1973) for her conceptualization of the two-person career, and what I explore further as the "career-dominated marriage" (Ortiz 1994a), Taylor and Hartley (1975) for their use of Papanek's concept, and Fowlkes (1987) for a discussion of certain two-person career issues.
2. There are roughly 14 road trips during the major league baseball season. The longest is often a 15-game trip (i.e., 15 games in 15 days), and the shortest is usually a 7-game trip (i.e., 7 games in 7 days). On a typical road trip, the airplane is chartered, but, on short trips, the team may take a commercial flight. A bus transports the ball club from the airport to the hotel. If the hotel is some distance from the ballpark, the team takes an early bus from the hotel to the ballpark, which is usually 3 hours prior to game time. The wives follow on a later bus, typically an hour before game time, or some wives may take a cab. After the game is over, the wives and the teammates usually return together on the same bus, or the wives will return on an early bus, or they may take a cab. Also, if the wife travels with the ball club during the entire season, and other wives make the family trip, she often returns with some of them, and they take a later bus or cab, or walk if the ballpark is close to their hotel.
3. The concept of the stranger was originally suggested by Simmel (1950), embellished by Schutz (1971, pp. 91–105), and elaborated on by others (Fontana and Frey 1983; Jansen 1980; Sway 1981; Wood 1934). In using a Simmelian interpretation, the traveling wife is placed in the role of stranger because, while she is important to her husband, she is not important to his team and will never be a member of his team. Consequently, given her peripheral or marginal status (indeed, often nonperson status), she is both "accepted" and "rejected" by the teammates (Fontana and Frey 1983, p. 309). The traveling wife is in the role of stranger because she, as Simmel (1950, p. 407) has suggested, "is near and far *at the same time.*"

REFERENCES

Cavan, Sherri. 1966. *Liquor License: An Ethnography of Bar Behavior.* Chicago: Aldine.

Fontana, Andrea and James Frey. 1983. "The Placekicker in Professional Football: Simmel's Stranger Revisited." *Qualitative Sociology* 6:308–321.

Fowlkes, Martha R. 1987. "The Myth of Merit and Male Professional Careers: The Roles of Wives." Pp. 347–360 in *Families and Work,* edited by Naomi

Gerstel and Harriet Engle Gross. Philadelphia, PA: Temple University Press.

Goffman, Erving. 1959. *The Presentation of Self in Everyday Life.* Garden City, NY: Anchor/Doubleday.

———. 1963. *Behavior in Public Places: Notes on the Social Organization of Gatherings.* New York: Free Press.

———. 1967. "The Nature of Deference and Demeanor." Pp. 47–95 in *Interaction Ritual: Essays on Face-to-Face Behavior.* Garden City, NY: Anchor/Doubleday.

Jansen, Sue Curry. 1980. "The Stranger as Seer or Voyeur: A Dilemma of the Peep-Show Theory of Knowledge." *Qualitative Sociology* 2:22–55.

Lutz, Michael A. 1993. "Athlete Opts to Cheer Birth and Misses Game." *San Francisco Examiner,* October 19: pp. A1, A14.

Ortiz, Steven M. 1994a. "When Happiness Ends and Coping Begins: The Private Pain of the Professional Athlete's Wife." Unpublished doctoral dissertation, Department of Sociology, University of California, Berkeley.

Papanek, Hanna. 1973. "Men, Women, and Work: Reflections on the Two-Person Career." *American Journal of Sociology* 78:852–872.

San Francisco Chronicle. 1993. "Oilers Dock New Daddy; He May Sue." October 19: pp. D1–2.

Simmel, Georg. 1950. "The Stranger." Pp. 402–408 in *The Sociology of Georg Simmel,* translated and edited by Kurt H. Wolff. New York: Free Press.

Sway, Marlene B. 1981. "Simmel's Concept of the Stranger and the Gypsies." *Social Science Journal* 18:41–50.

Taylor, Mary G. and Shirley Foster Hartley. 1975. "The Two-Person Career: A Classic Example." *Sociology of Work and Occupations* 2:354–372.

Wood, Margaret Mary. 1934. *The Stranger: A Study in Social Relationships.* New York: Columbia University Press.

Topics for Further Examination

CHAPTER 4

- Go to the main Web sites for Girl Scouts (http://www.gsusa.org) and Boy Scouts (http://www.scouting.org) and compare the two organizations. Are their programs similar or different? What similarities and differences do you observe in the Web sites themselves? What effect do these organizations have on gender socialization?
- Go to the U.S. Department of Education website, http://www.dol.gov/ and search for differences in men and women in higher education.
- Using InfoTrac (specify refereed journals only), search for "gender" and "socialization." Examine the titles of the first 20 articles and the journals they are in and observe the different ways that recent literature discusses gender socialization.

CHAPTER 5

- Find articles and Web sites that offer critiques of gender stereotypes in the mass media, popular culture, and consumer culture. For example, visit the following Web site: http://womensissues.about.com/cs/genderstereotypes/.
- Locate research on the impact of American media images of masculinity and femininity on the self-perceptions of women and men in non-Western cultures such as India, Thailand, and Kenya.

CHAPTER 6

- Look up research and Web sites on patterns of eating disorders and cosmetic surgery procedures among women and men in the United States today. Check out this Web site: www.pbs.org/wgbh/nova/thin.
- Explore the world's largest online library of books and articles on homosexuality: http://www.questia.com/Index.jsp?CRID=homosexuality &OFFID=se1

CHAPTER 7

- Check out the most recent research on women and work done by the Institute for Women's Policy Research at http://www.iwpr.org
- Using the Web, calculate a gender ratio of the top executives in a sample of the largest firms in this country.
- Use InfoTrac to check out the most recent research on gender and leisure or gender and sports.

CHAPTER 8

- Go to http://www.umbc.edu/wmst and check out their resource sites to find information on families (if you can't find something immediately, try the miscellaneous link).
- Check out the most recent research on gender and relationships using InfoTrac.

CHAPTER 9

- Visit http://www.mencanstoprape.org to see what men are doing to stop rape.
- Use InfoTrac to try to find the latest information on domestic violence and gender.

CHAPTER 10

Nothing Is Forever

The title of this chapter stands for the principle that change is inevitable. Like the ever-evolving patterns of the kaleidoscope, change is inherent in all of life's patterns. Anything can be changed and everything does change, from the cells in our bodies to global politics. There is no permanent pattern, no one way of experiencing or doing anything that lasts forever. This fact of life can be scary, but it can also be energizing. The mystery of life, like the wonder of the kaleidoscope, rests in not knowing precisely what will come next.

The readings in this chapter address the changing terrain of gender. In particular, they help us to understand a fundamental paradox of the gender order—it is simultaneously resistant and open to change. If one only takes a snapshot view of life, it may appear as though the current gender arrangements are relatively fixed. However, an expanded view of gender over time and across cultures reveals the well-researched fact that gender meanings and practices are as dynamic as any other aspect of life. Patterns of gender continuously undergo change, and they do so at every level of experience, from the individual to the global. Michael Schwalbe (2001) observes that there is both chance and pattern in the lives of individuals and in the bigger arena of social institutions. He makes the point that no matter how many rules there might be and no matter how much we know about a particular person or situation, "social life remains a swirl of contingencies out of which can emerge events that no one expects" (127). As a result, life, including its gendered dimensions, is full of possibilities.

Social constructionist theory is especially helpful in understanding the inevitability of change in the gender order. Recall that social constructionist research reveals the processes by which people create and maintain the institution of gender. It underscores the fact that gender is a human invention, not a biological absolute. Particular gender patterns keep going only as long as people share the same ideas about gender and keep doing masculinity and femininity in a routine, predictable fashion (Schwalbe, 2001; Johnson, 1997). Given that humans create gender, all gender patterns are subject to disruption and alteration by people who, individually and collectively, choose to invent and negotiate new ways of thinking about gender and doing gender. That is, gender is amenable to change at all levels of life, from the individual and interactional to the institutional and global.

Individuals Do Make a Difference At the micro level of daily interaction, individuals can, and do, destabilize the gender order. They do so by choosing to bend conventional gender rules or to transcend the rules altogether (Lorber, 1994). For example, women and men create partnerships based on equity, including shared carework and housework roles. An increasing number of individuals and couples have been changing the structure of work in relationship to family by creating home-based businesses (Worley and Vannoy, 2001). Others purposefully transgress the boundaries of sexual and gender identities by deliberately mixing appearance cues via makeup, clothing, hairstyle, and other modes of self-presentation (Lorber, 1994). The reading by Gillian A. Dunne illustrates the transformative potential of micro-level challenges to the gender status quo when individuals can conceptualize a different way of life and act on it. Dunne's article focuses on the ways in which lesbian mothers have rethought and reconfigured the two-parent family as they go about inventing and enacting new parenting roles.

Large-Scale Change At the macro or institutional level of the gender order, change comes about through large-scale forces and processes, both planned and unplanned. For example, major economic transitions, such as the Industrial Revolution, have fueled intense and extensive unplanned social transformations in every aspect of life. The Industrial Revolution was a turning point in world history, marking the transition from agricultural to industrial economies (Kivisto, 1998). As a consequence of this massive economic and technological transformation, far-reaching changes took place in all social institutions and relationships, including the social organization of work and family. Prior to industrialization, women's labor was essential to agricultural life. Women, men, and children worked side by side to grow food, make clothing, raise animals, and otherwise contribute to the family economy (Lorber, 2001).

As the Industrial Revolution got underway, productive work moved from the home into factories, and work came to be defined as valuable only if it produced a paycheck. Although important work was still done at home, housework did not produce a paycheck, and notably, women did this work almost exclusively. The negative outcome was that household labor transformed into an invisible and devalued, but still necessary, activity. Alongside these changes in the organization of work, religious, medical, and political leaders created and promoted a new gender ideology. This ideology, called the "doctrine of separate spheres," reinforced the new gendered division of labor into paid and unpaid work created by the Industrial Revolution. The doctrine was based on the belief that a woman's natural calling was to be a housewife and mother, while a man's responsibility was to follow wage work, embrace the breadwinner role, and participate in the political process (Godwin and Risman, 2001).

The profound changes in gender relations and the organization of work and family wrought by the Industrial Revolution continue to be a source of conflict for many women and men today. For example, although most married women with children now work outside the home, the doctrine of natural separate spheres continues to operate as an ideal against which "working women" who have children are often negatively evaluated.

Currently, globalization is a major force for change. The term globalization refers to the increasing interconnectedness of social, political, and economic activities worldwide (Held et al., 1999). To put it another way, transnational forces such as geopolitical conflicts, global markets, multinational corporations, transnational media, and the migration of labor now strongly influence what happens in specific countries and locales (Connell, 2000). For example, the international trading system—dominated by the United States, the European Union, and Japan—encompasses almost every nation in the world. Films and television programs, especially those produced in the West, circulate the globe, threatening to marginalize cultural and creative difference (Barber, 2002). The negative ecological consequences of globalization, such as the burning of tropical rain forests in countries such as Brazil and the consumption of gas-guzzling automobiles by Americans, reach into every nook and cranny of the Earth (Barber, 2002). English has rapidly become the dominant language of the global infrastructure and its common markets. Social movements such as feminism now engage women and men in almost every nation in the world (Shaw and Lee, 2001).

Robert Connell (2000), an important sociologist of gender studies (see Introduction), argues that globalization has created a world gender order. This world gender order, he states, has several interacting dimensions: (1) a gender division of labor in a "global factory" in which poor women and children provide cheap labor for transnational corporations owned by businessmen from the major economic powers (see Lee in Chapter 7), (2) the marginalization of women in international politics, and (3) the dominance of Western gender symbolism in transnational media.

However, globalization is not monolithic. There are countervailing forces challenging the homogenizing and hegemonic aspects of globalization. For example, indigenous cultures interact with global cultures to produce new cultural forms of art and music that challenge the dominance of Western popular culture (see Chapter 5). Globalization has also spawned transnational social movements, such as feminism, the slow food movement, and environmentalism, which have gone global to address worldwide problems of Western hegemony, global inequality, and human rights. The reading by Lionel Cantu offers a rich analysis of one global force—tourism—by which modern political and economic codependency between the United States and Mexico is transforming gender norms and sexual scripts among Mexicans in both countries.

Social Movements Large-scale change may also come about in a planned fashion. Social movements are prime examples of change that people deliberately and purposefully create. They are conscious, organized, and collective efforts to work toward cultural and institutional change, and share distinctive features including organization, consciousness, noninstitutionalized strategies (such as boycotts and protest marches), and prolonged duration (Kuumba, 2001). The United States has a long history of people joining together in organizations and movements to bring about justice and equality. For example, the labor union movement, socialist movement, civil rights movement, and gay and lesbian movement have been important vehicles for change that might not otherwise have happened.

One of the most durable and flexible social movements is feminism. Consider the fact that the feminist movement has already lasted for two centuries. At the opening of the nineteenth century, feminism emerged in the United States and Europe. By the early twentieth century, feminist organizations appeared in urban centers around the world. By the turn of the twenty-first century, feminism had grown into a transnational movement in which groups work at local and global levels to address militarism, global capitalism, racism, poverty, violence against women, economic autonomy for women and other issues of justice, rights, and peace (Shaw and Lee, 2001).

The reading by Allan Johnson examines planned change and the individual's role in making it happen. He offers an array of activities and forms of both ordinary and creative resistance that people can engage in to "plant the seeds of change" in themselves and their communities. Johnson makes the point that change in our gendered ways of life can be deliberately mapped out. He argues that we have the power to choose to change ourselves and our world. This is especially so if we are willing to forge relationships with others who recognize a need for change and are ready to get involved in developing modes of living together that do not rely on inequalities and destructive or exploitive patterns of living.

The Complexity of Change Not only is social change pervasive at micro and macro levels of life and a function of both planned and unplanned processes, it is also uneven and infinitely complex. For example, change doesn't unfold in a linear, predictable fashion. Change often takes people by surprise. Let's examine gender change again. Sometimes it is dramatically visible, marking the moment when gender relations have transformed. Consider the passage of the Nineteenth Amendment to the U.S. Constitution in 1920 that guaranteed women the right to vote. This one historic moment uplifted the public status of women. But more often, change consists of alterations in the fabric of gender relations that are not immediately visible to us, both in their determinants and their consequences.

For instance, we now know that a complex set of factors facilitated the entry of large numbers of single and married women into the paid work force and higher education in the second half of the twentieth century. Those factors included very broad economic, political, and technological developments that transformed the United States into an urban, industrial capitalist nation (Stone and McKee, 1998). Yet no one predicted the extent of change in gender attitudes and relations that would follow the entry of women into the work force. It is only after the fact that the many implications of the movement of women into the work force and higher education have been identified and assessed. For example, marital relationships in the United States have moved toward greater equity in response to the reality that most married women are no longer wholly dependent on their husbands' earnings. As married women have increasingly embraced paid work, their husbands have increasingly reconceptualized and rearranged their priorities so that they can devote more attention to parenthood (Goldscheider and Rogers, 2001).

Finally, it is important for us to recognize that change occurs even under the most rigid and oppressive of social circumstances. Sociological research has demonstrated how patriarchal traditions in seemingly rigid social institutions can be altered. For example, studies of the "forced" integration of women into previously all-male military academies, such as West Point and the Citadel, show that although women struggled against a powerful wall of male resistance, they have in the end shown that they can "do military masculinity" (Kimmel, 2000). The article by Shahin Gerami and Melodye Lehnerer explores the case of women's resistance to religious fundamentalism in Iran, a classic patriarchal nation. Gerami and Lehnerer uncovered covert strategies of resistance that Iranian women deploy to soften the impact of patriarchal restrictions and, in some instances, to undermine them.

Prisms of Gender and Change Returning briefly to the metaphor of the kaleidoscope, let us recall that the prism of gender interacts with a complex array of social prisms of difference and inequality, such as race and sexual orientation. The prisms produce ever-changing patterns at micro and macro levels of life. Our metaphor points to yet another important principle of changing gender arrangements. We can link gender change to alterations in other structural dimensions of society, such as race, class, and age. For example, as Americans have moved toward greater consciousness and enactment of gender equality, they have also come to greater consciousness about the roles that heterosexism (i.e., the institutionalization of heterosexuality as the only legitimate form of sexual expression) and homophobia (i.e., the fear and hatred of homosexuality) play in reinforcing rigid gender stereotypes and relationships (see Chapter 6). It has become clear to many seeking gender justice that it cannot be achieved without eliminating homophobia and the heterosexist framework of social institutions such as family and work.

Additionally, gender transformation in the United States is inextricably tied to race. This is true both historically and today. The first wave of feminism in the United States was an outgrowth of the antislavery movement, and it was the politics of racial justice that led to the second wave of feminism (Freedman, 2002). Racism, as well as ageism, classism, and other forms of oppression, had to be addressed by feminists because the struggle to achieve equal worth for women had to include all women. Anything less would mean failure.

The Inevitability of Change Collectively, the articles in this chapter invite the reader to ask "Why should I care about or get involved in promoting change in the gender status quo?" That's a good question. After all, why should one go to the trouble of departing from the standard package of gender practices and relationships? Change requires effort and entails some risk. On the other hand, the cost of "going with the flow" can be high. There are no safe places to hide from change. Even if we choose "not to rock the boat" by closing ourselves off to inner and outer awareness, change will find us.

There are two reasons for this fact of life. First, we cannot live in society without affecting others and in turn being affected by them. Each individual life intertwines with the lives of many other people, and our words and actions do have consequences, both helpful and harmful. Every step we take and every choice we make affect the quality of life of a multitude of people. If we choose to wear blinders to our connections to them, we run the risk of inadvertently diminishing their chances and our own chances of living fulfilling lives (Schwalbe, 2001). For example, when a person tells a demeaning joke about women, he or she may intend no harm. However, the (unintended) consequences are harmful. The joke reinforces negative stereotypes, and telling the joke gives other people permission to be disrespectful of women (Schwalbe, 2001).

Second, we can't escape broad, societal changes in gender relations. By definition, institutional and societal level change wraps its arms around all of us. Think about the widespread impact of laws such as the Equal Pay Act and Title VII, preventing discrimination against women and people of color, or consider how sexual harassment legislation has redefined and altered relationships in a wide array of organizational settings. Reflect on the enormous impact of the large numbers of women who have entered the workforce since the latter half of the twentieth century. The cumulative effect of the sheer numbers of women in the workforce has been revolutionary in its impact on gender relations in family, work, education, law, and other institutions and societal structures.

Given the inevitability of change in gender practices and relationships, it makes good sense to cultivate awareness of who we are and what our responsibilities to one another are. Without awareness, we cannot exercise control over our actions and their impact on others. Social forces shape us, but those forces change. Every transformation in societal patterns reverberates through our lives. Developing the "social literacy" to make sense of the changing links between our personal experience and the dynamics of social patterns can aid us in making informed, responsible choices (O'Brien, 1999).

The final reading in this chapter highlights the value of acquiring and applying sociological insight to create and alter social arrangements. It is the edited transcript of a discussion among three sociologists: Margaret Andersen, Anne Bowler, and Michael Kimmel. Andersen and her colleagues call attention to the complex, dynamic relationship between feminist theory and practice. In doing so, they address some of the critical issues in understanding the disjunctures between structural power and individual experience, and they underscore the importance of sociology to uncovering the connections between "the personal and the political." The authors end their discussion on an upbeat note, one that reminds us that social life is full of possibilities. Andersen says that the sociological perspective allows us to "see ways to create different institutions and different relationships." Society is not fixed. Gender patterns are not cast in stone. Nothing is forever. We can and do make a difference.

REFERENCES

Barber, Benjamin R. (2002). "Jihad vs. McWorld." In *McDonaldization: The reader,* G. Ritzer (Ed.), pp. 191–198. Thousand Oaks, CA: Pine Forge Press.

Connell, R. W. (2000). *The men and the boys.* Berkeley: University of California Press.

Freedman, Estelle. (2002). *No turning back: The history of feminism and the future of women.* New York: Ballantine Books.

Godwin, Frances K., and Barbara J. Risman. (2001). "Twentieth-century changes in economic work and family." In *Gender mosaics,* D. Vannoy (Ed.), pp. 134–144. Los Angeles: Roxbury Publishing Company.

Goldscheider, Frances K., and Michelle L. Rogers. (2001). "Gender and demographic reality." In *Gender mosaics,* D. Vannoy (Ed.), pp. 124–133. Los Angeles: Roxbury Publishing Company.

Held, David and Anthony McGrew, David Goldblatt, and Jonathan Perraton. (1999). *Global transformations.* Stanford, CA: Stanford University Press.

Johnson, Allan. (1997). *The gender knot.* Philadelphia: Temple University Press.

Kimmel, Michael. (2000). "Saving the males: The sociological implications of the Virginia Military Institute and the Citadel." *Gender & Society* (14)4:494–516.

Kivisto, Peter. (1998). *Key ideas in sociology.* Thousand Oaks, CA: Pine Forge Press.

Kuumba, M. Bahati. (2001). *Gender and social movements.* Walnut Creek, CA: Altamira Press.

Lorber, Judith. (2001). *Gender inequality: Feminist theories and politics.* Los Angeles: Roxbury Publishing Company.

———. (1994). *Paradoxes of gender.* New Haven, CT: Yale University Press.

O'Brien, Jodi. (1999). *Social prisms.* Thousand Oaks, CA: Pine Forge Press.

Schwalbe, Michael. (2001). *The sociologically examined life.* Mountain View, CA: Mayfield Publishing Company.

Shaw, Susan M. and Janet Lee. (2001). *Women's voices, feminist visions.* Mountain View, CA: Mayfield Publishing Company.

Stone, Linda and Nancy P. McKee. (1998). *Gender & culture in America.* Upper Saddle River, NJ: Prentice Hall.

Worley, Jennifer Campbell and Vannoy, Dana. (2001). "The challenge of integrating work and family life." In *Gender mosaics,* D. Vannoy (Ed.), pp. 165–173. Los Angeles: Roxbury Publishing Company.

49

Opting into Motherhood

Lesbians Blurring the Boundaries and Transforming the Meaning of Parenthood and Kinship

BY GILLIAN A. DUNNE

In this reading, Gillian Dunne analyzes data from a study called the Lesbian Household Project. The study employed qualitative and quantitative methods to illuminate the work and parenting experiences of 37 cohabiting lesbian couples with dependent children across England. Dunne notes that both the sensitive nature of the research topic and the invisibility of lesbians in the population make it difficult to claim that the sample is representative. Notwithstanding, Dunne's research into the lives of women parenting together reveals a world in which they challenge the connections between biological and social motherhood, and often put egalitarian ideals into practice.

1. Why does Dunne view heterosexuality as playing a central role in reproducing gender inequality?
2. Why do women parenting with women have a "head start" over heterosexual couples with respect to egalitarianism?
3. How do women parenting together challenge the connections between biological and social motherhood?

The extension of educational and employment opportunities for women, together with widening experience of the "plastic" nature of sexualities (Giddens 1992, 57), has enabled increasing numbers of Western women to construct independent identities and lifestyles beyond traditional marriage, motherhood, and indeed, heterosexuality (Dunne 1997). As contemporary women's identities expand to incorporate the expectations and activities that have been traditionally associated with masculinity, there has not been an equivalent shift of male identity, let alone practice, into the traditional domains of women. Exceptions not withstanding (Blaisure and Allen 1995; Doucet 1995; Ehrensaft 1987; VanEvery 1995), a distinctly asymmetrical division of labor remains the majority pattern (Berk 1985; Brannen and Moss 1991; Ferri and Smith 1996; Gregson and Lowe 1995; Hochschild 1989). The intransigent nature of the gender division of labor means that women continue to perform the bulk of domestic work and that mothers bear the brunt of the social and economic penalties associated with caring for children. Men's relative freedom from the time constraints and labor associated with the home and parenting enables them to be more single-minded in the pursuit of employment opportunities and retain their labor market advantages. . . .

While contemporary women begin to see the demands of motherhood as conflicting with their newly won bid for autonomy, there has been a recent shift in attitudes toward parenting among the lesbian population. A rising awareness of alternatives to heterosexual reproduction has led to the growing recognition that their sexuality does not preclude the possibility of lesbian and gay people having children. In Britain and in the United States, we are witnessing the early stages of a "gayby" boom, a situation wherein lesbian women and gay men are opting into parenthood in increasing numbers. According to Lewin, "The 'lesbian baby boom' and the growing visibility of lesbians who became mothers through donor insemination constitute the most dramatic and provocative challenge to traditional notions of both family and of the non-procreative nature of homosexuality" (1993, 19). In this article, I want to address this apparent contradiction between childlessness as resistance and lesbian motherhood as provocative challenge by showing that the mothering experiences that lesbians are opting into are qualitatively different from those that some women seek to avoid.

I take the view that sexuality is socially and materially constructed and that heterosexuality plays a central role in reproducing gender inequality (Dunne 1997, 1998d, 2000). The dominance of heterosexuality is the outcome of institutional processes that render alternatives undesirable and/or unimaginable (Dunne 1997; Rich 1984) and that construct gender difference and gender hierarchies (Butler 1990, 17; Rubin 1975). Consequently, there is a crucial relationship between gender and sexuality. . . .

I wish to support and extend Lewin's observations on single lesbian mothers by drawing on my work on lesbian couples who have become parents via donor insemination. I argue that an attentiveness to the gender dynamic of sexuality illuminates additional challenges that arise when women combine with women to rear children—the possibility of showing what can be achieved when gender difference as a fundamental structuring principle in interpersonal relationships is minimized (see Dunne 1997, 1998a). I suggest a complex and contradictory situation for lesbians who have opted into motherhood via donor insemination. By embracing motherhood, lesbians are making their lives "intelligible" to others—their quest to become parents is often enthusiastically supported by family and heterosexual friends. However, their sexuality both necessitates and facilitates the redefinition of the boundaries, meaning, and content of parenthood. When women parent together, the absence of the logic of polarization to inform gender scripts, and their parity in the gender hierarchy, means that, to borrow Juliet's words, "We have to make it up as we go along." Their similarities as women insist on high levels of reflexivity and enable the construction of more egalitarian approaches to financing and caring for children. In this way, some of the more negative social consequences of motherhood can be transformed. Although not unique in their achievements, nor assured of their success, women parenting with women have a head start over heterosexual couples because of their structural similarities and the way that egalitarianism is in the interests of both partners.[1]

★ ★ ★

PARENTING CIRCUMSTANCES

The sample includes 8 households where children were from a previous marriage, 1 household where the children were adopted, and 28 (75 percent) where they had been conceived by donor insemination. In the majority of households (60 percent), there was at least one child younger than five; and in 40 percent of households, coparents were also biological mothers of older, dependent, or nondependent children. The research revealed a fairly unique and important opportunity for women parenting together—the possibility of detaching motherhood from its biological roots through the experience of social motherhood. Interestingly, 15 women in the study expressed a long-standing desire to mother as a social experience but a strong reluctance to experience motherhood biologically. These women had often taken responsibility for siblings in their families of origin and for the children of others usually featured in their lives and occupational choices. This social-biological separation also meant that motherhood is not necessarily ruled out for women who have fertility problems. Parenting was depicted as jointly shared in 30 households (80 percent). As we will see in the three case studies, in contrast to men who share mothering (Ehrensaft 1987) yet remain happy with the identity of father, the singularity and exclusivity of the identity of mother represented a major problem for women parenting together.

[A]lmost all of the women who had experienced donor insemination organized this informally—they rarely used National Health or even private fertility services. Respondents tended to want to know the donor, and in 86 percent of households, this was the case. A wide range of reasons was given for this preference. A common feeling related to wanting to know that a good man, in terms of personal qualities, had a role in creating their child. Sometimes more specific ideas about biogenetic inheritance came up in discussions, and for Jewish women there was a preference for Jewish donors. Some employed the metaphor of adoption—the idea that children should have the option of knowing their biological father at some stage in the future. Commonly, donors were located through friendship networks or by advertising. Occasionally, they made use of the informal women's donor networks that exist in many British cities. When organized informally, children were always conceived by self- or partner insemination, and the majority became mothers in their current lesbian relationship.

Lesbian motherhood undermines a core signifier of heterosexuality and challenges heterosexual monopoly of and norms for parenting. The social hostility toward those parents and children who transgress the sanctity of heterosexual reproduction is such that the decision to become a mother by donor insemination can never be easily made. Typically, respondents described a lengthy period of soul-searching and planning preceding the arrival of children. For some, this process lasted as long as seven years. Unlike most women, they had to question their motives for wanting children, to critique dominant ideas about what constitutes a "good" mother and family, and to think about the implications of bringing up children in a wider society intolerant of difference. Informing this process was much research—reading the numerous self-help books that are available on lesbian parenting, watching videos on the topic, and attending discussion groups. I would suggest that lesbian parenting via donor insemination is the "reflexive project" par excellence described by Giddens (1992, 30). For respondents in partnerships, a central part of this process was the exploration of expectations in relation to parenting, for example, attitudes to discipline, schooling, and if and how far responsibilities would be shared. Key considerations related to employment situations. Respondents did not expect or desire a traditional division of labor, and thus timing was often influenced by their preference to integrate child care and income generation. In the meanwhile, potential donors were contacted. Respondents described a fairly lengthy process of negotiation with donors that focused on establishing a mutuality in parenting expectations and, if he was previously unknown to the couple, getting to know each other and developing confidence. While recognizing the generosity of potential donors, some were rejected because of personality clashes or concerns about motives, but more usually, rejection was because a donor wanted too much or too little involvement.

Men featured in the lives of most of the children, and it was not unusual for donors to have regular contact with their offspring (40 percent of households); in three households, fathers were actively co-parenting. This involvement was usually justified in terms of providing children with the opportunity to "normalize" their family arrangements by being able to talk to peers about doing things with father. Donors were usually gay men—and all male co-parents were gay. This preference appeared to be based on three main assumptions. First was the respondents' perceptions of gay men as representing more aware, acceptable, and positive forms of masculinity. Their desire to involve men (donors or other male friends) in the lives of children, particularly boys, was often described as being about counteracting dominant stereotypes of masculinity. Second, because of the particularities of gay men's lifestyles, respondents believed that they would be less likely to renege on agreements. Third, they thought that should a dispute arise, a heterosexual donor (particularly if he were married) had greater access to formal power to change arrangements in relation to access and custody. That none expressed any serious difficulties in relation to father and/or donor involvement attests to the value of the careful negotiation of expectations before the arrival of children. It also says much about the integrity and generosity of the men concerned, although it must be noted that most had preschool-age children, and conflicts of interests may come as the children mature.

In situations where children had been conceived in a previous marriage or heterosexual relationship, there was more diversity and conflict regarding fathers' involvement. In several cases, the father had unsuccessfully contested custody on the grounds of the mother's lesbianism. Indeed, two had appeared on daytime television arguing that their ex-wives' sexuality conflicted with their capacity to be good mothers. There were also examples of good relations between mothers and ex-husbands. While there were several examples of fathers having lost contact with their children, in most cases, respondents suggested that the child or children had more quality time with their fathers after divorce than before. Despite tensions and possible conflict between mothers and ex-husbands, these respondents suggested that they worked hard to maintain their children's relationships with the fathers. Thus, ironically, in this group as well as in the donor insemination group of parents, there are examples of highly productive models of cooperation between women and men in parenting.

The role of fathers and/or donors and other male friends in children's lives reminds us that lesbian parenting does not occur in a social vacuum. While generally hostile to the idea of the privatized nuclear family, respondents were keen to establish more extended family networks of friends and kin. Often, respondents described the arrival of children as bringing them closer to or helping repair difficult relations with their families of origin. Typically, they described a wide circle of friends (lesbian "aunties," gay "uncles," and heterosexual friends) and kin supporting their parenting.

I now want to illustrate some of these themes by drawing on the voices of respondents in three partnerships where parenting was shared and where men were involved.

VIVIEN AND CAY'S STORY

> We do feel lonely and unsupported and isolated at times, but we also feel very confident and excited about the way that we've carved out our family and the way that we go forward with it and the way that we parent. So although it's kind of a lonely path because there's not a lot of us to kind of reflect on each other, I don't see that as, oh, poor us. I see that more as, well, we're trying something out here and we've just got to get on with it. (Vivien)[2]

It was not uncommon to find a woman who had been married and had grown-up children who was starting over again with a partner who wanted to have children herself ($n = 4$). Women parenting together was understood as offering the opportunity to experience parenting in new and exciting ways that were tempered by the wisdom that comes from already having raised children. Cay and Vivien are fairly typical of these households. Vivien, age 44, has a grown-up son, Jo, who lives independently. Cay, age 32, is the biological mother of two boys, Frank, age four, and Steve, age two. When we first met, they had been living together for six years in a small terraced house in inner-city Birmingham. Cay, born in North America, is a self-employed illustrator of children's books who supplements her income by working as a cleaner. Vivien, of Irish-Greek descent, recently completed a degree and acquired her "first real job" as a probation officer. Cay told me that she had always wanted to have children and that her sexuality had not changed this desire. Vivien was enthusiastically supportive of the idea although she did not want to go through a pregnancy herself.

Like the vast majority of respondents, Vivien and Cay organized donor insemination informally. They had little difficulty in locating a willing donor—John, an old friend of Vivien's.

> Vivien: It worked out well. He's my oldest friend, and we've known each other since we were teenagers, and he has the same kind of coloring and stuff, he could be my brother in terms of coloring and looks. Originally we asked one of my brothers to donate, and he felt he would maybe want more of an involvement, more of a say in the children's lives, and we wanted somebody who would let us have the responsibility and would take on a sort of a kindly uncle role. And John agreed to do that.

The description "kindly uncle" was frequently used by respondents to describe what was a fairly limited yet enthusiastic relationship between a donor and his child or children. Respondents almost always wanted to retain responsibility for bringing up their children. Like most of the couples in the study, Vivien and Cay regard these responsibilities as shared.

> Cay: It can't be anything but joint I think. The way we've approached it is that if it's not totally agreeable between both of us, it couldn't have really gone forward, given the kind of relationship we have. We've seen other people, you know, where one parent has said, "Well, I want a child and that's it." But the other one says, "Yes, you can have one, but I don't want to have lots of responsibility." That's not our way.

When respondents described their parenting as jointly shared, they meant that each partner took an active role in the routine pleasures, stresses, and labor of child care. . . .

Like Vivien, respondents took great pleasure in child care, and this was reflected in their ordering of priorities. Cay suggested that because she and Vivien had joint responsibility for housework, they were less subject to the tyranny of maintaining high domestic standards—a sentiment reflected across the sample. This, together with their shared approach to doing tasks, she believed, gave them more fun time with their children—this was supported in the time-use data across the sample. For example, comparison of respondents' time use with trends for married parents with young children revealed that regardless of the employment status of married mothers, because they did the bulk of routine domestic work, it occupied far more of their time than child care, while the reverse was the case for respondents (Dunne 1998a).

Vivien and Cay described their roles before and after the arrival of children as interchangeable; earlier, Cay had been the main earner when Vivien was a student. Routinely, birth mothers and co-parents alike spoke of seeking integrated lives—valuing time with children, an identity from the formal workplace, and the ability to contribute financially. Within reason, they were prepared to experience a reduced standard of living to achieve the kind of quality of life desired. Thus, there was an unusually wide range of partner-employment strategies in the sample. Like Vivien and Cay, some took turns in who was the main earner, while others (a quarter of the sample) opted for half-time employment for both parents. Rather than the polarization of employment responsibilities that characterizes married couples' parenting experiences, particularly when children are young,[3] few households had extreme partner differences in employment hours, and being the birth mother was a poor predictor of employment hours. . . .

While Vivien and Cay describe themselves as the boys' mothers, in common with most respondents, they struggle over terminology to describe and symbolize that relationship. Because of the singularity and exclusivity of the label *mother* or *mum* and/or their feminist critique of the way the term can eclipse other important aspects of identity, respondents often preferred to encourage the use of first names, special nicknames, or the word *mother* borrowed from another language.

> Vivien: Yes, [we are the boys' mothers] absolutely, yes. Very much so.
>
> Interviewer: What do they call you?
>
> Vivien: By our names. . . . They very rarely use the word *mother.*
>
> Cay: In fact [Frank] never used the word *mother* until he started going to school, and then, hearing the other kids saying it, it was just a kind of copying thing.

Here we catch a glimpse of some of the everyday pressures toward social conformity and the dilemmas experienced by parents and children as they negotiate a world hostile to difference. Just as this motivated some to involve donors in their children's lives so that they had the option to pass as relatively "normal" in school, many respondents relented and used the term *mother* to describe the biological mother.

Cay's parents live abroad and are described as proud grandparents. Because John, the boys' father, is not out about his sexuality to his parents, they have no knowledge about the boys. Vivien's parents are dead, but her immediate kin are actively involved in supporting them.

> Vivien: My brothers are thrilled, though. My brothers treat the children as if they were their own kids. They don't separate them, you know, they don't see them as any less their kids. And their cousins that they're totally unrelated to just are their cousins, and in fact Tom looks like one of my cousins in Ireland. He doesn't look like any of Cay's.

However, the very positioning outside conventionality that enables the construction of more creative approaches to organizing parenting brings also the problem of lack of recognition and validation from the outside world. Vivien speaks for many in the study:

> I think we have to acknowledge that within this house we can sit down and we can talk about the equality that we feel and the experiences that we have and the confidence that we have in our relationship and in our parenting. But very little outside of this house tells us that those things that we're talking about tonight are actually true. . . . I think heterosexual friends that we have tend to probably see our relationship in their own terms. . . . I don't think they've got an insight into how much we really do work together. . . . You know, we have to work at it all the time, we have to forge links with the school, we have to forge links with this and forge links with that, we have to work hard at being good neighbors and making contact with the neighbors so that as the children come along they're not surprised and they can adjust. We're doing the work, we're doing the outreach, we're doing the education, and what we get back is the right to be ourselves, sort of, as long as we're careful.

Again, their experience underscores the difficulties associated with challenging the normative status of heterosexuality in relation to reproduction and the organization of parenting roles. Constantly, these pioneering women feel obliged to justify their alternative families and approaches to parenting to a wider society that cannot see beyond the constraints of heterosexuality and that is informed by media representations that vilify lesbian parents. Their struggle for validation was not confined to the heterosexual world.

> Vivien: Other lesbians I think may see its as trying to repeat some sort of heterosexual relationship, and that's not what we're trying

> to do. So we have to kind of justify it to our heterosexual friends and justify it to our lesbian friends.

The contradiction illuminated here between being a lesbian and being a mother serves to remind us that while it can be argued that assisted conception is an important expression of the ideologies supporting compulsory motherhood, it is less easy to apply this thinking to lesbian mothers. Within lesbian culture, the absence of children within a relationship does not constitute failure. In fact, research (Sullivan 1996) supports much of what respondents said about their decision to have children going against established societal norms, specifically those of the lesbian and gay community. Until recently, this community, particularly the radical or revolutionary wing, has been suspicious of motherhood because of fears of constraints on women's autonomy and the importation of oppressive family arrangements (see Green 1997).

THELMA AND LOUISE'S STORY

★ ★ ★

It was not unusual for both partners to have experienced biological motherhood as the result of donor insemination while in their relationship. At the time of first contact, four couples were in this situation (this number had risen to seven at the follow-up stage two years later). In these households, children were brought up as siblings, and parenting was equally shared. The experiences of Thelma and Louise are not atypical of mothers in this situation. They have been living together for seven years in an apartment that they own in inner-city Manchester. They have two daughters, Polly, age four, and Stef, age two. Thelma works in desktop publishing, and Louise is a teacher. Like many in the sample, Thelma and Louise operationalize shared parenting by reducing their paid employment to half-time. They both wanted to have children; their decisions about timing and who would go first were shaped by emotional and practical considerations. Thelma needed to build up sufficient clientele to enable self-employment from home, and Louise wanted to gain more secure employment. . . .

By the time Louise was pregnant, two years after Thelma, she was in a much stronger position at work, having undergone retraining. She had secured a permanent position in teaching and, after maternity leave, arranged a job share with a friend. Like women more generally, respondents' careers had rarely progressed in a planned linear manner. Instead, their job histories have a more organic quality (see Dunne 1997)—moving across occupations and in and out of education or training. However, in contrast to married women more generally, where the gender division of labor supports the anticipation of financial dependence on husbands when children are young (Mansfield and Collard 1988), an important consideration in the timing of the arrival of children for most biological mothers in this study was the achievement of certain employment aims that would enable greater financial security and allow time to enjoy the children. Their gender parity and this approach to paid employment meant that there were not major earning differentials between partners. . . .

After several miscarriages with an earlier donor, Thelma finally got pregnant. Again, they used their friendship networks to locate a donor who then took on a "kindly uncle" role.

> Louise: He was just living with a friend of ours, it was just brilliant.
>
> Thelma: Yeah, and ended up being a really good friend as well. . . . I got pregnant the first go really.
>
> Interviewer: And then did you have any views on how much involvement he should have?
>
> Thelma: I think we both wanted a known father and yes, if they wanted some involvement, that was fine. The clearly defined lines were, we're the parents of the

children—or of the child at that time—and so any kind of parenting decisions would always be ours.

Interviewer: And what will Polly call her donor?

Thelma: His name—and she calls him Daddy Paul. So I mean she doesn't ever really call him Daddy. Either she calls him Paul or Daddy Paul.

Louise: He is a bit like an uncle [to them both] she'd see now and again, you know, he'd be like this kindly uncle figure, who'd take her to the pics and take her to the zoo and that kind of thing. Give her treats.

They originally planned that Paul would be the donor for Louise; however, there were difficulties in conception, so a new donor was found. Hugh, a gay friend of Thelma's brother, who was temporarily living in England, agreed. While Thelma and Louise both wanted to experience motherhood biologically, they viewed parenting as shared, and this situation was legally recognized in their gaining of a joint parental responsibility order.

Louise: We don't just happen to have a relationship and happen to have two children. We always thought joint, that's why the court thing was important to us. They are sisters and I defy anybody to question that. That's very important to us and we also made it clear that if we ever split up, if I depart with Stef into the horizon and Thelma with Polly, that we have joint care for them.

Again, their interpretation of shared parenting brought them up against the limitations of language to describe a social mother's relationship to a child.

Thelma: They both call us Mum.

Louise: It started off that you were going to be Mum and I was going to be Louise, and then coming up to me giving birth to Stef, it just got a bit kind of funny, so we thought it's not really going to work any more because if they're sisters how come?—it just all didn't work, so now we're both Mums. And they just call us Mum.

Thelma: Stef says Mummy Louise or Mummy Thelma.

Louise: And Polly mostly calls us Louise and Thelma doesn't she?

Thelma: Yeah she does. She calls us both Mum when she wants to, but mostly she calls us by our names.

Louise: The last couple of years she's started calling me Mum.

Some of the immensity of the creative project in which lesbians engage is revealed in the tensions in the last two extracts and in the next. While they describe the children as having two mothers, Louise reminds us of the contingent nature of this. The rule of biological connection is unquestioned in the assumption that in the event of a breakup each will depart into the horizon with her own child. This next extract illustrates other practical difficulties faced by the couple as they engage with the wider society.

Louise: It's a lot easier now because we've both had a child. I don't think I had any role models in terms of being a nonbiological mum. There's a thing that if you want to be acknowledged as a parent, you just had to "come out." It's the only way to explain that you're a parent. And even that is a very hard way to explain you're a parent. My inner circle at work would know and it's funny—I nearly wrote it down one day—because it was just like some days I'd be a parent and some days I wasn't. So it would depend on what day of the week it was and who I was talking to. I think I made it harder for us by me not being called Mum [in the early stages]. Because as soon as people found out you weren't the mum, then they'd just—it was like "who the hell are you then?"

Such is the power of ideas about the singularity and the exclusivity of the identity of "Mum" in a social world structured by heterosexual norms that polarize parenting along lines of gender. Respondents had a store of both amusing and uncomfortable stories about other people's confusions about who was the mother of the child or children or the status of social mothers.

Again, the family has interesting and extensive kinship networks. The children have two fathers. Paul was not out to his elderly parents so they did not know about his child. However, Hugh, who comes to England several time a year to see them, had told his mother.

> Louise: I think Hugh was terrified of telling his mother—he's an only child—had a very close relationship with his mother and he was terrified of tell her. And she was absolutely delighted with it—"I'm the *children's* grandmother"—she's Stef's grandmother biologically, but she's also Polly's socially. So she's just been this incredible grandmother.
>
> Thelma: Paul's parents don't know. His parents are quite old, they're in their late eighties and they don't know he's gay and I don't think he'd ever tell them. So for him that one's a secret. But his sister knows.
>
> Louise: I think we'd be more worried by it, but I just guess by the time the kids are old enough—I think you've got to start coming out very confidently once you've kids, you can't be messing around really. And it would worry me I think if—if Paul explained to them [the children], that he's not been able to tell his parents. I'm just hoping that by the time it comes up, they won't be around any more.

This discussion illustrates several important themes that featured across the sample. First, respondents were keen to avoid keeping secrets from their children about their conception. Second, they articulated high levels of positivity about being lesbian[4]—this was seen as essential for supporting their children in their dealings with the outside world. Third, all expressed the desire to have their social bonds recognized by friends and kin as being equivalent to blood ties. Finally, kinship was calculated in a remarkable variety of ways. Kin appeared highly flexible in this, with countless examples like Hugh's mother. . . .

Without exception, respondents believed that they approached and experienced parenting in ways that were very different from the heterosexual norm. They were redefining the meaning and content of motherhood, extending its boundaries to incorporate the activities that are usually dichotomized as mother and father. Going against prevailing norms was never without difficulties and disappointments. In joint and individual interviews, respondents usually singled out the ability and commitment to communicate as crucial. They spoke of arrangements being constantly subject to negotiation and the need to check in regularly with each other so that routines that may lead to taking the other for granted could be rethought and sources of conflict discussed.

BONNIE AND CLAUDIA'S STORY

★ ★ ★

In three partnerships, donors were actively co-parenting from separate households—becoming a "junior partner in the parenting team" as one father described himself. In two cases, the father's parenting was legally recognized in a joint residency order. Bonnie, Claudia, and Philip share the care of Peter, age two. Bonnie and Claudia have lived together for nine years in a terraced house in inner-city Bristol. Bonnie, Peter's biological mother, works full-time in adult education, and Claudia has a half-time teaching post. They describe and contrast their feelings about wanting to have children:

> Claudia: Well, I think it was something that I was looking for when I was looking for a relationship. So I think it was a more immediate thing for me. You were interested in principle. And I knew the father—this is

Philip—although not with the view to having children. So you got to know him after we met really. And then the subject came up.

Bonnie: I think for you it had always been like a lifelong thing.

Claudia: I always wanted a baby. I wanted us to have about two.

Bonnie: She was just obsessed with babies, weren't you? Whereas, I wasn't really like that, I come from a big family and I like having lots of people around me. It was more for me that I didn't want to have *not* had children. It's different, because I didn't want to look back and think, Oh Christ, I didn't have any children. But I tend to get very caught up in whatever I'm doing, and I was busy doing my job and having this relationship and our friends. So in a way it was Claudia's enthusiasm and sense of urgency about it that actually pushed us to making a decision, taking some action. And the only reason I ended up having the baby was that Claudia had a whole series of fertility problems. We just always decided, didn't we, that if one of us had a problem the other one would.

Their experience illustrates another fairly unique advantage for women who want to become mothers in a lesbian relationship—if one partner has fertility problems, the other may agree to go through the pregnancy instead. There were three other examples of partners swapping for this reason, and several others expressed their willingness to do so. . . .

In their negotiations with Philip over the four years that preceded the birth of Peter, they came to the decision that he would be an actively involved father.

Claudia: Philip wanted a child, and he, I think, was also looking for a kind of extended family relationship, wasn't he?—with us and the children. But he also wants his freedom, I suppose, his lifestyle, a lot of which he needs not to have children around for. Yes, so it fits in the sense that what we get is time without Peter, to have a relationship that needs its own sort of nurturing and stuff, and he gets special time with Peter and a real bonding. I mean he's seen Peter every day since he's been born. So he has become part of the family, hasn't he?—in a sense, or we've become part of his. But we live in two separate homes. People sometimes don't realize that.

Claudia's words alert us to another underlying reason for respondents' confidence in fathers and/or donors retaining a more minor role in children's lives—routine child care does not usually fit in with the lifestyles of most men, gay or heterosexual. The masculine model of employment that governs ideas of job commitment and what constitutes a valuable worker is based on the assumption that employees are free from the constraints of child care.

After extended maternity leave, Bonnie returned to her successful career in adult education. At this point, Claudia, despite being the higher earner, reduced her employment hours to half-time so that she could become Peter's main caregiver. Men's superior earnings are often described by egalitarian-minded heterosexual couples as ruling out opportunities for shared parenting (Doucet 1995; Ehrensaft 1987). However, women parenting together, without access to ideologies that polarize parenting responsibilities, bring fresh insights to this impasse, which supports gender inequalities.

Bonnie: We started in a completely different place [from heterosexual couples]. I think we feel it's just much easier to be cooperative and to be more creative in the way that we share out paid work and domestic work, because that's how we look at it. We're constantly chatting about it, aren't we, over the weeks, and saying, "How does it feel now? Are you still thinking about staying on part-time?" and we've talked about what it would be like if I went part-time as well, and could we manage on less money?

> Claudia: Yes, and I think the thing that's part of the advantage is that in a conventional setup, although it may be easier to start with, everyone knowing what they are supposed to be doing, but the men don't know their children so they miss out. . . . . I'm having a balanced life really.
>
> Bonnie: I think that's why we've got the space to enjoy our child in a way that a lot of heterosexuals perhaps don't. It's so easy to fall in—the man earns slightly more so it makes sense for him to do the paid work, and women have babies anyway. Because we could potentially each have had the child it's all in the melting pot. Nothing is fixed.
>
> Claudia: And I don't think a lot of women [enjoy mothering]. They think they're going to, but they get isolated and devalued, and lose their self-confidence and self-esteem.

It was not unusual to find the higher earner in a partnership reducing her hours of employment to share care or become the main caregiver. In contradiction to the dictates of rational economic models, this was often justified on the grounds that a person in a higher paid or higher status occupation has more power and may be less penalized for time out than someone in a more marginal position (Dunne 1998a). I would argue that their rationale (like the part-time/part-time solution) can actually make good long-term financial sense. It also illuminates masculine assumptions in relation to value—the idea that market work is superior to caring.

As in the vast majority of households (Dunne 1998a), routine domestic work was fairly evenly divided between Claudia and Bonnie. Their guiding principle was that "neither should be running around after the other." . . .

Peter goes to a private nursery three days a week (the costs are shared with Philip), and the rest of his care is divided between Claudia, Bonnie, and Philip.

> Bonnie: Philip lives in the next street, and so he can just come round every day after work or pick Peter up from nursery and bring him back and do his tea, bath and things, and then we'll roll in about 6:30 or whenever, or sometimes one of us is here anyway.
>
> Claudia: Yes, we try to work that one of us is always at home, either with him or working at home. . . . Quite often there's days when we both have to commute, so Philip usually covers. . . . He's the only one of us who works locally and he's got a bleep [beeper] as part of his job and it's ideal because the nursery can call at any time if there's an emergency.
>
> Interviewer: It strikes me you've got the most ideal situation!
>
> Bonnie: Yes, we think so! [laughing] We're the envy of the mother and toddler group.

Their experience with Philip provides a radical alternative model of cooperative parenting between women and men, based on a consensual nonsexual relationship with a father who is interested in being actively involved in his child's life. In effect, Philip is prepared to engage in mothering,[5] and in doing so, he shares some of the social penalties associated with this activity—all three parents collaborate in balancing the demands of employment and child care, and the result is the lessening of its overall impact. . . .

Again, finding the right words to describe their parenting relations was difficult. Bonnie expresses a common feminist critique of the label *mummy,* which is hostile to ways that it can be employed to subsume other aspects of a woman's identity.

> Bonnie: I've always been quite keen that Peter should know what our names are anyway. I think there's something completely depersonalizing about the way women sit around and talk about a child's mummy as if she's got no identity. It's fine if there's a baby in the room and it's your child, but everyone will say, "Ask Mummy, tell Mummy." But you become this amorphous mummy to

> everybody. All women are sort of mummy, they don't have their own identity. So I've been quite keen that he should grow up knowing that people have roles and names, and that you should be able to distinguish between the two.

Yet, her radicalism is tempered by her recognition and desire to celebrate her special connection with the child, and she becomes swayed by arguments for the best interests of the child.

> Bonnie: But I also feel completely contradictory, that there is something very special emotionally about having your own mummy.
>
> Claudia: And then Philip had very strong feelings about it all, didn't he? He'd always been clear that he wanted to be Daddy, and while we went on holiday together last summer, he made it very clear that he thought that in some sense you needed to be recognized as Peter's mother, that that was important, an important thing in terms of what the relationship meant, and that it would be wrong to deny Bonnie that. . . . Yes, he [also thought] that Peter would, if we started him calling both of us Mummy, sooner or later he'd be ridiculed by some of the other children, and then he would have a terrible conflict of loyalties, does he go with the crowd or does he protect us? And that we shouldn't put him in that position. So we went for Mummy, Daddy and Claudia. And then he started calling me Mummy anyway. But now he calls me Addie. [laughter]

This Mummy, Daddy, and Claudia configuration that then evolved into Claudia being called Mummy or the nickname Addie is potentially very undermining of the co-mother. Other couples specifically avoided involving biological fathers to this extent because of such complications of status and role. Claudia's confidence in her relationship with Peter was affirmed through her experience of mothering as main caregiver and, hopefully, by their capacity to be aware of the issues, as the discussion above appears to indicate. Philip's desire for recognition as Daddy is at one level less problematic. He earns this validation through his active involvement in parenting, and because he is not attempting to share fatherhood with a partner, there are no additional complications in relation to exclusion. However, the gender dynamics of this are interesting. While much of the social aspect of Philip's parenting involves the activities of mothering, he is content with the identity of dad. Conversely, in common with the rest of the sample, rather than draw upon dominant polarized heterosexual frameworks—mother/father—respondents extend the meaning of motherhood to include so-called fathering activities such as breadwinning. This raises the wider question, What exactly is a father?

Once again, their parenting is supported by a complex network of kin who have been encouraged to recognize and act upon social as well as biological ties. As they map out the main people supporting their parenting, Bonnie and Claudia discuss the input of kin:

> Bonnie: That's my sister Holly and her partner Vickie, who is dyke as well, which is very nice, and they live round the corner as well. So in a sense they are part of our community, very much so, and Vickie was around for the birth. So they lead a different sort of lifestyle in the sense that they haven't got any children, so they're definitely sort of aunts that come in and do babysitting and things. They're sort of busy but they're important, and we promote the relationship actually, don't we?
>
> Interviewer: What about Philip's parents? Do they have any . . . ?
>
> Bonnie: Yes, there's Philip's mum and dad. They see him two or three times a year—it's only been a year and a half, but they've made a lot of effort. They came down just after his birth.

It is no simple act, however, for extended family to claim kinship ties in these nontraditional situations that require coming to terms with a relative's sexuality. While part of being lesbian and gay is about learning how to come out to self and others, I think we have given scant attention to the work involved when heterosexual family members, particularly elderly parents, claim kinship ties that require coming out on behalf of others. For Philip's parents, it was easier for them to explain his entry into fatherhood to other family members by inventing a complicated story about Philip and Bonnie being or having been lovers. . . .

As Claudia had been adopted, her family was used to the complexity of kinship relations.

> Claudia: [My family is] all interested and very supportive but there's no one nearby to pop in. . . . They all only see him about twice a year. Family get-togethers, isn't it?
>
> Bonnie: And you made an effort to go and visit and show Peter off.

Claudia's biological parents were described as treating Peter similarly to their other grandchildren, all of whom receive scant attention. Interestingly, in the case of her adoptive parents, in common with many other respondents, the arrival of children helped rebuild bridges after earlier estrangement over issues of sexuality.

> Claudia: Well, [my adoptive parents] have much more difficulty with me being a lesbian than my parents do. And they've virtually rejected me really. Not immediately when I came out but later on. And then [my adoptive mother], since she found out that I was trying to get pregnant, has been completely supportive. I think [my adoptive father] finds it more difficult.
>
> Interviewer: And she thinks of Peter as your son?
>
> Claudia: Yes. And she describes herself as his adoptive grandmother.

Bonnie's mother could see distinct advantages in her daughter's parenting arrangements:

> My mum is Peter's grandmother. She's very, very involved with Peter, totally supportive of this relationship, and thinks that—why hadn't anyone ever mentioned it before? It seems a great way to bring up children. Having brought seven children up without the help of my father, she now thinks it's wonderful not only to have a supportive woman partner but a father involved who lives up the road. It's great. Peter sees more of his father than most children probably do. So she's good.

Aside from a wide circle of friends, Bonnie, Claudia, and Philip had support from parents and siblings, with their son Peter looking forward to presents from four sets of grandparents.

CONCLUSION

These three stories illustrate many common themes that emerged across the sample, particularly the creativity and cooperation that appear to characterize much of the parenting experience of lesbian couples. I have focused on the involvement of fathers and/or donors and on the complexity of kinship to show how like and unlike these families are to other sorts of family formations. I could equally have looked at the important friendship networks that supported their parenting, the presence of lesbian aunties and heterosexual friends. Lesbian families are usually extended families, supported by elaborate networks of friends and kin.

In common with single lesbian mothers in the United States (Lewin 1993, 9), kin occupy an important place in respondents' accounts of their social interaction. My focus on couples in shared parenting situations reveals other interesting dimensions of kinship: the complexity of these relations and the importance respondents placed on having nonbiogenetic ties recognized and validated by family of origin. . . .

Regardless of whether parenting was shared, mothering was usually carried out in a context where mothers experienced a great deal of practical and emotional support from their partners, where routine domestic responsibilities were fairly evenly shared, and where there was a mutual recognition of a woman's right to an identity beyond the home. Beyond the confines of heterosexuality, they had greater scope to challenge the connections between biological and social motherhood and fatherhood. By deprivileging the biological as signifier of motherhood (although this appears to be contingent on the relationship remaining intact) and the capacity to mother, many were actively engaged in extending the meaning, content, and consequence of mothering to include both partners (or even fathers) on equal terms. . . .

They consequently have greater scope to operationalize their egalitarian ideals in relation to parenting. The high value they attached to nurturing, together with their desire to be fair to each other, meant that within reason they were prepared to experience a reduced standard of living (see Dunne 1998a). Their views about what constitutes shared parenting were less distorted by ideologies that dichotomized parenting along lines of gender in such a way that men can be seen and see themselves as involved fathers when they are largely absent from the home (Baxter and Western 1998, Ferri and Smith 1996). Consequently, their solution to the contradiction was to integrate mothering and breadwinning.

In their everyday lives of nurturing, housework, and breadwinning, respondents provide viable alternative models for parenting beyond heterosexuality. While our focus is on lesbian partners, anecdotal evidence suggests that lesbians are also founding parenting partnerships on the basis of friendship—with gay men or other lesbians. By finding a way around the reproductive limitations of their sexuality, they experience their position as gatekeepers between children and biological fathers in an unusual way. Ironically, we find examples of highly productive models of cooperation between women and men in bringing up children. Unhampered by the constraints of heterosexuality, they can choose to include men on the basis of the qualities they can bring into children's lives. It is no accident, I believe, that respondents usually chose to involve gay men. These men were seen as representing more acceptable forms of masculinity, and their sexuality barred them from some of the legal rights that have been extended to heterosexual fathers.

Their positioning outside conventionality and the similarities they share as women enable and indeed insist upon the redefinition of the meaning and content of motherhood. Thus, when choosing to opt into motherhood, they are anticipating something very different from the heterosexual norm. . . .

They challenge conventional wisdom by showing the viability of parenting beyond the confines of heterosexuality. Rather than being incorporated into the mainstream as honorary heterosexuals, by building bridges between the known and the unknown, their lives represent, I believe, a fundamental challenge to the foundation of the gender order.

NOTES

1. Both VanEvery (1995) and Ehrensaft (1987, 20) mention that women are the driving force in the quest to achieve and maintain egalitarianism. Both comment on the extent to which structural factors, such as men's superior earnings, and wider social expectations mediate success in this respect.
2. To maintain confidentiality, the names of participants and their children and their geographical location and occupations have been changed. To give some sense of their employment circumstances, I have assigned similar kinds of occupations.
3. While British mothers are more likely now than in the past to be employed full-time, it is mothers rather than fathers who balance the demands of paid work and child care. It is very unusual for mothers and fathers to have similar lengths of paid-work weeks, even when mothers are employed full-time (Dunne 1998a; Ferri and Smith 1996).
4. I was struck by the almost unanimous confidence of the sample in their sexuality—respondents saw their lesbian identity as a great source of advantage. Their identification as lesbian rather than gay was also evidence of their usually feminist inclinations. In a

previous life history study of lesbians who were generally not mothers (Dunne 1997), there were more examples of ambiguity in this respect. I suspect respondents' self-assurance is related to a combination of factors including historical period, being in fulfilling relationships, their achievement of motherhood, and the process of soul-searching that preceded this.

5. Silva (1996) draws a useful distinction between motherhood, a uniquely female experience, and mothering, which, although usually a female practice, can be performed by either gender.

REFERENCES

Baxter, J., and M. Western. 1998. Satisfaction with housework: Examining the paradox. *Sociology* 1:101-20.

Berk, S. F. 1985. *The gender factory: The appointment of work in American households.* New York: Plenum.

Blaisure, K., and K. Allen. 1995. Feminists and the ideology and practice of marital equality. *Journal of Marriage and the Family* 57:5-19.

Brannen, J., and P. Moss. 1991. *Managing mothers: Dual earner households after maternity leave.* London: Unwin Hyman.

Butler, J. 1990. *Gender trouble: Feminism and the subversion of identity.* New York: Routledge.

Doucet, A. 1995. Gender equality, gender difference and care. Ph.D. diss., Cambridge University, Cambridge, UK.

Dunne, G. A. 1997. *Lesbian lifestyles: Women's work and the politics of sexuality.* London: MacMillan.

———. 1998a. "Pioneers behind our own front doors": Towards new models in the organization of work in partnerships. *Work Employment and Society* 12(2):273-95.

———. 1998b. A passion for "sameness"? Sexuality and gender accountability. In *The new family?* edited by E. Silva and C. Smart. London: Sage.

———. 1998c. Opting into motherhood: Lesbian experience of work and family-life. London School of Economics, Gender Institute Discussion Paper Series 6.

———. 1998d. Add sexuality and stir: Towards a broader understanding of the gender dynamics of work and family life. In *Living "difference": Lesbian perspectives on work and family life,* edited by G. A. Dunne. New York: Haworth.

———. 1999. Balancing acts: On the salience of sexuality for understanding the gendering of work and family-life opportunities. In *Women and work: The age of post-feminism?* edited by L. Sperling and M. Owen. Aldershot, UK: Ashgate.

———. 2000. Lesbians as authentic workers? Institutional heterosexuality and the reproduction of gender inequalities. *Sexualities.* 3(2): 133-148.

Ehrensaft, D. 1987. *Parenting together: Men and women sharing the care of the children.* New York: Free Press.

Fenstermaker, S., C. West, and D. H. Zimmerman. 1991. Gender inequality: New conceptual terrain. In *Gender family and economy, the triple overlap,* edited by R. L. Blumberg. London: Sage.

Ferri, E., and K. Smith. 1996. *Parenting in the 1990s.* London: Family Policy Studies Center.

Gartrell, N., A. Banks, J., Hamilton, N., Reed, H. Bishop, and C. Rodas. 1999. The national lesbian family study 2: Interviews with mothers of toddlers. Forthcoming. *American Journal of Orthopsychiatry.* 69(3): 362-369.

Giddens, Anthony. 1992. *The transformation of intimacy.* Cambridge, MA: Polity.

Green, S. 1997. *Urban amazons: The politics of sexuality, gender and identity.* Basingstoke, UK: MacMillan.

Gregson, N., and M. Lowe. 1995. *Servicing the middle-classes: Class, gender and waged domestic labor.* London: Routledge.

Hochschild, A. R. 1989. *The second shift.* New York: Avon.

Lewin, E. 1993. *Lesbian mothers.* Ithaca, NY: Cornell University Press.

Mansfield, P., and J. Collard. 1988. *The beginning of the rest of your life: A portrait of newly wed marriage.* London: MacMillan.

McAllister, F., and L. Clarke. 1998. *Childless by choice: A study of childlessness in Britain.* London: Family Policy Studies Centre.

Rich, A. 1984. On compulsory heterosexuality and lesbian existence. In *Desire: The politics of sexuality,* edited by A. Snitow, C. Stansell, and S. Thompson. London: Virago.

Rubin, G. 1975. The traffic in women: Notes on the "political economy" of sex. In *Towards an anthropology of women,* edited by R. R. Reiter. London: Monthly Review Press.

Silva, E. 1996. The transformation of mothering. In *Good enough mothering?* edited by E. Silva and C. Smart. London: Routledge.

Sullivan, M. 1996. Rozzie and Harriet? Gender and family patterns of lesbian coparents. *Gender & Society* 10(6):747-67.

VanEvery, J. 1995. *Heterosexual women changing the family: Refusing to be a "wife."* London: Taylor Francis.

50

Unraveling the Gender Knot

BY ALLAN JOHNSON

This reading is valuable for the insight that Johnson offers into: (1) the intellectual and emotional obstacles that stand in the way of individuals understanding positive social change, and (2) how individuals can overcome those obstacles in order to contribute to change. He offers specific suggestions for becoming involved in unraveling the gender knot and moving our own lives, and the lives of those around us, toward equal worth and justice for all people. Johnson also addresses the value of finding the courage and taking the risks to plant the seeds of change.

1. What are the two myths about social change that get in the way of individuals understanding change and participating in it?
2. Why do oppressive systems, such as patriarchy, often seem stable?
3. Which of Johnson's suggestions for how to participate in positive change appeal to you and why?

What is the knot we want to unravel? In one sense, it is the complexity of patriarchy as a system—the tree, from its roots to the smallest outlying twig. It is misogyny and sexist ideology that keep women in their place. It is the organization of social life around core patriarchal principles of control and domination. It is the powerful dynamic of fear and control that keeps the patriarchal engine going. But the knot is also about our individual and collective paralysis around gender issues. It is everything that prevents us from seeing patriarchy and our participation in it clearly, from the denial that patriarchy even exists to false gender parallels, individualistic thinking, and cycles of blame and guilt. Stuck in this paralysis, we can't think or act to help undo the legacy of oppression.

To undo the patriarchal knot we have to undo the knot of our paralysis in the face of it. A good place to begin is with two powerful myths about how change happens and how we can contribute to it.

MYTH #1: "IT'S ALWAYS BEEN THIS WAY, AND IT ALWAYS WILL BE"

Given thousands of years of patriarchal history, it's easy to slide into the belief that things have always been this way. Even thousands of years, however, are a far cry from what "always" implies unless we ignore the more than 90 percent of humanity's time on Earth that preceded it. Given all the archaeological evidence pointing to the existence of goddess-based civilizations and the lack of evidence for perpetual patriarchy, there are plenty of reasons to doubt that life has always been organized around male dominance or any other form of oppression. . . . So, when it comes to human social life, the smart money should be on the idea that *nothing* has always been this way or any other.

This should suggest that nothing *will* be this way or any other, contrary to the notion that patriarchy is here to stay. If the only thing we can count on is change, then it's hard to see why we should believe for a minute that patriarchy or any other kind of social system is permanent. Reality is always in motion. Things may appear to stand still, but that's only because we have short attention spans, limited especially by the length of a human life. If we take the long view—the *really* long view—we can see that everything is in process all the time. Some would argue that everything *is* process, the space between one point and another, the movement from one thing toward another. What we may see as permanent end points—world capitalism, Western civilization, advanced technology, and so on—are actually temporary states on the way to other temporary states. Even ecologists, who used to talk about ecological balance, now speak of ecosystems as inherently unstable. Instead of always returning to some steady state after a period of disruption, ecosystems are, by nature, a continuing process of change from one arrangement to another and never go back to just where they were.

Social systems are also fluid. A society isn't some hulking *thing* that sits there forever as it is. Because a system only happens as people participate in it, it can't help but *be* a dynamic process of creation and recreation from one moment to the next. In something as simple as a man following the path of least resistance toward controlling conversations (and a woman letting him do it), the reality of patriarchy in that moment comes into being. This is how we *do* patriarchy, bit by bit, moment by moment. It is also how individuals can contribute to change—by choosing paths of *greater* resistance, as when men resist the urge toward control and women resist their own subordination. Since we can always choose paths of greater resistance or create new ones entirely, systems can only be as stable as the flow of human choice and creativity, which certainly isn't a recipe for permanence. In the short run, patriarchy may look stable and unchangeable. But the relentless process of social life never produces the exact same result twice in a row, because it's impossible for everyone to participate in any system in an unvarying and uniform way. Added to this are the dynamic interactions that go on among systems—between capitalism and the state, for example, or between families and the economy—that also produce powerful and unavoidable tensions, contradictions, and other currents of change. Ultimately, systems can't help but change, whether we see it or not.

Oppressive systems often *seem* stable because they limit our lives and imaginations so much that we can't see beyond them. But this masks a fundamental long-term instability caused by the dynamics of oppression itself. Any system organized around control is a losing proposition because it contradicts the essentially uncontrollable nature of reality and does such violence to basic human needs and values. As the last two centuries of feminist thought and action have begun to challenge the violence and break down the denial, patriarchy has become increasingly vulnerable. This is one reason why male resistance, backlash, and defensiveness are now so intense. . . .

Patriarchy is also destabilized as the illusion of masculine control breaks down. Corporate leaders alternate between arrogant optimism and panic, while governments lurch from one crisis to another, barely managing to stay in office, much less solving major social problems such as poverty, violence, health care, middle-class angst, and the excesses of global capitalism. Computer technology supposedly makes life and work more efficient, but it does so by chaining people to an escalating pace of work and giving them less rather than more control over their lives. The loss of control in pursuit of control is happening on a larger level, as well. As the patriarchal obsession with control deepens its grip on everything from governments and corporations to schools and religion, the overall degree of control actually becomes less, not more. The scale on which systems are out of control simply increases. The stakes are higher and the capacity for harm is greater, and together they fuel an upward spiral of worry, anxiety, and fear.

As the illusion of control becomes more apparent, men start doubting their ability to measure up to patriarchal standards of manhood. We have been here before. At the turn of the twentieth century, there was widespread white male panic in the United States about the "feminization" of society and the need to preserve masculine toughness. From the creation of the Boy Scouts to Teddy Roosevelt's Rough Riders, a public campaign tried to revitalize masculinity as a cultural basis for revitalizing a male-identified society and, with it, male privilege. A century later, the masculine backlash is again in full bloom. The warrior image has re-emerged as a dominant masculine ideal, from *Rambo, Diehard,* and *Under Siege* to right-wing militia groups to corporate takeovers to regional militarism to New Age Jungian archetypes in the new men's movement.[1]

Neither patriarchy nor any other system will last forever. Patriarchy is riddled with internal contradiction and strain. It is based on the false and self-defeating assumption that control is the answer to everything and that the pursuit of more control is always better than contenting ourselves with less. The transformation of patriarchy has been unfolding ever since it emerged seven thousand years ago, and it is going on still. We can't know what will replace it, but we can be confident that patriarchy will go, that it *is* going at every moment. It's only a matter of how quickly, by what means, and toward what alternatives, and whether each of us will do our part to make it happen sooner rather than later and with less rather than more human suffering in the process.

MYTH #2: THE MYTH OF NO EFFECT, AND GANDHI'S PARADOX

Whether we help change patriarchy depends on how we handle the belief that nothing we do can make a difference, that the system is too big and powerful for us to affect it. In one sense the complaint is valid: if we look at patriarchy as a whole, it's true that we aren't going to make it go away in our lifetime. But if changing the entire system through our own efforts is the standard against which we measure the ability to do something, then we've set ourselves up to feel powerless. It's not unreasonable to want to make a difference, but if we have to *see* the final result of what we do, then we can't be part of change that's too gradual and long term to allow that. We also can't be part of change that's so complex that we can't sort out our contribution from countless others that combine in ways we can never grasp. Problems like patriarchy are of just that sort, requiring complex and long-term change coupled with short-term work to soften some of its worst consequences. This means that if we're going to be part of the solution to such problems, we have to let go of the idea that change doesn't happen unless we're around to see it happen and that what we do matters only if we *make* it happen. In other words, if we free ourselves of the expectation of being in control of things, we free ourselves to act and participate in the kind of fundamental change that transforms social life.

To get free of the paralyzing myth that we cannot, individually, be effective, we have to change how we see ourselves in relation to a long-term, complex process of change. This begins by changing how we relate to time. Many changes can come about quickly enough for us to see them happen. When I was in college, for example, there was little talk about gender inequality as a social problem, whereas now there are women's studies programs all over the country. But a goal like ending gender oppression takes more than this and far more time than our short lives can encompass. If we're going to see ourselves as part of that kind of change, we can't use the human life span as a significant standard against which to measure progress. . . .

[W]e need to get clear about how our choices matter and how they don't. Gandhi once said that nothing we do as individuals matters, but that it's vitally important that we do it anyway. This touches on a powerful paradox in the relationship between society and individuals. In terms of the patriarchy-as-tree metaphor, no individual leaf on the tree matters; whether it lives or dies has no effect on much of anything. But collectively, the leaves are essential to the whole tree because they photosynthesize the sugar that feeds it. Without leaves, the tree dies. So, leaves both matter and they don't, just as we matter and we don't. What each of us does may not seem like much, because in important ways, it *isn't* much. But when many people do this work together, they can form a critical mass that is anything but insignificant, especially in the long run. If we're going to be part of a larger change process, we have to learn to live with this sometimes uncomfortable paradox rather than going back and forth between momentary illusions of potency and control and feelings of helpless despair and insignificance.

A related paradox is that we have to be willing to travel without knowing where we're going. We need faith to do what seems right without necessarily knowing the effect that will have. We have to think like pioneers who may know the *direction* they want to move in or what they would like to find, without knowing where they will wind up. Because they are going where they've never been before, they can't know whether they will ever arrive at anything they might consider a destination, much less what they had in mind when they first set out. If pioneers had to know their destination from the beginning, they would never go anywhere or discover anything. In similar ways, to seek out alternatives to patriarchy, it has to be enough to move *away* from social life organized around dominance and control and to move *toward* the certainty that alternatives are possible, even though we may not have a clear idea of what those are or ever experience them ourselves. It has to be enough to question how we think about and experience different forms of power, for example, how we see ourselves as gendered people, how oppression works and how we participate in it, and then open ourselves to experience what happens next. When we dare ask core questions about who we are and how the world works, things happen that we can't foresee; but they don't happen unless we *move,* if only in our minds. As pioneers, we discover what's possible only by first putting ourselves in motion, because we have to move in order to change our position—and hence our perspective—on where we are, where we've been, and where we *might* go. This is how alternatives begin to appear: to imagine how things might be, we first have to get past the idea that things will always be the way they are.

In relation to Gandhi's paradox, the myth of no effect obscures the role we can play in the long-term transformation of patriarchy. But the myth also blinds us to our own power in relation to other people. We may cling to the belief that there is nothing we can do precisely because we know how much power we do have and are afraid to use it because people may not like it. If we deny our power to affect people, then we don't have to worry about taking responsibility for how we use it or, more significant, how we don't. This reluctance to acknowledge and use power comes up in the simplest everyday situations, as when a group of friends starts laughing at a sexist joke and we have to decide whether to go along. It's a moment in a sea of countless such moments that constitutes the fabric of all kinds of oppressive systems. It is a crucial moment, because the group's seamless response to the joke reaffirms the normalcy and unproblematic nature of it and the sexism behind it. It takes only one person to tear the fabric of collusion and apparent consensus. . . .

Our power to affect other people isn't simply about making them feel uncomfortable. Systems shape the choices that people make primarily by providing paths of least resistance. We typically follow those paths because alternatives offer greater resistance or because we aren't even aware that alternatives exist. Whenever we openly choose a different path, however, we make it possible for people to see both the path of least resistance they're following and the possibility of choosing something else. This is both radical and simple. When most people get on an elevator, for example, they turn and face front without ever thinking why. We might think it's for purely practical reasons—the floor indicators and the door we'll exit through are at the front. But there's more going on than that, as we'd discover if we simply walked to the rear wall and stood facing it while everyone else faced front. The oddness of what we were doing would immediately be apparent to everyone, and would draw their attention and perhaps make them uncomfortable as they tried to figure out why we were doing that. Part of the discomfort is simply calling attention to the fact that we make choices when we enter social situations and that there are alternatives, something that paths of least resistance discourage us from considering. If the possibility of alternatives in situations as simple as where to stand in elevator cars can make people feel uncomfortable, imagine the potential for discomfort when the stakes are higher, as they certainly are when it comes to how people participate in oppressive systems like patriarchy.

If we choose different paths, we usually won't know if we affect other people, but it's safe to assume that we do. When people know that alternatives exist and witness other people choosing them, things become possible that weren't before. When we openly pass up a path of least resistance, we *increase* resistance for other people around that path because now they must reconcile their choice with what they've seen us do, something they didn't have to deal with before. There's no way to predict how this will play out in the long run, and certainly no good reason to think it won't make a difference.

The simple fact is that we affect one another all the time without knowing it. . . . This suggests that the simplest way to help others make different choices is to make them myself, and to do it openly so they can see what I'm doing. As I shift the patterns of my own participation in patriarchy, I make it easier for others to do so as well, *and harder for them not to.* Simply by setting an example—rather than trying to change them—I create the possibility of their participating in change in their own time and in their own way. In this way I can widen the circle of change without provoking the kind of defensiveness that perpetuates paths of least resistance and the oppressive systems they serve.

It's important to see that in doing this kind of work we don't have to go after people to change their minds. In fact, changing people's minds may play a relatively small part in changing systems like patriarchy. We won't succeed in turning diehard misogynists into practicing feminists. At most, we can shift the odds in favor of new paths that contradict core patriarchal values. We can introduce so many exceptions to patriarchal rules that the children or grandchildren of diehard misogynists will start to change their perception of which paths offer the least resistance. Research on men's changing attitudes toward the male provider role, for example, shows that most of the shift occurs *between* generations, not within them.[2] This suggests that rather than trying to change people, the most important thing we can do is contribute to the slow sea change of entire cultures so that patriarchal forms and values begin to lose their "obvious" legitimacy and normalcy and new forms emerge to challenge their privileged place in social life.

In science, this is how one paradigm replaces another.[3] For hundreds of years, for example, Europeans believed that the stars, planets, and sun revolved around Earth. But scientists such as Copernicus and Galileo found that too many of their astronomical observations were anomalies that didn't fit the prevailing paradigm: if the sun and planets revolved around Earth, then they wouldn't move as they did. As such observations accumulated, they made it increasingly difficult to hang on to an Earth-centered paradigm. Eventually the anomalies became so numerous that Copernicus offered a new paradigm, for which he, and later Galileo, were persecuted as heretics. Eventually, however, the evidence was so overwhelming that a new paradigm replaced the old one.

In similar ways, we can think of patriarchy as a system based on a paradigm that shapes how we think about gender and how we organize social life in relation to it. The patriarchal paradigm has been under attack for several centuries and the defense has been vigorous, with feminists widely regarded as heretics who practice the blasphemy of "male bashing." The patriarchal paradigm weakens in the face of mounting evidence that it doesn't work, and that it produces unacceptable consequences not only for women but, increasingly, for men as well. We help to weaken it by openly choosing alternative paths in our everyday lives and thereby providing living anomalies that don't fit the prevailing paradigm. By our example, we contradict patriarchal assumptions and their legitimacy over and over again. We add our choices and our lives to tip the scales toward new paradigms that don't revolve around control and oppression. We can't tip the scales overnight or by ourselves, and in that sense we don't amount to much. But on the other side of Gandhi's paradox, it is crucial where we "choose to place the stubborn ounces of [our] weight."[4] It is in such small and humble choices that patriarchy and the movement toward something better actually happen.

STUBBORN OUNCES: WHAT CAN WE DO?

★ ★ ★

What can we do about patriarchy that will make a difference? I don't have the answers, but I do have some suggestions.

Acknowledge That Patriarchy Exists

A key to the continued existence of every oppressive system is people being unaware of what's going on, because oppression contradicts so many basic human values that it invariably arouses opposition when people know about it. The Soviet Union and its East European satellites, for example, were riddled with contradictions that were so widely known among their people that the oppressive regimes fell apart with barely a whimper when given half a chance. An awareness of oppression compels people to speak out, breaking the silence on which continued oppression depends. This is why most oppressive cultures mask the reality of oppression by denying its existence, trivializing it, calling it something else, blaming it on those most victimized by it, or drawing attention away from it to other things. . . .

It's one thing to become aware and quite another to stay that way. The greatest challenge when we first become aware of a critical perspective on the world is simply to hang on to it. Every system's paths of least resistance invariably lead *away* from critical awareness of how the system works. Therefore, the easiest thing to do after reading a book like this is to forget about it. Maintaining a critical consciousness takes commitment and work; awareness is something we either maintain in the moment or we don't. And the only way to hang on to an awareness of patriarchy is to make paying attention to it an ongoing part of our lives.

Pay Attention

Understanding how patriarchy works and how we participate in it is essential for change. It's easy to have opinions; it takes work to know what we're talking about. The easiest place to begin is by reading, and making reading about patriarchy part of our lives. Unless we have the luxury of a personal teacher, we can't understand patriarchy without reading, just as we need to read about a foreign country before we travel there for the first time, or about a car before we try to work under the hood. Many people assume they already know what they need to know about gender since everyone has a gender, but they're usually wrong. Just as the last thing a fish would discover is water, the last thing we'll discover is society itself and something as pervasive as gender dynamics. We have to be open to the idea that what we think we know about gender is, if not wrong, so deeply shaped by patriarchy that it misses most of the truth. This is why feminists talk with one another and spend time reading one another's work—seeing things clearly is tricky business and hard work. This is also why people who are critical of the status quo are so often self-critical as well: they know how complex and elusive the truth really is and what a challenge it is to work toward it. People working for change are often accused of being orthodox and rigid, but in practice they are typically among the most self-critical people around. . . .

Reading, though, is only a beginning. At some point we have to look at ourselves and the world to see if we can identify what we're reading about. Once the phrase "paths of least resistance" entered my active vocabulary, for example, I started seeing them all over the place. Among other things, I started to see how easily I'm drawn to asserting control as a path of least resistance in all kinds of situations. Ask me a question, for example, and the easiest thing for me to do is offer an answer whether or not I know what I'm talking about. "Answering" is a more comfortable mode, an easier path, than admitting I don't know or have nothing to say.[5] The more aware I am of how powerful this path is, the more I can decide whether to go down it each time it presents itself. As a result, I listen more, think more, and talk less than I used to. . . .

Little Risks: Do Something

The more we pay attention to what's going on, the more we will see opportunities to do something about it. We don't have to mount an expedition to find those opportunities; they're all over the place, beginning in ourselves. As I became aware of how I gravitated toward controlling conversations, for example, I also realized how easily men dominate group meetings by controlling the agenda and interrupting, without women objecting to it. This pattern is especially striking in groups that are mostly female but in which most of the talking nonetheless comes from a few men. I would find myself sitting in meetings and suddenly the preponderance of male voices would jump out at me, an unmistakable hallmark of male privilege in full bloom. As I've seen what's going on, I've had to decide what to do about this little path of least resistance and my relation to it that leads me to follow it so readily. With some effort, I've tried out new ways of listening more and talking less. At times it's felt contrived and artificial, like telling myself to shut up for a while or even counting slowly to ten (or more) to give others a chance to step into the space afforded by silence. With time and practice, new paths have become easier to follow and I spend less time monitoring myself. But awareness is never automatic or permanent, for patriarchal paths of least resistance will be there to choose or not as long as patriarchy exists.

As we see more of what's going on, questions come up about what goes on at work, in the media, in families, in communities, in religion, in government, on the street, and at school—in short, just about everywhere. The questions don't come all at once (for which we can be grateful), although they sometimes come in a rush that can feel overwhelming. If we remind ourselves that it isn't up to us to do it all, however, we can see plenty of situations in which we can make a difference, sometimes in surprisingly simple ways. Consider the following possibilities:

- *Make noise be seen.* Stand up, volunteer, speak out, write letters, sign petitions, show up. Like every oppressive system, patriarchy feeds on silence. Don't collude in silence. . . .
- *Find little ways to withdraw support from paths of least resistance and people's choices to follow them, starting with ourselves.* It can be as simple as not laughing at a sexist joke or saying we don't think it's funny; or writing a letter to the editor objecting to sexism in the media. . . .
- *Dare to make people feel uncomfortable, beginning with ourselves.* At the next local school board meeting, for example, we can ask why principals and other administrators are almost always men (unless your system is an exception that proves the rule), while the teachers they control are mostly women. Consider asking the same thing about church, workplaces, or local government. . . .

It may seem that such actions don't amount to much until we stop for a moment and feel our resistance to doing them—our worry, for example, about how easily we could make people feel uncomfortable, including ourselves. If we take that resistance to action as a measure of power, then our potential to make a difference is plain to see. The potential for people to feel uncomfortable is a measure of the power for change inherent in such simple acts of not going along with the status quo.

Some will say that it isn't "nice" to make people uncomfortable, but oppressive systems like patriarchy do a lot more than make people feel uncomfortable, and it certainly isn't "nice" to allow them to continue unchallenged. Besides, discomfort is an unavoidable part of any meaningful process of education. We can't grow without being willing to challenge our assumptions and take ourselves to the edge of our competencies, where we're bound to feel uncomfortable. If we can't tolerate ambiguity, uncertainty, and discomfort, then we'll never go beneath the superficial appearance of things or learn or change anything of much value, including ourselves.

- *Openly choose and model alternative paths.* As we identify paths of least resistance—such as women being held responsible for child care and other domestic work—we can identify alternatives and then follow them openly so that other people can see what we're doing. Patriarchal paths become more visible when people choose alternatives, just as rules become more visible when someone breaks them. Modeling new paths creates tension in a system, which moves toward resolution. . . .
- *Actively promote change in how systems are organized around patriarchal values and male privilege.* There are almost endless possibilities here because social life is complicated and patriarchy is everywhere. We can, for example,

—Speak out for equality in the workplace.

—Promote diversity awareness and training.

—Support equal pay and promotion for women.

—Oppose the devaluing of women and the work they do, from the dead-end jobs most women are stuck in to the glass ceilings that keep women out of top positions.

—Support the well-being of mothers and children and defend women's right to control their bodies and their lives.

—Object to the punitive dismantling of welfare and attempts to limit women's access to reproductive health services.

—Speak out against violence and harassment against women wherever they occur, whether at home, at work, or on the street.

—Support government and private support services for women who are victimized by male violence.

—Volunteer at the local rape crisis center or battered women's shelter.

—Call for and support clear and effective sexual harassment policies in workplaces, unions, schools, professional associations, churches, and political parties, as well as public spaces such as parks, sidewalks, and malls.

—Join and support groups that intervene with and counsel violent men.

—Object to theaters and video stores that carry violent pornography. . . .

—Ask questions about how work, education, religion, family, and other areas of family life are shaped by core patriarchal values and principles. . . .

- *Because the persecution of gays and lesbians is a linchpin of patriarchy, support the right of women and men to love whomever they choose.* Raise awareness of homophobia and heterosexism. . . .
- *Because patriarchy is rooted in principles of domination and control, pay attention to racism and other forms of oppression that draw from those same roots. . . .*

[P]atriarchy isn't problematic just because it emphasizes *male* dominance, but because it promotes dominance and control as ends in themselves. In that sense, all forms of oppression draw support from common roots, and whatever we do that draws attention to those roots undermines *all* forms of oppression. If working against patriarchy is seen simply as enabling some women to get a bigger piece of the pie, then some women probably will "succeed" at the expense of others who are disadvantaged by race, class, ethnicity, and other characteristics. . . . [I]f we identify the core problem as *any* society organized around principles of control and domination, then changing *that* requires us to pay attention to all of the forms of oppression those principles promote. Whether we begin with race or gender or ethnicity or class, if we name the problem correctly, we'll wind up going in the same general direction.

- *Work with other people.* This is one of the most important principles of participating in social change. From expanding consciousness to taking risks, it makes all the difference in the world to be in the company of people who support what we are trying to do. We can read and talk about books and issues and just plain hang out with other people who want to understand and do something about patriarchy. Remember that the modern women's movement's roots were in consciousness-raising groups in which women did little more than sit around and talk about themselves and their lives and try to figure out what that had to do with living in patriarchy. It may not have looked like much at the time, but it laid the foundation for huge social movements. One way down this path is to share a book like this one with someone and then talk about it. Or ask around about local groups and organizations that focus on gender issues, and go find out what they're about and meet other people. . . . Make contact; connect to other people engaged in the same work; do whatever reminds us that we aren't alone in this.
- *Don't keep it to ourselves.* A corollary of looking for company is not to restrict our focus to the tight little circle of our own lives. It isn't enough to work out private solutions to social problems like patriarchy and other forms of oppression and keep them to ourselves. It isn't enough to clean up our own acts and then walk away, to find ways to avoid the worst consequences of patriarchy at home and inside ourselves and think that's taking responsibility. Patriarchy and oppression aren't personal problems and they can't be solved through personal solutions. At some point, taking responsibility means acting in a larger context, even if that means just letting one other person know what we're doing. It makes sense to start with ourselves; but it's equally important not to *end* with ourselves.

If all of this sounds overwhelming, remember again that we don't have to deal with everything. We don't have to set ourselves the impossible task of letting go of everything or transforming patriarchy or even ourselves. All we can do is what *we* can *manage* to do, secure in the knowledge that we're making it easier for other people—now and in the future—to see and do what *they* can do. So, rather than defeat ourselves before we start:

- *Think small, humble, and doable rather than large, heroic, and impossible.* Don't paralyze yourself with impossible expectations. It takes very little to make a difference. . . .
- *Don't let other people set the standard for us.* Start where we are and work from there. . . . set reasonable goals ("What small risk for change will I take *today?*"). As we get more experienced at taking risks, we can move up our lists. . . .

In the end, taking responsibility doesn't have to be about guilt and blame, about letting someone off the hook or being on the hook ourselves. It is simply to acknowledge our obligation to make a contribution to finding a way out of patriarchy, and to find constructive ways to act on that obligation. We don't have to do anything dramatic or Earth-shaking to help change happen. As powerful as patriarchy is, like all oppressive systems, it cannot stand the strain of lots of people doing something about it, beginning with the simplest act of speaking its name out loud.

★ ★ ★

NOTES

1. See James William Gibson, *Warrior Dreams: Violence and Manhood in Post-Vietnam America* (New York: Hill and Wang, 1994).
2. J. R. Wilkie, "Changes in U.S. Men's Attitudes Towards the Family Provider Role, 1972-1989." *Gender and Society* 7, no. 2 (1993): 261-279.
3. The classic statement of how this happens is by Thomas S. Kuhn, *The Structure of Scientific Revolutions* (Chicago: University of Chicago Press, 1970).
4. This is a line from a poem by Bonaro Overstreet that was given to me by a student many years ago. I have not been able to locate the source.
5. Or, as someone once said to me about a major corporation that valued creative thinking, "It's not OK to say you don't know the answer to a question here."

51

De Ambiente

Queer Tourism and the Shifting Boundaries of Mexican Male Sexualities

BY LIONEL CANTÚ

Cantú's article is a fascinating exploration of the impact of the growing global tourist industry on the lives and identities of gay Mexican men. He frames tourism as a form of migration and views it as an important dimension of globalization. In particular, Cantú's study examines the links between the rapid increase in queer tourism in Mexico and the "development and commodification" of Mexican gay culture, gay space, and the emergence of a Mexican gay and lesbian movement. Integral to this reading is an examination of the impact of American tourists' stereotypes of Mexican men on Mexican masculinities and sexualities.

1. How does the relationship between queer tourism and Mexican masculinities and sexualities manifest itself as both sexual colonization and liberation?
2. How is the Mexican definition of homosexuality different from the mainstream American definition of homosexuality?
3. Are there links between the arguments and findings of this reading and the article by Farrer on Chinese discos in Chapter 5? If so, what are they?

★ ★ ★

The purpose of this essay is to examine two sides of queer tourism "south of the border": the development of gay and lesbian tourism in Mexico and the effects of this industry on Mexican sexualities. I should state up front that I refer to "gay and lesbian tourism" as an identity-based industry and to "queer tourism" as a larger market that encompasses a multitude of identities, including both native and foreign heterosexuals, bisexuals, and transgenders. I argue that in the relationship between gay and lesbian tourism and Mexican sexualities, dimensions of both sexual colonization and liberation are at work. Furthermore, I assert that to understand Mexican sexualities, we must move away from one-dimensional cultural models and examine these sexualities from a more complex and materialist perspective that recognizes that culture, social relations, and identities are embedded in global processes.

Between 1997 and 2000, in Mexico and the United States, I collected oral histories from more than thirty men and ethnographic field data for a larger research project on the relationship between sexuality and migration among Mexican men.[1] In addition, I have collected and analyzed archival data on Mexican tourism in general and on gay and lesbian Mexican tourism in particular, such as travel guides, magazines, and material posted on Web sites. In my original project I focused on the experiences of men so as to examine closely the multiple intersecting dimensions, including gender, that shaped their lives.[2] Thus in this essay I am conscious of the multiple ways in which gender shapes gay and lesbian tourism in Mexico, but my ethnographic data are limited by the design of my original project, which focused on Mexican men who have sex with men (MSM).[3] My intent is not to reproduce lesbian invisibility with this essay but to acknowledge and address the limitations of this research.

Although my original research focused on migration in a stricter sense, I soon realized that tourism was not only an important factor in the lives of the men I interviewed but a form of migration itself (in a broader sense of the word). While my ancestry is Mexican, I myself am not; I am Chicano. Thus, although my purpose in Mexico was entirely academic, I was a visitor, an outsider—in a word, a tourist. Despite the voyeuristic tendencies of both, there is a difference between my roles as ethnographer and tourist that is relevant to this essay and my analysis. My gaze as an ethnographer was aimed at understanding the political economy of sexuality in Mexico as it differentially shapes the lives of men. I was not in Mexico on vacation; nonetheless, Mexicans often read me in the public spaces of the plazas, the bars, and the streets as a tourist. Thus the intersection of my ethnographic and tourist roles informs my analysis of queer tourism in Mexico.

DE LOS OTROS: OTHERNESS AND THE BOUNDARIES OF SEXUALITY

De los otros [of the others] is a Mexican phrase used by members of the dominant group to refer to "homosexuals" and thus to mark difference.[4] In Mexico the act of having homosexual relations, in and of itself, does not make one "homosexual"; if it did, then possibly everyone would be *de ambiente.* For although "homosexuality" is stigmatized, bisexuality is reportedly common among Mexican men.[5] . . .

According to the important work of Joseph Carrier, Stephen O. Murray, and others, homosexuality in Mexico is defined not by the biological sex of the participants but by the gendered roles that they perform in the sexual act.[6] In this model only the *pasivo* [passive] participant is marked as homosexual. However, it is not clear if this gendered construction applies only to men. The dearth of literature on Mexican female homosexuality should come as no surprise, given the relative invisibility of lesbian sexuality in academic discourse until recently.[7] The literature that does exist focuses not on sexual behavior but on public gender performance, and here too it seems that the nonnormative role, the "masculine" *marimacha,* is marked. In addition, anthropological reports of male homosexuality have asserted that a "gay" identity, as understood in the American context, does not exist in Mexico.[8] "Gay" identity and culture have been understood, therefore, as American constructions—alien to the Mexican social landscape. . . .

This essay examines the complexities of Mexican sexualities from a political-economic perspective, with a particular focus on queer tourism, to understand how these dimensions shape the sexual identities of Mexican men.

ENLACES/RUPTURAS FRONTERIZAS [BORDER LINKAGES/RUPTURES]

. . . In this section I highlight the political-economic links between the United States and Mexico that have given rise to gay and lesbian tourism in Mexico. In addition, I highlight the border ruptures created through "tolerance zones," sexual borderlands in which Mexican male sexualities are fixed even as they are transformed. . . .

The economic relationship between the United States and Mexico can perhaps best be described as "codependency." Economic ties between Mexico and the United States have been strengthened by formal economic bonds such as the General Agreement on Tariffs and Trade and the Organization for Economic Cooperation and Development (now the World Trade Organization) and by the signing of the North American Free Trade Agreement. In fact, Mexico is now the United States's second largest trading partner. However, these economic ties are only one dimension of the countries' complex relations. Mexico's long history as a labor source for American business has created transnational social and cultural links as well.

While both the migration and the more recent transnational literatures have largely ignored the sexual dimensions of the social and cultural links across borders, a growing number of scholars have begun to examine these issues. For instance, Pierrette Hondagneu-Sotelo and Matthew C. Gutmann have demonstrated the transformation of gender norms among Mexicans both in the United States and Mexico. Oliva M. Espín, Gloria González-López, and I have demonstrated that sexual norms, behaviors, and identities among Mexican men and women are also transformed through migration.[9] As a geopolitical boundary and militarized zone, however, the U.S.-Mexican border is difficult to cross if you are moving north and relatively easy if south. For complex reasons (including immigration laws and policies and the politics of gender and sexuality), the crossing is more difficult for Mexicans who might be branded "homosexual."[10] Thus, for many queer Mexican men and women, migrating to urban areas within Mexico has proved a better alternative.[11]

As in other developing countries, the latter half of the twentieth century was a period of increased urbanization in Mexico. . . . With the implementation of the Hundred Cities Program, part of the 1995–2000 National Program for Urban Development, the Mexican government tried to divert urban migration from congested metropolitan areas to other locations by refocusing development programs to other parts of the country.[12] Thus, by developing its tourism industry, Mexico hopes to draw labor from urban centers to new tourist development sites.[13] The socioeconomic changes brought by this development beg for scholarly attention, yet sexuality has been largely overlooked.[14] Thus part of my focus in this essay is the recognition that urbanization trends and projects are tied not only to the development of gay and lesbian communities throughout Mexico but to queer tourism.

Tourism has become an increasingly important sector of the Mexican economy. . . . While it is impossible to know definitively what proportions of Mexican tourists are gay, lesbian, or bisexual, certain factors point to the development of queer tourism in the country. These factors include the development and commodification of Mexican "gay" culture and space and the rise of a Mexican gay and lesbian movement.[15]

Gay bars are relatively new in Mexico, although they seem to be historically linked to urbanization and the development of *zonas de tolerancia* [tolerance zones, or red light districts] early in the twentieth century. Arising in the postrevolutionary period, *zonas de tolerancia* were conceived as a way to regulate spatially various forms of social deviance, including prostitution and homosexuality. The *zonas* were thus gendered and sexualized spaces where those who transgressed gender norms were located and where men could satisfy their more "licentious" desires. . . . The *zonas* of the border region proved particularly attractive to Americans during Prohibition. The spaces included areas where both male homosexual and transvestite bars were located and provided an escape from moral restraint for men who led public heterosexual lives. Once established, the *zonas* became legitimized spaces for "immoral" activity that attracted sexual tourism from north of the border, where morality was more closely policed.[16] By the mid-twentieth century the Mexican border towns were already firmly established as sites of sexual tourism for men on both sides of the border.[17]

Scholars have mentioned the growing popularity of the term *gay* as an identity label in Mexico among both men and women (who also use the term *lesbian*).[18] While the label is sometimes written as *gai,* it clearly refers to a sexual identity, culture, and movement similar in many ways to what one finds in the United States. Lumsden points to political-economic reasons for changes in identity constructions.[19] The combination of urbanization and industrialization, paralleled with the creation of *zonas de tolerancia,* in all probability provided the social spaces where sexual minorities could establish social networks and, to a degree, create community. This spatial segregation resulted in queer zones or ghettoes in some urban cases. In the nation's capital, for instance, a gay and lesbian community has lived in the *zona rosa* for some years.[20] In Puerto Vallarta the south side of the city has become the de facto "gay side," with bars, hotels, and other establishments that cater to a gay male—especially tourist—clientele. As in the United States, however, an entire city has become identified as a "gay space." Guadalajara has become known as the San Francisco of Mexico due to its gay and lesbian population. . . .

Another factor in the development of queer tourism in Mexico has been the slow but steady rise of a Mexican gay and lesbian movement. Lumsden asserts that Mexico's gay and lesbian movement has its roots in the student movement of 1968 that rocked Mexico and the rest of the world. . . . Since the 1970s gay and lesbian organizations have been created (and disbanded) throughout Mexico. Gay pride festivities are held in various Mexican cities and have become tourist attractions in themselves. In fact, Cancún hosted the International Gay Pride Festival in 2001. These examples illustrate that a "gay" (and lesbian) identity exists in Mexico; although not a clone of the American construction, it has many similarities to it.[21]

The rise of a "gay" identity is linked to the transnational ties forged by globalization between Mexico and the United States. My interviews with Mexican men in Guadalajara and migrants in the Los Angeles area were particularly useful in shedding light on this matter. Some informants had information about gay life in the United States from a number of sources, including magazines, newspapers, videos, the Internet, and travel. Middle- to upper-class Mexicans could travel to the United States (as well as other countries) and encounter the "gay lifestyle" firsthand. They then returned and shared their experiences. As one middle-class gay man in his mid-forties put it, "The majority were interested in going to the United States to experience it and change their social status, because traveling gave them a certain characteristic, the ability to say they were a very traveled, worldly person." This insight seems to be partly what gave rise in the 1980s to a new sexual identity label (still used to a certain degree, reportedly by lesbians as well as by gay men), *internacional.* Carrier reported use of this label by the "hipper" and younger homosexual men during his field research. The label that referred to men with a versatile sexual repertoire obviously has transborder connotations. Clearly, then, Mexican sexualities are being transformed through transnational processes and links, including tourism.

QUEER TOURISM IN MEXICO

★ ★ ★

While a growing body of literature has examined the role of tourism and globalization on sexuality in the Pacific Rim, how these phenomena have influenced Latin American countries has been largely ignored.[22] However, the rapid growth of gay and lesbian tourism in every part of the world—estimated at U.S. $17 billion by the International Gay and Lesbian Travel Association (IGLTA)—demands greater attention.[23] With its proximity to the United States and its affordability compared to other international sites, Mexico has become a desired destination for many gay and lesbian tourists. . . .

Although Mexico has long been a favored vacation spot among Americans, its growing popularity as a gay and lesbian tourist destination is due largely to specific marketing efforts. Founded in 1983, the IGLTA is an international body with member organizations throughout the world and increasing representation in Latin America, including Mexico.[24] Gay and lesbian cruise companies, such as Atlantis, RSVP, and Olivia Cruises and Resorts, have also contributed to Mexico's popularity as a destination.[25] In addition, a number of travel magazines and Web sites cater to gay and lesbian patrons, and a growing number of gay and lesbian travel agencies offer packages throughout Mexico. *Ferrari Guides Gay Mexico* lists more than forty businesses that offer travel arrangements throughout Mexico to a gay and lesbian clientele; not surprisingly, nearly all of them are based in major urban centers in the United States.[26]

Among the numerous international travel guides for gay and lesbian tourists, including comprehensive guides by Spartacus, Ferrari, Odysseus, and Damron, several focus on travel information for the queer tourist in Mexico. Three queer tourist guidebooks focus exclusively on Mexico and target a male audience: *Gay Mexico, Ferrari Guides Gay Mexico,* and *A Man's Guide to México and Central America.*[27] Each guide gives general information useful to any tourist (e.g., money-exchange information and maps) and information specifically for the queer tourist (e.g., bathhouse locations and helpful Spanish phrases for meeting men).

All three guides provide city-by-city information. *A Man's Guide to México and Central America* lists fourteen cities; *Gay Mexico,* twenty-five; and *Ferrari Guides Gay Mexico,* forty-three. While Mexico's urban centers (e.g., Mexico City, Guadalajara, and Monterrey) and mainstream tourist destinations (e.g., Cancún, Acapulco, and Los Cabos) are listed, the guides also list towns and cities "off the beaten path," especially for the queer tourist. These include León, the shoe capital of the world, and Oaxaca, with the neighboring villages. The distinction between these sites and gay tourist sites is supposedly one of "authenticity." But while the travel guides warn gay tourists of the dangers of crossing into these native grounds, they are in reality tourist sites, too. Thus "off the beaten path" means not that the sites are nontourist spaces but that they are more mainstream, catering to a "straighter" clientele.

An examination of these guides and tourist services suggests that the two most commonly represented sides of Mexico are the "just like home" and the exotic. As Richard D. Black (a.k.a. Ricardo) explains: "For Americans, Mexico is close, yet foreign. For any traveler, it's different yet has many of the comforts of home. It offers something for everyone! You're in for a great time!"[28] While both the "just like home" and the exotic representations emphasize the homoerotic aspects of different sites in Mexico, the former targets American tourists who want to vacation with all the gay comforts of home, while the latter attempts to attract those who seek erotic adventures unavailable in suburban American home life. Both representations speak to MacCannell's insight that

> the frontiers of world tourism are the same as the expansion of the modern consciousness with terminal destinations for each found throughout the colonial, ex-colonial, and future-colonial world where raw materials for industry and exotic flora, fauna, and peoples are found in conglomeration. The tourist world has also been established *beyond* the frontiers of existing society, or at least beyond the edges of the Third World. A *paradise* is a traditional type of tourist community, a kind of last resort, which has as its defining characteristic its location not merely outside the physical borders of urban industrial society, but just beyond the border of the peasant and plantation society as well.[29]

Yet for queer tourism there also exists a "border" tension between the lure of an exotic paradise and the dangers of homophobia in foreign lands. Here Mexico seems to represent a homosexual paradise free of the pressures of a modern "gay life style," where sexuality exists in its "raw" form yet where the dangers of an uncivilized heterosexual authority also threaten.

Gay and lesbian cruises target those who favor mediated adventures during which they can enjoy prefabricated representations of the exotic and always return to their "home away from home" either aboard ship or in a hotel. The cruise destinations tend to be either in the "Mexican Caribbean" on the east coast or in the Baja area on the west. Take, for instance, Atlantis's description of its services:

> Atlantis vacations are designed for the way we enjoy ourselves today. We created the concept of an all-gay resort vacation and are the leaders in all-gay charters of first class resorts and cruise ships. All at exotic locations, exclusively ours for these special weeks, with an emphasis on friendship and camaraderie. Places where you can always be yourself and always have fun. That's the way we play.[30]

The "just like home" approach to gay travel thus allows the "best of both worlds": one can "play" on exotic beaches, but under controlled conditions. A new *zona de tolerancia* is born: a queer space protected from the threat of cultural mismatch, including homophobia. Thus, in this borderland, the tourist can enjoy Mexico's pleasures under a controlled environment, free from the less "civilized" world *del otro lado* [on the other side] of tourist boundaries.

Compared to the cruise advertisements, gay guidebooks are more apt to give stereotypical representations of Mexican men. Consider David's description:

> Many Mexican men are often breathtaking in their beauty. They are sensual and often unabashedly sexual. Proud to be male, aware of their physical nature, they are often ready to give of themselves and sometimes receive in return. . . . The adventurous visitor may want to go farther afield in search of the men for whom Mexico is particularly famed: the butch *hombres* who would never walk into a place known to be gay, but who are ready to spring to attention when they catch a man's eye. These are men who must be pursued.[31]

This excerpt exemplifies a "colonial desire," which Robert J. C. Young defines as the dialectic of attraction and repulsion, to conquer (and be conquered by) the hypermasculine and sexually charged racial other.[32] The guidebooks reinforce the colonial message with advice on "rewarding" Mexican men for their services with gifts or money; they suggest that financial compensation for homosexual sex is a cultural norm.

Such representations are reminiscent of what is commonly referred to as "Spanish fantasy heritage."[33] From the end of the nineteenth to the mid-twentieth centuries the prevailing image of Mexicans (both in the United States and in Mexico) was of "gay caballeros" and "dark and lovely señoritas" lazily dancing the night away under Spanish tile roofs. This representation was used to sell a romantic, exotic image of California and Mexico to tourists early in the twentieth century. Contemporary gay tourist images seem either to play up the "Latin lover" image or to emphasize a more "savage" version of the *gay* caballero; both images abound in gay travel . . . guides. . . . The images on the covers of Black's guide and Córdova's . . . represent the "Latin lover" look, a light-complected (though tanned) young man at least partially clothed and waiting to give the queer male tourist a *bienvenida* in a romantic setting. The image on the cover of David's guide . . . is a darker mestizo with facial hair seen in an ambiguous setting and framed suggestively. . . . Not surprisingly, David's book contains more information for the traveler looking for experiences "off the beaten path." Beyond the stereotypes they depict of Latino masculinity, these images represent contradictions of internalized homophobia and the quest for the elusive "real" man among gay tourists themselves.

One of the ironies of this quest is that as more gay male tourists look for these exotic places, for virgin territory, "off the beaten path" of mainstream gay tourist sites, the sites become, in effect, conquered territory—gay tourist spots. This type of invasion is complicated, of course, by tensions between a certain level of sexual liberation, on the one hand, and sexual conquest, on the other.

That is, as gay and lesbian tourists become more common in an area, a certain level of normalization occurs. However, it is not clear to what extent the opening of more legitimized queer space is a positive effect for Mexico's queer population and to what extent this space is framed as American. The expansion of gay and lesbian tourism in Mexico is but one of the more visible manifestations of this queer manifest destiny. Another is the gay bars that now operate throughout Mexico, some with American-sounding names, like Relax (Acapulco), Blue City (Cancún), MN'MS (Ciudad Juarez), and the Door (Mexico City). Each of these bars is located in a site commonly visited by foreign gay tourists. While they are not technically or legally off-limits to queer Mexican nationals, only those of at least middle-class backgrounds and with more sophisticated tastes who might mix better with tourists are often found in them. *Zonas de tolerancia,* yes, but within limits.

How this tourism influences the lives and sexual identities of Mexican men and women remains to be researched, but Murray (as well as Black) reports that the hospitality industry is a common employer of Mexican gays and lesbians.[34] In addition, male prostitution (either as an occupation or as a "part-time" activity) is obviously linked to gay male tourism.[35] More recently, epidemiologists have become more concerned about tourism and the spread of HIV in the country.

GAY CABALLEROS AND PHALLIC DREAMS: LIFE IN THE SEXUAL BORDERLANDS

> In that coordination of and responsibility for the diverse efforts of the nation, tourism has its place, a place that is characterized by its diversification, for it is indifferent to nothing and affects all, the local as much as the foreign. . . . And that does not refer only to foreign tourism, those visitors from afar who discover new realities, even as they offer them.
>
> —HÉCTOR MANUEL ROMERO
> "NADA ES INDIFERENTE AL TURISMO"

As this quotation makes clear, everything and everyone is impacted by tourism. Mexico's gay *ambiente* has changed dramatically over the last decade. What was once an underground world of private parties in the homes of homosexual men has become a more public and therefore more visible phenomenon. Santiago, a former professor of drama in his forties who lives in Guadalajara, describes gay life in Mexico prior to its contemporary vogue:

> Twenty years ago I started in *el ambiente.* The *ambiente* in Guadalajara was something magical; it was sensational. There weren't any of the places that we have today. Before, the parties were at regular houses. The richest men of *el ambiente* would give you an invitation. They made invitations for everyone. Most of us in *el ambiente* knew each other, we were like a family. Some treated us well; if you were good-looking, you would end up going to bed with the most rich and famous. . . . As one of my friends of *el ambiente* says, "Before, being a *joto* was a privilege; now it is a vulgarity; anyone can be *gay.*"

Santiago's comments illustrate not only that class and physical appearance (in all likelihood race)[36] are factors in constructing the strata of Mexican homosexuality but, perhaps more important, that the boundaries that once constrained *los jotos* have been dramatically reshaped, giving rise to a "gay" identity that anyone can share. The creation of gay and lesbian bar space, which has largely replaced private parties, is the result of both the gay and lesbian movement, which has made gays and lesbians more visible in Mexico, and greater demand for gay bars, which is strengthened through foreign gay tourism. The bars range from upscale discos to humble cantinas, and, although most patrons were men, there were women as well at the establishments I visited in Guadalajara. The most salient feature of the bars is their class stratification: cantinas are open to practically anyone, and upscale discos cater to a wealthier clientele. Santiago's sense that "anyone can be *gay*" does not mean that social stratification no longer exists in Mexico's *ambiente.* On the contrary, "otherness" remains a marked facet of life among *los otros,* but its meaning is reformulated in a context of sexual commodification and queer tourism. Class remains an important marker, but there is greater mixing across classes, and also across male and female social worlds. than at private parties.

The commodification of Mexican gay spaces presents a complex set of factors in the lives of queer Mexican men. The spaces created allowed for the development of an identity and a community that served as the foundations of the gay and lesbian movement in Mexico. Queer tourists shape these spaces both through their contact with Mexicans and through the creation of new spaces to serve their needs. Contact provides an exchange of cultural and political information about issues of queerness that has an impact on men's lives.

Some men use this information as a rationale for migrating to the United States, which they construct as a more tolerant place. Armando, a thirty-two-year-old man from Jalisco, lives in the Los Angeles area, where he works in HIV services. In seminary, where he was studying to be a priest, he heard stories of gay life in the United States from friends (including other seminarians) who had been there, and later he himself visited the United States on missionary exchange programs. After a priest advised him to move to the north, where he could live more openly as a gay man, he emigrated.

Lalo, a thirty-three-year-old gay-identified man from Guadalajara, is a similar case. After his family rejected him for being gay, he moved to Puerto Vallarta, where he worked in hotels. Many of his coworkers were also gay, and they helped him by giving him an apartment. On their advice he then emigrated to southern California:

> The people in the hotel would tell me, "Go to the United States, it's beautiful, you make good money and there are a lot of homosexuals. You can hold [your lover's] hand and kiss in public and nothing happens." I thought it was an ideal world where homosexuals could be happy. [But I] learned that it wasn't true that homosexuals were free, that they could hold hands or that Americans liked Mexicans.

Soon after moving to California, he discovered not only that homophobia does exist in the United States but that he had to deal as well with racism, a tribulation that all immigrants face.[37] The fantasy Lalo had been told of gay life in the United States is a common one. Santiago refers to it as the *sueño-fálico,* or "phallic dream," in which queer men in Mexico envision the United States as a sexual utopia, an erotic land of milk and honey. The irony, of course, is that many American gay tourists have similar dreams when they visit Mexico's gay resorts. . . .

QUEER MEXICO?

Borders are both real and imagined. In the Western queer imaginary Mexico and its men are somehow locked in a spatiotemporal warp of macho desire. Mexico seems to represent a place fixed in time where "real" men can be found. The stereotype of the Mexican macho is alive and well in the imagination of American tourists. Far from cultural stagnation, however, Mexico has undergone profound changes in the last several decades. These changes have shaped the everyday experiences of Mexican men, including their meanings and identities of gender and sexuality. While anthropologists working in Mexico in the 1970s and 1980s asserted that "gay" identities did not exist as they are understood in an American context, this is no longer so. The boundaries of Mexican sexual identities are changing even as the spaces that produce them are remapped. The development of gay and lesbian tourism in the country has been a key factor in these changes.

The relationship between queer tourism and Mexican male sexualities is complex and multiply constituted, but in this essay I have highlighted some of these dimensions, especially as they are linked to a sexual political economy. . . .

Although the rise of gay and lesbian tourism in Mexico was not a planned outcome of the nation's tourist development project, it has caused important sociopolitical reverberations. It is in some ways ironic that those on the margins of Mexican society, *los otros,* and those on the margins of other nations, especially the United States, should come together under a nationalist project. However, life in these sexual borderlands has elements that are both liberalizing and oppressive. A central question is to what extent the Mexican nationalist project is willing to embrace not only its gay and lesbian tourists but, more important, its gay and lesbian citizens. The answer may lie in the demands not of Mexican gays and lesbians but of a queer market and the political economy of space.

NOTES

1. I conducted this research for my dissertation primarily in the greater Los Angeles and Guadalajara areas.
2. Although I had intended to interview both men and women. I decided to focus on men largely to simplify my research questions and to reduce the number of research variables. However, my decision was also a consequence of issues of access and my own identity as a gay Latino.
3. While gay tourists are much more visible than lesbian tourists to Mexico, Mexico is an increasingly popular destination for lesbian tourists. For instance, Olivia Cruises and Resorts offers a cruise of the Mexican Caribbean and a Club Med resort vacation on the Sonoro Bay of Mexico's western coast. The relatively small size of the lesbian tourist market is not peculiar to Mexico: the comparative thickness of travel guides for gay men and for lesbians alone reveals these differences.
4. Members of the dominant group are commonly referred to as *bugas* by queer Mexicans.
5. During my research in Mexico both the radio and the newspapers reported that bisexuality among Mexican men was a "social problem" due to its prevalence and its assumed link to the spread of HIV. Other scholars have reported on the prevalence of bisexuality, including Clark L. Taylor and Joseph Carrier. For discussions of the acceptance or the stigmatization of bisexuality in Mexico see Annick Prieur, *Mema's House, Mexico City: On Transvestites. Queens, and Machos* (Chicago: University of Chicago Press, 1998); and Murray, *Latin American Male Homosexualities.*
6. Joseph Carrier, *De Los Otros: Intimacy and Homosexuality among Mexican Men* (New York: Columbia University Press, 1995): Murray, *Latin American Male Homosexualities.*
7. See, e.g., Yvonne Yarbro-Bejarano, "Crossing the Border with Chabela Vargas: A Chicana Femme's Tribute," in *Sex and Sexuality in Latin America,* ed. Daniel Balderston and Donna J. Guy (New York: New York University Press, 1997), 33–43; and Norma Mogorevejo, *Un amor que se atrevió a decir su nombre* [A love that dare not speak its name] (Mexico City: Centro de Documentación y Archivo Historico Lesbico, 2000).
8. It is important to note that the focus of this literature is both the label and the identity of "gay." There are numerous historical and contemporary terms for homosexuality in Spanish; see Murray and Dynes, "Hispanic Homosexuals."
9. Pierrette Hondagneu-Sotelo, *Gendered Transitions: The Mexican Experience of Immigration* (Berkeley: University of California Press, 1994); Matthew C. Gutmann, *The Meanings of Macho: Being a Man in Mexico City* (Berkeley: University of California Press, 1996); Oliva M. Espín, *Women Crossing Boundaries: A Psychology of Immigration and Transformations of Sexuality* (New York: Routledge, 1999); Gloria González-López, "Beyond the Bed Sheets, beyond the Borders: Mexican Immigrant Women and Their Sex Lives" (Ph.D. diss., University of Southern California, 2000); Lionel Cantú, "Border Crossings: Mexican Men and the Sexuality of Migration" (Ph.D. diss., University of California, Irvine, 1999).
10. See Eithne Luibheid. "Racialized Immigrant Women's Sexualities: The Construction of Wives, Prostitutes, and Lesbians through U.S. Immigration" (Ph.D. diss, University of California, Berkeley, 1998); Luibheid. "'Looking like a Lesbian': The Organization of Sexual Monitoring at the United States–Mexican Border," *Journal of the History of Sexuality* 8 (1998): 477–507; and Cantú, "Border Crossings."
11. Alvaro Sanchez-Crispin and Alvaro Lopez-Lopez, "Gay Male Places of Mexico City," in *Queers in Space: Communities, Public Places, Sites of Resistance,* ed. Gordon Brent Ingram, Anne-Marie Bouthillette, and Yolanda Retter (Seattle: Bay, 1997), 197–212.
12. See www.sedesol.gob.mx/desuryvi/desurb/p100c.htm.

13. See Mary Lee Nolan and Sidney Nolan, "The Evolution of Tourism in Twentieth-Century Mexico," *Journal of the West* 27, no. 4 (1988): 14–26; and Michael J. Clancy, "Tourism and Development: Evidence from Mexico," *Annals of Tourism Research* 26 (1999): 1–20.
14. See Michelle E. Madsen Camacho, "The Politics of Progress: Constructing Paradise in Huatulco, Oaxaca" (Ph.D. diss., University of California, Irvine, 2000).
15. Although, to my knowledge, there are no statistics on the countries of origin of gay and lesbian Mexican tourists, one may assume (given the marketing tactics of the industry and Mexican tourism statistics) that their demographics reflect those of tourists in general; that is, a majority are from the United States.
16. Visiting "boys town" brothels on the Mexican side of the border remains a rite of passage for some young American men, who cross the border looking for mostly heterosexual adventure.
17. This phenomenon seems consistent with George Chauncey's discussion of homosexuality in pre–World War II New York, where marginal (racially segregated) areas became havens for different types of "deviance," including homosexuality, and a playground for the more well-to-do (*Gay New York: Gender, Urban Culture, and the Makings of the Gay Male World, 1890–1940* [New York: Basic, 1994]). However it is not clear how prevalent the *zonas* were outside border towns and the urban center of Mexico City.
18. See, e.g., Lumsden, *Homosexuality, Society, and the State in Mexico;* Murray, *Latin American Male Homosexualities;* and Prieur, *Mema's House, Mexico City.*
19. Lumsden, *Homosexuality, Society, and the State in Mexico.*
20. To my knowledge, the demographics of this community are undocumented. However, traditional Mexican culture makes it easier for unmarried men (as opposed to women) to live apart from their families, and economic conditions and the growing popularity of the area also affect its demographics.
21. Even in the United States the *gay* label fails to capture the numerous experiences and identities commonly grouped under it.
22. The Caribbean, particularly Cuba, is beginning to be studied. See, e.g., Julia Davidson, "Sex Tourism in Cuba," *Race and Class* 8 (1996): 39–49; and Ian Lumsden, *Machos, Maricones, and Gays: Cuba and Homosexuality* (Philadelphia: Temple University Press, 1996). For the Pacific Rim see, e.g., Thanh-Dam Truong, *Sex, Money, and Morality: Prostitution and Tourism in Southeast Asia* (Atlantic Highlands, NJ: Zed, 1990); and C. Michael Hall, "Gender and Economic Interests in Tourism Prostitution: The Nature, Development, and Implications of Sex Tourism in South-East Asia," in *Tourism, A Gender Analysis,* ed. Vivian Kinnaird and Derek Hall (New York: Wiley, 1994), 142–63.
23. International Gay and Lesbian Travel Association, 2000, www.iglta.com. There are conflicting estimates for this market. Tourism Industry Intelligence estimated the global market at U.S. $10 billion in 1994 (reported in Briavel Holcomb and Michael Luongo, "Gay Tourism in the United States," *Annals of Tourism Research* 23 [1996]: 711–13), and an industry survey by Community Marketing estimated the American gay and lesbian market alone at more than $47.3 billion ("Gay and Lesbian Travel Demographics," www.mark8ing.com/summary.html).
24. The IGLTA is represented in Argentina, Bolivia, Brazil, Chile, Colombia, Costa Rica, Ecuador, and Venezuela as well.
25. While Atlantis and RSVP advertise themselves as open to both men and women, their main market seems to be men; Olivia Cruises and Resorts targets a lesbian clientele.
26. Richard D. Black, *Ferrari Guides Gay Mexico: The Definitive Guide to Gay and Lesbian Mexico* (Phoenix: Ferrari, 1997).
27. Eduardo David, *Gay Mexico: The Men of Mexico* (Oakland, Calif.: Floating Lotus, 1998); Black, *Ferrari Guides Gay Mexico;* Señor Córdova, *A Man's Guide to Mexico and Central America* (Beverly Hills, Calif.: Centurion, 1999). While Black's guide does have some information for women, the guide is aimed mostly at men, no doubt because gay men are more visible than lesbians in Mexico.
28. Black, *Ferrari Guides Gay Mexico,* 14.
29. MacCannell, *Tourist,* 183.
30. Atlantis, www.atlantisevents.com.
31. David, *Gay Mexico,* 27–28.
32. Robert J. C. Young, *Colonial Desire: Hybridity in Theory, Culture, and Race* (London: Routledge, 1995).
33. See Carey McWilliams, *North from Mexico: The Spanish Speaking People of the United States* (New York: Praeger, 1948).
34. Murray, *Latin American Male Homosexualities;* Black, *Ferrari Guides Gay Mexico.* Michelle E. Madsen Camacho, who has conducted research on the Mexican tourist industry in Huatulco, Mexico, reports that businesses in the hospitality industry, particularly hotels, often desire gay men as workers due to the perception that gay men have a "higher" cultural aesthetic, a cultural capital, that serves the businesses' needs (pers. com., 1998). See also Madsen Camacho, "Politics of Progress."
35. See, e.g., Carrier, *De Los Otros;* Murray, *Latin American Male Homosexualities.*

36. Although racial differences and racism are not commonly discussed in the literature on Mexican homosexuality, those with European features tend to be privileged over those with African or Indian features.

37. See Luibheid, "Racialized Immigrant Women's Sexualities"; and Cantú, "Border Crossings."

52

Women's Agency and Household Diplomacy

Negotiating Fundamentalism

BY SHAHIN GERAMI AND MELODYE LEHNERER

Patriarchy is often misunderstood as a system of inequality in which all men oppress all women in every arena of life. In fact, patriarchy takes many forms, and other social hierarchies always mediate. Simply put, not all men have control over all women all the time. For example, in some patriarchies, age status places older women in a position of privilege in relationship to younger people. Additionally, patriarchies typically allow some authority to women to keep their loyalty, and women often find ways to subvert or resist oppression. This reading is a close examination of the strategies used by Iranian women to negotiate Islamic fundamentalism. The authors view their work as moving feminism beyond "victimization narratives" of women's lives.

1. Why did the authors choose a narrative style of representation?
2. What are the four strategies that Iranian women have developed to create agency?
3. Do you see any parallels between the strategies discussed in this reading and strategies employed by American women?

When evaluating the impact of Islamic orthodoxy on women's status, some scholars and many leaders of Islamic groups have disputed the usefulness of the fundamentalist label (Muadudi 1982; Piscatori 1994; Sidahmed and Ehteshami 1996). It is not the purpose of this article to enter into that debate. Nevertheless, to clarify our understanding of the term "fundamentalism," we refer to active contemporary movements organized against a perceived assault on the religious foundation of social order. We propose that despite the diversity of their format and ideology, two issues stand out in these movements: the role of the state and the conceptualization of gender (Gerami 1996; Klatch 1987; Lazarus-Yefeh 1988; Martin and Appleby 1994). A characteristic of these movements is the basic assumption that modernism or secularism is to blame for society's moral degeneration. To correct this moral degeneration, the ideal society, which is created following divine order, enforces female domesticity and modesty through a protected and private family with woman as the functionary and man as the gatekeeper. In short, fundamentalism stands as "a protest against the assault on patriarchal structural principles" (Riesebordt 1993, 202).

For fundamentalists, "women's behavior is regarded not only as being symptomatic of cosmic dislocation but as being its cause" (Hawley 1994, 27). Consequently, gender is often isolated as the social category used by fundamentalist movements' leaders to mobilize their supporters and discredit their opposition (Mazumdar 1994; Mumtaz 1994). Both Marshall (1984) and Zuhur (1992) have documented the practice of Islamic fundamentalist leaders who try to resurrect gender separation as the core of an ideal society. The Iranian experience is most illustrative of this tack (Papanek 1994); therefore, we focus on this country. For example, within Iran, serious attempts have been made to reinstitute *Shariat,* imposed veiling, and to restrict the movement of women in the public sphere, all signs of a return to a more domestic role for women (Gerami 1996).

From 1979 to 1995, Iran underwent major changes, among them a national uprising and the establishment of the Islamic Republic, an eight-year war with Iraq, an international economic embargo, a cultural revolution, and an overhaul of both the labor market and the civil service bureaucracy (Esposito 1990; Hooshang 1990; Keddie and Hooglund 1986). Pre-revolutionary Iranian discourse was imbued with gender symbolism (Gerami forthcoming). This symbolism thrived during the revolution and since has become the hallmark of the Republic's legitimization rhetoric.

Of importance, women have actively participated in the formulation of the Islamic gender discourse; they have responded to its oppressive force and have redesigned its contours (Paidar 1996). In addition, women's involvement has taken many shapes, as in their artistic and literary productions (A. Naficy 1994; H. Naficy 1994), their active advocacy and agency in the family courts (Mir-Hosseini 1993), their collective opposition to mandatory veiling (Tabari and Yeganeh 1982), and most significantly, their grassroots activism in the 1997 presidential election (Poya 1999; Rajaee 1999). All these acts, whether through open and direct negotiations, collective resistance, subversive techniques, individual co-optation, acquiescence, or collaboration, are illustrations of women's agency within a fundamentalist framework.

RESEARCH GOALS: PATRIARCHY, AGENCY, AND STRATEGIES OF TRANSFORMATION

Those forms of strategizing that women undertake when living within a set of concrete constraints such as fundamentalism have been defined by Kandiyoti as patriarchal bargains (1988, 275). Such strategizing varies over time and according to caste, class, and ethnicity. Most important, these patriarchal bargains exert a powerful shaping influence on women's gendered subjectivity and determine the nature of gender ideology in different contexts (1988, 275). In the narratives that follow, we identify four strategies women used to transform a fundamentalist framework—subversion, co-optation, acquiescence, and collaboration—and thus create analytical categories to expand our understanding of women's agency. In effect, our research goal is to move past victimization narratives and add to the growing body of feminist work that illustrates that the gender terrain in a fundamentalist context, such as Iran, is discursive rather than determined (deGroot 1996). . . .

In narratives of Iranian women, we show that some women collaborated with the regime by aligning their interest with those of the state's agenda, whereas others submitted to family demands and state authority. In contrast, there were women who manipulated both of these forces or played one against the other. Finally, we identify subversion as a strategy in which women undermined these forces. . . .

RESEARCH METHOD AND SAMPLE: THE NEED TO KNOW

The first author of this article, Gerami, is Iranian, was educated in both Iran and the United States, and annually visits family and friends in Iran. This feature of personal biography is perceived as an asset among feminist researchers because it merges personal interest with the "need to know" (see Reinharz 1992, 258-93). Between 1978 and 1995, Gerami made a series of trips to Iran to record women's responses to the Islamic state and, most important, their responses to the changes that took place between 1985 and 1995.

During this time, Gerami met with focus groups and conducted in-depth interviews. The format of the focus groups and the in-depth interviews was open-ended, allowing the respondents to set the research agenda. . . . The data from these qualitative sources led to the design and administration of a survey study reflective of lived experience (Gerami 1996). For example, one focus group composed mainly of professional women (two physicians, one lawyer, and three teachers) raised as an issue the impact of state-imposed restrictions relating to where one might work, dress codes, and hours allowed to work. These themes also gave direction to the in-depth interviews as well as the fieldwork that followed. Gerami observed more than 132 families in three cities: Tehran, Qum, and Tabriz. Eventually, she focused on 58 families that remained in Iran and with whom she had contact. All families in this study were middle class and all were more than two generations urban. Eight of these families were considered upper middle class, and 10 can be ranked as lower middle class for their localities. In addition, all of the families were headed by a man. Over time, three of the men became "martyrs" in the war with Iraq, and four died of other causes. Therefore, by 1995, seven women in this group were widowed.

Of the total, 32 women had at least a high school education. Of this group, 16 had two or more years of higher education. By 1995, six women had returned to college and four had obtained their college degrees. Eight women had degrees in such professions as medicine, law, and social work. The rest of the women, 26 in all, had less than a high school education.

Initially, 27 of the women were employed, 19 of whom were civil service employees. In 1985 (five years after the revolution and four years into the war with Iraq), many of these women were fired or forced to resign. Professional women were most affected. For example, with the implementation of *Shariat,* three women judges were banned from the judiciary. Similarly, four women employed by the Iranian National Radio and Television were purged as part of an ongoing state campaign to control these powerful communication mediums. Fifteen of the originally employed women were teachers—high school, elementary, and special education. None were fired, although all complained of incessant harassment. Ultimately, three factors influenced the way these women negotiated their gender roles: (1) social class, (2) family structure, and (3) degree of support for the fundamentalist regime.

COLLECTIVE STORIES

To best capture "the goals and intentions of these human actors" we have chosen a narrative style of representation (Bauer 1991, 1993; Richardson 1995). Narrative, as defined by

Laurel Richardson, is a valuable form of representation for sociologists. It is the primary way by which social actors, including sociologists, "make sense of their lives" through reference to a "sequence of events" such as birth, death, loss of a job, and war (Richardson 1995, 214). For example, the decision whether to resign or continue to work involved difficult family negotiations that took into consideration the impact of the war and the ongoing needs of the family. The narratives that follow capture these family negotiations as they evolved over time and present what Richardson refers to as "collective stories" (1995, 212).

Collective stories convey "an individual's story by narrativizing the experiences of the social category to which the individual belongs rather than by telling the particular individual's story" (Richardson 1995, 212). As the narratives unfold, the reader will note that bargaining with patriarchy took many forms and occurred in both the public and private spheres of life.

Islamist women respected the division of public and private as outlined in the Islamic discourse, and many chose to collaborate with the state. In contrast, secular women, many of whom identified with Islamic cultural and religious traditions, believed Islamic fundamentalism had created an arbitrary and unfair division of society along gendered lines. Their resolve to challenge this division often was hampered by conflicting pressures from family and state. The narratives that follow are divided into two major categories: Islamist and secular. Narratives include a brief description of each woman's social condition and her agency within that context. The expression of agency yielded four strategies that we label collaboration, acquiescence, co-optation, and subversion. Collaboration is defined as actively supporting state policies that are designed to reinstitute women's primary role as domestic using formal and oppressive tactics. Acquiescence is submitting to these state policies. Co-optation is manipulating these state policies, whereas subversion is undermining them.

NEGOTIATING WITH FAMILY AND STATE: ISLAMIST SOLUTIONS

In the sample of 58 women, 12 claimed strong support for the regime and publicly declared this support by wearing and practicing *Hijab Islami* (Gerami 1996, 133). *Hijab Islami* is defined as a commitment to a separation of social spheres (public/private) as proscribed in appearance and behavior codes (Gerami 1996). Three of the stories that follow are of women whose husbands or fathers had restricted their movements under the previous regime. For these women, serving the Ayatollah and the revolution allowed them the freedom to venture out for prayers, work, education, and volunteer activism (Hegland 1990). Such activities illustrate Zuhur's definition of "sex role expansion," which is linked to power in the public domain, both social and economic (1992, 93). Specifically, the activity of monitoring the behavior of others yielded these women social power as volunteer informers and economic power as paid enforcers. The closing narratives in this section illustrate varied degrees of success in regard to negotiation strategies.

Ameneh and Massy Collaborate: Enforcing *Hijab Islami*

The reforms brought on by the state prior to the revolution reflected upper- and middle-class interests and a move toward "modernism," or secularism, which promotes change and innovation at both the societal and individual level of experience (Inkeles 1983). Women from the lower middle class came from families that were less affected by such state policies, and complementary sex roles were taken for granted by the women from this class. They lacked opportunities for a higher education, and seeking employment was perceived as undermining their traditional bargain with patriarchy. This bargain, identified by Kandiyoti as a classic strategy, accommodated to varying degrees male authority in both the private and public spheres (1997, 273). Consequently, if these women sought to expand their sex roles, it had to be within the parameters of that particular bargain. The revolution offered such an avenue, giving them increased spatial mobility and control over others.

Ameneh was an ardent supporter of the regime who expressed this support by attending prayer meetings and doing volunteer work for the war effort. She and her two oldest daughters were active in their neighborhood *kommittehs* (committees). They served the cause through activities such as sewing uniforms and making provisions for the soldiers. Most important, they served the cause by keeping an eye on their neighbors. These supportive activities, domestic and limited to the private domain, yielded these women power over other women in their neighborhood and brought them in-kind benefits such as household and/or food items at reduced rates.

Massy, also a woman from the lower middle class, took a similar approach to Ameneh but chose to enter the public arena through paid labor. Women like Massy became supporters of state ideology. These women were referred to as "sisters"; they tended to enter the rank and file of the social control agencies of the state. Three of the 12 Islamist women in Gerami's survey were identified by self and others as "sisters." Sisters or female morality guards were always on the lookout for violators of *"Good Hijabi"* codes. They joined neighborhood *kommittehs,* spied on opposition groups, filled vacant government posts (resulting from purges), and reported on nonbelievers. Younger women loyalists were recruited to fill the positions of women who were fired. When employed, they helped enforce *hijab* and participated in the purging of women coworkers. Their loyalty to the regime often led to their being promoted over senior staff.

Massy's family background was modest and limited in terms of personal opportunities. She was fifth in a family of seven children. Her parents, along with her older brother, ran a small bakery. She had two years of middle school education, was married at the age of 16 to a plant worker, and had two children. She became active in the revolution and organized the first *kommitteh* in her neighborhood. She was an outspoken zealot when it came to observing *Hijab Islami.* Her neighbors complained of her interferences and her aggressive manners. These traits, while objectionable to other women in the community, brought her to the attention of a powerful cleric. He offered her a job as a sister, enforcing *Hijab Islami* in the municipal offices. There she would check women for their Islamic attire and admonish them for their indiscretions. While employed by the city, she obtained her high school equivalency. This credential led to her becoming a manager of a secretarial pool. In this administrative role, she maintained her totalitarian ways and consequently advanced the state's cause as well as her own.

Fatimeh Acquiesces: Submitting to Family Demands

Fatimeh, Ameneh's youngest daughter, was a civil service worker. To keep her job she was dependent on Ameneh to baby-sit. Ameneh was a reluctant babysitter and often complained about Fatimeh's children. She urged her, using *Quranic* verses, to stay home, care for her husband and children, and serve Islam. This pressure from her mother combined with a long commute to work, the burden of housework, and chronic health problems eventually led to Fatimeh's decision to quit her job. She did this in spite of the fact that her husband, a mechanic, could not provide a sufficient income for the family needs. Once she made this decision, Fatimeh joined her mother and older sisters in their volunteer activities. The combined pressures of state and family removed Fatimeh from the paid labor force but not from the public domain. She, like her mother and sisters, had a visible presence in the community.

Aisha Co-Opts: Manipulating State Educational Policy

Some women in this group of Islamist supporters had to overcome their families' opposition to their participating in public life. This was particularly true for the women from the socially and fiscally conservative bazaar families. The strategy they used was to co-opt the state's rhetoric embraced by their families. They consciously decided to conduct themselves according to "*Good Hijabi,*" thus ensuring family and state support. In addition, they were able to capitalize on an early parochial education, which facilitated their admission to state-run universities. University credentials eventually placed them in a position to replace purged women in the workforce. The combined effects of co-opting state power enhanced their "extradomestic pursuits" (Read and Bartkowski 2000, 405).

Aisha's family was composed of merchants in the steel trade. Her parents opposed pre-revolutionary public education and could afford to send their children to Islamic academies. Aisha had finished high school and ordinarily would have had to stay home and get married; however, the revolution offered her an avenue to public life. With the help of her mother and grandmother, she convinced her father that attending university was a legitimate public activity for a devout woman. The three women argued that under the Islamic Republic, universities were safe for women and allowed them to serve the revolution and Islam. This rationale made sense to her father because Islamic tradition supported women's education as an enhancement to raising Islamic families. Aisha did attend a state university and eventually took a position as a high school science teacher.

Zarin Subverts, Shirin Acquiesces: Contrasting Tactics of Shahid Widows

During the war and its aftermath, the state propaganda machine created a mass hysteria about martyrdom. Martyrs were defined as anyone whose death could be related to the revolution or the war. It was promoted as the highest spiritual, individual, and social status imagined. The *shahid* (martyr to the cause) was assured of immediate admission to heaven and eternal bliss. His family was guaranteed high social esteem, material benefits such as more rationed goods, and preferential treatment in educational and employment opportunities. Given these state-provided perks, men of all ages volunteered to serve and many died leaving behind a wife and family.

The Shahid Foundation, a state bureaucracy, was established to provide for the families of the martyrs. At the beginning of the war, the Shahid families were handsomely compensated, but as resources began to dwindle so did the Foundation's support of these families. While the Foundation provided some financial security for the widows, its procedures were intrusive and its staff authoritarian. The Foundation disapproved of working widows, especially if they had young children. The Foundation required the consent of the legal guardian in decisions related to children who, according to *Shariat,* would be the paternal grandfather or uncle in the absence of the father. In addition, the Foundation had a great deal of influence on the decision to remarry. The stories of Shirin and Zarin, sisters, illustrate contrasting negotiating strategies: acquiescence and subversion.

Shirin Acquiesces: Submitting to State Family Policy Shahids' widows were encouraged to marry the veterans of the war. These state-sanctioned marriages often were arranged by the local branch of the Foundation. Shirin, the second daughter in a family of seven, had minimum education and lived in the provincial capital of Qum. Her first husband, although a zealot, was a good provider and father. After his martyrdom, Shirin and her two young children became wards of the Shahid Foundation. She had no chance of employment and her widowed mother could only offer emotional support. Shirin's bargaining power was limited, causing her to yield to the pressure of a state-arranged marriage to a young war veteran. However, he was abusive to Shirin. This abuse was tolerated by family and friends because they thought his behavior was a consequence of war injuries. Shirin eventually had three children by this man. The cumulative effects of this state-arranged marriage prevented Shirin from caring for her first two children, who became the responsibility of her mother. Shirin's experience illustrates a woman whose strategy for survival was to succumb to the pressures of state and family. She had neither financial resources nor a strong extended family to negotiate an alternative strategy.

Zarin Subverts: Undermining State Family Policy In contrast, Shirin's older sister, Zarin, through her own agency, resisted victimization. She was the widow of a wealthy businessman with whom she had three daughters. Although she had less than a high school education, she was wise in the ways of the system. For example, as a Shahid widow she was entitled to a phone line, a scarce commodity during the war and one that often was traded without ever being installed. Under the claim that her teenage girls might call boys on the phone, the Foundation administrators had rejected her request. Her father-in-law, the legal guardian of the girls, supported the state's position. Through sheer persistence, Zarin prevailed and obtained a phone line, which she immediately sold. This knowledge of the system also allowed her to resist the Foundation's position on remarriage. Although a middle-age widow, she married a man who was not a veteran. Both factors would have led to state disapproval. To receive her widow's benefit and specifically to maintain the supervision of her daughters, she did not officially register her second marriage. Zarin had resources to resist and knowledge to subvert the state system.

NEGOTIATING WITH FAMILY AND STATE: SECULAR SOLUTIONS

The remaining 46 women in the sample did not support the regime and the changes taking place. For those employed, the work environment had become exceedingly hostile, characterized by ongoing harassment and state purges. Their resistance to this hostility took many forms, among them a commitment to remain in the workplace. The first four narratives that follow illustrate varying degrees of success in regard to this commitment. In contrast, for those women who remained at home, the changing conditions offered them unexpected opportunities to expand their family roles. We close our series of narratives discussing these opportunities.

Soyra and Mahin Acquiesce

Soyra: Submitting to Workplace Policy Soyra sought to earn money to keep her two sons safe from the military draft; however, her resolve to resist the state was not as strong as the state's ability to harass her. Soyra, a middle-level senior civil service employee resigned three years after the revolution. Her retreat to the private domain was precipitated by her dislike of the enforcement of *"Hijab Islami"* codes. She claimed that the dress and behavior codes at her workplace had become intolerable and that she could not stand the indignities any longer. For example, on one occasion, a government inspector had seen (from under her desk) that she was wearing sling-back shoes and had "written her up." In response to this incident, her supervisor warned her that she must wear "regulation shoes." After a series of these kinds of events, Soyra resigned from her job.

Atypically, her husband, Hossein, believed she was making a mistake, as did her sister-in-law, Parvin. Parvin, also a teacher, claimed that Soyra was "too tense about the regulations." What she should have done, according to her family, was "take a lengthy unpaid leave and bide her time until the conditions improved." Hossein, seeing her decision as a detriment to the family, could not prevail on her to change her mind. Through contacts and without financial support from Soyra, he did manage to place one of their sons in a secure military station, but the other son was sent to the front and wounded.

Mahin: Submitting to Family Demands Under the new regime, Mahin was transferred from her neighborhood teaching position to a new school in the outskirts of the city of Qum. Even though she was not happy about the commute, she did not want to resign. She once told Gerami, "Our principal is relaxed about new regulations. Most days we take a dish to school and have lunch together with all teachers. We eat and have a good time." In addition, Mahin did not want to stay home because "If I stayed home my mother-in-law would drop in to visit all the time. She meddles in our life and causes problems between my husband and me. She wants me to stay home." Mahin's husband, Ahmad, agreed with his mother. He used religious teachings to support his position, stating "Allah has made women gentle to be mothers and at home and that working outside exposed women to the rough elements of the society." He even took on a second job as a taxi driver so that he could be the sole provider for his family. This was a great source of pride among Iranian men. With these sorts of family pressures, Mahin resigned and soon started having children. By 1995, Mahin had three daughters and a son. Consequently, Ahmad was able to fulfill both his desire to have her at home and to provide the state with children. When she chose to return to her domestic responsibilities full time, Mahin clearly acquiesced to the pressures of family and state.

Nasrin, Shahla, and Ziba Co-Opt

Nasrin: Manipulating State Educational Policy The Republic's educational reform policy set quotas for women entering institutions of higher education. In addition, educational reform included exclusionary policies that affected both men and women. Specifically, men were banned from majoring in women's health care, thus giving women unprecedented dominance in medical schools and the health professions. Many women used these state policies to achieve three personal goals: (1) acquire a higher education, (2) pursue a career in women's health care, and (3) avoid the full-time responsibilities of wife and mother.

Nasrin and her husband, Mohammed, both came from affluent extended families in the provincial capital city of Qum. They married soon after the revolution. At that time, he had a high school education and she was finishing her midwife training in a special college of the State Medical School. The general understanding between Nasrin and Mohammed was that she would finish her education and as a couple they would negotiate her pursuit of a career. Immediately after graduating, she started working in a city hospital. In her free time and with the help of her parents, she established a private clinic. Even when the city was being bombed, Nasrin remained committed to work and women's health care.

Mohammed's career path was less clear. It was assumed that he would manage his father's financial dealings and his wholesale carpet trade. But with his father still active and very much in charge of his business dealings, Mohammed had a minor occupational role. As a result of his blocked advancement in his father's business, Mohammed became a "house husband." After three years of marriage, Nasrin and Mohammed had a daughter. He became the primary caretaker with occasional help from his mother. This was the only child they had due to Nasrin's career commitments. By pursuing a state-sanctioned education and career, Nasrin co-opted the state's intent to return women to their domestic roles.

Shahla: Manipulating State Family Policy During the cultural revolution, when the universities were closed, many families were left with the option of early marriage for their teenage daughters. This closure, coupled with a reduced minimum age of marriage, the haphazard distribution of contraceptive methods, and new financial incentives for additional children, reinforced the government's pro-natal policies. For a variety of reasons, many young women accommodated these policies even if it meant going against their families' wishes.

After finishing high school, Shahla was not admitted to any of the universities and colleges near her town of Tabriz. This did not bother her because she was not interested in acquiring a higher education. At 18, she wanted to get married and move out of her parents' home. Her mother, Sohila, did not want her to marry at such a young age. With her interests blocked, Sohila became involved in soliciting suitors who had the prospect of employment in the capital city of Tehran. Her motive for this selective screening was the hope that there would be better opportunities for Shahla in a metropolitan area. Shahla did marry a civil engineer. The couple moved often and eventually settled in the provincial capital of Shiraz. They had two children. Shahla co-opted the state and followed her own wishes for a family.

Ziba: Manipulating State Policy and Family Expectations Fundamentalist families from all classes believed that women could best serve the Islamic Republic by remaining at home. In contrast, many secular middle-class families encouraged their daughters to acquire a college education and pursue a career. Revolution and the war frustrated these secular family plans. Those daughters who preferred homemaking to employment manipulated the state's emphasis on marriage and motherhood to challenge family expectations.

Ziba and Ali were married during the early years of the revolution. He was a talented engineer moving up the ladder of a government ministry; she was a young girl with a high school education. Because her family valued education, it was assumed that Ziba should aspire to become a professional like her siblings. Her husband also believed she should seek higher education, but when the war started many of their plans were disrupted. Their house in southern Iran was destroyed by the Iraqi troops and Ziba had to move in with her parents in Tehran. As an engineer, Ali was constantly traveling to strategic locations to supervise the reconstruction of bombed facilities. He eventually became a prominent undersecretary for a state ministry.

Under these stressful conditions, Ziba had two daughters and completed her college education; she was poised to fulfill her family's expectations to begin a career. Ali, understanding his in-laws' sentiments, never openly opposed this assumption. Nevertheless, neither did he hide his desire for a large family, especially sons. Ziba's third child was a son. The effect of raising three children, running a household, and contending with Ali's long absences consumed Ziba's time. With these responsibilities and her lack of any experience in the workforce, Ziba chose not to pursue a career. As she stated, "I am not prepared for the hassle and challenges of work." Ziba had decided that a career was not practical or necessary. Ali's demanding government job provided high status and income and gave security to his family as well as Ziba's aging parents. In effect, Ziba saw her decision as a trade-off—her career aspirations for a secure living as a bureaucrat's wife.

Multiple Subversion Strategies

Mehri Subverts: Undermining State Draft Policy During the war, secular women were particularly concerned with protecting their draft-age sons. These women—teachers, clerks, and other working women—would say, "You know, I hate these people (the regime), but we have to think of our son." For these women, securing a safe future for their children was a family priority best met with their own continued involvement in the paid labor force.

Although the draft age was 18, the state had made it illegal for young men 15 and older to leave the country. This policy was rigidly enforced. In addition, host countries had tightened their visa requirements for Iranians, leading to the necessity of smuggling sons across the border into Turkey or Pakistan. Both destinations required traversing rugged and mountainous terrain. These extraordinary conditions imposed a heavy burden financially and emotionally on a family.

Mehri and her husband, Ali, illustrate a family responding to this situation. They had two sons, ages 15 and 13. Because of their opposition to the war, they had decided to smuggle them out of the country. Mehri's job as an accountant provided the money needed for the boys to exit through the southeastern border into Pakistan. Eventually, both boys entered a United Nation's refugee camp and immigrated to Sweden. Mehri and Ali remained in Iran. After 25 years of work, she retired. Mehri's resistance strategy ultimately protected her sons from becoming "martyrs to the cause."

Frabia and Shiba: Undermining State Distribution Policy Immediately after the war with Iraq, the government issued each family a ration book listing all members and their identity. Based on the booklet, families were issued coupons to receive scarce goods from sugar to gasoline. Procuring these goods meant having knowledge of what items were offered, when and where these items could be acquired, and the open market prices of these items. Procurement involved waiting in long lines and was mostly women's responsibility. This responsibility was quite burdensome to women with health problems or small children; it was insulting to women who believed it was "beneath their life station." Consequently, women in these social circumstances would sell their coupons at a reduced price to dealers or have the dealers shop for them in exchange for a portion of the goods. This practice set the stage for a black market that supported dealers, traders, and hawkers. Working and lower-middle-class women assumed the roles of dealers and traders in the black market; the men in this class assumed the role of hawkers. Frabia and Shiba, sisters, capitalized on this black market situation.

Frabia, the younger sister, was married to a civil service employee who was a heroin addict. According to her mother, she supported the family and his habit through her black market activities and was able to keep the family "respectable." Less dramatically, Shiba's participation supplemented the income of her husband, who was a tradesman. While the rationing system was in effect, these sisters established a network of clients and hawkers with whom they would trade, barter, and sell coupons and goods. They described their work as "fun" even though in reality it provided economic survival for their families. They referred to women who would not stand in lines as "fancy free and stomach empty." When the ration system in Iran was phased out, both sisters moved into more risky markets. For example, Shiba began trading in illegal currencies. Because of their ability to subvert the state rationing system, these women stand in stark contrast to the image of Iranian women confined to their homes and economically dependent on their husbands.

Roya, Rana, and Sara: Undermining State Lifestyle Restrictions Class position did not limit a woman's decision to enter the black market, it only influenced the nature of the operation. The revolution, the war with Iraq, and the international blockade had made it difficult to maintain the pre-revolutionary standard and style of living that upper-class women had attained (Sciolino 1997). Upscale boutiques were closed or ransacked by vigilante groups. Luxury items such as perfumes, cosmetics, china, crystal, and fashion merchandise were banned or just not imported. As a consequence, the fashion market went underground and into the homes of the wealthy. Upper-middle-class women who had an eye for bargains and good taste were able to use these skills to become fashion entrepreneurs.

Roya, Rana, and Sara were close friends. They formed a business in black market luxury items. To raise revenue for this business venture, they had to borrow money as well as sell personal items. For example, Roya sold the gold pieces and Persian rugs that were part of her trousseau. She was proud that she had raised the capital without her husband's help. In addition to raising their own funds, these women were able to capitalize on their class standing in two major ways. First, they knew shopping seasons abroad. Second, they had an extensive network of relatives outside of Iran who helped them obtain visa applications. They traveled to Turkey, Singapore, and Europe to purchase clothing and small household items that could be marketed in Tehran.

An illustration of the marketing savvy of these three women occurred in 1985. At this time, many families had lost loved ones in the war with Iraq. According to custom, women in mourning wear black for approximately six months. Fashion-conscious, upper-class mourners had difficulty finding "decent" mourning attire and *hijab.* At a handsome profit, Roya, Rana, and Sara provided high-quality black clothing to this class of mourner. In addition to knowing what would sell, these women knew how to sell. Upon arriving home from a shopping trip abroad, they would arrange for a "show" in their homes, which were in up-scale and fashionable neighborhoods. These shows were advertised by word of mouth, by means of fliers, or by private invitation. They were quite informal and conducted more like a tea party than an opportunity to engage in illegal transactions.

None of these upper-class entrepreneurs considered or used work-related terms to describe their activities, nor did they claim financial need. These activities were defined as being "fun" or "just a hobby." As Rana put it, "I enjoy shopping and bargain hunting and I have family abroad to visit." Nevertheless, the profits gained from these "social gatherings" were very real. They provided much-needed income to families that were especially affected by the revolution. For example, Rana's husband had been a colonel in the Shah's army and was for all purposes under house arrest. With the support of their families, these women resisted the state's attempt to ban their lifestyle.

DISCUSSION AND CONCLUSION

. . . The 1979 revolution and the Islamic Republic hegemonized a fundamentalist code advocating domesticity and modesty for women. Faced with oppressive state policies and fluctuating family pressures, Iranian women had to negotiate a sense of self at two levels of interaction: societal and familial.

Some collaborated with the state while others co-opted its ideology or subverted its demands. Their social class, education, and family's religious orientation informed women's strategies. . . . [N]one of the secular women in this group collaborated with the regime. They were mostly middle class and had more education. The existence of a mutual distrust between their family and local state machinery made collaboration a null strategy for these women. On the other hand, serving the new regime provided working-class women who had strong ties to the neighborhood Mosque an avenue to expand their gender roles. Unlike the other three strategies, collaboration posed no counterpressure for them between family demands and state policies.

In contrast, the strategy of acquiescence was used by women of both the working and middle class. Shirin, an Islamist, represents countless women in the South, resourceless and subject to oppressive family and state rules. Similarly, Fatimeh represents women who have to choose between a work environment that fails to support them and a family that needs their income. The secular examples retell the same tale of counterpressures from state and family and women's tactics to preserve their identity. Soyra faced cross-pressure from her secular family to co-opt or subvert and a fundamentalist state to quit or collaborate. She acquiesced to the state's demand. Mahin, on the other hand, gave in to a family that had co-opted the state's rhetoric to make her stay at home.

Using various co-optation strategies, four women from middle-class families manipulated state and family orientations to forge their identity. Aisha used the state's validation of her family's Islamic orientation to seek a higher education and employment. Ziba and Shahla used state gender policies to resist their secular families' demands, whereas Nasrin skillfully used the state's segregation policies to further her professional ambitions.

Women who used a subversion strategy came from different classes and educational levels. Among the secularists, Mehri, Roya, Rana, and Sara were upper middle class and savvy urbanites. Of the four, only Mehri had a college education. Zarin, the Islamist using subversion strategy, was upper middle class but within a minimum education. While Roya and her friends used their upper-middle-class fashionable network to subvert the system, Zarin used her Shahid husband's status to undermine the state's control as well as her father-in-law's authority. Shiba and Frabia, lower class and minimally educated, established a clientele of middle-class women and a network of black market operatives to bend the state's distribution policies.

Previous studies of power have analyzed and valorized men's negotiation tactics. In contrast, we have proposed analytical categories for documenting women's negotiation strategies. Women's strategies differ from men's because men have access to authority in the Weberian sense and women's access to power is often through indirect tactics. When men cooperate, women must contrive, and when men collaborate, women scheme or subvert. We want to expose this devaluation of women's empowerment. Therefore, by offering these categories, we acknowledge women's power and their innovative tactics. In the words of Comaroff, Iranian women "in their everyday production of goods and meaning, acquiesce yet protest, reproduce yet seek to transform their predicament" (1985, 1). We have demonstrated that in crafting their agency, Iranian women negotiate between the fundamentalist state's agenda and their family's patriarchal authority. . . .

REFERENCES

Amirahmadi, Hooshang. 1990. *Revolution and economic transition: The Iranian experience.* Albany: State University of New York Press.

Bauer, Janet. 1991. A long way home: Islam in the adaptation of Iranian women refugees in Turkey and West Germany. In *Iranian refugees and exiles since Khomeni,* edited by A. Fathi. Costa Mesa, CA: Masda.

———. 1993. Ma'ssoam's tale: The personal and political transformations of a young Iranian "feminist" and her ethnographer. *Feminist Studies* 19:519–48.

Comaroff, Jean. 1985. *Body of power spirit of resistance.* Chicago: University of Chicago Press.

deGroot, Joanna. 1996. Gender, discourse, and ideology in Iranian studies: Towards a new scholarship. In *Gendering the Middle East,* edited by D. Kandiyoti. Syracuse, NY. Syracuse University Press.

Esposito, John L. 1990. *The Iranian revolution: Its global impact.* Miami: Florida International University Press.

Gerami, Shahin. 1996. *Women and fundamentalism: Islam and Christianity.* New York: Garland Publishing, Inc.

———. Forthcoming. Mullahs, martyrs, and men: Conceptualizing masculinities in the Islamic republic of Iran. *Men and Masculinities.*

Hawley, John S. 1994. *Fundamentalism and gender.* Oxford, UK: Oxford University Press.

Inkeles, Alex. 1983. *Exploring individual modernity.* New York: Columbia University Press.

Kandiyoti, Deniz. 1988. Bargaining with patriarchy. *Gender & Society* 2:274–89.

———. 1997. Beyond Beijing: Obstacles and prospects for the Middle East. In *Muslim women and the politics of participation: Implementing the Beijing platform,* edited by Afkhami and Friedel. Syracuse, NY: Syracuse University Press.

Keddie, Nikki R., and Eric Hooglund. 1986. *The Iranian revolution and the Islamic republic.* Syracuse, NY: Syracuse University Press.

Klatch, Rebecca. 1987. *Women of the new right.* Philadelphia, PA: Temple University Press.

Lazarus-Yefeh, Hava. 1988. Contemporary fundamentalism: Judaism, Christianity, Islam. *Jerusalem Quarterly* 47:27–39.

Marshall, Susan. 1984. Paradoxes of change: Culture crisis, Islamic revival, and the reactivation of patriarchy. *Journal of Asian and African Studies* 19:1–17.

Martin, Marty, and R. Scott Appleby. 1994. *Accounting for fundamentalism: The dynamic character of movements.* Chicago: University of Chicago Press.

Mazumdar, Sucheta. 1994. Moving away from a secular vision? Nation and the cultural construction of Hindu India. In *Identity politics and women: Cultural reassertion and feminisms in international perspective,* edited by Valentine Moghadam. Boulder, CO: Westview.

Mir-Hosseini, Ziba. 1993. *Marriage on trial: A study of Islamic family law.* London: I. B. Tauris.

Muadudi, Abul A'la. 1982. Political theory of Islam. In *Islam in transition: Muslim perspectives,* edited by Donohue and J. L. Esposito. New York: Oxford University Press.

Mumtaz, Khawor. 1994. Identity politics and women: "Fundamentalism" and women in Pakistan. In *Identity politics and women: Cultural reassertion and feminisms in international perspective,* edited by Valentine Moghadam. Boulder, CO: Westview.

Naficy, Azar. 1994. Images of women in classical Persian literature and the contemporary Iranian novel. In *The eye of the storm: Women in post-revolutionary Iran,* edited by Afkhami and Friedl. Syracuse, NY. Syracuse University Press.

Naficy, Hamid. 1994. Veiled vision/powerful presence: Women in post-revolutionary Iranian cinema. In *The eye of the storm: Women in post-revolutionary Iran,* edited by Afkhami and Friedl. Syracuse, NY: Syracuse University Press.

Paidar, Parvin. 1996. Feminism and Islam in Iran. In *Gendering the Middle East,* edited by D. Kandiyoti. Syracuse, NY: Syracuse University Press.

Papanek, Hanna. 1994. The ideal woman and the ideal society: Control and autonomy in the construction of identity. In *Identity politics and women,* edited by V. Moghadam. Boulder, CO: Westview.

Piscatori, James. 1994. Accounting for Islamic fundamentalism. In *Accounting for fundamentalisms: The dynamic character of movements,* edited by M. Martin and R. S. Appleby. Chicago: University of Chicago Press.

Poya, Maryam. 1999. *Women, work and Islamism.* London: Zed Books.

Rajaee, Farhang. 1999. A thermidor of "Islamic yuppies?" Conflict and compromise in Iran's politics. *Middle East Journal* 53:217–20.

Reinharz, Shulamit. 1992. *Feminist methods in social research.* New York: Oxford University Press.

Richardson, Laurel. 1995. Narrative and sociology. In *Representation in ethnography,* edited by J. VanMaanen. Thousand Oaks, CA: Sage.

Riesebordt, Martin. 1993. *Pious passion: The emergence of modern fundamentalism in the United States and Iran.* Los Angeles: University of California Press.

Sciolino, Elaine. 1997. The channel under the chador. *The New York Times Magazine,* 4 May.

Sidahmed, Abdel Salam, and Anoushiravan Ehteshami. 1996. *Islamic fundamentalism.* Boulder, CO: Westview.

Zuhur, Sherifa. 1992. *Revealing reveiling: Islamist gender ideology in contemporary Egypt.* Albany: State University of New York Press.

53

Do We Still Need Feminist Theory?

A Conversation

BY MARGARET L. ANDERSEN, ANNE E. BOWLER, AND MICHAEL S. KIMMEL

The last reading in this chapter is the panel discussion of a conversation among three feminist scholars in sociology. Their focus is on the relationship between feminist theory and feminist practice. In the course of their discussion, they explore connections and disconnections between theory and practice, as well as the impact of recent widescale economic, political, and cultural changes on feminist perspectives and practices. Important to this reading is the discussion of group power versus individual feelings of power or powerlessness, and the value of sociological analysis of gender processes, patterns, and change.

1. Why is it important to distinguish between group power and individual agency?
2. What is the unique role of sociology in theorizing gender?
3. Why is the relationship of theory to practice an important issue for discussion and debate?

Margaret Andersen: "Do we still need feminist theory?" This is a question that could be answered by simply saying yes since, as sociologists, we know that sociological thought is always grounded in theoretical assumptions, however implicit or explicit. What are the major paradigms of feminist theory? In the early years of the women's movement, feminist theory could be neatly organized into three major paradigms: liberal feminism, socialist feminism and radical feminism; now we would add multiracial feminism and postmodernism. Do these paradigms still hold as orienting frameworks to understand contemporary gender relations and the status of women in society? Furthermore, what is the connection between feminist theory and feminist practice? These questions guide our discussion. First, we ask whether feminist theory must be tied to feminist practice or to a social movement.

Anne Bowler: My answer is a qualified no. There must be space for theory *qua* theory—not simply in feminist theory but also in other theoretical inquiries. At the same time, the best theoretical work both comes from and speaks to practical social problems, such as racial stratification, stratification in the paid work force and—a particularly pressing issue today—the problem of poverty and welfare reform.

To answer the question of the connection between feminist theory and social movements, we have to ask, "What are the viable movements of today?" Is there a feminist socialist movement? Does it make sense to still talk about socialist feminism in the absence of a feminist socialist movement? What does feminist theory have to offer other movements, such as the movement to preserve the right to legal and safe abortion? I do not see much evidence that people in these movements are looking to feminist theory as a guideline. In fact, I see a tremendous fissure between feminist theory and feminist practice.

Michael Kimmel: My answer is a qualified yes. Speaking as a man, I ask what does feminist theory have to do with my day-to-day interactions with colleagues, staff, and students? There feminist theory is most assuredly connected to feminist practice. I try to maintain a kind of feminist practice and ongoing self-reflection about feminist practice while working in the institutional context in which I find myself.

Margaret Andersen: My answer is that feminist theory *is* linked to the feminist movement and vice versa. That is not to say that there are not times when there is a fissure between them, but most feminist academic work has stemmed from the women's movement. I wonder if we are thinking about theory with a "big T" and movement with a "big M." The women's movement has become more institutionalized than it was when the early categories of feminist theory—liberal feminism, socialist feminism and radical feminism—were developed. Now, the women's movement has multiple meanings, including our everyday practice as feminists in formal and bureaucratic organizations and in less formal places, such as families and communities. Theory and practice need to be linked; how sociologists think about feminist theory really originates from feminist politics.

Audience: Feminist studies in sociology are now grounded in more specific empirical results. Feminist scholarship has turned to examining the social construction of gender and race, and we have developed studies of the specific experiences of particular groups of women. Unlike in the past, we see more changing historical patterns, rather than grand theory, so making broad claims, such as that women's status is all the result of patriarchy or all the result of capitalism, seems inappropriate. We can no longer see "women" as an undifferentiated category in non-specific historical ages.

Michael Kimmel: I think this point is really important. From my perspective, I see those early exhilarating feminist texts as speaking about women as a single, unified category. Although this has been criticized since, it actually defines radical feminism and distinguishes radical feminism from other feminist theories. Thus, the very thing that defined radical feminism was the positing of women as a single unitary category in which all women's experience, though somewhat modified by various things, was essentially the same. What historians and sociologists have done is explode the myth of women's unitary experience, meaning that radical feminism can no longer be used as it was twenty-five years ago to explain the position of *all* women.

Margaret Andersen: What worries me about giving up those old categories, particularly about losing sight of socialist feminism, is that even with all of the current attention to race, class and gender in feminist scholarship, the analysis of class has virtually disappeared. It is somewhat surprising how rarely you see class analyzed seriously in gender scholarship and how seldom the word capitalism even appears in feminist scholarship. Yet, were I to list the current phenomena that seem to be most affecting women's lives, the list would include global capitalism, increasing class and race polarization, globalization of technology, immigration, and other large macro-social forces that are reflected in some of the empirical literature, but are not informing in a grander way how we are conceptualizing gender. Although we have done much work on race, class has tended to be less visible—even in race, class, and gender studies.

Audience: Why has class disappeared in feminist theory? With regard to the decline of socialist feminism, one of the forces that has led to its decline is the movement of theorizing into the academy. Many feminists who were writing feminist theory early on have moved into the realm of practice. This wasn't just a retreat, but the mobilization to create gender studies and women's studies programs within universities means a lot of energy has gone into activism within the academy. Given the relative privilege of university women and the risk to their careers of taking more radical stances, socialist feminism has declined.

Michael Kimmel: I would add that the decline of the left—not simply the end of the Vietnam War, but also the fall of the Berlin Wall and the collapse of the Soviet Union—means that there are alternate poles around which theorizing is being done. The discrediting of the anti-capitalist critique has also dovetailed with the rise of "girl power" and the association of feminism with individual agency. Thus, there is an absence of class analysis because we have witnessed the collapse of class critique and we celebrate the hyper-individualism of girls' agency.

Anne Bowler: This is also tied to the rise of consumerism.

Michael Kimmel: Exactly, but where is agency when one is shopping at Victoria's Secret and buying up Madonna CDs? This is especially interesting in light of the recent interest in globalization because the global division of labor is also a *global gender* division of labor. Class is reemerging on a global scale in ways that socialist feminists could not have dreamed of in the 1970s. It seems to me that globalization is going to shift socialist feminist analysis again. It will provide openings as well as fissures in the edifice in which a new socialist feminist critique can take place.

Anne Bowler: Shifting the focus a bit, I'd like to ask Michael how the various themes that emerge from studying the experiences of men and masculinity contribute to the development of feminist theory.

Michael Kimmel: Let me start by saying that I speak as a man. While I know this audience sees and understands the significance of that fact, it is less clear to me that my students see its significance. A few years ago I was teaching a Sociology of Gender course which I usually teach every other semester. In the semester when I was not teaching it, the woman who did asked me to give a guest lecture in the class. When I walked into the class as a guest lecturer, one of the students looked up and said, "Oh, finally an objective opinion." In a way, I can say far more radical things than either Anne or Maggie—for example, in speaking about systematic inequality. Students perceive that, because you are a woman, you are biased; but, when I say the same thing, they want to know if it is going to be on the test.

Part of what feminism and gender studies have done is to examine precisely the mechanisms by which women are inscribed as gendered and men are inscribed as ungendered—and, presumably, objective and universally generalizable. One of the first themes in the feminist critiques of the sciences has been to demonstrate that this objectivity is a gendered myth. Now that there are some men doing critical gender studies, the first task that we have is to decenter men or decenter masculinity as the unexamined norm against which all events are measured. I also think that feminist theory has made gender visible both as a category of identity and as a system of inequality, positing questions of difference and questions of inequality. Yet, there are obstacles for engaging men in feminist theory. Why aren't there more men who are thinking about gender? What is the resistance that men face to doing this?

The critique of feminism as first articulated in the 1970s was that women's experience as individuals and the aggregate structural distribution of power were similar—that is, symmetrical. Women were not in power. This is really easy to measure. Empirically, you could look at the distributions of corporate boards, college boards, trustees, parliaments and houses of legislature. Women as a group were simply not in power. Furthermore, women didn't feel powerful individually, so there was symmetry between women's individual experience of powerlessness and women's aggregate structural powerlessness. We then applied this thinking to men's lives. Thus, our analysis went like this: Men are in power. Look at all those boards and legislature. Men must feel powerful and, therefore, it is time for men to give up the power. And women would say that was a great idea. But, once said, men's hands would go up and say, "What are you talking about? I don't have any power. My kids boss me around, my wife bosses me around, my boss bosses me around. I'm completely powerless."

It occurs to me that we failed to understand that, although men were as a group in power, men individually did not feel powerful. Part of this is the illusion that we are always looking one step above us in the hierarchy—who we do not have power over—and at the same time forgetting who we *do* have power over.

My argument is that feminist theory, as we originally comprehended it, failed to resonate with men was because it assumed the same symmetry in men's lives that we observed and talked about in women's lives. Aggregate power in the world does not translate to individual men feeling powerful. Unless we attend to that feeling of powerlessness among many men, along with structural power, we will continue to leave men out of gender studies. This is also a tremendous point of entry for men into this discussion. What are the structures of power that leave men, as well as women, feeling powerless despite overwhelming gender inequality at the structural level?

Audience: Women are experiencing a similar disjunction. For example, contemporary students tell us: "I was raised by a feminist mother, I am personally empowered. I have all the feelings of power in my life, but you're telling me that I'm not powerful." Young women do not want to hear this. There is the personal disconnection between their feeling individually powerful and their fury that they are in fact structurally disempowered. We need to have more discussion about this disjunction between personal feelings and structural power.

Margaret Andersen: What we are pointing to is the unique role of sociology in theorizing gender, as reflected in theorizing about the connection between social structure and identity. With much of feminist theory now coming from the humanities, the focus on social structure may be getting lost. What role does sociology have in building feminist theory now?

Anne Bowler: Sociology contributes an understanding of the reality of social structure. For example, sociology includes the ability to apprehend identity structures—race, class, gender, sexuality—as social facts, to use Durkheim's term. These are not just textual manifestations; they are real. It is nothing new to point out that much of feminist theory in the past ten or fifteen years has come from poststructuralism and postmodernism. Both theoretical discussions have made contributions, but have very severe limitations.

The major contribution of a poststructural approach to gender has been to demonstrate how the very category of woman or man is historically and culturally constructed—or, I should say, women and men. By this, I mean to place a special emphasis on how these categories change throughout history and in different social locations.

Similarly, the major contribution of a postmodernist approach to gender has been, like postmodernist theory more generally, to challenge long-held ideals of objectivity—in particular the idea that there can be a single "Truth"—and to challenge the evolutionary approach to social change that emphasizes the ideal (some might say myth) of infinite progress. Also, like poststructuralist theory, postmodernism challenges the degree to which popularly accepted categories (for example, in the area of sexuality the presumed binary opposition of heterosexual and homosexual or masculine and feminine) are fixed and timeless.

At the same time the limitations of these theoretical trends is that, in their most extreme forms, social structure and the lived experience of individuals is ignored or even erased. For example, the major focus of a poststructuralist approach lies in its emphasis on texts, in particular, the different meanings by which texts can be decoded, reinterpreted, excavated for heretofore hidden meanings. A notable and highly problematic tendency coming out of this emphasis on text, however, has been to minimize the significance of social and material factors. In its most extreme forms, society simply becomes another "text."

The limitation of postmodernist theory is, in my judgment, even more problematic. It is one thing to claim that the idea of a single, monolithic "Truth" is questionable but the tendency here lies in the unwitting adoption of a kind of radical relativism wherein truth is ephemeral or even a myth. This denies the lived experience of women and men, especially, I would argue, poor women and women of color—in other words, some of those very women who are experiencing the effects of welfare reform and its impact on poverty.

A second major problem with postmodernism lies in its attempt to revel in the realm of the popular as part of a rejection of the modernist emphasis on so-called "high culture." The tendency here is to privilege cultural forms and practices as sources of opposition to systems of social stratification and social control. It has become something of a maxim that culture—specifically, popular culture—equals "resistance," but watching daytime television, as an example, is not automatically an act of rebellion against dominant, hegemonic norms.

Margaret Andersen: The panel is in agreement about the sociological contribution to the analysis of the connection between, to use old language, the personal and the political. However, what may be getting lost is one of the major insights of postmodernism and poststructuralism—the focus on social construction. Patricia Yancey Martin once incisively observed in a conversation, "All postmodernism involves is literary critics discovering symbolic interaction." Her insight makes sense to me because I think that the task of feminist theory is to see how systems of representation, identity, and structure are linked. Ideology mediates how we experience social structure and see ourselves represented *by* it and *in* it. What is uniquely sociological in feminist work is connecting the level of individual and collective experience to concepts such as power and class that have largely disappeared in some feminist writing.

Anne Bowler: This is precisely the reason I did not simply invoke social structure but also social facts. Class, like gender, is a social fact and how it gets produced and reproduced is through social practices or discourses (a word I dread using because it has become so overused and abused that it has come to mean everything and nothing). The late French philosopher Michel Foucault employed the term "discourses" to demonstrate how systems of power operate not only through the spoken or written word but also through practices and institutions.[1] Consider the example of gender in the paid labor force. Gendered relations are produced and reproduced not just through what is said or how one is heard but by such things as gender segregation, the gender wage gap, and the presence or absence of job ladders and sexual harassment policies. Attention to factors such as these constitutes a major force behind the analytical shift in sociology from gender norms to gendered institutions.[2] This is why I think the concept of social facts is so important.

Michael Kimmel: We have to be careful about overemphasizing the question of identity. Doing so risks psychologizing feminist theory and downplaying institutions, organizations, structures, discourses, and so forth. I recall the first texts on men that came out in the mid-1970s, texts that largely came out of a psychological framework. They mirrored the texts that were being written about women at the time. There would be a chapter on work, a chapter on family, a chapter on sexuality and then, at the end, there would be a chapter on Black men and gay men. It was if all those other chapters—on work, family and the like—assumed they were about straight white men, and that Black men or gay men were "problem" masculinities that had to be explained only in terms of difference from the norm.

Now, multicultural feminism has exploded that construction and you could not do a book around men or women organized in this same way—as if all men were alike. What has happened in the university is that these diverse identities have become so balkanized and enfranchised in academic studies and centers that men rarely talk to each other across their differences. There is an over-focus on identity and the under-analysis of how these identities interpenetrate or complicate each other. We need to study how they are related to each other and how they are not.

Margaret Andersen: This problem has troubled me on a number of fronts. Identity is a term that has a more particular meaning for me as a sociologist than it may for people in other disciplines. For sociologists, identity means how one sees oneself in relation to others in the context of a social structure. Yet, as I talk with people in literary studies and in ethnic studies, I hear identity being used to refer to everything. Identity has become such a big term that it means almost nothing. Work on identity, as insightful as it might be about diverse experiences, is rarely analytical about the structures of stratification and power. In many cases, identity is being used just as Michael suggested: highly individualistic. Thus, you lose a picture of the total structural context in which identities are formed.

If you use the experiential location of, for example, African-American women, to reveal the operation of systems of domination and subordination and how racism operates then you reveal social structure. And, you do not always have to have everyone at the table to see this because, if you reveal how social structure operates, you can then locate different groups within this structure. For example, analyzing the structural bases of race, class, and gender can also reveal how other structures of oppression—sexuality, age, nationality, religion, and others—are part of the same system. Although exactly how identities and experiences are created across these different experiences differs and is not exactly the same, the processes are similar and overlapping.

Michael Kimmel: Another institutional arena where we need to look is political discourse. We are confronted with a popular culture that is increasingly demanding simple binary answers—yes/no, for/against—to every issue. Every television program I have been on in the past few years wants to have the opposing view. Now, if you have a show about the Holocaust, they also have to have a Holocaust denier.

Also, on a cultural note, more of our students have read *Men Are From Mars, Women Are From Venus.* This is the number one best seller on college campuses. It is, in fact, the number one best selling self-help book of all time. More students have read that than will ever read careful and deft empirical analyses of gender, class, and race.

This is a failure of nerve in some ways on our part. We resist the impulse to speak boldly and broadly and paint a large canvas. Instead, when we are interviewed in the public media, we spend a lot of time backing away, always arguing that something is more complicated than the media make it. The absence of sociologists on the public stage has to do with that kind of hesitation and qualification. I don't think we've learned yet—and we need to learn—how to speak boldly and still remain true to what we know.

Margaret Andersen: It is hard enough to teach undergraduate students over the course of a semester to think in other than individualistic terms, much less presenting such ideas in a fifteen second sound bite. Yet, coming back to the earlier part of our conversation about the significance of class and power as structured relationships, we can draw on the rich array of empirical work that we now have in gender scholarship. We can benefit from the empirical phase of feminist scholarship that we have produced and use these studies to present feminist conclusions to the public.

Audience: There is no reason for us not to be bold in our statements about what we know. We know that class matters, we know that race matters, we know that gender matters, we know that ethnicity matters. And, as complicated as it may be, we have to say that in as straightforward manner as we can.

Anne Bowler: Based on the last statements that have been made, I would say that as feminist theorists we have a two-pronged challenge. One is to take what is a wealth of empirical research that did not exist twenty years ago, synthesize it, and make bold statements. And the other is to continue this empirical research, but also to think about how feminist research has ignored the experiences of certain women. Had feminist theory been centered from the start in the experience of women of color and the experience of working-class women and women living in poverty our analyses and questions would have been very different.

Michael Kimmel: When sociologists do really good empirical work, we can identify processes that speak beyond limited empirical studies of sex stereotyping and tokenism.

This is so much of what we try to do in our work. I remind us that that there is a precedent for using what we do know. One of the things about feminist theory—I opened with this and I want to close with this—is that this has always remained central to me: To remember to whom I am accountable. It is a privilege being a university professor in the late twentieth and early twenty-first centuries, living the lives we are able to live. We should recognize how that privilege was attained and recognize the people to whom I feel we are accountable—those who are not so privileged. I think we are accountable to those who are working for a world where gender, race, and class are *not* the best predictors of life chances. We know that these are overwhelmingly, the most powerful predictors of life chances. That's the bad news we keep giving our students. I want to live in a world in which they are not. That means that my work has to be political.

Audience: One of my students said that, when she really grasped the sociological perspective, she realized she would always be poor. What do I say to her?

Margaret Andersen: I think it is probably better to know what is happening to you than not. Within sociology, you can also see ways to create different institutions and different relationships. That is the beauty of the social constructionist argument because society is not just fixed. Things can change. We must teach our students that as well.

NOTES

1. See Foucault, *The Archeology of Knowledge.* Translated by A. M. Sheridan Smith. New York: Harper Colophon, 1969, 1971/1976 and "Politics and the Study of Discourse," *Ideology and Consciousness* 3 (1978): 7–26.
2. Acker, Joan. "Gendered Institutions: From Sex Roles to Gendered Institutions," *Contemporary Sociology* 21 (September 1992): 565–569.

Topics for Further Examination

CHAPTER TEN

- Look up research on men who participated in the first and second waves of feminism in the United States. Check out the following Web sites: http://www.feminist.com/men.htm and http://www.nomas.org.
- Browse websites on women's organizations and gender issues such as: http://www.iwpr.org, http://www.feminist.org, http://www.un.org/womenwatch/ and http://www.wilpf.org.
- Locate articles on the impact of globalization on human rights. Go to the following Web site: http://www.pbs.org/globalization/home.

EPILOGUE

Possibilities

This book began with the metaphor of the kaleidoscope to help understand the complex and dynamic nature of gender. Viewed kaleidoscopically, gender is not static. Gender patterns are social constructions, reconstructed as they intersect with multiple and changing social prisms such as race, ethnicity, culture, class, and sexual orientation. In concluding, we want to emphasize the dynamic nature of gender, underscoring the possibilities that the future holds. No one can predict the future; therefore, we will illustrate the dynamic nature of gender using stories of how changing gender patterns shaped our lives and lives of many other people.

Reflecting on the course of our lives, we are struck by the depth and breadth of changes that have occurred in American culture, institutions, and social relationships since we were young girls in the 1950s. Many of these changes have been positive; patterns of oppression have been reduced and the opportunity and power to participate meaningfully in America's social institutions have been extended to more people. We benefited from and participated in bringing about changes in the genderscape.

The cultural climate of the 1950s forged our early lives. Though romanticized in film and TV, that decade was in fact a deeply troubled time of blatant racism, sexism, and other forms of oppression. Civil rights had not yet been extended to people of color, women's rights were negligible, gay and lesbian Americans were largely closeted, poverty was ignored, political dissent was strongly discouraged, child abuse went unacknowledged or hidden, and an atomic war seemed ready to break out at any moment.

We didn't learn about the women's movement then, even though there had been significant organized social movements for gender change beginning almost 100 years before we were born, culminating in the right to vote in 1921. The second wave of feminism and subsequent women's movements began when we were very young, after World War II, inspired by books such as Simone de Beauvoir's *The Second Sex* (1952, first published in France in 1949) and Betty Friedan's *The Feminine Mystique* (1963). Although not widely recognized, the United Nations Charter of 1945 affirmed the equal rights of men and women and established the U.N. Commission on the Status of Women a few years later (Schneir, 1994). However, in many ways women were still second-class citizens in the 1950s.

We grew up in families not unlike those of many white Americans of the 1950s. Our parents held traditional views of proper roles for women and men. They tried to live up to those roles, and yet, like many, they often failed or fell short. They suffered with their failures in silence, behind the closed doors of the nuclear family of that era. For men and women, marriage and children came with the gender territory. Social sanctions in society maintained this territory. For example, people called women who didn't marry "old maids" pejoratively, and looked at couples who didn't bear children suspiciously. Also, there were limited reproductive control options, and abortion, although widespread, was illegal. This left most women and many men with few options except for getting married and having children.

Jessie Bernard (1975) described men's and women's roles in white, middle-class families of that time as being destructive to adults and children alike. She spoke of "the work intoxicated father" and the "pathogenic mother" as the end result of the efforts to fulfill the cultural imperative for a traditional family. Bernard observed that middle-class, white family roles at that time were stressful and unsatisfying. Men detached from their families as they struggled to earn enough for the household, while women shouldered the sole responsibility of raising perfect children and keeping a perfect home. Family and gender researchers (e.g., Bernard, 1972; Rapp, 1983; Coontz, 1992; Schwartz, 1994; Coontz, 1997) found that this arrangement of distinct and separate roles did not foster full and loving relationships between men and women or parents and children. And, of course, by now you can guess that most people could not achieve this "perfect family."

However, television reflected the image of the happy family behind the white picket fence and for decades it stood as an ideal, even for those who failed to meet it. Marriages were supposed to be happy, but often were not. Getting out of a conflict-ridden or abusive marriage was very difficult. There were divorces, but the courts only granted these if they decided there were "appropriate violations" of the marriage contract (Weitzman, 1985). As a result, many marriages persisted, even when there was unbearable alienation, violence, and abuse.

There was little recognition of domestic violence. Marital rape and rape in general were not taken seriously, legally or socially. For example, the police, when called to a domestic conflict, often ignored pleas of beaten or raped women, and the rules and procedures that guided police work made interventions almost meaningless. There was a great deal of resistance to changes in legislation relative to domestic violence and rape, including marital rape. The words of one U.S. Senator captured the general attitude at that time; he said, "if you can't rape your wife, who can you rape" (Russell, 1982). Thus, domestic violence and rape reforms took many years to implement.

As you can imagine, women in our mothers' cohort had few choices and opportunities. We noted in Chapter 7 that the "best" occupations most women could aspire to were limited to clerical or secretarial jobs, teaching, and nursing. Salaries were low and women had a difficult time living independently, both financially and socially. There was no such legal concept as workplace sexual harassment and equal pay for equal work was rarely considered. Married women who did work outside the home often had fragmented work lives, defined by their primary responsibility of caring for children and husbands.

The situation for white women was bad, but it was much worse for women of color and immigrant women. For example, most African-American women worked outside the home, but were confined to the lowest paying and most degrading of jobs such as domestic work. These jobs had even fewer protections against sexual harassment and workplace inequalities. Men of racial minority groups and immigrant men also endured considerable inequalities in the workplace as well as other domains of life.

Post–World War II saw a considerable increase in the number of people entering college; however, almost all new students were white men taking advantage of the G.I. Bill. Considerable gender segregation in higher education persisted throughout the 1950s, with many colleges and universities denying admittance to women and racial minorities. The proportion of women in higher education increased only slightly during the 1950s, from 31.6% in 1950 to only 35.9% in 1959 (U.S. Department of Education, 2000). Some of the most exclusive colleges and universities maintained gender segregated spaces even after women were admitted. For example, although women were admitted to Harvard Law School in 1950, they were denied access to the only eating space at the Law school until 1970 (Deckard, 1979). Harvard law school even limited the days on which women could ask questions in classes (Deckard, 1979). Money was another resource denied to many women; there were few scholarships for women because administrators and faculty felt "men needed the money more" (Deckard, 1979:130).

The decades of the 1960s and 1970s brought about the awakening of political consciousness and action by Americans from many walks of life. Early in this period, the civil rights movement resulted in the dismantling of many legal barriers to participation in American life for African-Americans. In the 1960s, the second wave of feminism picked up steam, spawning several organized political movements as described in Chapter 2. These social movements focused attention on the social disadvantages faced by women and brought about considerable change, including greater legal, economic, political, educational, and familial equality. Other social movements emerged out of this culture of change: gay and lesbian rights, anti-war and peace, environmental, and children's rights movements.

We were fortunate to enter early adulthood during this time of positive social change. For example, Title IX opened up avenues in education previously closed to our mothers; attempts at pay equity made our labor somewhat more valuable than that of our mothers' labor; the naming and litigation of sexual harassment, marital rape, and other forms of gender violence made our lives somewhat safer than our mothers' lives; and the women's movement empowered us and helped us to understand how we could contribute to a more just world. Our lives took us down different pathways, yet many of the same social change forces touched us deeply, and eventually our lives intersected.

Kay joined the ranks of one of the first waves of college-bound baby boomers in the mid-1960s. Her undergraduate years coincided with the civil rights movement, the height of the Vietnam War, the development of a strong political left, and the emergence of countercultural life styles. She found her intellectual passion in sociology, a home for her experimental self in the counterculture, and a political focus in the antiwar movement. With B.A. in hand, Kay entered the work world at the very moment that the first of a series of economic recessions set in. She bounced from one, unsatisfying, low-paying, female-type job to another and quickly found herself at an intellectual and emotional crossroads. Kay chose a pathway that few women had gone down, graduate school.

In 1971, Kay joined the ranks of a cutting-edge cohort of graduate students in departments of sociology across the United States. Although dominated by white men, this cohort contained more working-class white women and people of color than earlier generations. They moved through graduate school at the same time that the second wave of feminism spawned organized movements for gender equality around the world. Kay's consciousness expanded to embrace feminism. She took one of the first gender courses, Sex Roles, to be offered in any American institution of higher education. Feminism opened up a world of choices never before available. Empowered and exhilarated, Kay chose a nontraditional life course, as did many of the women in her graduate school cohort. Women postponed marriage and children. Others chose singlehood or cohabitating relationships. Still others chose child-free marriage. All pursued careers.

Of course change is never smooth and even-handed. Although the women's movement was well under way during Kay's graduate school years, women students and professors had regular encounters with sexism both in and out of the classroom. Sexual harassment was built into the everyday educational experiences of women who pursued a Ph.D. Also, as the women of Kay's cohort entered the professional world of teaching and research, barriers to hiring, tenure, and promotion would prove to be part of their ongoing struggle for respect, security, and equality.

Unlike Kay, Joan was a "good girl." She went to a secretarial school and pursued a gender-type job as a secretary. She became dissatisfied because she was not receiving raises or being paid as well as others who had 2-year college degrees, so she went to a community college to earn that degree. One course in sociology was all Joan needed and she was a student of understanding social processes. She married and had two children, but continued on in school, moving from community college to university-level education. In trying to understand the patterns of daily life, gender became a major explanatory framework for her. The isolation of women during childrearing years and the lack of institutions to support child rearing developed into major interests, as well as the effects of work patterns on men and women. Joan received her Ph.D. while rearing children. Her consciousness raising occurred more informally, via books and one-on-one conversations with friends, often while caring for children.

Entry into professional sociology, while at first off-putting given the white, male dominance of the field, became a path to connect ideas with action and to form meaningful relationships with an array of women. It was through a relatively new organization, Sociologists for Women in Society, that Joan met Kay and many others who taught and worked to improve the situation for women in society.

Together with another sociologist, we started a regional chapter of Sociologists for Women in Society. We became involved in Women's Studies on our respective campuses, conducted research on women in the arts, work, family, education, and women's bodily and emotional experiences. Today, we continue the journey toward a world that is more just and humane, a world in which people can achieve their potential unimpeded by the social prisms of difference and inequality described in this book.

Although it may seem that most of the work toward gender equality has been accomplished, as individuals and as members of a global society, we have more work to do. For example, women still do not have equal pay for equal work; glass ceilings and sticky floors continue to keep women from high paying jobs; only a few countries offer gender equitable parental leaves; most women work a double day, at paid labor and in the home; violence against women remains a serious problem; heterosexist and homophobic beliefs and behaviors maintain restrictive gender patterns while oppressing gay men and lesbians; racism continues to degrade and diminish the lives of women and men of color; and hegemonic masculinity limits the life experiences of most men.

Although there is still more to be accomplished, the good news is that in a very short period of time—our lifetimes—much positive change has occurred in gender relations. A third wave of feminism has emerged among young people (Baumgardner and Richards, 2000) and many of the movements that Barbara Ryan discussed in Chapter 2 remain strong. Clearly, more change is on the way. What possibilities for change do *you* see in your future? How might you make a difference?

REFERENCES

Baumgardner, Jennifer and Amy Richards. (2000). *Manifesta: Young women, feminism, and the future.* New York: Farrar, Straus and Giroux.

Bernard, Jessie. (1972). *The future of marriage.* New York: World Publishing.

———. (1975). The bitter fruits of extreme sex-role socialization. From *Women, wives, mothers.* Chicago: Aldine Publishing Company.

Coontz, Stephanie. (1992). *The way we never were: Americans and the nostalgia trap.* New York: Basic Books.

———. (1997). *The way we really are: Coming to terms with America's changing families.* New York: Basic Books.

de Beauvoir, Simone. (1952). *The second sex.* New York: Knopf.

Deckard, Barbara Sinclair. (1979). *The women's movement: Political, socioeconomic, and psychological issues.* New York: Harper & Row.

Friedan, Betty. (1963). *The feminine mystique.* New York: W. W. Norton.

Rapp, Rayna. (1982). "Family and class in contemporary America." In *Rethinking the family: Some feminist questions,* B. Thorne, with M. Yalom (Eds.), pp. 168–187. New York: Longman.

Russell, Diana E. H. (1982). *Rape in marriage.* New York: Macmillan.

Schneir, Miriam. (1994). *Feminism in our time: The essential writings, World War II to the present.* New York: Vintage Books.

Schwartz, Pepper. (1994). *Peer marriage: How love between equals really works.* New York: Free Press.

U. S. Dept. of Education. (2000). Degree-Granting Postsecondary: Enrollment. Table 173. Downloaded 9/30/02.

Weitzman, Lenore J. (1985). *The divorce revolution: The unexpected social and economic consequences for women and children in America.* New York: Free Press.

CREDITS

This page constitutes an extension of the copyright page. We have made every effort to trace the ownership of all copyrighted material and to secure permission from copyright holders. In the event of any question arising as to the use of any material, we will be pleased to make the necessary corrections in future printings. Thanks are due to the following authors, publishers, and agents for permission to use the material indicated.

Chapter 1

1. Copyright © 1998 from "Who's on Top?" in *My Gender Workbook* by Kate Bornstein, pp. 35–71. Reproduced by permission of Routledge, Inc., part of The Taylor & Francis Group.
2. Sharon E. Preves, "Sexing the Intersexed" (2001). From *Signs: Journal of Women in Culture and Society* 27(2), pp. 523–526. Reprinted with permission of the University of Chicago Press.
3. Robert Sapolsky, "The Trouble with Testosterone." Reprinted with the permission of Scribner, an imprint of Simon & Schuster Adult Publishing Group, from *Trouble with Testosterone: and Other Essays on the Biology of the Human Predicament* by Robert M. Sapolsky, pp. 149–159. Copyright © 1997 by Robert M. Sapolsky.
4. Betsy Lucal, "What It Means to Be Gendered Me" *Gender & Society* 13(6): 781–797. Copyright © 1999 by Sage Publications. Reprinted by permission of Sage Publications, Inc.
5. Serena Nanda, "Multiple Genders among North American Indians" (2000). In *Gender Diversity: Crosscultural Variations*. Copyright © 2000 by Waveland Press. Used with permission from the author.

Chapter 2

6. Peggy McIntosh, "White Privilege and Male Privilege: Unpacking the Invisible Knapsack." Copyright © 1998 by Peggy McIntosh. Used by permission of the author and Wellesley Centers for Women.
7. Patricia Hill Collins, "Toward a New Vision: Race, Class and Gender as Categories of Analysis and Connection" (1993). Copyright © 1993 *Race, Class, and Gender*. Reproduced with permission.
8. Yen L. Espiritu, "Race, Gender, Class in the Lives of Asian Americans" (1997). Copyright © 1997 *Race, Class, and Gender*. Reproduced with permission.
9. Alfredo Mirandé, "'Macho': Contemporary Conceptions" (1998). In *Hombres y Machos* (Boulder, CO: Westview Press), pp. 63–70.
10. Barbara Ryan, "Identity Politics in the Women's Movement" (2001). In *Identity Politics in the Women's Movement*, pp. 1–16. Used by permission of New York University Press.

Chapter 3

11. John Balaban, translator, "Girl Without a Sex" from *Spring Essence: The Poetry of Ho Xuan Huong*. Copyright © 2000 by John Balaban. Reprinted with the permission of Copper Canyon Press, P.O. Box 271, Port Townsend, WA 98368-0271.
12. Christine Helliwell, "It's Only a Penis: Rape, Feminism, and Difference" (2000). In *Signs: Journal of Women in Culture and Society* 25(3), pp. 789–816. Reprinted with permission of the University of Chicago Press.
13. Wairimũ Ngarũiya Njambi and William E. O'Brien, "Revisiting 'woman-woman marriage'" (2000). *NWSA Journal* 12(1), pp. 1–23. Reprinted with permission of Indiana University Press.
14. Maria Alexandra Lepowsky, "Gender and Power" from *Fruit of the Motherland*. Copyright © 1993 by Columbia University Press. Reprinted with the permission of the publisher.
15. Sally Cole, "Maria, a Portuguese Fisherwoman" in *Women of the Praia: Work and Lives in a Portuguese Coastal Community*. Copyright © 1991 by Princeton University Press. Reprinted by permission of Princeton University Press.

Chapter 4

16. C. Shawn McGuffey and B. Lindsay Rich, "Playing in the Gender Transgression Zone: Race, Class, and Hegemonic Masculinity in Middle Childhood" *Gender & Society* 13(5), pp. 608–627. Copyright © 1999 by Sage Publications. Reprinted by permission of Sage Publications, Inc.
17. Dorothy E. Smith, "Schooling for Inequality" (1999). *Signs: Journal of Women in Culture and Society* 5(4), pp. 1147–1151. Reprinted with permission of the University of Chicago Press.
18. Reprinted with permission of Bernice R. Sandler from www.bernicesandler.com. Sandler is a Senior Scholar in Residence at the Women's Research and Education Institute, consults extensively with institutions and others about women's equity, including sexual harassment, discrimination, and the chilly climate. She has given over 2000 presentations, written many articles, and serves as an expert witness in discrimination cases. Sandler can be contacted at: Bernice R. Sandler Senior Scholar, Women's Research and Education Institute, 1350 Connecticut Avenue, Suite 850, Washington, D.C. 20036 Phone: 202 833 3331 Fax: 202 785 5605 E-mail: sandler@bernicesandler.com website: www.bernicesandler.com.
19. Ellen Goodman, "College Gender Gap Stirs Old Bias" (August 4, 2002*) The Boston Globe*. Copyright © 2002, The Washington Post Writers Group. Reprinted with permission.
20. Ann Arnett Ferguson, "Bad Boys: Public Schools in the Making of Black Masculinity (2000). In *Bad Boys*, pp. 169–194. Copyright © 2000 by University of Michigan Press. Reproduced with permission.
21. Elizabeth Gilbert, "My Life as a Man" (August 2001). *GQ*, pp. 150–155, 189–190. Copyright © 2001 by Elizabeth Gilbert. Reprinted with permission of The Wylie Agency, Inc.
22. Leora Tanenbaum, "Slut! Growing Up Female with a Bad Reputation" (2000). In *Slut!*, pp. 1–7, 237–239. Reprinted with permission from Seven Stories Press.

Chapter 5

23. Adie Nelson, "The Pink Dragon Is Female: Halloween Costumes and Gender Markers" (2000). *Psychology of Women Quarterly* 24: 137–144. Reprinted by permission of Blackwell Publishing.
24. Reprinted with the permission of The Free Press, a division of Simon & Schuster Adult Publishing Group, from *Can't Buy My Love: How Advertising Changes the Way We Think and Feel* by Jean Kilbourne. Copyright © 1999 by Jean Kilbourne. Previously published as *Deadly Persuasion*.
25. Jacqueline Urla and Alan C. Swedlund, "The Anthropometry of Barbie: Unsettling Ideals of the Feminine Body in Popular Culture" (1995). In *Deviant Bodies: Critical Perspectives on Difference in Science and Popular Culture*, ed. by Jennifer Terry and Jacqueline Urla, pp. 277–313. Reprinted by permission of Indiana University Press.

26. Rana A. Emerson, "'Where My Girls At?' Negotiating Black Womanhood in Music Videos" *Gender & Society* 16(1): 115–135. Copyright © 2002 by Sage Publications. Reprinted by permission of Sage Publications, Inc.
27. James Farrer, "Disco 'Super-Culture': Consuming Foreign Sex in the Chinese Disco" (1999). *Sexualities* 2(2): 147–166. Reprinted by permission of Sage Publications, UK.

Chapter 6

28. Paula Black and Ursula Sharma, "Men Are Real, Women Are 'Made Up': Beauty Therapy and the Construction of Femininity" (2001). In *The Sociological Review*, pp. 100–116. Reprinted by permission of Blackwell Publishers.
29. Fatema Mernissi, "Size 6: The Western Woman's Harem." Reprinted with permission of Atria Books, an imprint of Simon & Schuster Adult Publishing Group, from *Scheherazade Goes West* by Fatema Mernissi. Copyright © 2001 by Fatema Mernissi.
30. Nicola Gavey, Kathryn McPhillips, and Marion Doherty, "'If It's Not On, It's Not On'—Or Is It? Discursive Constraints on Women's Condom Use" *Gender & Society* 15(6): 917–934. Copyright © 2001 by Sage Publications. Reprinted by permission of Sage Publications, Inc.
31. Pamela Block, "Sexuality, Fertility, and Danger: Twentieth Century Images of Women with Cognitive Disabilities (2000). *Sexuality and Disability* 18(4), pp. 239–254. Reprinted by permission of Kluwer Academic/Plenum Publishers.
32. Jennifer Lois, "Peaks and Valleys: The Gendered Emotional Culture of Edgework" *Gender & Society* 15(3): 381–406. Copyright © 2001 by Sage Publications. Reprinted by permission of Sage Publications, Inc.

Chapter 7

33. Amy S. Wharton, "Feminism at Work" *Annals of the American Academy* 471: 167–182. Copyright © 2000 by Sage Publications. Reprinted by permission of Sage Publications, Inc.
34. Peter Levin, "Gendering the Market: Temporality, Work and Gender on a National Futures Exchange" *Work and Occupations* 28(1): 112–130. Copyright © 2001 by Sage Publications. Reprinted by permission of Sage Publications, Inc.
35. Ching Kwan Lee, from *Gender and the South China Miracle: Two Worlds of Factory Women*, pp. 1–9, 160–161. Copyright © 1998 by The Regents of the University of California. Reprinted with the permission of The University of California Press.
36. Toni M. Calasanti and Kathleen F. Selvin, "Gender, Social Inequalities, and Retirement Income" (2001) from *Gender, Social Inequalities, and Aging*, pp. 93–120 (AltaMira Press). Reprinted with permission of Rowman & Littlefield Publishers, Inc.
37. Gai Ingham Berlage, "Marketing and Publicity Images of Women's Professional Basketball Players from 1977 to 2001" (2004). Written for this volume.
38. Alexis J. Walker, "Couples Watching Television: Gender, Power, and the Remote Control." From *Journal of Marriage and the Family* 58: 813–823. Copyright © 1996 by the National Council on Family Relations, 2989 Central Ave. NE, Suite 550, Minneapolis, MN 55421. Reprinted by permission.

Chapter 8

39. Karen Walker, "Men, Women, and Friendship: What They Say and What They Do" *Gender & Society* 8(2): 246–265. Copyright © 1994 by Sage Publications. Reprinted by permission of Sage Publications, Inc.
40. Kathleen Gerson, "Moral Dilemmas, Moral Strategies, and the Transformation of Gender: Lessons from Two Generations of Work and Family Change," in *Gender & Society* 16(1): 8–28. Copyright © 2002 by Sage Publications. Reprinted by permission of Sage Publications, Inc.
41. Gracia Clark, "Mothering, Work, and Gender in Urban Asante Ideology and Practice." Reproduced by permission of the author and the American Anthropological Association from *American Anthropologist* 101(4), pp. 717–729. Not for sale or further reproduction.
42. Maxine P. Atkinson and Stephen P. Blackwelder, "Fathering in the 20th Century." From *Journal of Marriage and the Family* 55: 975–986. Copyright © 1993 by the National Council on Family Relations, 2989 Central Ave. NE, Suite 550, Minneapolis, MN 55421. Reprinted by permission.
43. Karen D. Pyke, "Class-Based Masculinities: The Interdependence of Gender, Class, and Interpersonal Power" *Gender and Society* 10(5): 527–549. Copyright © 1996 by Sage Publications. Reprinted by permission of Sage Publications, Inc.

Chapter 9

44. Julia Hall, "It Hurts to Be a Girl: Growing Up Poor, White, and Female" *Gender & Society* 14(5): 630–643. Copyright © 2000 by Sage Publications. Reprinted by permission of Sage Publications, Inc.
45. Scott Straus, "Escape from Animal House: Frat Boy Tells All" (1996). *On the Issues*, Winter, pp. 26–28. Used with permission.
46. Reprinted by permission of Transaction Publishers. "Sexual Trafficking in Women: International Political Economy and the Politics of Sex" by Andrea Marie Bertone in *Gender Issues* 18(1). Copyright © 2000 by Transaction Publishers.
47. Beth A. Quinn, "Sexual Harassment and Masculinity: The Power and Meaning of 'Girl Watching'" *Gender & Society* 16(3): 386–402. Copyright © 2002 by Sage Publications. Reprinted by permission of Sage Publications, Inc.
48. Steven M. Ortiz, "Traveling with the Ball Club: A Code of Conduct for Wives Only" (1997) *Symbolic Interaction* 20(3): 225–249. Copyright © 1997 by JAI Press. Reprinted by permission of University of California Press.

Chapter 10

49. Gillian Dunne, "Opting into Motherhood: Lesbians Blurring the Boundaries and Transforming the Meaning of Parenthood and Kinship" *Gender & Society* 14(1): 11–35. Copyright © 2000 by Sage Publications. Reprinted by permission of Sage Publications, Inc.
50. The edited chapter "Unraveling the Gender Knot" and corresponding notes appears in *The Gender Knot: Unraveling Our Patriarchal Legacy*, by Allan G. Johnson, pp. 232–253, 288. Reprinted by permission of Temple University Press and the Author. Copyright © 1997 by Allan G. Johnson. All rights reserved.
51. Lionel Cantú, "De Ambiente: Queer Tourism and the Shifting Boundaries of Mexican Male Sexualities," in *GLQ: A Journal of Lesbian and Gay Studies*, Vol. 8, No. 1. Copyright © 2002 by Duke University Press. All rights reserved. Used by permission of the publisher.
52. Shahin Gerami and Melodye Lehnerer, "Women's Agency and Household Diplomacy: Negotiating Fundamentalism" *Gender & Society* 15(4): 556–573. Copyright © 2001 by Sage Publications. Reprinted by permission of Sage Publications, Inc.
53. Margaret L. Andersen, Anne E. Bowler, and Michael Kimmel, "Do We Still Need Feminist Theory? A Conversation." Written for this book.